MATH REVIEW

By

Charles, Chien-Chung, Feng

This book aims at illustrating fundamental concepts in basic math. While the examples, tips, and problems are designed to be relevant to real-life scenarios, some of them are not meant to reflect the practical situation exactly. Reliance of the information is strictly at your own risk.

Every effort has been made to ensure the correctness and accuracy of the contents. You are welcome to send any corrections, comments, and suggestions to charlesfengmath@gmail.com.

PREFACE

The purpose of this book

This book provides a comprehensive and practical review of middle to high school mathematics. While its main purpose is to elevate mathematical skills of middle and high school students, it is also an indispensable guide to various math standardized tests including but, not limited to SAT.

In any sports and athletics, learning, practicing, and mastering the basic skills are the foundations of high performance and achievements. The same is true for mathematics. Middle and high school mathematics encompases a broad range of basic math concepts and is the foundation for future advanced math learning. When solving practice problems in this book, first think through all the basic mathematical concepts you have learnt. Try to derive the solution through analysis and logical deduction from these basic concepts. Repeat this process over and over again until a final solution is reached. Only consult the solution provided after persistent trials and errors instead of jumping into reading the solution right away. This process is time consuming, but it is an essential step to understanding basic concepts in depth and solidifying a firm foundation. You might even derive new concepts on your own through this process.

In this book, various theorems and formulas are explained in a detailed step-by-step fashion. The most basic concepts are the building blocks where a final solution is derived from. A variety of visuals and illustrations are also used to aid understanding. Learning to solve problems through this process enhances analytical and logical thinking ability which prepares students to solve any type of problems.

A special chapter, Context Problems, was compiled to illustrate the steps to solve math problems we often encounter in our daily lives and the thinking process behind it. The problems encompass areas such as food, clothing, housing, transportation, schooling, entertainment, health, finance, sports, science, and much more. Learning how to solve these problems not only helps students get ready for standardized tests, but also equips the students with essential mathematical skills for life.

To help students prepare for standardized tests in the most efficient and effective manner, this book provides an innovative "concept check" section up front. Through concept check questions, students gain insights into their strengths and weaknesses in terms of understanding fundamental concepts. With this insight, students can then better strategize study and allocate appropriate time to each chapter. If students do well in the concept check problems for a particular chapter, students can even skip the entire chapter all together. With more than 1000 examples, practice questions, and detailed explanations, this book provides the invaluable training students need to home in on each fundamental concept in order to achieve high scores in standardized tests.

How to read this book

1. Before reading a chapter, first complete the "concept check" questions for that chapter. For each question that you get wrong, follow the instructions to find solution and detailed explanations in the chapter itself.
2. The visual "thought bullet points" in the explanation guide you through the step-by-step thought process behind each solution. Focus on where you got stuck initially and carefully tease out your mistake. Once you fully understand the explanations, it is also very important that you finish all the exercises in that chapter to solidify the concept. When you get stuck, do not look at the answer right away. Try to think through all the mathematical concepts you know. Pay special attention to the pieces of information that were missed initially. Following this process is essential to enhance problem-solving skills.
3. To maximize efficiency, skip sections you know well and focus on the chapters with concepts you are less familiar with. This is why working on concept check first is important.
4. When you complete all the chapters, take the tests at the end. They provide you with experience of what the exam is like and what to expect.

THE FLOW-CHART OF READING THIS BOOK

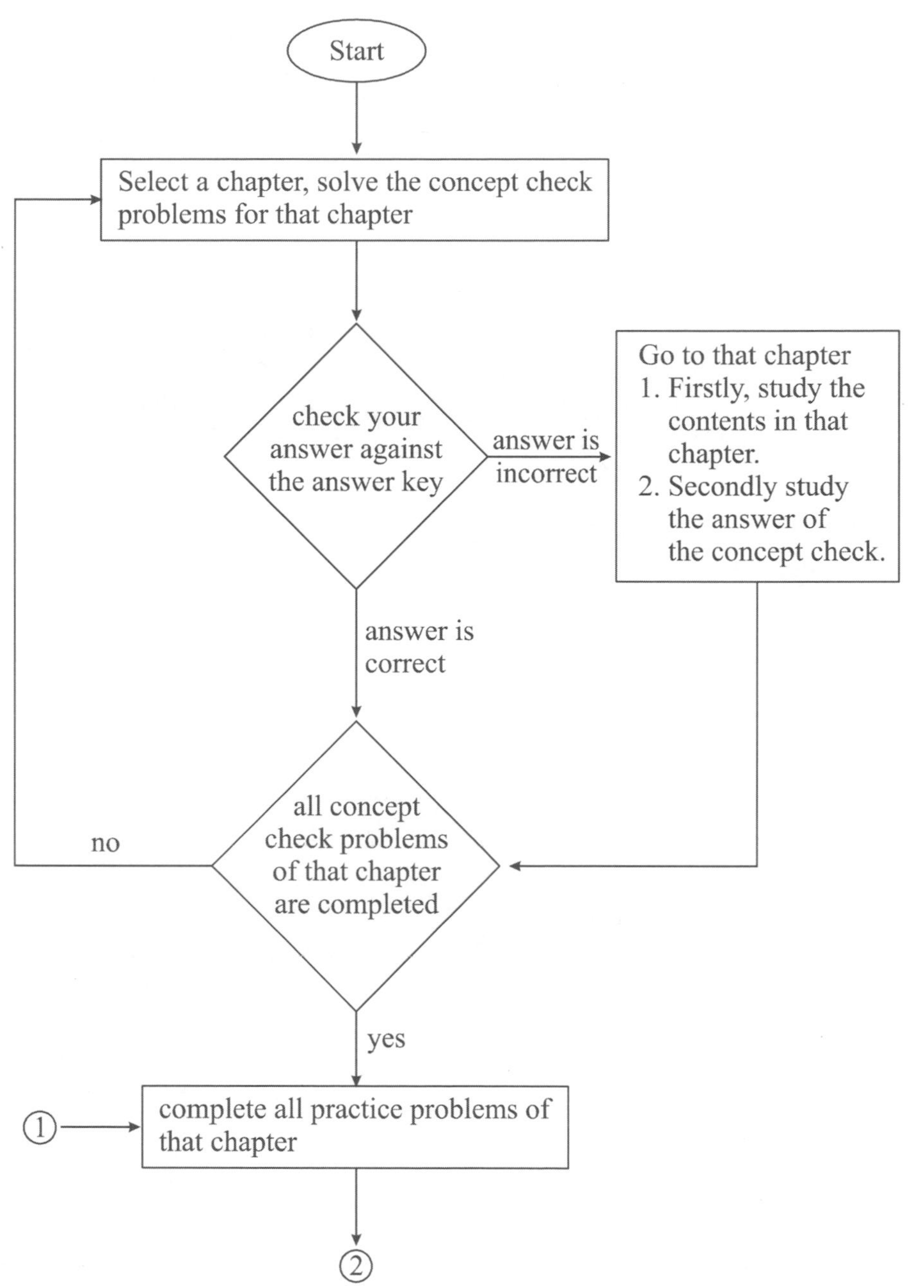

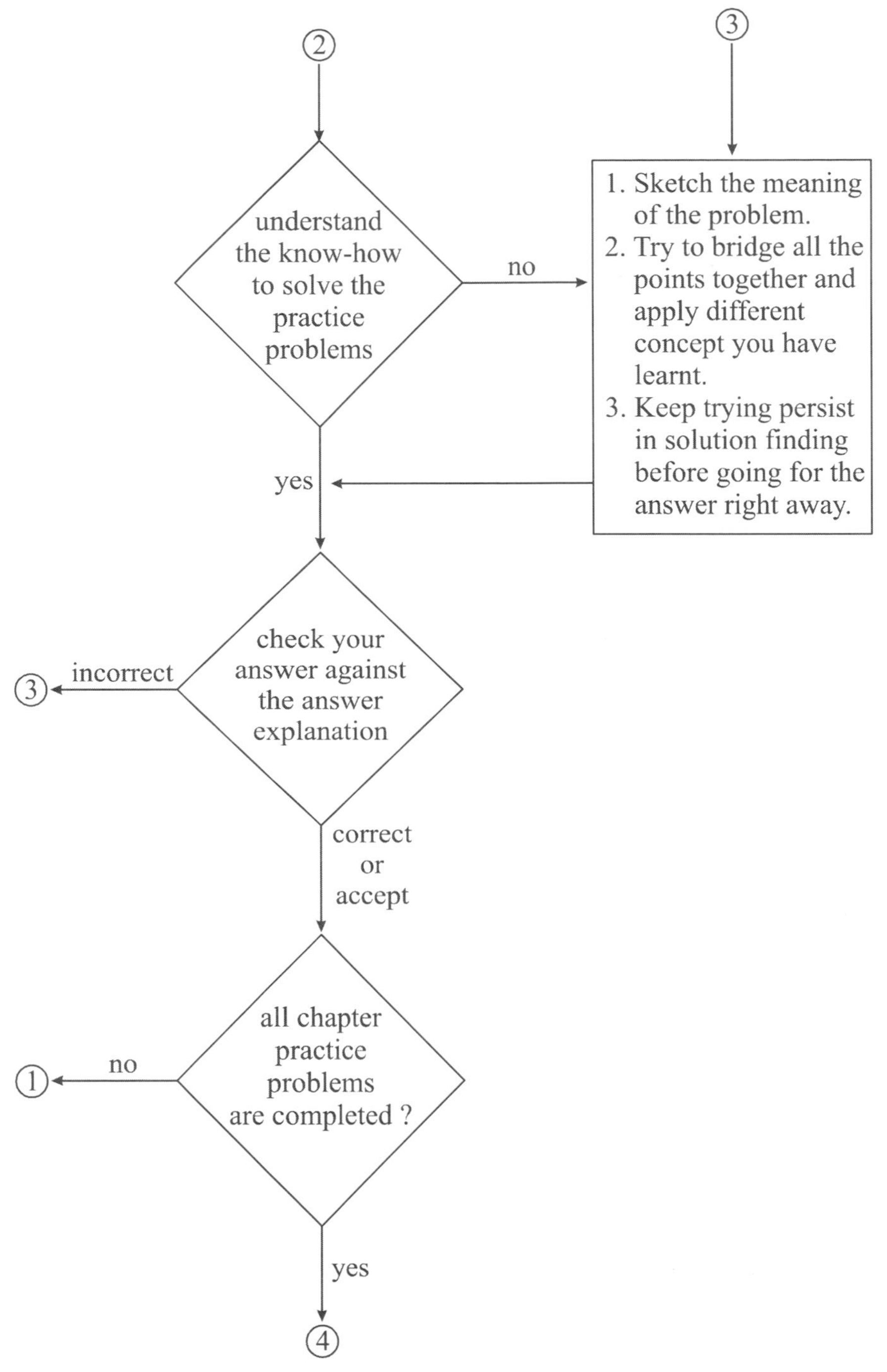

②
③
understand the know-how to solve the practice problems
no
1. Sketch the meaning of the problem.
2. Try to bridge all the points together and apply different concept you have learnt.
3. Keep trying persist in solution finding before going for the answer right away.
yes
check your answer against the answer explanation
incorrect
③
correct or accept
all chapter practice problems are completed ?
no
①
yes
④

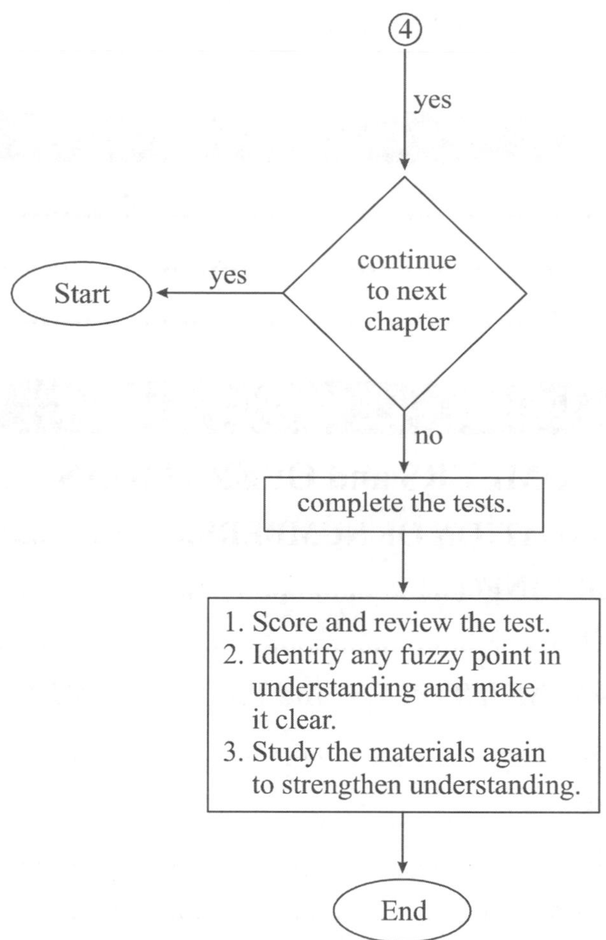

Acknowledgements

I would like to thank my wife Shirley for her constant support throughout this journey and assisting in data collection, editing, and file managements. I would also like to give thanks to my children and grandchildren, Ken, Caroline, Stefan, Katy, Brennan, Connor, Yohan, Jens for content review and feedback. They are as important to this book getting done as I was. Thank you all so much. Finally, I would greatly appreciate any corrections, comments, and suggestions from you, my readers. I hope you benefit from reading this book as much as I did writing it. Thank you!

CONTENTS

PART(I):PREFACE

PART(II):MATH LEVEL1/LEVEL II REVIEW

PART (III) Model Test

Concept Check
Chapter 1
Numbers and Operations

1-01. Which of the following numbers is irrational? (Rational Numbers)

(A) $\frac{7}{3}$

(B) $-\sqrt{9}$

(C) $\sqrt{3}$

(D) $3.\dot{3}$ (3.33333333…)

1-02. Which of the following numbers is rational? (Rational Numbers)

(A) $\sqrt{2}$

(B) e (2.71828…)

(C) π (3.14159…)

(D) $-\sqrt{\frac{1}{9}}$

1-03. P, Q, M, and R lie on the same number line in that order. Point M is the midpoint of $\overline{PR}$ and $PQ=\frac{1}{2}QR$, which of the following is the distance between point M and Q, in terms of PR? (Number Line)

(A) $\frac{1}{6}PR$

(B) $\frac{1}{3}PR$

(C) $\frac{1}{4}PR$

(D) $\frac{1}{2}PR$

1-04. If $x-8$ is an odd integer, x could be which of the following? (Odd/Even Number)

(A) −8

(B) 0

(C) 1

(D) 8

1-05. If k is an odd number, which of the following must be an even number? (Odd/Even Numbers)

(A) $k+4$

(B) $k-4$

(C) $k\times4$

(D) $k\div4$

1-06. What is the G.C.F. (Greatest Common Factor) of 252 and 126? (GCF)

(A) 42

(B) 48

(C) 52

(D) 126

1-07. Which of the following is the L.C.M. (Least Common Multiple) of 186 and

72? (L.C.M)

(A) 372

(B) 794

(C) 1568

(D) 2232

1-08. If the sum of n consecution integers is 64, what is the greatest value of n? (Consecutive integers)

(A) 64

(B) 126

(C) 128

(D) 130

1-09. Which of the following operations yields the highest value? (Fraction Operations)

(A) $(\frac{3}{8}+\frac{1}{4})+\frac{1}{3}$

(B) $(\frac{2}{3}\times\frac{1}{5})-\frac{1}{3}$

(C) $(\frac{1}{2}\div\frac{1}{3})\times\frac{1}{3}$

(D) $(2\frac{1}{2}-\frac{3}{4})\div\frac{1}{3}$

1-10. In Jens' high school, 3/5 of the students take the school bus to school. If in this school, 3/5 of the students are boys and only 1/5 of them do not take the school bus to school, what is the fraction of girls who do not take the school bus to school? (Two-way of Classification)

(A) $\frac{6}{25}$

(B) $\frac{7}{25}$

(C) $\frac{8}{25}$

(D) $\frac{9}{25}$

1-11. Convert 0.12% to fraction. (Decimal Conversion)

(A) $\frac{12}{100}$

(B) $\frac{3}{25}$

(C) $\frac{12}{10000}$

(D) $\frac{3}{2500}$

1-12. What is the result of $1234\div234$ (Rounded to Hundredths)? (Rounded)

(A) 5.28

(B) 5.3

(C) 5.27

(D) 5.274

1-13. If the light from a certain planet in the galaxy reaches Earth in a year, how far is it from Earth? (The speed of light is about 300,000 km/s) (Scientific Notation)

1-14. What is the scientific notation of 0.00000032? (Scientific Notation)

(A) 0.32×10^{6}

(B) 0.32×10^{-6}

(C) 3.2×10^{7}

(D) 3.2×10^{-7}

1-15. The highest score of the mid-term math exam in Katy's class is 92, and the lowest score is 64. If x represents the score of any student in the class, which of the following is true? (Absolute Value)

(A) $|x-78|\le 14$

(B) $|x-78|<14$

(C) $|x-78|\ge 14$

(D) $|x-78|>14$

1-16. Find the result of

$$\frac{(3+2i)-(4-5i)(1+i)}{2-i}=$$

(Complex Number I)

(A) $1+2i$

(B) $1-2i$

(C) $|1+2i|$

(D) -3

1-17. What is the rectangular (Standard) form of $4(\cos\frac{\pi}{3}+i\sin\frac{\pi}{3})$?

(Complex II)

1-18. find $(1-i)^8$ (Complex II)

1-19. What is the polar form of $3+3i$? (Complex II)

1-20. If $z_1=4e^{i\frac{3\pi}{4}}$ and $z_2=e^{i\frac{4}{3}\pi}$, what is the value of $z_1 \cdot z_2$? (complex II)

1-21. What is the trigonometric from of $\frac{\sqrt{3}}{2}-\frac{\sqrt{3}}{2}i$? (Complex II)

Concept Check
Chapter 2
Sets, Sequence, Defined Operations, Patterns and Logic

2-01. If a set is defined as $x = \{x \mid 20 < x < 40, x \text{ is a multiple of } 3\}$, how many members are there in the set? (Set)

(A) 5
(B) 6
(C) 7
(D) 8

2-02. If $P = \{x \mid x$ represents all non-negative numbers$\}$, and $N = \{x \mid x$ represents all non-positive numbers$\}$, what are $P \cup N$ and $P \cap N$? (Set)

2-03. If $P = \{1, 3, 5, 7\}$, and CuP$\cup P = \{1, 2, 3, 4, 5, 6, 7, 8\}$, what are the values in (CuP)? (Set)

2-04. How many subsets does the set $\{a, b, c, d\}$ have at maximum? (Set)

(A) 4
(B) 8
(C) 15
(D) 16

2-05. $\{1, 2\} \subset S \subseteq C\{1, 2, 3, 4\}$ How many sets of S satisfy the expression above? (Set)

(A) 1
(B) 2
(C) 3
(D) 4

2-06. 1, 3, 5, 7, …

Based on the sequence shown above, what is the value of the 13th term? (Sequence)

(A) 13
(B) 18
(C) 22
(D) 25

2-07. 2, 4, 8, 16, 32, …

In the list above, what is the sum of the first 10 terms? (Sequence)

2-08. 8, 4, 2, 1, $\frac{1}{2}$, $\frac{1}{4}$, …

The sequence shown above has infinite number of terms, what is the sum of this sequence? (Sequence)

(A) 15
(B) $15\frac{1}{2}$
(C) 16
(D) $16\frac{1}{4}$

2-09. Let the operation $\boxed{\not}$ be defined by $p \boxed{\not} q = \frac{p}{q}$ for all positive numbers. If $8 \boxed{\not} (4 \boxed{\not} x) = 4$, what is the value of x? (Defined Operation)

2-10. 1100110011001100110011…

Based on the bit pattern shown above, what is the value of the 121th bit? (Pattern)

2-11. $-5, -4, -2, -1, 1, 2, 4, 5, \ldots$

Based on the number sequence shown above, what is the number after 5? (pattern)

2-12. Given $T_{n+2} = 2T_n + T_{n+1}$ where T_n represents the term at nth position, if $T_1 = 2$, $T_2 = 5$, what is the value of T_5? (Pattern)

2-13. What is the value of A and B in the following truth table? (Logic)

P	Q	P and Q ($P \wedge Q$)	P or Q ($P \vee Q$)
T	T	T	T
T	F	F	T
F	T	A	T
F	F	F	B

T = Truth, F =False

$A =$

$B =$

2-14. Given the statement "If it rains, the ground gets wet", which of the following statements is true? (Logic)

(A) If it doesn't rain, the ground doesn't get wet.

(B) If it is cloudy, then the ground will get wet.

(C) If the ground is dry, it didn't rain.

(D) If the ground is not wet, it is raining.

2-15. Little league baseball teams A, B, C, D, and E are competing for the championship in the annual playoff. 5 students try to predict the outcomes.

Allen's prediction: team C gets the 2nd place; team A gets the 3rd place.

Betty's prediction: team A gets the championship; team E gets the 4th place.

Charles's prediction: team B gets the 3rd place; team D gets the 4th place.

David's prediction: team C gets the 2nd place; team E gets the 4th place.

Eric's prediction: team D gets the championship; team B gets the second place.

It turns out that exactly half of each student's prediction is correct, which team got the championship? (Logic)

(A) team A

(B) team B

(C) team C

(D)team D

Concept Check
Chapter 3
Average, Ratio, Rate, Proportion, and Percent

3-01. April's previous 4 math exam scores are 92, 86, 90, 88. If she expects the average score of 90 for her 5 math exams, what should the fifth score be? (Average)

3-02.

	10	9	8	7	6	5	4	3	2	1	0
Katy	1	3		2	1		1	2			
Tom		1	1		3	2	2		1		
Jim				2		1	4		1		2

The table above shows the archery scores of three people. The column represents the score each shot obtains while the number in the cell represents the number of shots with that score for a particular person. What is the average score of these 3 people per shot? (Average)

3-03. "Golden ratio" is when the ratio of the sum of two quantities to the larger one is about 1.6 (it is an irrational number 1.6180339887…)
The ratio of Mary's height to the length of her legs is golden ratio. If the length of her upper body is 24.75 inches, how long are her legs?

3-04. If Yohan filled up his car with 7 gallons of gasoline and it costed him \$28. The car can be driven for exactly 140 miles with this amount of gasoline.

I. How many miles can the car drive per gallon?
II. How many gallons of gasoline are needed per mile?
III. How many gallons of gasoline can be bought with one dollar?
IV. How much does one gallon of gasoline cost (in dollars)?
V. How many miles can the car drive per dollar?
VI. How much does it cost for the car to drive for one mile (in dollars)?

3-05. Two cups of coffee cost \$6.8. How much does it cost to buy 5 cups of the same coffee? (Rate)

(A) 15
(B) 16
(C) 17
(D) 18
(E) 19

3-06. P people can complete a job in D days, how many days are needed to do the same job for $P+1$ people? (Rate)

(A) $\frac{PD}{P+1}$

(B) $\frac{(P+1)}{PD}$

(C) $\frac{(P+1)D}{P}$

(D) $\frac{(D+1)P}{D}$

(E) $\frac{D+1}{P}$

3-07. Alex has 34 fluid ounces of a 28% alcohol solution. How many ounces of 58% alcohol solution must be added into Alex's solution to make it a 32% alcohol solution?

(Rounded to hundredth) (Proportion)

3-08. The price of a certain overcoat is \$90. During promotion, the price is reduced to \$78. What is the percentage of price reduction? (Proportion)

(A) 11.1%

(B) 13.3%

(C) 15.5%

(D) 17.7%

(E) 19.9%

3-09. There are d dollars in a drawer. Bonnie took 5 dollars away. In terms of d, what is the percentage of money that is still left in the drawer?

(A) $\frac{100(d-5)}{d}\%$

(B) $100(a-5)\%$

(C) $\frac{d-5}{100}\%$

(D) $\frac{100d}{d-5}\%$

(E) $\frac{d-5}{100d}\%$

Concept Check
Chapter 4
Fundamentals of Algebra

4-01. The difference between Tad's age (T) and Steven's age (S) is 3 years less than twice Tad's age. Which of the following expressions best fit the above statement? (Expression and Operations)

(A) $T - S = 2T - 3$

(B) $S - 2T = 2T - 3$

(C) $|S - T| = 2T - 3$

(D) $|2ST - S| = T - 3$

4-02. There are n people who are planning to come to Tim's weekend party. Each of them pitched in to share the total cost of d dollars. However, two more guests are coming and would like to pitch in as well, how much will each person who already paid be refunded? (Expression and Operation)

(A) $\frac{d}{n+2}$

(B) $\frac{d}{n-2}$

(C) $d(\frac{1}{n} - \frac{1}{n+2})$

(D) $d(\frac{1}{n+2} - \frac{1}{n})$

4-03. Simplify $-5(a+2b)+\frac{1}{2}(4a-8b)$

(Expression and Operations)

4-04. Simplify $2(x^2-1)-[3(x^3-x^2+\frac{2}{3}y)$
$-(x^2+y^2)+3]+\frac{2}{3}$

(Expression and Operations)

4-05. $(a-b)(a^2+ab+b^2)=$

(Expression and Operations)

4-06. $(4x^4+x^3-3x^2+5x-1)\div(x^2+x+1)=?$

(Expression and Operations)

4-07. Factor the expression by pulling out common factor or factors: $(a+b)^3(a-2b)-(a+b)(a-2b)^3=?$

(Factoring)

4-08. Factor the expression by grouping the common terms: $x^3-2x^2-x+2=?$

(Factoring)

4-09. Factor the expression by using cross-multiplication: $2x^2-5x-3=?$

(Factoring)

4-10. Factor the expression by using formula: $16x^2 - 81 = ?$ (Factoring)

4-11. Given equation $7x + 3 = 12$, what is the value of x? (Linear Equation)

4-12. Given linear equation $3x + a = 5$, what is the value of x in terms of a? (Linear Equation)

4-13. Solve $\begin{cases} 2x + 3y = -1 \\ -3x + y = 7 \end{cases}$.

(System Equation)

4-14. Solve $\frac{-2x+3}{3} \geq 3$. (Inequality)

4-15. Solve $\begin{cases} 2x + 3 > 2 \\ 3x - 5 < 1 \end{cases}$.

(System Inequality)

4-16. Solve $|3x - 2| \leq 5$. (Inequality)

4-17. Solve the inequality $|x - 2| \geq 3$ and use number line to show the range. (Inequality)

4-18. Solve $\begin{cases} x + 3 \geq 3 \\ 2x - 1 < 1 \end{cases}$. (System inequality)

4-19. Solve $3x^2 - 4x + 1 = 0$ by using formula. (Quadratic Equation)

4-20. Given $f(x) = x^2 + 3x + 1$, $g(x) = 2x + 1$, what is the value of $f(g(2))$? (Function)

4-21. Given $f(x) = x^2 + 2x$ and $g(x) = x - 2$, what is the sum of $3f(x) + 2g(x)$? (Function I)

4-22. Find the domain and range of $f(x) = \frac{1}{3} + \sqrt{x}$. (Function I)

4-23. Find the inverse function of $f(x) = 3x - 1$. (Function I)

4-24. Mr. Martin bought a new car that costed \$35,000. The salesman told him that this car depreciates by 10% each year. Which of the following expressions best models the depreciation? (Modeling)

(A) $D = 35{,}000 \times (1 - 0.1)^x$

(B) $D = 35{,}000 \times (0.9)(x)$

(C) $D = 35{,}000 \times 0.1^x$

(D) $D = 35{,}000 \times 0.1(x)$

Concept Check
Chapter 5
Polynomials

5-01. If $P(x)=x^3-2x^2+3x-6$ can be divided by $x+k$, what is the value of k? (Factor theorem)

(A) -1

(B) -2

(C) -3

(D) -4

5-02. Find the remainder of $x^{15}+2x^{14}-x^{13}+3x^2-2x+3$ when it is divided by $x-1$. (Remainder)

(A) 0

(B) 2

(C) 4

(D) 6

5-03. $P(x)=x^3+ax^2+bx-2$ has two factors $(x+1)(x-2)$, what is the value of $a+b$? (Polynomial)

5-04. Solve $x^5+3x^4+2x^3+6x^2-3x-9=0$ by factoring. (Higher-degree equations)

5-05. Solve $x^3-27=0$ by formula. (Higher-degree equations)

5-06. Solve $4x^4+8x^2-3=0$ by quadratic formula. (Higher-degree equation)

5-07. Solve $4x^4+8x^2+3=0$ by substitution. (Higher-degree equations)

5-08. Simplify $\dfrac{(x^{-6})(y^5)(x^4y^{-1})}{(x^{-2})y^4}+3(a^{-3}b^{-2}c)\times(a^3b^2c^{-1})$. (Exponent)

5-09. If $4^{x+1}+(22)(2)^{x-1}-3=0$, what is the value of x? (Exponent)

(A) -1

(B) -2

(C) -3

(D) -4

5-10. Simply $\dfrac{\sqrt{2}+\sqrt{5}}{\sqrt{3}-1}$. (Radical)

5-11. Solve $\sqrt{x-1}-\sqrt{4x}+2=0$. (Radical)

5-12. Solve $\dfrac{2x-3}{x-1}+\dfrac{3x-2}{x+1}=1$. (Rational equation)

5-13. Solve $\dfrac{x}{x-2}-\dfrac{2}{x}-\dfrac{3}{x(x-2)}=0$. (Rational Equation)

Concept Check
Chapter 6
Coordinate and Analytical Geometry

6-01. Point $P_1(2,3)$ and $P_2(5,-4)$ are on xy-plane, what are the coordinates of the midpoint of segment $\overline{P_1P_2}$? (Coordinate plane)

6-02. In question 6-01 above, what is the distance between $P_1(2,3)$ and $P_2(5,-4)$? (Distance)

6-03. $A(1,3)$ and $B(-3,-5)$ are the starting and ending points of segment $\overline{AB}$ respectively. What is the slope of $\overline{AB}$? (Straight Lines)

6-04. A straight line has the equation $2x+3y+4=0$. What is its slope-intercept form? (Algebraic form of straight lines)

6-05. Given point $A(3,2)$ and point $B(4,-2)$, what is the algebraic form of segment $\overline{AB}$? (Algebraic form of straight line)

6-06. The slope of line ℓ is $-\frac{1}{3}$ and y-intercept is 3. What is the algebraic form of line ℓ? (Algebraic form of straight line)

6-07. If line L: $y=\frac{2}{3}x+5$ is parallel to line M: $y=mx+b$, what is the value of m? (Straight lines)

6-08. In the question above, if line ℓ is perpendicular to line M, what is the value of m? (Straight lines)

6-09. $\begin{cases} 5x+3y=6 \\ 15x+9y=24 \end{cases}$

How many solutions are there for the systems of equations above? (Straight line)

6-10. $\begin{cases} 2x-3y=7 \\ 4x-6y=14 \end{cases}$

How many solutions are there for the systems of equation above? (Straight line)

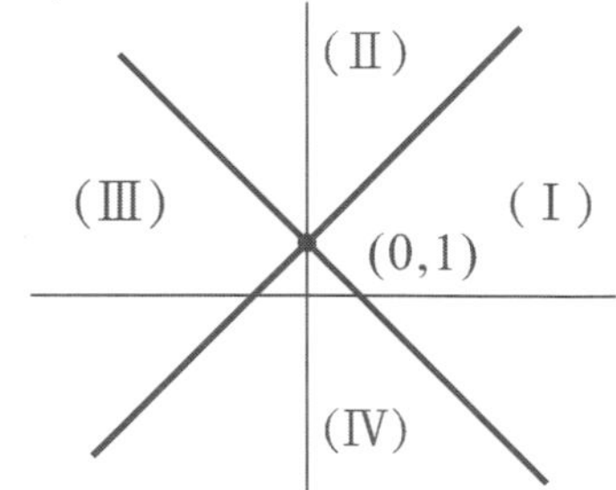

6-11. $\begin{cases} x+y>2 \\ 3x+2y>-3 \end{cases}$

Draw the graph solution for the systems of inequalities above. (Solved by Graphs)

6-12. Point $A(-3, 2)$ lies on the graph $f(x)$. If the graph $f(x)$ is shifted to the right by 2 units and down by 3 units, what is the new coordinates of point A? (Before And After Images)

(A) $(-5, 5)$
(B) $(-5, -1)$
(C) $(-1, -1)$
(D) $(-1, -5)$

6-13. The graph of $f(x)=x^2+2x-5$ was shifted right by 3 units and then up by 4 units, which of the following is the function of the new graph? (Before And After Images)

(A) $f(x)=(x+2)^2-4$
(B) $f(x)=(x-2)^2-2$
(C) $f(x)=(x-4)^2-4$
(D) $f(x)=(x-4)^2+4$

6-14. Find the concave, vertex, symmetric equation, focus, and intersect points with x and y axes of the parabola $f(x)=2x^2+4x-5$ and graph it. (Parabola)

6-15. The parabola $f(x)=-2x^2+4x+6$ was shifted to the right by 2 units and up by 4 units, what is the coordinates of the vertex after the shift? (Before and After Images)

6-16. What is the maximum value of $f(x)=-x^2+6x+8$? (Parabola)

6-17. Find the equation (general form) of an ellipse with (3, 2) as its center, major axis of 8 units, and minor axis of 6 units. (Ellipse)

6-18. What are the asymptotes, foci, and length of transverse and conjugate axes of the hyperbola $\frac{x^2}{16}-\frac{y^2}{9}=1$?

6-19. Find the rectangular coordinates of point $(2, \frac{\pi}{3})$? (Polar coordinates)

6-20. Convent the polar equation $r=\frac{3}{3-2\cos\theta}$ to rectangular coordinates? (Polar coordinates)

Concept Check
Chapter 7
Data Presentation and Basic Statistics

7-01.

Grade	Number of Students	Percentage
A^+	44	6.8%
A	80	12.4%
A^-	104	16.1%
B^+	120	18.5%
B	113	17.5%
B^-	95	k%
C^+	42	6.5%
C	35	5.4%
C^-	14	2.1%

The table above shows the grades of a math midterm exam in West High School. What percentage (k%) of the students has grade B^-?

And what percentage of the whole school has grade B and up?

(Table)

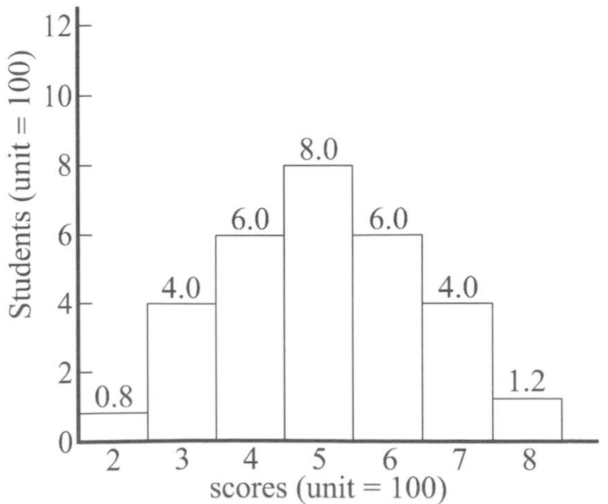

7-02. The histogram on the left shows the scores of Math Competition in North Gate district, 2019.

(1) Based on the histogram above, what is the average score in this math competition?

(2) What percent of the students scored over 600?

(Histogram)

Sales performance Of Ma's Company (January ~ June, 2019)

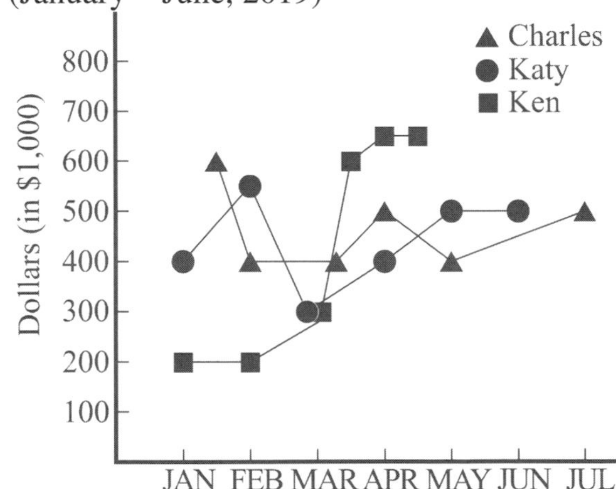

7-03. Based on the information presented in the line-graph above, answer the following questions:

(1) Who has the most sales revenue?

(2) Who has the biggest sales revenue range?

(Line graph)

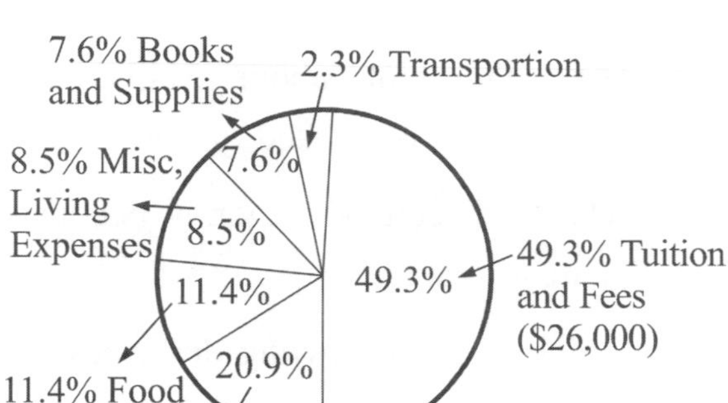

7-04. The Annual cost of attending a particular four-year college is shown above.

1. What is the ratio of Tuition Fees to Misc. and Living Expenses?
2. How much does a student need to pay for food in this 4-year college?

(Pie Chart)

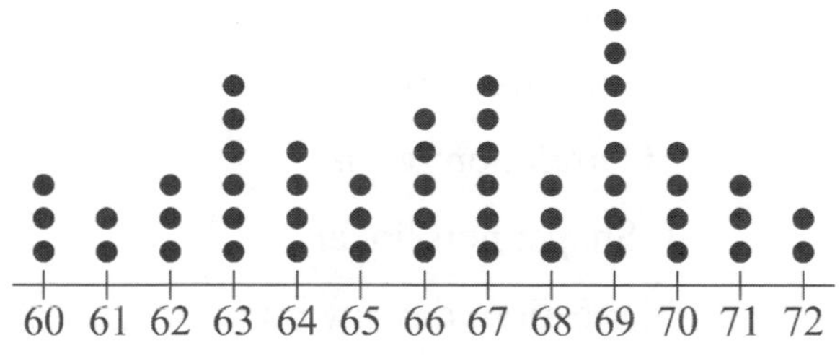

Height in inches

7-05. The dot-plot above shows the height of Stephen and Paul's classmates.

(1) What is the mean of the heights (in inches)?

(2) What is the median?

(3) What is the mode?

(4) What is the range?

(Dot Plot)

7-06. A sorted data list is shown below. Draw a stem-leaf-plot of this list, and find the mean, median, mode, and range.

11, 12, 12, 14, 15, 15, 15, 20, 22, 33, 35, 36, 42, 48, 53, 55, 62, 63, 65, 66, 72, 74, 77

(Stem-Leaf-Plot)

7-07. Data list: 30, 46, 52, 58, 60, 64, 66, 70, 72, 75, 75, 78, 80, 80, 83, 84, 85, 85, 85, 85, 85, 88, 88, 90, 92, 95, 100

The data list above shows the Math midterm exam scores of Mr. Jackson's Math class. What is the mean, median, mode, range, interquartile range, Q_1 (first quartile), Q_3 (third quartile), and outliers if any?

(Basic statistics)

average scores (points)

1 2 3 4 5 6

1 2 3 4 5 6 7 8

Professional training (hours) in a workday

7-08. The scatterplot above shows the average scores of 16 games vs. the professional training hours in a workday for a male basketball player.

1. Draw the line of best fit.
2. Find the function of the line of best fit.
3. Predict the average points he would get if there is no professional training at all?
4. Estimate the average points he would get if he receives 7 hours of professional training in a workday?

(The line of best fie)

7-09. Below lists the heights of Peter's classmates: 58, 60, 60, 62, 63, 64, 65, 66, 66, 67, 67, 67, 67, 68, 69, 70, 72, 73, 73, 74.

If peter's height is 64 inches, what percentile does he fall in?

If Peter's twin-brother George is 70 inches tall, what percentile does George fall in?

(Percentile)

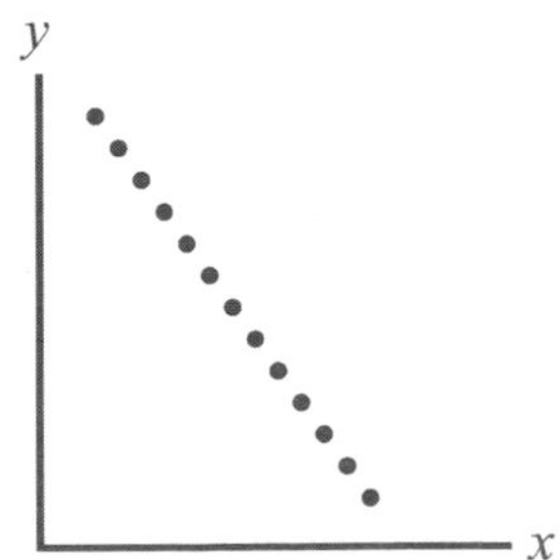

7-10. The scatterplot above shows the relationship of x and y, $y = f(x)$. When x becomes bigger, y becomes smaller.

Which of the following describes the relationship graph above?

(A) Shape: liner
Direction: downward
Strength: strong
Correlation: week

(B) Shape: liner
Direction: downward
Strength: strong
Correlation: strong

(C) Shape: non-linear
Direction: downward
Strength: strong
Correlation: weak

(D) Shape: non-linear
Direction: downward
Strength: strong
Correlation: strong

(Correlationship)

7-11. Ms. Philips will select 5 students from 32 students in her class using the system sampling method to participate in an inter-school Mathematics competition. Students are numbered from 1 to 32 and the systematic sampling interval is 10.

If student number 4 is selected by the teacher, what is the number of the fifth student selected (the last student)? (Data Sampling)

7-12. A Spanish teacher gave 5 student volunteers Spanish-learning CDs in hope to improve their Spanish test scores. As a result, all five students performed really well on the test. Therefore, the teacher concluded that the CDs are very helpful. However, this conclusion cannot be established due to which of the following?

(A) Not the right people
(B) Not the right time
(C) Not the right CDs
(D) Not the right place

(Data Sampling)

7-13. A certain company wants to know the popularity of its products in S-city. They collected 1,200 valid sample questionnaires and 800 of them indicated that the customers love the products. At 95% confidence level, what is the confidence interval? (Confidence Interval)

Concept Check
Chapter 8
Counting and Probability

8-01. By rolling 2 fair dice, how often can we expect the sum of the dice to be below 4?, and how often can we expect the points of the two dice are equal ? (Counting)

8-02. On the menu of a particular restaurant, main courses include steak, lamb chops, seabass, pork chops, and lobster. Appetizers include salad, tomato soup, and clam chowder. Desserts include cheese cake, coffee, ice cream, and black tea. If April has to choose one and only one main course, appetizer, and dessert for her dinner, how many selections does she have? (Counting)

8-03. There are 6 people in the meeting room. Each person must shake hands with another person once. How many different ways of handshaking are there?
(Counting)

8-04. There are three red balls, 4 white balls, and 5 yellow balls in the bag. If Shirley takes out one ball from the bag, what is the probability that the ball is white?
(Probability)

8-05. 5 players from an eight-player team will be selected to participate in 5 friendly tennis games. Currently there are 3 top players among the 8 players. The coach decided to arrange the top players arbitrarily in the first, third and fifth match. The rest of the players are scheduled to the second and the fourth match. How many different arrangement are there? (combination)

8-06. There is a long shelf on the counter that can hold 6 cups. Jack has 3 cups of coffee, 2 cups of coke, and 1 cup of water.
How many ways are there for Jack to arrange these 6 cups on the shelf ?
(Permutation)

8-07. Four couples Aa, Bb, Cc, and Dd would like to be seated at a round table (capitalized letter represents male while lowercase letter represents female).

How many different seating arrangements are there if people with the same gender must not be seated together? (circular permutation)

8-08. The cups produced by a certain company are being packaged into boxes. Each box contains 10 cups. During quality inspection, 3 cups are taken out of each box. If 2 or more cups have bad quality, the whole box will be eliminated. If there is a box which contains 4 bad cups, what is the probability that this box will be eliminated? (probability)

8-09. Jens is a good baseball player, his bat average is 0.25. If the coach assigns him to bat 6 times in a game, what is the probability that he will get exactly 2 hits? (Repeating Experiments)

8-10. According to statistics, the winning rate for a basketball player in each game is $\frac{1}{3}$ and losing rate is $\frac{2}{3}$. Ryne and Kim negotiated that in each game, if a particular player wins, Ryne gives $12 to Kim. Otherwise, Kim gives $4 to Ryne. What is the probability that Kim wins $45 from Ryne after 8 games? (Probability)

Concept Check
Chapter 9
Plane and Solid Geometry

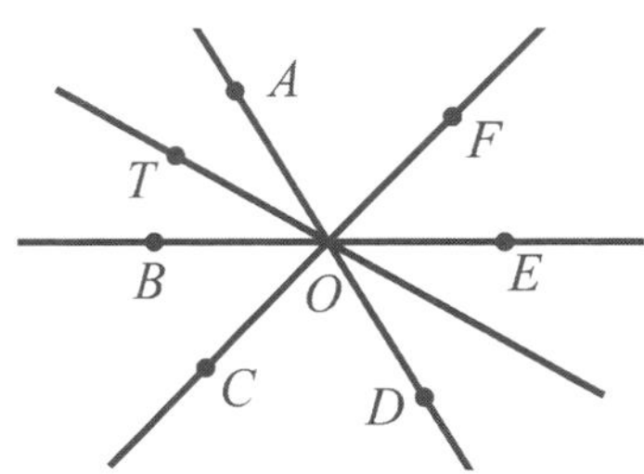

9-01. In the figure above, $\overrightarrow{OT}$ bisects $\angle AOB$, $\overrightarrow{OF}$ bisects $\angle AOE$, if $\angle BOT = (2x-8)°$, and $\angle FOE = (3x+3)°$, what is the value of x?

(Angle Measurement)

9-02. In a triangle $\triangle ABC$, if $2(\angle B + \angle C) = \angle A$ what is the measure of $\angle A$?

(Angle Measurement)

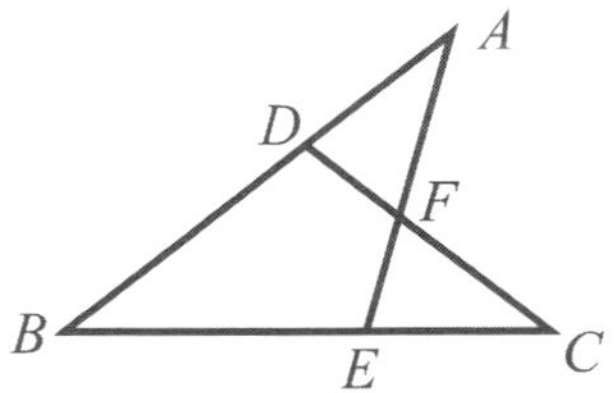

9-03. In the figure above, $\angle ABC = 35°$, $\angle BCD = 30°$, $\angle EFC = 40°$, what is the measure of $\angle BAE$?

(Congruent of Triangle)

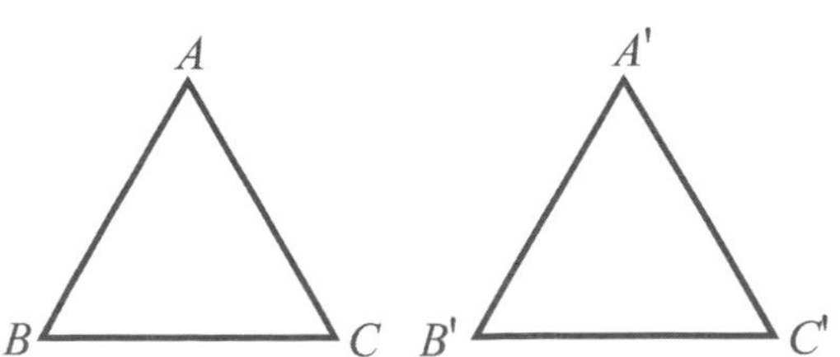

9-04. $\triangle ABC$ and $\triangle A'B'C'$ are shown above, which of the following does not prove $\triangle ABC \cong \triangle A'B'C'$?

(A) $\overline{AB} = \overline{A'B'}$, $\overline{BC} = \overline{B'C'}$, and $\overline{AC} = \overline{A'C'}$ (S.S.S)

(B) $\overline{AB} = \overline{A'B'}$, $\overline{AC} = \overline{A'C'}$, and $\angle A \cong \angle A'$ (S.A.S)

(C) $\overline{AB} = \overline{A'B'}$, $\overline{AC} = \overline{A'C'}$, and $\angle B \cong \angle B'$ (S.S.A)

(D) $\overline{AB} = \overline{A'B'}$, $\angle B = \angle B'$, and $\angle C = \angle C'$ (A.A.S)

9-05. Explain whether the following sets of triangles are congruent or not.

(Congruent of Triangle)

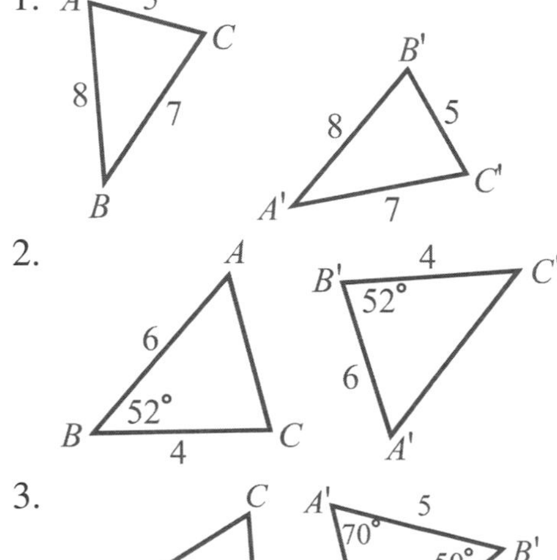

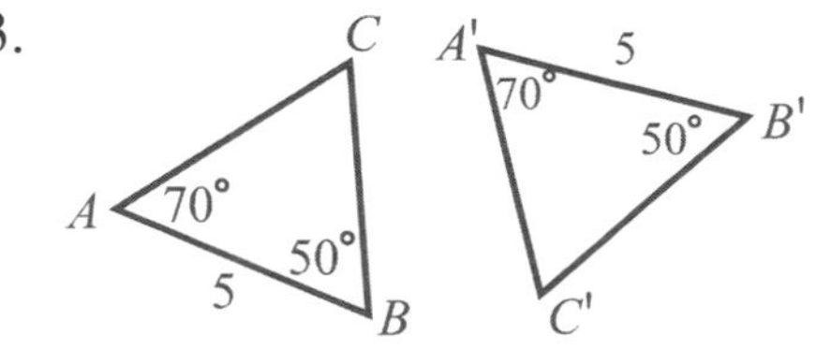

4.

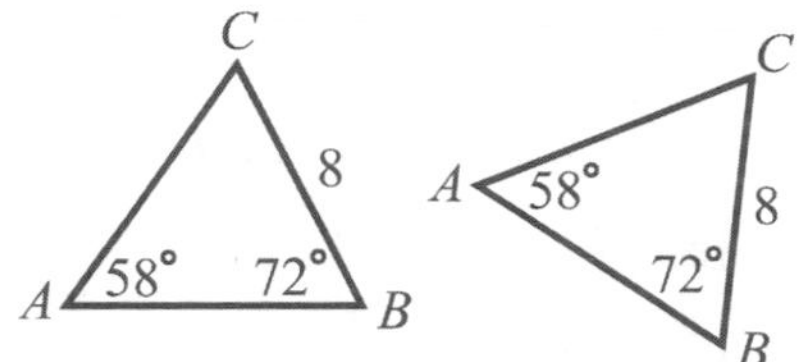

5.

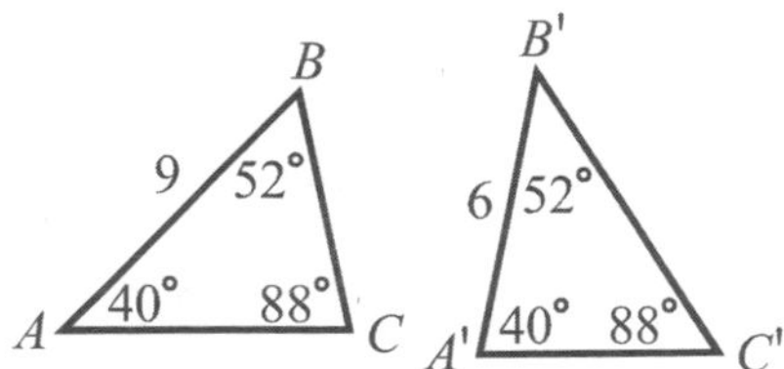

6.

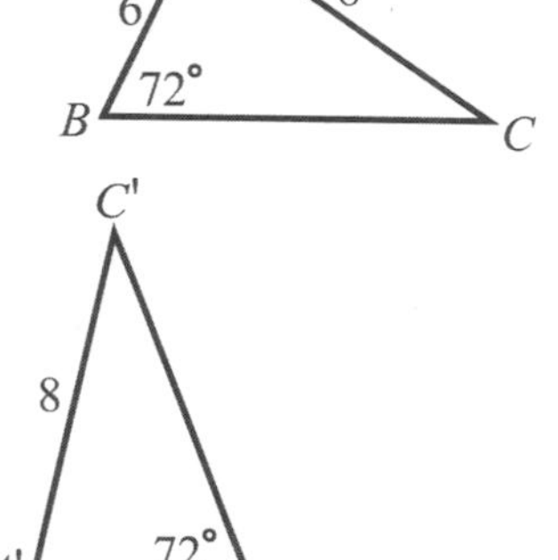

Note: Figure not drawn to scale

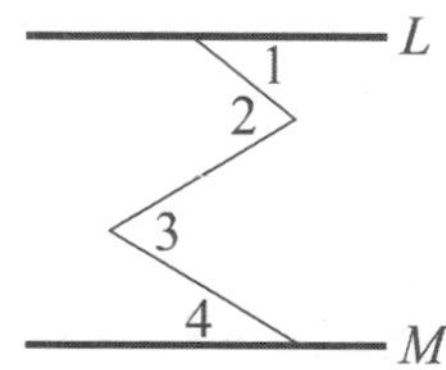

9-06. In the figure above, $L \parallel M$, $\angle 1 = 30°$, $\angle 4 = 40°$, $\angle 2 = 68°$, what is the measure of $\angle 3$?
(Parallel lines)

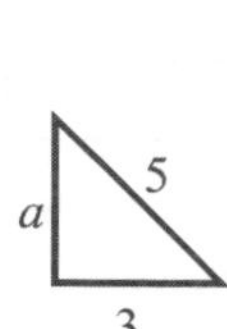

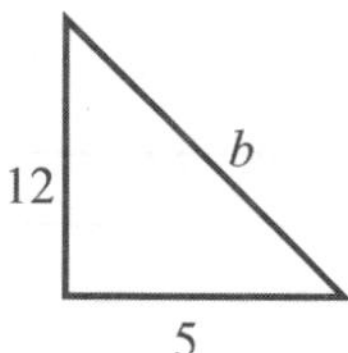

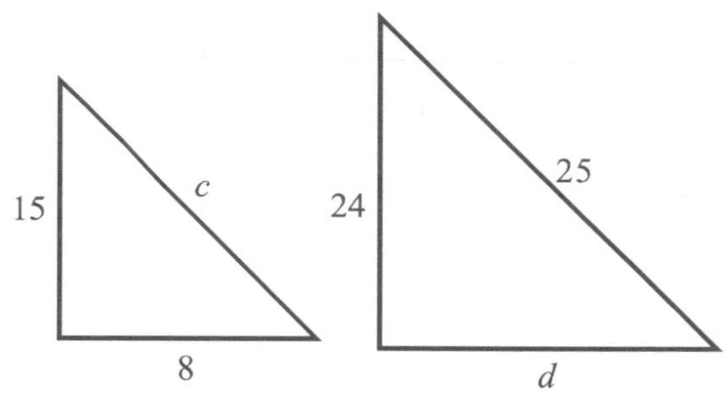

9-07. According to Pythagorean triplets, what is the value of *a*, *b*, *c*, and *d*?
(Pythagorean Theorem)

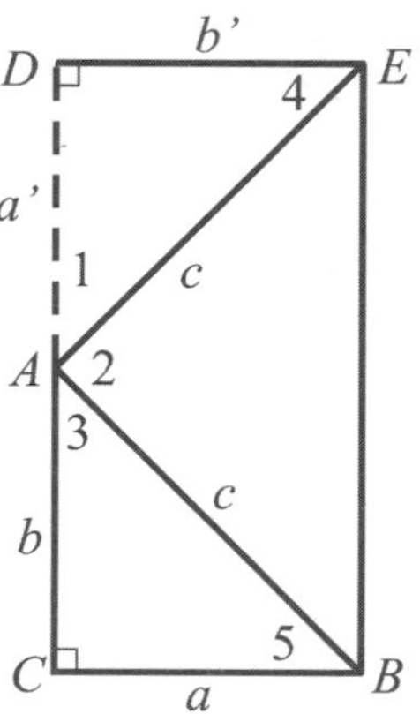

9-08. Here is one way to prove Pythagorean Theorem, please fill-in the reasons for $a^2 + b^2 = c^2$

A. Extend $\overline{CA}$ to D, let $\overline{AD} = a'$
let $\overline{DE} \perp \overline{AD}$, and $\overline{DE} = b'$
$\triangle ADE \cong \triangle ACB$ Reason:______

B. $\angle 2 = 90°$ Reason:______

C. Trapezoid $DEBC$

$$= \triangle ADE + \triangle AEB + \triangle ACB$$

$$\therefore \ \frac{1}{2}(b' + a)(b + a')$$

$$= \frac{1}{2}a'b' + \frac{1}{2}c \cdot c' + \frac{1}{2}ab$$

$$\frac{1}{2}(a+b)^2 = \frac{1}{2}ab + \frac{1}{2}ab + \frac{1}{2}c^2$$

$$(a+b)^2 = c^2 + 2ab$$

$$a^2 + b^2 + 2ab = c^2 + 2ab$$
$$\therefore \quad a^2 + b^2 = c^2$$

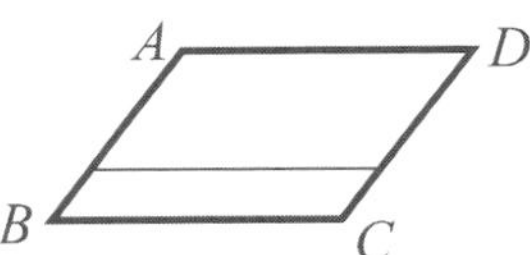

9-09. In the figure above, which of the following does not prove that $\square ABCD$ is a parallelogram?

(Parallelogram)

(A) $\overline{AD} = \overline{DC}$ and $\overline{AD} \parallel \overline{BC}$

(B) $\overline{AD} = \overline{BC}$ and $\overline{AB} = \overline{DC}$

(C) $\overline{AD} = \overline{BC}$ and $\overline{AD} \parallel \overline{BC}$

(D) $\overline{AD} \parallel \overline{BC}$ and $\overline{AB} \parallel \overline{DC}$

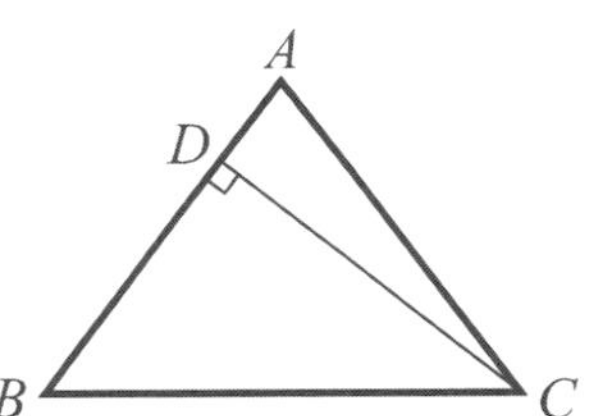

9-10. In the figure, $\overline{AB} = \overline{AC} = 20$, $\overline{BC} = 32$, $\overline{CD} \perp \overline{AB}$, what is the length of $\overline{CD}$?

(Triangle)

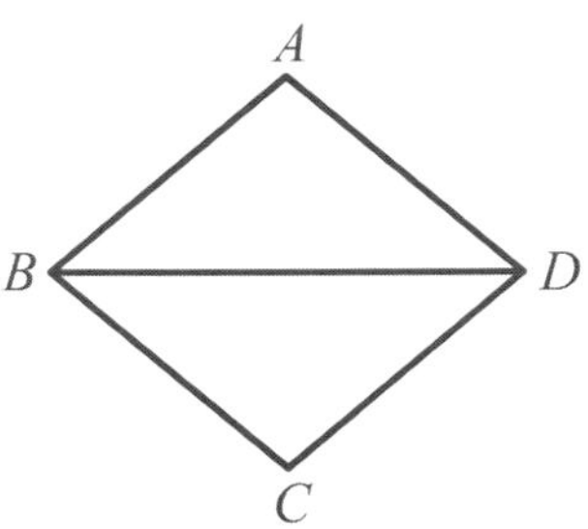

(Not drawn to scale)

9-11. The figure of a rhombus is shown above, If $\overline{BD} = 6$ and $\overline{AD} = 5$, what is the measure of the area of the rhombus *ABCD* above?

(Rhombus)

9-12. Put a check mark in the cell where the statement in each row under the first column is true for the shape specified in each column.

Name	Parallelogram	Rectangle	Square	Rhombus	Kite
Figures					
Two pairs of parallel sides					
Opposite sides are of equal length					
Opposite angles are equal					
4 angles are equal					
4 sides are equal					
Two adjacent sides are of equal length					
The diagonals bisect each other					
The diagonals are perpendicular to each other					
The diagonals are equal of length					

9-13. 1. What is the sum of the internal angles of a 5-sided polygon (pentagon)?

2. What is the sum of the external angles of a 5-sided polygon? (Pentagon)

3. What is the measure of an internal angle of a regular pentagon? (Pentagon)

4. What is the sum of the internal angles of a six sided polygon (Hexagon)?

5. What is the measure of any angle of a regular hexagon? (Polygon)

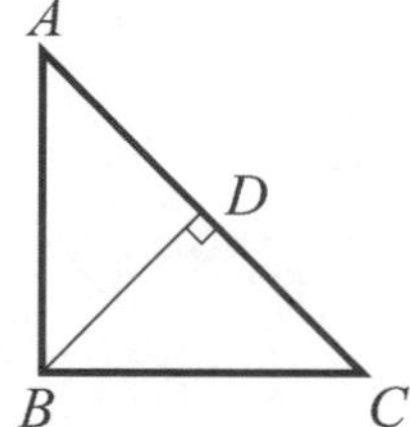

9-14. In the figure above, *ABC* is a right angle, $\overline{BD} \perp \overline{AC}$, $AB = 10$, $AD = 8$. What is the length of $\overline{BC}$? (Right Triangle)

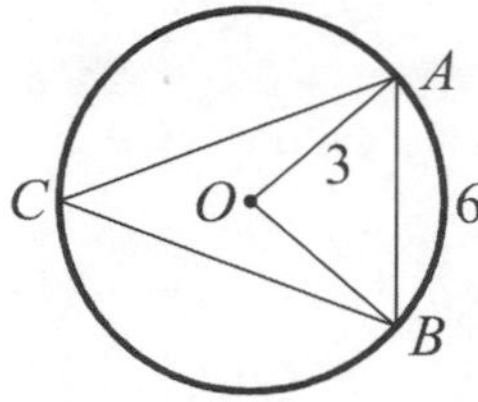

9-15. Circle *O* is shown above. If the radius = 3 and the length of $\overset{\frown}{AB}$ is 6, what is the measure of $\angle AOB$ and $\angle ACB$? (Circle)

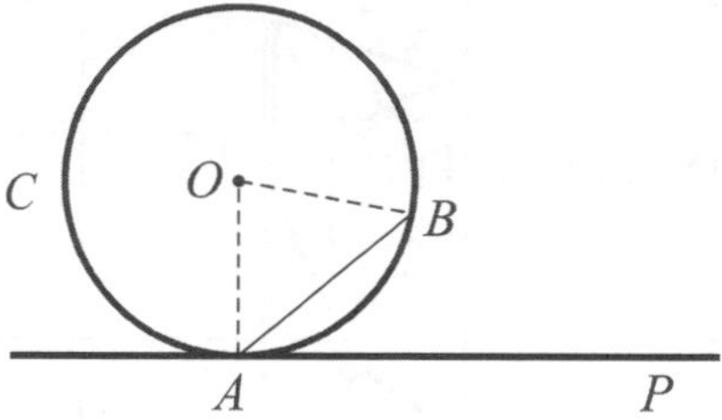

9-16. Circle O is shown above. If $\angle AOB = 80°$, what is the measure of $\angle PAB$? (Circle)

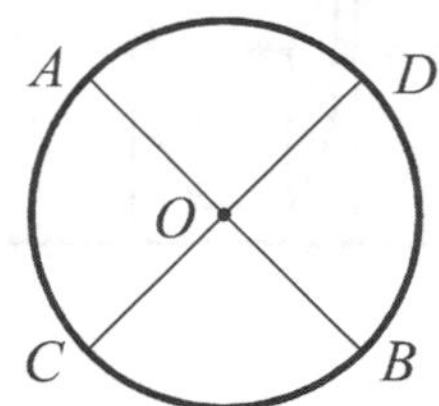

9-17. Circle *O* with center at *O* is shown above. If the measure of $\angle AOC = 60°$ and the radius = 4, what is the length of $\overset{\frown}{AC} + \overset{\frown}{BD}$? (Circle)

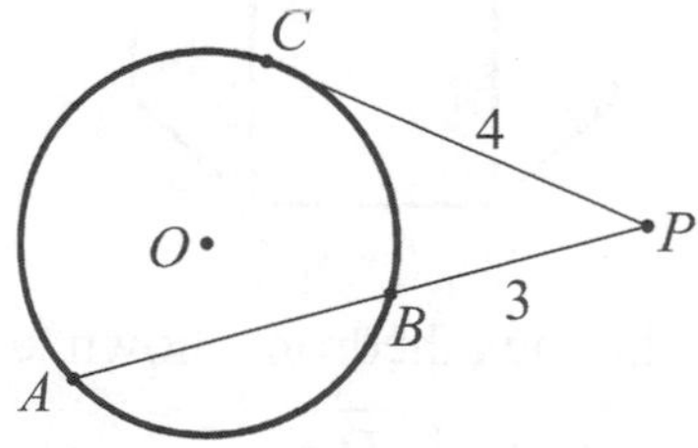

9-18. In the figure above, $\overline{PC}$ is tangent to circle O at point *C*, if $\overline{PC} = 4$, and $\overline{PB} = 3$, what is the length of $\overline{PA}$? (Circle)

9-19. Please fill in the cells in this table

Term	Figures	Vertices	Edges	Faces
tetrahedron				
Pentahedron				
Hexahedron				

(Prisms)

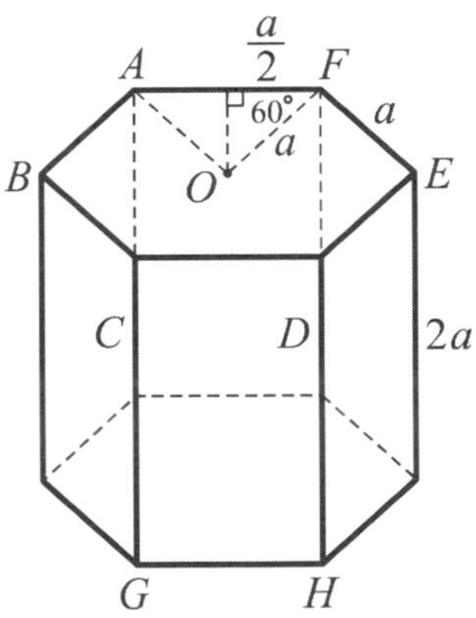

9-20. In the hexahedron shown above, if $\overline{AF} = a$ and $\overline{DH} = 2a$, what is the area and the volume of this hexahedron in terms of a? (Prisms)

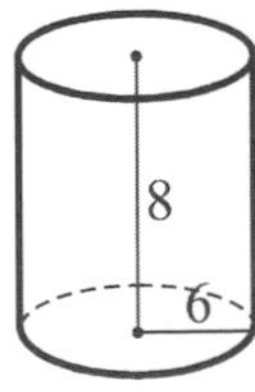

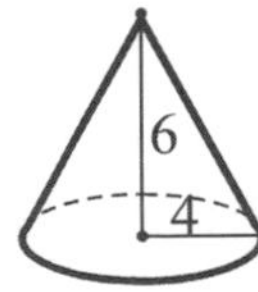

9-21. Two cylinders and one cone are shown above, what are the ratios of their volumes?
(Cylinder/Cone)

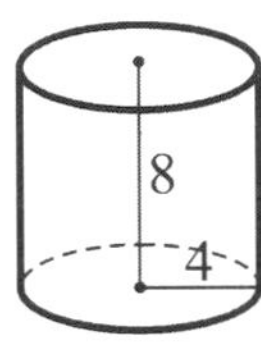

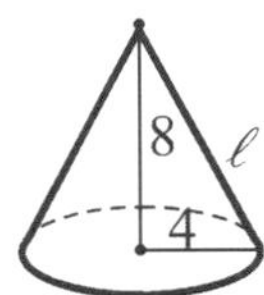

9-22. A cylinder and a cone are shown above, what is the ratio of their surface area?
(Cylinder/Cone)

Concept Check
Chapter 10
Trigonometry

10-01. Which of the following is equivalent to $-750°$? (Negative angle)

(A) 750°

(B) 660°

(C) 330°

(D) 30°

10-02. What is the measurement in degrees that is equivalent to 15 radians? (Radians)

(A) $\frac{\pi}{12}$

(B) $\frac{12}{\pi}$

(C) $\frac{120}{\pi}$

(D) $\frac{1200}{\pi}$

(E) $\frac{2700}{\pi}$

10-03. If the radius of a circle is 6 inches, how many radians is a central angle that has an arc of 3π inches?

10-04. $\cos 300° + \cos 120° = ?$ (Operations)

(A) $\frac{\sqrt{3}}{2}$

(B) $\frac{\sqrt{2}}{3}$

(C) 0

(D) 1

(E) -1

10-05. $\sin(\frac{\pi}{2}+\theta)+\sin(\frac{\pi}{2}-\theta)+\cos(\frac{\pi}{2}+\theta)$ $+\cos(\frac{\pi}{2}-\theta)+\sin(\pi+\theta)+$ $\cos(\pi-\theta)=?$ (Operation)

(A) 0

(B) $\sin\theta$

(C) $\cos\theta$

(D) $\sin\theta-\cos\theta$

(E) $\cos\theta-\sin\theta$

10-06. Evaluate $\frac{\cos 315° + \tan 240°}{(\cot 210°)(\sin 300°)}$.

(Operation)

(A) $\frac{\sqrt{2}}{3}$

(B) $\frac{2\sqrt{2}}{3}$

(C) $-\frac{\sqrt{2}+2\sqrt{3}}{3}$

(D) $\frac{\sqrt{3}+2\sqrt{2}}{3}$

(E) $\frac{\sqrt{2}+2\sqrt{3}}{3}$

10-07. Given $\csc\theta = -\frac{5}{3}$ and $\pi \le \theta \le \frac{3}{2}\pi$, what is the value of $\cot\theta$? (Trigonometric identities and quadrants)

(A) $\frac{4}{3}$

(B) $-\frac{4}{3}$

(C) $\frac{3}{4}$

(D) $-\frac{3}{4}$

(E) None of above

10-08. If $\sin\theta = \frac{3}{5}$ and $\cot\theta < 0$, then $\cos\theta = ?$ (Trigonometric identities and quadrants)

(A) $-\frac{4}{5}$

(B) $\frac{4}{3}$

(C) $-\frac{3}{4}$

(D) $\frac{3}{4}$

(E) None of above

10-09. Evaluate $\dfrac{\frac{1}{\csc x} + \frac{1}{\sec x}}{(\cot x)(\sin x)} = ?$ (operation)

(A) $\sin x$

(B) $\cos x$

(C) $\sin x + 1$

(D) $\tan x + 1$

(E) $\cot x + 1$

10-10. What is the period of $y = \frac{1}{3}\sin(2x + \frac{\pi}{6}) - 2$? (Graph of sine)

(A) $\frac{\pi}{4}$

(B) $\frac{\pi}{3}$

(C) $\frac{\pi}{2}$

(D) π

(E) 2π

10-11. What is the amplitude of $y = \frac{1}{3}\sin(2x + \frac{\pi}{6}) - 2$? (Graph of sine)

(A) $\frac{1}{3}$

(B) $\frac{2}{3}$

(C) $\frac{1}{2}$

(D) $\frac{3}{2}$

(E) 3

10-12. What is the phase shift of $y = \frac{1}{2}\cos(2x + \frac{\pi}{6}) - 2$? (Graph of cosine)

(A) $-\frac{\pi}{12}$

(B) $\frac{\pi}{12}$

(C) $-\frac{\pi}{6}$

(D) $\frac{\pi}{6}$

(E) $\frac{\pi}{3}$

10-13. What is the middle line of $y=\frac{1}{2}\cos(2x+\frac{\pi}{6})-2$? (Middle line)

(A) $y=-\frac{1}{3}$

(B) $y=-3$

(C) $y=-2$

(D) $y=2$

(E) $y=\frac{2}{3}$

10-14. What is the value of $\tan^{-1}(\tan(-2))$? (Inverse trigonometric function)

(A) 2

(B) −2

(C) 1

(D) −1

(E) None of the above

10-15. What is the value of $\cos(\tan^{-1}(-\frac{4}{3}))$? (Inverse trigonometric function)

(A) $-\frac{3}{4}$

(B) $-\frac{3}{5}$

(C) $\frac{3}{4}$

(D) $\frac{3}{5}$

(E) None of above

10-16. $\tan(\sin^{-1}(-\frac{1}{3}))=$? (Inverse trigonometric function)

(A) $-\frac{1}{3}$

(B) $-\frac{\sqrt{2}}{3}$

(C) $\frac{\sqrt{2}}{3}$

(D) $-\frac{\sqrt{2}}{4}$

(E) $\frac{\sqrt{2}}{4}$

10-17. $\tan(\tan^{-1}(-2))=$?

(A) −2

(B) 2

(C) $-\frac{1}{2}$

(D) $\frac{1}{2}$

(E) None of above

10-18. Given $\tan(\theta+\frac{\pi}{6})=\cot(\frac{\pi}{3})$ and $\theta<\frac{\pi}{2}$, what is the value of $\cos\theta$? (Trigonometric equation)

(A) 0

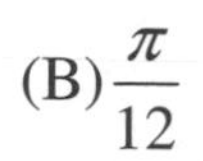

(B) $\frac{1}{2}$

(C) 1

(D) $\frac{\sqrt{2}}{2}$

(E) $\frac{\sqrt{3}}{2}$

10-19. Given $\pi < x < \frac{3}{2}\pi$, solve the equation $2\cot x \sin x + 1 = 0$ (Trigonometric equation)

(A) –60°

(B) 60°

(C) 120°

(D) 300°

(E) None of above

10-20. Given $0 \le \theta \le 2\pi$, how many values of θ satisfy the equation. $2\sec^2\theta + 7\tan\theta + 1 = 0$? (Trigonometric equation)

(A) one

(B) two

(C) three

(D) four

(E) five

10-21. In $\triangle ABC$, $\angle B = 45°$, $b = 3$, and $\angle C = 30°$, what is the value of side C? (Theorem of sine)

(A) $\frac{\sqrt{2}}{2}$

(B) $\sqrt{2}$

(C) $\frac{\sqrt{2}}{2}$

(D) $\frac{3\sqrt{2}}{2}$

(E) $\frac{2\sqrt{3}}{3}$

10-22. In triangle ABC, $a = 6$, $b = 3$, and $c = 5$, what is the value of $\cos B$? (Theorem of cosine)

(A) $\frac{3}{5}$

(B) $\frac{5}{8}$

(C) $\frac{5}{11}$

(D) $\frac{8}{13}$

(E) $\frac{13}{15}$

10-23. What is the measurement of the largest angle in a triangle whose sides have a ratio of 6:5:3? (Theorem of cosine)

10-24. In triangle ABC, $\angle B = 40°$ and $\angle C = 20°$, what is the ratio of a to b? (Theorem of sine)

10-25. If $\triangle ABC$ has the sides $a = 3$, $b = 4$, $c = 5$, what is the area of triangle ABC? (Heron's Formula)

Concept Check
Chapter 12
Function (II)

12-01. Which of the following functions or relations is odd, even, or both even and odd? (Odd/Even Function)

(1) $y = x$

(2) $x^2 + y^2 = 1$

(3) $x - 2y + 1 = 0$

(4) $x - y = 2$

(5) $f(x) = x^5 - 1$

12-02. Which of the following graphs could be the graph of $P(x) = -7x^5 + x^4 - 3x + 1$? (Polynomial Function)

(A)

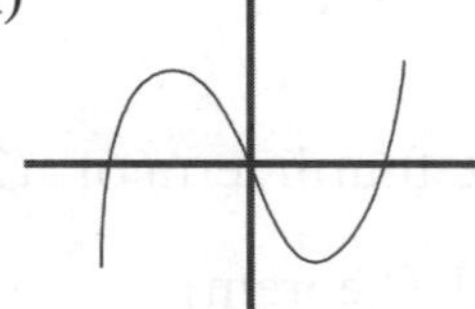

(B)

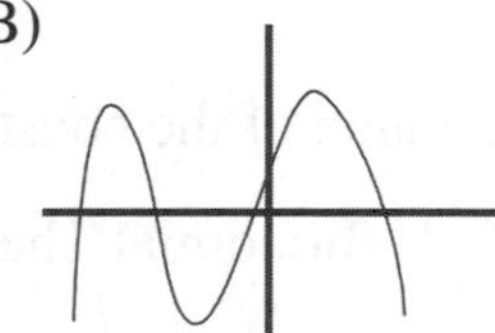

(C)

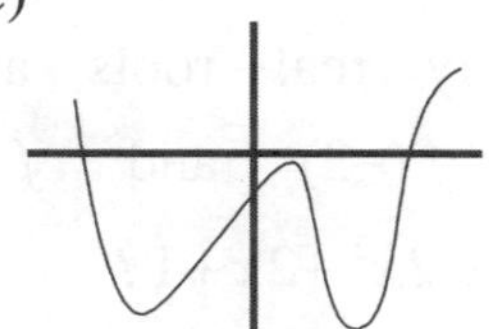

(D)

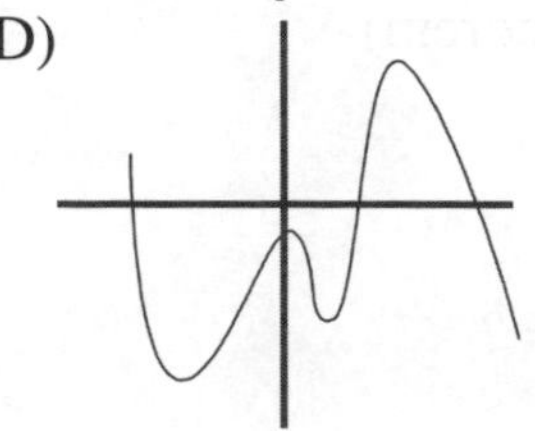

12-03. If $P(x) = x^3 - 2x^2 + 3x - 6$ can be divided by $x + k$, what is the value of k? (Polynomial Function)

(A) 1

(B) 2

(C) −1

(D) −2

12-04. Find the remainder of $x^{15} + 2x^{14} - x^{13} + 3x^2 - 2x + 3$ when it is divided by $x - 1$. (Remainder Theorem)

(A) 0

(B) 4

(C) 6

(D) 8

12-05. If $P(x) = x^3 + ax^2 + bx - 2$ has two factors $(x+1)$ and $(x-2)$, what is the value of $a + b$?
(Remainder Theorem)

(A) 3

(B) −3

(C) 2

(D) −2

12-06. $P(x) = x^5 - 3x^4 + 5x^3 + 5x^2 + 4x - 12$

If the sum of all roots for $P(x)$ is a

and the product of all roots for $P(x)$ is b, what is the value of $a \times b$? (Properties of Roots)

(A) -36

(B) 36

(C) -12

(D) 12

12-07. If the roots of $P(x) = 2x^2 + 5x + 3 = 0$ are α and β, what is the value of $\frac{1}{\alpha^2} + \frac{1}{\beta^2}$? (Properties of Roots)

(A) $\frac{1}{36}$

(B) 36

(C) $\frac{9}{13}$

(D) $\frac{13}{9}$

12-08. Find the possible rational zeros of $f(x) = x^5 + 2x^4 + 3x^3 - 2x^2 - 3x - 2$. (Zeroes)

12-09. If $f(x) = x^3 + 7x^2 + 7x - 15 = 0$ have 3 roots, α, β, and γ, what is the value of $\frac{\alpha\beta\gamma}{\alpha + \beta + \gamma}$? (Zeroes)

12-10. Find the real roots and conjugate roots of $x^3 - 2x^2 + 2x - 1 = 0$. (Conjugate Roots)

12-11. How many positive and negative zeroes are there for polynomial $P(x) = 5x^5 - 4x^4 + 3x^3 + 2x^2 - x + 1$? (Zeroes)

12-12. What is the solution of $P(x) = 2x^3 + 9x^2 + 13x + 6 \geq 0$? Sketch the graph of $P(x) > 0$. (Polynomial Inequality)

12-13. Use graph to show $P(x) = (x-1)^2 [\frac{(x-2)(x+3)}{x-1}] \geq 0$. (Rational Inequality)

12-14. Solve $\frac{(x+1)(x-2)}{(x-1)(x-3)} \geq 0$ $(x \neq 1, x \neq 3)$. (Rational Inequality)

12-15. What is the fourth term of $(2x - y)^5$? (Bionomial Theorem)

12-16. What is the value of the constant term of $(x - \frac{1}{x})^{12}$? (Bionomial Theorem)

12-17. How many real roots are there between $f(-2)$ and $f(2)$ for $P(x) = x^3 + 2x^2 + 2x + 1$? (Root Theorem)

12-18. 1. $2^{\log_2 x} = ?$

2. $\log_2 {2^3} = ?$

3. $\log_2^{3^2} = ?$

4. $\log_8^{2\times 3}$?

5. $\log_2^{\frac{2}{3}} = ?$

6. $\ln^e = ?$

7. $\frac{\log_2^5}{\log_2^3} = ?$

8. $\log_{2^3}^{x^3} = ?$

9.Solve $\log^{(2x+1)} + \log^{(x-2)} = \log^{10}$.

(Logarithm)

12-19. Solve $\ln^{(x+1)} + \ln^{(2x-3)} = \ln^{4x}$.

(Exponential Equation)

12-20. Solve $\log^{(2x+1)} + \log^{(x-2)} = \log^{7}$.

(Logarithmic Equation)

12-21. Sketch the graph of following rational function $f(x) = \dfrac{x^4 + x^3 - 7x^2 - x + 6}{x+3}$.

Indicate on the graph if there are any holes of the function.

(Rational Function)

12-22. What is the graph for parametric equation $x = \sqrt{p} - 3$ and $y = p - 3$?

(Parametric Equation)

12-23. What is the graph for parameters equations $\sin\theta = x$, $\cos\theta = y$?

(Parametric Equation)

12-24. Graph following parametric functions $\begin{cases} x = 2w \\ y = 4w^2 - 8w \end{cases}$.

(Parametric Equation)

12-25. $f(x)$ is a piecewise function formed by a parabola $y = (x-2)^2 - 2$ and a straight line $y = 2x + 1$. Find:

I. Domain

II. Range

III. Graphs

(Piecewise Function)

Concept Check
Chapter 13
Vector

13-01. What is the resultant of $2\vec{a}+3\vec{b}$ if $\vec{a}=(2,3)$ and $\vec{b}=(4,5)$?

13-02. Given $\overrightarrow{AB}=(6,3)$ and point $A=(2,4)$, what is the coordinates of point B?

13-03. Find the norm and unit vector of $\vec{a}=(3,2)$.

13-04. If $|\vec{a}|=2$, $|\vec{b}|=3$, the measure of the angle between $\vec{a}$ and $\vec{b}$ is $60°$, what is the value of $|\vec{a}+3\vec{b}|$?

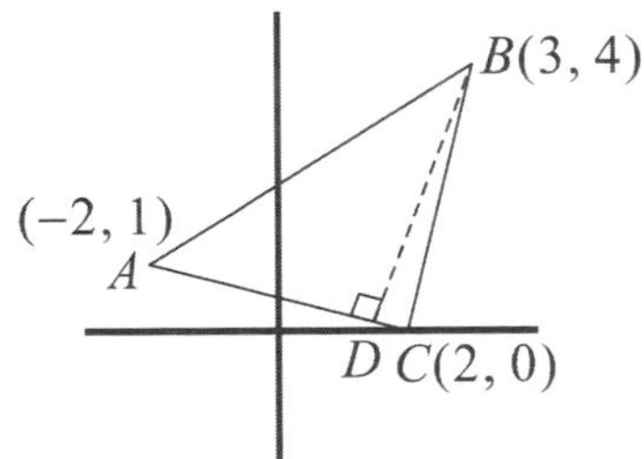

13-05. $\triangle ABC$ is shown above. What is the length of normal projection from $\overrightarrow{AB}$ onto $\overrightarrow{AC}$?

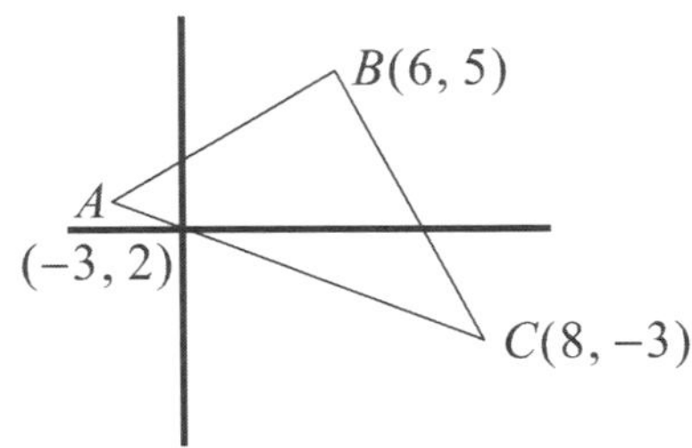

13-06. Triangle ABC is shown above. Use the method of vector to find the area of $\triangle ABC$.

13-07. Find the distance between point $A(-1,3,2)$ and point $B(3,-1,2)$.

13-08. A straight line ℓ in a three-dimensional space passes thorough point $A(2,1,3)$ and point $B(3,-2,1)$. What is the equation of ℓ in parametric form and in proportional form?

13-09. Find the distance from a point (1, 1, 2) to line $\frac{x+1}{2}=\frac{y-1}{-2}=\frac{z+2}{1}$.

13-10. Given that the normal vector of plane E is $\vec{n}=(1,2,3)$ and that point $(1,-1,2)$ is on plane E, what is the equation of plane E?

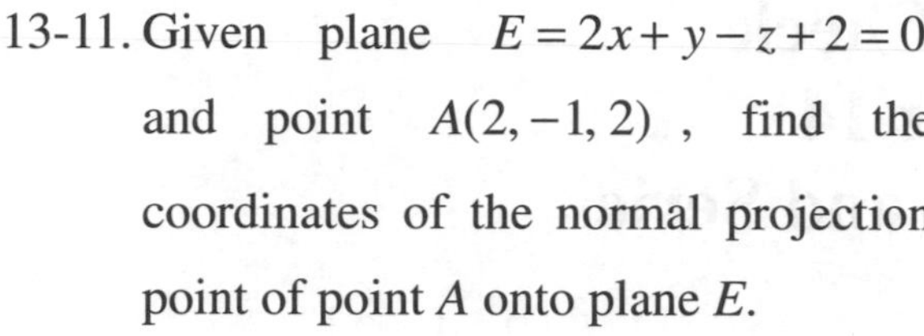

13-11. Given plane $E = 2x + y - z + 2 = 0$ and point $A(2, -1, 2)$, find the coordinates of the normal projection point of point A onto plane E.

13-12. Given plane $E = x + 2y - z + 3 = 0$ and point $A(1, 3, -2)$, what is the distance between point A and plane E?

13-13. Given plane $E_1 = x - 2y + z + 2 = 0$ and plane $E_2 : 2x - 4y + 2z + 8 = 0$ what is the distance between these two parallel planes?

13-14. Given plane $E_1 : x + y - z + 3 = 0$ and plane $E_2 : 2x + y - z + 1 = 0$, if θ is the angle between plane E_1 and plane E_2, what is the value of $\cos\theta$?

Concept Check
Chapter 14
Matrix, Limit, and Series

14-01. Given $A=\begin{bmatrix}1 & 4 & 7\\ 2 & 5 & 8\\ 3 & 6 & 9\end{bmatrix}\begin{matrix}-\text{row}1\\ -\text{row}2\\ -\text{row}3\end{matrix}$, if we exchange row 1 and row 2 to form matrix $B=\begin{bmatrix}2 & 5 & 8\\ 1 & 4 & 7\\ 3 & 6 & 9\end{bmatrix}$, which of the following is correct?

(A) $A=B$

(B) $\det(A)=\det(B)$

(C) $\det(A)=-\det(B)$

(D) $A>B$

14-02. Given $A=\begin{bmatrix}1 & 4 & 7\\ 2 & 5 & 8\\ 3 & 6 & 9\end{bmatrix}$ and $B=\begin{bmatrix}1 & 2 & 3\\ 4 & 5 & 8\\ 7 & 6 & 9\end{bmatrix}$, which of the following is true?

(A) $\det A=\det(B)$

(B) $\det A>\det(B)$

(C) $\det A=-\det(B)$

(D) $\det A<\det(B)$

14-03. Given $A=\begin{bmatrix}-1 & 2\\ 0 & 3\end{bmatrix}$ and $B=\begin{bmatrix}-1 & 3\\ 2 & 1\end{bmatrix}$, what is $3A+2B$?

14-04. Given $A=\begin{bmatrix}2 & 1 & 2\\ 1 & 0 & 2\end{bmatrix}$, $B=\begin{bmatrix}1 & 0 & -2\\ -2 & 1 & 1\\ 1 & -1 & 0\end{bmatrix}$, and $C=\begin{bmatrix}2 & 3 & -2\\ 1 & 0 & 1\end{bmatrix}$, what is $3AB-2CB$?

14-05. Continue from question 4 above, does $3AB-2BC$ make sense? Why?

14-06. Find the inverse matrix of $A=\begin{bmatrix}2 & -1\\ 1 & 0\end{bmatrix}$.

14-07. If A, B, and C are order 2 square matrices and if $AB=\begin{bmatrix}-1 & 2\\ -1 & -2\end{bmatrix}$, $AC=\begin{bmatrix}4 & -1\\ -2 & -1\end{bmatrix}$, and $B+C=\begin{bmatrix}0 & -1\\ 3 & 3\end{bmatrix}$, what is matrix A?

14-08. Use Cramer's method to solve the system equations $\begin{cases}x+y+z=2\\ 2x-y+3z=9\\ x-3y-z=2\end{cases}$.

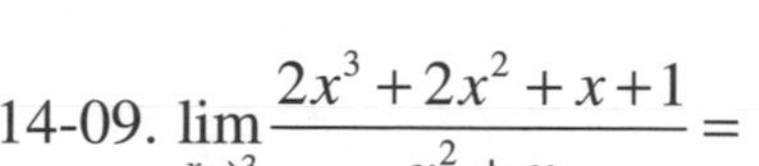

14-09. $\lim_{x \to 3} \frac{2x^3 + 2x^2 + x + 1}{x^2 + x} =$

14-10. $\lim_{x \to 0} \frac{x^2 + x}{x} =$

14-11. $\lim_{x \to 1} \frac{\sqrt{x+1} + \sqrt{x}}{\sqrt{x^2+1} + 1} =$

14-12. If $\lim_{x \to 1} \frac{a\sqrt{x+3} + b}{x - 1} = 2$, what is the value of a?

14-13. $\sum_{n=1}^{100} n =$

14-14. $\sum_{n=1}^{\infty} 2^n =$

14-15. $\sum_{n=1}^{\infty} (\frac{1}{2})^n$

14-16. $\sum_{n=1}^{\infty} \frac{4^n - 2^n - 1^n}{8^n} =$

Answer Key
Chapter 1

1-01. (C) (P.53)

1-02. (D) (P.53)

1-03. (A) (P.55)

1-04. (C) (P.56)

1-05. (C) (P.57)

1-06. (D) (H.C.F) (P.60)

1-07. (D) (L.C.M) (P.60)

1-08. (C) (P.60)

1-09. (D) (P.62)

1-10. (B) (P.64)

1-11. (D) (P.66)

1-12. (C) (P.67)

1-13. 9.5×10^{12} km (P.68)

1-14. (D) (P.68)

1-15. (A) (P.70)

1-16. (D) (P.72)

1-17. $8+8\sqrt{3}i$ (level II) (P.75)

1-18. 16 (level II) (P.76)

1-19. $3\sqrt{2}e^{\frac{\pi}{4}i}$ (level II) (P.77)

1-20. $12e^{\frac{\pi}{12}i}$ (level II) (P.77)

1-21. $\frac{\sqrt{6}}{2}(\cos\frac{7}{4}\pi+i\sin\frac{7}{4}\pi)$ (P.77)

Answer Key
Chapter 2

2-01 (C) (P.93)

2-02 $P \cup N = \{x \mid x \text{ are real numbers}\}$
$P \cap N = \{0\}$ (P.93)

2-03 $CuP = \{2, 4, 6, 8\}$ (P.93)

2-04 (D) (P.94)

2-05 (C) (P.94)

2-06 (D) (P.95)

2-07 2046 (P.97)

2-08 (C) (P.97)

2-09 2 (P.98)

2-10 1 (P.100)

2-11 7 (P.100)

2-12 37 (P.100)

2-13 $A \to F$ (P.104)
$B \to F$

2-14 (C) (P.104)

2-15 (C) (P.107)

Answer Key
Chapter 3

3-01 94 (P.115)

3-02 5.30 (P.116)

3-03 41.25 inches (P.117)

3-04 I. 20 miles/gallon (P.118)
II. 0.05 gallons/mile
III. 0.25 gallons/dollar
IV. 4 dollars/gallon
V 5 miles/dollar
VI $\frac{1}{5}=0.2$ dollars/mile

3-05 (C) (P.119)

3-06 (A) (P.119)

3-07 5.23 ounces (P.121)

3-08 (B) (P.123)

3-09 (A) (P.124)

Answer Key
Chapter 4

4-01 (C) (P.140)

4-02 (C) (P.141)

4-03 $-3a-14b$ (P.143)

4-04 $-3x^3+6x^2+y^2-2y-4\frac{1}{3}$ (P.143)

4-05 a^3-b^3 (P.143)

4-06 Quotient $=4x^2-3x-4$ (P.143)
Remainder $=12x+3$

4-07 $3b(a+b)(a-2b)(2a-b)$ (P.145)

4-08 $(x-1)(x+1)(x-2)$ (P.146)

4-09 $(2x+1)(x-3)$ (P.146)

4-10 $(4x-9)(4x+9)$ (P.146)

4-11 $\frac{9}{7}$ (P.148)

4-12 $\frac{(5-a)}{3}$ (P.149)

4-13 $x=-2$, $y=1$ (P.151)

4-14 $x\le-3$ (P.156)

4-15 $-\frac{1}{2}<x<2$ (P.156)

4-16 $-1\le x\le\frac{7}{3}$ (P.157)

4-17

(P.157)

4-18 $0\le x<1$ (P.158)

4-19 $x=1$, $\frac{1}{3}$ (P.162)

4-20 41 (P.170)

4-21 $3x^2+8x-4$ (P.170)

4-22 domain: $x\ge0$ (P.170)
Range: $y\ge\frac{1}{3}$

4-23 $\frac{x}{3}+\frac{1}{3}$ (P.170)

4-24 (A) (P.172)

Answer Key
Chapter 5

5-01 (B) (P.183)

5-02 (D) (P.183)

5-03 −3 (P.184)

5-04 $x = -3,\ -1, 1$ (P.186)

5-05 $x = 3$ (P.187)

5-06 $x = 1.35$ (P.187)

5-07 no real root (P.188)

5-08 4 (P.191)

5-09 (B) (P.191)

5-10 $\dfrac{(\sqrt{3}+1)(\sqrt{2}+\sqrt{5})}{2}$ (P.196)

5-11 $x = 1$ or $x = \dfrac{25}{9}$ (P.197)

5-12 $x = 0$, or $x = \dfrac{3}{2}$ (P.197)

5-13 $x = 1$ (P.198)

Answer Key
Chapter 6

6-01 $(\frac{7}{2}, -\frac{1}{2})$ (P.206)

6-02 $\sqrt{58}$ (P.207)

6-03 2 (P.209)

6-04 $y = -\frac{2}{3}x - \frac{4}{3}$ (P.212)

6-05 $y = -4(x-3)+2$ (P.212)

6-06 $y = -\frac{1}{3}x + 3$ (P.213)

6-07 $\frac{2}{3}$ (P.219)

6-08 $-\frac{3}{2}$ (P.219)

6-09 No solution (P.219)

6-10 Infinite number of solutions (P.219)

6-11 (P.220)

$x + y > 2$

$3x + 2y > -3$

6-12 (C) (P.224)

6-13 (B) (P.225)

6-14 (1) concave upward (P.237)

(2) Vertex $(-1, -7)$

(3) symmetric equation: $x = -1$

(4) Focus: $(-1, -\frac{55}{8})$

(5) Intersected points with x-axis: $(-2.87, 0)$, $(0.87, 0)$ with y-axis:

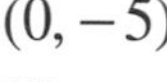
$(0, -5)$

(6)

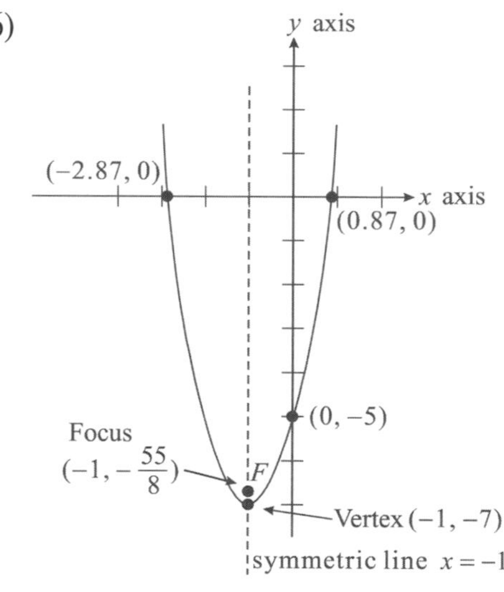

6-15 $(3, 12)$ (P.238)

6-16 17 (P.239)

6-17 $9x^2 + 16y^2 - 54x - 64y + 1 = 0$ (P.244)

6-18 (1) asymptotes: $y = \pm\frac{3}{4}x$ (P.246)

(2) foci: $(5, 0)$, $(-5, 0)$

(3) transverse: 8

(4) conjugate: 6

6-19 $(1, \sqrt{3})$ (P.248)

6-20 $5x^2 + 9y^2 - 12x - 9 = 0$ (P.249)

Answer Key
Chapter 7

7-01 14.7%，71.3% (P.264)

7-02 (1) Average = 452.17 (P.267)

(2) 17%

7-03 (1) Charles (P.269)

(2) Ken

7-04 (1) 5.8 : 1 (P.270)

(2) 24,048

7-05 (1) Median: 66.19 inches (P.277)

(2) Median: 66.5

(3) Mode: 69

(4) Range: 12 inches

7-06 (1) Mean = 39.87 (P.278)

(2) Median = 36

(3) Mode = 15

(4) Range = 66

7-07 (1) Mean = 75.96 (P.280)

(2) Median = 80

(3) Mode = 85

(4) Range = 70

(5) IQR = $85-66=19$

(6) $Q_1=66$

(7) $Q_3=85$

(8) lower outlier = 30

7-08 (P.283)

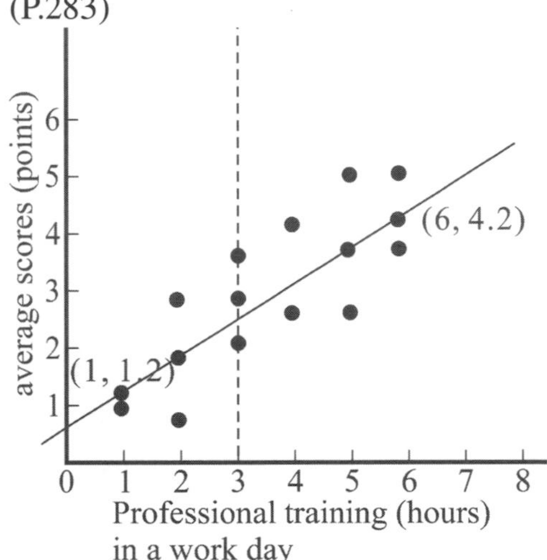

(1) See graph above for the line of best fit.

(2) $y=0.6x+0.6$

(3) 0.6 points

(4) 4.8 points

7-09 (1) percentile = 29% (P.288)

(2) 78%

7-10 (B) (P.293)

7-11 12 (P.294)

7-12 (A) (P.295)

7-13 [0.64, 0.70] (P.299)

Answer Key
Chapter 8

8-01 8 (P.313)

8-02 60 (P.313)

8-03 15 (P.315)

8-04 $\frac{1}{3}$ (P.317)

8-05 120 (P.320)

8-06 60 (P.322)

8-07 144 (P.325)

8-08 $\frac{1}{3}$ (P.330)

8-09 0.3 (P.331)

8-10 0.02 (P.334)

Answer Key
Chapter 9

9-01 19 (P.345)

9-02 120° (P.348)

9-03 75° (P.350)

9-04 (C) (P.351)

9-05 (1) Congruent, S.S.S (P.351)
(2) Congruent, S.A.S (P.351)
(3) Congruent, A.S.A (P.351)
(4) Congruent, A.A.S (P.351)
(5) Not Congruent, A.A.A (P.351)
(6) Not Congruent, S.S.A (P.351)

9-06 78° (P.352)

9-07 $a=4$ (P.352)
$b=13$
$c=17$
$d=7$

9-08 see (P.352)
(1) $\triangle ADE \cong \triangle ACB$ (S.A.S)
(2) $\angle 3 = \angle 4$,
$m\angle 2 = 180° - m(\angle 1 + \angle 4) = 90°$

9-09 (A) (P.362)

9-10 19.2 (P.363)

9-11 24 (P.363)

9-12 See table on page (P.366)

9-13 (1) 540° (P.366)
(2) 360°
(3) 108°
(4) 720°
(5) 120°

9-14 7.5 (P.371)

9-15 $\angle AOB = 2$ (Radian) (P.375)
$\angle ACB = 1$ (Radian)

9-16 40° (P.375)

9-17 $\frac{8}{3}\pi$ (P.375)

9-18 $\frac{16}{3}$ (P.375)

9-19 (P.384)

Term	Figures	Vertices	Edges	Faces
Tetrahedron		6	9	5
Pentahedron		10	15	7
Hexahedron		12	18	8

9-20 area $= 3(\sqrt{3}+4)a^2$ (P.385)

Volume $= 3\sqrt{3}a^3$

9-21 72 : 9 : 8 (P.391)

9-22 96 : 51.76 (P.392)

Answer Key
Chapter 10

10-01 (C) (P.406)

10-02 (E) (P.408)

10-03 $\frac{\pi}{2}$ (P.408)

10-04 (C) (P.413)

10-05 (D) (P.413)

10-06 (C) (P.415)

10-07 (A) (P.415)

10-08 (A) (P.416)

10-09 (D) (P.416)

10-10 (D) (P.423)

10-11 (A) (P.424)

10-12 (A) (P.424)

10-13 (C) (P.424)

10-14 (E) (P.426)

10-15 (D) (P.426)

10-16 (D) (P.427)

10-17 (A) (P.427)

10-18 (C) (P.434)

10-19 (E) (P.434)

10-20 (D) (P.435)

10-21 (D) (P.436)

10-22 (E) (P.436)

10-23 $\cos^{-1}(-\frac{1}{15})$ (P.436)

10-24 1.36 (P.437)

10-25 6 (P.437)

Answer Key
Chapter 12

12-01 (1) $y = x \Rightarrow$ (Odd) (P.478)
(2) $x^2 + y^2 = 1$ (Even)
(3) $x - 2y + 1 = 0$ (Neither odd nor even)
(4) $x - y = 2$ (Neither odd nor even)
(5) $f(x) = x^5 - 1$ (Neither odd nor even)

12-02 (D) (P.481)

12-03 (D) (P.481)

12-04 (C) (P.482)

12-05 (B) (P.482)

12-06 (B) (P.484)

12-07 (D) (P.484)

12-08 no rational zeros (P.486)

12-09 $-\frac{15}{7}$ (P.487)

12-10 Real Root $x = 1$ (P.489)
conjugate Root $x = \frac{1 \pm \sqrt{3}i}{2}$

12-11 Could have 4, 2, 0, positive zeros and 1 negative zero (P.490)

12-12 $\{-2 < x < -\frac{3}{2}\} \cup \{-1 < x\}$ (P.491)

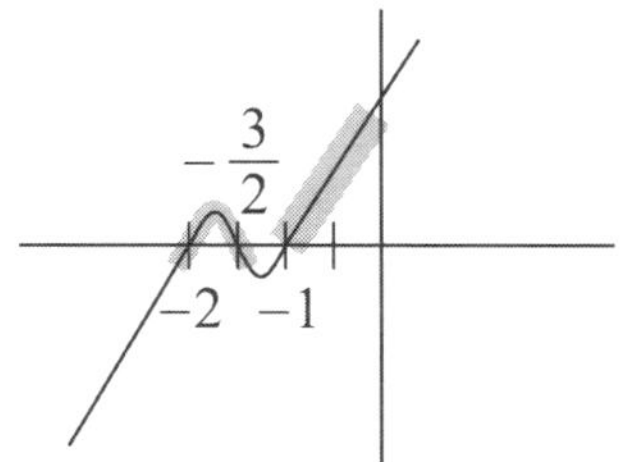

12-13 $\{-3 \le x < 1\} \cup \{2 \le x\}$ (P.492)

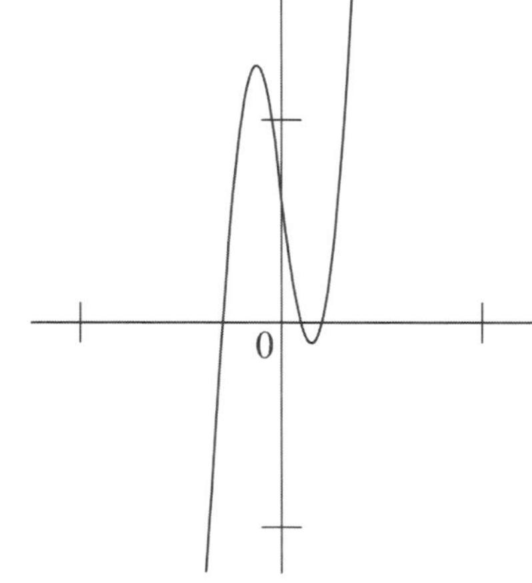

12-14 $\{x \le 1\} \cup \{1 < x \le 2\} \cup \{3 < x\}$ (P.493)

12-15 $-40x^2y^3$ (P.495)

12-16 924 (P.495)

12-17 one or three (P.495)

12-18 1.x (P.501)

2.3

3. $2\log_2^3$

4. $\log_8^2+\log_8^3$

5. $1-\log_2^3$

6.1

7. $\log_3^5$

8. $\log_2^x$

9. $x=3$

12-19 $x=3$ (P.502)

12-20 $x=3$ (P.502)

12-21 $f(x)=\dfrac{x^4+x^3-7x^2-x+6}{x+3}$ (P.508)

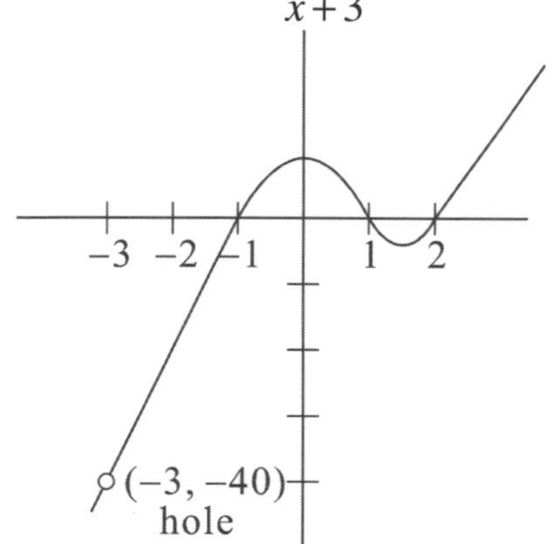

12-22 parabola (P.511)

12-23 circle (P.511)

12-24 (P.512)

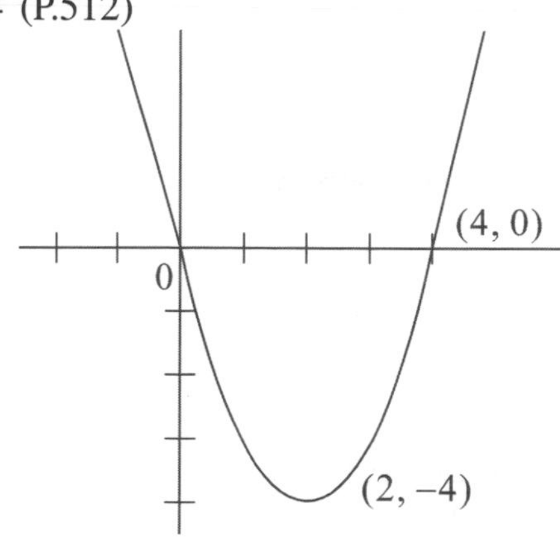

It is a parabola with the vertex at $(2,-4)$

12-25 (P.514)

I. Domain: All real numbers,

II. Range: all real numbers

III. Graph:

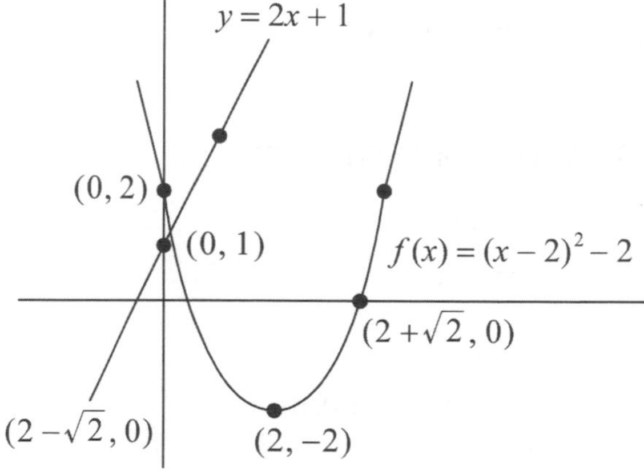

Answer Key
Chapter 13

13-01 (16, 21) (P.525)

13-02 (8, 7) (P.525)

13-03 $\frac{(3, 2)}{\sqrt{13}}$ (P.525)

13-04 10.15 (P.529)

13-05 $\sqrt{17}$ (P.530)

13-06 39 (P.532)

13-07 $4\sqrt{2}$ (P.535)

13-08 parametric Form: (P.535)

$x = 2 + t$

$y = 1 - 3t$

$z = 3 - 2t$

proportional Form:

$\frac{x-2}{1} = \frac{y-1}{-3} = \frac{z-3}{-2}$

13-09 3.60 (P.538)

13-10 E: $x + 2y + 3z - 5 = 0$ (P.539)

13-11 $P(1, -\frac{3}{2}, \frac{5}{2})$ (P.541)

13-12 $D = 4.90$ (P.543)

13-13 $D = \frac{\sqrt{6}}{3}$ (P.544)

13-14 ±0.94 (P.545)

Answer Key
Chapter 14

14-01 (C) (P.561)

14-02 (A) (P.561)

14-03 $\begin{bmatrix} -5 & 12 \\ 4 & 11 \end{bmatrix}$ (P.564)

14-04 $\begin{bmatrix} 18 & -13 & -7 \\ 5 & -4 & -2 \end{bmatrix}$ (P.565)

14-05 It is undefined. (P.566)

14-06 $A^{-1} = \begin{bmatrix} 0 & 1 \\ -1 & 2 \end{bmatrix}$ (P.566)

14-07 $A = \begin{bmatrix} 2 & 1 \\ 0 & -1 \end{bmatrix}$ (P.567)

14-08 $x = 1$ (P.575)
$y = -1$
$z = 2$

14-09 $\frac{19}{3}$ (P.590)

14-10 1 (P.590)

14-11 1 (P.590)

14-12 $a = 8$ (P.591)

14-13 5050 (P.597)

14-14 Goes to infinity (P.598)

14-15 1 (P.598)

14-16 $\frac{11}{21}$ (P.598)

CHAPTER 1
NUMBERS and OPERATIONS

1-1 CLASSIFICATION OF NUMBERS

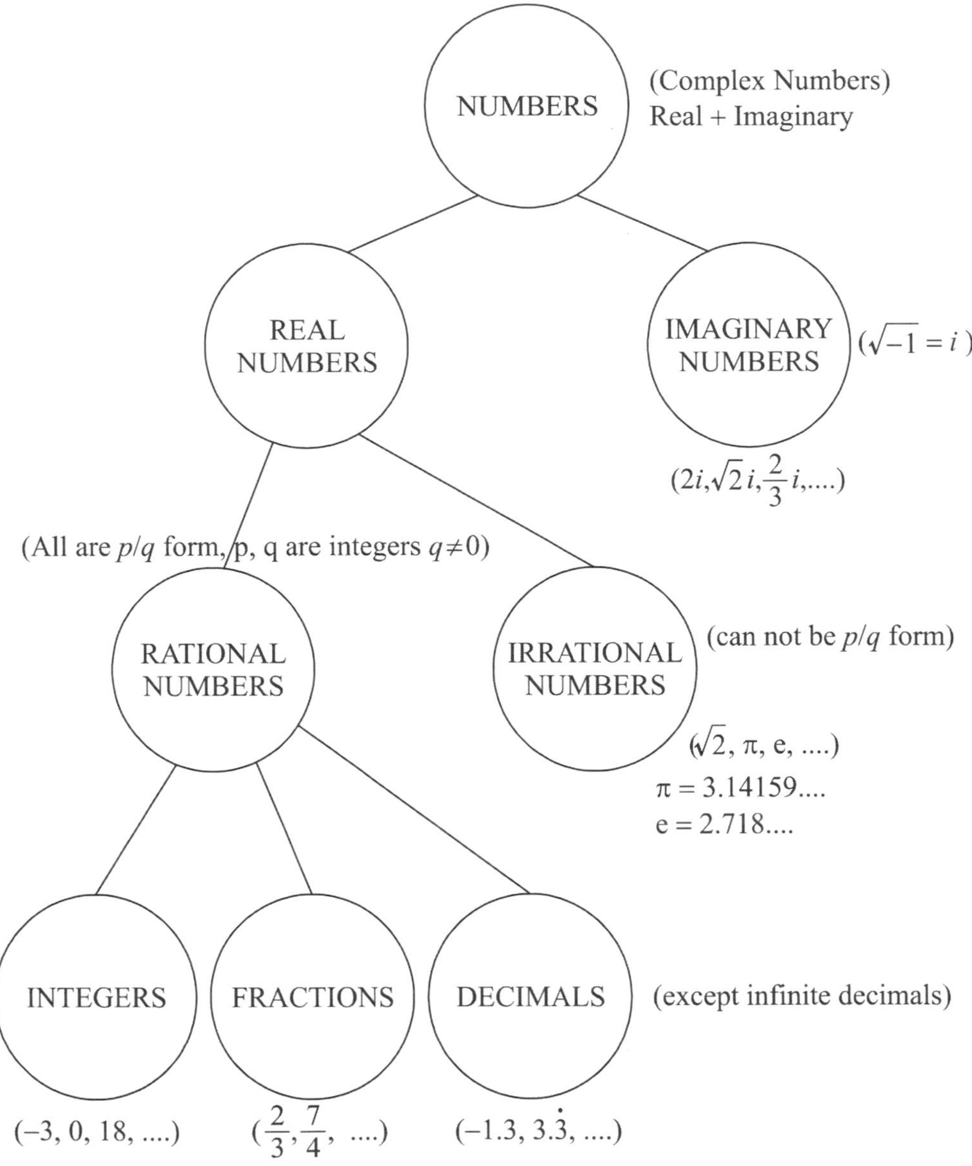

- **Complex Numbers:**

  Real part + imaginary part Such as $3+2i$, $\sqrt{2}+\sqrt{3}i$, ...

- **Rational Numbers:**

  Any real number which can be re-written as p/g form, where p, q are integers, and $q \neq 0$, is a rational number.

Such as: -8, $3\frac{1}{2}$, 3.58, $\sqrt{9}$, $-3.\dot{3}$, ...,

can be converted to $\frac{p}{q}$

forms such as

$-\frac{8}{1}$, $\frac{7}{2}$, $\frac{358}{1000}$, $-\frac{10}{3}$, ...

★Tip

(let $0.\dot{3}=x$, then $3.\dot{3}=10x$)

$10x-x=3.\dot{3}-0.\dot{3}=3$

$\Rightarrow 9x=3$. $x=\frac{3}{9}$

$\Rightarrow \frac{1}{3}$ $\quad -3.\dot{3}=-(3+\frac{1}{3})=-(\frac{10}{3})$

Answer to Concept Check 1-01

Irrational Number means that the number can't be converted to the form of p/q.

So,

(A) $\frac{7}{3}$, $P=7$, $q=3$, rational

(B) $-\sqrt{9} \Rightarrow -3 \Rightarrow -\frac{3}{1}$, $P=-3, q=1$, rational

(C) $\sqrt{3} \Rightarrow 1.7320508076...$

Can not be converted to p/q form so, it is an irrational number

(D) $3.\dot{3} \Rightarrow 3\frac{3}{10-1} \Rightarrow 3\frac{3}{9} \Rightarrow 3\frac{1}{3} \Rightarrow \frac{10}{3}$ $\quad P=10$, $q=3$ rational number

Answer is (C)

Answer to Concept Check 1-02

(A) $\sqrt{2} \Rightarrow 1.4142135624$ ....irrational

(B) $e \Rightarrow 2.71828$ ....irrational

(C) $\pi \Rightarrow 3.14159$ ....irrational

(D) $-\sqrt{\frac{1}{9}} \Rightarrow -\frac{1}{3}$ ....rational

Answer is (D)

1-2 NUMBER LINE

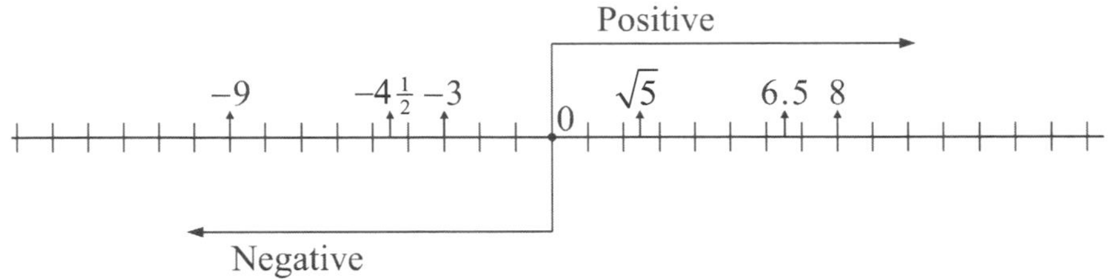

- Every real number can be positioned on the number line. The positive numbers are positioned on the right side of "zero", and the negative numbers are on the left side of "zero".

  Number "o" is neither " positive " nor "negative".

◆Example:

In the figure above, point A, B, and C lie on the same number line in that order, and all the tick marks are equally spaced. What is the value of B?

The distance from point A to point C is $11-(-5)=16$ units. And there are only 8 tick marks, so, each tick mark is equal to 2 units.

From point A to point B, there are only 2 tick marks, so they are $2\times2=4$ units apart.

The value of point B is $-5+4=-1$.

◆Example:

The Value of point A can be represented as $A(x)$, where x is the Value of point A.

There are three points $P(3)$, $Q(-2)$, and $R(x)$, if $QR=2PQ$. What value could be x?

$R'(x)$ $Q(-2)$ $P(3)$ $R(x)$

Since $PQ=|-2-3|=|-5|=5$

$QR=2\times5=10$ $|x-(-2)|=10$ $|x+2|=10$

$x+2=10$ or $x+2=-10$

x could be 8, or -12

> ★Tip
>
> $\overline{PQ}$ represents the segment from point P to point Q.
>
> PQ represents the length of segment $\overline{PQ}$

Answer to Concept Check 1-03

$$QM = PM - PQ$$
$$= \frac{1}{2}PR - \frac{1}{3}PR \qquad (PQ + 2PQ = PR)$$
$$= (\frac{1}{2} - \frac{1}{3})PR = (\frac{3-2}{6})PR$$
$$= \frac{1}{6}PR$$

Answer is (A)

1-3 INTEGERS

- An integer could be in the form of a fraction (like 3/1 $\Rightarrow$ 3), or in the form of a decimal (such as 3.0 $\Rightarrow$ 3)
- Whole Number: {0, 1, 2, 3, 4, ……..} all are non-negative integers.
- Natural Numbers: also called counting numbers: {1, 2, 3, 4, 5, ……..} may include zero "o", There is no universal agreement.
- Odd Numbers: Such as – 5, – 3, – 1, 1, 7, ……
- Even Number: Such as – 4, – 2, "0", 8, 12, 14, ……
  "0" is neither positive nor negative, but it is an even number.
- Rules

Addition and Subtraction

even $\pm$ even $\Rightarrow$ even even $\pm$ odd $\Rightarrow$ odd odd $\pm$ odd $\Rightarrow$ even

Multiplication

$\text{even} \times \text{even} \Rightarrow \text{even}$
$\text{even} \times \text{odd} \Rightarrow \text{even}$
$\text{odd} \times \text{odd} \Rightarrow \text{odd}$

Division

There is no rule for division.

The result could be any number, even it could be not an integer

◆Example:

Given 3 and 7

then $3+7=10$ (even)

$3-7=-4$ (even)

$3\times 7=21$ (odd)

$3\div 7=\frac{3}{7}$ (not an integer)

◆Example:

Given 3 and 8

then $3+8=11$ (odd)

$3-8=-5$ (odd)

$3\times 8=24$ (even)

$3\div 8=\frac{3}{8}$ (not an integer)

★Tip

A. '1' is a multiple of '0'. (False)
Since 1/0 is undefined.

B. '1' is a factor of '0' (True)
Since $0/1=0 \Rightarrow Q+0$
(Quatient$=0$)(Remainder$=0$)

C. '0' is a multiple of '1'. (True)
Since $0/1=0$.

D. '0' is a factor of '1'. (False)
Since 1/0 is undefined.

Answer to Concept Check 1-04

$x-8$ is an odd number, then

(A) $-8-8=-16$ it is an even number

(B) $0-8=-8$ it is also an even number

(C) $1-8=-7$ it is an odd number

(D) $8-8=0$ it is an even number

Answer is (C)

Answer to Concept Check 1-05

(A) $k+4$ such as $3+4=7$ odd

(B) $k-4$ such as $3-4=-1$ odd

(C) $k\times4$ such as $3\times4=12$ even

(D) $k\div4$ such as $\frac{3}{4}$ not an integer

Answer is (C)

1-4 FACTORS, MULTIPLES, and PRIME NUMBERS

A. Factors

All the numbers that can be evenly divided by a certain number are(is) called the factors of this number.

B. Multiples

If a number can be evenly divided by another number, then this number is a multiple of another number.

Given two numbers a and b, if

$\frac{a}{b}=Q+"0"$, then "a" is a multiple of "b", "b" is a factor of "a".

• Properties of "1" and "0"

A. "1" is a factor of any integer, and any integer is the multiple of "1".

B. "0" is a multiple but, not a factor of any non-zero integer, any non-zero integer is a factor of "0".

For instance,

$\frac{5}{1}=5+0$ evenly divisible 5, so, "1" is a factor of "5" then, "5" is a multiple of "1".

C. Prime Number

An integer greater than "1" and only has two factors "1" and itself, it is called "prime number".

• The first 10 prime number are: 2, 3, 5, 7, 11, 13, 17, 19, 23, 29

• All prime numbers are odd but "2".

• "1" is not a prime number. Since it is not greater than "1".

D. Prime Factoring

◆Example:

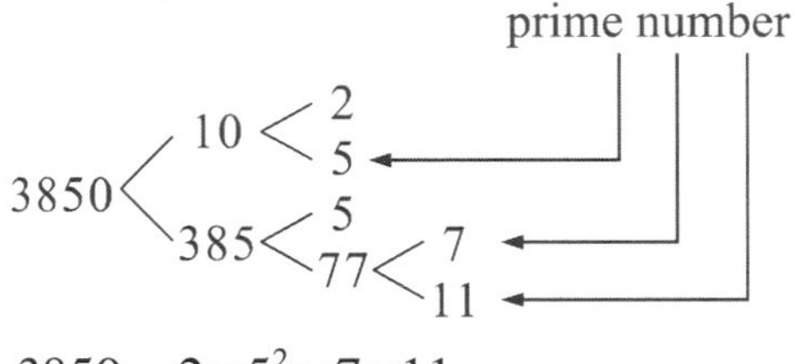

$3850 = 2\times5^2\times7\times11$

E. The Number of Factors

If $p=a^m\times b^n$ (a, b are prime numbers)

then total number of factors of $p=(m+1)\times(n+1)$

Example: Find the number of factors of 72.

72 < 2, 36 < 2, 18 < 2, 9 < 3, 3

$72=2\times2\times2\times3\times3=2^3\times3^2$

The number of factors $=(3+1)\times(2+1)=4\times3=12$

$2^3=2^0,\ 2^1,\ 2^2$, and $2^3=1, 2, 4, 8$

$3^2=3^0,\ 3^1$, and $3^2=1, 3, 9$

3

4

	1	3	9
1	1	3	9
2	2	6	18
4	4	12	36
8	8	24	72

⇒ 12 number ⇒ 12 factors ⇒ $(3+1)\times(2+1)$, $4\times3=12$

F. G.C.F: Greatest Common Factor

◆Example:

For 8 has the factors of 1, 2, 4, 8. For 12 has the factors of 1, 2, 3, 4, 6, 12

The common factors are: 1, 2, 4

The greatest factor of the common factors is "4". So, the G.C.F of 8 and 12 is "4".

• One way to get G.C.F. (Greatest Common Factor)

◆Example:

G.C.F of 140 and 3850

$$140 \left\langle \begin{matrix} 10 < \begin{matrix} 2 \\ 5 \end{matrix} \\ 14 < \begin{matrix} 2 \\ 7 \end{matrix} \end{matrix} \right. = 2^2 \times 5 \times 7$$

$3850 = 2 \times 5^2 \times 7 \times 11$

$140 = \boxed{2} \times 2 \times \boxed{5} \times \boxed{7}$

$3850 = \boxed{2} \times 5 \times \boxed{5} \times \boxed{7} \times 11$

$\therefore$ G.C.F of 140 and 3850 is $2 \times 5 \times 7 = 70$

G. Common Multiple

Example: "20" is a multiple of "2"

"20" is also a multiple of "5"

Therefore "20" is a common multiples of "2" and "5".

H. Least Common multiple L.C.M.

The numbers of common multiple are unlimited, but there is only one "least common multiple". It is so called "LCM".

• How to find L.C.M.

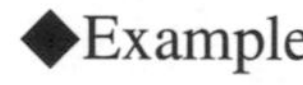 **Step 1.** Find the prime factors of this number

◆Example

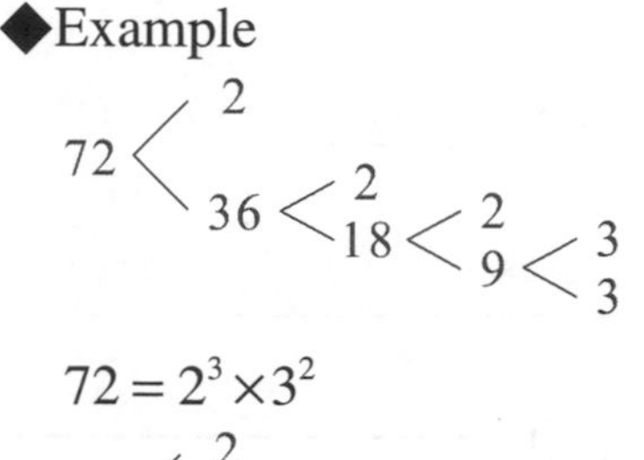

$$72 \left\langle \begin{matrix} 2 \\ 36 < \begin{matrix} 2 \\ 18 < \begin{matrix} 2 \\ 9 < \begin{matrix} 3 \\ 3 \end{matrix} \end{matrix} \end{matrix} \end{matrix} \right.$$

$72 = 2^3 \times 3^2$

$$88 \left\langle \begin{matrix} 2 \\ 44 < \begin{matrix} 2 \\ 22 < \begin{matrix} 2 \\ 11 \end{matrix} \end{matrix} \end{matrix} \right.$$

$88 = 2^3 \times 11$

 **Step 2.** Multiply all prime factors. (The common factors treated as one term)

L.C.M. $= 2^3 \times 3^2 \times 11$

Answer to Concept Check 1-06

The prime factoring of 252

$$252 \to ②,\ 126 \to 2,\ 63 \to ③,\ 21 \to ③,\ ⑦$$

$252 = ② \times 2 \times ③ \times ③ \times ⑦$

$$126 \to ②,\ 63 \to ③,\ 21 \to ③,\ ⑦$$

$126 = ② \times ③ \times ③ \times ⑦$

Common Factors: $2\times3\times3\times7$

G.C.F (greatest common factor) is $2\times3\times3\times7=126$

Answer is (D)

Answer to Concept Check 1-07

Prime Factoring

$$186 \to 2,\ 93 \to 3,\ 31$$ ② × ③ × 31

$$72 \to 2,\ 36 \to 2,\ 18 \to 2,\ 9 \to 3,\ 3$$ ② × 2 × 2 × ③ × 3

L.C.M. = ② × ③ × 31 × 2 × 2 × 3 = $2\times2\times2\times3\times3\times31 = 2,232$

Answer is (D)

Answer to Concept Check 1-08

~~−63~~, ~~−62~~, ⋯, ‘0’, ~~1~~, ⋯, ~~63~~, 64

$63+63+1+1=128$

(−63) (+63)

Answer is (C)

1-5 FRACTION

What is a fraction?

- The form of rational numbers represented by $\frac{p}{q}$

(When p, q are integer and $q \neq 0$), such as, $\frac{5}{1}, \frac{2}{5}, \frac{8}{3}, \ldots\ldots$

- The form of Division $\frac{6}{2}$

- A part of a whole

$\frac{3}{8}$ represents 3 parts of 8 equal parts it is usually read, “three eighths”

- Ratio, Rate

Such as, the ratio is 3 to 4.

Can be written as $\frac{3}{4}$.

*The denominator is never “zero”.

For every 8 students, there are 3 girls.

Girls are $\frac{3}{8}$ of students.

A. Addition and Subtraction of Fraction

◆Example:

* The denominators must be the same

$\frac{2}{3}+\frac{3}{4}$

Change $\frac{2}{3} \Rightarrow \frac{2\times 4}{3\times 4}=\frac{8}{12}$ ($3\times 4=12$ is the L.C.M. of 3 and 4)

Change $\frac{3}{4} \Rightarrow \frac{3\times 3}{4\times 3}=\frac{9}{12}$

Then $\frac{2}{3}+\frac{3}{4}=\frac{8}{12}+\frac{9}{12}=\frac{17}{12}$

B. Multiplication and Division

◆Example:

$\frac{2}{3}\times\frac{3}{4}=\frac{2\times 3}{3\times 4}=\frac{6}{12}=\frac{1}{2}$

numerator × numerator
⇒ new numerator
denominator × denominator
⇒ new denominator

$\frac{2}{3}\div\frac{3}{4}=\frac{2}{3}\times\frac{4}{3}=\frac{2\times4}{3\times3}=\frac{8}{9}$

$\frac{4}{3}$ is the reciprocol of $\frac{3}{4}$

- Proper fraction:

Such as: $\frac{3}{4}$ (numerator is less than denominator)

- improper fraction

Such as: $\frac{4}{3}$ (numerator is greater than denominator)

- mixed number

It is made of a whole number and a fraction

Such as: $3+\frac{2}{3} \Rightarrow 3\frac{2}{3} \Rightarrow \frac{3\times3+2}{3} \Rightarrow \frac{11}{3}$

- lowest terms(simpliest form)

Such as: $\frac{16}{24} \Rightarrow \frac{2\times\cancel{8}}{3\times\cancel{8}} \Rightarrow \frac{2}{3}$

Answer to Concept Check 1-09

(A) $(\frac{3}{8}+\frac{1}{4})+\frac{1}{3}=\frac{3+2}{8}+\frac{1}{3}=\frac{5\times3+8\times1}{24}=\frac{15+8}{24}=\frac{23}{24}$

(B) $(\frac{2}{3}\times\frac{1}{5})-\frac{1}{3}=\frac{2}{15}-\frac{1}{3}=\frac{2-5}{15}=-\frac{3}{15}=-\frac{1}{5}$

(C) $(\frac{1}{2}\div\frac{1}{3})\times\frac{1}{3}=(\frac{1}{2}\times\frac{3}{1})\times\frac{1}{3}\cdot=\frac{1}{2}\times\frac{\cancel{3}^{1}}{1}\times\frac{1}{\cancel{3}_{1}}=\frac{1}{2}$

(D) $(2\frac{1}{2}-\frac{3}{4})\div\frac{1}{3}=(\frac{5}{2}-\frac{3}{4})\div\frac{1}{3}=(\frac{10}{4}-\frac{3}{4})\times\frac{3}{1}=\frac{7}{4}\times\frac{3}{1}=\frac{21}{4}$

$\frac{21}{4}>\frac{23}{24}>\frac{1}{2}>-\frac{1}{5}$

So, (D) the greatest Value

Answer is (D)

- Two-way Classification

Use a table or a chart to show the classifications based on different categories.

◆Example:

In North New point High School, boys are 5/9 of all students and 3/4 of them are 16 years old or older. If 3/5 of all girls' age are greater or equal to 16, What fraction of all students is the girls whose age are less than 16 years old?

Two steps to solve this kind of problem.

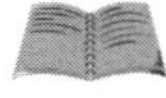 **Step 1.** Draw a two dimensional table and fill-in the information which the question already provided or hinted.

	boys	girls	total
age ≥ 16	A	D	G
age < 16	B	E	H
Total	C	F	I

 **Step 2.** According to the question,

$$C=\frac{5}{9},\quad F=1-\frac{5}{9}=\frac{4}{9}$$

$$A=\frac{5}{9}\times\frac{3}{4}=\frac{15}{36}=\frac{5}{12}$$

$$B=\frac{5}{9}-\frac{5}{12}=\frac{20-15}{36}=\frac{5}{36}$$

$$D=\frac{4}{9}\times\frac{3}{5}=\frac{12}{45}=\frac{4}{15}$$

$$E=\frac{4}{9}-\frac{4}{15}=\frac{20-12}{45}=\frac{8}{45}\text{ (answer)}$$

$$G=A+D=\frac{5}{12}+\frac{4}{15}=\frac{25+16}{60}=\frac{41}{60}$$

$$H=B+E=\frac{5}{36}+\frac{8}{45}=I-G=1-\frac{41}{60}=\frac{19}{60}$$

$$I=1$$

	boys	girls	total
age ≥ 16	$\frac{5}{12}$	$\frac{4}{15}$	$\frac{41}{60}$
age < 16	$\frac{5}{36}$	$\frac{8}{45}$	$\frac{19}{60}$
Total	$\frac{5}{9}$	$\frac{4}{9}$	1

The number of girls whose age are less than 16 years old is 8/45 of the number of all students.

If the question is: 'What is the fraction of the number of girls who are younger than 16 in all girls?' then the answer is

$$\frac{8}{45} \div \frac{4}{9} = \frac{8}{25} \times \frac{9}{4} = \frac{2}{5}$$

Answer to Concept Check 1-10

Step 1. draw a two-way tables and fill in the information provided.

	boys	girls	total
take the school bus	A	B	$\frac{3}{5}$
Do not take the school bus	$\frac{1}{5}$ of $\frac{3}{5}$	C?	E
	$\frac{3}{5}$	D	1

Step 2.

	boys	girls	total
take the school bus	$A = \frac{12}{25}$	$\frac{3}{25}$	$\frac{3}{5}$
Not take the school bus	$\frac{1}{5} \times \frac{3}{5} = \frac{3}{25}$	$\frac{2}{5} - \frac{3}{25} = \frac{7}{25}$	$\frac{2}{5}$
total	$\frac{3}{5}$	$\frac{2}{5}$	1

Step 3. the answer is $\frac{7}{25}$

$$A = \frac{4}{5} \times \frac{3}{5} = \frac{12}{25}$$

$$B = \frac{3}{5} - \frac{12}{25} = \frac{15-12}{25} = \frac{3}{25}$$

$$D = \frac{2}{5}$$

$$E = \frac{2}{5}$$

Answer is (B)

1-6 DECIMALS

A. Representation

Any real number can be represented by two parts, one is the integer part and the other one is the decimal part which follows a decimal point.

Such as: 1 2 3 4 5• 6 7 8 9
→ decimal point

1 2 3 4 5• 6 7 8 9 Can be re-written as

$1\times10^4+2\times10^3+3\times10^2+4\times10^1+5\times10^0+6\times10^{-1}+7\times10^{-2}+8\times10^{-3}+9\times10^{-4}$

	10^4	10^3	10^2	10^1	10^0	10^{-1}	10^{-2}	10^{-3}	10^{-4}
So	1	2	3	4	5	6	7	8	9
	Ten Thousands	Thousands	Hundreds	Tens	Ones	Tenths	Hundredths	Thousandths	Ten thousendths

B. Decimal and Fraction Conversion

 **Step 1.** count the number of decimal places that follow the decimal point.

For instance: 0.75, $n=2$

 **Step 2.** Rewrite the decimal as a fraction with a denominator of 1×10^n and a numerator of decimal which is without decimal point.

For instance: 0.75, $\frac{75}{1\times10^2}=\frac{75}{100}$

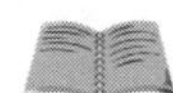 **Step 3.** Simplify the fraction

For instance: $\frac{75}{100}\Rightarrow\frac{3}{4}$

◆Examples:

$0.75=\frac{75}{100}=\frac{3}{4}$,

2 decimal 2 zeroes

$0.0625=\frac{625}{10000}$

4 decimal 4 zeroes

C. Percent and Fraction Conversion

◆For example:

$$25\% = 25\times\frac{1}{100} = \frac{25}{100} = 0.25$$

then $0.25 = \frac{25}{100} = \frac{1}{4}$

Caution: $0.25\% = 0.25\times\frac{1}{100} = \frac{25}{100}\times\frac{1}{100} = \frac{25}{10000} = \frac{1}{400}$

D. Fraction to Decimal Conversion

◆For Example:

such as: $\frac{3}{7}$

Just use simple division: (3) divided by denominator (7)

$3 \div 7 = 0.4285714286\ldots\ldots$

Answer to Concept Check 1-11

$$0.12\% = \frac{12}{100}\times\frac{1}{100} = \frac{\overset{3}{\cancel{\overset{6}{\cancel{12}}}}}{\underset{2500}{\cancel{\underset{5000}{\cancel{10000}}}}} = \frac{3}{2500}$$

Answer is (D)

E. Rounded Off

It depends upon the next digit after the designated rounded digit.

If the digit is greater or equal to "5" then the designated digit must add 1 to it, otherwise remain un-changed.

such as:

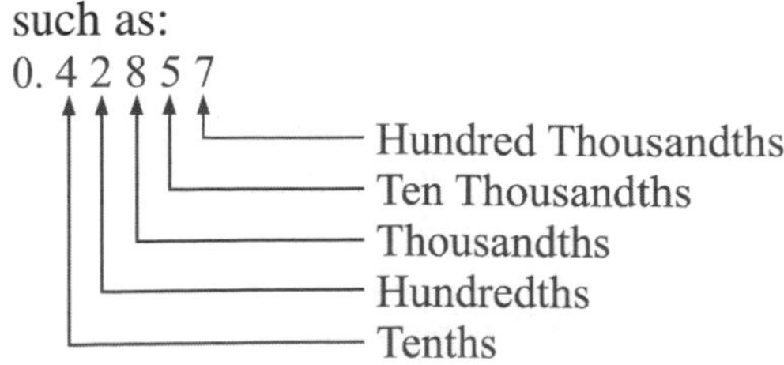

if rounded off at thousandths, then

0.42857 will be changed to 0.429 (8 + 1)

Answer to Concept Check 1-12

$1234 \div 234 = 5.2735........$

- 5 — Ten thousandths
- 3 — Thousandths
- 7 — Hundredths
- 2 — Tenths

the digit in thousandths is 3, so hundredths remained unchanged, $5.2735 \Rightarrow 5.27$

all the digits after hundredths are truncated

Answer is (C)

F. Scientific Notation

◆Example:

- $1,234 = 1.234 \times 10^3$
- $123, 456, 789 = 1.23456789 \times 10^8$
- $900, 000, 000 = 9 \times 10^8$
- $0.00123456 = 1.23456 \times 10^{-3}$
- $0.000000009 = 9 \times 10^{-9}$
- $0.000010009 = 1.0009 \times 10^{-5}$

◆Example:

The average of a certain animal cell's diameter is about 50 microns.

The average diameter of a certain bacteria is about 6 microns, the average diameter of a sort of virus is about 50 nanometers. How many times is the volumn of this animal's cell to the volume of this bacteria, and to the volume of this sort of virus?

(50 microns = 50×10^{-6} meters)

The volume of an animal cell is about $\frac{4}{3}\pi(25\times10^{-6})^3$, and the volume of a bacteria is about $\frac{4}{3}\pi(3\times10^{-6})^3$

$$\frac{\text{Volume of animal's cell}}{\text{Volume of a bacteria}} = \frac{\frac{4}{3}\pi(25\times10^{-6})^3}{\frac{4}{3}\pi(3\times10^{-6})^3} = (\frac{25}{3})^3 \doteq 8^3 \doteq 512 \text{ times}$$

The volume of a virus is about $\frac{4}{3}\pi(25\times10^{-9})^3$

$$\frac{\text{Volume of animal's cell}}{\text{Volume of virus}} = \frac{\frac{4}{3}\pi(25\times10^{-6})^3}{\frac{4}{3}\pi(25\times10^{-9})^3} = (\frac{10^{-6}}{10^{-9}})^3 = (10^3)^3 = 10^9 = 1{,}000{,}000{,}000$$

$\Rightarrow$ a billion times

Answer to Concept Check 1-13

1 year $= 365\times24\times60\times60 = 31{,}536{,}000$ seconds

The speed of the light is 300,000 km/s $= 3.0\times10^5$ km/s

The distance of 1 light-year is $31{,}536{,}000\times3.0\times10^5 = 94{,}608\times10^8$ km

$\doteq 95{,}000\times10^8$ km $\doteq 9.5\times10^{12}$

scientific notation $= 9.5\times10^{12}$ km

Answer to Concept Check 1-14

$0.00000032 \Rightarrow 3.2\times10^{-7}$

Answer is (D)

1-7 ABSOLUTE VALUE

The absolute value is denoted as $|x|$, actually it is the distance from point x to the original point "o", so, it is always positive value no matter what the value of x is.

If $|x| = a$, a is always positive,

x could be a, or x could be $-a$

◆Example:

If $|x|=3$, means, $x=3$, or $x=-3$

If $|x|>3$, means, $x>3$, or $x<-3$

If $|x|<3$, means, $-3<x<3$

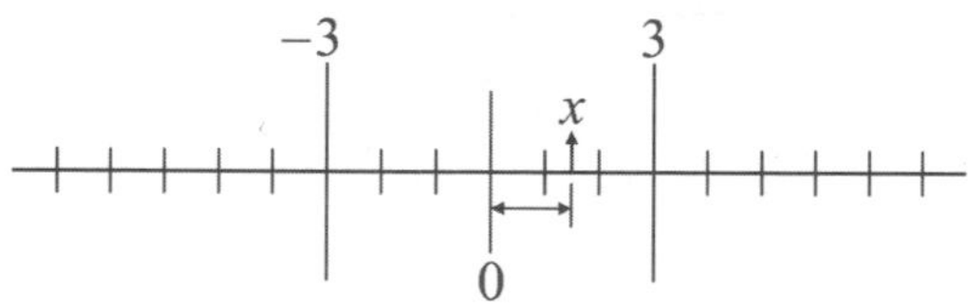

The distance from x to zero is less than 3, so, it could be any point between 3 and –3.

If $|x-3|=5$

means: $x-3=5$, $x=8$

or $x-3=-5$, $x=-5+3=-2$

If $|x-3|>5$

means: $x-3>5$ $x>8$

or $x-3<-5$, $x<-2$

If $|x-3|<5$

$x-3<5$ and $x-3>-5$

$-2<x<8$

◆Example:

If $|x|\le 5$, and $|y|\le 2$, what is the least value of $x-y$?

For $|x|\le 5$ means $-5\le x\le 5$

For $|y|\le 2$ means $-2\le y\le 2$

The least value of $x-y$ means

The smallest value of x minus the largest value of y.

Such that, the least value of $x-y$ is equal to the smallest value of $x=-5$ minus the biggest value of $y=2$, it is $-5-2=-7$.

- How to use the absolute value to express the values within an interval.

◆Example:

If the numbers are within the interval between 70 and 150, $70<x<150$, we can see that the distances from any point in this interval to the midpoint is less than the half of the interval. So, The midpoint of the interval 70 and 150 is $\frac{150+70}{2}=110$, we may use $|x-110|<40$ to describe all the values of x within the interval between 70 and 150.

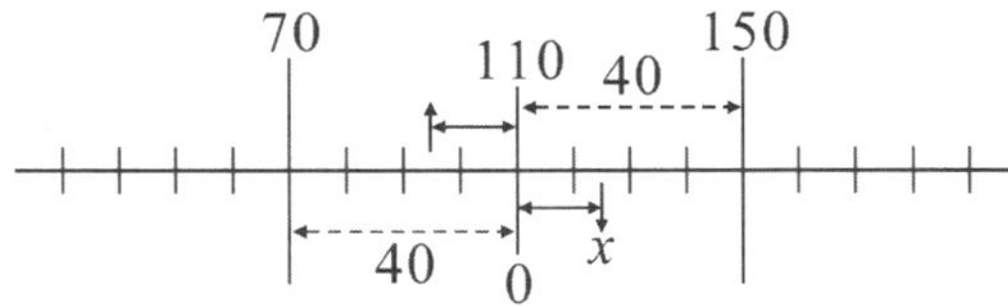

$|x-110|<150-110 \Rightarrow |x-110|<40$

Answer to Concept Check 1-15

mid-point
64 x 92

The score at the midpoint is $\frac{64+92}{2}=78$, The distance from point (92) to the midpoint is

$92-78=14$

So, any point x can be represented by

$|x-78|\le 14$

Answer is (A)

1-8 COMPLEX NUMBER (I)

Standard Form $a+bi$

a is real part

bi is imaginary part

"i" is defined as $\sqrt{-1}$, such as $\sqrt{-3}=\sqrt{3}\cdot\sqrt{-1}=\sqrt{3}i$

a is any real number, could be "o"

such as: $0+3i \Rightarrow 3i$ (only imaginary part)

b could be zero too, that $3+0i$ has only real the part.

- The properties of i:

  $i^1=\sqrt{-1}=i$
  $i^2=\sqrt{-1}\cdot\sqrt{-1}=-1$

$i^3 = i^2 \cdot i = -i$

$i^4 = i^3 \cdot i = -i^2 = 1$

therefore, $i^7 = i^4 \cdot i^3 = -i$

$i^{17} = i^{16} \cdot i = (i^4)^4 \cdot i = 1^4 \cdot i = i$

$i^{26} = (i^4)^6 \cdot i^2 = i^2 = -1$

• The operation of complex numbers

1. Addition and subtraction

perform real part and imaginary part separately

such as: $(5+3i)+(-3+2i)=(5-3)+(3+2)i=2+5i$

such as: $(\sqrt{3}+\sqrt{2}i)-(\sqrt{2}+\sqrt{3}i)=(\sqrt{3}-\sqrt{2})+(\sqrt{2}-\sqrt{3})i$

2. Multiplication

The same rules as real number operation only remember to simplify the results.

such as: $(5+3i)\times(-3+2i)=-15+10i+-9i+6i^2=-15+10i-9i-6=-21+i$

3. Division

Conjugate: $a+bi$ is the conjugate of $a-bi$

◆Example:

Use conjugate to do the division of complex numbers.

$(2+3i)\div(5+6i)$

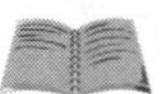 **Step 1.** change it to fraction form

$$(2+3i)\div(5+6i)=\frac{2+3i}{5+6i}$$

 **Step 2.** multiply the conjugate of $(5+6i)$ to both denominator and numerator

$$\frac{(2+3i)(5-6i)}{(5+6i)(5-6i)}=\frac{10-12i+15i-18i^2}{(5)^2-(6i)^2}=\frac{10+3i-18(-1)}{25-36i^2}=\frac{28+3i}{25+36}$$

$$=\frac{28+3i}{61}$$

◆Example 1:

One of the root of $x^2+ax+1=0$ is $2+i$, what is the value of a?

If the root of $x^2+ax+1=0$ is $2+i$, there must be a conjugate root $2-i$

So, $(2+i)+(2-i)=-a$

$\therefore\ a=-4$

◆Example 2:

What is the Value of $|2+3i|$?

$$|2+3i|=\sqrt{2^2+3^2}=\sqrt{4+9}=\sqrt{13}$$

◆Example 3:

Find the standard form of $\dfrac{-1}{3+i}$

$$\frac{-1}{3+i}=\frac{(-1)(3-i)}{(3+i)(3-i)}=\frac{-3+i}{9-i^2}=\frac{-3+i}{10}=\frac{-3}{10}+\frac{i}{10}=-\frac{3}{10}+\frac{1}{10}i$$

◆Example 4:

What is the Value of i^{52}?

$$i^{52}=i^{13\times 4}=(i^4)^{13}=(1)^{13}=1$$

Answer to Concept Check 1-16

$$\frac{(3+2i)-(4-5i)(1+i)}{2-i}=\frac{3+2i-(4+4i-5i-5i^2)}{2-i}=\frac{3+2i-4-4i+5i+5i^2}{2-i}$$

$$=\frac{3+2i-4-4i+5i-5}{2-i}=\frac{-6+3i}{2-i}=\frac{(-6+3i)(2+i)}{(2-i)(2+i)}$$

$$=\frac{-12-6i+6i+3i^2}{4-i^2}=\frac{-15}{5}=-3$$

Answer is (D)

1-9 COMPLEX NUMBER (II)

A. Graphical Representation

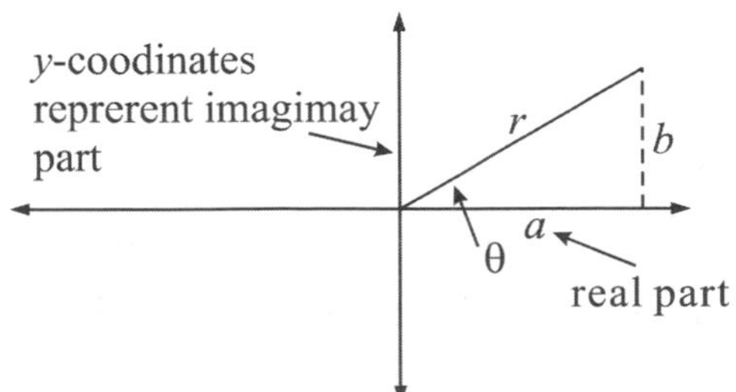

Modulus is the distance from the origin denoted as "r".

So, $r=\sqrt{a^2+b^2}$

θ: angle between modulus and x-axis,

$0 \le \theta \le 2\pi$

Usually use a character 'z' to represent a complex number.

$z=a+bi$, $|z|=\sqrt{a^2+b^2} \Rightarrow r$

$a=r\cos\theta$, $b=r\sin\theta$

$z=r\cos\theta+ri\sin\theta=r(\cos\theta+i\sin\theta)$

B. De Moivre's Formula

If $z=r(\cos\theta+i\sin\theta)$, and $z'=r'(\cos\theta'+i\sin\theta')$

Then $z\cdot z'=rr'[\cos(\theta+\theta')+i\sin(\theta+\theta')]$

$\dfrac{z}{z'}=\dfrac{r}{r'}[\cos(\theta-\theta')+i\sin(\theta-\theta')]$

$z^n=r^n(\cos n\theta+i\sin n\theta)$

proof: $z\cdot z'=r(\cos\theta+i\sin\theta)\cdot r'(\cos\theta'+i\sin\theta')$

$$=rr'(\cos\theta\cos\theta'+i\sin\theta'\cos\theta+i\sin\theta\cos\theta'+i^2\cdot\sin\theta\sin\theta')$$

$$=rr'[\cos\theta\cos\theta'-\sin\theta\sin\theta'+i(\sin\theta\cos\theta'+\sin\theta'\cos\theta)]$$

$$=rr'[\cos(\theta+\theta')+i\sin(\theta+\theta')]$$

$$\frac{z}{z'}=\frac{r(\cos\theta+i\sin\theta)}{r'(\cos\theta'+i\sin\theta')}=\frac{r}{r'}\cdot\frac{(\cos\theta+i\sin\theta)(\cos\theta'-i\sin\theta')}{(\cos\theta'+i\sin\theta')(\cos\theta'-i\sin\theta')}$$

$$=\frac{r}{r'}\frac{\cos\theta\cos\theta'+i\sin\theta\cos\theta'-i\cos\theta\sin\theta'-i^2\sin\theta\sin\theta'}{\cos^2\theta'-i^2\sin^2\theta'}$$

$$=\frac{r}{r'}\frac{\cos\theta\cos\theta'+\sin\theta\sin\theta'+i(\sin\theta\cos\theta'-\cos\theta\sin\theta')}{\cos^2\theta+\sin^2\theta'}$$

$$=\frac{r}{r'}=\cos(\theta-\theta')+i\sin(\theta-\theta')$$

$z^n=r^n(\cos n\theta+i\sin n\theta)$

When $n=1$ $z^1=r(\cos\theta+i\sin\theta)$ holds

When $n=2$ $z^2=r^2(\cos 2\theta+i\sin 2\theta)$

$z^2=z\cdot z=r(\cos\theta+i\sin\theta)\cdot r(\cos\theta+i\sin\theta)$

$=r^2[(\cos^2\theta-\sin^2\theta)+i(\sin\theta\cos\theta+\cos\theta\sin\theta)]$

$=r^2[\cos(\theta+\theta)+i\sin(\theta+\theta)]$

$=r^2[\cos(2\theta)+i\sin(2\theta)]$ (holds)

When $n=3$, $z^3=z^2\cdot z=r^2[\cos(2\theta)+i\sin(2\theta)]\cdot r$

$(\cos\theta+i\sin\theta)=r^3[(\cos(2\theta+\theta)+i\sin(2\theta+\theta)]$

$z^3=r^3(\cos 3\theta+i\sin 3\theta)$ holds

If z^k holds $z^k=r^k(\cos k\theta+i\sin k\theta)$

$$\begin{aligned} z^{k+1} &= z^k\cdot z=r^k(\cos k\theta+i\sin k\theta)\cdot r(\cos\theta+i\sin\theta) \\ &= r^{k+1}(\cos k\theta\cdot\cos\theta+i\cos k\theta\sin\theta+i\sin k\theta\cdot\cos\theta+i^2\sin k\theta\cdot\sin\theta) \\ &= r^{k+1}[(\cos k\theta\cos\theta-\sin k\theta\sin\theta)+i(\cos k\theta\sin\theta+\sin k\theta\cos\theta)] \\ &= r^{k+1}[\cos(k\theta+\theta)+i\sin(k\theta+\theta)] \\ &= r^{k+1}[\cos(k+1)\theta+i\sin(k+1)\theta] \end{aligned}$$

So, $z^n=\cos n\theta+i\sin n\theta$ (holds)

C. The Properties of Complex Number

1. $a+bi=c+di$, then $a=c$, and $b=d$
2. $a+bi=0$, $a=0$ and $b=0$
3. If conjugate of z_1 and z_2 are represented by $\overline{z_1}$, and $\overline{z_2}$ respectively

then 1. $\overline{z_1}+\overline{z_2}=\overline{z_1+z_2}$

2. $\overline{z_1}-\overline{z_2}=\overline{z_1-z_2}$

3. $\overline{z_1}\cdot\overline{z_2}=\overline{z_1z_2}$

4. $z\overline{z}=|z|^2$

Proof: $\overline{z_1}+\overline{z_2}=\overline{z_1+z_2}$

(1) $z_1=a+bi$, $z_2=c+di$

$\overline{z_1}=a-bi$, $\overline{z_2}=c-di$

$\overline{z_1}+\overline{z_2}=(a+c)-(b+d)i$

$\overline{z_1}-\overline{z_2}=(a-c)-(b-d)i$

$\overline{z_1+z_2}=\overline{(a+bi)+(c+di)}=\overline{(a+c)+(b+d)i}=(a+c)-(b+d)i=\overline{z_1}+\overline{z_2}$

(2) $\overline{z_1}-\overline{z_2}=\overline{z_1-z_2}$

$\overline{z_1}-\overline{z_2}=(a-bi)-(c-di)=(a-c)-i(b-d)$

$\overline{z_1-z_2}=\overline{(a+bi)-(c+di)}=\overline{(a-c)+(b-d)i}=(a-c)-i(b-d)$

$\overline{z_1}-\overline{z_2}=\overline{z_1-z_2}$

(3) $\overline{z_1}\cdot\overline{z_2}=(a-bi)(c-di)=ac-(bc+ad)i-bdi^2=(ac+bd)-(bc+ad)i=\overline{z_1\cdot z_2}$

(4) $z\overline{z}=(a+bi)(a-bi)=a^2+b^2=|z|^2$

◆Example 1: If $z_1 = 2+3i$, $z_2 = 1-i$, what is $z_1 + z_2 = ?$

$z_1 + z_2 = (2+3i)+(1-i) = (2+1)+i(3-1) = 3+2i$

◆Example 2: If $z_1 = 1+i$, and $z_2 = 1-i$

what is $z_1 - z_2$?

$z_1 - z_2 = (1+i)-(1-i) = (1-1)+[1-(-1)]i = 0+2i = 2i$

◆Example 3: If $z_1 = 1+i$, and $z_2 = 1-i$, what is $z_1 \cdot z_2$?

$z_1 \cdot z_2 = (1+i)(1-i) = 1-i^2 = 1+1 = 2$

◆Example 4: If $z_1 = 2+3i$ and $z_2 = 3+2i$, what is the Value of $\frac{z_1}{z_2}$?

$$\frac{z_1}{z_2} = \frac{2+3i}{3+2i} = \frac{(2+3i)(3-2i)}{(3+2i)(3-2i)} = \frac{6-4i+9i-6i^2}{3^2-4i^2}$$

$$= \frac{12+5i}{9+4} = \frac{12+5i}{13}$$

◆Example 5: If $z_1 = 2+3i$, $z_2 = 1-i$, what is the Value of $\overline{z_1} + \overline{z_2}$?

($\overline{z_1}$ and $\overline{z_2}$ are the conjugates of z_1 and z_2 respectively)

According to the formula

$\overline{z_1} + \overline{z_2} = \overline{z_1 + z_2} = \overline{(2+3i)+(1-i)} = \overline{3+2i} = 3-2i$

Another way:

$\overline{z_1} = 2-3i$, and $\overline{z_2} = 1+i$, then

$\overline{z_1} + \overline{z_2} = (2-3i)+(1+i) = 3-2i = \overline{z_1 + z_2}$

◆Example 6: If $z_1 = 1+2i$, and $z_2 = 2+i$, what is the value of $\overline{z_1} \cdot \overline{z_2}$?

$z_1 = 1+2i$, $\overline{z_1} = 1-2i$; $z_2 = 2+i$, $\overline{z_2} = 2-i$

$\overline{z_1} \cdot \overline{z_2} = (1-2i)\cdot(2-i) = (2-i-4i+2i^2) = (2-5i-2) = -5i$

Answer to Concept Check 1-17

$$x = 4\cos\frac{\pi}{3} = 4\times\frac{1}{2} = 2$$

$$y = 4\sin\frac{\pi}{3} = 4\times\frac{\sqrt{3}}{2} = 2\sqrt{3}$$

$$\therefore\ 4(\cos\frac{\pi}{3} + i\sin\frac{\pi}{3}) = 4(2+2\sqrt{3}i) = 8+8\sqrt{3}i$$

◆Example 7: $(1+i)^{10}=[\sqrt{2}(\frac{1}{\sqrt{2}}+i\frac{1}{\sqrt{2}})]^{10}=[\sqrt{2}(\cos\frac{\pi}{4}+i\sin\frac{\pi}{4})]^{10}$

$=(\sqrt{2})^{10}(\cos\frac{10\pi}{4}+i\sin\frac{10\pi}{4})=2^5\cdot(0+i)=2^5 i$

Other solution:

$(1+i)^{10}=[(1+i)^2]^5=(1+2i+i^2)^5=(2i)^5=2^5 i^5=2^5 i$

Answer to Concept Check 1-18

Since $a+bi=\sqrt{a^2+b^2}(\cos\theta+i\sin\theta)$

$1-i=\sqrt{2}(\frac{1}{\sqrt{2}}-\frac{1}{\sqrt{2}}i)$

$\cos\theta=\frac{1}{\sqrt{2}}$, $\sin\theta=-\frac{1}{\sqrt{2}}$

θ in quadrant IV

$\theta=2\pi-\frac{\pi}{4}=\frac{7}{4}\pi$

$1-i=\sqrt{2}(\cos\frac{7}{4}\pi+i\sin\frac{7}{4}\pi)$

$(1-i)^8=(\sqrt{2})^8[\cos\frac{7}{4}(8)\pi+i\sin(\frac{7}{4})\cdot(8)\pi]$ De Moivre's Formula

$=16(\cos 14\pi+i\sin 14\pi)=16(\cos 0°+i\sin 0°)=16(1+0i)=16$

(+)1

θ

(−)

$\sqrt{2}$

1

D. Polar Form of Complex Number

The standard polar from is $re^{i\theta}$ where r is distance from origin. (modulus)

For instance, the standard from of $5+2i$

$r=\sqrt{5^2+2^2}=\sqrt{29}$

So, the polar form is

$z=re^{i\theta}=\sqrt{29}e^{i\tan^{-1}\frac{2}{5}}$

$z_1\cdot z_2=r_1\cdot r_2 e^{i(\theta_1+\theta_2)}$

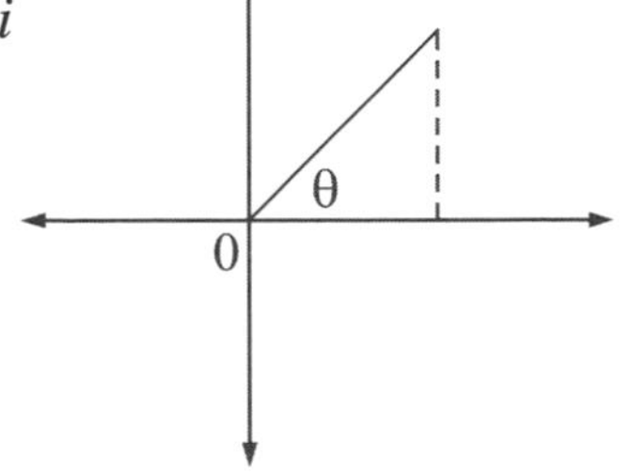

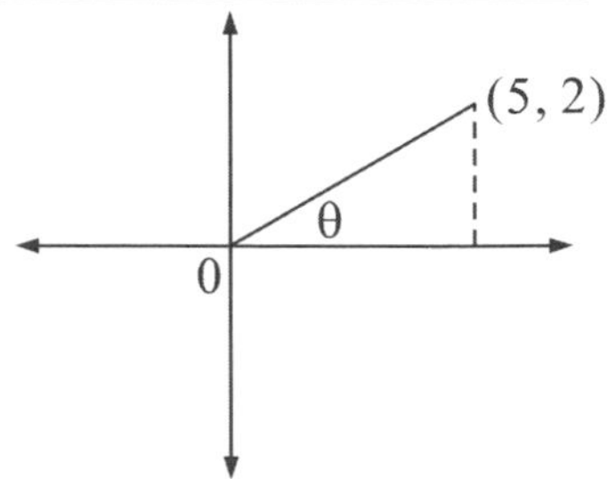

Answer to Concept Check 1-19

the polar from of $3+3i$

$\because$ The standard form of polar form is $re^{i\theta}$, $r=\sqrt{3^2+3^2}=3\sqrt{2}$,

$\tan\theta=\frac{3}{3}=1$, $\theta=\frac{\pi}{4}$

$3+3i=3\sqrt{2}e^{\frac{\pi}{4}i}$

Answer to Concept Check 1-20

$z_1=4e^{i\frac{3}{4}\pi}$, $z_2=3e^{i\frac{4}{3}\pi}$

$r_1=4$ $\quad r_2=3$, $z=z_1\cdot z_2=re^{i\theta}=r_1\cdot r_2e^{i(\theta_1+\theta_2)}$

$z_1\cdot z_2=r_1\cdot r_2e^{i(\theta_1+\theta_2)}=(3\cdot 4)e^{i(\frac{3}{4}\pi+\frac{4}{3}\pi)}=12e^{i\frac{9+16}{12}\pi}=12e^{i\frac{25}{12}\pi}=12e^{i(\frac{24}{12}\pi+\frac{1}{12}\pi)}$

$=12e^{i\frac{1}{12}\pi}=12e^{\frac{\pi}{12}i}$

Answer to Concept Check 1-21

$z=a+bi=|z|(\cos\theta+i\sin\theta)$

$|z|=\sqrt{a^2+b^2}$, $\cos\theta=\frac{a}{|z|}$, $\sin\theta=\frac{b}{|z|}$

$\frac{\sqrt{3}}{2}-\frac{\sqrt{3}}{2}i=\frac{\sqrt{6}}{2}(\cos\theta+i\sin\theta)$

So, $\cos\theta=\frac{\frac{\sqrt{3}}{2}}{\frac{\sqrt{6}}{2}}=\frac{\sqrt{3}}{\sqrt{6}}$, $\sin\theta=\frac{-\frac{\sqrt{3}}{2}}{\frac{\sqrt{6}}{2}}=-\frac{\sqrt{3}}{\sqrt{6}}=-\frac{1}{\sqrt{2}}$

So, θ is in quadrant IV $\theta=2\pi-\frac{\pi}{4}=\frac{7}{4}\pi$

$\frac{\sqrt{3}}{2}-\frac{\sqrt{3}}{2}i=\frac{\sqrt{6}}{2}(\cos\frac{7}{4}\pi+i\sin\frac{7}{4}\pi)$

CHAPTER 1
PRACTICE

1. If $11 \le t \le 17$, $13 \le s \le 21$, and $2 \le u \le 8$, what is the largest possible value of $\frac{t-s}{u}$?

(A) 2

(B) 2.8

(C) 4

(D) 5.2

2. If the sum of n consecutive integers is 33, what is the greatest value of n?

(A) 65

(B) 66

(C) 67

(D) 68

3. If one pound of tiger shrimp costs $6.99, how much does 0.79 pound of this tiger shrimp cost, (rounded to the nearest cent).

4. The decimal form of 1/7 is an infinite decimal, what is the value of the 100th digit after decimal point?

5. If $P/2$ is an integer, which of the following must be true ?

I. P is even

II. $(2P+5)\div 2$ is odd

III. $(5P+2)\div 2$ is odd

(A) I only

(B) II only

(C) III only

(D) I and III.

6. If $p \times q \times r = 84$, Where p, q, and r are different integers, and $p > q > r > 1$, what is the greatest possible value of p ?

(A) 14

(B) 18

(C) 21

(D) 22

7. $2\frac{2}{3} \div \frac{4}{5} + \frac{3}{4}\left(\frac{\frac{1}{3}+\frac{1}{2}}{\frac{1}{5}-\frac{1}{4}}\right) =$

8. If $10 \le p \le 15$, $12 < t \le 21$, and $3 < s < 7$, where p, t, and s are integers, what is the smallest value of $\frac{p-t}{s}$?

(A) $-\frac{11}{6}$

(B) $-\frac{5}{7}$

(C) $\frac{5}{7}$

(D) $\frac{2}{3}$

9. If $x=\frac{1}{3}$ and $a=\frac{1}{2}$, what is the value of $\frac{a}{x}+\frac{x}{a}=?$

(A) $\frac{6}{13}$

(B) $\frac{13}{6}$

(C) $\frac{1}{2}$

(D) 2

10. The scores of the final term math examination, in Mr. Black's class, was between 62 and 98. If S stands for the scores of the students, which of the following represents all possible values of S?

(A) $|S-80|\le 18$

(B) $|S-80|\ge 18$

(C) $|18-S|\le 80$

(D) $|S-18|\ge 80$

11. If $o<m<1$, which of the following expressions must be true?

(A) $m^2>m>\sqrt{m}$

(B) $m^2>\sqrt{m}>m$

(C) $\sqrt{m}>m^2>m$

(D) $\sqrt{m}>m>m^2$

12. If $-1<n<1$, which of the following could be true?

(A) $n^3>n^2>n$

(B) $n^2>n>n^3$

(C) $n^2>n^3>n$

(D) $n^3>n>n^2$

13. Which of the following is equivalent to $\frac{1+i}{1-i}+\frac{2-i}{2+i}$?

(A) $\frac{2}{3}-i$

(B) $\frac{2}{3}+i$

(C) $\frac{3}{5}+\frac{i}{5}$

(D) $\frac{3}{5}-\frac{i}{5}$

14. April and Joyce together buy 0.89 pound of beef. If a pound of the beef costs \$5.79, how much will each person pay on average? (rounded to cents)

15. On a number line, given point $A(-6)$ and point $B(2)$, if $P(x)$ is a point on the number line, and $\overline{AP}=3\overline{BP}$, What is the value(s) of x ?

(A) 2

(B) 4

(C) 6

(D) None of above

16. How many factors does $P=5\times7\times11\times13$ have, including, P ?

(A) 4

(B) 8

(C) 15

(D) 16

17. If a, b, c, and d are all different integers greater than 3, which of the following is the greatest?

(A) $(a+b)c+d$

(B) $a+b(c+d)$

(C) $a+(bc+d)$

(D) $(a+b)(c+d)$

18. In a sequence each term after the first term is half of proceeding term plus 2. If S is the first term of the sequence and $S \neq 0$, what is the fraction of the third term to the second term?

(A) $\frac{3}{4}S$

(B) $\frac{4+S}{8+S}$

(C) $\frac{12+S}{8+2S}$

(D) $\frac{8+2S}{12+S}$

19. Faith participated in school's a three-day field trip. Her mother prepared some pocket money for her. On the first day, she spent 3/20 of the pocket money, and on the second day she spent $6 less than what she spent on the first, on the third day, after she spent $16, there are still 3/5 of the pocket money left, how much was the pocket money?

20. In a certain Chess Chub, $\frac{3}{9}$ of the members wear glasses. And $\frac{4}{9}$ of the members are female. If $\frac{1}{5}$ of the female members wear glasses, what fraction of the number of male who do not wear glasses?

(A) $\frac{11}{45}$

(B) $\frac{14}{45}$

(C) $\frac{11}{25}$

(D) $\frac{14}{25}$

21. If a positive integer K can be represented as $K=m^3 \times n^6$, where m and n are distinct prime numbers. How many factors does K have, including K?

(A) 28

(B) 32

(C) 36

(D) 40

22. Cynthia's ATM password is made with the prime factoring of $6a20=2^b \times 3^c \times d \times 17$. The password represented by 'a b c d'. What is the password ?

23. The price of apples is from \$1.45 per pound to \$5.15 per pound. If p represents the price of apple per pound, which of the following inequalities best represents their range ?
(A) $|x-1.85| \le 3.3$
(B) $|x-3.3| \le 1.85$
(C) $|x-1.85| \ge 3.3$
(D) $|x-3.3| \ge 1.85$

24. If the sum of a list of consecutive integers from – 32 to an integer k is 102, what is the value of k?
(A) 35
(B) 36
(C) 37
(D) 38
(E) 39

25. What is the trigonometric form of $\frac{\sqrt{3}}{2}-\frac{\sqrt{3}}{2}i$?

26. If $z_1=4(\cos\frac{3}{2}\pi+i\sin\frac{3}{2}\pi)$, and $z_2=3(\cos\frac{7}{4}\pi+i\sin\frac{7}{4}\pi)$ find $z_1 \cdot z_2$ and $\frac{z_1}{z_2}$ in trigonometric and (in) rectangular form.

27. What is the polar form of $3+3i$?

28. If $z_1=4e^{i\frac{2\pi}{4}}$, $z_2=3e^{i\frac{4}{3}\pi}$, What is the Value of z_1z_2?

29. There are 20 students in Cynthia's class, she likes to offer each of her students one cheese cake, and she just like to buy 20 cheese cakes.
It happens that a certain cake shop offers 3 kinds of discounts.
A. Buy 3 get one free
B. 20% off
C. Increase the number of purchase by 20%
If one cheese cake costs \$2.5, which of the discount plan should she take and how much is saved from the most expensive plan ?

30. If the solutions of $|ax+1| \ge b$ are $x \le -2$ or $x \ge 8$, a, b are real numbers, what is the value of $a+b$?

CHAPTER 1
ANSWERS and EXPLANATIONS

1. (A)

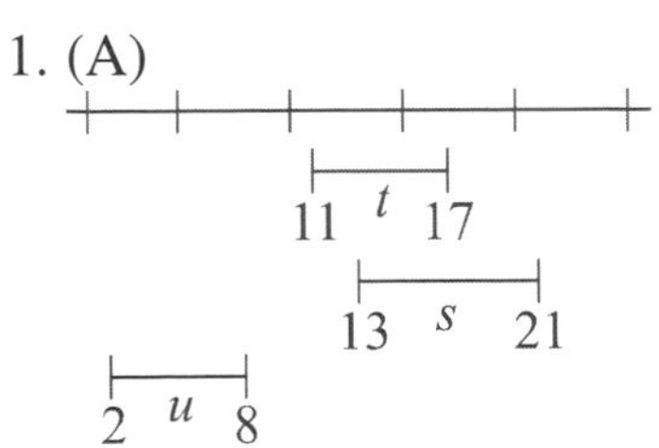

The greatest value of $\frac{t-s}{u}$, when $t-s$ is the greatest and u is the smallest value.

The greatest for $t-s$ is $17-13=4$

The smallest value of u is 2

The greatest value of $\frac{t-s}{u}$ is $\frac{4}{2}=2$

Answer is (A)

2. (B)

The consecutive list could be

A. $33=(\cancel{-15})+(\cancel{-14})+(\cancel{-13})+\cdots$
$+(\cancel{13})+(\cancel{14})+(\cancel{15})+16+17$,

$n=15+15+1+2=33$

also could be

B. $33=(\cancel{-9})+(\cancel{-8})+(\cancel{-7})+\cdots$
$+\cancel{7}+\cancel{8}+\cancel{9}+(10)+(11)+(12)$

$n=9+9+1+3=22$

also could be

C. $33=(\cancel{-32})+(\cancel{-31})+(\cancel{-30})+\cdots$
$+(\cancel{30})+(\cancel{31})+(\cancel{32})+33$

So, the list of case c has the greatest member of items

So, $n=32+32+1+1=66$
(‘0’) (‘33’)

Answer is (B)

3. 5.52

1 pound costs \$6.99

0.79 pound costs $0.79\times 6.99=5.5221$

the digit after cent is “2”

So, the result is \$5.52

4. 8

$1\div 7=0.\underline{142857}\underline{142857}14........$

For every 6-digits is a repeating group.

$100\div 6=16\cdots 4$

We have 16 repeating groups and the 4th digit of the 17th group is ‘8’.

Answer is 8.

5. (A)

I. $\frac{P}{2}$ is an integer, P is divisible by “2”

P must be even, True

II. $2P+5 \Rightarrow$ even + odd $\Rightarrow$ odd
$\frac{(2P+5)}{2} \Rightarrow$ not an integer (False)

III. $5P+2 \Rightarrow$ even + even $\Rightarrow$ even
$\frac{(5P+2)}{2} \Rightarrow$ it could be odd or even

Such as $\frac{(5\times 100+2)}{2}=251$ odd

Such as $\frac{(5\times10+2)}{2}=26$ even

Only I is true

Answer is (A)

6. (A)

$84=2\times2\times3\times7=2\times3\times14$

The greatest value of P is 14

Answer is (A)

7. $\frac{55}{6}$

$$2\frac{2}{3}\div\frac{4}{5}+\frac{3}{4}\left(\frac{\frac{1}{3}+\frac{1}{2}}{\frac{1}{5}-\frac{1}{4}}\right)=\frac{8}{3}\div\frac{4}{5}+\frac{3}{4}\left(\frac{\frac{5}{6}}{\frac{-1}{20}}\right)$$
$$=\frac{8}{3}\times\frac{5}{4}+\frac{3}{4}\bullet\frac{-100}{6}=\frac{10}{3}+\frac{-100}{8}$$
$$=\frac{80-300}{24}=\frac{-55}{6}$$

8. (A)

The smallest value of fraction $\frac{p-t}{s}$ is the largest value of denominator of s and the smallest value of numerator of $p-t$.

The largest of 's' is '6' $(3<s<7)$

The smallest value of "$p-t$" is

$|0-2|=-11$

So, The smallest value of $\frac{p-t}{s}$

$\frac{p-t}{s}=\frac{-11}{6}$

The answer is (A)

9. (B)

$$\frac{a}{x}+\frac{x}{a}=\frac{\frac{1}{2}}{\frac{1}{3}}+\frac{\frac{1}{3}}{\frac{1}{2}}=\frac{3}{2}+\frac{2}{3}=\frac{9}{6}+\frac{4}{6}=\frac{13}{6}$$

Answer is (B)

10. (A)

$\frac{(62+98)}{2}=80$, the differences of all scores from the midpoint score "80" is

$|S-80|\le18$

Answer is (A)

11. (D)

suppose $m=\frac{1}{3}$, for $0<\frac{1}{3}<1$

$m^2=(\frac{1}{3})^2=\frac{1}{9}$, $\sqrt{m}=\sqrt{\frac{1}{3}}=\frac{\sqrt{3}}{3}=\frac{1.712}{3}$

since $\frac{1.712}{3}>\frac{1}{3}>\frac{1}{9}$, so, $\sqrt{m}>m>m^2$

Answer is (D)

12. (C)

If $n=\frac{1}{3}$ $n^3=\frac{1}{27}$, $n^2=\frac{1}{9}$

so $n>n^2>n^3$

if $n=-\frac{1}{3}$ $n^3=-\frac{1}{27}$, $n^2=\frac{1}{9}$

$n^2>n^3>n$

(A) $n^3>n^2>n$ not true

(B) $n^2>n>n^3$ not true

(C) $n^2>n^3>n$ true

(D) $n^3>n>n^2$ not true

Answer is (C)

13. (C)

$$\frac{1+i}{1-i}+\frac{2-i}{2+i}=\frac{(1+i)^2}{(1-i)(1+i)}+\frac{(2-i)^2}{(2+i)(2-i)}$$
$$=\frac{(1+i)^2}{1-i^2}+\frac{(2-i)^2}{4-i^2}$$
$$=\frac{(1+i)^2}{2}+\frac{(2-i)^2}{5}$$
$$=\frac{1+2i+i^2}{2}+\frac{4-4i+i^2}{5}$$
$$=\frac{1+2i-1}{2}+\frac{4-4i-1}{5}$$
$$=\frac{2i}{2}+\frac{3-4i}{5}$$
$$=i+\frac{3}{5}-\frac{4}{5}i$$
$$=\frac{i}{5}+\frac{3}{5}=\frac{3}{5}+\frac{i}{5}$$

Answer is (C)

14. 2.58

5.79 (dollars) × 0.89 (pound) = 5.1531

$\frac{5.153}{2}=2.5765$ ←rounded

each pays $2.58

15. (C)

$x=0$, or $x=6$

case 1:

(−6) P(x) (2)

A 3 1 B

$|2-(-6)|=8$

$8\div(3+1)=2$

each segment is equal to 2 units.

so, $P(x)$ is at $-6+3\times2=0$

$x=0$

case 2:

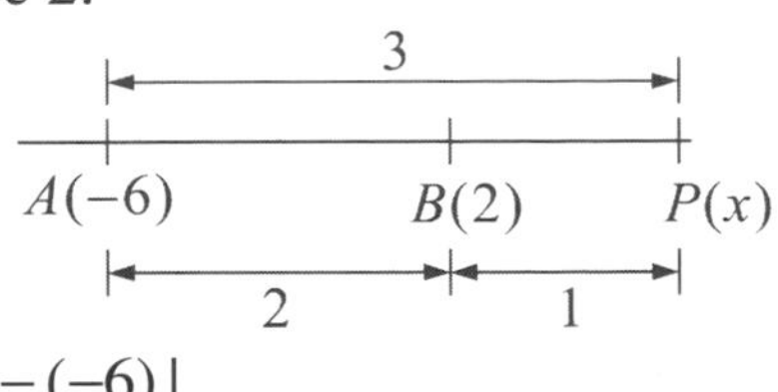

$\frac{|2-(-6)|}{2}=4$ (each unit)

$P(x)$ at $2+4=6$

$\therefore\ x=6$

16. $P=5\times7\times11\times13$

Positive factors are 5, 7, 11, and 13

For factor "5" We have

1 factor: 5, just one

2 factors: 5×7, 5×11, 5×13, three

3 factors: $5\times7\times11$, $5\times7\times13$, $5\times11\times13$, three

4 factors: $5\times7\times11\times13$, one

total $=1+3+3+1=8$

For factor 7 we have

1 factors:7, just one

2 factors: 7×11, 7×13, two

3 factors: $7\times11\times13$, just one

4 factors: none

total $=1+2+1=4$

For factor 11 we have

1 factors: 11, just one

2 factors: 11×13, just one

total $=1+1=2$

For factor 13 we have

1 factors: 13, just one

2 factors: none

total = 1

we have positive factors are 8 + 4 + 2 + 1 = 15

but we also have factor "1"add it up.

15 + 1 = 16 factors.

other solution: $P=5\times7\times11\times13$

Total factors $=(1+1)\times(1+1)\times(1+1)\times(1+1)$

$=2^4=16$

Answer is (D)

17. (D)

(A) $(a+b)c+d=ac+bc+d$

(B) $a+b(c+d)=a+bc+bd$

(C) $a+(bc+d)=a+bc+d$

(D) $(a+b)(c+d)=ac+ad+bc+bd$

D is the greatest

Answer is (D)

18. (C)

First term: S

Second term is $2+\frac{S}{2}$

Third term is

$$2+\frac{2+\frac{S}{2}}{2}=\frac{4+2+\frac{S}{2}}{2}=\frac{6+\frac{S}{2}}{2}$$

$$=\frac{\frac{12+S}{2}}{2}=\frac{12+S}{4}$$

$$\therefore \frac{\frac{12+S}{4}}{2+\frac{S}{2}}=\frac{\frac{12+S}{4}}{\frac{4+S}{2}}=\frac{12+S}{2(4+S)}=\frac{12+S}{8+2S}$$

Answer is (D)

19.

1st day | 2nd day | 3rd day

3/20 | 3/20-6 | 16 | $\frac{3}{5}$

$$\frac{3}{20}+(\frac{3}{20}-\$6)+\$16=1-\frac{3}{5}$$

$$\frac{6}{20}+\$10=\frac{2}{5}$$

$$\$10=\frac{8}{20}-\frac{6}{20}=\frac{2}{20}$$

$$\therefore \text{ pocket money}=\frac{10}{\frac{2}{20}}=\$100$$

20. (D)

Step 1.

Draw a two way classification

	Male	Female	Total
Wearing	A	(1/5) C	3/9
Not Wearing	B?	D	F
Total	E	4/9	G

Step 2.

$G=1$

$$C=\frac{1}{5}\times\frac{4}{9}=\frac{4}{45}$$

$$A=\frac{3}{9}-C=\frac{3}{9}-\frac{4}{45}=\frac{15-4}{45}=\frac{11}{45}$$

$$E=1-\frac{4}{9}=\frac{5}{9}$$

$$B=E-A=\frac{5}{9}-\frac{11}{45}=\frac{25-11}{45}=\frac{14}{45}$$

$$\frac{B}{E}=\frac{\frac{14}{45}}{\frac{5}{9}}=\frac{14\times\cancel{9}^{1}}{\cancel{45}_{5}\times5}=\frac{14}{25}$$

Answer is (D)

21. (D)

The number of factors are

$(3+1)\cdot(6+1)=4\times 7=28$

22. 6 a 2 0 is a multiple of 17.

(1) When $a=0$ $602 \div 17 = 35.41$ (indivisible)

$a=1$ $612 \div 17 = 36$ (divisible)

so, $6120 \div 17 = 360$

(2)

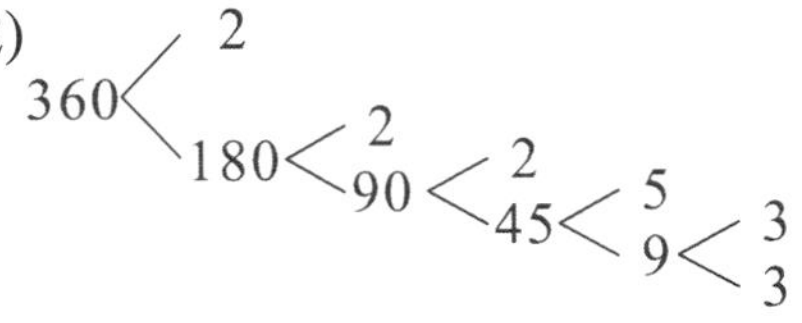

$360 = 2^3 \times 3^2 \times 5^1$

$6120 = 2^3 \times 3^2 \times 5^1 \times 17$

$a=1,\ b=3,\ c=2,\ d=1$

password$=1321$

23. (B)

The price of apples are from \$1.45 per pound to \$5.15 per pound

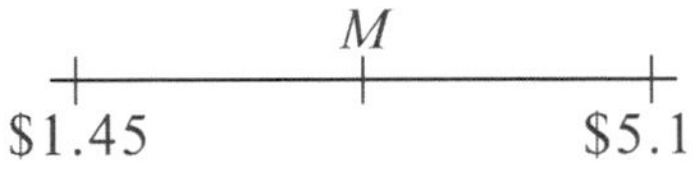

The middle price $M = \dfrac{1.45+5.15}{2} = 3.3$

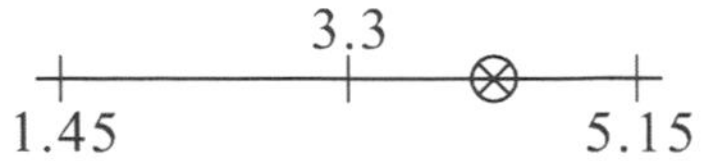

So, the value of × from 3.3 is always less than $5.15-3.3=1.85$

The inequality should be $|x-3.3| \le 1.85$

means $1.45 \le x \le 5.15$

Answer is (B)

24. (A)

~~-32~~, −31, −30, …….. 0, 1, 2,

…… ~~32~~, 33, 34, 35

$33+34+35=102$

$k=35$

Answer is (A)

25. Asking: trigonometric form?

Since $r=\sqrt{x^2+y^2}=\sqrt{\dfrac{3}{4}+\dfrac{3}{4}}=\dfrac{\sqrt{6}}{2}$

$\left(\tan\theta = \dfrac{-\frac{\sqrt{3}}{2}}{\frac{\sqrt{3}}{2}} = -1\right)$

since $\left(-\dfrac{\sqrt{3}}{2}, \dfrac{\sqrt{3}}{2}\right)$ is in quadrant II

so, $\theta = 180° - 45° = 135°$

$\left(\text{or } \pi - \dfrac{\pi}{4} = \dfrac{3\pi}{4}\right)$

The trigonometric form is

$\dfrac{\sqrt{6}}{2}\left(\cos\dfrac{3\pi}{4} + i\sin\dfrac{3\pi}{4}\right)$

26. $z_1 \cdot z_2 = (4)(3)[\cos(\dfrac{3}{2}\pi + \dfrac{7}{4}\pi)$

$+ i\sin(\dfrac{3}{2}\pi + \dfrac{7}{4}\pi)$

$= 12\,[\cos\dfrac{13}{4}\pi + i\sin(\dfrac{13}{4}\pi)]$

Since $o \le \theta < 2\pi$,

$\cos\dfrac{13}{4}\pi = \cos(\dfrac{12}{4}\pi + \dfrac{1}{4}\pi)$

$= \cos\dfrac{1}{4}\pi = \dfrac{\sqrt{2}}{2}$,

$\sin\dfrac{13}{4}\pi = \sin\dfrac{1}{4}\pi = \dfrac{\sqrt{2}}{2}$

$$z_1 \bullet z_2 = 12(\cos\frac{1}{4}\pi + i\sin\frac{1}{4}\pi)$$
$$= 6(\sqrt{2} + i\sqrt{2})$$
$$\frac{z_1}{z_2} = \frac{4}{3}[\cos(\frac{3}{2} - \frac{7}{4})\pi + i\sin(\frac{3}{2} - \frac{7}{4})\pi]$$
$$= \frac{4}{3}[\cos(-\frac{\pi}{4}) + i\sin(-\frac{\pi}{4})]$$
$$= \frac{4}{3}(\frac{\sqrt{2}}{2} - \frac{\sqrt{2}}{2}i) = \frac{2\sqrt{2}}{3}(1-i)$$

27. $r = \sqrt{3^2 + 3^2} = 3\sqrt{2}$, $\tan\theta = \frac{3}{3} = 1$

θ is in quadrant I, $\theta = \frac{\pi}{4}$

Thus $3 + 3i = 3\sqrt{2}e^{\frac{\pi}{4}i}$

28. $z_1 \bullet z_2 = r_1 r_2 e^{i(\frac{2\pi}{4} + \frac{4}{3}\pi)} = (3 \bullet 4)e^{i(\frac{6+16}{12})\pi}$

$= 12e^{\frac{22}{12}\pi i} = 12e^{\frac{11}{6}\pi i}$

29. Take plan A, save 4.17

Plan A: 20÷(3+1)=5

5×(2.5×3)=37.5

Plan B: 20×(1−0.2)×2.5=40

Plan C: 20÷1.2×2.5=41.67

41.67−37.5=4.17

30. $\frac{4}{3}$

get midpoint of

-2, and 8, $\frac{-2+8}{2} = 3$

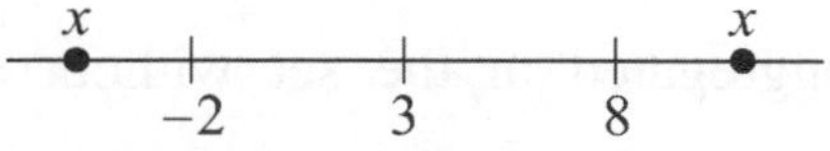

For $x \leq -2$ or $x \geq 8$

So, $|x-3| \geq 5$

Divided by 3 $|\frac{1}{3}x - 1| \geq \frac{5}{3}$

$|-\frac{1}{3}x + 1| \geq \frac{5}{3}$

$\therefore\ a = -\frac{1}{3}\quad b = \frac{5}{3}\quad a + b = \frac{4}{3}$

CHAPTER 2
SETS, SEQUENCES, DEFINED OPERATIONS, PATTERNS, and LOGIC

2-1 SETS

A set is a composition of aggregated objects that have certain relationship among them

A. The Description of a set

- List all elements

  for instance, $S=\{3,5,6,8\}$

- Described by variables

  for instance, $S=\{x \mid x \geq 10\}$

  means in set S, all elements are equal to or greater than 10.

- Described by graph

  $x=\{x \mid x \geq 2\}$

  −2 −1 0 1 2

B. The basic properties of a set

- Certainty

  $S=\{a,b,c\}$

  Set S contains 3 elements of a, b, and c. There is no other element in set S.

- Diversity

  $S=\{a,b,c\}$ where $a \neq b \neq c$

  If $S=\{a,a,b,b,b,c,c\}$, then only $S=\{a,b,c\}$ counts.

- No particular order

  $S=\{1,100,2,13,-5\}$

  Set S contains 5 elements 1, 100, 2, 13, −5 that aggregated in the set without any particular order.

- Classification

1. Limited Set

   $S = \{0, 1, 2, 3\}$ The number of elements is fixed.

   S contains only 4 elements

2. Unlimited Set

   $S = \{x \mid x > 0\}$

   S contains all the numbers greater than '0'.

3. Empty Set

   $S = \{\ \}$ or $S = \phi$ (Greek letter pronounced as "Phi")

   There is no element in the set. But, if $S = \{\phi\}$, it means there is an element, called 'empty' in set S. So, S is not an empty set.

C. Unions and Intersection

- Unions

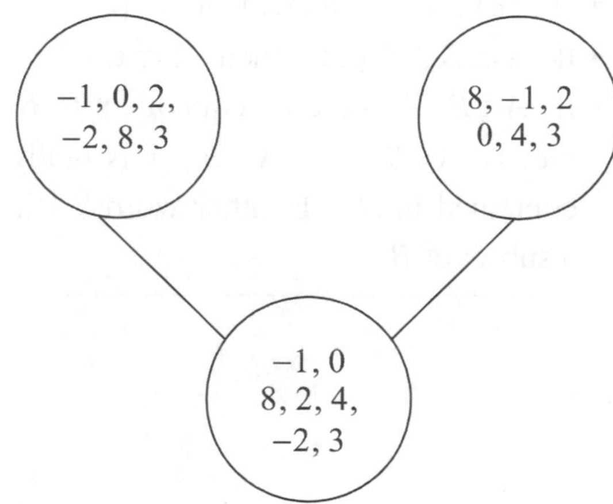

Select elements from all sets to form the new set, keep just one copy for duplicated elements

Set $A = \{-1, 0, 2, -2, 8, 3\}$

Set $B = \{8, -1, 2, 0, 4, 3\}$

$A \cup B = \{-1, 0, 8, 2, 4, -2, 3\}$

Ordering does not matter

- Intersection

1, 2, 4, 7 | 3 5 | 0, −1, −2, 6

Both set A and set B contains the same elements 3, 5

$A \cap B = \{3, 5\}$

Set $A = \{1, \ 2, \ 3, \ 4, \ 5, \ 7\}$

Set $B = \{0, -1, -2, \ 3, \ 5, \ 6\}$

$A \cap B = \{3, 5\}$

D. Equal

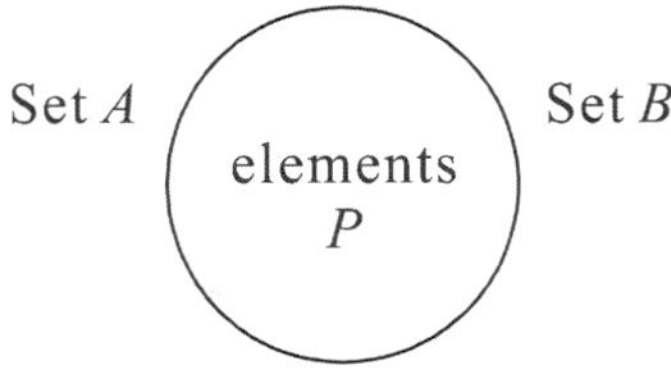

Every element in Set A is also in Set B.

$P \in A$ and also $P \in B$

$\therefore \ A = B$

> ★Tips
> - ϕ is a null set, or an empty set,
> - $\phi \subseteq \{\phi\}$ ϕ is a subset of $\{\phi\}$,
> - $\phi \in \{\phi\}$, ϕ is an element of $\{\phi\}$,
> - If $A \subseteq B$, $B \subseteq C$, then $A \subseteq C$
> - If $A \subseteq B$, A could be equivalent to B, means $A = B$, but $A \subset B$, A is really contained in B. In other words, A is a subset of B.

◆Example:

$A = \{a, b, c, d\}$, $B = \{a, b, c, d\}$

Then $A = B$

If $A = \{a, b, c, d\}$, $B = \{a, b\}$

A contains B but not equal to B and B is a real subset of A

E. Complement

$U = A \cup CuA$

$\phi = A \cap CuA$

$A = Cu(CuA)$

Set U

U CuA A

complement of set A

> ★Tips
> The notation for complement of set A can be CuA, A', or $\overline{A}$.

◆Example:

$U = \{1, 2, 3, 4, 5, 6, 7\}$, $P = \{2, 3, 5, 7\}$

$U = P \cup CuP$ $\because$ $CuP = \{1, 4, 6\}$

U

1, 4, 6

2, 3, 5, 7

$\phi=\{2, 3, 5, 7\}\cap\{1, 4, 6\}$

$\{2, 3, 5, 7\}=Cu(Cup)=Cu(1, 4, 6\}=\{2, 3, 5, 7\}$

F. The relations between sets

• Subsets

Set A

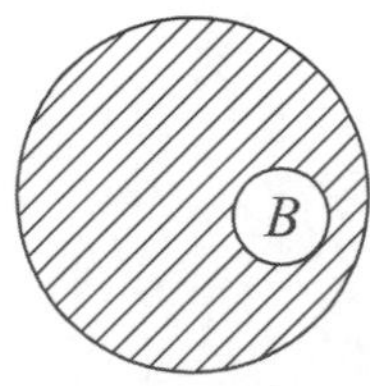

B is a subset of Set A

Denoted $B \subset A$, B is a real subset of A.

G. Subsets

◆Example 1:

$\{1, 2\}$

The subsets are $\{1\}$, $\{2\}$, $\{1, 2\}$, ϕ.

There are 4 subsets of $\{1, 2\}$.

$\{a, b, c\}$ can have $\{a\}$, $\{b\}$, $\{c\}$, $\{a, b\}$, $\{a, c\}$, $\{b, c\}$, $\{a, b, c\}$ and ϕ, 8 subsets.

If there are n elements in a set, the number of subsets is 2^n.

Set $\{1, 2, 3, 4\}$ has $4^2 = 16$ subsets.

◆Example 2:

Set $A=\{x|x\leq 12\}$, $B=\{x|x\geq b\}$, if $A\cap B=\phi$, what is the value of b?

$A = \{x \mid x \leq 12\}$ B

12 b (somewhere greater than 12)

So, $b>12$

◆Example 3:

Set $A=\{x|1<x<9\}$, $B=\{x|a<x<3a+1\}$

if $A\cap B=\{x|3<x<5\}$, what is the description of B ?

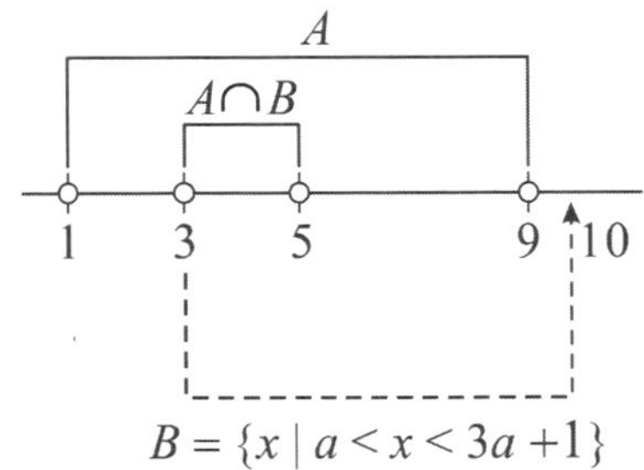

$\therefore$ $a=3$, $3a+1=10$

$B=\{x|3<x<10\}$

◆Example 4:

Set $A=\{a+2, a+3\}$, $B=\{b-1, b+1\}$

if $A\cup B=\{0, 2, 5, 6\}$, what is the value of $a+b$?

$A=\{a+2, a+3\}$

the relation between a_1 and a_2

$a_2-a_1=1$, we find $a+2=5$, $a+3=6$

So, $a=3$

$A\cup B=\{0, 2, 5, 6\}$

$B=\{b-1, b+1\}$

$b_2-b_1=b+1-(b-1)=2$

$\therefore$ So, $b_1=0$, $b_2=2$, so, $b=1$

$a+b=3+1=4$

◆Example 5:

$U=\{1, 2, 3, 4, 5, 6, 7, 8\}$

$A=\{1, 2, 3, 4\}$, $B=\{3, 4, 5, 6\}$

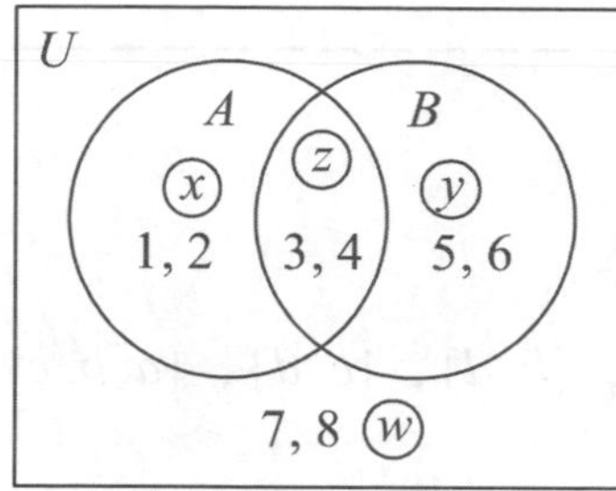

$X=\{1, 2\}=A\cap(CuB)=\{1, 2, 3, 4\}\cap\{1, 2, 7, 8\}$

$Y=\{5, 6\}=B\cap(CuA)=\{3, 4, 5, 6\}\cap\{5, 6, 7, 8\}$

$Z=\{3, 4\}=A\cap B=\{1, 2, 3, 4\}\cap\{3, 4, 5, 6\}$

$W=\{7, 8\}=Cu(A\cup B)=Cu\{1, 2, 3, 4, 5, 6\}=\{7, 8\}=(CuA)\cap(CuB)=\{5, 6, 7, 8)\cap\{1, 2, 7, 8\}$

$=\{7, 8\}$

Answer to Concept Check 2-01

The multiples of 3 between 20 and 40 are 21, 24, 27, 30, 33, 36, 39. The are 7 mumbers.

Answer is (C)

Answer to Concept Check 2-02

$P=\{x\,|\,x$ are all non-negative numbers$\}$

Means all numbers $x\geq 0$

$N=\{x\,|\,x$ are all non-positive numbers$\}$

Means all numbers $x\leq 0$

$P\cup N=\{x\,|\,x$ are all real number$\}$

means all real numbers

$P\cap N=\{0\}$ or "0" is the intersection of set P and set N.

Answer to Concept Check 2-03

$CuP\cup P=\{1, 2, 3, 4, 5, 6, 7, 8\}$

If $P=\{1, 3, 5, 7\}$

So, $CuP=\{2, 4, 6, 8\}$

Answer to Concept Check 2-04

Set $\{a, b, c, d\}$ has 4 elements

So, the maximum subset it has $2^4 = 16$ subsets

such as $\{a\}$, $\{b\}$, $\{c\}$, $\{d\}$, $\{a,b\}$, $\{a,c\}$, $\{a,d\}$, $\{b,c\}$, $\{b,d\}$, $\{c,d\}$, $\{a,b,c\}$, $\{a,b,d\}$, $\{a,c,d\}$, $\{b,c,d\}$, $\{a,b,c,d\}$, ϕ

Answer is (D)

Answer to Concept Check 2-05

$\{1, 2\} \subset S \subseteq \{1, 2, 3, 4\}$, S must be $\{1, 2, \cdots\cdots\}$, but not more than $\{1, 2, 3, 4\}$

So, S must be one of $\{1, 2, 3\}$, $\{1, 2, 4\}$, $\{1, 2, 3, 4\}$

Answer is (C)

2-2 SEQUENCE

A. Arithmetic sequence

The differences of each term from its previous term is always the same.

◆Example:

3, 5, 7, 9, 11, 13, 15

2 2 2 2 2 2

"2" is the common difference

"3" is the first term

"15" is the last term

The pattern of the sequence is

$T_n = a + (n-1) \cdot d$

Where a is the first term,

T_n is the nth term

d is the difference

$S_n = a + (a+d) + (a+2d) + \cdots\cdots + [(a+(n-2)d]$

a ⇒ the first term

d ⇒ the common difference

S_u ⇒ the sum of terms

Since Average $= \dfrac{a+[a+(n-1)\cdot d]}{2} = \dfrac{2a+(n-1)\cdot d}{2}$

$S_n = \dfrac{2a+(n-1)d}{2} \times n = \dfrac{n}{2}[2a+(n-1)\cdot d]$

◆Example:

$S_n = 2+4+6+8+10+12 = \dfrac{6}{2}[2\cdot 2+(6-1)\cdot 2] = 3(4+10) = 42$

$[a = 2, d = 2, \text{and } n = 6]$

◆Example:

If in an arithmetic sequence $a=5$, $d=2$, what is the value of the 14th term and the sum of the first 24 terms ?

The list of the sequence is 5, 7, 9, 11, ……

The 14th term $= a+(14-1)\cdot d = 5+13\times 2 = 31$

The sum of the first 24 terms

$S_{24} = \dfrac{24}{2}[2\times 5+(24-1)\times 2] = 12[10+23\times 2] = 672$

Answer to Concept Check 2-06

1, 3, 5, 7, ⋯ it is an arithmetic sequence, $a=1$, $d=3-1=2$

The 13th term is $a+(13-1)\cdot 2 = 1+12\cdot 2 = 1+24 = 25$

Answer is (D)

B. Geometric Sequence

- Constant Ratio Between Terms

If a ⇒ the first term

r ⇒ constant ratio

ar^{n-1} ⇒ the last term of n terms

S_u ⇒ the sum of the sequence for n terms

The pattern of the sequence is $T_n = ar^{n-1}$

$S_n = a + ar + ar^2 + ar^3 + \cdots + ar^{n-1} \cdots\cdots①$

$r \cdot S_n = ar + ar^2 + ar^3 + ar^4 + \cdots + ar^{n-1} + ar^n \cdots\cdots②$

②–① $r \cdot S_n - S_n = ar^n - a$

$S_n(r-1) = a(r^n - 1)$

$$S_n = \frac{a(r^n-1)}{r-1} \text{ or } S_n = \frac{a(1-r^n)}{1-r}$$

◆Example:

$S_n = 2+4+8+16+32+64+128$

$a=2,\ r=2,\ n=7$

$S_7 = \frac{a(r^n-1)}{r-1} = \frac{2(2^7-1)}{2-1} = 2(128-1) = 2\times127 = 254$

◆Example:

$S_n = 1+3+9+27+81$

$a=1,\ r=3,\ n=5$

$S_5 = \frac{1\cdot[3^5-1]}{3-1} = \frac{(243-1)}{2} = \frac{242}{2} = 121$

$S_n = \frac{a(r^n-1)}{r-1} = \frac{a(1-r^n)}{1-r}$

if $0<r<1$, and $n \to \infty$ (infinite)

$r^n \Rightarrow 0$

So, $S_n = \frac{a(1-0)}{1-r} = \frac{a}{1-r}$

◆Example:

$$S_n = 1+\frac{1}{2}+\frac{1}{4}+\frac{1}{8}+\frac{1}{16}+\cdots+\frac{1}{2^{n-1}}$$

$$a=1,\quad r=\frac{1}{2},\quad n\to\infty$$

So, $S_n = \dfrac{1}{1-\frac{1}{2}} = \dfrac{1}{\frac{1}{2}} = 2$

Answer to Concept Check 2-07

2, 4, 8, ⋯

It is a geometric sequence

$$a=2,\quad r=\frac{4}{2}=2,\quad T=10$$

$$\text{Sum}=\frac{a(1-r^n)}{1-r}=\frac{2(1-2^{10})}{1-2}=\frac{2-2048}{-1}=\frac{-2046}{-1}=2046$$

Answer to Concept Check 2-08

The sequence of 8, 4, 2, 1, ⋯is an infinite geometric sequence,

$$a=8,\quad r=\frac{4}{8}=\frac{1}{2}<1$$

$$\text{Sum}=\frac{a}{1-r}=\frac{8}{1-\frac{1}{2}}=\frac{8}{\frac{1}{2}}=16$$

Answer is (C)

2-3 DEFINED OPERATIONS

Designed Models for Math Operations

◆Example 1:

Let the operation ⊕ be defined for all numbers by $a\oplus b=(a+b)(a-b)$, if $9\oplus u=0$, what is the value of u?

$\because\ 9 \oplus u = (9+u)(9-u) = 9^2 - u^2 = 0$

$\therefore\ u^2 = 9,\ \therefore\ u = \pm 3$

◆Example 2:

Let the operation $\boxed{-}$ be defined by $a \boxed{-} b = \dfrac{a-b}{a+b}$. If $m \boxed{-} n = 5$.

What is the value of $\dfrac{m}{n}$?

$\because\ m \boxed{-} n = \dfrac{m-n}{m+n} = 5$

$\therefore\ 5m + 5n = m - n,\ \therefore\ 4m = -6n$

$\therefore\ m = \dfrac{-6}{4} n = -\dfrac{3}{2} n$

$\dfrac{m}{n} = \dfrac{-\dfrac{3}{2}n}{n} = -\dfrac{3}{2}$

◆Example 3:

Let the operation "!" be defined by $n! = n(n-1)(n-2)(n-3)\cdots 3 \cdot 2 \cdot 1$, and let the operation nCr be defined by $nCr = \dfrac{n!}{(n-r)!r!}$, when n, and r are positive integer, what is the value of $4C_2$?

$\therefore\ 4C_2 = \dfrac{4!}{(4-2)!2!} = \dfrac{4 \cdot 3 \cdot 2 \cdot 1}{(2 \cdot 1) \cdot (2 \cdot 1)} = 3 \times 2 \times 1 = 6$

Answer to Concept Check 2-09

$8 \boxed{/} (4 \boxed{/} x) = \dfrac{8}{\dfrac{4}{x}} = 2x$

$2x = 4,\ x = 2$

2-4 PATTERNS

A list of numbers that follow a certain rule or a group of rules

◆Examples:

(1) 3, 5, 4, 6, 5, 7, 6 ⋯

Based on the list of numbers shown above, what is the value of the 8th term ?

3, 5, 4, 6, 5, 7, 6, ?

2 −1 2 −1 2 −1 [2]

According to the rule of the pattern the 8th term is $6+2=8$

(2) 1, 2, 4, 7, 11, 16, ⋯

Based on the list above, what is the value of the 8th term

1, 2, 4, 7, 11, 16, 22, (29) → 8 th term

1 2 3 4 5 6 7

The 8th term $=22+7=29$

(3) 1, 2, 4, 7, 8, 10, 13, 14, ⋯

Based on the sequence shown above, what is the value of the 9th term ?

1, 2, 4, 7, 8, 10, 13, 14, [16] → 9 th term

1 2 3 1 2 3 1 2

The 9th term $=14+2=16$

(4) 110011001100110011 ⋯

The bit pattern shown above, what is the value of the 19th bit ?

1100 1100 1100 1100 11 [0]0 19th bit

$19 \div 4 = 4 \cdots 3$

The value of the 19th is "0"

(5) 0, 3, 8, 15, 24, □, ⋯

Based on the list of numbers shown above, what is the value of the 6th term?

0, 3, 8, 15, 24, [35]

3 5 7 9 11

(6) The data list : 1, 2, 3, 5, 8, 13, 21, ⋯

$T_3 = T_1 + T_2$, $T_4 = T_2 + T_3$, ⋯

$T_5 = 8 = T_3 + T_4 = 3+5$

each number is obtained by adding up two previous numbers.

This data set is called 'Fibonacci Sequence'

(7) If $T_1 = 1$, $T_2 = 3$, $T_3 = 7$, $\cdots$

$T_{n+1} = 2T_n + 1$, what is the value of T_8 ?

$T_8 = 2T_7 + 1$, $T_7 = 2T_6 + 1$, $T_6 = 2T_5 + 1$, $T_5 = 2T_4 + 1$, $T_4 = 2T_3 + 1$, so, $T_4 = 15$, $T_5 = 31$, $T_6 = 63$, $T_7 = 127$, $T_8 = 255$

(8) 1, 3, 7, 15, 31, 63, $\boxed{127}$

2 4 8 16 32 64

$63 + 64 = 127$

Answer to Concept Check 2-10

$1100110011001100110011001100110011\cdots$

The list-pattern repeats for every four bits.

The 121st list should be $121 \div 4 = 30 \cdots 1$

So, the 121st bit is "1"

Answer to Concept Check 2-11

The list is

$-5, -4, -2, -1, 1, 2, 4, 5, \cdots$

+1 +2 +1 +2 +1 +2 +1 (+2)

so, $5 + 2 = 7$, $T_9 = 7$

Answer to Concept Check 2-12

Since $T_5 = 2T_3 + T_4$, $T_4 = 2T_2 + T_3$, $T_3 = 2T_1 + T_2$, $T_1 = 2$, $T_2 = 5$, $T_3 = 2 \times 2 + 5 = 9$

$\therefore$ $T_4 = 2T_2 + T_3 = 2 \times 5 + 9 = 19$

$T_5 = 2T_3 + T_4 = 2 \times 9 + 19 = 37$

2-5 LOGIC

A. Terminologies

- Statement: It is a sentence to describe a fact or event. It may be true or false
- Negation: It is opposite of the statement

P	$\sim P$
True	False
False	True

- Conjunction: P and Q denoted as $P \wedge Q$
- Disjunction: P and Q denoted as $P \vee Q$

B. Truth Table

P	Q	P and Q ($P \wedge Q$)	P and Q ($P \vee Q$)
T	T	T	T
T	F	F	T
F	T	F	T
F	F	F	F

◆Example:

One of the rides in a theme park says, "If your age is greater than or equal to 15 years or your height is greater than or equal to 5 ft, then you are permitted to take this ride". The Truth Table will be counted as a follows.

P	Q	Age ≥ 15 year and height ≥ 5 ft ($P \wedge Q$)	Age ≥ 15 year or height ≥ 5 ft ($P \vee Q$)
T	T	T (OK)	T (ok)
T	F	F (Not OK)	T (ok)
F	T	F (Not OK)	T (ok)
F	F	F (Not OK)	F (Not OK)
		(AND)	(OR)

P: Age ≥ 15 years, Q: height ≥ 5 feet

- If-then statements.

  It is a conditional statement.

  If some thing happens, then some other thing happens.

  If "P", then "Q", P is a hypothesis and Q is a conclusion.

C. Conditional Statement

Statements	Form	Sample	Conclusion
Conditional	$P \to Q$	If it rains, then the ground gets wet	T
Converse:	$Q \to P$	If the ground gets wet, then it rains	F(Not sure)
Inverse:	$\sim P \to \sim Q$	If it doesn't rain, then the ground does not get wet	F(Not sure)
Contrapositive	$\sim Q \to \sim P$	If the ground does not get wet, then it does not rain	T

> Conditional statement = Contrapositive statement
> If P then Q = If $\sim Q$ then $\sim P$

◆For example:

"If it is perfect, then it is beautiful" is true.

P = it is perfect, Q = it is beautiful

Statements	Form	Sample	Conclusion
Conditional	$P \to Q$	If it is perfect, then it is beautiful	True
Converse:	$Q \to P$	If it is beautiful, then it is perfect	False
Inverse:	$\sim P \to \sim Q$	If it is not perfect, then it is not beautiful	False
Contrapositive	$\sim Q \to \sim P$	If it is not beautiful, then it is not perfect	True

> The conditional statement is always logically equivalent to the contrapositive statement

◆Example 1:

If the statement: "All handbags sold at Store *A* are over \$100" is true. Which of the following statements must be true?

(A) Handbags under \$100 are not sold by Store *A*.

(B) One hand bag is priced at \$90.

(C) All bags over \$100 are sold by Store *A*

(D) Among all merchandise sold by Store *A*, the ones over \$100 munt be handbags

(E) All merchandise sold at Store *A* are over \$100.

Conditional statement: if P = all handbags sold at Store A, then Q = over \$100 is true.

In this case the contrapositive statement is:

Not over \$100, Not sold by Store A. The answer is (A)

◆Example 2:

- **Conditional statement:**

If it is a frog, then it is an amphibian

- **Converse statement:**

If it is an amphibian, then it is a frog

(False)

- **Inverse statement:**

If it is not a frog, then it is not an amphibian,

(False)

- **Contrapositive statement:**

If it is not an amphibian then it is not a frog. (true)

◆Example 3:

If the statement 'If he is born in the U.S.A, then he is a citizen of the U.S.A' is true. Which of the following statements must be true?

(A) He is not born in the U.S.A, then he is not a citizen of the U.S.A.

(B) He is a citizen of the U.S.A then he is born in the U.S.A.

(C) He is not a citizen of the U.S.A. then he is not born in the U.S.A.

Conditional statement:

P = born in U.S.A. Then Q = citizen of U.S.A.

Contrapositive statement:

$\sim Q$ then $\sim P$, "not a citizen of U.S.A then not born in U.S.A."

Answer is (C)

Answer to Concept Check 2-13

Case A, $F \wedge T$, is False, so $A \to F$

Case B, $F \vee T$, is False, so $B \to F$

Answer to Concept Check 2-14

The original statement: "If it rains, then the ground gets wet"

1. In if – then statement, only contrapositive statement is logically equivalent to the original statement.
2. Contrapositive statement

   If the ground doesn't get wet, then it doesn't rain.

Answer is (C)

D. Logical reasoning

It is a kind of reasonable and consistent thinking through inductive or deductive ways to get to the conclusion.

◆Example 1:

Who is who ?

The triples sat in a row, the eldest brother always tells the truth, the second brother sometimes tells the truth and the youngest brother always tells a lie.

The person sitting in the first place said "the person sitting in the second chair is the eldest brother", the person sitting in the second chair said "I am the third youngest brother",

The person sitting in the last chair said "the person sitting in the second chair is the second brother"

Who is who?

Analysis:

Brother 1	Brother 2	Brother 3

1. Brother 1 said "Brother 2 is the eldest brother", infer that he is not the eldest brother. If he would be the eldest brother, he would have said I am the eldest brother.
2. Brother 2 said "I am the youngest brother", definitely. he is not the eldest brother.
3. Neither of first place nor second are the eldest brother, so the person sitting in the third chair is the eldest brother, and he said brother 2 is the second old brother. The second brother is sitting in the second chair, and finally, the person sitting in the first place is the youngest brother.

◆Example2:

The color of headscarf ?

In an ancient country, the penalty was very cruel. Once three criminals were caught, named criminal A, criminal B, and criminal C. respectively. The king was very angry and sentenced these three criminals to death.

The king asked them to line up in accordance with ABC's order, A is at the forefront, B at the middle, C is at the end. B can see A's head and C can see both of A'S and B'S head. The king said that I will cover your eyes and put a color headscarf on each of your head. If anyone can guess out the color of the headscarf on his head, he can be released from his sin. The king said: I have only five headscarves here, three of them are black, two of them are white.

When the three criminals' eye masks were opened at the same time, the last C did not dare to guess the color of his headscarf. B did not dare to guess the color of his headscarf either. Then criminal A said that 'I dare to guess the color of my headscarf'. As a result, his guess is right. And the king pardon his sin.

What color is his headscarf?

When their eye maskers were opened simultaneously,

	A	B	C
(1)	W	W	B
(2)	B	W	?
(3)	W	B	?
(4)	B	B	?

1. There are 3 cases Criminal C dare not to guess the color of his headscarf
   In case 1, C definitely knows that he wear black headscarf. Therefore, it must be one of (2), (3), or (4) cases.
2. Criminal B referred that if criminal A had wore white color headscarf ,then he must wear black color, for there are only two white headscarves, unfortunately, A did not wear white color headscarf.
3. The situation made criminal B not able to guess the color of his headscarf. Therefore criminal A knew for sure that his headscarf color was "Black".

◆Example3:

> Bubble sort using the methods of comparison and exchanging to re-arrange a chaotic data set to an ordered ascending or descending ordered data list, is so called.
> Bubble sort which compares the two adjacent data setting the smaller on the top. Through some paths, the chaotic data set become an arranged data set. It is somewhat like the bubble rises from the bottom to the surface, so it is called the bubble sort.

- Bubble Sort

$16 < 36$, $13 < 36$, $17 < 36$, $27 < 36$, $25 < 36$, $18 < 36$, $32 < 36$, $01 < 36$, $05 < 36$

Original	Path 1	Path 2	Path 3	Path 4	Path 5	Path 6	Path 7	Path 8	Path 9
16	16	13	13	13	13	13	13	01	01 no more exchange
36	13	16	16	16	16	16	01	05	05
13	17	17	17	17	17	01	05	13	13
17	27	25	18	18	01	05	16	16	16
27	25	18	25	01	05	17	17	17	17
25	18	27	01	05	18	18	18	18	18
18	32	01	05	25	25	25	05	25	25
32	01	05	27	27	27	27	27	27	27
01	05	32	32	32	32	32	32	32	33
05	36	36	36	36	36	36	36	36	36

Answer to Concept Check 2-15

Deduced from a set of facts to the conclusion

Allen		Betty		Charles		David		Eric	
C_2	A_3	A_1	E_4	B_3	D_5	C_2	E_4	D_1	B_2

Step 1. Suppose C_2 is true, than E_4 is "False"

Step 2. From Betty A_1 is "True"

Step 3. From Eric D_1 is 'False', ' B_2 is True'

Step 4. From Allen C_2 is "False", contradiction to the original prediction

Step 5. Suppose C_2 is "False", then A_3 is "True"

Step 6. B_3 is "False", from Charles D_5 is "True"

Step 7. A_1 is "False", from betty E_4 is "True"

Step 8. D_1 is "False", from Eric B_2 is "True"

Step 9. B_2, A_3, E_4, D_5, C_1 must be "True"

Team C won the championship

Answer is (C)

CHAPTER 2
PRACTICE

1. In a sequence, the first term is -15, and each term after the first term is 7 more than the previous term, what is the 77th term?

(A) 514

(B) 515

(C) 516

(D) 517

$5,\ -10,\ 20,\ -40,\ \cdots\cdots$

2. In the sequence above, what is the value of the 10th term?

(A) 2560

(B) -2560

(C) 5120

(D) -5120

3. Let operation $\boxed{\times}$ be defined by $x\boxed{\times}\,y=3x^2+\frac{y}{4}$ for all real numbers of x and y. If $2\boxed{\times}\,k=14$, what is the value of k ?

(A) 8

(B) 10

(C) 12

(D) 14

4. Let the operation E be defined as $x\,\mathrm{E}\,y=2x-y$ for all values of x and y. If $a\,\mathrm{E}\,b=4$, what is the value of $\frac{b}{a}$?

(A) 2

(B) 4

(C) $2-\frac{4}{a}$

(D) $2+\frac{4}{a}$

5. What is the next number in the following pattern ?

2, 1, 4, 4, 6, 7, 8, 10, 10, 13, 12, 16, 14 ?

(A) 19

(B) 22

(C) 25

(D) 27

6. $1,\ -3,\ 9,\ -27,\ 81,\ -243,\ 929,\ \cdots$

In the sequence above, the first term of the sequence is 1 and each term after the first term is -3 times the previous term immediately before it. How many terms of the first 100 terms are greater than 500 ?

(A) 47

(B) 49

(C) 51

(D) 53

7. In the sequence $a_n = a_{n-1} + a_{n-2}$, where $n > 2$, if $a_6 = 13$, $a_7 = 21$, what is a_3?

(A) 1

(B) 2

(C) 3

(D) 4

8. + + + + ……

In the figure above, each equilateral triangle is formed by connecting the midpoint of the three sides of the previous triangle. What is the sum of the areas of those infinite triangles ?

(A) $2A$

(B) $\frac{7}{3}A$

(C) $\frac{5}{3}A$

(D) $\frac{4}{3}A$

9. 4, 3, −1, −4, −3, 1, 4, 3, −1, −4, ⋯

The sequence is shown above, the first term is 4 and the second term is 3, each term after the second term can be obtained by subtracting the previous term from the term before it. What is the value of the 71th term?

(A) −4

(B) −3

(C) 3

(D) 4

10. −3, −2, −1, 0, 1, 2, −3, −2, −1, 0, ⋯

The sequence shown above is formed by repeating the six number to the infinity. What is the sum of the first 150 terms of the sequence ?

(A) 0

(B) −3

(C) −35

(D) −75

TEAM	GAME	WON	LOSE	TIE
A	60	42	13	5
B	60			4
C	60	31	23	6
D	60	23	34	3

11. There are four little baseball teams in Cherry County. Each team has 20 games for the other three teams. The results of the games are as shown in the table above. However, the data was accidentally defaced. What are the correct data ?

12. Let $x \boxed{*} y$ be defined by x^2y, and $x \boxed{+} y$ be defined by $x^2 + y$ for all values x and y. What is the value of $2 \boxed{+} (3 \boxed{*} 4)$?

(A) 40

(B) 48

(C) 52

(D) 60

13. The operation $x \otimes y$ is defined by $x \otimes y = \frac{x^2 - y^2}{2xy}$. If $m \otimes 1 = \frac{3}{4}$, which of the following could be a possible value of m ?

(A) $-\frac{1}{2}$

(B) $-\frac{1}{3}$

(C) $-\frac{1}{4}$

(D) $-\frac{1}{6}$

14. What is the 125th digit on the right of the Decimal point in the decimal form of $\frac{6}{11}$?

(A) 3

(B) 4

(C) 5

(D) 6

15. If $p \ominus q = \frac{p+q}{p-q}$, for all values of p and q, $p \neq q$. What is the value of $(\frac{p \ominus q}{p})$ $(\frac{p \ominus q}{q})$, given $p=3$ and $q=2$?

16. At North Elementary School in a certain year, all teachers wear with glasses. If the statement is true which of the following statements is also true?

(A) People with glasses in North Elementary School must be teachers.

(B) People without glasses must not be teachers in this school.

(C) There are more teachers with glasses in this school than other schools.

(D) David is not the teacher of this school, so, he does not wear glasses.

17. In Noelle's class, every student must take at least one foreign language class to fulfill second language requirements.

8 students in her class take Japanese, 12 students take German, and 18 students take Spanish. Among them, 4 students take both Japanese and German, 6 students, take both German and Spanish, and 6 students take both Japanese and Spanish.

If there are 24 students total in Noelle's class, how many of them take all 3 languages classes ?

18. $A=\{0, 2, 3, 5\}$, $B=\{0, 1, 2, 3, 4\}$

$S=\{0, 1, 2, 3, 4, 5, 6, 7\}$

What is $Cs(A \cup B)$?

(A) ϕ

(B) $\{0\}$

(C) $\{6, 7\}$

(D) $\{0, 1, 2, 3, 4, 5, 6, 7\}$

19. Set $A=\{k+3, k+4\}$, $B=\{t-2, t+1\}$ if $A \cup B=\{0, 3, 5, 6\}$, what is the value of $k+t$?

(A) 4

(B) 6

(C) 8

(D) 10

20. Charles has drawn 13 spades and 13 hearts, from a deck of poker cards. And ask the audience to shuffle the cards and arrange the cards face down into two columns. Charles predicts that "The spade-cards in the long column is k cards more than the heart-cards in the short column." When audience opens the cards then the prediction is true. If the two columns are arranged to be 17 and 9, what is the number of k ?

(A) 2

(B) 3

(C) 4

(D) 5

21. Connor and Jens play games, there are a total of four boxes placed on a table, one of which contains treasures and the other three boxes are empty. Before the game starts, Connor first chooses one box and then Jens chooses another box. The host opened two boxes without treasures and asked Connor whether he likes to exchange Jens' box. Connor always wants to exchange Jens' box.

After playing it many times, most of the time Connor gets the treasures, why?

CHAPTER 2
ANSWERS and EXPLANATIONS

1. $a=-15,\ d=7$

$T_{77}=a+(d)(77-1)$

$=-15+(77-1)(7)$

$=-15+532=517$

Answer is (D)

2. $T_{10}=5\cdot r^{n-1}=5\cdot(-2)^{10-1}$

$=5\cdot(-2)^9$

$=-2560$

Answer is (B)

3. $2\boxtimes k=3(2)^2+\frac{k}{4}=12+\frac{k}{4}=14$

$\therefore\ \frac{k}{4}=2\quad k=8$

Answer is (A)

4. $a \text{ E } b=2a-b=4$

$2a-b=4,\ b=2a-4$

$\therefore\ \frac{b}{a}=\frac{2a-4}{a}=2-\frac{4}{a}$

Answer is (C)

5. 2, 1, 4, 4, 6, 7, 8, 10, 10, 13, 12, 16, 14, $\boxtimes$ ...

+3 +3 +3 +3 +3 +3

2, 1, 4, 4, 6, 7, 8, 10, 10, 13, 12, 16, 14, □

+2 +2 +2 +2 +2 +2

$x=19$

Answer is (A)

6. The list 1, −3, 9, −27, 81, −243, 929, ⋯

The value of the term greater than 500 start from T_7 929. So, there are $100-6=94$, half of them are greater than 500, and half of them are negative.

So there are $94\div 2=47$

Answer is (A)

7. $a_7=a_6+a_5,\ 21=13+a_5$

$a_5=8$

$a_6=a_5+a_4,\ 13=8+a_4$

$a_4=5$

$a_5=a_4+a_3,\ 8=5+a_3$

$a_3=8-5=3$

Answer is (C)

8. Sum $=A+\frac{A}{4}+\frac{A}{16}+\cdots$

$a=A,\ r=\frac{1}{4}$

$S=\frac{A}{1-\frac{1}{4}}=\frac{4}{3}A$

Answer is (D)

9. repeating for every 6 digits.

$71\div 6=11\cdots\cdots 5$

one group is 4, 3, −1, −4, −3, 1, the fifth digrt is −3, Answer is (B)

10. $150 \div 6 = 25 \cdots\cdots 0$

The sum of one repeating group is -3 there are 25 groups, so, $25\times(-3) = -75$

Answers is (D)

TEAM	GAME	WON	LOSE	TIE
A	60	42	13	5
B	60			4
C	60	31	23	6
D	60	23	34	3

11.(1) The logic is that the number of total 'won' is equal to the number of total 'LOSE'.

So, $42 + W + 31 + 23$

$= 13 + L + 23 + 34$

$96 + W = 70 + L$, $L = W + 26$

(2) The total games for team B is 60,

$W + L + 4 = 60$, $W + L = 56$

So, $W + W + 26 = 56$, $2W = 30$,

$W = 15$

(3) $L = 15 + 26 = 41$

12. $3 \boxed{*} 4 = 3^2 \cdot 4 = 36$

$2 \boxed{+} (3 \boxed{*} 4) = 2^2 \boxed{+} 36 = 2^2 + 36 = 40$

Answer is (A)

13. $x \otimes y = \dfrac{x^2 - y^2}{2xy}$

so, $m \otimes 1 = \dfrac{m^2 - 1}{2m} = \dfrac{3}{4}$

$\therefore\ 6m = 4m^2 - 4$

$4m^2 - 6m - 4 = 0$

$(2m - 4)(2m + 1) = 0$

$\therefore\ m = \dfrac{2}{4} = \dfrac{1}{2}$

or $m = -\dfrac{1}{2}$

Answer is (A)

14. $6 \div 11 = 0.545454\cdots\cdots$

$125 \div 2 = 62.5$

It means that the 125th digit after decimal point is at the beginning of at the 63rd group, it is "5".

Answer is (C)

15. $p \ominus q = \dfrac{p+q}{p-q}$, given $p = 3$, $q = 2$

so $\left(\dfrac{p \ominus q}{p}\right) \ominus \left(\dfrac{p \ominus q}{q}\right)$

$= \left(\dfrac{\frac{3+2}{3-2}}{3}\right) \ominus \left(\dfrac{\frac{3+2}{3-2}}{2}\right)$

$= \left(\dfrac{5}{3}\right) \ominus \left(\dfrac{5}{2}\right) = \dfrac{\frac{5}{3} + \frac{5}{2}}{\frac{5}{3} - \frac{5}{2}} = -5$

16. Key: IF-Then Statement

The only true statement is contrapositive statement

Original Statement:

All teachers wear glasses. The contrapositive statement: 'If people do not wear glasses they are not the teacher of the school' so, Answer is (B).

17.

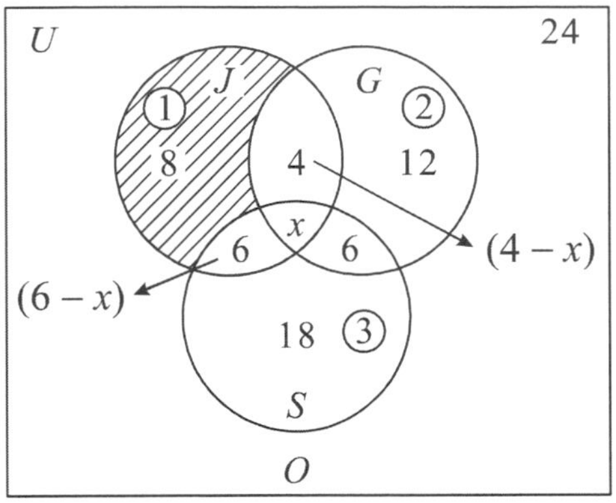

U 24
J ① 8
G ② 12
4, x, 6, 6
(6 − x) (4 − x)
18 ③ S
O

(1) $8-(6-x)-(4-x)-x$

$=8-6+x-4+x-x$

$=-2+x$

(2) $12-(6-x)-(4-x)-x$

$=12-6+x-4+x-x$

$=2+x$

(3) $18-(6-x)-(6-x)-x$

$=18-6+x-6+x-x$

$=6+x$

$24-0=(-2+x)+(2+x)+(6+x)$

$+(6-x)+(6-x)+(4-x)+x$

$=6+3x+16-3x+x=22+x$

$\therefore\ x=2$

Answer: 2 students take all 3 languages

18. (C)

$S=\{0, 1, 2, 3, 4, 5, 6, 7\}$

$A\cup B=\{0, 2, 3, 5\}\cup\{0, 1, 2, 3, 4\}$

$=\{0, 1, 2, 3, 4, 5\}$

$Cs(A\cup B)=\{6, 7\}$

Answer is (C)

19. (A)

Since $A\cup B=\{0, 3, 5, 6\}$

$k+3$ must be 5, $k+4$ must be 6

so, $k+3=5$, $k=2$

$t-2$ must be 0, $t+1$ must be 3

$\therefore\ t-2=0$, $t=2$

$\therefore\ k+t=2+2=4$

20. (C)

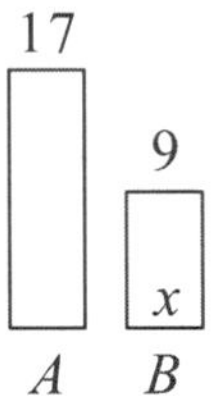

17 9
x
A B

Assume there are x hearts in column B, then there are $(13-x)$ hearts in column A. The spade card in column A is $17-(13-x)=4+x$

so, $k=(4+x)-x=4$

21. At the beginning of the game, Connor only had $\frac{1}{4}$ chance to get the treasures, and the other side had $\frac{3}{4}$ chance to have the treasure-box. When the host open 2 boxes without the treasure, Jens' box is most likely to have the treasures, because of on the Jens' side with $\frac{3}{4}$ odds.

CHAPTER 3
AVERAGE, RATIO, RATE, PROPORTION, and PERCENT

3-1 AVERAGE

$$\text{Average} = \frac{\text{total values}}{\text{number of elements}}$$

◆Example:

24 pieces of candy are distributed to 6 students. How many pieces of candy can be distributed to each student ?

Each student receives (on average) $= \frac{24}{6} = 4$ (pieces of candy)

Weighted Average

$$= \frac{\text{element 1}\times\text{weight 1}+\text{element 2}\times\text{weight 2}+\cdots+\text{element } n\times\text{weight } n}{\text{element 1}+\text{element 2}+\text{element 3}+\cdots+\text{element } n}$$

◆Example:

Archery scores

points	10	9	8	7	6	5
Arrows	2	3	2	1	1	1

The table above shows Alice's archery competition scores.

What was her average score ?

$$\text{Average} = \frac{2\times10+3\times9+2\times8+1\times7+1\times6+1\times5}{10} = \frac{81}{10} = 8.1$$

 Answer to Concept Check 3-01

$5\times90-(92+86+90+88)=94$

The fifth math score must be 94

Answer to Concept Check 3-02

$$A=\frac{1\times10+4\times9+8\times1+7\times4+6\times4+5\times3+4\times7+3\times2+2\times2+0\times2}{30}$$

$$=5.30$$

3-2 RATIO

A comparison of two quantities

Three forms

- Colon: 5 : 4

In Connor's class, the ratio of the number of boys to the number of girls is 5 : 4.

- To: 5 to 4

In Connor's class, the ratio of the number of boys to the number of girls is 5 to 4.

- Fraction: $\frac{5}{4}$

In Connor's class, the number of boys is $\frac{5}{4}$ the number of girls.

◆Example:

A white cat and a black cat cooperated to catch mice, and they caught a total of 45 mice. The ratio of the number of mice caught by the white to the number of mice caught by the black cat is 5 : 4. How many mice did the white cat catch ?

Method 1:

The ratio is 5 : 4.

It also means the white cat caught $\frac{5}{5+4}=\frac{5}{9}$ of the total number of mice.

So, white cat caught $\frac{5}{9}\times45=25$

Method 2:

The ratio is 5 : 4, rewrite it as a fraction $\frac{5}{4}$, and it also equals $\frac{5k}{4k}$,

So, $5k+4k=45$, $9k=45$, $k=5$

The mice that white cat caught

$5\times(5)=25$

Answer to Concept Check 3-03

Assume the length of her leg is ℓ.

so, $\dfrac{24.75+\ell}{\ell}=1.6 \quad 24.75=1.6\ell-\ell=0.6\ell$

$\therefore\ \ell=\dfrac{24.75}{0.6}=41.25$ (inches)

3-3 RATE

A rate is similar to a ratio, but it is a comparison of different type of unit.

For example: $\dfrac{240\text{ miles}}{4\text{ hours}}$ (different units), $\dfrac{\$5.0}{3\text{ ice cream}}$, $\dfrac{20\text{ hours}}{5\text{ workers}}$

↗ different units

$\dfrac{240\text{ miles}}{4\text{ hours}}=60\text{ miles / hour}$

↑ unit rate

“per” mile means a unit rate

$\dfrac{28\text{ dollars}}{140\text{ miles}}$ ← The unit in PER is always a divisor or a denominator

Answer to Concept Check 3-04

Miles	gasoline	cost
140	7	\$28

I. per gallon: $\dfrac{\text{miles}}{\text{gallon}} = \dfrac{140}{7} = 20$ miles / gallon

II. per mile: $\dfrac{\text{gallon}}{\text{mile}} = \dfrac{7}{140} = \dfrac{1}{20} = 0.05$ gallon / mile

III. per dollar: $\dfrac{\text{gallon}}{\text{dollar}} = \dfrac{7}{28} = \dfrac{1}{4} = 0.25$ gallon / dollar

IV. per gallon: $\dfrac{\text{dollar}}{\text{gallon}} = \dfrac{28}{7} = 4$ dollars / gallon

V. per dollar: $\dfrac{\text{mile}}{\text{dollar}} = \dfrac{140}{28} = 5$ miles / dollar

VI. per mile: $\dfrac{\text{dollar}}{\text{mile}} = \dfrac{28}{140} = \dfrac{1}{5} = 0.2$ dollars / mile

◆Example:

Brennan runs a 5,000-meter race, in the first 3,000 meters, he uses 15 minutes. At this rate, how long does he need in minutes to finish the rest of the race ?

(A) 8

(B) 9

(C) 10

(D) 11

(E) 12

$\dfrac{3000}{15} = 200$ meter/minute (For the first 15 minutes)

$5000 - 3000 = 2000$ (The rest distance of the race)

$\dfrac{2000}{200} = 10$ minutes

Answer is (C)

Answer to Concept Check 3-05

$\frac{6.8}{2}=\frac{x}{5}$ $\therefore$ $x=\frac{6.8}{2}\times 5=17$

Answer is (C)

Answer to Concept Check 3-06

The total workload is $P\times D$ people-hours one more person into the team it takes $\frac{P\times D}{P+1}$ to complete the job.

Answer is (A)

3-4 PROPORTION

$\frac{a}{b}=\frac{c}{d}$

Equal ratios

◆Example 1:

Ruby's sister earns \$3 for each \$2 Ruby earns. If she earned \$120 last week, how much did Ruby earn last week ?

The ratio of Ruby's sister's wage to Ruby's wage is 3 : 2, and assume Ruby's wage is x,

$\frac{3}{2} \times \frac{120}{x}$

$3x=240$ $\therefore$ $x=80$

Ruby earned \$80 last week

◆Example 2:

On a certain map, 1 inch represents 5 miles, if the direct distance from city A to city B shows 3.5 inches on the map, how far is the real distance from city A to city B ?

$\frac{1}{5} \times \frac{3.5}{x}$

$x=17.5$

The distance between city A and city B is 17.5 miles

◆Example 3:

A certain salt solution 12 ounces contains $\frac{1}{4}$ salt, how much salt must be added to the solution to make it contain $\frac{1}{3}$ salt ?

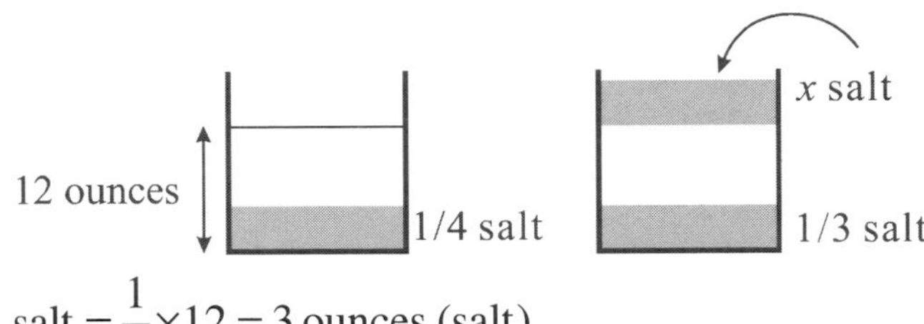

salt $=\frac{1}{4}\times 12=3$ ounces (salt)

salt ↘

so, $\frac{x+3}{12+x}=\frac{1}{3}$

solution

$3x+9=12+x$

$\therefore\ 2x=3,\ x=\frac{3}{2}$

◆Example 4:

A tray in a certain oven can bake 16 cookies at one time. If $\frac{1}{4}$ pound of flour can make 32 cookies, how many times of repeating baking is needed to complete baking the cookies that are made with 1 pound of flour ?

(A) 8

(B) 9

(C) 10

(D) 11

(E) 12

How many cookies are made with one pound of flour ?

$$\frac{\frac{1}{4}}{32}=\frac{1}{x}$$

$x=128,\ 128\div 16=8$ repeating times

Answer is (A)

Answer to Concept Check 3-07

Alex has 34 ounces (fluid ounce) of a 28% alcohol solution. How many ounces of 58% alcohol solution must be added into Alex's solution to makes a 32% alcohol solution ? (Rounded to hundredth)

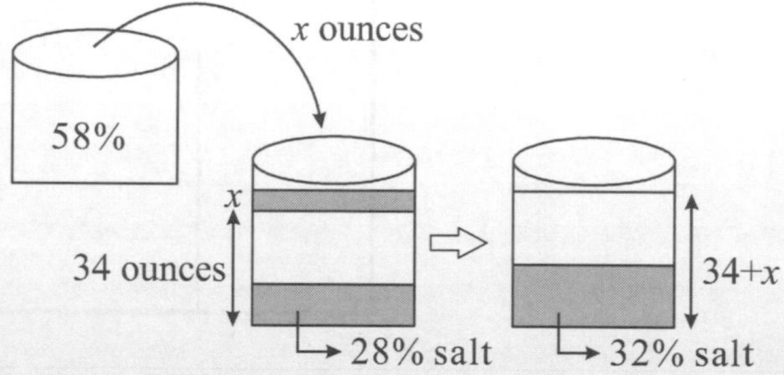

Step 1. Pour x ounces of 58% alcohol solution into Alex's solution.

Step 2. How much alcohol in total $34\times0.28+0.58x$

Step 3. How many ounces of new solution ?

$34+x$

Step 4. percent $=\dfrac{\text{part}}{\text{whole}}$

$$32\%=\frac{34\times0.28+0.58x}{34+x}\times100\%$$

same as: $0.32=\dfrac{34\times0.28+0.58x}{34+x}$

Step 5. $0.32\times34+0.32x=0.28\times34+0.58x$

$0.26x=0.04\times34 \quad \therefore \; x=\dfrac{0.04\times34}{0.26}=5.23$

so, 5.23 ounces of 58% alcohol solution must be added into Alex's solution.

3-5 PERCENTAGE (PERCENT)

A. Percent:

$$\frac{21}{100}\times100\%=21\%$$

→ Denominator = 100

B. Change fraction or decimal to percentage

(1) Fraction

$\frac{4}{5}=\frac{4\times 20}{5\times 20}=\frac{80}{100}=80\%$

$\frac{5}{4}=\frac{5\times 25}{4\times 25}=\frac{125}{100}=125\%$

(2) Decimal

$0.325=32.5\%$

$0.3=30\%$

$3.32=332\%$

$\frac{1}{2}=50\%$	$\frac{1}{3}=33.33\%$	$\frac{1}{4}=25\%$	$\frac{1}{5}=20\%$
	$\frac{2}{3}=66.67\%$	$\frac{2}{4}=50\%$	$\frac{2}{5}=40\%$
		$\frac{3}{4}=75\%$	$\frac{3}{5}=60\%$
			$\frac{4}{5}=80\%$

C. Percentage

$$\text{percent} =\frac{\text{part}}{\text{whole}}\times 100\%$$

◆Example:

There are 22 apples in a basket, 12 of them are red apples, what is the percentage of the apples in the basket that are red apples ?

$\text{Percent} =\frac{12}{22}\times 100\% = 54.55\%$

◆Example:

5 is 20% of a number. What is the value of the number ?

$\text{percent} =\frac{\text{part}}{\text{whole}}\times 100\%$

$\therefore\ \text{Whole}=\frac{\text{part}}{\text{percent}}$

$\text{Number} =\frac{5}{20\%}=\frac{5}{\frac{1}{5}}=25$

D. Increasing & Decreasing

Amount of increase = Original (whole) × the % of increase

Amount of decrease = original (whole) × the % of decrease

◆Example:

Amount of increase $=20\times\frac{10}{100}=2$

or Amount of increase $=20\times 0.1=2$

The sales volume of a house real estate company's transactions this month increased by 10% compared with the previous month. If the company's sales last month were 20 million dollars, how much were sold this month ?

$20\times(1+0.1)=22$ million dollars

$$\text{Percent Increase} = \frac{\text{New Amount} - \text{Old Amount}}{\text{Old Amount}}\times 100\%$$

$$\text{Percent Decrease} = \frac{\text{Old Amount} - \text{New Amount}}{\text{Old Amount}}\times 100\%$$

◆Example:

David's salary was \$80,000 last year, for some reason, it decreased to \$70,000, how much percentage was decreased ?

Percent Decreased $=\frac{80,000-70,000}{80,000}\times 100\% = 12.5\%$

◆Example:

The original price of a jacket is \$72, when it is on sale, the tag price is \$50, how much percentage is the discount ? (Rounded to hundredth)

Percent Decreased $=\frac{\text{Odd Amount} - \text{New Amount}}{\text{Old Amount}}\times 100\%$

Percent Decreased $=\frac{72-50}{72}\times 100\% = \frac{22}{72}\times 100\% = 30.56\%$ (Rounded to hundredth)

Answer to Concept Check 3-08

$\frac{90-78}{90}=0.13$

percentage decrease $=\frac{90-78}{90}\times 100\% = 13\%$

Answer is (B)

Answer to Concept Check 3-09

$\frac{d-5}{d}\times 100\%$

Answer is (A)

E. Some typical applications

(1) Commissions

◆Example 1:

Charles is a used car salesman. He receives 2% commission on his total sale. If during a period of time, his sales total is \$360,000, how much is his commission ?

part = whole × %

$\therefore$ Commission $=360,000\times 0.02=7,200$

(2) Sales Tax

◆Example 2:

Frank buys a new television for a price of \$1,200. How much in total will he pay if the sales tax 9% is added to this bill ?

sales tax $=1200\times 0.09=108$

Total amount $=1,200+108=1,308$

(3) Discount

◆Example 3:

Joyce is shopping for a pair of new jeans. The original price is \$35. If she gets a 35% discount, how much does she pay ?

She pays 100−35=65 (%)

35×65%=22.75

Other way:

percent decrease $=\frac{\text{Old Amount}-\text{New Amount}}{\text{Old Amount}}\times 100\%$

$$0.35 = \frac{35 - x}{35}$$

$$0.35 \times 35 = 35 - x$$

$$\therefore \; x = 35 - 0.35 \times 35 = 22.75$$

(4) Interest

◆Example 4:

How much interest will Mary pay for a \$150,000 loan with an 6% annual interest rate for 10 years ?

Interest $= p \times r \times t$

principle (p) rate (r) time (t)

Interest $= 150,000 \times 0.06 \times 10 = 90,000$

CHAPTER 3 PRACTICE

1. Change 0.2% to the form of fraction.

2. Change $\frac{1}{2}\%$ to the form of decimal.

3. What percentage is $\frac{5}{6}$? (Rounded to hundredth)

4. 6 is 5% of a number. What is the value of the number ?
   (A) 30
   (B) 60
   (C) 90
   (D) 120
   (E) 150

5. The price of a certain suit of clothes is \$90. For some reason, the selling price went down to \$78, what is the percentage of the price decrease ?
   (A) 11.1%
   (B) 13.3%
   (C) 15.5%
   (D) 17.7%
   (E) 19.9%

6. Stefan has a 30 ounces salt solution which contains 15% of salt. How many ounces of salt must be add to the solution to make a 25% salt solution ?
   (A) 2.8
   (B) 3.4
   (C) 4
   (D) 5.2
   (E) 6.4

7. Michael and his family went out to eat at a French restaurant. The bill for food was \$78, because it was a special day, they got a 10% discount. If the sales tax was 9%, and the tip was 18%, how much in total did Michael's family pay ?
   (A) 79.32
   (B) 82.48
   (C) 85.02
   (D) 89.16
   (E) 94.32

8. There are d dollars in a drawer. Agnes took 5 dollars away, in terms of d, what percent of money still left in the drawer ?
   (A) $\frac{100(d-5)}{d}\%$
   (B) $100(d-5)\%$
   (C) $\frac{d-5}{100}\%$
   (D) $\frac{100d}{d-5}\%$
   (E) $\frac{d-5}{100d}\%$

9. **Table of Weight In Alex's Class**

Weight	120	125	130	135	140	145	150	155	160
Students	2	1	3	4	6	3	4	2	1

The above table shows the weight of the students in Alex's class, Find the average weight.

10. If $5:4=x:16$, then $x=$

(A) 20

(B) 22

(C) 25

(D) 28

(E) 30

11. The ratio of 7 to 2 is equal to the ratio of 28 to what number ?

(A) 4

(B) 8

(C) 28

(D) 56

(E) 72

12. The ratios of Andrew's, Brain's, and Charles' salaries are 5 : 3 : 2. If the total amount of their salaries is $180,000 annually, what is the amount of the least salary among the three people ?

(A) $36,000

(B) $42,000

(C) $54,000

(D) $65,000

(E) $83,000

13. If the ratio of x to y is 5 to 6, which of the following could be true ?

(A) $x=1,\ \ y=\frac{5}{6}$

(B) $x=2,\ \ y=\frac{5}{3}$

(C) $x=3,\ \ y=\frac{15}{6}$

(D) $x=4,\ \ y=\frac{6}{5}$

(E) $x=5,\ \ y=6$

14. What percent of 5 is 8 ?

(A) 80%

(B) 90%

(C) 120%

(D) 140%

(E) 160%

15. The prices of the same products of companies A and B are adjusted to be the same for some reason. The price of the product of company A is actually increase to $\frac{1}{6}$ of the original, while the price of the product of company B is reduced by $\frac{1}{5}$. What is the ratio of

their original prices ?

(A) 11:3

(B) 17:13

(C) 24:35

(D) 27:37

(E) 31:23

16. Shirley filled the gas of 8 gallons for \$44.72, and she can drive her car for exactly 176 miles. How many miles does \$1 of gas drive ? (Rounded to tenth)

(A) 3.94

(B) 5.87

(C) 6.32

(D) 7.05

(E) 7.81

17. On Anne's map, $\frac{1}{2}$ inch represents 8 miles, if the direct distance from city A to city B shows 4.2 inches, what is the real direct distance from city A to city B in miles ?

(A) 63.2

(B) 64.2

(C) 65.2

(D) 66.2

(E) 67.2

18. $\frac{1}{5}\% =$

(A) 0.2

(B) 0.02

(C) 0.002

(D) 0.0002

(E) 0.00002

19. 5 is 25% of a number, what is the value of the number ?

(A) 35

(B) 30

(C) 25

(D) 20

(E) 15

20. What percent of 58 is 8 ?

(A) 7.25%

(B) 8.3%

(C) 13.8%

(D) 14.5%

(E) 16%

21. If $\frac{x}{y}=\frac{1}{3}$ and $\frac{y}{z}=\frac{2}{5}$, then $\frac{x}{z}=$

(A) $\frac{2}{15}$

(B) $\frac{1}{5}$

(C) $\frac{2}{3}$

(D) $\frac{1}{3}$

(E) $\frac{3}{5}$

22. If 10^3 cubic meters of sea water can extract 0.03 gram of salt, how many grams of salt can be refined from 10^8 cubic meters of sea water ?

(A) 1,000
(B) 2,000
(C) 3,000
(D) 4,000
(E) 5,000

23. The price of gasoline is d dollars for 7 gallons and each gallon can drive m miles. In terms of m and d, what is the cost, in dollars, of the gasoline to drive for 1 mile ?

(A) $\frac{7d}{m}$
(B) $\frac{md}{7}$
(C) $\frac{d}{7m}$
(D) $\frac{7d}{m}$
(E) $7m \cdot d$

24. In a certain university, 34% are freshmen, and among them 48% are girls; 28% are sophomores and 45% are girls; 22% are juniors, and among them 42% are girls; 16% are seniors, and among them 40% are girls. What is the percentage of boys in the university ?

(A) 44.56%
(B) 50.56%
(C) 55.44%
(D) 60.44%
(E) 65.44%

25. Points P, Q, R, S, and T lie on a line in that order. If $PT : PR = 3 : 1$, $PT : PQ = 4 : 1$, and $PT : ST = 5 : 1$, what is the value of $QS : PR$?

(A) 21 : 17
(B) 33 : 17
(C) 27 : 20
(D) 33 : 20
(E) cannot determined from the information given

26. Sam has 5 gallons of a 30% alcohol solution. How many quarts of an 80% alcohol solution is needed to be added to Sam's original alcohol to create a 50% alcohol solution ?

(A) 3.33
(B) 13.32
(C) 18.56
(D) 22.32
(E) 26.64

27. The age ratio of ken, Caroline, and Connor is 6:5:2 this year. Five years ago, the age ratio was 11:9:3. So what is the age ratio of the three people five

years later ?

(A) 4:3:1

(B) 6:4:1

(C) 11:9:3

(D) 13:11:5

(E) 15:13:11

28. If $x:y=2:5$, what is the value of $(x+y):(8x+y)$?

(A) $\frac{1}{5}$

(B) $\frac{2}{5}$

(C) $\frac{1}{3}$

(D) $\frac{1}{4}$

(E) $\frac{7}{15}$

CHAPTER 3
ANSWERS and EXPLANATIONS

1. $\frac{1}{500}$

Key: $0.2\% = \frac{0.2}{100} = \frac{2}{1000} = \frac{1}{500}$

2. 0.005

Key: $\frac{1}{2}\% = \frac{\frac{1}{2}}{100} = \frac{1}{200} = 0.005$

3. 83.33%

Key: $\frac{5}{6} = 0.83333 = 83.33\overline{3}\% = 83.33\%$

(rounded to hundredth)

4. (D)

Key: percent $= \frac{\text{part}}{\text{whole}} \times 100\%$

$\therefore\ 0.05 = \frac{6}{\text{whole}}$

$\therefore$ whole $= \frac{6}{0.05} = 120$

Answer is (D)

5. (B)

decrease% $= \frac{\text{original} - \text{New}}{\text{original}} \times 100\%$

$\% = \frac{90-78}{90} \times 100\% = 13.3\%$

Answers is (B)

6. (C)

salt x

x

30 ounces

30 ounces

15% salt

25% salt

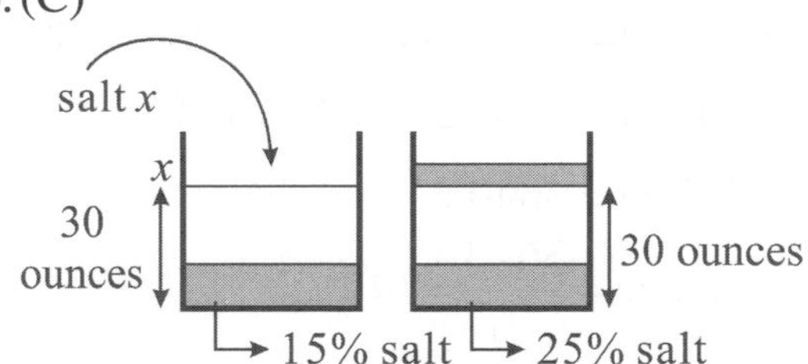

Assume x ounces salt added to the solution then, $\frac{30\times0.15+x}{30+x} = 0.25$

$30+x$ → total

cross multiplication

$0.25\times(30+x) = 30\times0.15+x$

$30\times0.25+0.25x = 30\times0.15+x$

$\therefore\ 0.75x = 7.5-4.5$

$\therefore\ x = \frac{3}{0.75} = 4$

Answer is (C)

7. (D)

Key: (1) bill $= 78\times0.9 = 70.2$

(2) sales tax $= 70.2\times0.09 = 6.32$

(3) tips $= 70.2\times0.18 = 12.64$

$\therefore$ payment $= 70.2+6.32+12.64$

$= 89.16$

Answer is (D)

8. (A)

Key: took 5 dollars away means $d-5$ dollars left in drawer percent

$= \frac{d-5}{d} \times 100\% = \frac{100(d-5)}{d}\%$

Answer is (A)

9. 140

Total $= 120\times 2 + 125\times 1 + 130\times 3 + 135\times 4$

$+140\times 6 + 145\times 3 + 150\times 4 + 155\times 2$

$+160\times 1 = 3640$

Average $= \dfrac{3640}{26} = 140$

10. (A)

$\dfrac{5}{4} = \dfrac{x}{16}$

$x = \dfrac{5\times 16}{4} = 20$

Answer is (A)

11. (B)

Key: $\dfrac{7}{2} = \dfrac{28}{x}$

28 is 4 times of 7 so, x must be 4 times of

2 $\quad 2\times 4 = 8$

Answer is (B)

12. (A)

Key: $A:B:C = 5:3:2$

imply $A = 5k$, $B = 3k$, $C = 2k$

$\therefore\ 5k + 3k + 2k = 180{,}000$

$\therefore\ k = 18{,}000$

The least is $2k \quad x\times 18{,}000 = 36{,}000$

Answer is (A)

13. (E)

Key: $\dfrac{x}{y} = \dfrac{5}{6}$

(A) $x = 1,\ \ y = \dfrac{5}{6}$,

Since $y = \dfrac{x}{\frac{5}{6}}$,

$x = 1,\ \ y = \dfrac{6}{5}$ (Not True)

(B) $x = 2,\ \ y = \dfrac{5}{3}$,

$x = 2,\ \ y = \dfrac{12}{5}$ (Not True)

(C) $x = 3,\ \ y = \dfrac{15}{6}$,

$x = 3,\ \ y = \dfrac{3}{\frac{5}{6}} = \dfrac{18}{5}$ (Not True)

(D) $x = 4,\ \ y = \dfrac{6}{5}$,

$x = 4,\ \ y = \dfrac{4}{\frac{5}{6}} = \dfrac{24}{5}$ (Not True)

(E) $x = 5,\ \ y = 6$,

$x = 5,\ \ y = \dfrac{5}{\frac{5}{6}} = \dfrac{30}{5} = 6$ (True)

Answers is (E)

14. (E)

Key: percerht $= \dfrac{\text{part}}{\text{whole}} \times 100\%$

$\therefore$ percent $= \dfrac{8}{5}\times 100\% = 1.6\times 100\%$

$= 160\%$

Answer is (E)

15. (C)

Assume the price now is P, the original price of company A' product is PA, and company B' product is PB.

then $P=PA(1+\frac{1}{6})=PB(1-\frac{1}{5})$

So, $\frac{7}{6}PA=\frac{4}{5}PB$, $\frac{PA}{PB}=\frac{\frac{4}{5}}{\frac{7}{6}}=\frac{24}{35}$

Answer is (C)

16. (A)

Key: $\frac{\text{miles}}{\text{dollar}}=\frac{176}{44.72}=3.94$

Answer is (A)

17. (E)

Key: $\frac{\frac{1}{2}}{8}=\frac{4.2}{x}$

$\frac{1}{2}x=33.6 \quad \therefore \quad x=67.2$

Answer is (E)

18. (C)

$\frac{1}{5}\%=\frac{\frac{1}{5}}{100}=\frac{1}{500}=0.002$

Answer is (C)

19. (D)

Key: $0.25=\frac{\text{part}}{\text{whole}}=\frac{5}{x}$

$\therefore \quad x=\frac{5}{0.25}=20$

Answer is (D)

20. (C)

Key: percent $=\frac{8}{58}=0.1379=13.8\%$

Answer is (C)

21. (A)

Key: $\frac{x}{y} \times \frac{1}{3} \quad \therefore \quad y=3x$

$\frac{y}{z}=\frac{3x}{z}=\frac{2}{5} \quad \therefore \quad 15x=2z$

$\therefore \quad \frac{x}{z}=\frac{2}{15}$

Answer is (A)

22. (C)

Key: $\frac{10^3}{0.03}=\frac{10^8}{x}$

$\therefore \quad x=\frac{10^8\times0.03}{10^3}=10^{8-3}\times0.03$

$10^5\times0.03=3,000$ grams

Answer is (C)

23. (C)

Key: $\frac{\$}{\text{mile}}$ is how much for 1 mile

Since 1 gallon of gasoline can drive the car for *m* miles.

7 gallons of gasoline can drive the car for 7 m miles,

so, for 1 mile we need $\frac{d}{7m}$ dollars

Answer is (C)

24. (C)

Key: (1) Freshmen boys

$=0.34\times(1-0.48)=0.1768$

(2)Sophomore boys

$=0.28\times(1-0.45)=0.154$

(3) Junior boys

$=0.22\times(1-0.42)=0.1276$

(4) Senior boys

$=0.16\times(1-0.4)=0.096$

So, boys in the university are

$0.1768+0.154+0.1276+0.096$

$=0.5544$

Answer is (C)

25. (D)

Step 1: Understand and Analyze the problem, draw a figure of a number line, to make it clear to read.

$$\frac{PT}{PR}=\frac{3}{1},\ \frac{PT}{PQ}=\frac{4}{1},\ \frac{PT}{ST}=\frac{5}{1}$$

Step 2: Asking $QS : PR$

Step 3: Key:

The relationship of $\overline{QS}$ and $\overline{PR}$

From the figure you will find

$QS=QT-ST$

$=\frac{3}{4}PT-\frac{1}{5}PT=\frac{11}{20}PT$

$(PQ=\frac{1}{4}PT)$

$PR=\frac{1}{3}PT$

So, we can find the ratio of QS : PR

Step 4: Operations

$$\frac{QS}{PR}=\frac{\frac{11}{20}PT}{\frac{1}{3}PT}=\frac{33}{20}$$

Answer is (D)

26. 13.32

Step 1: Understand the problem and draw a figure to help analyzing the situation

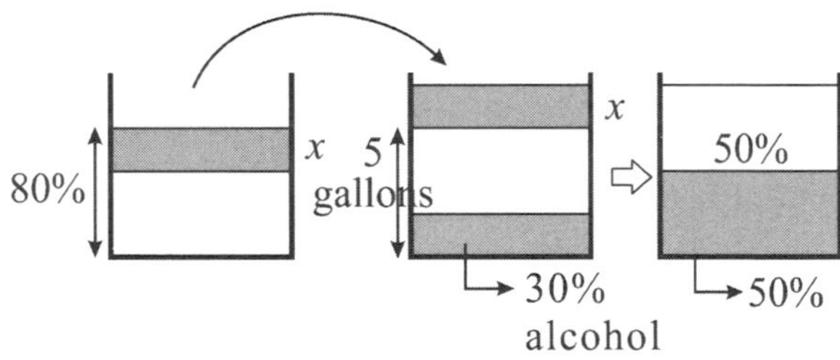

Step 2: Asking : how many "quarts"

Step 3: Key: percent

$$=\frac{\text{alcohol}}{\text{alcohol}+\text{solution}}$$

Step 4: operations

assume add x gallons of 80% alcohol into Sam's solution,

then, $\frac{0.8x+5\times0.3}{5+x}=0.5$

$\therefore\ 5(0.5)+0.5x=0.8x+5\times0.3$

$\therefore\ 2.5+0.5x=0.8x+1.5$

$\therefore\ 0.3x=1,\ \frac{1}{0.3}=3.33$ gallons

1 gallons = 4 quarts,

3.33 gallons 3.33×4 =13.32 quarts

Answer is (B)

27. (D)

Assume the ages of these three people are 6K, 5K, and 2K.

$(6K-5):(5K-5):(2K-5)=11:9:3$

$\therefore\ \frac{5K-5}{2K-5}=\frac{9}{3}$

$15K-15=18K-45$

$3K=30,\ K=10$

So, 5 years later, ken$=6\times10+5=65$

Caroline$=5\times10+5=55$

Connor$=2\times10+5=25$

$65:55:25=13:11:5$

Answer is (D)

28. (C)

Let $x=2k,\ y=5k$

So, $x+y=2k+5k=7k$

$8x+y=8(2k)+5k=21k$

$\therefore\ \frac{x+y}{8x+y}=\frac{7k}{21k}=\frac{1}{3}$

Answer is (C)s

CHAPTER 4
FUNDAMENTALS OF ALGEBRA

4-1 EXPRESSIONS and OPERATIONS

A. Terminologies

Variable

A variable is an unknown numeric value usually represented by a letter such as x and y in an equation. Sometimes the value of a variable varies within a designated range.

For instance, $x > 3$, x can represent any number which is greater than 3. For instance, it could be 3.2, $3\frac{2}{3}$, $5\frac{1}{2}$, 7.8787, …

Term

A term is formed by a number, a variable, or numbers and variables multiplied together in an expression. Terms are the units operated by mathematical operations.

For example,

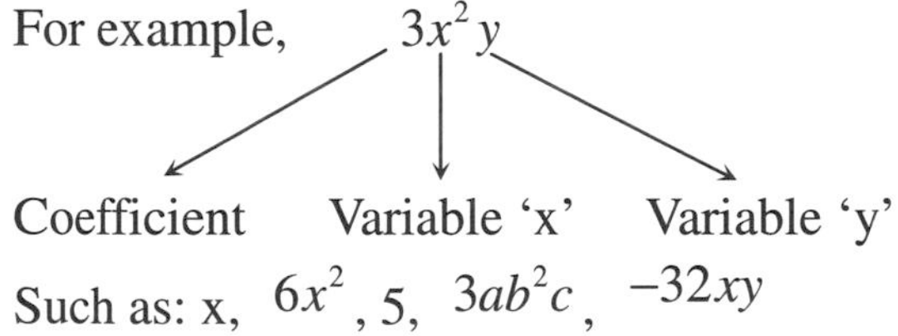

Such as: x, $6x^2$, 5, $3ab^2c$, $-32xy$

Constant

A number on its own is referred to as a constant. A constant term is a term that consists of a number only.

For example, in an expression $2x+4$, 4 is a constant term.

Coefficients

Coefficient is a number that is being multiplied by a variable in a term.

For example, 1 in x, 6 in $6x^2$, 3 in $3ab^2c$, and -32 in $-32xy$ are coefficients.

Like Terms

Like terms are terms that have the same formation in terms of variables and expressions, only the coefficient may be different.

For example: x, $3x$, $-5x$, ... are like terms

$\frac{1}{2}x^2y$, $3.5x^2y$, $120x^2y$, ... are a like terms

Algebraic Expression

An algebraic expression is an expression consisting of one or more terms connected by mathematical operations (such as addition, subtraction, multiplication, and division).

For example:

$3x+5y$, $3x-5y+7x^2$, $\frac{x^2+1}{y}$, ...

→ connected by the operation of addition or subtraction

Monomial

Monomial is an algebraic expression with just one term.

For example: $3x^2$

Binomial

Binomial is an algebraic expression with two terms.

For example: $3x^2-3y^2$, x^2y+2y^2

Trinomial

Trinomial is an algebraic expression with three terms.

For example: $ab^2+2ab+c^2$, $x^2+2xy+y^2$, ...

Polynomial

Polynomial is an algebraic expression with a combination of terms. The variables must have non-negative whole number power.

For example: $x^2+2xy+y^2+5x+4y+6$

↑ Constant term (the 6)

B. Keywords

The following table demonstrates some keywords we often encounter in context problems that indicate mathematic operations. How these keywords are used in context problem and how to represent them as algebraic equations are illustrated in the example section below.

Math Operators	Keywords in Context
Equal (=)	Same as, is, was, has, equivalent to, ……..
Addition (+)	Plus, sum, increase, more than, total, older than, and, ……..
Subtraction (−)	Decrease, difference, less than, subtract, fewer, younger than, ……..
Multiplication (×)	Times, product, twice, triple, ……..
Division (÷)	Divide, quotient, for every, per, ……..

Greater than (>)	Greater than or equal to (≥)	Less than (<)	Less than or equal to (≤)
is more than, is greater than, ⋮ ⋮	at least, is not less than, minimum value, ⋮ ⋮	is less than, below, ⋮ ⋮	maximum value, at most, is not greater than, is not older than, ⋮ ⋮

- Some examples

(1)Is, was

Adam's age is 5 times Brain's age $A=5B$

Charles' networth is twice as much as David's 5 years ago.

(2)Has

Edward has the same number of apples as Frank does.

$E = F$

(3)Equivalent to

Three plus five is equivalent to eight.

$3+5=8$

(4)Same as

Grace's math score is the same as Harry's score.

$G = H$

(5)Sum

The sum of 5 and x is y.

$y = 5 + x$

(6)Increase

Jack's salary was increased by 5%.

$J = J \times (1+0.05)$

(7)More than

Katy's score is 5 points more than Michael's score.

$k = M + 5$

(8)Older than

Peter is 2 years older than his brother.

$P = 2 + B$

(9)Difference

The difference between 5 and 3 is 2.

$2 = 5 - 3$

The difference in average income between city A and B is 30k.

$30000 = | A - B |$

(10)Decrease

Comparing to last year, the price index is decreased by 2%.

$$2\% = \frac{\text{last year's price} - \text{this year's price}}{\text{last year's price}} \times 100\%$$

(11)Less than

Four is four less than eight.

$4=8-4$

(12)fewer than

William has 3 dollars fewer than Elia's.

(13)Times

Agnes has 8 more marbles than 3 times the number of marbles Bonnie has.

$A=3B+8$

(14)Difference

The difference between Ted's age and Steven's age is 3 years.

$|T-S|=3$

(15) Half of Joe's bats are blue.

$\frac{1}{2}J=\text{Blue}$

(16) For

Caroline earned 5 dollars for every 2 dollars Janet earned.

$\frac{c}{J}=\frac{5}{2}$, or $c:J=5:2$

(17)Is greater than

5 is greater than 2

$5>2$

(18)Is less than

5 is less than 7

$5<7$

Answer to Concept Check 4-01

Let Tad's age be T, and Steven's ages be S. The difference between Ted's age and Steven's age is the absolute value of $T-S$ or $S-T$ $\Rightarrow$ $|T-S|$ or $|S-T|$

Since the difference is 3 years less than twice Ted's age: $2T-3$

$\Rightarrow |T-S|=|S-T|=2T-3$

Therefore the correct answer is (C)

 Answer to Concept Check 4-02

n people share d dollars is $\frac{d}{n}$. Increase 2 people, the share is $\frac{d}{n+2}$.

The original people will pay $\frac{d}{n}-\frac{d}{n+2}$.

That is $d(\frac{1}{n}-\frac{1}{n+2})$.

The correct answer is (C)

C. Operations

• Addition and Subtraction

Just Combine Like Terms

Example:

$$-5(a+2b)+\frac{1}{2}(4a-8b)=-5a-10b+2a-4b=-3a-14b$$

Example:

$$2(x^2-1)-[3(x^3-x^2+\frac{2}{3}y)-(x^2+y^2)+3]+\frac{2}{3}$$

$$=2x^2-2-[3x^3-3x^2+2y-x^2-y^2+3]+\frac{2}{3}$$

$$=2x^2-2-[3x^3-4x^2+2y-y^2+3]+\frac{2}{3}$$

$$=2x^2-\frac{4}{3}-3x^3+4x^2-2y+y^2-3$$

$$=-3x^3+6x^2+y^2-2y-\frac{13}{3}$$

• Multiplication

It is the same as numbers multiplication. Just remember to Combine like terms as the last step

1. use distribution law.

Example:

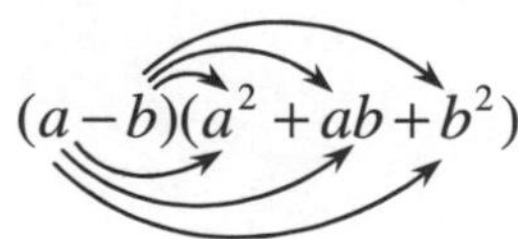
$(a-b)(a^2+ab+b^2)$

$$= a \cdot a^2 + a \cdot ab + a \cdot b^2 - a^2b - ab^2 - b^3$$
$$= a^3 + \cancel{a^2b} + \cancel{ab^2} - \cancel{a^2b} - \cancel{ab^2} - b^3$$
$$= a^3 - b^3$$

For Example:

$$(x^3 - 2x^2 - 3)(3x+1) - (2-3x^2)(5x+1)$$
$$= (x^3 - 2x^2 - 3) \cdot 3x + (x^3 - 2x^2 - 3) \cdot 1 - (2-3x^2) \cdot 5x - (2-3x^2)$$
$$= 3x^4 - 6x^3 - 9x + x^3 - 2x^2 - 3 - 10x + 15x^3 - 2 + 3x^2$$
$$= 3x^4 + 10x^3 + x^2 - 19x - 5$$

2. Use Foil method for typical binomials multiplication. (first outer inner last)

inner last

$(a+b)\ (c+d)$

first outer

Example: $(3x-y)(x-3y)$

First: $(3x)(x) = 3x^2$

Outer: $(3x)(-3y) = -9xy$

Inner: $(-y)(x) = -xy$

Last: $(-y)(-3y) = 3y^2$

Add first/outer/inner/last all together.

$(3x-y)(x-3y) = 3x^2 - 10xy + 3y^2$

• Division

Term division is the reverse of Multiplication from the highest order to the lowest.

◆Example:

$$(4x^4 + x^3 - 3x^2 + 5x - 1) \div (x^2 + x + 1)$$

$$\begin{array}{rr|l}
 & 4x^4 + x^3 - 3x^2 + 5x - 1 & x^2 + x + 1 \\
-) & 4x^4 + 4x^3 + 4x^2 & \\
\hline
 & -3x^3 - 7x^2 + 5x - 1 & 4x^2 - 3x - 4 \\
-) & -3x^3 - 3x^2 - 3x & \\
\hline
 & -4x^2 + 8x - 1 & \\
-) & -4x^2 - 4x - 4 & \\
\hline
\text{Remainder} \longrightarrow & 12x + 3 &
\end{array}$$

-4 → For elimination of the third term

$-3x$ → For elimination of the second term

$4x^2$ → For elimination of the first term

so, $(4x^4 + x^3 + 5x - 1) \div (x^2 + x + 1) = 4x^2 - 3x - 4 + \dfrac{3(4x+1)}{x^2+x+1}$

Answer to Concept Check 4-03

Simplify $-5(a+2b)+\frac{1}{2}(4a-8b)=-5a-10b+2a-4b=-3a-14b$

Answer to Concept Check 4-04

Simplify $2(x^2-1)-[3(x^3-x^2+\frac{2}{3}y)-(x^2+y^2)+3]+\frac{2}{3}$

$$=2x^2-2-[3x^3-3x^2+2y-x^2-y^2+3]+\frac{2}{3}$$

$$=2x^2-2-3x^3+3x^2-2y+x^2+y^2-3+\frac{2}{3}$$

$$=-3x^3+6x^2+y^2-2y-5+\frac{2}{3}$$

$$=-3x^3+6x^2+y^2-2y-4\frac{1}{3}$$

Answer to Concept Check 4-05

Multiplication of $(a-b)(a^2+ab+b^2)=a^3+\cancel{a^2b}+\cancel{ab^2}-\cancel{a^2b}-\cancel{ab^2}-b^3=a^3-b^3$

Answer to Concept Check 4-06

$(4x^4+x^3-3x^2+5x-1)\div(x^2+x+1)$

$$\begin{array}{rl|l}
 & 4x^4+x^3-3x^2+5x-1 & x^2+x+1 \\
-) & 4x^4+4x^3+4x^2 & 4x^2-3x-4 \longrightarrow \text{quotient} \\
\hline
 & -3x^3-7x^2+5x-1 & \\
-) & -3x^3-3x^2-3x & \\
\hline
 & -4x^2+8x-1 & \\
-) & -4x^2-4x-4 & \\
\hline
 & 12x+3 & \longrightarrow \text{remainder}
\end{array}$$

$(4x^4+x^3-3x^2+5x-1)\div(x^2+x+1)=(4x^2-3x-4)+\dfrac{12x+3}{x^2+x+1}$

4-2 FACTORING

Factoring is the reverse operation of multiplication. The purpose of factoring is to make the polynomials' expressions easier to work with.

There are some common methods for factorings.

A. By pulling out the common factors from each term.

Example: $P = 6x^2y + 3xy - 4xy^2 + 2x^2y^2$

The common factor is xy,

$P = xy(6x + 3 - 4y + 2xy) = (6x - 4y + 2xy + 3)(xy)$

B. By grouping the terms with potential common factor.

Sometimes trials and errors are needed to find common factors.

Example: $P = 2x^3 - 6x^2y + 4xy^2 + x^2y - 3xy^2 + 2y^3 = 2x^3 + x^2y - 6x^2y - 3xy^2 + 4xy^2 + 2y^3$

$P = x^2(2x + y) - 3xy(2x + y) + 2y^2(2x + y)$

$= (2x + y)(x^2 - 3xy + 2y^2)$

$= (2x + y)(x - 2y)(x - y)$

C. By using the formulas

- $(a + b)^2 = a^2 + 2ab + b^2$

  Example: $4x^2 + 12xy + 9y^2 = (2x)^2 + (2)(2x)(3y) + (3y)^2 = (2x + 3y)^2$

- $(a - b)^2 = a^2 - 2ab + b^2$

  Example: $4x^2 - 12xy + 9y^2 = (2x)^2 - (2)(2x)(3y) + (3y)^2 = (2x - 3y)^2$

D. By using the difference of squares $a^2 - b^2 = (a - b)(a + b)$

Example: $9x^2 - 4y^2 = (3x)^2 - (2y)^2 = (3x - 2y)(3x + 2y)$

E. By using the sum and difference of cubes.

$a^3 + b^3 = (a + b)(a^2 - ab + b^2)$

Example: $27x^3 + 8 = (3x)^3 + 2^3 = (3x + 2)((3x)^2 - (3x)(2) + (2)^2) = (3x + 2)(9x^2 - 6x + 4)$

$a^3 - b^3 = (a - b)(a^2 + ab + b^2)$

Example: $8x^3 - \frac{1}{27} = (2x)^3 - (\frac{1}{3})^3 = (2x - \frac{1}{3})[(2x)^2 + (2x)(\frac{1}{3}) + (\frac{1}{3})^2]$

$= (2x - \frac{1}{3})(4x^2 + \frac{2}{3}x + \frac{1}{9})$

F. By using cross-multiplication of the quadratic polynomials

For $(ax+b)(cx+d)=acx^2+adx+bcx+bd$

$$=acx^2+(ad+bc)x+bd$$

$a \quad b$

$c \quad d$

$ad+bc$ (Find middle term)

Example: x^2+3x+2

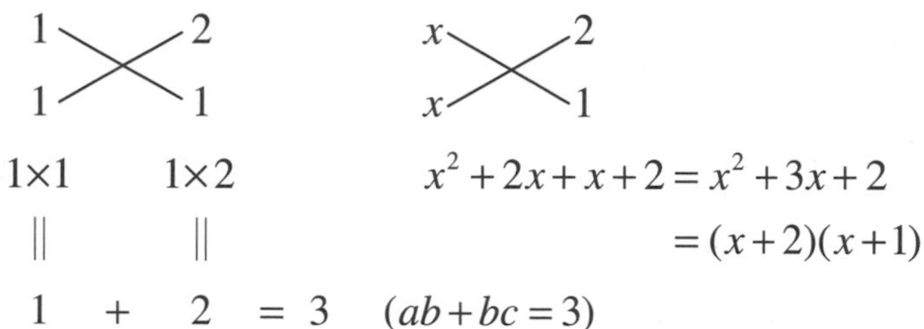

$1\times1 \quad 1\times2$

$1 + 2 = 3 \quad (ab+bc=3)$

$x^2+2x+x+2=x^2+3x+2$

$=(x+2)(x+1)$

So, $x^2+3x+2=(x+2)(x+1)$

Example: $x^2-xy-2y^2=(x-2y)(x+y)$

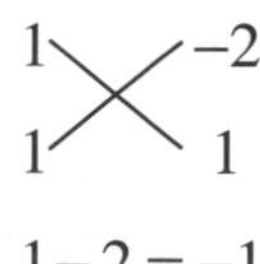

$1-2=-1$

G. By using the difference of perfect square.

Example: factoring the expression

$$P=x^2+4x-9$$
$$=x^2+2(x)\cdot2+2^2-2^2-9$$
$$=(x+2)^2-13$$
$$=(x+2)^2-(\sqrt{13})^2$$
$$=(x+2-\sqrt{13})(x+2+\sqrt{13})$$

Answer to Concept Check 4-07

$(a+b)^3(a-2b)-(a+b)(a-2b)^3=(a+b)[(a+b)^2(a-2b)-(a-2b)^3]$

$=(a+b)(a-2b)[(a+b)^2-(a-2b)^2]=(a+b)(a-2b)(a+b+a-2b)(a+b-a+2b)$

$=(a+b)(a-2b)(2a-b)(3b)=3b\cdot(a+b)\cdot(a-2b)\cdot(2a-b)$

Answer to Concept Check 4-08

$x^3 - 2x^2 - x + 2$

$= x^2(x-2) - (x-2)$

$= (x^2 - 1)(x-2)$

$= (x-1)(x+1)(x-2)$

Answer to Concept Check 4-09

$2x^2 - 5x - 3 = (2x+1)(x-3)$

$$\begin{array}{ccc} 2x & \searrow\!\!\!\!\swarrow & 1 \\ x & & -3 \\ \hline \end{array}$$

$-6x + x = -5x$

Answer to Concept Check 4-10

$16x^2 - 81$

$= (4x)^2 - 9^2$

$= (4x-9)(4x+9)$

4-3 LINEAR EQUATION

Standard Form.

$ax + b = 0$

- b → Constant
- x → Variable
- a → Coefficient

- The steps to solve a linear equation

Step 1. Simplify the contents in the parenthesis.

Step 2. Combine the like terms.

Step 3. Combine the variable terms at one side of the equal symbol and move all other terms to the other side including constants and other variable (variables) if any.

 **Step 4.** Simplify the result.

 **Step 5.** Plug in your result to the original equation to double check your answer.

◆Examples:

1. $3x-4=8$

$$\begin{array}{r} 3x-4=8 \\ +4\ \ +4 \\ \hline 3x\ \ \ =12 \end{array}$$

$$\frac{3x}{3}=\frac{12}{3}$$

$x=4$

Check $3(4)-4=8$ (O.K.)

2. $2x+3=11$

$$\begin{array}{r} 2x+3=11 \\ -3\ \ -3 \\ \hline 2x\ \ \ =8 \end{array}$$

$$\frac{2x}{2}=\frac{8}{2}$$

$x=4$

Check $2(4)+3=11$ (O.K.)

> ★TIPS
> Shifting terms to the other side of the equal sign is the same as adding or subtracting the same value on both sides.

◆Example:

$3x-4=8$ Shift to the right of equal sign

$3x=8+4$ Caution: change sign

$3x=12$

$x=4$

◆Example:

$2x+3=11$

$2x=11-3$ change sign

$2x=8$

$x=4$

◆Example:

$$ax+b=c$$

$$ax=c-b$$

$$x=\frac{c-b}{a}$$

◆Example:

$$2a+2x=5$$

$$x=\frac{5-2a}{2}$$

◆Example:

$$6x+3-a=4x+7-b$$

$$6x-4x=7-b-(3-a)$$

$$2x=7-b-3+a$$

$$2x=4+a-b$$

$$\therefore\ x=\frac{4+a-b}{2}$$

Answer to Concept Check 4-11

$$7x+3=12$$

$$7x+3-3=12-3$$

$$7x=9$$

$$x=\frac{9}{7}$$

substitute to $7x+3=12$

$$7\left(\frac{9}{7}\right)+3=9+3=12$$

left side = right side

so, $x=\frac{9}{7}$ is the answer

$$\begin{array}{rl} & 7x+3=12 \\ -) & \quad\ \ 3\quad 3 \\ \hline & 7x\quad = 9 \\ \div 7) & \\ \hline & x=\frac{9}{7} \end{array}$$

Answer to Concept Check 4-12

$3x=5-a$

$x=\dfrac{5-a}{3}$

substitute to $3x+a=5$

left side $3(\dfrac{5-a}{3})+a=5-a+a=5$ right side.

so, $x=\dfrac{5-a}{3}$ is the answer

$$\begin{array}{rl} & 3x+a=5 \\ -) & \quad\ \ a=a \\ \hline \div 3)\ 3x & =5-a \\ \hline x & =\dfrac{5-a}{3} \end{array}$$

4-4 SYSTEM EQUATIONS (SIMULTANEOUS EQUATIONS)

An equation with two or more than two variables cannot be solved by itself individually. They have to be put together and solved simultaneous way. It is called system equations or simultaneous equations. Generally, there are two simultaneously ways to solve the system equations.

A. Elimination

Eliminate one variable by addition or subtraction or both.

Example: $\begin{cases} 3x+5y=2\cdots\cdots① \\ 2x-y=1\cdots\cdots② \end{cases}$

> ★Tips:
> Multiply 5 to both side of equation ②

Suppose you like to eliminate variable y,

First ①+②×5

You get $\begin{cases} 3x+5y=2\cdots\cdots① \\ 10x-5y=5\cdots\cdots③ \end{cases}$

Then, ①+③ $13x+0y=7$, so $x=\dfrac{7}{13}$

Substitute it back into equation ②

You get $2\times\dfrac{7}{13}-y=1$

$\therefore\ y=\dfrac{14}{13}-1=\dfrac{1}{13}$

The solution is $\begin{cases} x = \dfrac{7}{13} \\ y = \dfrac{1}{13} \end{cases}$

Check them with original equations.

For: $3x+5y=2$ substitute $x=\dfrac{7}{13}$ and $y=\dfrac{1}{13}$

you get $3\times\dfrac{7}{13}+5\times\dfrac{1}{13}=\dfrac{21}{13}+\dfrac{5}{13}=\dfrac{26}{13}=2$ (Checked)

For: $2x-y=1$

$2\cdot(\dfrac{7}{13})-\dfrac{1}{13}=\dfrac{14}{13}-\dfrac{1}{13}=\dfrac{13}{13}=1$ (Checked)

B. Substitution

Example: $\begin{cases} 3x+5y=2\cdots\cdots① \\ 2x-y=1\cdots\cdots② \end{cases}$

From ② $y=2x-1$, substitute it into ①

Then you get $3x+5(2x-1)=2$

$3x+10x-5=2$

$13x=7$

$x=\dfrac{7}{13}$ substitute it back into equation ②

you get $y=2\cdot(\dfrac{7}{13})-1=\dfrac{14}{13}-\dfrac{13}{13}=\dfrac{1}{13}$

The solution is $\begin{cases} x = \dfrac{7}{13} \\ y = \dfrac{1}{13} \end{cases}$

Answer to Concept Check 4-13

$\begin{cases} 2x+3y=-1 \cdots\cdots ① \\ -3x+y=7 \cdots\cdots ② \end{cases}$

① − ②×3 (use elimination method)

$$\begin{array}{r} 2x+3y=-1 \\ -)\ -9x+3y=21 \\ \hline 11x=-22 \end{array}$$

$x=-2$

Substitute to ①

$-4+3y=-1,\ 3y=3,\ y=1$

Check them with the original system equations

$-3(-2)+1=7$ checked.

Answer: $\begin{cases} x=-2 \\ y=1 \end{cases}$

4-5 LINEAR INEQUALITY

Inequality Signs:

- Not equal $A \neq B$
- Less than $A < B$
- Not Less than $A \nless B$
- Greater than $A > B$
- Not Greater than $A \ngtr B$
- Less than or equal to $A \leq B$
- Greater than or equal to $A \geq B$
- Equal to $A = B$

A. How to solve a linear inequality?

It is the same as solving the equations.

★Caution:

When multiplying or dividing negative value(s), the inequality sign has to be flipped.

◆Example:

$x > y$

multiply both sides by 5, you got

$5x > 5y$, OK

Then Divide both sides by 3, you got

$\frac{5x}{3} > \frac{5y}{3}$

Multiply or divide a positive value on both sides, inequality sign remains unchanged.

But for a negative value, the inequality sign has to be flipped.

◆Example:

$x > y$

Multiply both, by -5,

Then, $-5x < -5y$

Divide both, by -5

Then $\frac{-5x}{-5} < \frac{-5y}{-5}$

$x > y$ Inequality sign is flipped.

◆Example:

$2x - 3 > 8$

Add 3 to both sides

$2x - 3 > 8$

$2x - 3 + 3 > 8 + 3$

$2x > 11$

$\therefore\ x > \frac{11}{2}$

or $2x > 8 + 3$

$2x > 11$

$x > \frac{11}{2}$

or multiply -2 by both sides (caution: flip the inequality sign)

$(2x - 3)(-2) < 8(-2)$

$-4x+6<-16$

$-4x<-22$

$\dfrac{-4x}{-4}>\dfrac{-22}{-4}$

$x>\dfrac{11}{2}$

B. How to show the range of inequality ?

> Use number line to indicate the range of an inequality statement.

"○" it means not including the value

"●" it means including the value

◆Example:

$x>5$

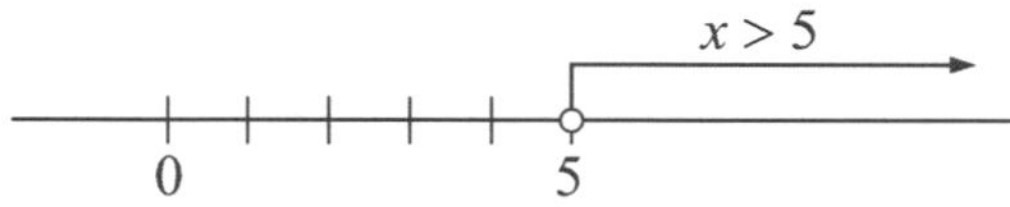

"●" means including

$x \le 5$

5

◆Example:

$\dfrac{-2x+3}{3}\ge 2$

$-2x+3\ge 6$

$-2x\ge 3$

Divided by -2

$\dfrac{-2x}{-2}\le\dfrac{3}{-2}$ (flip the sign)

$\therefore\ x\le -\dfrac{3}{2}$

Represented by number line

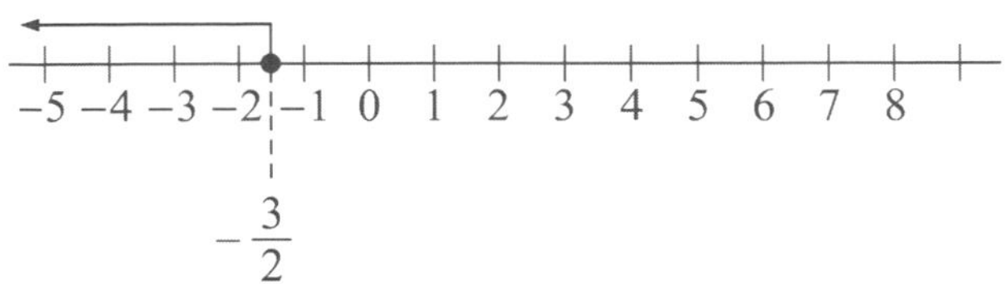

◆Example:

$|x| > 2$

This means: $x > 2$, if x positive

Or $x < -2$, if x negative

For example $x = -8$

Then $|-8| > 2$

◆Example:

$|x+5| \geq 2$

There are two possibilities:

A. $x+5 \geq 2$, $x \geq -3$

B. $x+5 \leq -2$, $x \leq -7$

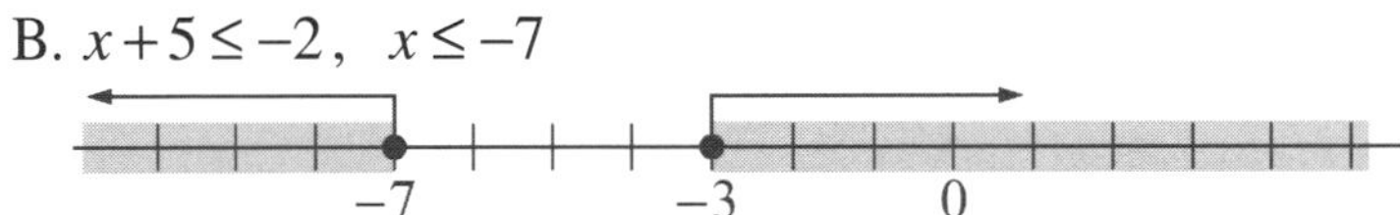

4-6 SYSTEM INEQUALITIES

- The steps to solve system inequalities.

 **Step 1.** Solve the inequality individually.

 **Step 2.** Find the intersection between them.

◆Example:

$$\begin{cases} 2x+3 \geq 2 \cdots\cdots ① \\ 3x-5 \leq 1 \cdots\cdots ② \end{cases}$$

For ① $2x+3 \geq 2$

$2x \geq -1$

$x \geq -\frac{1}{2}$

For ② $3x-5<1$

$3x\leq 6$

$x\leq 2$

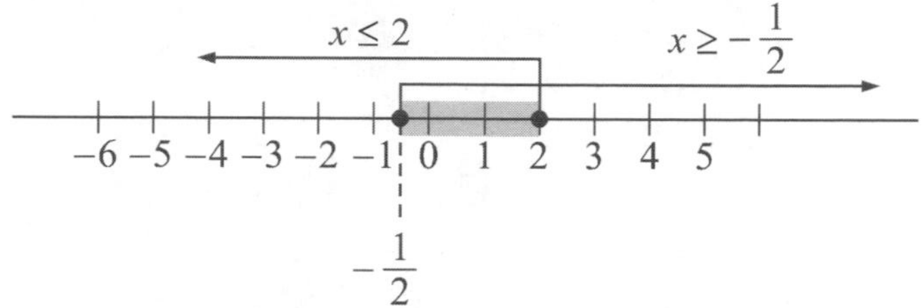

So, $-\frac{1}{2}\leq x\leq 2$

◆Example:

$$\begin{cases} \frac{3x}{5}+2\geq 6-\frac{x}{3} \cdots\cdots ① \\ \frac{2x+1}{3}-x<3 \cdots\cdots ② \end{cases}$$

For ① $\frac{3x}{5}+2\geq 6-\frac{x}{3}$

For the purpose to eliminate the denominator of both sides, multiple the L.C.M. $3\times 5=15$ by both sides.

$9x+30\geq 90-5x$

$14x\geq 60,\ \ x\geq\frac{30}{7}$

For ② $\frac{2x+1}{3}-x<3$

Multiply 3 to both sides

$2x+1-3x<9$

$-x+1<9$

$-x<8$

$x>-8$

So, the common solution to both inequalities is $x\geq\frac{30}{7}$

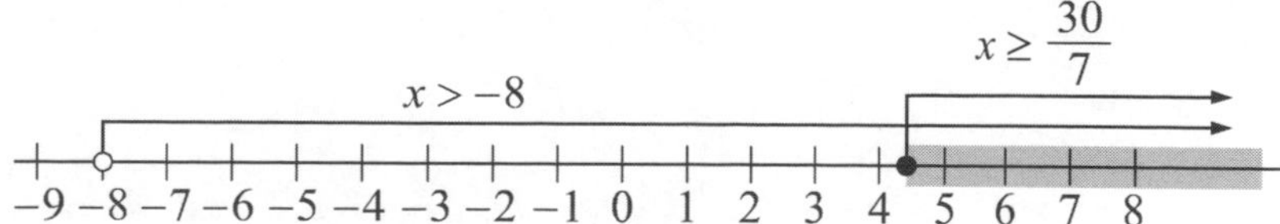

Answer to Concept Check 4-14

$\dfrac{-2x+3}{3} \geq 3$

$-2x+3 \geq 9$, $-2x \geq 6$, $-x \geq 3$

$x \leq -3$

$x \leq -3$

$-3 \quad 0$

Answer to Concept Check 4-15

Key: $\begin{cases} 2x+3>2 \cdots\cdots ① \\ 3x-5<1 \cdots\cdots ② \end{cases}$

Solving it individually and then find the intersection of these 2 statements.

$2x+3>2$, $2x>-1$, $x>-\dfrac{1}{2}$

$3x-5<1$, $3x<6$, $x<2$

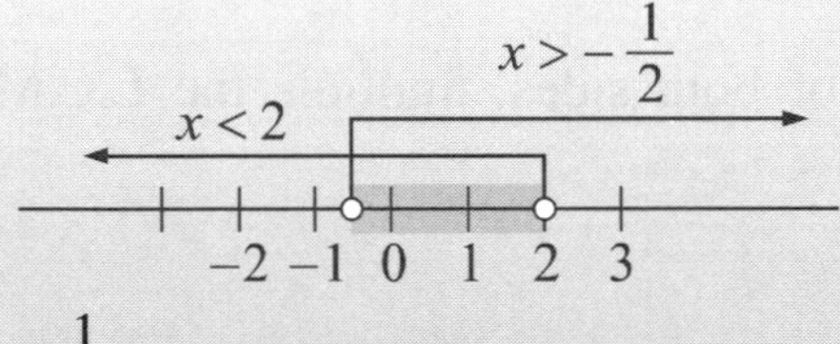

$-\dfrac{1}{2}<x<2$

Answer to Concept Check 4-16

$|3x-2| \le 5$

Key: $3x-2 \le 5 \cdots ①$

$3x-2 \ge -5 \cdots ②$

Operations:

For ① $3x-2 \le 5$

$3x \le 7$

$x \le \frac{7}{3}$

For ② $3x-2 \ge -5$

$3x \ge -3$

$\therefore\ x \ge -1$

$-1 \le x \le \frac{7}{3}$

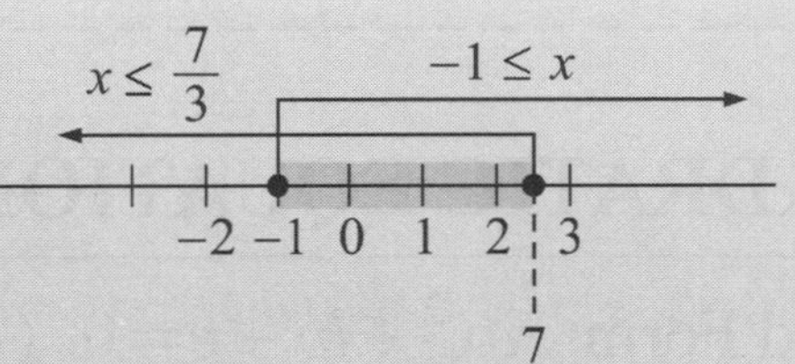

Answer to Concept Check 4-17

$|x-2| \ge 3$

Key: $x-2 \ge 3 \cdots ①$

$x-2 \le -3 \cdots ②$

Operation

so, for ①, $x-2 \ge 3$, it means $x \ge 5$

for ②, $x-2 \le -3$, it means $x \le -1$

Use number line to show the range.

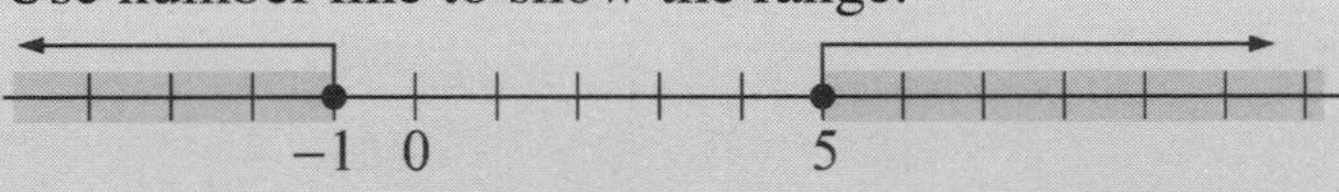

Answer to Concept Check 4-18

solve $\begin{cases} x+3\geq 3\cdots① \\ 2x-1<1\cdots② \end{cases}$

For ① $x+3\geq 3,\ x\geq 0$

For ② $2x-1<1,\ 2x<2,\ x<1$

$x\geq 0$

$x<1$

−2 −1 0 1 2

$0\leq x<1$

4-7 QUADRATIC EQUATIONS

• Standard Form: $ax^2+bx+c=0\ \ (a\neq 0)$
Any equation which can be rewritten to the form of $ax^2+bx+c=0$ is a quadratic equation.

A. How to solve quadratic equations

1. Solve by factoring

◆Example: solve $2x^2+5x+2=0$

By crossing-multiplication

$$\begin{matrix} 2 & \times & 1 \\ 1 & & 2 \end{matrix}$$

$4+1=5$ (The coefficient of the mid-term)

$2x^2+5x+2=(2x+1)(x+2)=0$

So, $x=-\dfrac{1}{2}$, or $x=-2$

◆Example: solve $3x^2+5x-2=0$

By crossing-multiplication

$$\begin{matrix} 3 & \times & -1 \\ 1 & & 2 \end{matrix}$$

$6-1=5$

$3x^2+5x-2=(3x-1)(x+2)=0$

$x=\dfrac{1}{3}$ or $x=-2$

2. Solve by rewriting the original equation to the form of $ax^2+bx+c=0$

◆Example: $x^4-3x^2-4=0$

Let $x^2=u$, then rewrite the equation to

$u^2-3u-4=0$

$(u-4)(u+1)=0$

$u=4$ or $u=-1$

then substitute back,

$x^2=4$, $x=\pm 2$

$x^2=-1$, x has no solution

The roots for this equation are $x=2$, or $x=-2$

3. solve by squaring

Example: $x^2-4x-5=0$

$x^2-4x+4-4-5=0$

$(x-2)^2-9=0$

$(x-2)^2-3^2=0$

$[(x-2)-3][(x-2)+3]=0$

$(x-5)(x+1)=0$

$x=5$ or $x=-1$

Example: $x^2+x-6=0$

$$x^2+x-6=x^2+x+(\frac{1}{2})^2-(\frac{1}{2})^2-6=(x+\frac{1}{2})^2-\frac{1+24}{4}$$

$$=(x+\frac{1}{2})^2-(\frac{5}{2})^2$$

$$=[(x+\frac{1}{2})+\frac{5}{2}][(x+\frac{1}{2})-\frac{5}{2}]=(x+\frac{1}{2}+\frac{5}{2})(x+\frac{1}{2}-\frac{5}{2})$$

$$=(x+\frac{6}{2})(x-\frac{4}{2})=(x+3)(x-2)=0$$

$\therefore$ $x=-3$ or $x=2$

4. Solve by formula

Example: solve $x^2-4x-5=0$

The roots of quadratic equation

$$x_1=\frac{-b+\sqrt{b^2-4ac}}{2a},\quad x_2=\frac{-b-\sqrt{b^2-4ac}}{2a}$$

In this example: $a=1$, $b=-4$, $c=-5$

$$x_1=\frac{4+\sqrt{16+20}}{2}=\frac{4+\sqrt{36}}{2}=\frac{4+6}{2}=5$$

$$x_2=\frac{4-\sqrt{16+20}}{2}=\frac{4-\sqrt{36}}{2}=\frac{4-6}{2}=-1$$

5. Solve standard form of quadratic equation

$$ax^2+bx+c=0$$

$$x^2+\frac{b}{a}x+\frac{c}{a}=0$$

$$x^2+\frac{b}{a}x+(\frac{b}{2a})^2-(\frac{b}{2a})^2+\frac{c}{a}=0$$

$$(x+\frac{b}{2a})^2-(\frac{b^2}{4a^2}-\frac{c}{a})=0$$

$$(x+\frac{b}{2a})^2-(\frac{b^2-4ac}{4a^2})=0$$

$$(x+\frac{b}{2a})^2-[\sqrt{\frac{b^2-4ac}{4a^2}}]^2=0$$

$$(x+\frac{b}{2a}-\sqrt{\frac{b^2-4ac}{4a^2}})(x+\frac{b}{2a}+\sqrt{\frac{b^2-4ac}{4a^2}})=0$$

$$(x+\frac{b-\sqrt{b^2-4ac}}{2a})(x+\frac{b+\sqrt{b^2-4ac}}{2a})=0$$

$$x=\frac{-b+\sqrt{b^2-4ac}}{2a} \quad \text{or} \quad x=\frac{-b-\sqrt{b^2-4ac}}{2a}$$

$$x=\frac{-b\pm\sqrt{b^2-4ac}}{2a}$$

Notice the relation of two roots.

$$x_1=\frac{-b+\sqrt{b^2-4ac}}{2a}, \quad x_2=\frac{-b-\sqrt{b^2-4ac}}{2a}$$

So, $\boxed{x_1+x_2=\frac{-2b}{2a}=-\frac{b}{a}}$

$$(x_1)(x_2)=(\frac{-b}{2a}+\frac{\sqrt{b^2-4ac}}{2a})(\frac{-b}{2a}-\frac{\sqrt{b^2-4ac}}{2a})$$

$$=(\frac{-b}{2a})^2-(\frac{\sqrt{b^2-4ac}}{2a})^2$$

$$=\frac{b^2}{4a^2}-\frac{b^2-4ac}{4a^2}=\frac{b^2-b^2+4ac}{4a^2}$$

$$=\frac{4ac}{4a^2}$$

$$=\frac{c}{a}$$

Example: Solve $2x^2+3x+1=0$

$a=2,\ b=3,\ c=1$

$$x=\frac{-3\pm\sqrt{3^2-4\times2\times1}}{2\times2}=\frac{-3\pm\sqrt{9-8}}{4}=\frac{-3\pm1}{4}$$

$x=-1$, or $x=-\frac{1}{2}$

Check: when $x=-1$

$2(-1)^2+3(-1)+1=2-3+1=0$

$2(-\frac{1}{2})^2+3(-\frac{1}{2})+1=\frac{1}{2}-\frac{3}{2}+1=0$

(Checked)

B. Discriminant

Since the roots of a quadratic equation are $x=\dfrac{-b\pm\sqrt{b^2-4ac}}{2a}$,

the value of b^2-4ac can be used to determine the nature of the roots.

1. If $b^2-4ac>0$, then

   $\dfrac{-b\pm\sqrt{b^2-4ac}}{2a}$ are two real numbers,

   There are two unequal real roots.

2. If $b^2-4ac=0$

   $x_1=x_2=\dfrac{-b}{2a}$

   There is only one root (also called double root)

3. If $b^2-4ac<0$

   There is no real root.

◆Example 1:

$2x^2+3x+1=0$

$a=2,\ b=3$, and $c=1$

$b^2-4ac=(3)^2-4(2)(1)=9-8=1>0$

so, there are two real roots.

◆Example 2:

$2x^2 - x + 8 = 0$

$a = 2$, $b = -1$, and $c = 8$

$b^2 - 4ac = (-1)^2 - (4)(2)(8) = 1 - 64 = -63 < 0$

There is no real root.

◆Example 3:

$2x^2 - 8x + 8 = 0$

$a = 2$, $b = -8$, and $c = 8$

$b^2 - 4ac = (-8)^2 - (4)(2)(8) = 0$

There is only one root (double root).

Answer to Concept Check 4-19

$3x^2 - 4x + 1 = 0$

$a = 3$, $b = -4$, $c = 1$

$$x = \frac{-(-4) \pm \sqrt{(-4)^2 - 4(3)(1)}}{2\times 3} = \frac{4 \pm \sqrt{16-12}}{6} = \frac{4 \pm \sqrt{4}}{6} = \frac{4 \pm 2}{6}$$

$$x_1 = \frac{6}{6} = 1, \quad x_2 = \frac{2}{6} = \frac{1}{3}$$

check: substitute $x = 1$ to the equation

$3(1)^2 - 4(1) + 1 = 3 - 4 + 1 = 0$ O.K.

$3(\frac{1}{3})^2 - 4(\frac{1}{3}) + 1 = \frac{1}{3} - \frac{4}{3} + 1 = -\frac{3}{3} + 1 = 0$ O.K.

4-8 FUNCTIONS (I)

A. What is a relation ?

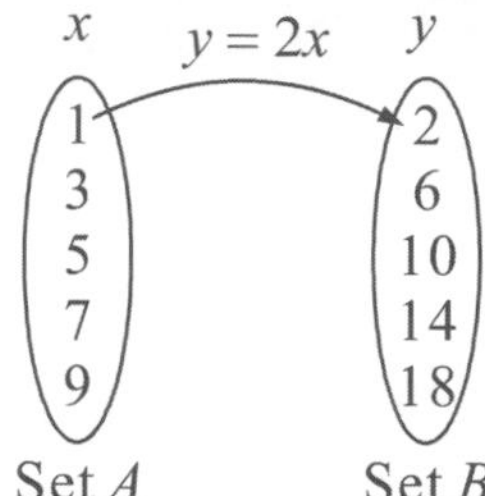

$y = 2x$ is a relation between set A and set B.

B. What is a variation ?

The value of a variable varies according to the variation of another variable.

• Direct variation

$y = kx$ (k is a constant)

◆Example: $y = 2x$, $k = 2$

x	-2	-1	0	1	2	3	4	5	6
y	-4	-2	0	2	4	6	8	10	12

• inverse variation

$y = \frac{k}{x}$ (k is a constant, $x \neq 0$)

◆Example: $k = 2$, $y = \frac{2}{x}$

x	-2	-1	2	3	4	5	6	···
y	-1	-2	1	$\frac{2}{3}$	$\frac{1}{2}$	$\frac{2}{5}$	$\frac{1}{3}$	···

C. What is a function ?

If an element in set x has "one and only one" corresponding element in set y, then this kind of relation or variation is so called "Function".

It is denoted as: $y = f(x)$

◆Example:

$y = f(x) = 2x$

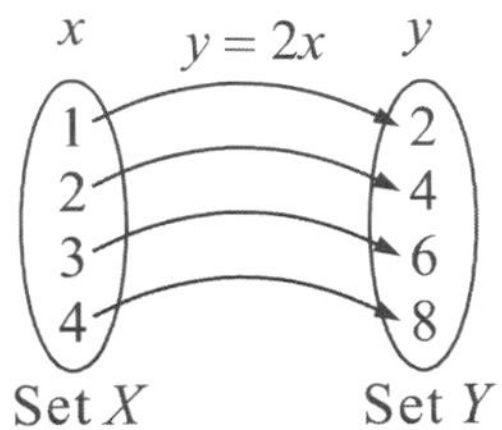

But $x = y^2$

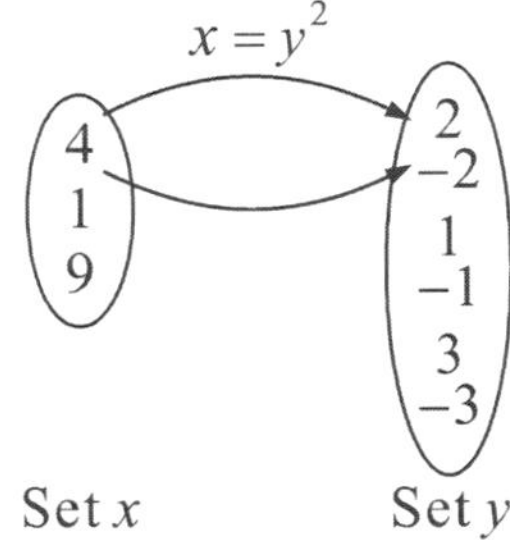

Because one element in set x (say 4) could have more than one $(2, -2)$ elements corresponding in set y, in this case there are two of them: 2, -2, therefore $x = y^2$ is not a function !

It is just a kind of relation.

D. Inverse Functions

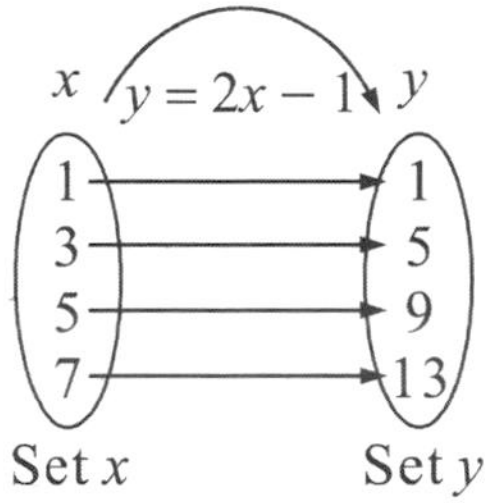

$f(x) = 2x - 1$

Each element in set x has one and only one corresponding element in set y, so $y = 2x + 1$ is a function.

If we can inversely map each element in set y to the element and only the element in set x, then we calls this kind of relationship as inverse function.

It is denoted as: $f^{-1}(x)$

1.How to find the inverse function ?

◆Example:

$y = f(x) = 2x - 1$

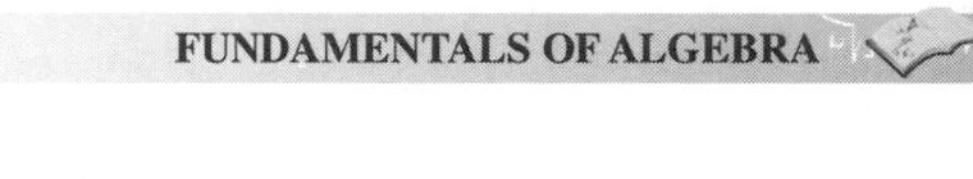

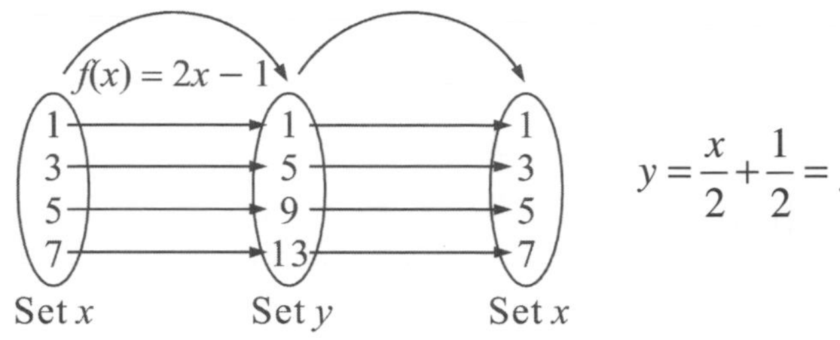

$$y=\frac{x}{2}+\frac{1}{2}=f^{-1}(x)$$

Step 1. Find the value of x in terms of y.

$$y=2x-1$$

$$2x=y+1$$

$$x=\frac{y}{2}+\frac{1}{2}$$

Because x, y are just two variables

we can exchange the variables to get another function.

 **Step 2.** $y=\frac{x}{2}+\frac{1}{2}$, it is the inverse function of $y=2x-1$

> Caution: The inverse function of a function may not be a true "function" ! If in the inverse function, one x value could result in more than one y values, it is not a (true) function.

◆Example:

Find the inverse function of $f(x)=3x+1$

Step 1. $y=3x+1$

$$3x=y-1$$

$$x=\frac{y}{3}-\frac{1}{3}$$

 **Step 2.** Exchange the name of variables

$$y=\frac{x}{3}-\frac{1}{3}$$

$$f^{-1}(x)=\frac{x}{3}-\frac{1}{3}$$

Any element in set y can only have one element in set x corresponding to it.

So $f^{-1}(x)$ is a function.

E. Domain and Range

Domain

A set of values that define the function.

For example:

x	−3	−2	−1	1	2	3	4
y	9	4	1	1	4	9	16

The domain of the function $f(x)$ is defined by the table above.

$f(x):\{-3,-2,-1,1,2,3,4\}$

Range

Range is a set of all output values of a function. In other words, it is the set of resulting values from the function operations given the set of domain values.

The range of the function are:

Set A: $\{1, 4, 9, 16\}$

◆Example:

Find the domain and range of the function $f(x)=3x+1$

x	$-\infty$	⋯	−3	−2	−1	0	1	2	2.5	⋯	∞
y	$-\infty$	⋯	−8	−5	−2	1	4	7	8.5	⋯	∞

Domain: $\{x \mid -\infty \text{ to } \infty\}$

all real numbers

Range: $\{x \mid -\infty \text{ to } \infty\}$

◆Example:

Find the domain and the range of the function.

$$f(x)=\sqrt{3x-1}+\frac{1}{x-3}$$

(1) $\sqrt{3x-1}$, so $3x-1$ must ≥ 0

$3x-1\geq 0 \qquad x\geq \frac{1}{3}$

(2) $\frac{1}{x-3}$, $x-3\neq 0$, $x\neq 3$

Therefore, the domain are $x\geq\frac{1}{3}$, and $x\neq 3$.

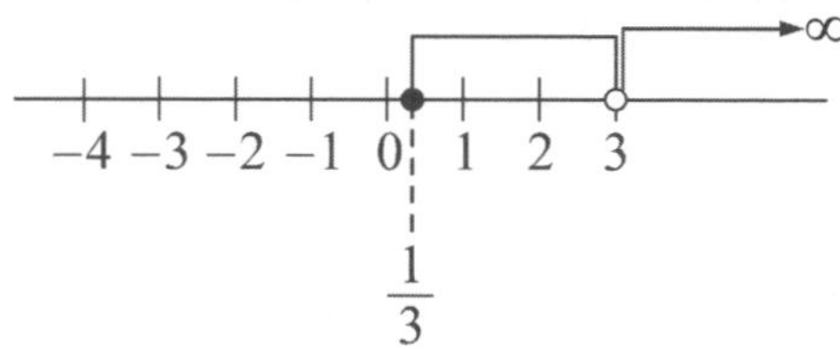

When $x=\frac{1}{3}$, the value of $f(x)$ is the smallest

$$f(\frac{1}{3})=0+\frac{1}{\frac{1}{3}-3}=-\frac{3}{8}$$

so, the range is $\{-\frac{3}{8}, 3\}$ and $\{3, \infty\}$.

> **★TIPS:**
> The range of a function is the domain of inverse function, if the inverse function is also a function.

◆Example:

Find the domain and range of $f(x)=\frac{x}{x^2-x+1}$.

Domain:

$$x^2-x+1=(x^2-x+\frac{1}{4})-\frac{1}{4}+1=(x-\frac{1}{2})^2+\frac{3}{4}$$

There is no value of x can make the denominator be zero.

So, $x=\{-\infty, \infty\}$ which means x could be all real numbers.

Range: Let $f(x)=\frac{x}{x^2-x+1}=y$

then $yx^2-yx+y=x \Rightarrow yx^2-(y+1)x+y=0$

Since x is real, so the discriminant

$\Delta=b^2-4ac\geq 0$

$[-(y+1)]^2-4\cdot y\cdot y=y^2+2y+1-4y^2=-3y^2+2y+1\geq 0$

$\therefore\ 3y^2-2y-1\leq 0\quad (3y+1)(y-1)\leq 0$

$-\frac{1}{3}$ 0 1

$-\frac{1}{3}\leq y\leq 1$

$-\frac{1}{3}\leq f(x)\leq 1$

F. The Operations with Functions

The same as algebraic operation.

Example: if $f(x)=-3x+2$, $g(x)=5x-1$

$f(x)+g(x)=(f+g)(x)=-3x+2+5x-1=2x+1$

$f(x)-g(x)=(f-g)(x)=-3x+2-5x+1=-8x+3$

$f(x)\cdot g(x)=(f\cdot g)(x)=(-3x+2)(5x-1)$

$=-15x^2+13x-2$

$\frac{f(x)}{g(x)}=\frac{f}{g}(x)=\frac{-3x+2}{5x-1}\quad (g(x)\neq 0)$

★TIPs:
$(f\cdot g)(x)=f(x)\cdot g(x)$
$(f\circ g)(x)=f(g(x))$

Composite function

When the variable of a function is also a function, It is called a composite function.

It is denoted as $(f\circ g)(x)$ or $f(g(x))$.

◆Example 1:

If $g(x)=2x+1$, $f(x)=x^2-1$

$(f\circ g)(x)=[g(x)]^2-1=(2x+1)^2-1$

Find the value of $(f\circ g)(1)$

Firstly, $g(1)=2(1)+1=3$

then, $f(3)=3^2-1=8$

Putting it together $(f\circ g)(1)=(2\cdot 1+1)^2-1=3^2-1=8$

◆Example 2:

if $g(x)=x+2$, $f(x)=x^2+1$

what is the value of $(f\circ g)(2)$

$g(2)=2+2=4$, $f(4)=4^2+1=16+1=17$

Alternatively, we get the answer from $(f\circ g)(2)$ directly:

$(f\circ g)(2)=(2+2)^2+1=16+1=17$.

◆Example 3:

If $f(x)=x+1$, $g(x)=2x+1$ what is the value of $f(g(2))$?

$g(2)=2\cdot 2+1=5$

$f(5)=5+1=6$

$\therefore\ f(g(2))=6$

◆Example 4:

If $f(x)=-2x+5$, which of the following is equal to $f(-3x)$?

(A) $-6x-5$

(B) $6x+5$

(C) $6x-5$

(D) $6x^2-15$

Since $f(-3x)=-2(-3x)+5=6x+5$

Answer is (B)

◆Example 5:

If $f(-3x)=6x+5$, what is $f(x)$?

Let $-3x=x'$, $x=-\dfrac{x'}{3}$

so, $f(x')=6(-\dfrac{x'}{3})+5=-2x'+5$

change x' to x, then $f(x)=-2x+5$ (x or x' are just variables)

◆Example 6:

If $f(x-2)=x^2-4x+6$, what is $f(x)$?

Let $x-2=x'$, then $x=x'+2$

$f(x')=(x'+2)^2-4(x'+2)+6$

$=x'^2+4x'+4-4x'-8+6$

$=x'^2+2$

change x' to x, $f(x)=x^2+2$

Answer to Concept Check 4-20

$f(x)=x^2+3x+1,\quad g(x)=2x+1$

$g(2)=2(2)+1=5$

$f(g(2))=5^2+3\cdot 5+1=25+15+1=41$

Answer to Concept Check 4-21

$f(x)=x^2+2x,\quad g(x)=x-2$

$\therefore\ 3f(x)+2g(x)=3(x^2+2x)+2(x-2)=3x^2+6x+2x-4=3x^2+8x-4$

Answer to Concept Check 4-22

$f(x)=\dfrac{1}{3}+\sqrt{x}$

$x\geq 0$

$f(x)\geq \dfrac{1}{3}$

Domain $0\leq x$, Range: $\dfrac{1}{3}\leq y$

Answer to Concept Check 4-23

1. $f(x)=3x-1$

Step 1. Let $y=3x-1$

Step 2. $3x=y+1,\quad x=\dfrac{y}{3}+\dfrac{1}{3}$

Step 3. exchange x,y variables $y=\dfrac{x}{3}+\dfrac{1}{3}$ (x, y are just symbols of variables)

Step 4. $f^{-1}(x)=\dfrac{x}{3}+\dfrac{1}{3}$

★Tips:

How to identify invalid domain for a function?

1. A domain value is invalid if it causes the function to contain a fraction with zero denominator.

   For instance, in the function $f(x)=\dfrac{3x+4}{x-1}$, when $x=1$, the $f(x)$ is undefined, therefore, $x=1$ must be eliminated from the domain.

2. A domain value is invalid when it causes the function to contain imaginary number.

   $f(x)=5\sqrt{x-1}$, when $x<1$, then $f(x)$ is a imaginary number which must be eliminated from the domain.

4-9 MODELING

Modeling is the process of creating an algebraic expression that represents the relationship between various variables.

◆Example:

The model for calculating the weekly pay for a salesperson:

$P(\text{pay}) = S(\text{sales})\times 0.02(2\%\ \text{rate}) + 200(\text{Basic pay})$

A. The Standard Form of a linear model

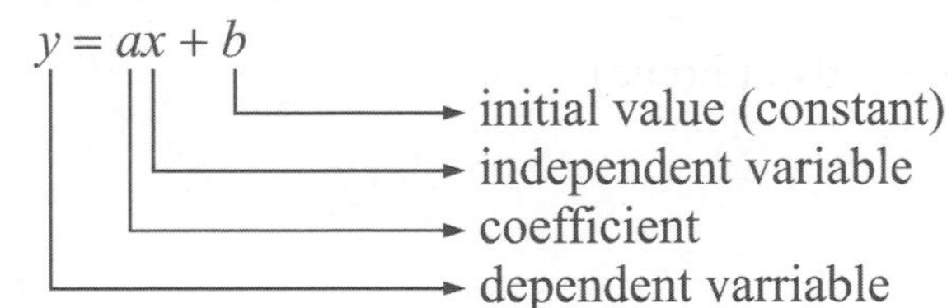

Initial value: The start value of y when $x=0$

Independent value: x varies by itself.

Coefficient: the multiple of x

Dependent Variable: y varies according to the changes of x.

B. How to find the independent and dependent variables in a context ?

As the wording suggests, if the value of a variable varies independently of other variable, it is referred to as independent variable. Usually in an experiment, it's the variable you manipulate. Dependent variable is the variable whose value changes based on the changes of the independent variables.

◆Example:

Generally speaking, on a mountain, for every 1,000 feet of elevation, the temperature drops by about 5.4° Fahrenheit. In other words, the change in elevation affects the change in temperature. On the other hand, when temperature changes, the height does not change, Therefore height is the independent variable, and temperature is dependent variable.

$T(\text{temperature}) = -\dfrac{h(\text{height})}{1000} \times 5.4 + T_G$ (Temperature on the bottom of the mountain)

$T \Rightarrow y$, dependent variable

$5.4 \Rightarrow a$, coefficient

$-\dfrac{h}{1000} \Rightarrow x$, independent variable

$T_G \Rightarrow b$ initial value

$y = ax + b$

◆Example:

The amount of candy to be prepared for party is based on how many guests who are expected to come. But the number of guests will not be affected by the amount of candy prepared. Therefore, guests is the independent variable while candy is the dependent variable.

$c(\text{candy}) = -2 \times p$ (people coming) $+ H$ (candy at home)

$c \Rightarrow y$, dependent variable

$-2 \Rightarrow n$, coefficient

$p \Rightarrow x$, independent variable

$H \Rightarrow$ initial value (candy in home)

Answer to Concept Check 4-24

The value of the first year $D = 35{,}000 \times (1 - 0.1) = 35{,}000 \times 0.9$

The value of the second year $D = 35{,}000 \times 0.9 \times (1 - 0.1) = 35{,}000 \times 0.9 \times 0.9$

The value of x years $D = 35{,}000 \times \underbrace{0.9 \times 0.9 \times \cdots \times 0.9}_{x} = 35{,}000 \times (0.9)^x$

Answer is (A)

CHAPTER 4
PRACTICE

1. Roman landscaping company charges its customers P dollars for every work-hour and b dollars for every square foot of land. Which of the following expressions best represents the cost of n workers, h hours, and f square feet of land ?

(A) $C = n \cdot h \cdot p + f \cdot b$

(B) $C = h \cdot p + n \cdot h + fb$

(C) $C = (fh + n)P$

(D) $C = f \cdot b + (n + h)P$

2. The formula of mark-up is $m = \dfrac{P-C}{C}$, what is the cost C in terms of P and m ?

(A) $C = \dfrac{P}{m+1}$

(B) $C = \dfrac{m}{P+1}$

(C) $C = \dfrac{1}{P+m}$

(D) $C = \dfrac{1}{P-m}$

> **★Tips**
> Mark-up: retailer adds a certain amount of money to the price of product to cover cost overhead and profit.

3. $\begin{cases} x + ay = 3 \\ 3x + y = 7 \end{cases}$

In the system equations above, a is constant and x and y are variables. What value of a causes the system equations to have no solution ?

(A) $\dfrac{1}{3}$

(B) $\dfrac{2}{3}$

(C) $\dfrac{4}{21}$

(D) $\dfrac{5}{21}$

4. The price for 1 pound of honey crisp apples is 3 times the price of 1 pound of red apples. If Katy bought 2 pounds of honey crisp apples and 4 pounds of red apples and paid \$15, how much would she pay for 1 pound of red apple ?

(A) \$1.5

(B) \$2

(C) \$2.5

(D) \$3

5. What are the solutions to $2x^2 - 8x + 5 = 0$.

(A) $-2 + \dfrac{\sqrt{6}}{2}$

(B) $2 + \dfrac{\sqrt{6}}{2}$

(C) $\dfrac{3}{2} + \dfrac{\sqrt{6}}{2}$

(D) $-\dfrac{3}{2} + \dfrac{\sqrt{6}}{2}$

6. If a and b both satisfy equation $2x^2-5x-3=0$. What is the value of $a+b$?

(A) $\frac{5}{2}$

(B) $\frac{7}{2}$

(C) 4

(D) 5

7. For the function $f(x)=kx^2-6$, k is a constant and $f(2)=2$. What is the value of $f(-2)$?

(A) 2

(B) -2

(C) 4

(D) -4

8. If $g(x)=-2x+2$, which of the following is equivalent to $g(-2x)$?

(A) $-6x-2$

(B) $-6x+2$

(C) $4x-2$

(D) $4x+2$

9. $\begin{cases} \frac{1}{4}x=\frac{1}{5}y \\ 0.3x-0.2y+2=12 \end{cases}$

10.

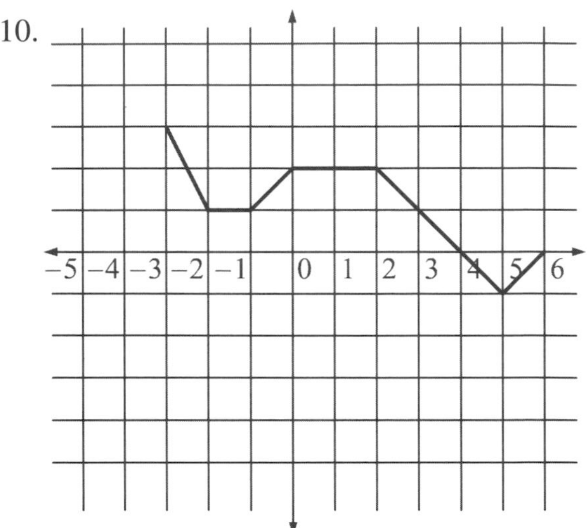

The graph of function $f(x)$ shown above, which of the following is (are) equal to 2 ?

(I) $f(\frac{3}{2})$

(II) $f(\frac{-2}{3})$

(III) $f(-\frac{3}{2})$

(A) I only

(B) II only

(C) III only

(D) I and III only

11. $S=\frac{1}{2}(vt+at^2)$

The equation above represents the distance S, in feet, of a object after moving in a steady condition with an initial velocity of v feet per second and an acceleration a feet/sec^2. Which of the following represents a in terms of S, v, and t ?

(A) $a = S - vt$

(B) $a = \dfrac{S - vt}{t^2}$

(C) $a = \dfrac{2S}{t^2} - \dfrac{v}{t}$

(D) $a = \dfrac{t^2}{S} + \dfrac{t}{v}$

12. If the solutions of the system equations $\begin{cases} 3x+2y=7 \\ 3x+ay=4a+1 \end{cases}$ are $x=\alpha$, $y=\beta$, and $\alpha+\beta=3$, what is the value of a ?

13. Which of the following represents the solution set for the inequality $\frac{2}{3}(x-\frac{4}{7})>-5$?

(A) $x>-\dfrac{97}{14}$

(B) $x<\dfrac{97}{14}$

(C) $x>-\dfrac{14}{97}$

(D) $x<\dfrac{14}{97}$

14. Which of the following is an equivalent form of the function $f(x)=(x-2)(x+3)$ and has a minimum value of $f(x)$ appears as a constant or a coefficient ?

(A) $f(x)=x^2-6$

(B) $f(x)=x^2+x+6$

(C) $f(x)=(x+\frac{1}{2})^2-\frac{25}{4}$

(D) $f(x)=(x-\frac{1}{2})^2-\frac{25}{4}$

15. A group of children try to distribute their candies from trick-or-treat. If each child receives 8 candies, then there are 6 candies left over. If each child receives 9 candies then there are shortage of 34 candies. How many children are in this group ?

(A) 34

(B) 36

(C) 38

(D) 40

16. If $g(x)=3x+1$ and $p(g(2))=10$ which of the following could be the function of $p(x)$?

(A) $x+10$

(B) $x-10$

(C) $x+3$

(D) $x-3$

17. If $\frac{1}{4}\leq 3-\frac{a}{3}\leq\frac{6}{5}$, what is the maximum possible value of a ?

18. $4x(2x+3)+3(2x+3)=ax^2+bx+c$ for all values of x, what is the value of $a+b+c$?

(A) 29

(B) 31

(C) 33

(D) 35

19. $\begin{cases} 4x+3y=2 \\ -x+2y=5 \end{cases}$

If p and q are the two roots of the system equations above, and p and q that satisfy the equation of $px+qy=5$, what is the value of $15-3p+6q$?

(A) 18

(B) 24

(C) 30

(D) 38

20. A sample of iron is heated to 1200℃ and then placed to cool. The temperature of the iron is given by the formula $T=1200-10t-0.5t^2$, where T is the temperature of the iron in degree Celsius and t is the time in second after the start of the cooling process.
After how many seconds will the temperature of the sample of iron be 800 ℃ ?

21. $\begin{cases} y-\dfrac{3}{a}x\le 2 \\ \dfrac{1}{a}x-\dfrac{1}{3}y\le -1 \end{cases}$

If the system of inequalities shown above has no solution, what is the value of a ?

(A) $\dfrac{1}{2}$

(B) $-\dfrac{1}{2}$

(C) There is no value of a that results in no solution

(D) There are infinitely many values of a except 0 that result in no solution

22. A math review book was so popular that the bookseller reduced the original price from $30 to $25 to reduce the burden on students, hoping to still breaks even on the book. The bookseller break even when its total revenue is equal to total cost of producing the book. Assume the cost C, in dollars, of producing the book is $C=n\times 20+2000$, how many more of the books does the bookseller need to sell at the new price to break even ?

(A) 150

(B) 200

(C) 250

(D) 300

CHAPTER 4
ANSWERS and EXPLANATIONS

1. Work one hour per worker = $p.

   $f \times b \Rightarrow$ material fee

   workers hours

   $n \times h \times p$

   = total cost for labours

   total cost $= nhp + fb$

   Answer is (A)

2. $m = \dfrac{p-c}{c}$

   $mc = p - c$

   $(m+1)c = p$

   $\therefore\ c = \dfrac{p}{m+1}$

   Answer is (A)

3. When system equations have no solution, their coefficients are proportional.

   $\dfrac{1}{3} = \dfrac{a}{1} \qquad a = \dfrac{1}{3}$

   Answer is (A)

4. $C = 3R$

   $2C + 4R = 15$

   $2 \times 3R + 4R = 15$

   $10R = 15$

   $R = 1.5$

   Answer is (A)

5. $a = 2,\ b = -8$, and $c = 5$

   $x = \dfrac{8 \pm \sqrt{64-40}}{4} = \dfrac{8 \pm \sqrt{24}}{4} = 2 \pm \dfrac{\sqrt{6}}{2}$

   Answer is (B)

6. $a + b = \dfrac{5}{2},\ ab = -\dfrac{3}{2}$

   Answer is (A)

7. $f(2) = 4k - 6 = 2$

   $k = 2$

   $f(-2) = 2x^2 - 6 = 2(-2)^2 - 6 = 2$

   Answer is (A)

8. $g(x) = -2x + 2$

   $g(-2x) = -2(-2x) + 2 = 4x + 2$

   Answer is (D)

9. $\dfrac{1}{4}x = \dfrac{1}{5}y,\ x = \dfrac{4}{5}y,$

   substitute, $0.3(\dfrac{4}{5}y) - 0.2y - 10 = 0$

   $\dfrac{1.2}{5}y - \dfrac{1}{5}y - 10 = 0,\ \dfrac{0.2}{5}y - 10 = 0,$

   $y = 250$

   $x = \dfrac{4}{5} \times 250 = 200$

10. $f(\frac{3}{2})=2$, $f(-\frac{2}{3})=-1.3$, $f(-\frac{3}{2})=-1$

so, only (I)

Answer is (A)

11. $S=\frac{1}{2}(vt+at^2)$

$2s=vt+at^2$

$2s-vt=at^2$

$a=\frac{2s-vt}{t^2}=\frac{2s}{t^2}-\frac{v}{t}$

Answer is (C)

12. $\begin{cases}3x+2y=7\cdots\cdots① \\ 3x+ay=4a+1\cdots\cdots②\end{cases}$

①−② $y(2-a)=7-4a-1=6-4a$

$y=\frac{6-4a}{2-a}$

$\because\ x=\alpha=3-\beta=3-\frac{6-4a}{2-a}$

$=\frac{6-3a-6+4a}{2-a}=\frac{a}{2-a}$

substitute, $3x+2y=7$,

$3\cdot\frac{a}{2-a}+2\cdot\frac{6-4a}{2-a}=\frac{3a+12-8a}{2-a}$

$=\frac{-5a+12}{2-a}=7$

$\therefore\ -5a+12=14-7a$, $2a=2$, $a=1$

13. (A)

$\frac{2}{3}(x-\frac{4}{7})>-5$

$\frac{3}{2}\times[\frac{2}{3}(x-\frac{4}{7})]>(\frac{3}{2})(-5)$

$x-\frac{4}{7}>-\frac{15}{2}$, $x>-\frac{15}{2}+\frac{4}{7}>\frac{-105}{14}+\frac{8}{14}$

$x>\frac{-97}{14}$

Answer is (A)

14. $f(x)=(x-2)(x+3)=x^2+x-6$

$=x^2+x+(\frac{1}{2})^2-(\frac{1}{2})^2-6$

$=(x+\frac{1}{2})^2-\frac{1}{4}-6$

$=(x+\frac{1}{2})^2-\frac{25}{4}$

Answer is (C)

15. Let the number of children $=n$

Let the number of candies $=c$

then $\begin{cases}c-8\cdot n=6\cdots① \\ 9n-c=34\cdots②\end{cases}$

①+② $n=40$

Answer is (D)

16. $g(x)=3x+1$, $g(2)=6+1=7$

(A) $p(x)=x+10$, $p(7)=7+10=17\neq 10$

(B) $p(x)=x-10$, $p(7)=7-10=-3\neq 10$

(C) $p(x)=x+3$, $p(7)=7+3=10$ (True)

Answer is (C)

17. $\frac{1}{4}\leq 3-\frac{a}{3}\leq\frac{6}{5}$

For $\frac{1}{4}\leq 3-\frac{a}{3}\ \Rightarrow\ \frac{1}{4}-3\leq-\frac{a}{3}$

$\Rightarrow\ 3-36\leq-4a$

$-33\leq-4a$, $33\geq 4a$, $\therefore\ a\leq\frac{33}{4}$

For $3-\frac{a}{3}\leq\frac{6}{5}\ \Rightarrow\ 45-5a\leq 18$

$\Rightarrow\ -5a \le -27$

$5a \le 27,\ a \ge \frac{27}{5}$

Maximum value of a is $\frac{33}{4}$

18. $4x(2x+3)+3(2x+3)$

$=8x^2+12x+6x+9$

$=8x^2+18x+9$

$a=8,\ b=18$, and $c=9$

$a+b+c=8+18+9=35$

Answer is (D)

19. $\begin{cases} 4x+3y=2 \cdots\cdots ① \\ -x+2y=5 \cdots\cdots ② \end{cases}$

①+②×4 $11y=22,\ y=2$

substitute, $-x=5-4,\ x=-1$

Since $x=-1$ and $y=2$ satisfy the equation of $px+qy=5,\ -p+2q=5$

$15-3(p-2q)=15-3(-5)=30$

Answer is (C)

20. $T=1200-10t-0.5t^2$

$800=1200-10t-0.5t^2$

$0.5t^2+10t-400=0$

$t^2+20t-800=0$

$(t+40)(t-20)=0$

$t=-40,\ t=20$

After 20 seconds.

21. $\begin{cases} y-\frac{3}{a}x \le 2 \\ \frac{1}{a}x-\frac{1}{3}y \le -1 \end{cases} \Rightarrow \begin{cases} -\frac{3}{a}x+y \le 2 \\ \frac{1}{a}x-\frac{1}{3}y \le -1 \end{cases}$

Since $\frac{-\frac{3}{\not a}}{\frac{1}{\not a}}=\frac{1}{-\frac{1}{3}}\text{(must)}\ne\frac{2}{-1}$

$\Rightarrow\ -3=-3\ne-2$

The ratio of x is the same as the ratio of y, not equal to the ratio of the constant term. So, the value of a can be any value except zero.

Answer is (D)

22. (B)

Cost = sales revenue

Old price: $C=20 \bullet n+2000=30 \bullet n$

$10 \bullet n=2000$

$n=200$

New price: $C=20 \bullet n+2000=25 \bullet n$

$5n=2000$

$n=400$

$400-200=200$

200 books needed to sell.

Answer is (B)

CHAPTER 5
POLYNOMIAL, EXPONENT, RADICAL, and RATIONAL EQUATIONS

Polynomials are algebraic expressions composed of variables, coefficients, and operations, all the exponents must be non-negative integers. This chapter discusses about Higher Degree Polynomials, Exponential, Radical, and Rational Equations.

5-1 POLYNOMIAL

Remainder Theorem and Factor Theorem both play very important roles in the manipulation of polynomials.

A. Remainder Theorem

> If $p(x)$ is a polynomial in x, $p(a)$ is the remainder when $f(x)$ is divided by $(x-a)$.

$P(x)$ is divided by $(x-a)$, where a is a constant, R is an algebraic expression.

$P(x)$ can be expressed as

$P(x)=(x-a)Q(x)+R$

So, substitute a to x, then

$P(a)=R$

◆Example 1:

$P(x)=3x^2-2x+1$ devided by $x-2$ using long division is shown below.

$$
\begin{array}{rl}
3x^2-2x+1 & \big|\ \underline{x-2} \\
-)\ \underline{3x^2-6x\quad} & \big|\ 3x+4 \\
4x+1 & \\
-)\ \underline{4x-8} & \\
9 &
\end{array}
$$

$$\frac{3x^2-2x+1}{x-2}=(3x+4)+\frac{9}{x-2}$$

Based on Remainder Theorem.

$P(2)=3(2)^2-2(2)+1=3\times4-4+1=12-4+1=9$

◆Example 2:

$P(x)=x^3+2x^2-x+5$ is divided by $x-2$

$$
\begin{array}{rl}
x^3+2x^2-x+5 & \big|\ x-2 \\
-)\ x^3-2x^2 & \ \ x^2+4x+7 \\
\hline
4x^2-x & \\
-)\ 4x^2-8x & \\
\hline
7x+\ 5 & \\
-)\ 7x-14 & \\
\hline
19 &
\end{array}
$$

$P(x)=(x-2)(x^2+4x+7)+19$

Based on Remainder Theorem.

$P(2)=2^3+2(2)^2-(2)+5=8+8-2+5=14+5=19$

◆Example 3:

The remainders of $P(x)$ divided by $x-1$ and $x+1$ are 3, and -5 respectively. What is the remainder of $P(x)$ divided by $(x-1)(x+1)$?

Set $P(x)=(x-1)(x+1)Q(x)+ax+b$ (Where $ax+b$ is the reminder)

So, $P(1)=a+b=3$, $P(-1)=-a+b=-5$

Therefore, $2a=8$, $a=4$, $b=-1$

The remainder is $4x-1$

B. Factor Theorem

In Remainder theorem

$P(x)=(x-a)Q(x)+R(x)$

If $(x-a)$ is a factor of $P(x)$

then there will be no remainder $R(x)$, in other words $R(x)=0$

$P(x)=(x-a)Q(x)+0$ no remainder

$P(a)=(a-a)\cdot Q(a)=0$

◆Example 1:

$P(x)=2x^3-x^2+4x-4$

If $(x-2)$ is a factor of $P(x)$

then $P(2)=2(2)^3-2^2-4(2)-4=16-4-8-4=0$

◆Example 2:

$P(x)=x^5-6x^4+10x^3-7x^2+7x+15$

Is $(x-3)$ a factor of $P(x)$?

$P(3)=3^5-6\times3^4+10\times3^3-7\times3^2+7\times3+15$

$=243-486+270-63+21+15$

$=0$

So, $x-3$ is a factor of $P(x)$.

◆Example 3:

$P(x)=x^3-2x^2+x-2$

$=x^2(x-2)+(x-2)$

$=(x-2)(x^2+1)$

$P(2)=(2-2)(2^2+1)=0$

So, $P(x)=x^3-2x^2+x-2$, has a factor of $x-2$.

◆Example 4:

$P(x)=x^3+7x^2+7x-15$

If $P(x)$ has a factor of $x-1$, then $P(1)=1+7+7-15=0$

If $P(x)$ has a factor of $x+3$, then $P(-3)=(-3)^3+7(-3)^2+7(-3)-15$

$=-27+63-21-15=0$

If $P(x)$ has a factor of $x+5$, than $P(-5)=(-5)^3+7(-5)^2+7(-5)-15$

$=-125+175-35-15=0$

◆Example 5:

If $P(x)=x^3+ax^2+bx-15$ has two factors: $(x-1)$ and $(x+3)$. What is the value of $a-b$?

$P(1)=1+a+b-15=0$, $a+b=14$

$P(-3)=(-3)^3+9a-3b-15=0$

$-27+9a-3b-15=0$

divide both sides by 3,

$-9+3a-b-5=0$

$3a-b=14$

$$\begin{cases} a+b=14\cdots① \\ 3a-b=14\cdots② \end{cases}$$

① + ② $4a=28$ $a=7$

substitute into ① $b=7$

$a-b=7-7=0$

Answer to Concept Check 5-01

$P(-k)=(-k)^3-2(-k)^2+3(-k)-6=0$

$-k^3-2k^2-3k-6=0$

$-k^2(k+2)-3(k+2)=0$

$(-1)(k+2)(k^2+3)=0$

$k=-2$, $k=\pm\sqrt{3}i$

$\therefore\ k=-2$

Answer is (B)

Answer to Concept Check 5-02

$P(x)=x^{15}+2x^{14}-x^{13}+3x^2-2x+3$

$P(1)=1^{15}+2(1)^{14}-(1)^{13}+3(1)^2-2(1)+3=1+2-1+3-2+3=6$

Answer is (D)

Answer to Concept Check 5-03

$P(x)=x^3+ax^2+bx-2$

has 2 factors $(x+1)(x-2)$

$P(-1)=(-1)^3+a(-1)^2+b(-1)-2=-1+a-b-2=0$

$\therefore\ a-b=3$

$P(2)=2^3+2^2\cdot a+2b-2=8+4a+2b-2=6+4a+2b=0$

$\therefore\ 4a+2b=-6$

$2a+b=-3$

From $P(-1)$ and $P(2)$

$a-b=3\cdots(1)$

$2a+b=-3\cdots(2)$

$(1)+(2)\quad 3a=0\quad \therefore\ a=0$

substitute into (1), yield $0-b=3$, $b=-3$

$\therefore\ a+b=0+(-3)=-3$

C. Higher-Degree Equations

1. Solved by factoring

◆Example:

Solve $x^4+3x^3+5x^2+7x+2=0$

If it has a factor, by factor theorem, it must be one of $(x+1)$, $(x-1)$, $(x+2)$, $(x-2)$.

Why?

Since $p(x)=x^4+3x^3+5x^2+7x+2=(x+a)(x^3+bx^2+cx+d)$,

(where, *a*, *b*, *c*, and *d* are real numbers)

$=x^4+bx^3+cx^2+dx+ax^3+abx^2+acx+ad$

$=x^4+(a+b)x^3+(c+ab)x^2+(d+ac)x+ad$

Therefore, $a+b=3$, $c+ab=5$, $d+ac=7$, and $ad=2$.

From $a\cdot d=2$, a must be a factor of 2.

So, a must be one of 1, -1, 2, or -2.

Let's try $a=-2$, $x+a=x+2$

Split terms and group them.

$P(x)=x^4+2x^3+x^3+2x^2+3x^2+6x+x+2$

$=x^3(x+2)+x^2(x+2)+3x(x+2)+(x+2)$

$=(x+2)(x^3+x^2+3x+1)$

There is no more factor of (x^3+x^2+3x+1) ; so, $x^4+3x^3+5x^2+7x+2=0$

$P(x)$ has only one real root $x=-2$.

◆Example:

Solve $x^3-2x^2-x+2=0$

Therefore there are $x^2(x-2)-(x-2)=0$

$(x-2)(x^2-1)=0$

$(x-2)(x-1)(x+1)=0$

therefore three roots 1, 2, -1

2.By formula

◆Example:

Solve $16x^4-81=0$

$16x^4-81=(4x^2-9)(4x^2+9)=0$

$(2x-3)(2x+3)(4x^2+9)=0$

$2x-3=0 \quad x=\frac{3}{2}$

$2x+3=0 \quad x=-\frac{3}{2}$

This polynomial has 2 real roots

3.Soved by quadratic format

$3x^4-2x^2-1=0$

Let $a=x^2$.

$3x^4-2x^2-1=3a^2-2a-1=(3a+1)(a-1)=0$

$a=-\frac{1}{3}$ or $a=1$

$x^2=-\frac{1}{3}$ or $x^2=1$

$$\begin{array}{ccc} 3 & \times & 1 \\ 1 & & -1 \\ \hline -3 & + & 1=-2 \end{array}$$

Since $x^2 = -\frac{1}{3}$ has no real solution

Therefore $x=1$ or -1.

4.Solved by factor theorem

◆Example:

Solve $x^5 - 5x^3 - x^2 + 7x - 2 = 0$

Try $x=1$, $P(1) = 1-5-1+7-2=0$

So, $x=1$ is a root.

Try $x=-1$, $P(-1) = -1+5-1-7-2=-6 \neq 0$

So, -1 is not a root.

Try $x=2$, $P(2) = 2^5 - 5(2)^3 - 2^2 + 7(2) - 2 = 32-40-4+14-2 = 0$

2 is another root.

Try $x=-2$, $P(-2) = (-2)^5 - 5(-2)^3 - (-2)^2 + 7(-2) - 2$

$= -32+40-4-14-2$

$= -12 \neq 0$

Because the constant is -2, you only need try 4 times, that are $+1$, -1, $+2$, -2

$x^5 - 5x^3 - x^2 + 7x - 2 = 0$

The polynomial has 2 real roots 1, 2

Answer to Concept Check 5-04

$P(x) = x^4(x+3) + 2x^2(x+3) - 3(x+3)$

$= (x+3)(x^4 + 2x^2 - 3)$

$= (x+3)[(x^2)^2 + 2x^2 - 3]$

$= (x+3)(x^2+3)(x^2-1)$

$$\begin{array}{ccc} 1 & \times & -3 \\ 1 & & 1 \\ \hline 1 & - & 3 = -2 \end{array}$$

let $P(x) = 0$

$(x^2+3)(x+3)(x^2-1) = 0$

$\therefore\ (x^2+3)(x+3)(x+1)(x-1) = 0$

The real number roots are $x=-3$, or $x=-1$, $x=1$.

Answer to Concept Check 5-05

$x^3 - 27 = (x^3 - 3^3)$

since $(a^3 - b^3) = (a-b)(a^2 + ab + b^2)$

$\therefore\ (x^3 - 3^3) = (x-3)(x^2 + 3x + 9) = 0$

$x - 3 = 0 \quad \therefore x = 3$, or $x^2 + 3x + 9 = 0$

$$x = \frac{-3 \pm \sqrt{9 - 4(1)(9)}}{2} = \frac{-3 \pm \sqrt{-27}}{2}$$

$$= -\frac{3}{2} \pm \frac{3\sqrt{3}}{2}i \quad \text{(not real roots)}$$

so, $x = 3$

> ★Tips
>
> $a^3 - b^3 = (a-b)(a^2 + ab + b^2)$
>
> $a^3 + b^3 = (a+b)(a^2 - ab + b^2)$

Answer to Concept Check 5-06

let $A = x^2$, then $4x^4 + 8x^2 - 3 = 0$

Let $4A^2 + 8A - 3 = 0$

$$A = \frac{-(8) \pm \sqrt{8^2 - 4(4)(-3)}}{2(4)} = \frac{-8 \pm \sqrt{64 + 48}}{8}$$

$$= -1 \pm \frac{\sqrt{112}}{8} = -1 \pm \frac{4\sqrt{7}}{8}$$

$$= -1 \pm \frac{\sqrt{7}}{2}$$

$$x^2 = \frac{-1 \pm \sqrt{7}}{2} = -\frac{1}{2} \pm \frac{\sqrt{7}}{2} = 0.5 \pm 1.32$$

$x^2 = 1.82$, or $x^2 = -0.82$ (meaningless)

Therefore $x = \sqrt{1.82} = 1.35$

Answer to Concept Check 5-07

$P(x)=0$

$4x^4+8x^2+3=(2x^2)^2+4(2x^2)+3=0$

Let $z=2x^2$,

then $P(x)=Q(z)=z^2+4z+3=(z+1)(z+3)$

Let $Q(z)=0$

$z=-1$, or $z=-3$

$2x^2=-1$, $2x^2=-3$

$\therefore\ x^2=-\dfrac{1}{2}$

$x=\pm\sqrt{\dfrac{1}{2}}i$ not real roots

When $2x^2=-3$,

$x^2=\dfrac{-3}{2}$

$x=\sqrt{\dfrac{3}{2}}i$ not real roots

So, there is no real root.

5-2 EXPONENT

Exponent is a mathematical symbol that indicates a number or an expression multiplied by itself multiple times.

For instance, a number multiplies itself by 5 times $a\times a\times a\times a\times a$ can be expressed as a^5 where a is the base of the exponent and 5 is the exponent.

A. Laws of Exponents

If a, b are positive real number.

1. $a^m \cdot a^n = a^{m+n}$
2. $a^m \div a^n = a^{m-n}$
3. $(a^m)^n = a^{mn} = (a^n)^m$
4. $(ab)^m = a^m \cdot b^m$

5. $(\frac{a}{b})^m = \frac{a^m}{b^m}$

6. $a^{-m} = \frac{1}{a^m} = (\frac{1}{a})^m$

7. $a^{\frac{n}{m}} = \sqrt[m]{a^n}$

8. $a^0 = 1 \quad (a \neq 0)$

B. Simplification of Exponents

◆Example 1:

$a^5 \bullet a^3 \bullet a^1 = a^{5+3+1} = a^9$

◆Example 2:

$a^5 \div a^3 \div a^1 = a^{5-3} \div a^1 = a^2 \div a^1 = a^{2-1} = a$

◆Example 3:

$(a^3)^2 + 3a^5 + 2a^6 = a^{3\times 2} + 3a^5 + 2a^6 = a^6 + 3a^5 + 2a^6 = (1+2)a^6 + 3a^5 = 3a^6 + 3a^5$

$= 3a^5(a+1)$

◆Example 4:

$$\frac{(xy)^5 \bullet y^3 \bullet x^{-2}}{x^2 \bullet y^{-2} \bullet (\frac{1}{x})^3} = \frac{(x^5y^5)(y^3)(x^{-2})}{(x^2)(y^{-2}) \bullet (\frac{1^3}{x^3})} = \frac{(x^5y^5)(y^3)(x^{-2})}{(x^2)(y^{-2})(x^{-3})} = \frac{x^{5-2}y^{5+3}}{x^{2-3}y^{-2}}$$

$$= \frac{x^3y^8}{x^{-1}y^{-2}} = x^{3-(-1)}y^{8-(-2)} = x^{3+1}y^{8+2} = x^4y^{10}$$

C. Exponential Equation

◆Example 1:

If $5^{2x} - (22)(5)^x - 75 = 0$, what is the value of x ?

Let $5^x = u$, then the equation change to $u^2 - 22u - 75 = 0 \quad (u-25)(u+3) = 0$

$u = 25, \quad u = -3$

$5^x = 25$, but $5^x \neq -3$ (exponent cannot be negative)

Since $5^x = 5^2$ therefore $x = 2$

◆Example 2:

If $\frac{(3)^{x^2-1}}{(3)^{-x+1}}=1$, what is the value of x ?

First let's simplify $\frac{(3)^{x^2-1}}{(3)^{-x+1}}=\frac{(3)^{x^2}}{(3^{-x})\cdot 3^1\cdot 3^1}=(3)^{x^2}\cdot(3)^x\cdot(3)^{-2}=1=3^0$

$(3)^{(x^2+x-2)}=3^0$

Therefore $x^2+x-2=0$ $(x+2)(x-1)=0$

$x=-2$ or $x=1$

◆Example 3:

If \$10,000 is invested at an annual interest rate of 1.5% compounded quarterly, what is the balance after 10 years ?

$B=10,000\times(1+\frac{0.015}{4})^{4(10)}=10,000\times(1.00375)^{40}=11,615$

◆Example 4:

A certain bacteria doubles in number every 10 minutes, if there are 10^6 bacteria at the beginning, what is the amount of bacteria in an hour ?

$B(t)=10^6\cdot(2)^{\frac{t}{10}}$

$B(60)=10^6\cdot 2^{\frac{60}{10}}=10^6\cdot 2^6=64\times10^6$

◆Example 5:

$$\begin{cases}2^{x+1}+(3)^{2y+1}=247\\(16)^{\frac{x}{4}}+9^y=83\end{cases}$$

A system of exponential equations are shown above, what are the values of x and y ?

$(16)^{\frac{x}{4}}+9^y=(2^4)^{\frac{x}{4}}+9^y=2^x+9^y=(2)^x+(3)^{2y}$

$$\begin{cases}2(2^x)+3(3^{2y})=247\cdots①\\2^x+3^{2y}=83\cdots②\end{cases}$$

①–②×2 $3(3^{2y})-2(3^{2y})=247-166$

$(3-2)(3^{2y})=81$

$3^{2y}=3^4$

$\therefore\ 2y=4,\ \ y=2$

substitute into $2^x+3^4=83 \quad 2^x+81=83$

$2^x=2 \quad \therefore\ x=1$

Answer: $x=1$, and $y=2$

Answer to Concept Check 5-08

Simplify

$\dfrac{(x^{-6})(y^5)(x^4y^{-1})}{(x^{-2})y^4}+3(a^{-3}b^{-2}c)(a^3b^2c^{-1})$

$=(x^{-6+4+2})(y^{5-1-4})+3(a^{-3+3}b^{-2+2}c^{1-1})$

$=x^0y^0+3a^0b^0c^0$

$=(1)(1)+3(1)(1)(1)$

$=1+3$

$=4$

Answer to Concept Check 5-09

$4^{x+1}+(22)(2^{x-1})-3=2^{2(x+1)}+(\dfrac{22}{2})(2^x)-3$

$=(4)(2^{2x})+(\dfrac{22}{2})(2^x)-3=0$

$(8)(2^{2x})+(22)2^x-6=0$

It $u=2^x$, $8u^2+(22)u-6=0$

$(8u-2)(u+3)=0$

$u=\dfrac{2}{8}=\dfrac{1}{4}$, or $u=-3$

$2^x=\dfrac{1}{4} \quad 2^x=2^{-2} \quad x=-2$

$2^x\neq-3$ (exponent can not be negative)

Answer (B)

5-3 RADICAL

- Radical is represented by the symbol $\sqrt[n]{x}$. It is read "*x* radical *n*." and it represents the "nth root of *x*" where *n* is referred to as the index, and *x* is the radicand. If a multiplied by itself n times equals *x*, then x radical *n* equals *a*.

- The properties of Radical.

  1. In $\sqrt[n]{x}$, if n is an even integer, x must be a positive number, $\sqrt{-16}$ is invalid.
  2. The difference between x^2 and $\sqrt{x}$.

     In this case of $x^2 = 9$, $x = 3$ or $x = -3$ (there are two roots).

     But in the case of $\sqrt{x} = 3$, x can only be 9 (there is one root only).
  3. All radical form can be treated as a fractional exponents. Such as $\sqrt[3]{3^2}$ can be converted to $3^{\frac{2}{3}}$
  4. Rationalization the Denominator

- Using perfect square, cube, or any other nth power to make the denominator without radical.

◆Example:

Simplify $\sqrt{\dfrac{2}{3}}$

$$\sqrt{\frac{2}{3}\times\frac{3}{3}} = \sqrt{\frac{6}{3^2}} = \frac{\sqrt{6}}{3}$$

◆Example:

Simplify $\dfrac{\sqrt{10}+\sqrt{3}}{\sqrt{2}}$

$$\frac{\sqrt{10}+\sqrt{3}}{\sqrt{2}}\times\frac{\sqrt{2}}{\sqrt{2}} = \frac{\sqrt{20}+\sqrt{6}}{\sqrt{2^2}} = \frac{\sqrt{5\times4}+\sqrt{6}}{2} = \frac{2\sqrt{5}+\sqrt{6}}{2}$$

- Conjugates

Using multiplication format $a^2 - b^2 = (a-b)(a+b)$, multiply the conjugate to make the denominator rationalized.

◆Example:

simplify $\dfrac{\sqrt{3}+\sqrt{2}}{\sqrt{2}-1}$

$$\frac{(\sqrt{3}+\sqrt{2})(\sqrt{2}+1)}{(\sqrt{2}-1)(\sqrt{2}+1)}=\frac{\sqrt{6}+(\sqrt{2})^2+\sqrt{3}+\sqrt{2}}{2-1}=2+\sqrt{2}+\sqrt{3}+\sqrt{6}$$

◆Example:

Simplify $\sqrt{18}+27\sqrt{\dfrac{1}{3}}+\dfrac{4}{\sqrt{3}-\sqrt{2}}=3\sqrt{2}+(27)(\sqrt{\dfrac{1}{3}})+\dfrac{(4)(\sqrt{3}+\sqrt{2})}{(\sqrt{3}-\sqrt{2})(\sqrt{3}+\sqrt{2})}$

$$=3\sqrt{2}+(27)\frac{\sqrt{1}(\sqrt{3})}{(\sqrt{3})(\sqrt{3})}+\frac{4\sqrt{3}+4\sqrt{2}}{3-2}$$

$$=3\sqrt{2}+\frac{(27)(\sqrt{3})}{3}+4\sqrt{3}+4\sqrt{2}$$

$$=3\sqrt{2}+9\sqrt{3}+4\sqrt{3}+4\sqrt{2}$$

$$=7\sqrt{2}+13\sqrt{3}$$

• Radical Equations

Radical Equations.
A radical equation is an equation where the variable is under a radical sign.

◆Example:

Solve $\sqrt{x-1}=2$

$(\sqrt{x-1})^2=2^2$

$x-1=4$, $x=5$

Substitute 5 into the original equation

$\sqrt{5-1}=\sqrt{4}=2$ (checked)

◆Example:

$\sqrt{x-1}+\sqrt{x}=2$

Isolate one radical to either side of the equation

$\sqrt{x-1} = 2 - \sqrt{x}$

Square both side.

$x - 1 = (2 - \sqrt{x})^2 = 4 - 4\sqrt{x} + x$

$-1 = 4 - 4\sqrt{x}$

$4\sqrt{x} = 5 \quad \sqrt{x} = \frac{5}{4}$

$\therefore \quad x = (\frac{5}{4})^2 = \frac{25}{16}$

Substitute $x = \frac{25}{16}$ into the original equation.

$\sqrt{\frac{25}{16} - 1} + \sqrt{\frac{25}{16}} = \sqrt{\frac{9}{16}} + \sqrt{\frac{25}{16}} = \frac{3}{4} + \frac{5}{4} = \frac{8}{4} = 2$ (checked)

5-4 RATIONAL EQUATIONS

A rational equation is an equation that contains at least one fraction whose numerator and denominator are polynomials. The standard form of a rational equations is $\frac{p(x)}{q(x)} = 0$, where $q(x) \neq 0$.

1. Simplify equations.
2. Cross-multiply to eliminate the denominator.
3. Solve the equation using the general approach.
4. Be sure to recheck the roots.

◆Example 1:

Solve $\frac{x-1}{x+1} = 3$

$x - 1 = 3x + 3$

$2x = -4 \quad x = -2$

Recheck: substitute $x = -2$ into the original equation $\frac{-2-1}{-2+1} = \frac{-3}{-1} = 3$ (checked)

◆Example 2:

$$\frac{2x+3}{x+1} + \frac{x+4}{x-1} = 1 = \frac{(2x+3)(x-1) + (x+1)(x+4)}{(x+1)(x-1)} = \frac{(2x^2 - 2x + 3x - 3) + (x^2 + 4x + x + 4)}{x^2 - 1}$$

$$=\frac{(2x^2+x-3)+(x^2+5x+4)}{x^2-1}=\frac{3x^2+6x+1}{x^2-1}=1$$

$3x^2+6x+1=x^2-1$

$2x^2+6x+2=0 \quad x^2+3x+1=0$

$x=\frac{-3\pm\sqrt{9-4}}{2}=\frac{-3\pm\sqrt{5}}{2}$

Recheck, substitute $x=\frac{-3\pm\sqrt{5}}{2}$ into original equations.

$$\frac{(2)(\frac{-3+\sqrt{5}}{2})+3}{\frac{-3+\sqrt{5}}{2}+1}+\frac{\frac{-3+\sqrt{5}}{2}+4}{\frac{-3+\sqrt{5}}{2}-1}$$

$$=\frac{-3+\sqrt{5}+3}{\frac{-3+\sqrt{5}+2}{2}}+\frac{\frac{-3+\sqrt{5}+8}{2}}{\frac{-3+\sqrt{5}-2}{2}}$$

$$=\frac{2\sqrt{5}}{-1+\sqrt{5}}+\frac{5+\sqrt{5}}{-5+\sqrt{5}}$$

$$=\frac{(2\sqrt{5})(-5+\sqrt{5})+(5+\sqrt{5})(-1+\sqrt{5})}{(-1+\sqrt{5})(-5+\sqrt{5})}$$

$$=\frac{-10\sqrt{5}+10-5+5\sqrt{5}-\sqrt{5}+5}{(5-\sqrt{5}-5\sqrt{5}+5)}$$

$=\frac{10-6\sqrt{5}}{10-6\sqrt{5}}=1$ (checked).

$x=\frac{-3-\sqrt{5}}{2}$ (also can be checked.)

Answer: $x=\frac{-3\pm\sqrt{5}}{2}$

◆Example 3:

Solve $\frac{x-5}{x+6}=3$

Multiply both sides by $x+6$

Yield $x-5=3(x+6)=3x+18$

$2x=-23 \quad \therefore \; x=-\frac{23}{2}$

Caution: Be sure to check and eliminate "extraneous roots". An extraneous root is a solution to an equation that turns out to be incorrect when substituting it into the original solution.

Substitute $x=-\frac{23}{2}$ into the original equation $\frac{-\frac{23}{2}-5}{-\frac{23}{2}+6}=\frac{\frac{-33}{2}}{\frac{-11}{2}}=3$. (checked)

Answer: $\frac{x-5}{x+6}=3$, the root is $x=-\frac{23}{2}$.

◆Example 4:

Solve $\frac{3x^2+7x+2}{x^2+3x+2}=2$

$$\frac{\cancel{(x+2)}(3x+1)}{\cancel{(x+2)}(x+1)}=2$$

$$\frac{3x+1}{x+1}=2$$

$3x+1=2x+2$

$x=1$

Substitute it into $P(x)$

$\frac{3(1)^2+7(1)+2}{(1)^2+3(1)+2}=\frac{3+7+2}{1+3+2}=\frac{12}{6}=2$ (checked)

So, $x=1$

Answer to Concept Check 5-10

$$\frac{\sqrt{2}+\sqrt{5}}{\sqrt{3}-1}=\frac{(\sqrt{2}+\sqrt{5})(\sqrt{3}+1)}{(\sqrt{3}-1)(\sqrt{3}+1)}$$

$$=\frac{\sqrt{6}+\sqrt{2}+\sqrt{15}+\sqrt{5}}{(\sqrt{3})^2-1}$$

$$=\frac{\sqrt{2}(\sqrt{3}+1)+\sqrt{5}(\sqrt{3}+1)}{3-1}$$

$$=\frac{(\sqrt{3}+1)(\sqrt{2}+\sqrt{5})}{2}$$

Answer to Concept Check 5-11

$\sqrt{x-1}-\sqrt{4x}+2=0$

$x-1=(\sqrt{4x}-2)^2=(\sqrt{4x})^2-4\sqrt{4x}+4=4x-8\sqrt{x}+4$

$8\sqrt{x}=3x+5$

$64x=(3x+5)^2=9x^2+30x+25$

$9x^2-34x+25=0$

$(9x-25)(x-1)=0$

$\therefore\ x=\frac{25}{9}$ or $x=1$

$$\begin{array}{cc} 9 & -25 \\ 1 & -\ 1 \\ \hline -25 & -9=-34 \end{array}$$

check $\sqrt{\frac{25}{9}-1}-\sqrt{\frac{4\times 25}{9}}+2=\sqrt{\frac{16}{9}}-\sqrt{\frac{100}{9}}+2$

$=\frac{4}{3}-\frac{10}{3}+2=\frac{-6}{3}+2=0$ checked.

$\sqrt{1-1}-\sqrt{4\times 1}+2=0-2+2=0$ checked

Answers are x = 1 or x = $\frac{25}{9}$

Answer to Concept Check 5-12

$\frac{2x-3}{x-1}+\frac{3x-2}{x+1}=1$

$(2x-3)(x+1)+(3x-2)(x-1)=(x-1)(x+1)=x^2-1$

$2x^2-3x+2x-3+3x^2-2x-3x+2=x^2-1$

$5x^2-6x-1=x^2-1$

$4x^2-6x=0,\ x(4x-6)=0$

$x=0$ or $4x=6,\ x=\frac{3}{2}$

Substitute them into one of the original equation.

For $x=0$, left side $=\frac{-3}{-1}+\frac{-2}{1}=3-2=1$ checked

For $x=\frac{3}{2}$, left side $=\frac{0}{\frac{1}{2}}+\frac{\frac{9}{2}-2}{\frac{5}{2}}=0+\frac{\frac{5}{2}}{\frac{5}{2}}=1$ checked

Answer to Concept Check 5-13

$$\frac{x}{x-2}-\frac{2}{x}-\frac{3}{x(x-2)}=0$$

multiply both sides by $x(x-2)$.

$x^2-2(x-2)-3=0$

$x^2-2x+4-3=0$

$x^2-2x+1=0$

$(x-1)^2=0$, $x=1$

Substitute into the original equation

$\frac{1}{1-2}-\frac{2}{1}-\frac{3}{1(1-2)}=-1-2+3=0$ checked

CHAPTER 5
PRACTICE

1. Which of the following is equivalent to $x^{\frac{3}{5}}$?

(A) $\sqrt[3]{x^5}$

(B) $\sqrt[5]{x^3}$

(C) $\sqrt[5]{\sqrt[3]{x}}$

(D) $\sqrt[3]{\sqrt[5]{x}}$

2. If $a=x^{-\frac{2}{3}}$, where $a>0$, $x>0$, which of the following equations gives x in terms of a ?

(A) $x=\dfrac{1}{\sqrt{a^3}}$

(B) $x=\dfrac{1}{\sqrt[3]{a^2}}$

(C) $x=\sqrt{a^3}$

(D) $x=a$

3. If $\sqrt{5x}=4.8$, $x=$

4. The equation $\dfrac{32x^2+8x-21}{kx-3}=8x+8+\dfrac{3}{kx-3}$ is true for all values of $x\neq\dfrac{3}{k}$, where k is a constant. Find the value of k.

5. If $(x+1)^{\frac{2}{3}}=16$, what is the value of x ?

(A) 63

(B) 64

(C) 65

(D) 66

6. If $g(x)=pr^x$ and if $g(0)=5$ and $g(1)=30$, Find $g(2)$.

(A) 180

(B) 220

(C) 260

(D) 300

7. If $f(x)=3^{-x}$ and $g(x)=27\cdot3^{-x}$, which of the following is true ?

(A) $g(x)=f(x-3)$

(B) $g(x)=f(x+3)$

(C) $g(x)=f(x)-3$

(D) none of them

8. Which is the solution to $(8^x)(2^4)=(\frac{1}{2})^x$?

(A) -2

(B) -1

(C) $-\dfrac{1}{12}$

(D) 1

9. If $\dfrac{x^2-x-6}{x^2-4x+3}=\dfrac{4}{3}$, find x

(A) -10

(B) -2

(C) 2

(D) 10

10. For a polynomial $p(x)$, if $p(2)=-3$, which of the following must be true ?

(A) $x+3$ is a factor of $p(x)$

(B) $x-3$ is a factor of $p(x)$

(C) $x-2$ is a factor of $p(x)$

(D) if $p(x)$ is divided by $x-2$, the remainder is -3

11. Given the polynomial $4x^4-2x^2+6x+k$, where k is a constant, what would k be when $\frac{4x^4-2x^2+6x+k}{x-2}$ has no remainder ?

(A) -48

(B) -68

(C) 48

(D) 68

12. If $\frac{a^{x^2}}{a^{y^2}}=a^{24}$, $a>1$ and $x-y=3$, what is the value of $x+y$?

(A) 3

(B) -3

(C) 8

(D) -8

13. For the radical equation of $\sqrt{3x^2-2}-k=0$ where $x>0$, and $k=5$, which of the following could be the value of x ?

(A) 1

(B) 2

(C) 3

(D) 4

14. If $2x-2y=3$, what is the value of $\frac{2^x}{2^y}$?

(A) $2^{\frac{3}{2}}$

(B) $2^{\frac{2}{3}}$

(C) 2^3

(D) Can not be determined

15. If $14^k=64$, what is the value of $14^{\frac{k}{2}+1}$?

(A) 33

(B) 112

(C) 256

(D) 448

16. Solve $5x-\sqrt{x}-1=0$

(A) 0.13

(B) 1.33

(C) 2.57

(D) Can not be determined

17. If the equation $\frac{3x^2+1}{2-x^2}=-3$ is true, which of the following is also true ?

(A) The equation has no solution.

(B) The equation has exactly one solution.

(C) The equation has exactly two solutions.

(D) The equation has infinitely many solutions.

18. If $27^{3-x}=81$, what is the value of x ?

(A) $\frac{3}{5}$

(B) $\frac{5}{3}$

(C) $\frac{3}{7}$

(D) $\frac{7}{3}$

19. The half-life of a radioactive substance is 80 years. How many kg of such substance would remain after 300 years given that there is 80 ky right now?

(A) 5.95

(B) 10.23

(C) 15.35

(D) 20.15

20. If $x^8=8888$, $\frac{x^7}{y}=22$, what is the value of xy ?

(A) 404

(B)444

(C) 85547

(D) 171094

21. Solve $\frac{2^{x^2+5}}{2^x}=128$

(A) $x=2$ or $x=-1$

(B) $x=-2$ or $x=-1$

(C) $x=2$ or $x=1$

(D) $x=-2$ or $x=-1$

22. If $s=6+\frac{5}{s}$, what is the value of $s^2+\frac{25}{s^2}$?

(A) 40

(B) 46

(C) 50

(D) 51

23. Simplify $\frac{x^5y^3z^2}{x^2yz^{-2}}$

(A) x^7y^4

(B) $x^3y^2z^4$

(C) x^3y^2

(D) $x^7y^4z^4$

24. If $3^t+3^t+3^t=3^5$, then $t=$

25. The remainders of $p(x)$ divided by $x+1$, and $x-2$ are 7 and 19 respectively, what is the remainder when $p(x)$ is divided by $(x+1)(x-2)$?

(A) 26

(B) $x+26$

(C) $4x-3$

(D) $4x+11$

CHAPTER 5
ANSWERS and EXPLANATIONS

1. $x^{\frac{3}{5}} \Rightarrow \sqrt[5]{x^3}$

   Answer is (B)

2. (A)

   $a = x^{-\frac{2}{3}}$, $a^{-\frac{3}{2}} = x^{(-\frac{2}{3})(-\frac{3}{2})} = x$

   $x = \dfrac{1}{a^{\frac{3}{2}}} = \dfrac{1}{\sqrt{a^3}}$

   Answer is (A)

3. $5x = (48)^2$

   $x = \dfrac{(4.8)^2}{5} = 4.6$

4. Since $32x^2 + 8x - 21$

   $= (kx-3)(8x+8)+3$

   $= 8kx^2 + 8kx - 24x - 24 + 3$

   $= 8kx^2 + 8x(k-3) - 21$

   So, $32 = 8k$, $k - 3 = 1$

   $\therefore\ k = 4$

5. $(x+1)^{\frac{2}{3}} = 16$

   $x + 1 = 16^{\frac{3}{2}} = (4^2)^{\frac{3}{2}} = 4^3 = 64$

   $\therefore\ x = 64 - 1 = 63$

   Answer is (A)

6. $g(0) = pr^0 = 5 \quad \therefore p = 5$

   $g(1) = pr^1 = 5 \cdot r^1 = 30$

   $r = \dfrac{30}{5} = 6$

   $g(2) = 5 \cdot 6^2 = 180$

   Answer is (A)

7. $g(x) = 3^3 \cdot 3^{-x} = 3^{3-x} = 3^{-(x-3)}$

   since $f(x-3) = 3^{-(x-3)}$

   $g(x) = f(x-3) = 3^{-(x-3)}$

   Answer is (A)

8. $(8^x)(2^4) = 2^{3x} \cdot 2^4 = 2^{3x+4} = (\frac{1}{2})^x = 2^{-x}$

   $\therefore\ 3x + 4 = -x$, $4x = -4$

   $\therefore\ x = -1$

   Answer is (B)

9. $\dfrac{x^2 - x - 6}{x^2 - 4x + 3} = \dfrac{\cancel{(x-3)}(x+2)}{\cancel{(x-3)}(x-1)}$

   $= \dfrac{x+2}{x-1} = \dfrac{4}{3}$

   $4x - 4 = 3x + 6$

   $\therefore\ x = 10$

   Substitute $x=10$ into the original equation $\dfrac{10^2-10-6}{10^2-40+3} = \dfrac{84}{63} = \dfrac{(21)(4)}{(21)(3)} = \dfrac{4}{3}$

   (checked)

   Answer is (D)

10. According to the Remainder Theorem, $p(2) = -3$ means that the remainder is

-3 when $p(x)$ is divided by $x-2$

Answer is (D)

11. (B)

$$\begin{array}{r|l} 4x^4+0-2x^2+6x+k & x-2 \\ \cline{2-2} 4x^4-8x^3 & 4x^3+8x^2+14x+34 \\ \hline 8x^3-2x^2+6x+k & \\ 8x^3-16x^2 & \\ \hline 14x^2+6x+k & \\ 14x^2-28x & \\ \hline 34x+k & \\ 34x-68 & \end{array}$$

$k=-68$

Answer is (B)

12. $\frac{a^{x^2}}{a^{y^2}}=a^{x^2-y^2}=a^{24}$

$\therefore\ (x+y)(x-y)=24$

$(x+y)\cdot(3)=24$

$\therefore\ x+y=8$

Answer is (C)

13. $\sqrt{3x^2-2}-k=0$, if $k=5$

$\sqrt{3x^2-2}-5=0$

$\sqrt{3x^2-2}=5,\ 3x^2-2=25$

$3x^2=27,\ x^2=9,\ x=\pm3$

Since $x>0$

so, $x=3$

Answer is (C)

14. $2x-2y=3$

$x=\frac{3}{2}+y,\ 2^x=2^{\frac{3}{2}+y}$

$$\frac{2^x}{2^y}=\frac{2^{\frac{3}{2}+y}}{2^y}=\frac{2^{\frac{3}{2}}\cdot2^y}{2^y}=2^{\frac{3}{2}}$$

Answer is (A)

15. $14^k=64$

$14^{\frac{k}{2}+1}=\sqrt{14^k}\cdot14$

$=\sqrt{64}\cdot14$

$=8\cdot14$

$=112$

Answer is (B)

16. $5x-\sqrt{x}-1=0$

$\sqrt{x}=5x-1$

$x=(5x-1)^2=25x^2-10x+1$

$25x^2-11x+1=0$

$$x=\frac{11\pm\sqrt{11^2-4(25)(1)}}{2\times25}$$

$$=\frac{11\pm\sqrt{21}}{50}$$

$$=\frac{11\pm4.58}{50}$$

$$x=\frac{15.58}{50}=0.31\ \text{or}\ \frac{6.42}{50}=0.13$$

Answer is (A)

17. Since $\cancel{3x^2}+1=-6+\cancel{3x^2}$

so, the equation has no solution.

Answer is (A)

18. $27^{3-x}=81$

$(3^3)^{3-x}=3^{(9-3x)}=81=3^4$

$9-3x=4$

$3x=5$

$x=\frac{5}{3}$

Answer is (B)

19. $w=80(0.5)^{\frac{300}{80}}=5.95$

Answer is (A)

20. $\frac{x^7}{y}=\frac{x^8}{xy}=22$

$\therefore\ xy=\frac{x^8}{22}=\frac{8888}{22}=404$

Answer is (A)

21. $\frac{2^{x^2+5}}{2^x}=128$

$2^{x^2-x+5}=2^7$

$\therefore\ x^2-x+5=7$

$x^2-x-2=0$

$(x-2)(x+1)=0$

$x=2$ or $x=-1$

Substitute x into the original equation

$x=2$

$2^{4+5-2}=2^7=128$ (checked.)

$x=-1$

$2^{(-1)^2+1+5}=2^7=128$ (checked.)

Answer is (A)

22. Since $s=6+\frac{5}{s}$, $s-\frac{5}{s}=6$

$(s-\frac{5}{s})^2=s^2+\frac{25}{s^2}-10=36$

$s^2+\frac{25}{s^2}=46$

Answer is (B)

23. $\frac{x^5y^3z^2}{x^2yz^{-2}}=x^{5-2}y^{3-1}z^{2-(-2)}=x^3y^2z^4$

Answer is (B)

24. $3^t+3^t+3^t=3\cdot 3^t=3^{t+1}$

Since $3^{t+1}=3^5$

$\therefore\ t+1=5$

$\therefore\ t=4$

25. Let $p(x)=(x+1)(x-2)Qx+ax+b$

$p(-1)=-a+b=7$

$p(2)=2a+b=19$

$\therefore\ 3a=12,\ a=4,\ b=11$

The remainder is $4x+11$

Answer is (D)

CHAPTER 6
COORDINATE and ANALYTICAL GEOMETRY

6-1 XY-COORDINATE PLANE

XY-coordinate plane is a grid composed of two perpendicular axes; the horizontal axis is often referred to as the x-axis, while the vertical axis is often referred to as the y-axis. There are four quadrants on an XY-coordinate plane. Quadrant I, Quadrant II, Quadrant III, and Quadrant IV. Below graph displays the signs of x and y values in each quadrant.

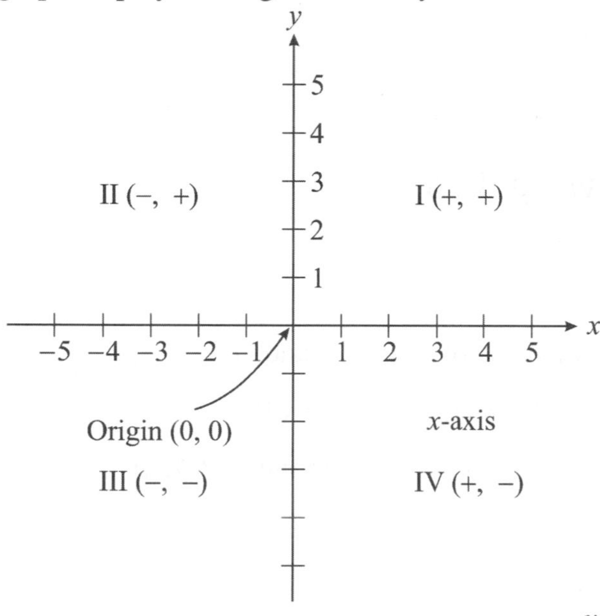

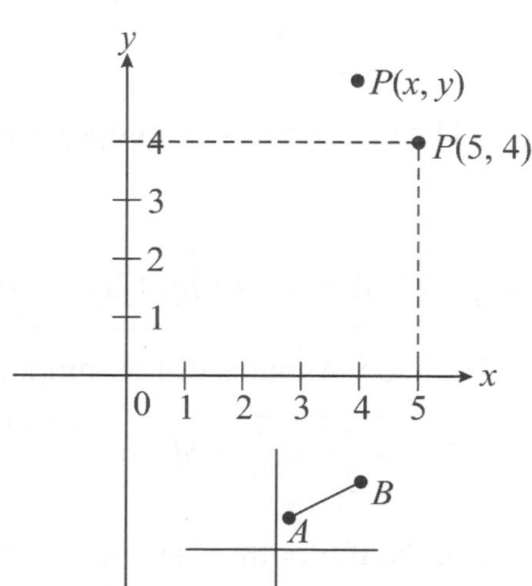

A. Point

A position in xy-plane represented by x and y coordinates is denoted by $P(x, y)$.

B. Segment

The line between two points is denoted as $\overline{AB}$.

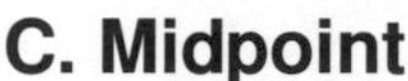

C. Midpoint

The coordinates of midpoint

$$M_x = \frac{x_1 + x_2}{2}, \quad M_y = \frac{y_1 + y_2}{2}$$

For example: Find the midpoint of $P(5, 3)$ and $P(3, 1)$.

$m_x = \dfrac{5+3}{2} = 4$, $m_y = \dfrac{3+1}{2} = 2$

Midpoint is $(4, 2)$

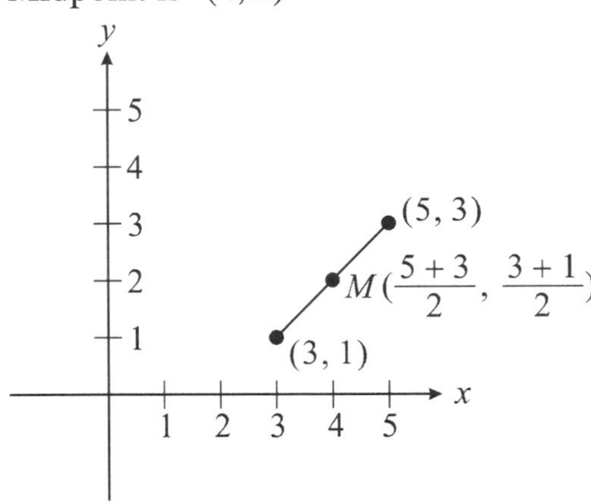

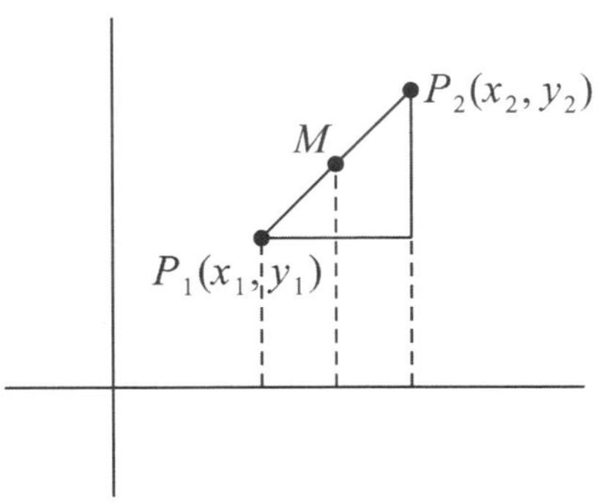

D. Distance from two points

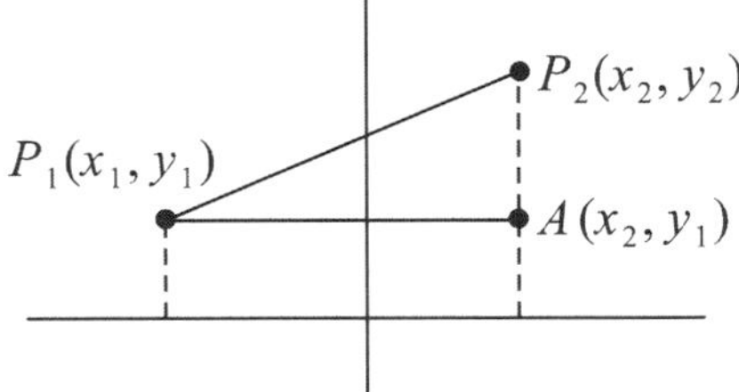

$\overline{P_1P_2}^2 = \overline{P_1A}^2 + \overline{P_2A}^2 = (x_2 - x_1)^2 + (y_2 - y_1)^2$

$|\overline{P_1P_2}| = \sqrt{(x_2 - x_1)^2 + (y_2 - y_1)^2}$

For example :Find the distance of $P_1(3, 2)$ and $P_2(2, -1)$.

$|\overline{P_1P_2}| = \sqrt{(2-3)^2 + (-1-2)^2} = \sqrt{(-1)^2 + 3^2} = \sqrt{10}$

For example: Find the midpoint of $P(-3, 1)$ and $P(4, 2)$.

$m_x = \dfrac{-3+4}{2} = \dfrac{1}{2}$

$m_y = \dfrac{1+2}{2} = \dfrac{3}{2}$

The coordinates of midpoint is $(\dfrac{1}{2}, \dfrac{3}{2})$.

Answer to Concept Check 6-01

If the coordinates of the midpoint are (M_x, M_y),

$M_x = \dfrac{2+5}{2} = \dfrac{7}{2}$, $M_y = \dfrac{3+(-4)}{2} = -\dfrac{1}{2}$

So, the midpoint is at $(\dfrac{7}{2}, -\dfrac{1}{2})$.

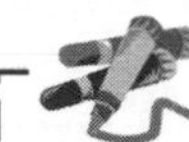

Answer to Concept Check 6-02

$$|\overline{P_1P_2}| = \sqrt{(5-2)^2+(-4-3)^2}$$
$$= \sqrt{3^2+7^2}$$
$$= \sqrt{9+49}$$
$$= \sqrt{58}$$

6-2 STRAIGHT LINE

A. Slope

Slope is a number that describes steepness and direction of a line. The greater the absolute value of a slope, the greater the steepness. The positive and negative sign of a slope indicates the direction.

In below pictures, slope of line ℓ is determined by the angle between the line (ℓ) and the x-axis. Slope is often denoted as m.

Positive slope--the line goes uphill from left to right.

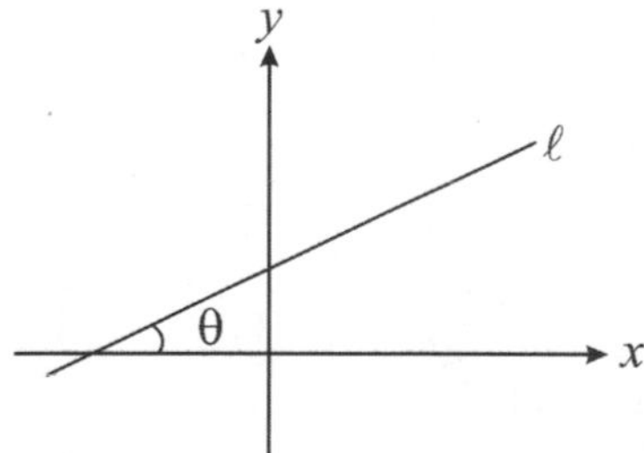

Negative slope--the line goes is downhill from left to right.

(x_1, y_1)

(x_2, y_2)

θ

$$\text{Slope} = \tan\theta = \frac{y_1 - y_2}{x_1 - x_2} \quad \text{or Slope} = \tan\theta = \frac{y_2 - y_1}{x_2 - x_1}$$

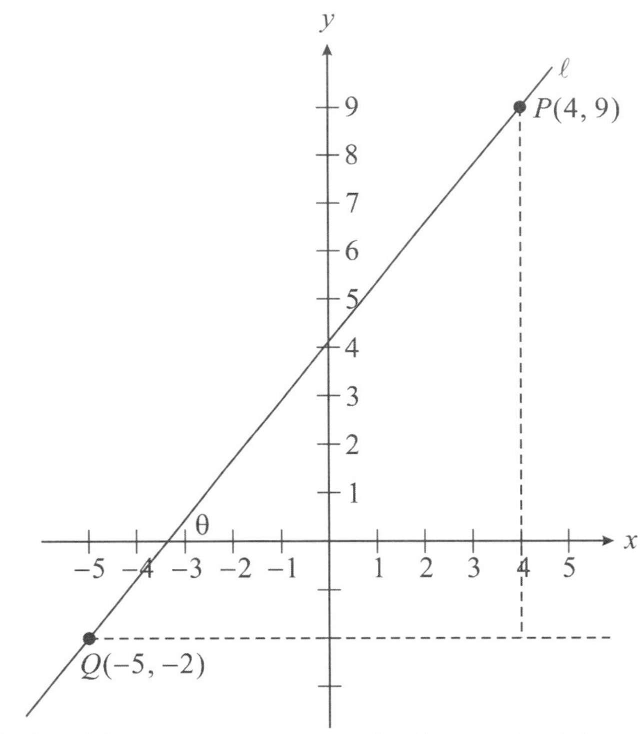

Slope of $l = \frac{9-(-2)}{4-(-5)} = \frac{11}{9}$ or Slope of $l = \frac{-2-9}{(-5)-4} = \frac{-11}{-9} = \frac{11}{9}$

When two lines are parallel. $m_1 = m_2$ (m: slope).

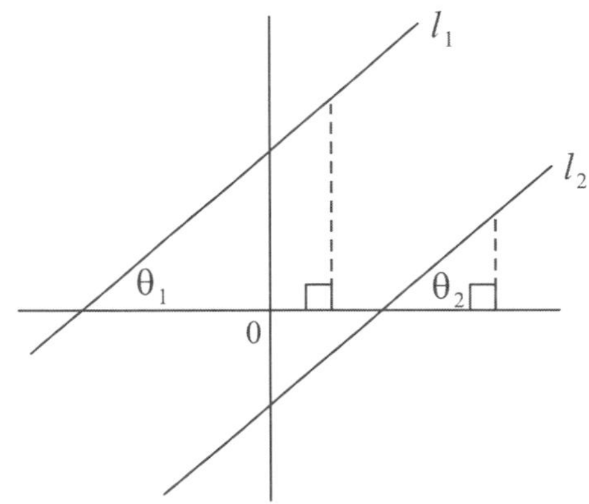

$m_1 = \tan\theta_1$, $m_2 = \tan\theta_2$

$m_1 = m_2$: $\tan\theta_1 = \tan\theta_2$, $\theta_1 = \theta_2$

$\therefore\ l_1 \parallel l_2$

When two lines are perpendicular to each other, the slope of line 1 multiplied by the slope of line 2 is equivalent to -1.

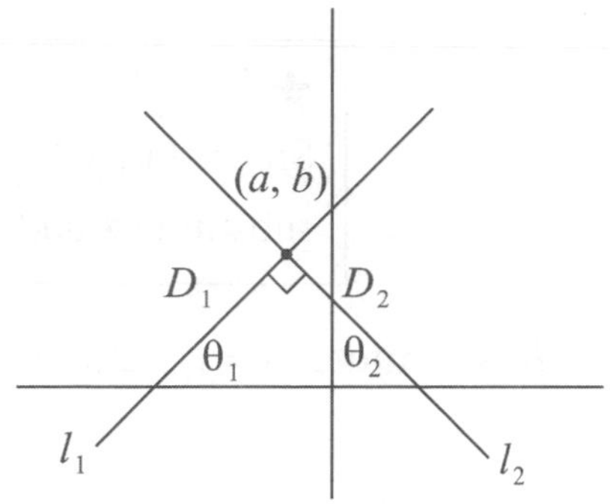

If $l_1 \perp l_2$, then $m_1 \times m_2 = -1$.

$$m_1 = \tan\theta_1 = \frac{D_2}{D_1}, \quad m_2 = \tan\theta_2 = -\frac{D_1}{D_2}$$

$$\therefore \quad m_1 \times m_2 = \tan\theta_1 \times \tan\theta_2 = \frac{D_2}{D_1} \times -\frac{D_1}{D_2} = -1$$

 Answer to Concept Check 6-03

The slope of $\overline{AB}$ is $\dfrac{-5-3}{-3-1} = \dfrac{-8}{-4} = 2$.

6-3 THE ALGEBRAIC FORMS OF STRAIGHT LINES

The equation of a straight line can be represented in multiple forms:

A. Standard Form:

$$Ax + By + C = 0$$

B. Slope-Intercept Form:

$$y = mx + b$$

m is the slope, b is the y intercept.

Since m and b are defined, then the straight line is defined.

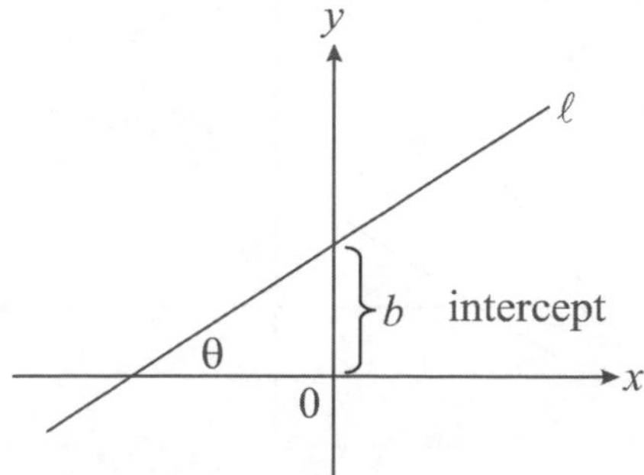

C: Two-Point Form

$p(x_1, y_1)$, $q(x_2, y_2)$

$$y = \frac{y_1 - y_2}{x_1 - x_2}(x - x_1) + y_1$$

> ★TIP
> Since (x_1, y_1) is a point on the line, we can substitute x and y with x_1 and y_1

Proof: $y = mx + b = \frac{y_1 - y_2}{x_1 - x_2}x + b$, substitute $y = y_1$, and $x = x_1$

$$b = y_1 - \frac{y_1 - y_2}{x_1 - x_2}x_1$$

$$y = \frac{y_1 - y_2}{x_1 - x_2}x + y_1 - \frac{y_1 - y_2}{x_1 - x_2} \times x_1 = \frac{y_1 - y_2}{x_1 - x_2}(x - x_1) + y_1$$

D: Point-Slope Form

$$y = m(x - x_1) + y_1$$

Proof: $y = mx + b$

$y_1 = mx_1 + b \quad b = y_1 - mx_1$

$y = mx + y_1 - mx_1 = m(x - x_1) + y_1$

E: Intercepts-Form

$$\frac{x}{a} + \frac{y}{b} = 1$$

(a is the intercept with x-axis, and b is the intercept with y-axis)

Proof: by slope-intercept form

$y = mx + b \quad m = -\frac{b}{a}$

$y = -\frac{b}{a}x + b$

$\therefore \frac{x}{a} + \frac{y}{b} = 1$

◆Example 1:

What is the equation of a straight line which has the slope $\frac{2}{3}$ and the y-intercept is 6 ?

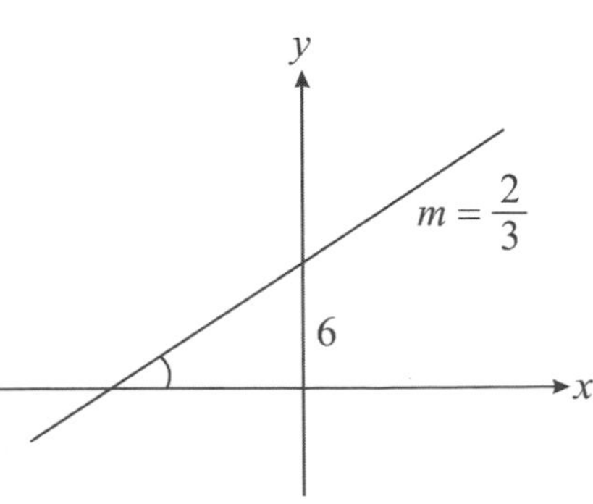

The form of a straight line is $y = \frac{2}{3}x + 6$

◆Example 2:

Line ℓ passes through $P(6, 3)$ and point $Q(2, -3)$, what is the equation of line ℓ ?

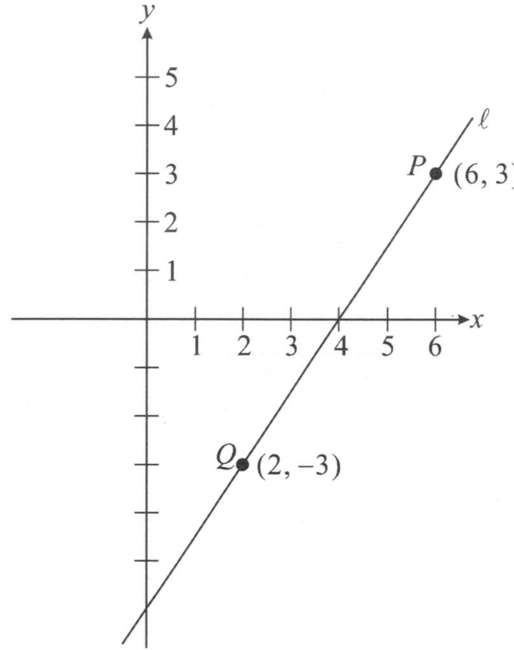

According to the straight line (Two-point) form, it is $y = \dfrac{y_1 - y_2}{x_1 - x_2}(x - x_1) + y_1$

$$
\begin{aligned}
y &= \frac{3-(-3)}{6-2}(x-6)+3 \\
&= \frac{6}{4}(x-6)+3 \\
&= \frac{3}{2}(x-6)+3 \\
&= \frac{3}{2}x-9+3 \\
&= \frac{3}{2}x-6
\end{aligned}
$$

Solution 2: Using slope-intercept form $y = mx + b$, $y = \dfrac{3-(-3)}{6-2}x + b = \dfrac{6}{4}x + b = \dfrac{3}{2}x + b$

Because the line passes though $(6, 3)$, substitute x and y with $(6, 3)$.

$$3 = \frac{3}{2}(6) + b = 9 + b$$

$$b = -6$$

$$y = \frac{3}{2}x - 6$$

◆Example 3:

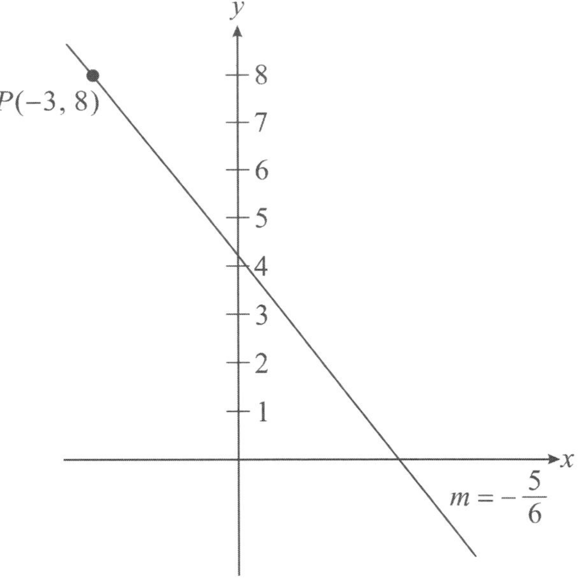

From point-slope form

$y = m(x - x_1) + y_1$

$y = -\frac{5}{6}[x - (-3)] + 8 = -\frac{5}{6}x - \frac{5}{2} + 8$

$= -\frac{5}{6}x + \frac{11}{2}$

Solution 2: slope-intercept

$y = mx + b$

$y = -\frac{5}{6}x + b$, substitute x and y with $(-3, 8)$

$8 = -\frac{5}{6}(-3) + b = \frac{5}{2} + b$

$\therefore\ b = 8 - \frac{5}{2} = \frac{11}{2}$

$\therefore\ y = -\frac{5}{6}x + \frac{11}{2}$

Answer to Concept Check 6-04

The general form of the straight line is $2x + 3y + 4 = 0$

$3y = -2x - 4$, $y = -\frac{2}{3}x - \frac{4}{3}$

slope $= -\frac{2}{3}$, y-intercept $= -\frac{4}{3}$

Answer to Concept Check 6-05

The two-point form is $y = \frac{y_1 - y_2}{x_1 - x_2}(x - x_1) + y_1$

$\therefore\ y = \frac{2 - (-2)}{3 - 4}(x - 3) + 2 = -4(x - 3) + 2$

Answer to Concept Check 6-06

slope-intercept form is $y = mx + b$, $m = -\frac{1}{3}$, $b = 3$

So, the form of the straight is $y = -\frac{1}{3}x + 3$.

6-4 LINEAR SYSTEM EQUATIONS SOLVED BY GRAPHS

Coordinates and analytic geometry are often used to solve linear equations. Below are some examples.

◆Example 1:

Solve system equations

$$\begin{cases} 3x - 2y = -1 \cdots ① \\ 2x + 3y = 7 \cdots ② \end{cases}$$

from ① $-2y = -3x - 1$, $y = \frac{3}{2}x + \frac{1}{2}$

from ② $3y = -2x + 7$, $y = -\frac{2}{3}x + \frac{7}{3}$

Use Slope-Intercept Form

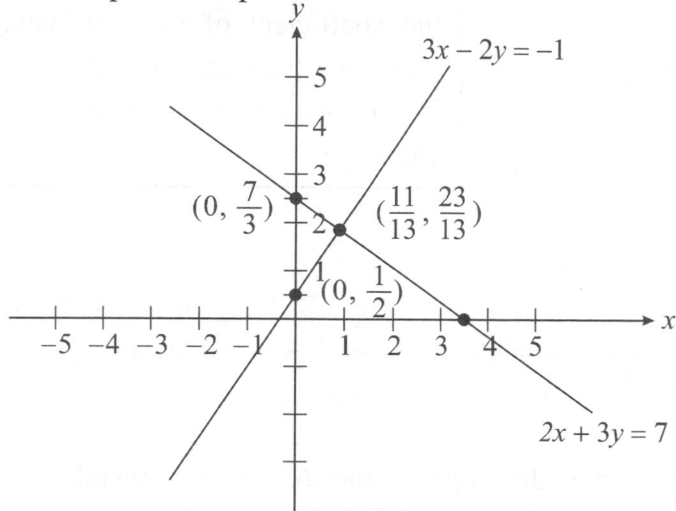

The intersection point is $(\frac{11}{13}, \frac{23}{13})$. There is one intersection point for this system of equations, that means there is one solution to this system of equations.

◆Example 2:

solve system equations

$$\begin{cases} 3x+4y=1\cdots① \\ x-y=-2\cdots② \end{cases}$$

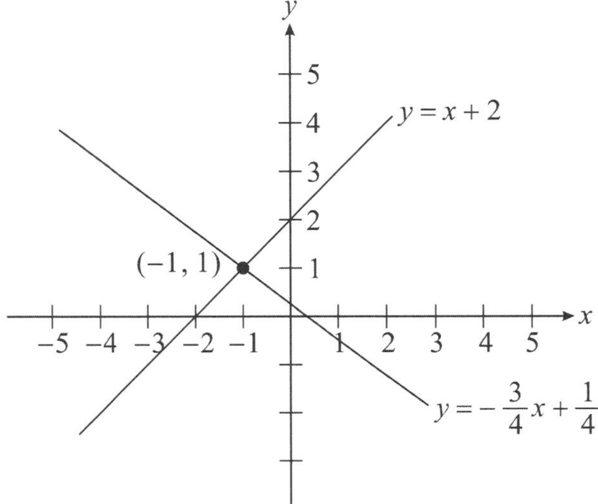

From ① $4y=-3x+1$, $y=-\frac{3}{4}x+\frac{1}{4}$

From ② $-y=-x-2$, $y=x+2$

The intersection point is $(-1,1)$.

"One intersection point means one solution"

> **★TIP**
> When the coefficient of the variables are proportional, but not the coefficient of the constants; these two lines are parallel; there is no solution to this system of equations.

◆Example 3:

$$\begin{cases} 3x+4y=1\cdots① \\ 6x+8y=3\cdots② \end{cases}$$

From ① $4y=-3x+1$, $y=-\frac{3}{4}x+\frac{1}{4}$

From ② $8y=-6x+3$, $y=-\frac{3}{4}x+\frac{3}{8}$

The slope of line $3x+4y=1$ is the same as the slope of line $6x+8y=3$, therefore two lines are parallel. There is no intersection. There system equations have no solution.

Such as, $\frac{3}{6}=\frac{4}{8}\neq\frac{1}{3}$

These two lines are parallel to each other.

◆Example 4:

$$\begin{cases}3x+4y=1\cdots① \\ 6x+8y=2\cdots②\end{cases}$$

From ① $4y=-3x+1$, $y=-\frac{3}{4}x+\frac{1}{4}$

From ② $8y=-6x+2$, $y=-\frac{3}{4}x+\frac{1}{4}$

Two lines are overlaid.

> ★Tip
>
> When two lines are overlaid, so, there are many solutions.
>
> When the coefficients of variables and constant are proportional, there are many solutions.
>
> These two lines are in fact the same line.

◆Example 5:

This example illustrates system of equations with one solution.

Solve system equations:

A. $\begin{cases}2x+3y=3 \\ x-y=-3\end{cases}$

$$\begin{cases}y=\frac{2}{-3}x+1\cdots① \\ y=x+3\cdots②\end{cases}$$

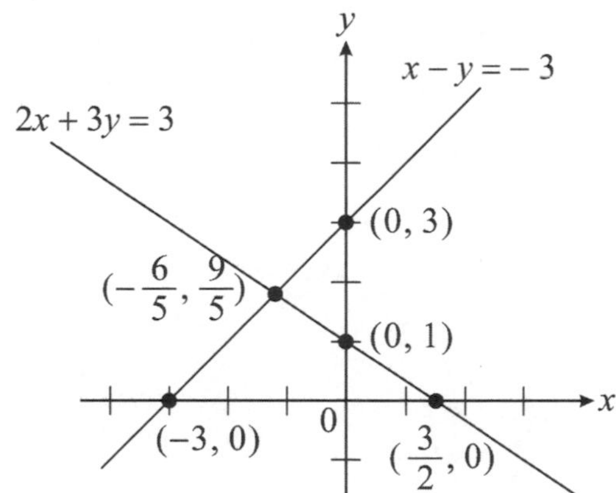

②−①

$x+\frac{2}{3}x+2=0$

$\frac{5}{3}x=-2$

$x=-\frac{6}{5}$ (substitute to ②)

$y=-\frac{6}{5}+3=\frac{9}{5}$

There is a solution $\begin{cases} x=-\frac{6}{5} \\ y=\frac{9}{5} \end{cases}$

◆Example 6:

This example illustrates system of equations with no solution.

B. $\begin{cases} 3x+2y=8\cdots① \\ 9x+6y=12\cdots② \end{cases}$

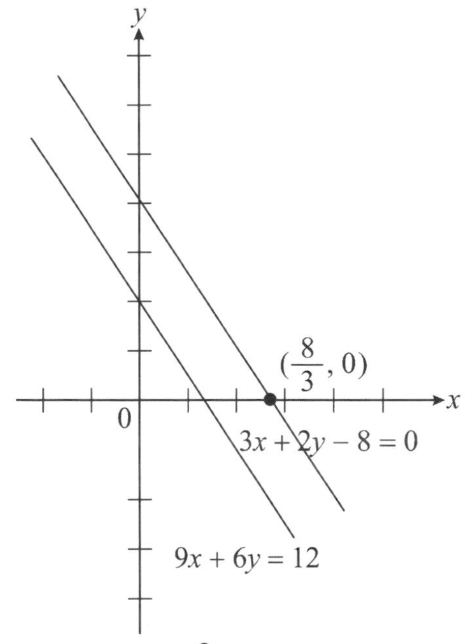

from ① $y=-\frac{3}{2}x+4$

when $x=0$, $y=4$

when $y=0$, $x=\frac{8}{3}$

from ② $y=\frac{-9}{6}x+\frac{12}{6}=-\frac{3}{2}x+2$

when $x=0$ $y=2$

$y=0$ $x=\frac{4}{3}$

They are paralleled to each other. (no solution)

From the coefficient of the equations

$$\frac{3}{9}=\frac{2}{6}=\frac{1}{3}\neq\frac{8}{12}$$

The ratios of coefficients of x and y are the same, but the ratio of constants is not the same as the ratio of coefficients. Therefore there is no solution to this system of equations.

◆Example 7:

This example illustrates system of equations with infinite number of solutions.

$$\begin{cases} 3x+2y=8\cdots① \\ 9x+6y=24\cdots② \end{cases}$$

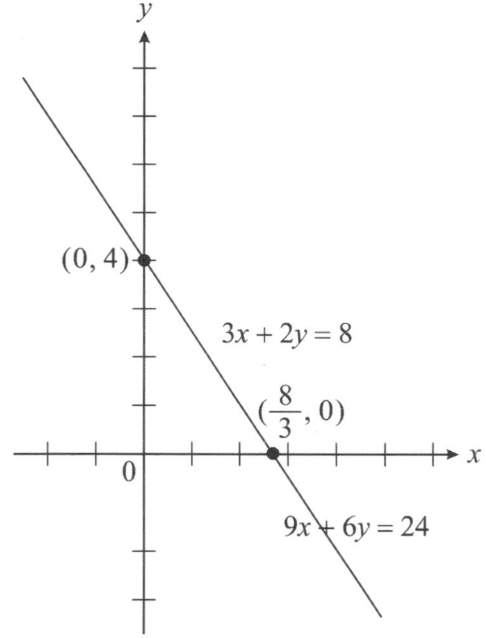

These two lines are actually the same.

There are infinite numbers of solutions to this system of equations.

How to solve linear system inequalities by graphs.

◆Example 1:

$$\begin{cases} 3x-2y>-2 \\ x+y<1 \end{cases}$$

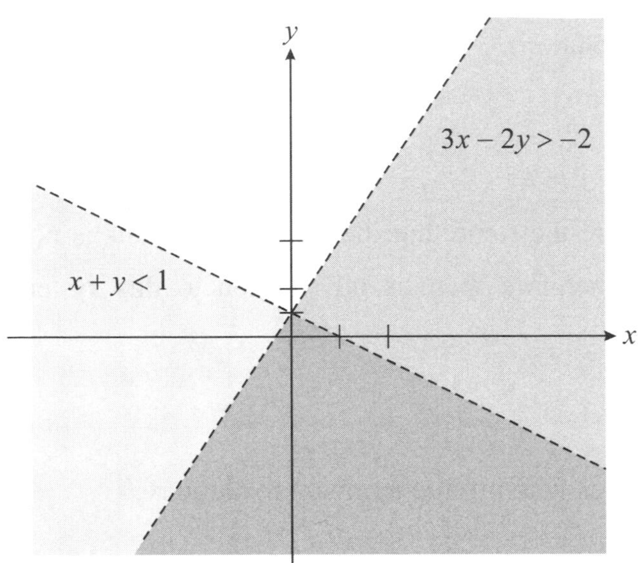

Ps: If the equation has greater/less than and equal relationship such as $3x-2y\geq-1$, it means the solutions include the points on the line itself. It is represented by a dark line, otherwise, dotted line. When you change $3x-2y>-1$ to $y<\frac{3}{2}x+\frac{1}{2}$, it is obvious that the y-values of all point in the region are lower than the y-values of all correspoulding point on the line.

◆Example:

Graph $\begin{cases}3x+4y<1\\x-y>-2\end{cases}$

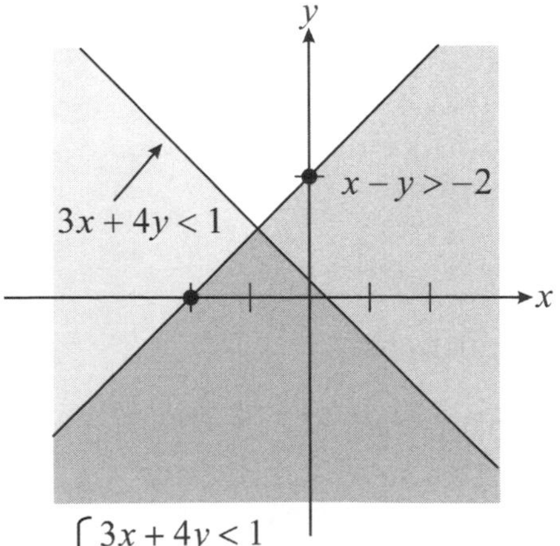

y

x − y > −2

3x + 4y < 1

x

$\begin{cases}3x+4y<1\\x-y>-2\end{cases}$

Answer to Concept Check 6-07

Since $L//M$, $m=\frac{2}{3}$.

Answer to Concept Check 6-08

Since $L \perp M$,

$m\times\frac{2}{3}=-1$, $m=-\frac{3}{2}$.

Answer to Concept Check 6-09

$\frac{5}{15}=\frac{3}{9}\neq\frac{6}{24}$

The two lines are parallel, therefore there is no solution to the system of equations.

Answer to Concept Check 6-10

$\frac{2}{4}=\frac{-3}{-6}=\frac{7}{14}$

The two lines are overlaid, so, then are infinite number of solutions.

Answer to Concept Check 6-11

$$\begin{cases} x+y>2 \\ 3x+2y>-3 \end{cases}$$

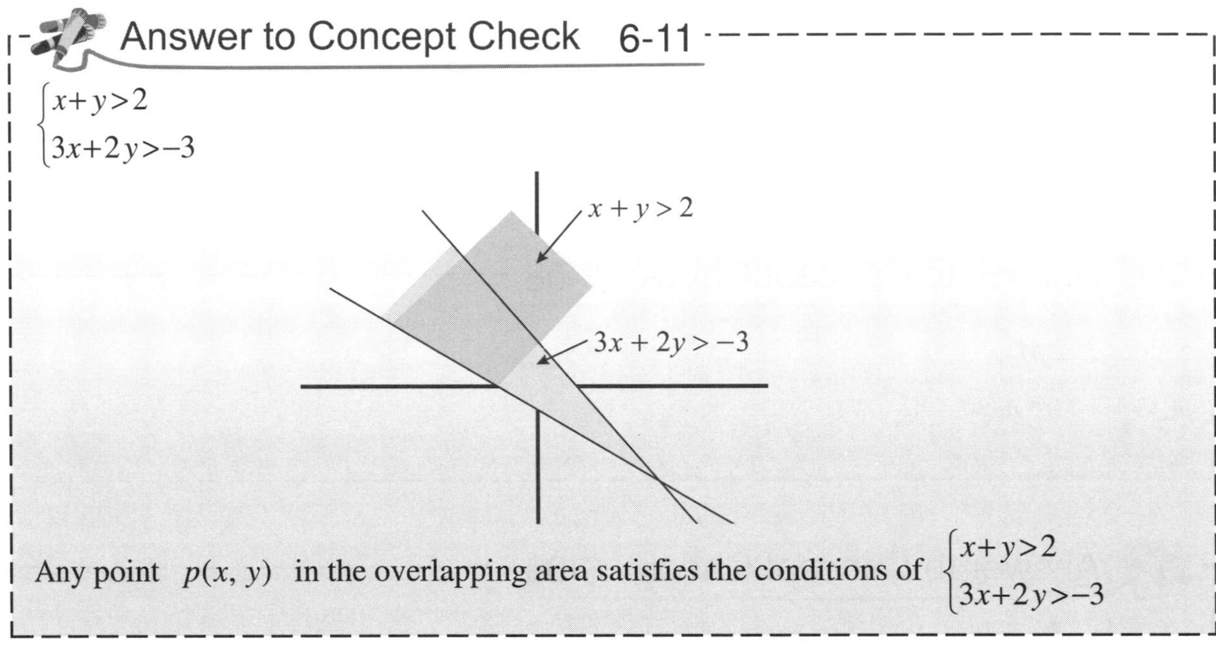

Any point $p(x, y)$ in the overlapping area satisfies the conditions of $\begin{cases} x+y>2 \\ 3x+2y>-3 \end{cases}$

6-5 BEFORE and AFTER IMAGE

Coordinates and analytic geometry can be used to represent image translation, reflection, scaling, and rotation. Below are some examples.

- Image left and right shift

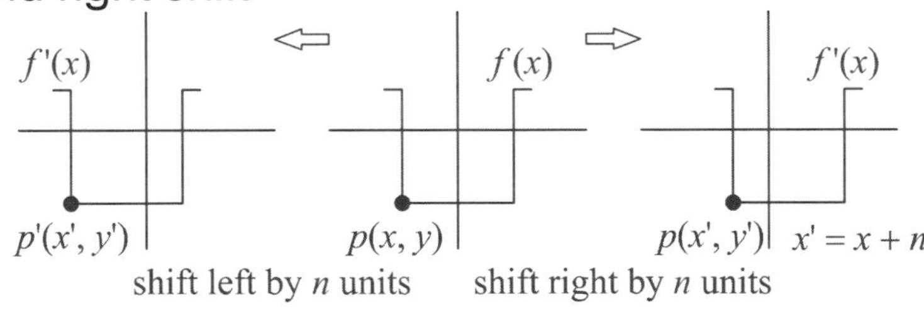

Shift the figure to left 3 units

then $x' = x - 3$

Shift the figure to right 3 units,

then $x' = x + 3$

If $P(-4, -4)$ shifts to right 3 units

$P'(-4+3, -4) \Rightarrow P(-1, -4)$

If $P(-4, -4)$ is shifted to left 3 units,

then $P(-4-3, -4) \Rightarrow P(-7, -4)$

If the graph of $f(x)$ is shifted right by n units, the function becomes $f(x-n)$; on the other hand, if it is shifted left by n units the function changed to $f(x+n)$.

- Image up and down shift

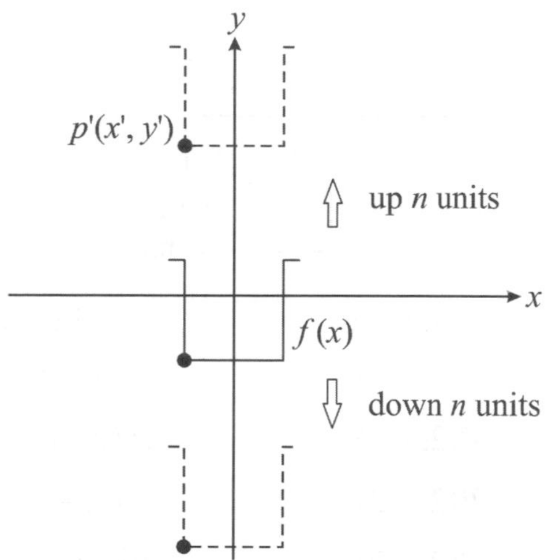

Shift $P(x)$ upwards n units, $P(x, y)$ will be changed to $P(x, y+n)$

Shift $P(x)$ downwards n units, $P(x, y)$ will be changed to $P(x, y-n)$

For example, when $P(-5,-3)$ is shifted up for 6 units, then $P(-5,-3)$ will be $P(-5,-3+6) \Rightarrow P(-5,3)$

On the other hand, when it is shifted downwards for 8 units, then $P(-5,-3)$ will be $P(-5,-3-8) \Rightarrow P(-5,-11)$

For function $f(x)$, if it is shifted upward(s) n units, it will be changed to $f(x)+n$, if shifted downward n units, it will be changed to $f(x)-n$.

◆More examples:

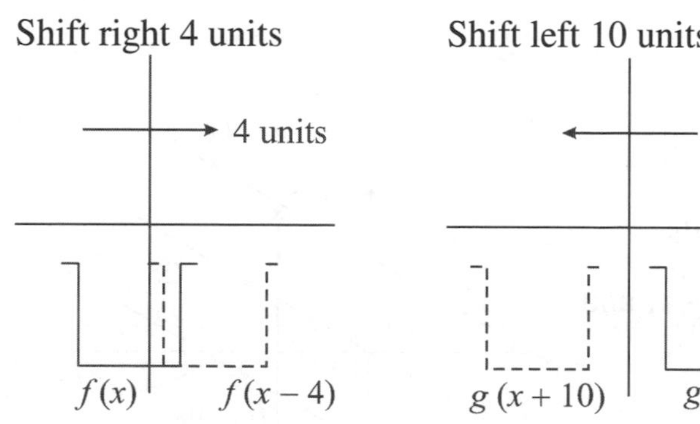

★Tip

$f(x \pm k)$ means every independent variable of $f(x)$ is shifted by $\pm k$ units.

$f(x) \pm k$ means dependent variable (y) is shifted by $\pm k$ units.

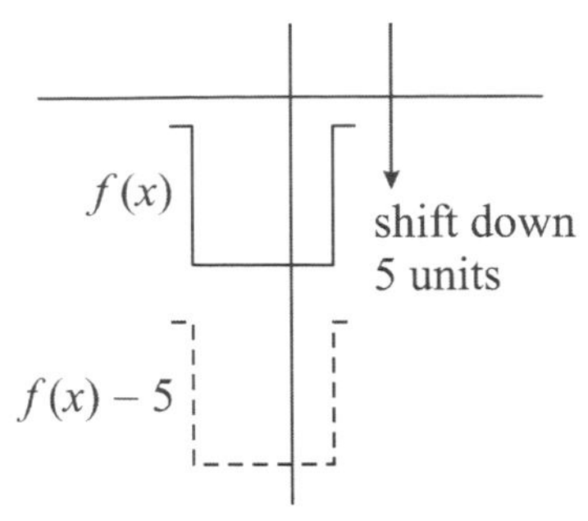

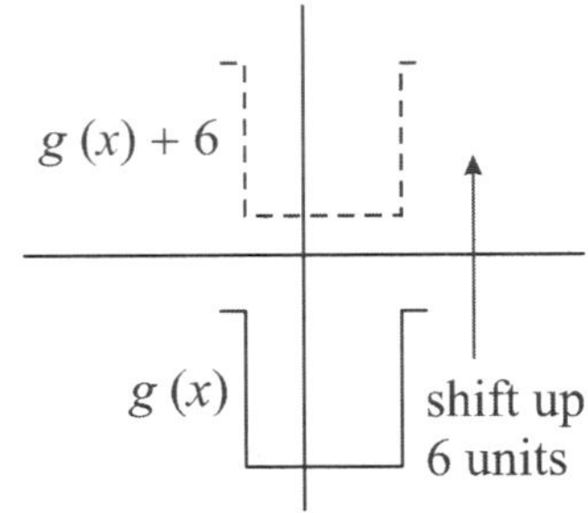

	pre-image	after-image
1. $f(x)-3$	$P(2,-2)$	$P'(2,-5)$
2. $f(x)+3$	$P(2,-2)$	$P'(2,1)$
3. $f(x-3)$	$P(2,-2)$	$P'(5,-2)$
4. $f(x+3)$	$P(2,-2)$	$P'(-1,-2)$
5. $f(-x)$	$P(2,-2)$	$P'(-2,-2)$
6. $-f(x)$	$P(2,-2)$	$P'(2,2)$
7. $2f(x)$	$P(2,-2)$	$P'(2,-4)$
8. $\frac{1}{2}f(x)$	$P(2,-2)$	$P'(2,-1)$
9. $3f(x+1)-2$	$P(2,-2)$	$P'(1,-8)$

The function says:

(A) shift left by 1 unit $P'(1,-2)$

(B) 3 times the height $P'(1,-6)$

(C) shift down by 2 units $P'(1,-8)$

◆Example 1:

The graph on the right shows line ℓ was shifted to the right for 6, units, and shifted up for 4 units. What is the new coordinates of point $(2,\frac{23}{5})$, and the new function of $f(x)=\frac{4}{5}x+3$?

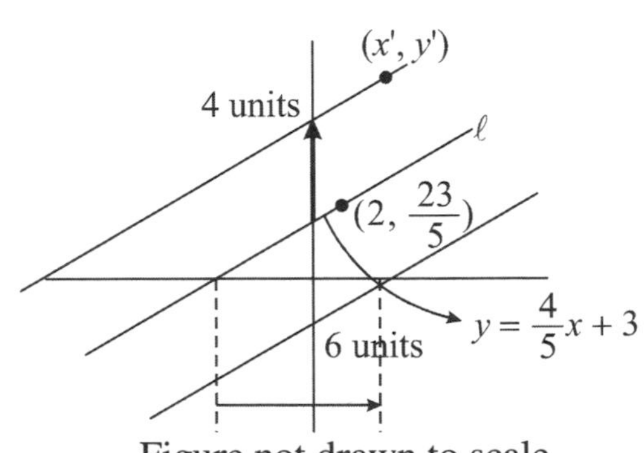

Figure not drawn to scale.

Set new point as (x', y')

$x'=x+6$, $y'=y+4$ (shift right and up)

So, $P(2,\frac{23}{5}) \Rightarrow P'(2+6,\frac{23}{5}+4) \Rightarrow P(8,\frac{43}{5})$

The function $y=\frac{4}{5}x+3$ is change to new function.

$y'-4=\frac{4}{5}(x'-6)+3 \Rightarrow y'=\frac{4}{5}(x'-6)+3+4 \Rightarrow y'=\frac{4}{5}(x'-6)+7$

Since y' and x' each represents a variable,

$y'=\frac{4}{5}(x'-6)+7 \Rightarrow y=\frac{4}{5}(x-6)+7$

◆Example 2:

The function $f(x)=x^2+x+1$ is shifted to the right by 3 units, what is the new function ?

When $f(x)$ is shifted to the right by 3 units, the new function is $f(x-3)$,

Plug $(x-3)$ into the original function $f(x)=x^2+x+1$ to obtain a new function:

$f'(x)=(x-3)^2+(x-3)+1=x^2-6x+9+x-3+1=x^2-5x+7$

Similarly, if the function is shifted to the left by 2 units, then $f(x)$ becomes

$f'(x)=(x+2)^2+(x+2)+1=x^2+4x+4+x+2+1=x^2+5x+7$

◆Example 3:

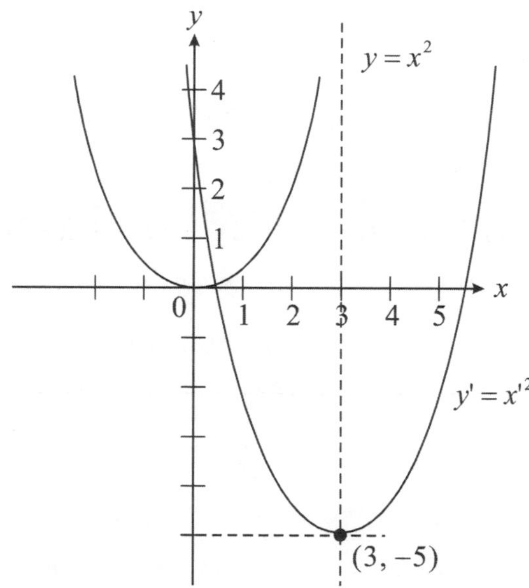

The parabola $y=x^2$ showed above, is shifted to the right by 3 units and shifts down by 5 units, what is the new function?

$x'=x+3$, $x=x'-3$

$y'=y-5$, $y=y'+5$

Substitute to original function $y=x^2$, $y'+5=(x'-3)^2$

$y' = x'^2 - 6x' + 9 - 5 = x'^2 - 6x' + 4$

$\therefore\ y = x^2 - 6x + 4$ (new function)

◆Example 4:

Shift $f(x) = 2x^2 + 4x + 10$ downwards by 4 units and rightward by 3 units, what is the new function.

Shift rightwards by for 3 units: $x' = x+3,\ x = x'-3$

Shift downwards for 4 units: $y' = y-4,\ y = y'+4$

$\therefore\ f(x') = y'+4 = 2(x'-3)^2 + 4(x'-3) + 10$

$= 2(x'-6x'+9) + 4x' - 12 + 10$

$= 2x'^2 - 12x' + 18 + 4x' - 2$

$= 2x'^2 - 8x' + 16$

$y' = 2x'^2 - 8x' + 16 - 4$

$= 2x'^2 - 8x' + 12$

Change x' to x, y' to y, $y = 2x^2 - 8x + 12$

Answer to Concept Check 6-12

1. $f(x)$ is shifted to the right by 2 units, the after image of point $(-3, 2)$ is $P(-3+2, 2) \Rightarrow P(-1, 2)$
2. $f(x)$ is shifted downwards by 3 units, the after image of $P(-1, 2)$ is $P(-1, 2-3)$ $\Rightarrow P(-1, -1)$

Answer is (C)

Answer to Concept Check 6-13

$f(x)=x^2+2x-5$ is shift to the right by 3 units.

$f'(x)=f(x-3)=(x-3)^2+2(x-3)-5$

$=x^2-6x+9+2x-6-5=x^2-4x-2$

The function is shifted up by 4 units, $y'=y+4$, $y=y'-4$

$y'-4=x'^2-4x'-2$

$y'=x'^2-4x'+2$

exchange $y'=y$, and $x'=x$ (all of them are variables)

$\therefore\ y=x^2-4x+2=(x^2-4x+4)-4+2=(x-2)^2-2$

Answer is (B)

6-6 QUADRATIC FUNCTION (PARABOLA)

- A parabola is a curve composed of a set of points whose distances from a fixed line and a fixed point are always the same.
- Fixed point: also called focus.
- Fixed line: also called directrix.

Below graph shows a parabola with focus $F(o, p)$ and vertex is at origin (0, 0) and directrix $y=-p$.

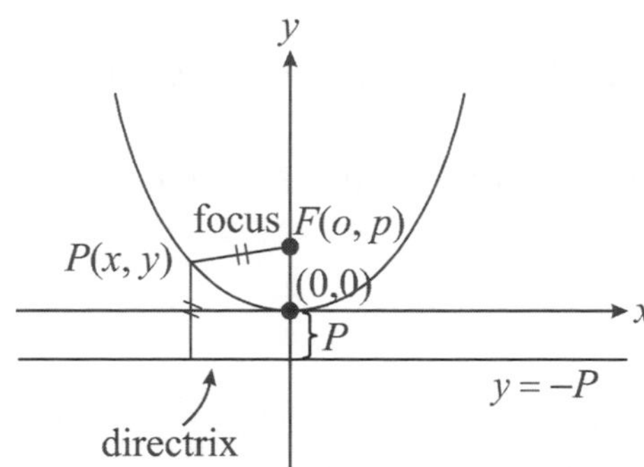

Set the coordinates of focus F as (O, P), and the distance from the origin point to the directrix is P.

Then, the distance from point (x, y) to the directrix is $y+p$. The distance from point $P(x, y)$ to the focus is $\sqrt{(x-0)^2+(y-p)^2}$.

According to the definition of parabola $y+p=\sqrt{(x-0)^2+(y-p)^2}$.

$(y+p)^2 = x^2 + (y-p)^2 = x^2 + y^2 - 2py + p^2$

$\cancel{y^2} + 2py + \cancel{p^2} = x^2 + \cancel{y^2} - 2py + \cancel{p^2}$

$\therefore\ 4py = x^2 \quad \therefore\ y = \dfrac{1}{4p}x^2$ (vertex is at the origin)

When the vertex of the parabola is shifted to (h, k), then the function of the parabola is changed to $y = (x-h)^2 + k$.

> Vertex Form of Parabola $y = (x-h)^2 + k$
> where (h, k) is the coordinates of vertex
> $y = \dfrac{1}{4p}x^2$ (p is the distance from focus to vertex), when vertex is at (0, 0)

• Horizontal graph of parabola

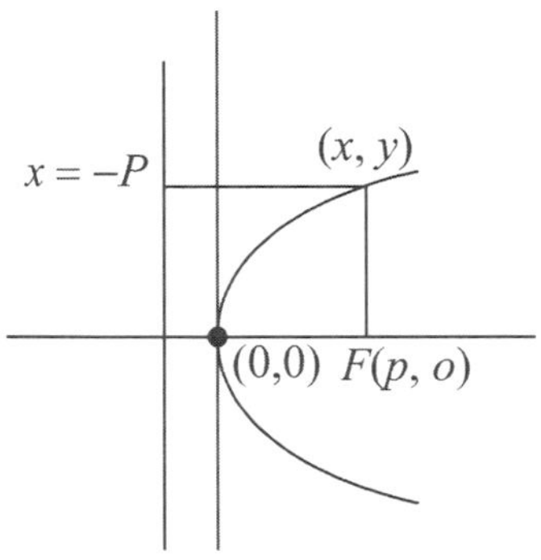

> Vertex Form of Parabola $x = (y-k)^2 + h$ vertex at (h, k)
> $x = \dfrac{1}{4p}y^2$, vertex at (0, 0) and focus at (p, o)

◆Example:

Find the focus and directrix of the parabola given by $y = x^2$.

Vertex form of parabola (Vertex at (0, 0))

$y = \dfrac{1}{4p}x^2 \quad \therefore\ \dfrac{1}{4p} = 1 \quad \therefore\ p = \dfrac{1}{4}$

Focus is at: $(0, \frac{1}{4})$

Directrix is $y = -\frac{1}{4}$

• The properties of parabola

The graphs of all quadratic functions are parabolas. Standard Form: $ax^2 + bx + c$

Since the vertex form of parabola is $y = (x-h)^2 + k$, it can be rewritten as $y = x^2 - 2hx + h^2 + k \Rightarrow ax^2 + bx + c$, where a indicates the property of the concave of the parabola, $b = -2h$, $c = h^2 + k$. Therefore, all quadratic functions are parabola.

1. $a > 0$ the parabola opens upwards

   $a < 0$ the parabola opens downwards

   $a = 0$ not a parabola

   For example, $y = x^2$ concaves upwards while $y = -x^2$ concaves downwards.

2. $|a|$: the bigger the magnitude of a, the narrower the concave of the parabola.

   $|a|$: the smaller the magnitude of a, the broader the concave of the parabola.

   For example: $y = \frac{1}{3}x^2$ is broader than $y = x^2$.

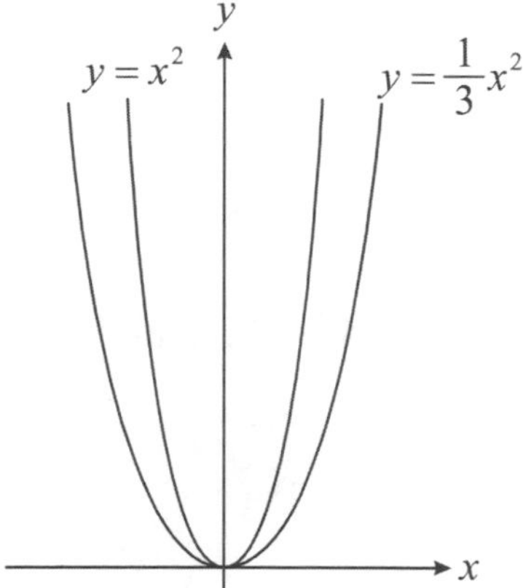

3. Symmetry line $x = -\frac{b}{2a}$ is parallel to y-axis

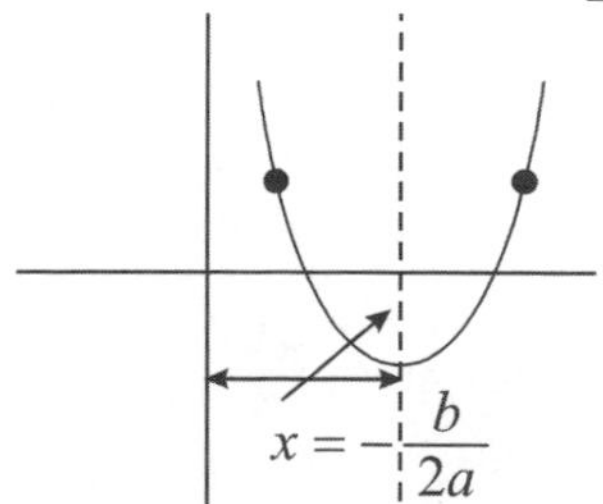

4. Focus of $y=\frac{1}{4p}x^2$ (p is the distance from focus to vertex, and vertex is at (0, 0))

$a=\frac{1}{4p}$, $p=\frac{1}{4a}$

coordinates of Focus are $(-\frac{b}{2a}, k\pm\frac{1}{4a})$ (where k is the y-coordinate of vertex)

◆Example 1:

The graph of $y=x^2$

x	⋯	–5	–4	–3	–2	–1	0	1	2	3	4	5	⋯
y	⋯	25	16	9	4	1	0	1	4	9	16	25	⋯

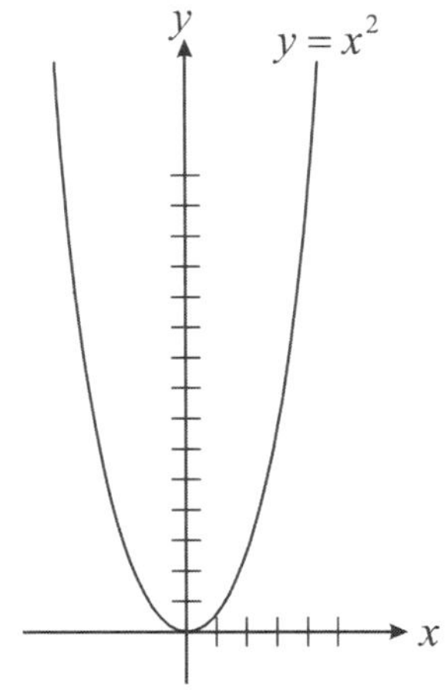

◆Example 2:

The vertex is at (h, k) and the axis of symmetry is $x=h$.

For example, the vertex of parabola $y=(x+4)^2-5$ is at $(-4,-5)$, and the axis of symmetry is $x=-4$.

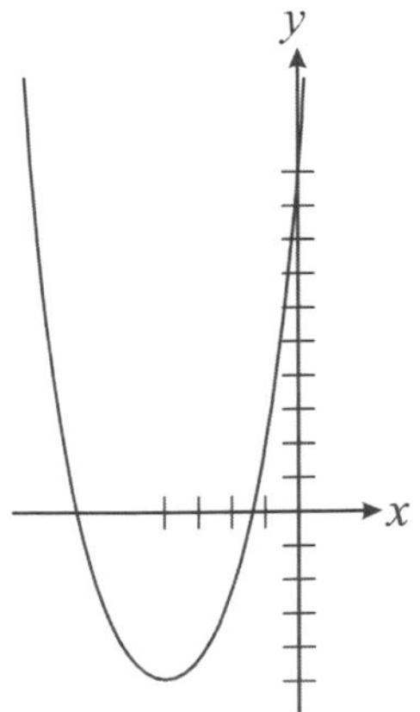

The graph of a quadratic function is a parabola, it is usually written in the form of $y=(x-h)^2+k$ where (h, k) is the vertex of the parabola.

If $h=0$ and $k=0$, the vertex of the parabola is at the origin.

◆Example 3:

Graph $y=-(x-3)^2-4$

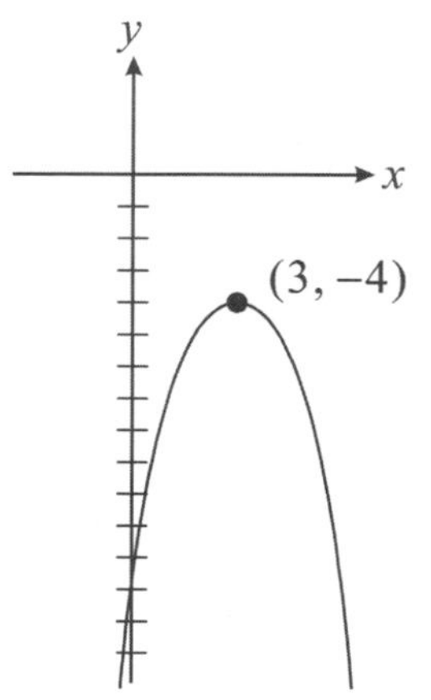

Vertex at $(3, -4)$

• The graphs of quadratic functions

The standard form of quadratic functions

$f(x)=ax^2+bx+c$, $a \neq 0$

Since the graph of a quadratic function (equation) is a parabola.
So, it is usually written in the form of $y=(x-h)^2+k$ (Vertex Form).

where (h, k) is the coordinates of the vertex of the parabola.

If $h=0$ and $k=0$, the vertex of the parabola is at the origin.

1. Proof: Vertex is at $(-\frac{b}{2a}, -\frac{b^2-4ac}{4a})$

Since $y=ax^2+bx+c=a[(x^2+\frac{b}{a}x+(\frac{b}{2a})^2-(\frac{b}{2a})^2+\frac{c}{a}]$

$=a[(x+\frac{b}{2a})^2-\frac{b^2-4ac}{4a^2}]$

$=a(x+\frac{b}{2a})^2-\frac{b^2-4ac}{4a}$

Therefore, vertex is at $(-\frac{b}{2a}, -\frac{b^2-4ac}{4a})$

2. $a > 0$, concave up, $a < 0$, concave down

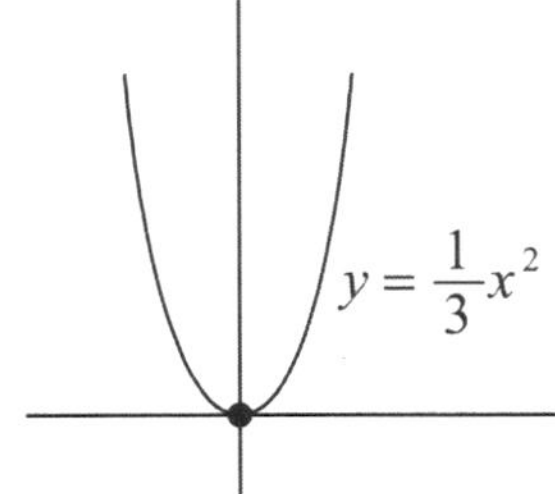

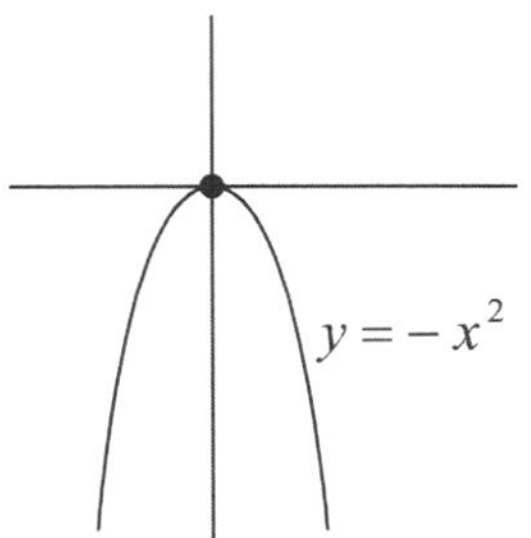

3. The smaller the magnitude of a, the broader the parabola.

Such as: $y = \frac{1}{3}x^2$

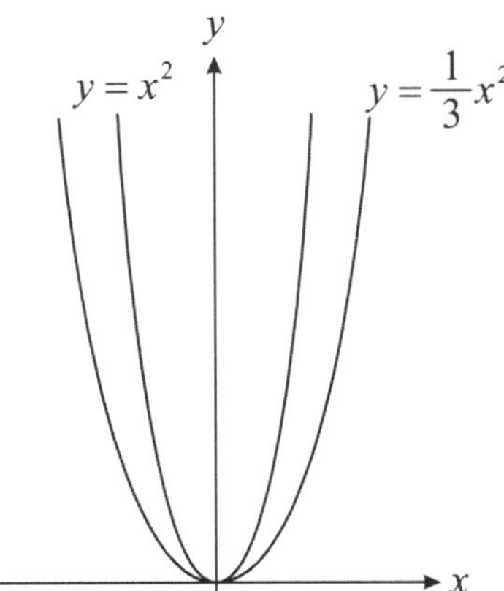

Vertex Form vertex at (h, k)
$y = (x-h)^2 + k$

◆Example:

Which of the following graphs could be the graph of $y = x^2 + 2x + c$?

(c is an integer)

(A)

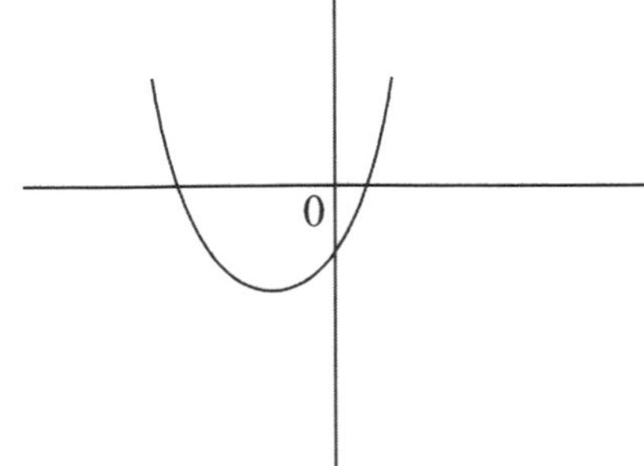

(B)

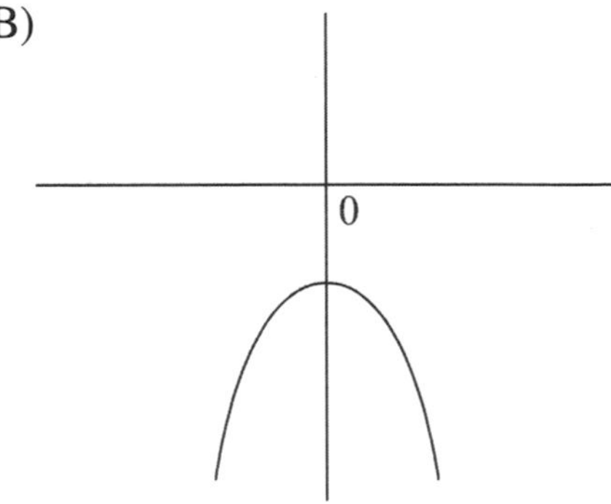

(C)

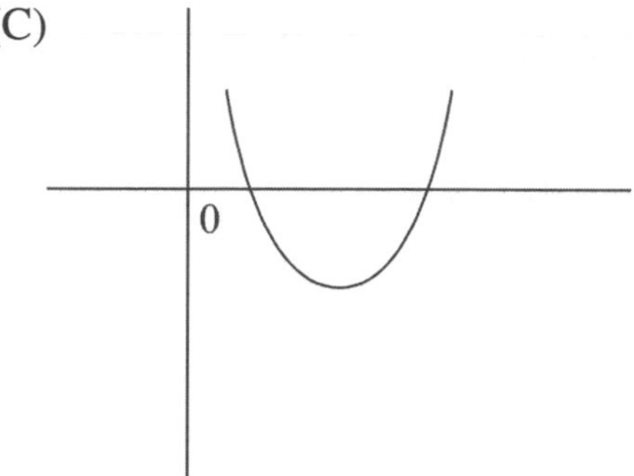

(D)

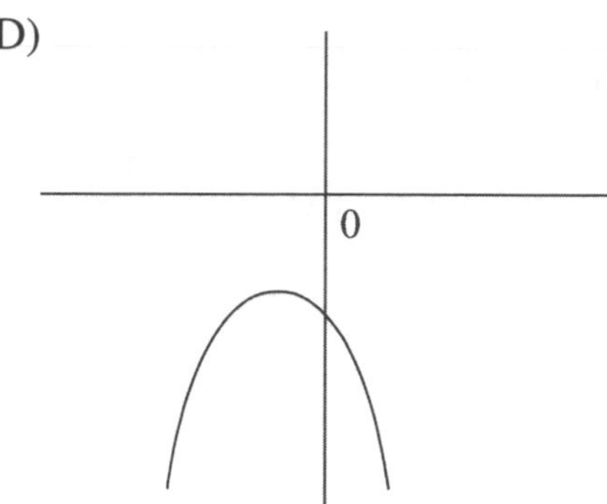

Because $a=+1$

1. (B) and (D) concave downward, can not be the graph
2. $y=x^2+2x+1-1+c=(x+1)^2+(c-1)$, vertex at $(-1, y)$

In c, vertex has positive x value therefore (C) is not correct. So, the answer is (A).

◆Example 1:

The function of a parabola is $f(x)=2x^2+3x+1$. Find its direction of concave, vertex, symmetric equation, focus coordinates, intersection with x, y axes, and graph.

$a=2,\ b=3,\ c=1$

1. $a=2$, the graph concave upwards
2. Vertex is at $(-\frac{b}{2a}, -\frac{b^2-4ac}{4a})$

   $\therefore$ Vertex is at $(-\frac{3}{2\times 2}, -\frac{9-(4)(2)(+1)}{4\times 2}) \Rightarrow (-\frac{3}{4}, -\frac{1}{8})$
3. Symmetric equation is $x=-\frac{b}{2a} \Rightarrow -\frac{3}{4}$.
4. The distance from focus to vertex is $\frac{1}{4p}=a \quad \therefore\ p=\frac{1}{4a} \quad \therefore\ p=\frac{1}{8}$

   Since the graph concaves upwards, the focus is above vertex by $\frac{1}{8}$,

   There fore, focus at $(\frac{1}{8}-\frac{1}{8})=0$, the coordinates of focus $=(-\frac{3}{4}, 0)$.
5. Intersection with x and y axes:

   $f(x)=y=2x^2+3x+1$, when $x=0$, $y=1$

   intersection with y-axis at $(0, 1)$.

When $y=0$, $x=\dfrac{-b\pm\sqrt{b^2-4ac}}{2a}$ $\Rightarrow$ $\dfrac{-3\pm\sqrt{9-4(2)(1)}}{2\times2}$ $\Rightarrow$ $-\dfrac{3}{4}\pm\dfrac{1}{4}$

intersection with x-axis at $(-\dfrac{3}{4}-\dfrac{1}{4},0)$ and $(-\dfrac{3}{4}+\dfrac{1}{4},0)$

$\Rightarrow$ $(-1,0)$ and $(-\dfrac{1}{2},0)$

6. Graph

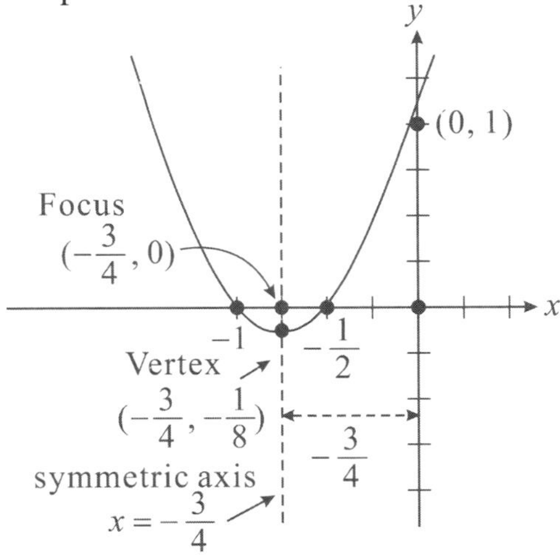

◆Example 2:

The parabola $f(x)=x^2+2x-3$

Find its direction of concave, vertex, symmetric equation, focus, intersection with x, y axes, and graph.

1. Concave: upwards $(a=1)$
2. Vertex: $f(x)=x^2+2x-3=x^2+2x+1-1-3=(x+1)^2-4$

   Vertex is at $(-1,-4)$
3. Symmetric equation: $x=-1$
4. Focus: since $\dfrac{1}{4p}=1$, $p=\dfrac{1}{4}$

   Distance between focus and vertex is $p \Rightarrow \dfrac{1}{4}$

   Since vertex at $(-1,-4)$, and concaves upward, so, focus is at $(-1,-4+\dfrac{1}{4})$

   $\Rightarrow$ $(-1,-3\dfrac{3}{4})$

5. Intersection with x, y axes.

   When $y=0$, $x^2+2x-3=(x+3)(x-1)=0$

   $\therefore$ $x=-3$, or $x=1$

   $x=0$ $y=0+2(0)-3=-3$.

   Intersection with x-axis at $(-3,0)$, $(1,0)$

   Intersection with y-axis at $(0,-3)$

6. Graph

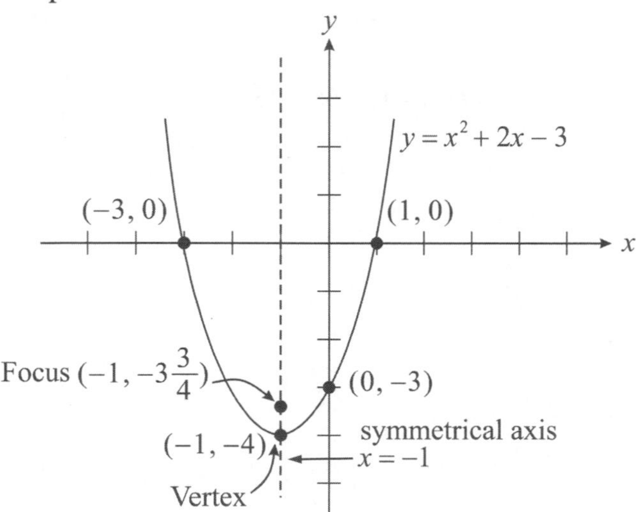

◆Example 3:

Find the direction of concave, vertex, focus, symmetrical axis, the intersection with x, y axes, and graph of the parabola $f(x)=x^2-6x+13$.

1. Concave: upwards ($a=1$)
2. Vertex: rewrite the function

   $f(x)=x^2-6x+13=(x^2-6x+9-9+13)=(x-3)^2+4$

   Vertex is at $(3,4)$.
3. Focus: Distance from vertex $\dfrac{1}{4p}=1$, $p=\dfrac{1}{4}$

   since vertex at $(3,4)$, and concaves upward so, focus is at $(3,4+\frac{1}{4}) \Rightarrow (3,4\frac{1}{4})$
4. Symmetric axis: $x=3$
5. Intersection with x, y, axes:

   When $x=0$, $y=13$

When $y=0$, $x=\frac{6\pm\sqrt{36-52}}{2}=\frac{6\pm\sqrt{-16}}{2}$

Intersection with y-axis at $(0,13)$

There is no intersection with x-axis

6. Graph

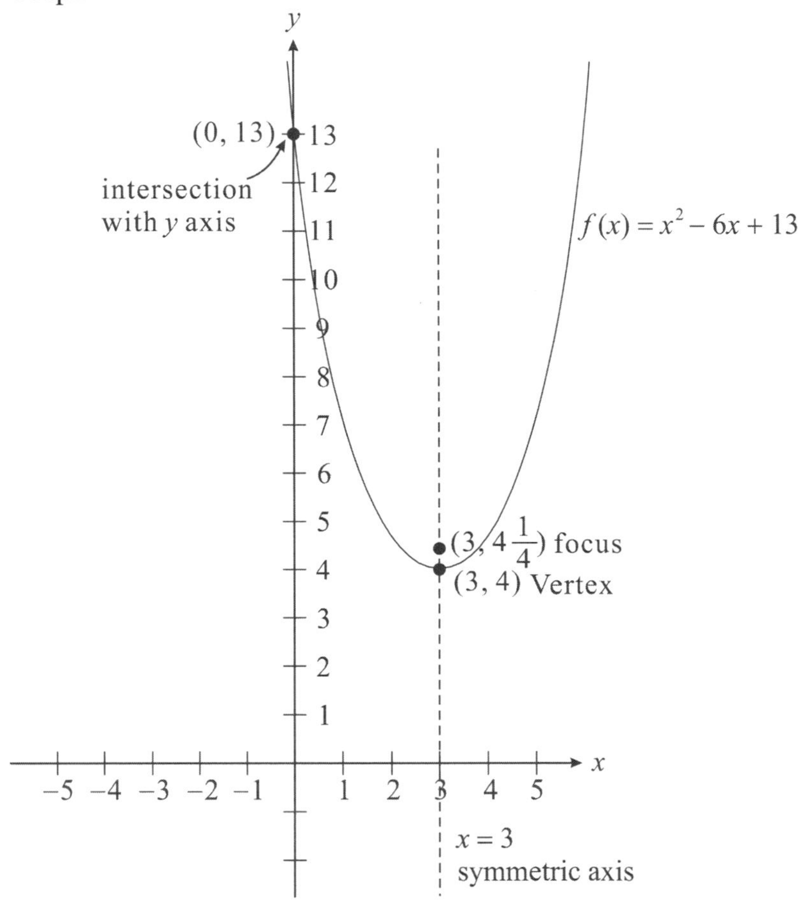

◆Example 4:

The function of a parabola is $f(x)=x^2-2x+5$. If we shift it to the left by 4 units and then shift it down by 7 units. Finally, transform it to $f'(x)=J(x)=-f(2x)$. Find the function of $J(x)$ and the graph of $J(x)$.

1. Shift $f(x)$ left by 4 units

$f(x)=x^2-2x+5$, shift left by 4 units

$g(x)=f(x+4)=(x+4)^2-2(x+4)+5=x^2+8x+16-2x-8+5=x^2+6x+13$

$g(x)$ to vertex form $g(x)=x^2+6x+3^2-3^2+13=(x+3)^2+4$

Vertex is at $(-3,4)$.

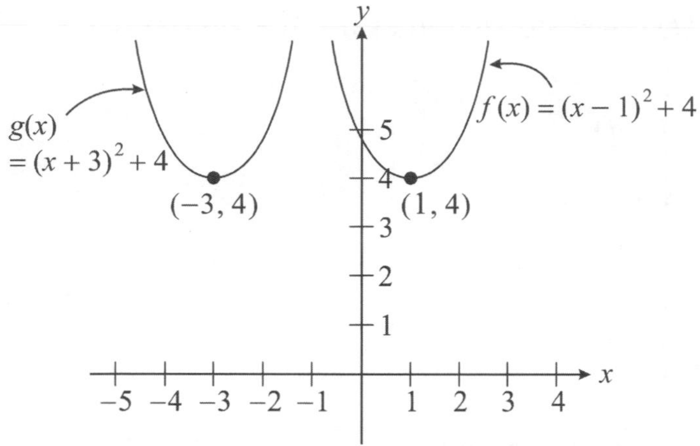

2. Shift down by 7 units:

$h(x)=g(x)-7=(x+3)^2+4-7=(x+3)^2-3$

Vertex is at $(-3,-3)$.

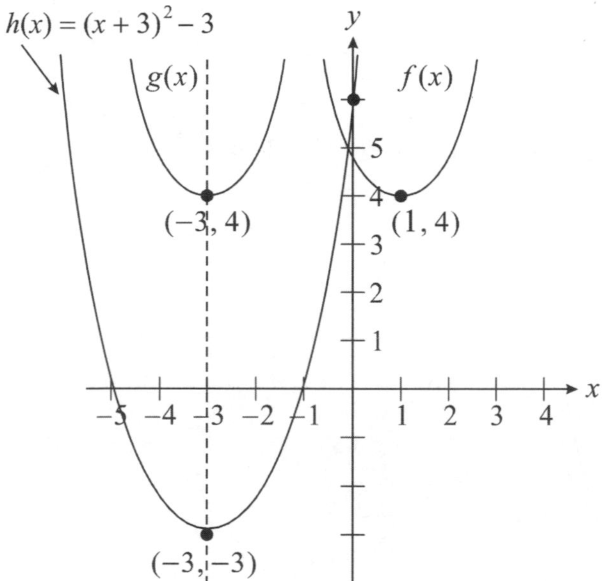

3. Flip $h(x)$ to $I(x)$ $\Rightarrow$ $I(x)=-h(x)=-x^2-6x-6=-((x+3)^2-3)=-(x+3)^2+3$

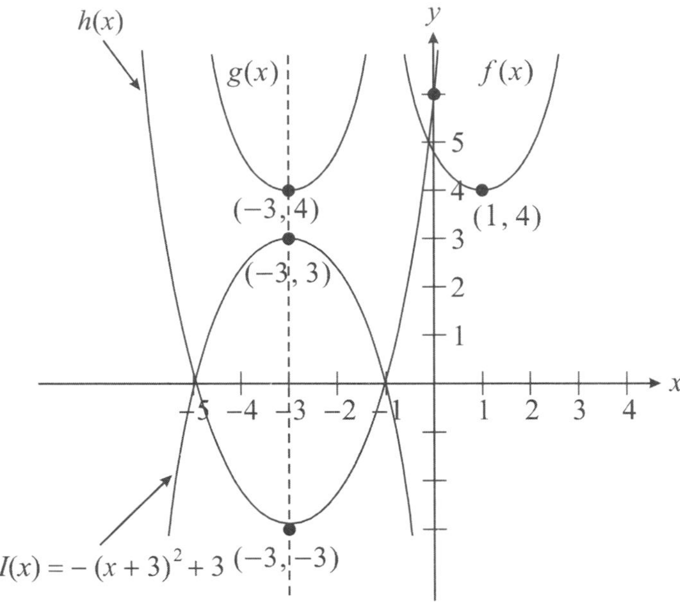

4. Transform $I(x)$ to $J(x)$

$$J(x)=I(2x)=-(2x)^2-6(2x)-6=-4x^2-12x-6=-4[x^2+3x+(\frac{3}{2})^2]+4(\frac{3}{2})^2-6$$

$$=-4(x+\frac{3}{2})^2+3$$

Vertex is at $(-\frac{3}{2}, 3)$

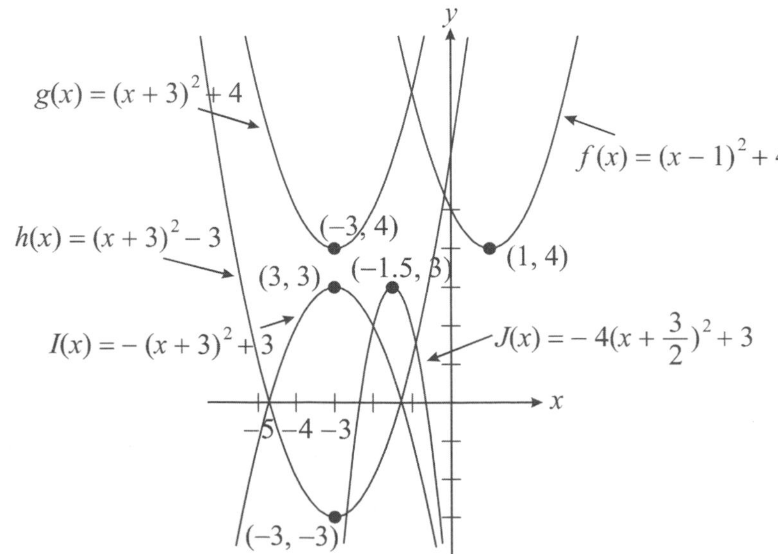

Answer to Concept Check 6-14

$f(x)=2x^2+4x-5=2(x^2+2x+1)-2-5=2(x+1)^2-7$

1. Concaves upwards.
2. Vertex is at $(-1,-7)$.
3. Symmetric equation $x=-1$
4. Focus: The distance from the focus to the vertex is p, $\frac{1}{4p}=a=2 \quad \therefore \quad p=\frac{1}{8}$

   $-7+\frac{1}{8}=-6\frac{7}{8} \Rightarrow -\frac{55}{8}$

   Focus at $(-1,-\frac{55}{8})$
5. Intersection point with x and y axes

   When $y=0 \quad 2x^2+4x-5=0$

   $x=\frac{-b\pm\sqrt{b^2-4ac}}{2a}=\frac{-4\pm\sqrt{16+40}}{4}=\frac{-4\pm\sqrt{56}}{4}=-1\pm\frac{1}{2}\sqrt{14}$

   When $x=0$: $\quad y=2(0)^2+4(0)-5=-5$

   intersection with y-axis $(0,-5)$ and with x-axis is at $(-1\pm\frac{1}{2}\sqrt{14},0)$.

   $\Rightarrow (-2.87,0)$, $(0.87,0)$
6. The graph of $y=2x^2+4x-5$:

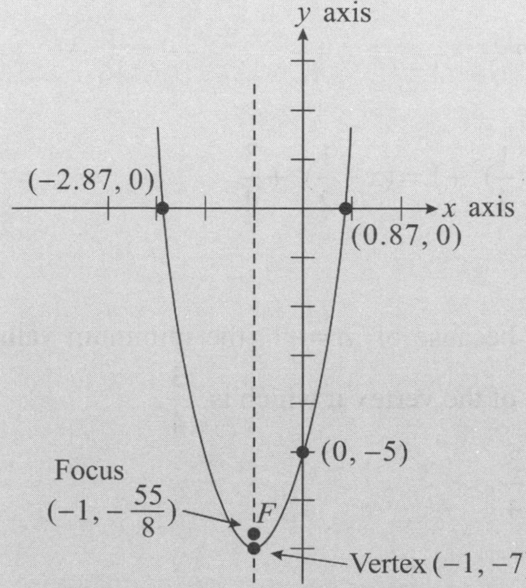

Answer to Concept Check 6-15

$f(x) = -2x^2 + 4x + 6$

1. Shift to the right by 2 units:

$$f'(x) = f(x-2) = -2(x-2)^2 + 4(x-2) + 6$$
$$= -2(x^2 - 4x + 4) + 4(x-2) + 6$$
$$= -2x^2 + 8x - 8 + 4x - 8 + 6$$
$$= -2x^2 + 12x - 10$$

2. Shift up by 4 units:

$$f''(x) = f'(x) + 4 = -2x^2 + 12x - 10 + 4 = -2x^2 + 12x - 6$$

3. Rewrite to vertex form:

$$-2x^2 + 12x - 6 = -2(x^2 - 6x + 9 - 9) - 6 = -2(x-3)^2 + 18 - 6 = -2(x-3)^2 + 12$$

4. Vertex is at $(3, 12)$

- The Maximum and minimum value of Quadratic function.

The graph of any quadratic function is a parabola. The y coordinate value of the vertex of a parabola is the minimum value of the function if the parabola concaves upwards, or is the maximum value if concave downwards.

◆Example 1:

What is the minimum value of $f(x) = x^2 + x + 1$?

 **Step 1.** Find the vertex of $f(x)$:

$$f(x) = x^2 + x + 1 = x^2 + x + (\frac{1}{2})^2 - (\frac{1}{2})^2 + 1 = (x + \frac{1}{2})^2 + \frac{3}{4}$$

Vertex is at $(-\frac{1}{2}, \frac{3}{4})$

 **Step 2.** The parabola concaves upwards, because of $a = +1$, the minimum value of the function is the y-coordinate of the vertex it which is $\frac{3}{4}$

 **Step 3.** Therefore the minimum value is $\frac{3}{4}$.

◆Example 2:

What is the maximum value of $f(x) = -x^2 + x + 1$?

 **Step 1.** Find the vertex of the parabola.

$$f(x)=-[x^2-x+(\frac{1}{2})^2-(\frac{1}{2})^2-1]=-(x-\frac{1}{2})^2+\frac{1}{4}+1$$

 **Step 2.** Vertex at is $(\frac{1}{2},\frac{5}{4})$

Because of the parabola concaves downwards.

 **Step 3.** the maximum value is $\frac{5}{4}$

Answer to Concept Check 6-16

 **Step 1.** Find the vertex of the parabola

$$f(x)=-x^2+6x+8=-[(x^2-6x+9)-9-8]$$

$$=-[(x-3)^2-17]=-(x-3)^2+17$$

 **Step 2.** Vertex is at $(3,17)$.

 **Step 3.** Maximum value is 17. (concave downwards)

◆Example 3:

What is the maximum value of $f(x)=-x^2+4x+6$?

$f(x)=-x^2+4x+6=-(x^2-4x+4)+4+6=-(x-2)^2+10$,

Since $a<0$, the parabola concaves downwards

$f(2)=10$ is maximum value.

◆Example 4:

What is the minimum value of $f(x)=x^2+8x-6$?

$f(x)=x^2+8x-6=x^2+8x+16-16-6=(x+4)^2-22$

When $x=-4$, $f(x)=-22$ is the minimum value, since concave upward.

◆Example 5:

Shifting $f(x)=x^2+4x+1$ rightwards by R units and upwards by U units, yields new function of $f(x)=x^2-4x+8$, what is the values of R and U?

$f(x)=x^2+4x+1=x^2+4x+4-4+1=(x+2)^2-3$

$f(x)=x^2-4x+8=x^2-4x+4-4+8=(x-2)^2+4$

Vertex of the original function is at $(-2, 3)$

The function of parabola is $y=(x+2)^2-3$

After shifting: $y=(x+2-R)^2+(-3+U)=(x-2)^2+4$

$\cancel{x}+2-R=\cancel{x}-2$, $-3+U=4$, $R=4$, $U=7$

6-7 CIRCLE

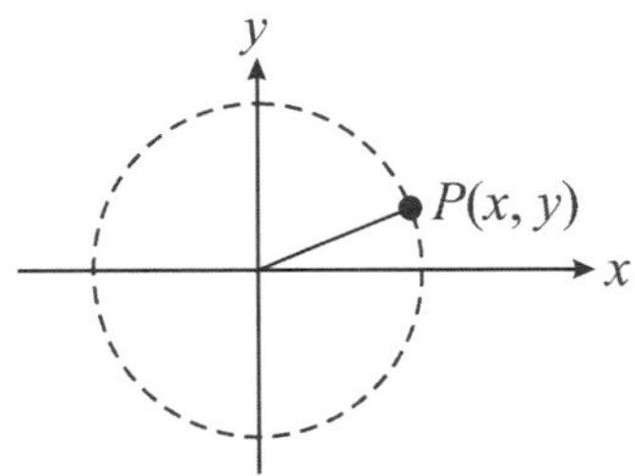

- A. circle is a curve in a plane generated by a set of points such that the distance from any one point to *a* fixed point the center is always the same.

  As the figure above let us set, the fixed distance as *r*, there $\sqrt{(x-0)^2+(y-0)^2}=r$

  $x^2+y^2=r^2$ center of the circle at origin.

  If we shift the center to $(3,-2)$, then $\sqrt{(x-3)^2+(y+2)^2}=r$

  $(x-3)^2+(y+2)^2=r^2$

  (1) Center become $(3,-2)$

  (2) Radius is *r*.

◆Example:

What are the center and radius of the circle

$36x^2-48x+36y^2+36y-119=0$

Divided both sides by 36:

$$x^2-\frac{48}{36}x+y^2+y-\frac{119}{36}=0$$

$$x^2-\frac{4}{3}x+y^2+y-\frac{119}{36}=0$$

$$x^2-\frac{4}{3}x+\frac{4}{9}-\frac{4}{9}+y^2+y+\frac{1}{4}-\frac{1}{4}-\frac{119}{36}=0$$

$$(x-\frac{2}{3})^2+(y+\frac{1}{2})^2-\frac{4}{9}-\frac{1}{4}-\frac{119}{36}=0$$

$(x-\frac{2}{3})^2+(y+\frac{1}{2})^2-\frac{16+9+119}{36}=0$

$(x-\frac{2}{3})^2+(y+\frac{1}{2})^2-\frac{144}{36}=0$

$(x-\frac{2}{3})^2+(y+\frac{1}{2})^2=\frac{144}{36}=(\frac{12}{6})^2=2^2$

Center is at $(\frac{2}{3},-\frac{1}{2})$, Radius is 2.

6-8 ELLIPSE

- An ellipse is formed by a set of points in a plane such that the sum of distances from any one point to the two distinct fixed points (foci) is the same.

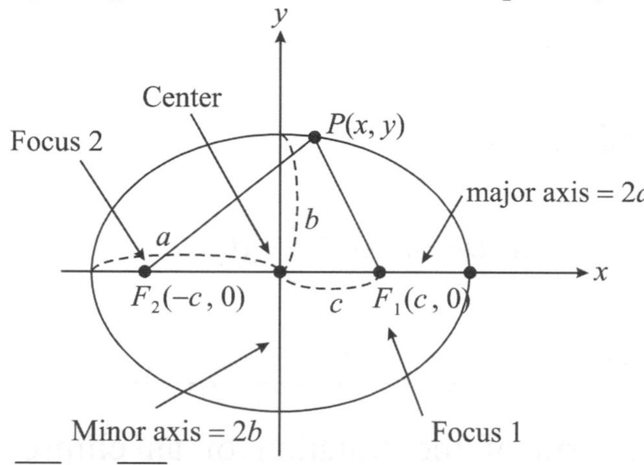

> ★Tip
> When $p(x, y)$ is at $(a, 0)$
> Due to the fact in ellipse, the sum of distances from any point to two foci is the same.
> $k=2a$ for any point $p(x, y)$

$|\overline{PF_1}|+|\overline{PF_2}|=k$

$PF_1=\sqrt{(x-c)^2+(y-0)^2}$

$PF_2=\sqrt{(x+c)^2+(y-0)^2}$

$\sqrt{(x-c)^2+y^2}+\sqrt{(x+c)^2+y^2}=k$

$k=2a$

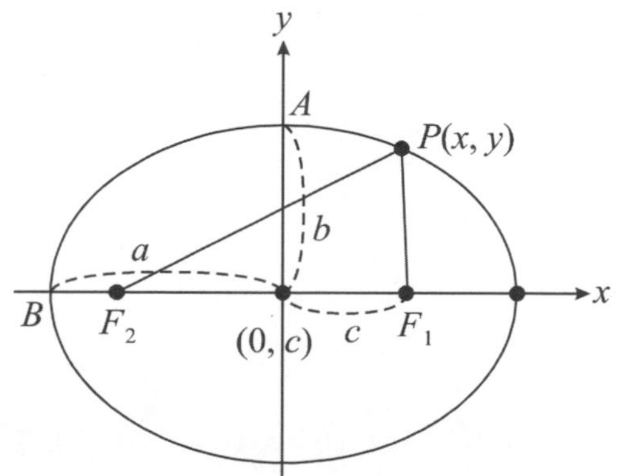

$PF_1 + PF_2 = 2a$

$\therefore\ AF_1 + AF_2 = PF_1 + PF_2 = BF_1 + BF_2 = 2a = k$ (By Definition)

So, $\sqrt{(x-c)^2 + y^2} + \sqrt{(x+c)^2 + y^2} = 2a$

$\sqrt{(x-c)^2 + y^2} = 2a - \sqrt{(x+c)^2 + y^2}$

Squaring both sides,

$(x-c)^2 + y^2 = 4a^2 - 4a\sqrt{(x+c)^2 + y^2} + (x+c)^2 + y^2$

$\cancel{x^2} - 2cx + \cancel{c^2} + \cancel{y^2} = 4a^2 + \cancel{x^2} + 2xc + c^2 + \cancel{y^2} - 4a\sqrt{(x+c)^2 + y^2}$

$4a^2 + 4xc = 4a\sqrt{(x+c)^2 + y^2}$

$a^2 + xc = a\sqrt{(x+c)^2 + y^2}$

Squaring both sides,

$a^4 + 2a^2xc + x^2c^2 = a^2 \cdot [(x+c)^2 + y^2]$

$a^4 + 2a^2xc + x^2c^2 = a^2(x^2 + 2xc + c^2 + y^2)$

$a^4 + \cancel{2a^2xc} + x^2c^2 = a^2x^2 + \cancel{2a^2xc} + a^2c^2 + a^2y^2$

$(a^2 - c^2)x^2 + a^2y^2 = a^4 - a^2c^2$

$\therefore\ AF_1 + AF_2 = 2a \quad \therefore\ AF_1 = AF_2 (\triangle AOF_1 \cong \triangle AOF_2),\ AF_1 = a$

$b^2 + c^2 = a^2$

So, $b^2x^2 + a^2y^2 = a^2(a^2 - c^2) = a^2b^2$

$\frac{x^2}{a^2} + \frac{y^2}{b^2} = 1$ is the standard form of the equation of an ellipse with center at (0, 0):

If major axis is vertical and minor axis is horizontal, then the equation is

$\frac{y^2}{a^2} + \frac{x^2}{b^2} = 1$

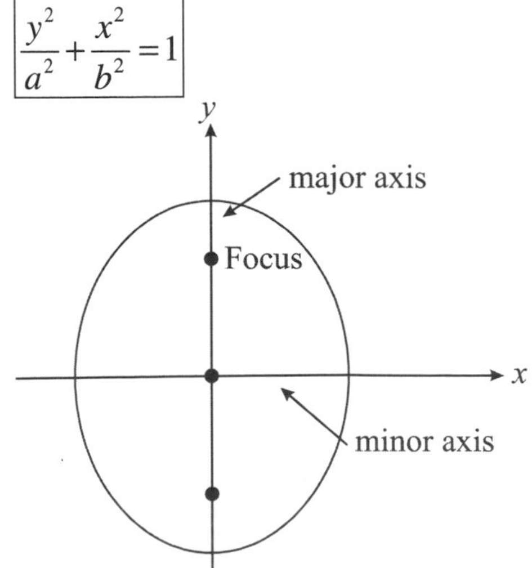

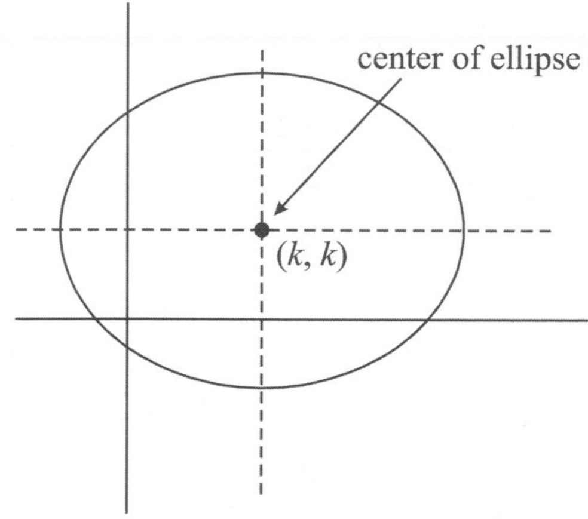

The standard equation of ellipse

$\dfrac{(x-h)^2}{a^2}+\dfrac{(y-k)^2}{b^2}=1$, where (h,k) is the coordinates of the center.

◆Example 1:

Find the center of an ellipse with the equation $9x^2+4y^2+36x-16y+16=0$.

$9(x^2+4x+4)+4(y^2-4y+4)-52+16=0$

$9(x+2)^2+4(y-2)^2=36$

$\dfrac{(x+2)^2}{4}+\dfrac{(y-2)^2}{9}=1$

The center is $(-2, 2)$.

◆Example 2:

What is the standard form and general form of an ellipse with the center at $(3,-2)$ and its major axis and minor axis are equal to 4 and 6 respectively ?

Standard form: $\dfrac{(x-3)^2}{2^2}+\dfrac{(y+2)^2}{3^2}=1$

General form: $9(x-3)^2+4(y+2)^2=36$

$9(x^2-6x+9)+4(y^2+4y+4)=36$

$9x^2-54x+81+4y^2+16y+16-36=0$

$9x^2+4y^2-54x+16y+61=0$

Answer to Concept Check 6-17

The standard form of the ellipse is $\frac{(x-3)^2}{(\frac{8}{2})^2}+\frac{(y-2)^2}{(\frac{6}{2})^2}=1$, $\frac{(x-3)^2}{4^2}+\frac{(y-2)^2}{3^2}=1$

$9(x-3)^2+16(y-2)^2=9\times16$

$9(x^2-6x+9)+16(y^2-4y+4)=9\times16$

$9x^2-54x+9^2+16y^2-64y+64=144$

$9x^2-54x+16y^2-64y+145=144$

$9x^2+16y^2-54x-64y+1=0$

6-9 HYPERBOLA

- A hyperbola is formed by a set of points in a plane such that the difference of distances from any one point to two distinct fixed points foci is constant.

  The standard form of the equation for hyperbola with center at original point is

  $\frac{x^2}{a^2}-\frac{y^2}{b^2}=1$ Transverse axis (horizontal)

  $\frac{y^2}{a^2}-\frac{x^2}{b^2}=1$ Transverse axis (vertical)

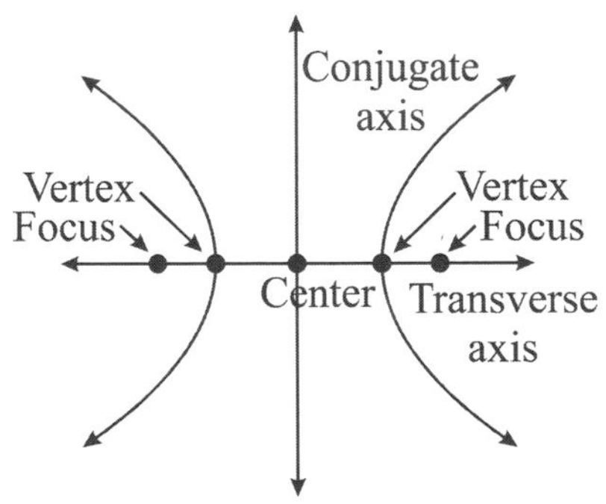

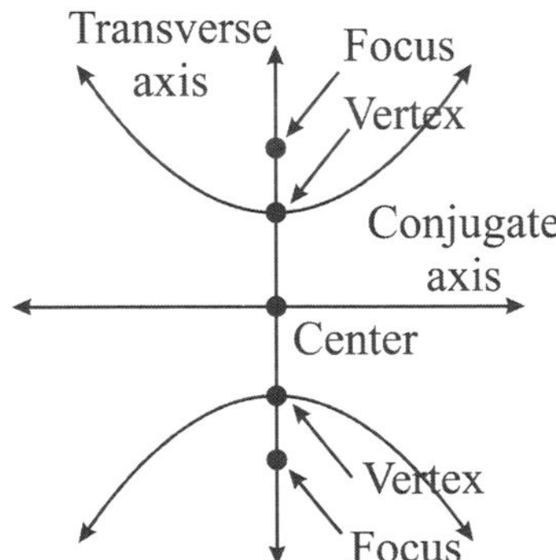

$|PF_1-PF_2|=k$ (constant)

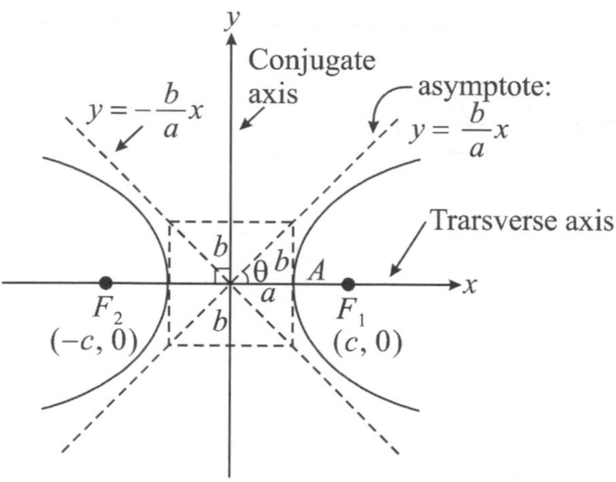

Since the distance from point $A(x, y)$ (a point on the locus) to F_2, minus the distance from point A to F_1 is $|AF_1 - AF_2| = 2a$ (Let the constant be $2a$)

$=|\sqrt{(x+c)^2+y^2}-\sqrt{(x-c)^2+y^2}|$

$|\sqrt{(x+c)^2+y^2}|=2a+|\sqrt{(x-c)^2+y^2}|$

$(x+c)^2+y^2=4a^2+4a\sqrt{(x-c)^2+y^2}+(x-c)^2+y^2$

$x^2+2cx+c^2+y^2-4a^2-[x^2-2cx+c^2+y^2]=4a\sqrt{(x-c)^2+y^2}$

$\cancel{x^2}+2cx+\cancel{c^2}+\cancel{y^2}-4a^2-\cancel{x^2}+2cx-\cancel{c^2}-\cancel{y^2}=4a\sqrt{(x-c)^2+y^2}$

$4cx-4a^2=4a\sqrt{(x-c)^2+y^2}$

$(cx-a^2)=a\sqrt{(x-c)^2+y^2}$

$(cx-a^2)^2=a^2[(x-c)^2+y^2]$

$c^2x^2-2\cancel{a^2}cx+a^4=a^2[x^2-2cx+c^2+y^2]=a^2x^2-\cancel{2a^2cx}+a^2c^2+a^2y^2$

$(c^2-a^2)x^2-a^2y^2=a^2(c^2-a^2)$

Let $c^2-a^2=b^2$

$b^2x^2-a^2y^2=a^2b^2$

$\frac{x^2}{a^2}-\frac{y^2}{b^2}=1$ Transverse axis (horizontal)

$\frac{y^2}{a^2}-\frac{x^2}{b^2}=1$ Transverse axis (vertical)

If the center of Hyperbola is at (h, k) then the equation is $\frac{(x-h)^2}{a^2}-\frac{(y-k)^2}{b^2}=1$

◆Example 1:

What are the asymptotes and foci of the hyperbola $\frac{x^2}{9}-\frac{y^2}{4}=1$?

$a=\sqrt{9}=3$, $b=\sqrt{4}=2$

$c^2=9+4=13$, $c=\sqrt{13}$

asymptotes: $y=\pm\frac{b}{a}x$, $y=\pm\frac{2}{3}x$

$(\frac{x}{3}+\frac{y}{2})(\frac{x}{2}-\frac{y}{3})=0$

$3x-2y=0$, $2x+3y=0$.

foci: $(c,0)$ and $(-c,0)$.

$(\sqrt{13},0)$ and $(-\sqrt{13},0)$

◆Example 2:

What is the center of the hyperbola $4x^2-9y^2-8x-18y-41=0$?

$4x^2-8x+4-9y^2-18y-9-41+5=0$

$4(x-1)^2-9(y+1)^2=36$

$\frac{(x-1)^2}{9}-\frac{(y+1)^2}{4}=1$

Center at $(1,-1)$

Answer to Concept Check 6-18

$\frac{x^2}{16}-\frac{y^2}{9}=1$, $a=4$, $b=3$,

So, $c=\sqrt{a^2+b^2}=\sqrt{16+9}=5$

Asymptotes: $y=\pm\frac{b}{a}x=\pm\frac{3}{4}x$

Foci are $(c,0)$ and $(-c,0)$.

So, they are at $(5,0)$ and $(-5,0)$.

The length of transverse axis $=2a=2\times4=8$.

The length of conjugate axis $=2b=2\times3=6$.

6-10 POLAR COORDINATES

The position of *a* point are determined by the distance r (from the origin) and the angle θ formed by the r and the x-axis.

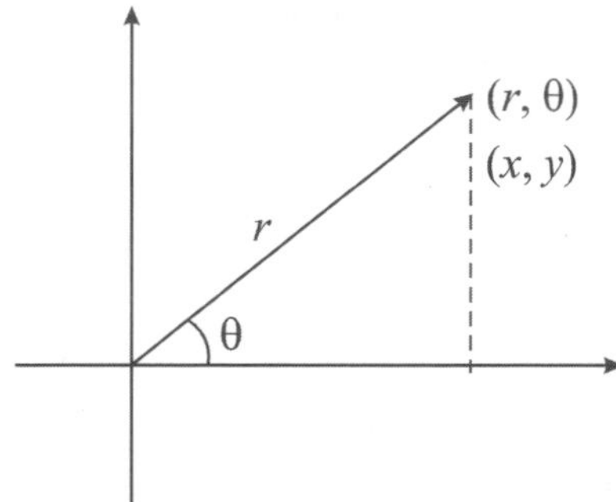

$x = r\cos\theta \quad y = r\sin\theta$

$\tan\theta = \dfrac{y}{x} \quad r^2 = x^2 + y^2$

◆Example 1:

If the coordinates of a polar point are $[3, \dfrac{\pi}{2}]$, what are its rectangular coordinates ?

$x = r\cos\theta = 3\cos\theta = 3\cos\dfrac{\pi}{2} = 0$

$y = r\sin\theta = 3\sin\theta = 3\ \sin\dfrac{\pi}{2} = 3$

The rectangular coordinates are $(0, 3)$

◆Example 2:

$(x, y) = (\sqrt{3}, -1)$, what are its polar coordinates ?

$x = \sqrt{3}, \quad y = -1 \quad x^2 + y^2 = 3 + 1 = 4 \quad r = 2$

$\theta = \tan^{-1}\dfrac{-1}{\sqrt{3}}$

$(\sqrt{3}, -1) = (2, \tan^{-1}\dfrac{-1}{\sqrt{3}}) = (2, -\dfrac{\pi}{6})$

◆Example 3:

What is the graph of the polar equation $r = 3$?

$\sqrt{x^2 + y^2} = 3$

$x^2 + y^2 = 9$ (It is a circle)

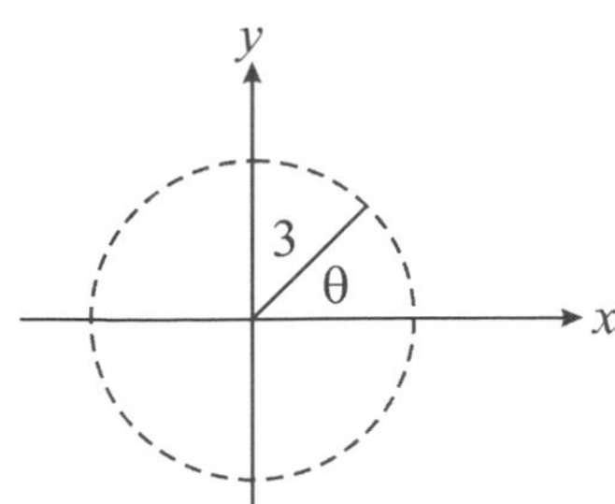

◆Example 4:

Convert the polar equation

$r=\dfrac{4}{3-2\sin\theta}$ to x-y equation.

$3r-2r\sin\theta=4$

$3(\sqrt{x^2+y^2})-2y=4$

$\sqrt{x^2+y^2}=\dfrac{4+2y}{3}$

$x^2+y^2=\dfrac{16+16y+4y^2}{9}$

$9x^2+9y^2=16+16y+4y^2$

$9x^2+5y^2-16y-16=0$

◆Example 5:

Convert $3x-6y+5=0$ to polar from.

$3r\cos\theta-6r\sin\theta+5=0$

$r(3\cos\theta-6\sin\theta)=-5$

$r=\dfrac{-5}{3\cos\theta-6\sin\theta}$

Answer to Concept Check 6-19

$r=2,\ \theta=\dfrac{\pi}{3} \Rightarrow 60°$

$x=2\cos\dfrac{\pi}{3}=2\cdot\dfrac{1}{2}=1$

$y=2\sin\dfrac{\pi}{3}=2\cdot\dfrac{\sqrt{3}}{2}=\sqrt{3}$

$(1,\sqrt{3})$

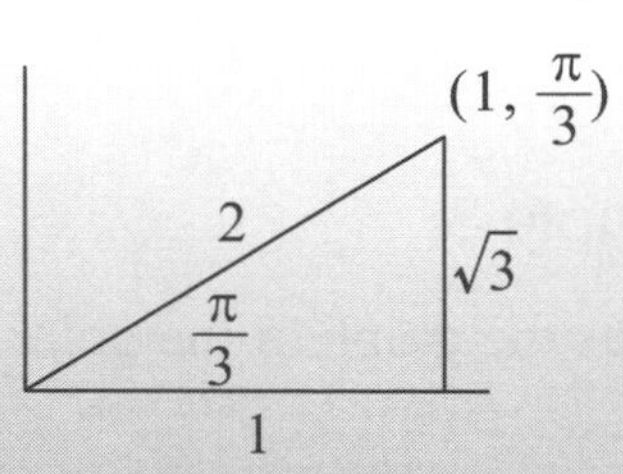

Answer to Concept Check 6-20

$$r = \frac{3}{3-2\cos\theta} = \frac{3}{3-2\cdot\frac{x}{r}} = \frac{3}{\frac{3r-2x}{r}}$$

$\therefore\ 3r-2x=3,$

$3\sqrt{x^2+y^2}=3+2x$

$9(x^2+y^2)=(3+2x)^2=9+12x+4x^2$

$9x^2-4x^2+9y^2-12x-9=0$

$5x^2+9y^2-12x-9=0$

CHATPER 6 PRACTICE

1. If line ℓ passes through the points $A(0,-2)$ and $(-4,0)$. If we flip it over the x-axis, what is the slope of the new line ?

(A) $\frac{1}{2}$

(B) $-\frac{1}{2}$

(C) 2

(D) -2

2. The equation of a circle is given by $x^2+y^2-6x+4y=12$, what is its area ?

(A) 25π

(B) 64π

(C) 144π

(D) 256π

3. In xy-plane, a parabola intersects the x-axis at $(3,0)$ and $(-3,0)$, what is the function of the parabola ?

(A) $y=3x^2+6x+9$

(B) $y=3x^2-6x+9$

(C) $y=x^2-9$

(D) $y=x^2+9$

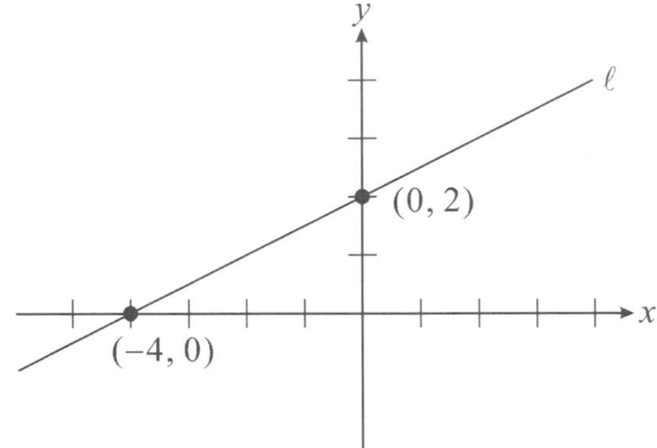

4. In the figure above, the graph of line ℓ is shown, which of the following could be the equation of line ℓ ?

(A) $y=-\frac{1}{2}x+2$

(B) $y=\frac{1}{2}x+2$

(C) $y=\frac{1}{2}x-2$

(D) $y=-\frac{1}{2}x-2$

5. In the xy-plane, line ℓ passes three points $(0,0)$, $(-3,k)$, and $(k,-12)$. What is the value of k ?

(A) -6

(B) -8

(C) -12

(D) -16

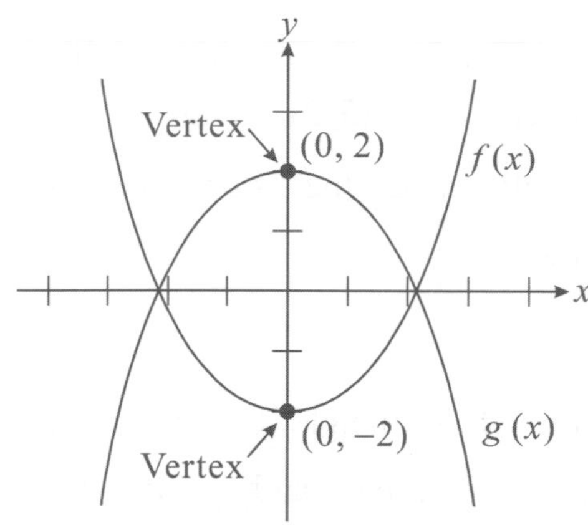

Note: Figure not drawn to scale

6. In the figure above, parabola $f(x)=4x^2-2$ and parabola $g(x)=-4x^2+2$ are symmetric. How many values of x make $f(x)+g(x)=0$?

(A) 2

(B) 3

(C) 4

(D) infinite number of x.

7. In xy-plane, line ℓ passes through points $(-4, 0)$ and point $(0, 3)$, and line m passes through point $(0, 7)$ and point $(k, -3)$. These two are perpendicular to each other. What is the value of k ?

(A) 6.5

(B) −6.5

(C) 7.5

(D) −7.5

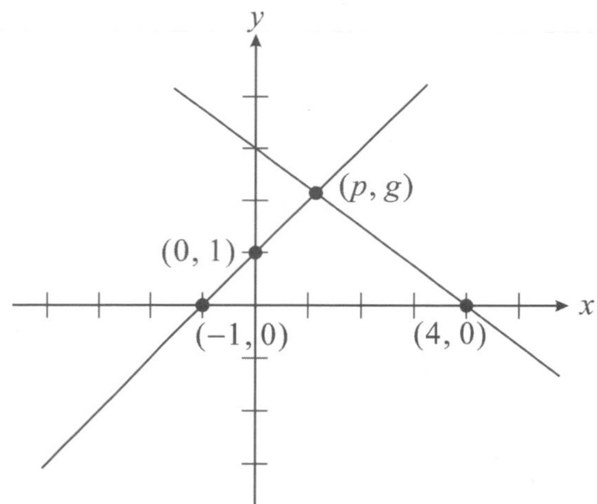

8. A line in the xy-plane has slope $-\frac{3}{4}$ and the intersection with the x-axis is $(4, 0)$. A second line passes through the points $(-1, 0)$, and $(0, 1)$. If the two lines intersect at the point (p, q), what is the value of $p-q$?

(A) −1

(B) 0

(C) $-\frac{8}{7}$

(D) $\frac{8}{7}$

9. If k is a real number such that $-1<k<1$, which of the following could be the graph of the equation $y=ky+kx+x-5$?

(A)

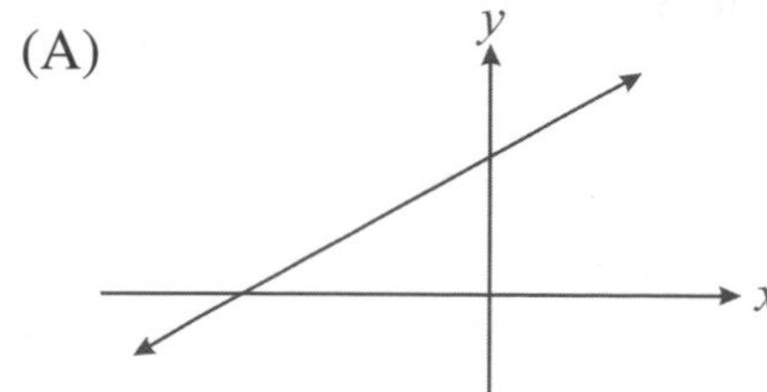

(B)

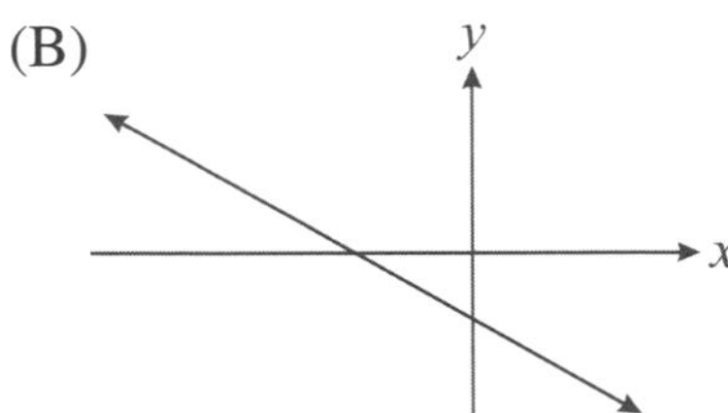

(C)

(D)

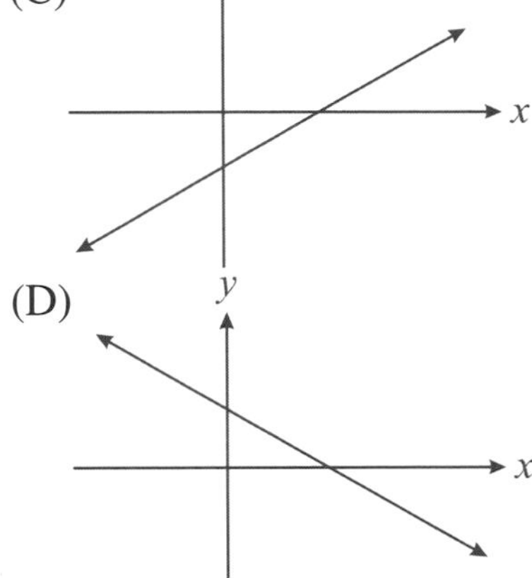

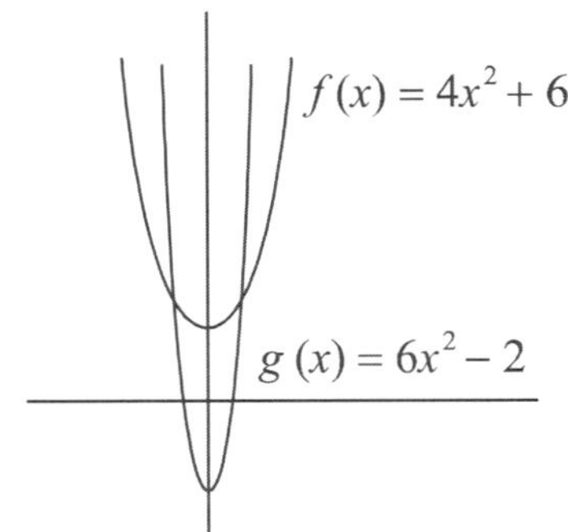

Note: Figure not drawn to scale

10. The parabola $f(x) = 4x^2 + 6$ and the parabola $g(x) = 6x^2 - 2$ are graphed in the xy-plane shown above. They intersect at point (p, q) and $(-p, q)$, what is the value p ?

(A) 0

(B) 1

(C) 2

(D) 3

11. Find the equation of a hyperbola whose two Foci are (5, 2) and (−3, 2) respectively, and the slope of one asymptote is $\frac{3}{4}$.

12. The area of a triangle formed by the intercepts of x-axis and y-axis of a straight line is 12. If the line passes point $(3, -2)$ and the length of these two intercepts are rational number, what is the equation of the line ?

13. A line in xy-plane has the intercepts with x-axis and y-axis at point $(a, 0)$ and point $(0, b)$ respectively. If $a > 0$, $b > 0$, and $a \neq b$, which of the following statements must be true ?

(A) The slope is positive

(B) The slope is negative

(C)The slope could be either positive or negative

(D)There is nothing to do with the slope

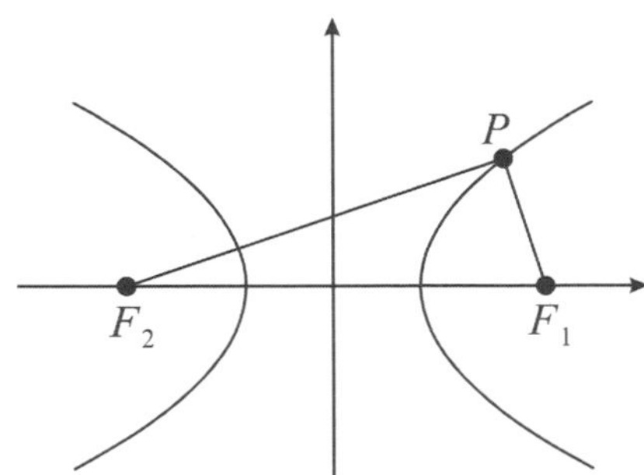

Note: Figure not drawn to scale.

14. The figure above shows the hypobola $\frac{x^2}{4}-\frac{y^2}{9}=1$. Point p lies on this hypobola in quadrant I. If F_1 and F_2 are two foci, and $\overline{PF_1}=\overline{PF_2}=1:3$, what is the perimeeter of ΔF_2PF_1 ?

15. Which of the following is the equation of line ℓ that is perpendicular to line $y=-\frac{2}{3}x+1$ and passes through the point $(1, 3)$?

(A) $y=\frac{3}{2}x+\frac{3}{2}$

(B) $y=\frac{2}{3}x+\frac{2}{3}$

(C) $y=\frac{3}{2}x-\frac{3}{2}$

(D) $y=\frac{2}{3}x+\frac{2}{3}$

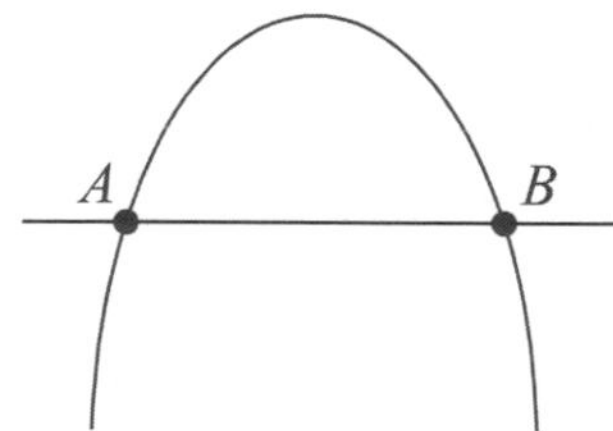

Note: Figure not drown to sacle.

16. The figure above shows a parabola with the equation of $y=ax^2+bx+c$ and the graph intercepts with x-axis at $A(2, 0)$, $B(6, 0)$, which of the following values is positive ?

(A) $b-c$

(B) $a-b$

(C) a

(D) $a+c$

17. The graph (not shown) of $y=ax^2+bx+c$ passes point $(-1, 0)$ and $(5, 0)$ and intercepts with the straight line $y=3x+3$ at its vertex, what is the value of $a+b+c$?

(A) 2

(B) 3

(C) 8

(D) 9

18. Line m in the xy-plane contains points from each of Quadrants I II, and III, but no points from Quadrant IV which of the following must be true ?

(A) The slope of line ℓ could be positive or negative.

(B) The slope of line ℓ is negative.

(C) The slope of line ℓ is positive.

(D) The slope of line ℓ is undefined.

19. Find the equation of an ellipse whose one focus is at $(-7, 3)$, the length of the major axis is 16 and the minor axis is collinear with $x=-1$.

20. Move a parabola left 4 units and up 5 units, so that the vertex of the parabola is at a new point $(4, 3)$. If the new

function of the parabola is now $y=x^2-8x+19$, which of the following is the original function of this parabola ?

(A) $y=x^2+5$

(B) $y=x^2-5$

(C) $y=x^2-16x+62$

(D) $y=x^2+16x-64$

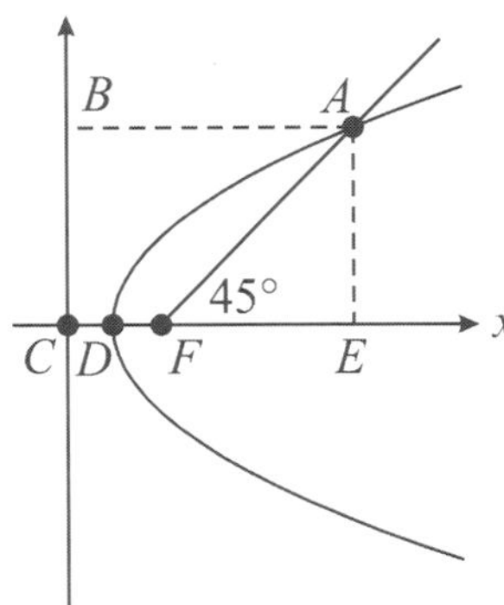

Note: Figure not drawn to scale

21. The figure above shows the graph of the orbit of a comet. In space, the orbit of a comet is a parabola and it runs with the sun as the focus. $\overline{AF}$ is the line segment which connects the comet and the sun. If the angle between $\overline{AF}$ and x-axis (the symmetric axis of the parabola) is 45° and the distance of $\overline{AF}$ is $4au$, where au is a astronomical unit, how many au is the shortest distance between the comet and the sun ?

(A) $(2-\sqrt{2})au$

(B) $(4-3\sqrt{2})au$

(C) $(2+2\sqrt{2})au$

(D) $2\sqrt{2}au$

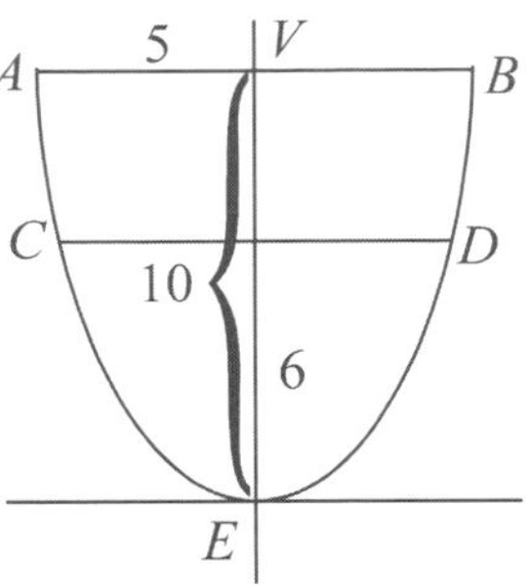

Note: Figure not drawn to scale.

22. The figure above shows a side view of a cup of wine that is not filled to the top. The inner edge of the cup is of a parabola shape. Given that the inner depth of the cup VE is 10 inches, the depth of the wine is 6 inches, and the inner width of the cup on the top AB is 5 inches, what is the inner width of the wine line $\overline{CD}$?

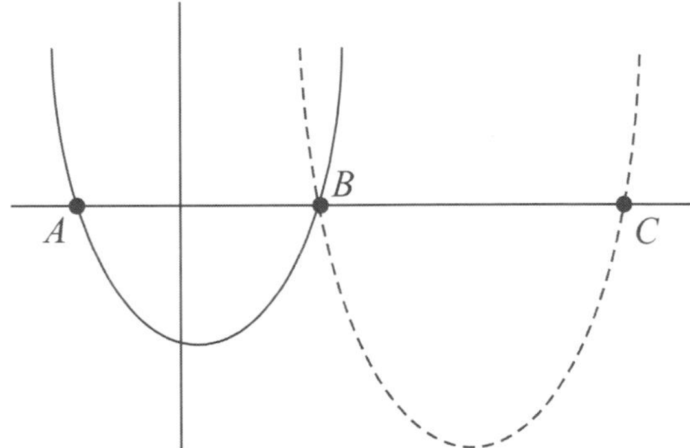

23. The figure above shows the graph of a parabola $y=x^2-6x+8$ which intercepts the x-axis at point A and point B.

If the graph is shifted 7 units to the right and k units down, the new graph intercepts with x-axis at point B and point C. What is the value of k ?

24. What are the coordinates of the foci of the ellipse with the equation $\frac{(x-3)^2}{9}+\frac{(y+2)^2}{4}=1$?

25. A line $y=x+\sqrt{5}$ is tangent to the ellipse $\frac{x^2}{a^2}+\frac{y^2}{4}=1$, what is the value of a?

26. What are the Asymptotes of the Hyperbola $\frac{x^2}{16}-\frac{y^2}{9}=1$?

27. Find the faci of $\frac{y^2}{25}-\frac{x^2}{4}=1$.

28. Find the center of the hyperbola $4x^2+9y^2+8x+36y-68=0$

29. If the rectangular coordinates are $(1,1)$, then what are its polar coordinates ?

(A) $(\frac{1}{2},\frac{\pi}{4})$

(B) $(\frac{1}{2},\frac{\pi}{2})$

(C) $(\sqrt{2},\frac{\pi}{4})$

(D) $(\sqrt{2},-\frac{\pi}{4})$

30. What is the rectangular form of the polar equation $r=\frac{3}{4-2\cos\theta}$?

31. Convert the rectangular equation $-2x+3y+4=0$ to polar form:

(A) $r=\frac{3}{4\cos\theta+\sin\theta}$

(B) $r=\frac{2}{3\cos\theta-2\sin\theta}$

(C) $r=\frac{4}{2\cos\theta+3\sin\theta}$

(D) $r=\frac{-4}{-2\cos\theta+3\sin\theta}$

32. Convert the rectangular form $4x^2+4y^2-9y=0$ to polar form.

(A) $r=\frac{3}{2}\sin\theta+\cos\theta$

(B) $r=\frac{9}{4}\cos\theta$

(C) $r=\frac{3}{2}\sin\theta$

(D) $r=\frac{9}{4}\sin\theta$

CHATPER 6
ANSWERS and EXPLANATIONS

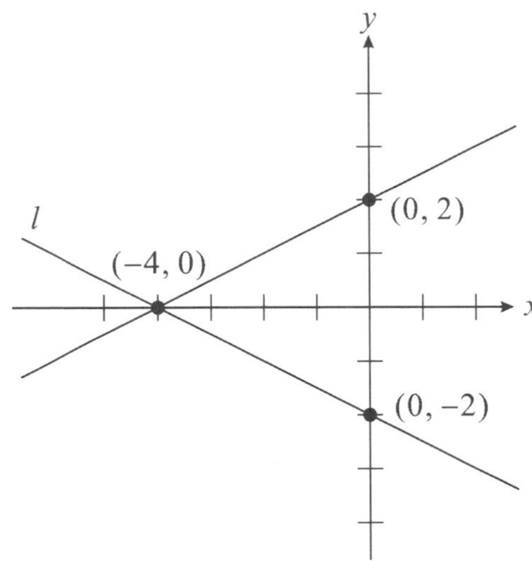

1.Base on slope-intercept form $y = mx + b$,

The original equation is

$$y = \frac{0-(2)}{-4-(0)}x + b = \frac{-2}{-4}x + b = \frac{1}{2}x + b$$

l passes through $(-4, 0)$,

$0 = \frac{1}{2} \times (-4) + b$, $b = 2$, the equation is

$$y = \frac{1}{2}x + 2$$

The new line must intercept with y-axis at (0, 2), so the slope is $\frac{-2-0}{0-(-4)} = \frac{-2}{4}$

$= -\frac{1}{2}$ or, since it's flipped over x-axis, the slope is the same but reversed sign.

Answer is (A).

2. $x^2 - 6x + 9 + y^2 + 4y + 4 - 13 = 12$

$(x-3)^2 + (y+2)^2 = 25 = 5^2$

area $= \pi r^2 = \pi \cdot 25 = 25\pi$

Answer is (A).

3. Assume the function of the parabola is $y = (x+a)^2 + b$, where $(-a, b)$ is the vertex of the parabola.

Since this parabola intersects x-axis at $(3, 0)$ and $(-3, 0)$, plug them into the equation above, we get

$(a+3)^2 + b = 0$, and

$(a-3)^2 + b = 0$

So, $a^2 + 6a + 9 + b = 0 \cdots ①$

$a^2 - 6a + 9 + b = 0 \cdots ②$

①−② $12a = 0$, $a = 0$

substitute to ① $9 + b = 0$ $b = -9$

The function of the parabola is

$y = x^2 - 9$

The answer is (C).

4. Based on slope-intercept form, the equation of line is

$y = mx + b$, then $2 = (m)(0) + b$, $b = 2$

and $0 = (m)(-4) + 2$ $\therefore$ $-4m = -2$,

$m = \frac{-2}{-4} = \frac{1}{2}$

The equation of ℓ is $y = \frac{1}{2}x + 2$

The answer is (B).

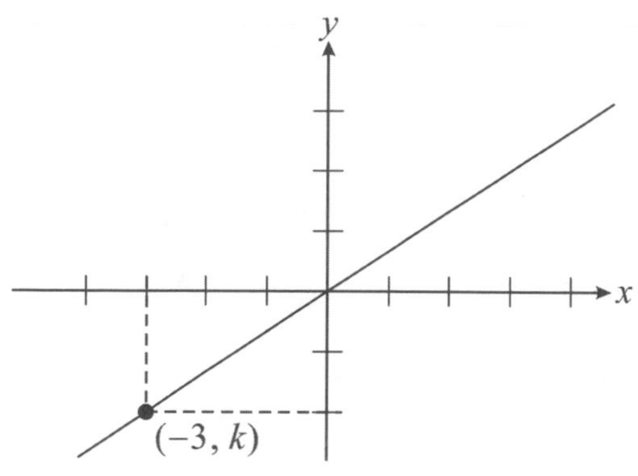

5. Assume line ℓ $y=mx+b$

since it passes $(0,0)$, $b=0$ since it passes $(-3,k)$, $y=mx$, $k=-3m$, $y=-\frac{k}{3}x$

Since it passes, $(k_1,-12)$,

$-12=-\frac{k}{3}(k)$ $k^2=36$ $\therefore$ $k=\pm 6$

Answer is (A).

6. Since 2 parabolas are totally symmetry, any value of y in $f(x)$, there is a opposite value of y in $g(x)$, $g(x)=-f(x)$, $f(x)+g(x)=0$

Answer is (D).

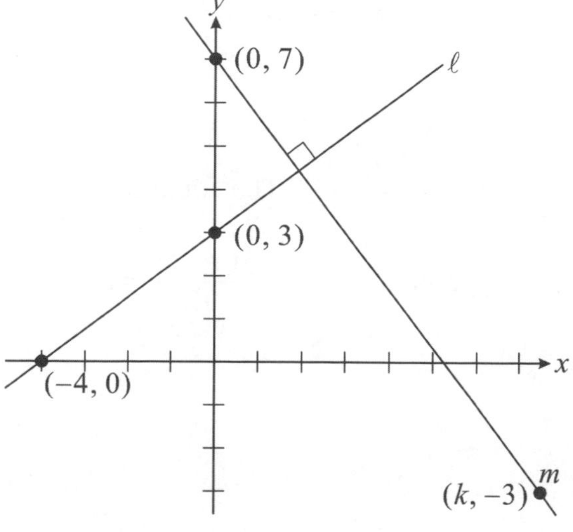

7. The slope of line ℓ: $\frac{3-0}{0-(-4)}=\frac{3}{4}$

Since they are perpendicular,

So, $m_1\times m_2=-1$ $\frac{3}{4}\times m_2=-1$

$\therefore$ $m_2=-\frac{4}{3}$

The equation of line m is: $y=-\frac{4}{3}x+b$

Since it passes through $(0,7):7=0+b$

$b=7$

$y=-\frac{4}{3}x+7$,

Because it passes through $(k,-3)$:

$-3=-\frac{4}{3}(k)+7$ $-\frac{4}{3}k=-10$

$\therefore$ $k=\frac{30}{4}=\frac{15}{2}$

Answers (C).

8. The equation of the first line is

$y=-\frac{3}{4}x+b$, when $x=4$, $y=0$

so $0=-\frac{3}{4}(4)+b$ $\therefore$ $b=3$

$\therefore$ $y=-\frac{3}{4}x+3\cdots①$

The equation of the second line is $y=mx+b$,

$y=\frac{1-0}{0-(-1)}x+b$

$y=x+b$, when $x=-1$, $y=0$,

$0=-1+b$

$\therefore$ $b=1$, $y=x+1\cdots②$

When they intersect, $-\frac{3}{4}x+3=x+1$

$\therefore \ \frac{7}{4}x = 2$

$\therefore \ x = \frac{8}{7} = p, \ y = \frac{8}{7} + 1 = \frac{15}{7} = q$

$p - q = \frac{8}{7} - \frac{15}{7} = -\frac{7}{7} = -1$

Answer is (A).

9. (C)

$y = ky + kx + x - 5$

$y - ky = (k+1)x - 5$

$y(1-k) = (k+1)x - 5$

$y = \frac{k+1}{1-k}x - 5$

From the slope-intercept form

$m = \frac{k+1}{1-k}, \ b = -5$

Since $-1 < k < 1$, $k+1$ is always positive and $1-k$ is also positive.

Therefore m is positive, and the intercept with y axis is negative.

So, the graph of (C) is the answer.

10. Since they interset at point (p, q), so, the value of y coordinate must be equal.

$f(x) = 4x^2 + 6 = g(x) = 6x^2 - 2$

$4x^2 + 6 = 6x^2 - 2$

$\therefore \ 2x^2 = 8, \ x^2 = 4, \ x = \pm 2$

The x-coordinate is 2, or -2,

Answer is (C)

11. Draw a figure of the hyperbola below.

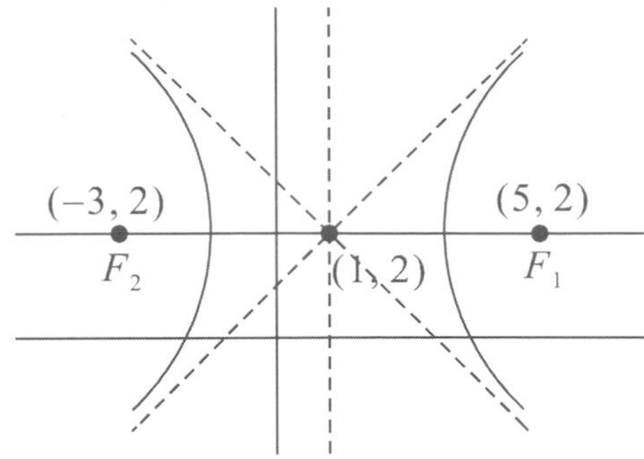

(1) $F_1(5, 2), \ F_2(-3, 2)$

$2c = 5 - (-3) = 8, \ c = 4$ (Let c be the distance from center to the focus)

(2) Center is at (1, 2)

hyperbola: $\frac{(x-1)^2}{a^2} - \frac{(y-2)^2}{b^2} = 1$

(3) Two asymptotes

$[\frac{(x-1)}{a}] \pm [\frac{(y-2)}{b}] = 0,$

$a(y-2) \pm b(x-1) = 0$

$ay - 2a + bx - b = 0, \ ay = -bx + 2a + b,$

$y = -\frac{b}{a}x + \frac{2a+b}{a}$

Slope $= -\frac{b}{a} = \frac{3}{4}, \ 4b = -3a, \ b = -\frac{3}{4}a$

(4) $a^2 + b^2 = c^2 = 16,$

$a^2 + (-\frac{3}{4}a)^2 = \frac{25}{16}a^2 = 16,$

$a^2 = \frac{16^2}{5^2}, \ a = \frac{16}{5}$

$b = -\frac{3}{4} \cdot \frac{16}{5} = \frac{12}{5}$, The equation of hyperbola $\frac{(x-1)^2}{(\frac{16}{5})^2} - \frac{(y-2)^2}{(\frac{12}{5})^2} = 1$

12. Set the intercepts are a and b, the equation is $\frac{x}{a}+\frac{y}{b}=1$ (intercepts-form).

Since the area of the triangle is 12.

So $\frac{1}{2}|ab|=12$, $ab=24$ or $ab=-24$

Since line passes $(3,-2)$:

(1) $\begin{cases} ab=24 \\ \frac{3}{a}+\frac{-2}{b}=1 \end{cases}$ (2) $\begin{cases} ab=-24 \\ \frac{3}{a}+\frac{-2}{b}=1 \end{cases}$

From (1) $\frac{3}{\frac{24}{b}}+\frac{-2}{b}=1 \Rightarrow \frac{3b}{24}-\frac{2}{b}=1$

$\Rightarrow \frac{b}{8}-\frac{2}{b}=1 \Rightarrow b-\frac{16}{b}=8$

$\Rightarrow b^2-16=8b$, $b^2-8b-16=0$

$\Rightarrow b=\frac{8\pm\sqrt{64+64}}{2}$

$\Rightarrow 4\pm4\sqrt{2}$ (irrational)

From (2) $\frac{3}{\frac{-24}{b}}+\frac{-2}{b}=1 \Rightarrow -\frac{b}{8}-\frac{2}{b}=1$

$\Rightarrow b^2+8b+16=0 \Rightarrow (b+4)^2=0$

$b=-4$, $a=6$

Equation: $\frac{x}{6}-\frac{y}{4}=1$

13. If $a>0$, $b>0$, $a\neq b$, the figure should look like the graphs below

Therefore, the slope must be negative.

Answer is (B).

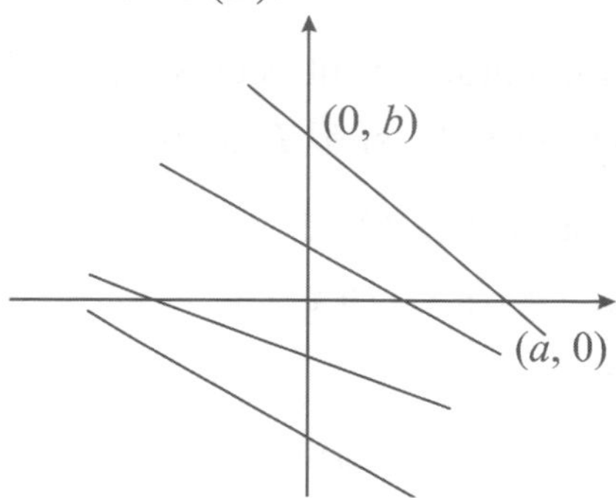

14. Since $\frac{x^2}{4}-\frac{y^2}{9}=1$,

so, $a=\sqrt{4}=2$, $b=\sqrt{9}=3$, $c=\sqrt{13}$.

Two Foci are $(\sqrt{13},0)$ and $(-\sqrt{13},0)$.

Since $\overline{PF_1}:\overline{PF_2}=1:3$,

Let $PF_1=k$, $PF_2=3k$

by definition $|\overline{PF_1}-\overline{PF_2}|=2a=2\times2=4$.

$3k-k=2k=4$

$k=2$

The perimeter of

$\Delta F_2PF_1 = PF_1+PF_2+F_1F_2$

$=2+3\times2+2\sqrt{13} = 8+2\sqrt{13}$.

15. Since line ℓ is perpendicular to line $y=-\frac{2}{3}x+1$, so its slope is m such that

$(-\frac{2}{3})(m)=-1$, $m=\frac{-1}{-\frac{2}{3}}=\frac{3}{2}$

line ℓ $y=mx+b$, $y=\frac{3}{2}x+b$

Since it passes through $(1,3)$,

$3=(\frac{3}{2})(1)+b$, $b=\frac{3}{2}$

$y=\frac{3}{2}x+\frac{3}{2}$, the answer is (A)

16. (A)

Since the parabola intercepts with x-axis at point $A(2,0)$ and $B(6,0)$

the equation should be

$ax^2+bc+c=a(x-2)(x-6)$

$\Rightarrow\ y=a(x^2-8x+12)$

$\Rightarrow\ y=ax^2-8ax+12a$

Since it concaves downwards, a is negative.

$b=-8a$ is positive.

$c=12a$ is negative.

(A) $b-c$ positive value minus negative is positive

(B) $a-b$ negative minus negative could be negative, could be positive.

(C) a a is negative.

(D) $a+c$ negative plus negative is negative.

Answer is (A).

17. (C)

Since the graph of $y=ax^2+bx+c$ passes through $(-1,0)$ and $(5,0)$, the vertex lies on the symmetrical axis which is $x=\dfrac{-1+5}{2}=2$.

Line $y=3x+3$ passes through the vertex, so, $y=3(2)+3=9$.

The coordinates of the vertex is $(2,9)$,

The equation of the parabola is

$y=a(x-2)^2+9=a(x^2-4x+4)+9$

$=ax^2-4ax+4a+9$

Substitute $(-1,0)$ into the equation

$0=a-4a\cdot(-1)+4a+9=a+4a+4a+9$

$=9a+9$

$\therefore\ a=-1$

So, $y=-x^2+4x+5$.

$a=-1$, $b=4$, and $c=5$

$a+b+c=-1+4+5=8$

Answer is (C).

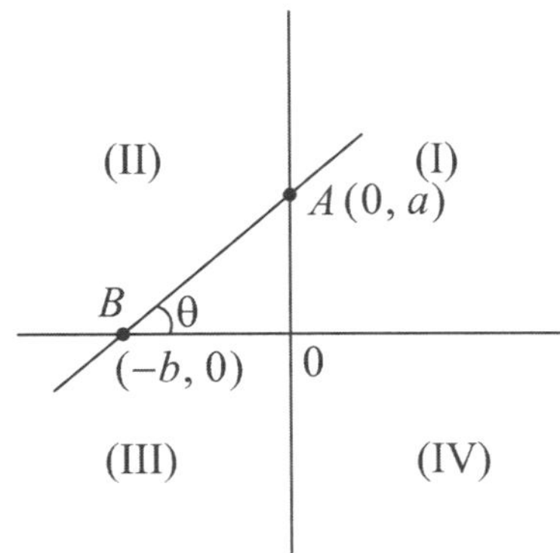

18. Let a, b be positive number, since line l contains points from quadrants I, II, and III but not IV.

Slope $=\dfrac{a-0}{0-(-b)}=\dfrac{a}{b}\Rightarrow$ positive

Answer (C).

19. Graph the ellipse below:

Since minor axis is collinear with $x=-1$, and the Y-coordinate of focus is 3, the center is at $(-1,3)$.

$c=-1-(-7)=6$, F_1 at $(5,3)$

$2a=16$, $a=8$

$a^2=b^2+c^2$, $8^2=b^2+6^2$, $b^2=8^2-6^2=28$

Equation: $\frac{(x+1)^2}{64}+\frac{(y-3)^2}{28}=1$

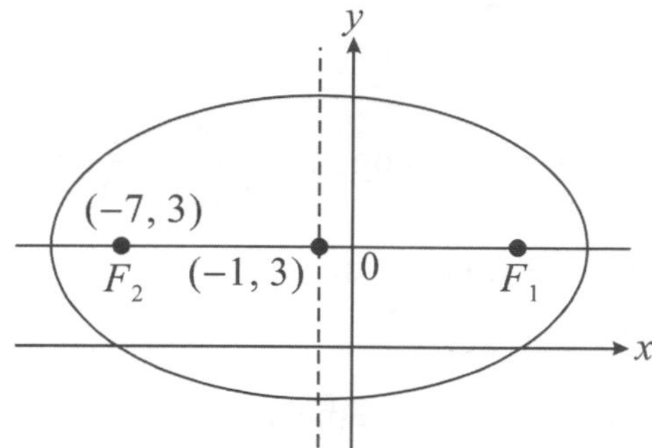

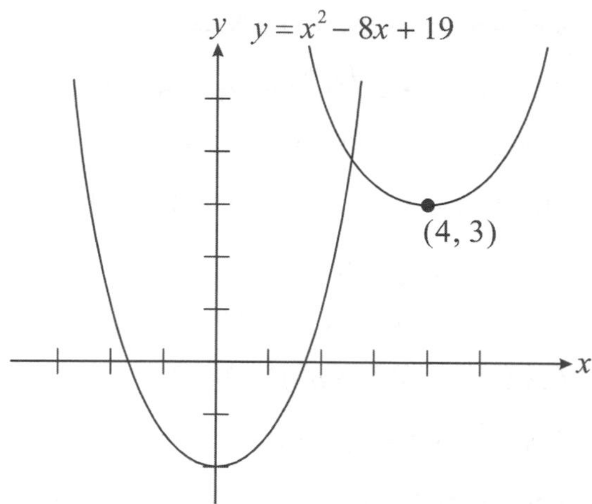

20. The function $y=x^2-8x+19$ was translated by shifted the original figure 4 units left and up 5 units.

(1) $y=x^2-8x+4^2-4^2+19=(x-4)^2+3$

Vertex is at (4, 3).

(2) Shift down 5 units:

$y'=(x-4)^2+3-5=(x-4)^2-2$

Shift right 4 units:

$y=(x-4-4)^2-2=(x-8)^2-2$

$=x^2-16x+62$

Note: Figure not drawn to scale

21. When the comet runs to D, $\overline{DF}$ is the shortest distance.

Draw $\overline{AE}\perp\overline{CE}$, $\overline{AB}\perp\overline{BC}$

$\overline{AB}=\overline{CE}=\overline{CF}+\overline{FE}$

$FE=4au\times\frac{\sqrt{2}}{2}=2\sqrt{2}au$

Since $\overline{AB}=\overline{AF}=4au$

$\therefore\ 4au=CF+2\sqrt{2}au$,

$CF=4au-2\sqrt{2}au=(4-2\sqrt{2})au$

$DF=DC$

$DF=\frac{1}{2}CF=(2-\sqrt{2})au$

Answer is (A)

22. 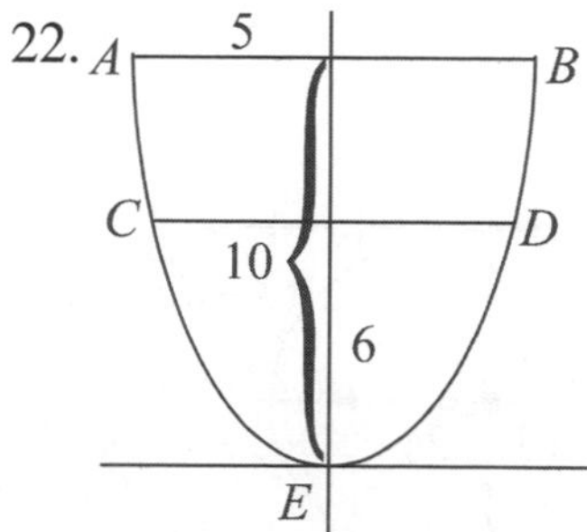

Let the equation of the parabola be $y=\frac{1}{4p}x^2$, then $(10=\frac{1}{4p}(2.5)^2)$

$40p=6.25$

$\therefore\ p=\frac{6.25}{40}$

substitute, $6=\frac{x^2}{4\times\frac{6.25}{40}}$

$\therefore\ x^2=6\times\frac{6.25}{10}$, $x=\pm1.94$

$\therefore\ CD=2\times1.94=3.28$ inches

23.(1) $y=x^2-6x+8=(x-2)(x-4)$, the graph intercepts the x-axis at point $(2, 0)$ and $(4, 0)$. $y=x^2-6x+8=(x-3)^2-9+8=(x-3)^2-1$, the vertex is at $(3, -1)$.

(2) If it shifts right for 7 units, the function changes to $y=(x-3-7)^2-1=(x-10)^2-1$

If it shifts down for k units, then the function changes to $y=(x-10)^2-1-k$.

(3) Point $B(4, 0)$ satisfies the new function $0=(4-10)^2-1-k$, $k=35$.

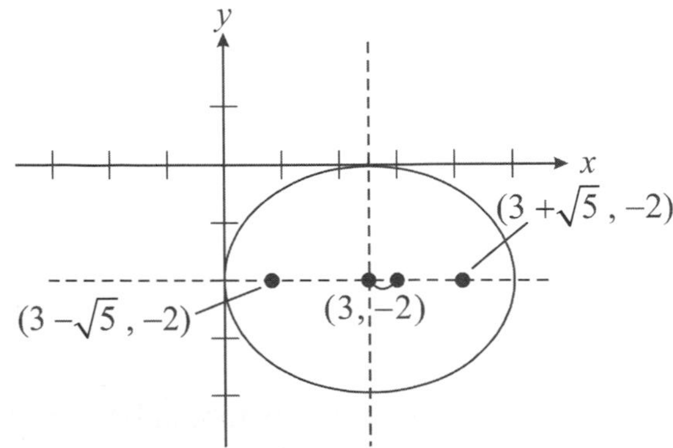

24. The center is at $(3, -2)$

Since $c^2=a^2-b^2$, $c^2=9-4=5$

$\therefore\ c=\sqrt{5}$

The coordinates of foci are $(3+\sqrt{5}, -2)$ $(3-\sqrt{5}, -2)$.

25. $y=x+\sqrt{5}$ is tangent to $\frac{x^2}{a^2}+\frac{y^2}{4}=1$

re-write the equation

$4x^2+a^2y^2=4a^2$

the y-coordinate is the same.

$4x^2+a^2(x+\sqrt{5})^2=4a^2$

$4x^2+a^2(x^2+2\cdot\sqrt{5}x+5)=4a^2$

$4x^2+a^2x^2+2\sqrt{5}a^2x+5a^2-4a^2=0$

$(4+a^2)x^2+2\sqrt{5}a^2x+a^2=0$

Since $y=x+\sqrt{5}$ is tangent to ellipse

There is only one root.

Therefore, $b^2-4ac=0$

$(2\sqrt{5}a^2)^2-4(4+a^2)(a^2)=0$

$20a^4-16a^2-4a^4=0$

$16a^4-16a^2=0$

$a^2(16a^2-16)=0$

$a^2=0$, $a^2=\frac{16}{16}=1\,(a\neq0)$

$\therefore\ a=1$

26. $\frac{x^2}{16}-\frac{y^2}{9}=1$

$\frac{x^2}{4^2}-\frac{y^2}{3^2}=1$

$y=\pm\frac{b}{a}x$

$y=\pm\frac{3}{4}x$

27. $\frac{y^2}{25}-\frac{x^2}{4}=1$ transvere axis is vertical

$a=5$, $b=2$

$c^2=25+4=29$, $c=\pm\sqrt{29}$

faci: $(0, \sqrt{29})$, $(0, -\sqrt{29})$

28. $4x^2+8x+9y^2+36y-68=0$

$4(x+1)^2-9(y-2)^2-4+36-68=0$

$4(x+1)^2-9(y-2)^2=36$

$\frac{(x+1)^2}{9}-\frac{(y-2)^2}{4}=1$

Center is at $(-1, 2)$.

29. $r^2=1^2+1^2=2 \quad r=\sqrt{2}$

$\tan\theta=\frac{y}{x}=\frac{1}{1}=1$

$\theta=\tan^{-1}1=\frac{\pi}{4}$

Its polar coordinates are $(\sqrt{2}, \frac{\pi}{4})$.

Answer is (C).

30. $4r-2r\cos\theta=3$

$4\sqrt{x^2+y^2}-2x=3$

$\sqrt{x^2+y^2}=\frac{2x+3}{4}$

$x^2+y^2=\frac{(2x+3)^2}{16}=\frac{4x^2+12x+9}{16}$

$16x^2+16y^2-4x^2-12x-9=0$

$12x^2+16y^2-12x-9=0$

31. $-2r\cos\theta+3r\sin\theta+4=0$

$r(-2\cos\theta+3\sin\theta)=-4$

$r=\frac{-4}{-2\cos\theta+3\cos\theta}$

Answer (D).

32. $4(x^2+y^2)-9\,r\sin\theta=0$

$4r^2=9r\sin\theta$

$4r=9\sin\theta$

$r=\frac{9}{4}\sin\theta$

Answer (D).

CHAPTER 7
DATA PRESENTATION and BASIC STATISTICS

7-1 DATA PRESENTATION

Data presentation is a structured way to show the characteristics, distribution, quantities, and trends of the data.

A. Table

◆Example:

MEMBERS IN CHESS CLUB OF NORTH HIGH SCHOOL

	Freshman	Sophomore	Junior	Senior	Total
Male	12	22	28	14	76
Female	7	10	11	10	38
Total	19	32	39	24	114

qualitative variables | quantitative variables | quantitative variables | quantitative variables | quantitative variables | quantitative variables

Answer to Concept Check 7-01

- The number of students who get B- is 95, and the total number of students in the high school is 647.

  $95 \div 647 = 14.7\%$

- The number of students who received the grades *b* and up is

  $44 + 80 + 104 + 120 + 113 = 461$

  $461 \div 647 = 0.713$

  Grade B and up is 71.3%

B. Bar Chart

Using the length of a bar graph to represent the quantity of an item.

◆Example:

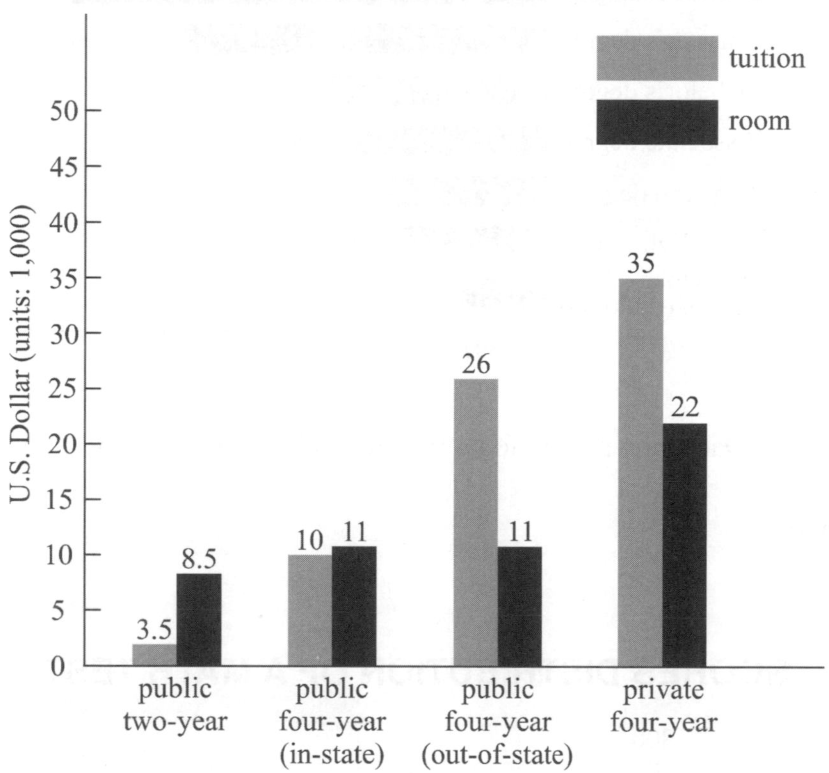

Data shown is fictional for illustration purpose only.

According to the graph above. The public four-year (out-of-state) is what percent increase of the public four-year (in-state) ?

The cost for public four-year (out-of-state) is $26+11=37 \Rightarrow \$37{,}000$, and the cost for public four-year (in-state) is $10+11=21 \Rightarrow \$21{,}000$.

Percent increase $= \dfrac{37{,}000-21{,}000}{21{,}000} \times 100\% = 76\%$

◆Example:

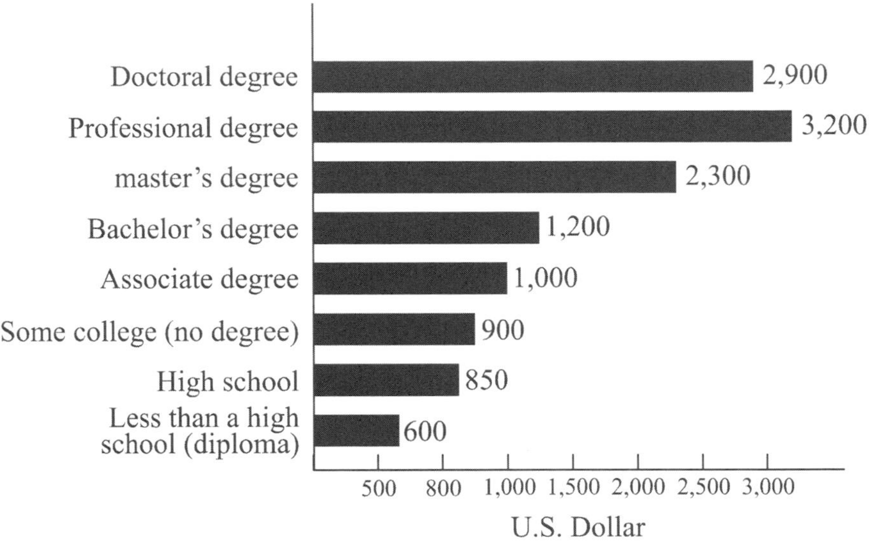

Data shown is fictional for illustration purpose only.

C. Histogram

◆Example:

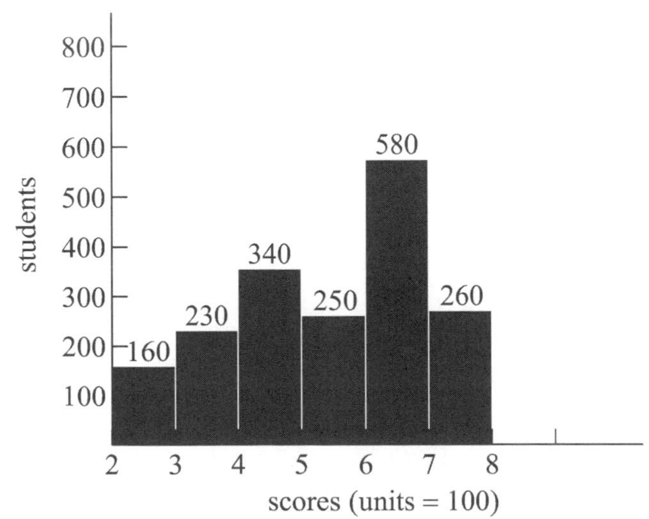

* The range of a histogram bar includes the beginning value excludes the ending value. For instance, the range of 3-4 contains all values greater than 3, but doesn't include the value 4.

* The total value of a histogram bar is calculated by taking the average multiplied by the value of y-axis. For instance, the value of range 4-5 is $340\times\frac{4+5}{2}=1,530$

Answer to Concept Check 7-02

$$\text{Average score} = \frac{\text{Total scores}}{\text{Total number of students}}$$

1. Total score $=[\frac{0+199}{2}\times0.8+\frac{200+299}{2}\times4+\frac{300+399}{2}\times6+\frac{400+499}{2}\times8$
$+\frac{500+599}{2}\times6+\frac{600+699}{2}\times4+\frac{700+799}{2}\times1.2]+\times100=1,356,500$

   $\text{Average}=\frac{1,356,500}{3000}=452.17$

2. $\frac{520}{3000}=0.1733=17.33\%$

D. Line Graph

◆Example:

THE STOCK PRICE OF COMPANY A

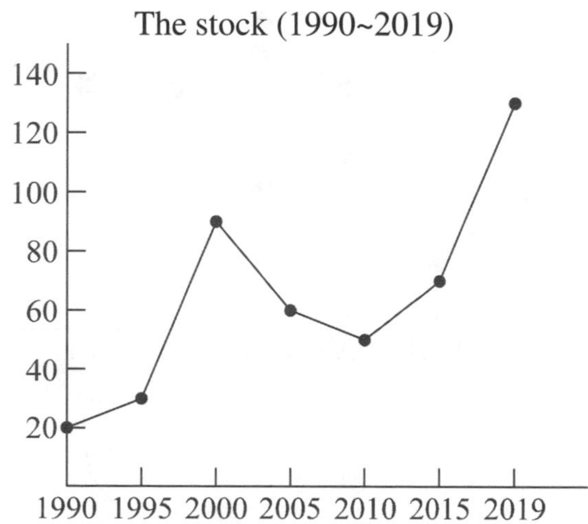

1.Which of the 5-year periods sees the biggest increase in price ?

2.The stock price in 2019 is what percent increase of that in 1990 ?

From 1990~1995: $\frac{30-20}{20}\times100\%=50\%$

From 1995~2000: $\dfrac{90-30}{50}=1.2 \Rightarrow 120\%$ increased

From 2000~2005: decreased

From 2005~2010: decreased

From 2010~2015: $\dfrac{70-50}{50}\times 100\% = 40\%$

From 2015~2019: $\dfrac{130-70}{70}\times 100\% = 85.7\%$

1. The 5-year period of 1995~2000 saw the biggest increase.
2. The stock price in 2019 is about \$130, and the stock price in 1990 is about \$20. The percentage raise $=\dfrac{130-20}{20}\times 100\% = 550\%$

E. Multiple line Graph

◆Example:

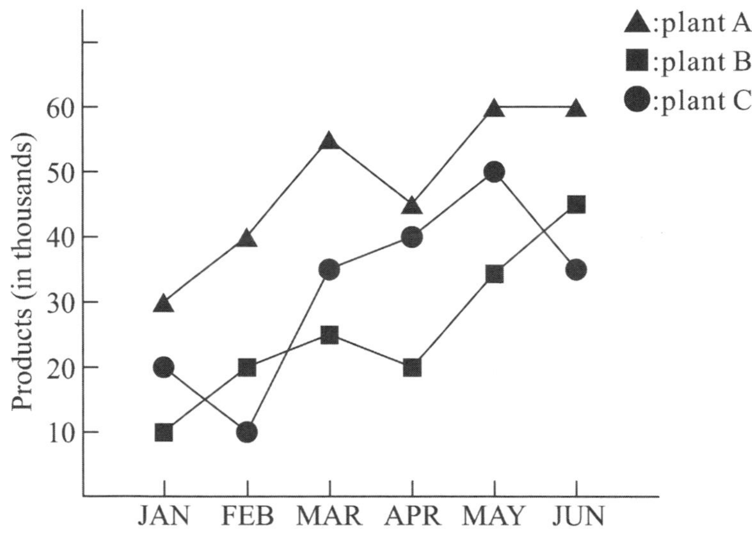

Which plant had the biggest **productivity** percentage increase between Feb and March ?

	Feb	Mach	Increase
Plant A	40	55	$\frac{55-40}{40}\times 100\% = 37.5\%$
Plant B	20	25	$\frac{25-20}{20} = 25\%$
Plant C	12	36	$\frac{36-12}{12} = 200\%$

Plant C has the biggest percentage increase.

Answer to Concept Check 7-03

1. ▲ Charles $= 600+400+400+500+400+500 = 2,800$

   ● Katy $= 400+550+300+400+500+500 = 2,650$

   ■ Ken: $200+200+300+600+650+650 = 2,600$

   Charles bas the biggest sales revenue.

2. ▲ Charles $600-400 = 200$

   ● Katy $500-300 = 200$

   ● Ken $650-200 = 450$

   Ken has the biggest sales revenue range.

F. Pie Chart

DEGREES HELD BY FACULTY MEMBERS IN *W* UNIVERSITY

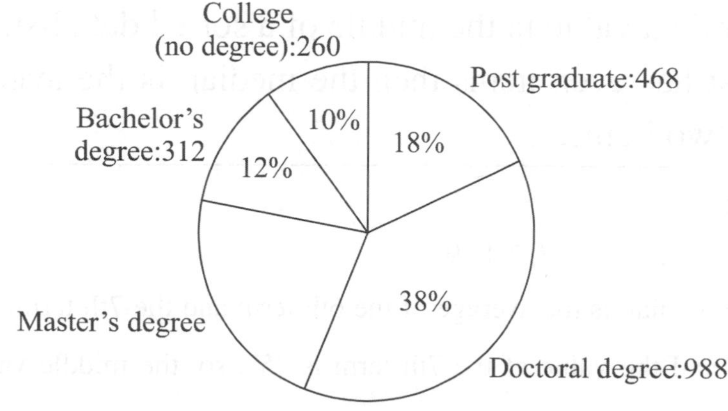

1. What percentage of faculty member holds master's degree ?

   $100\% - 12\% - 10\% - 18\% - 38\% = 22\%$

2. How many faculties holds master's degree ?
   Since $260 \div 0.1 = 2,600$, $2,600 \times 0.22 = 572$

Answer to Concept Check 7-04

1. Tuition fee $= 49.3\%$, living expense $= 8.5\%$
   $49.3 : 8.5 = 5.8 : 1$
2. $26,000 \div 0.493 = 52,738$
   $52,738 \times 0.114 = 6,012$
   $6,012 \times 4\,(\text{years}) = 24,048$

7-2 CENTRAL TENDENCY

Mean (arithmetic average)
The average value of whole data list.

◆Example:

If the math scores in Jack's class are:

68, 72, 72, 84, 88, 90, 90, 90, 98

$$\text{Mean} = \frac{68+72+72+84+88+90+90+90+98}{9} = \frac{752}{9} = 83.6$$

Median
The item or data value in the middle of a sorted data list.
If a data list has even items then the median is the average value of the middle two items.

For an Example:

Data list: 2, 3, 3, 4, 4, 4, 5, 6, 7, 7, 8, 9

12 items, so, the median is the average of the 6th term and the 7th term. The value of the 6th term is "4", and the value of the 7th term is "5", so, the middle value is $\frac{4+5}{2} = 4.5$

The median is therefore 4.5 of the data list above.

• Odd terms of the data list.

In odd terms, the median of a data set is just the middle term.

Example: 2, 3, 3, 4, 4, 4, 5, 6, 7, 7, 8, there are 11 terms, the median is the middle term $\frac{11+1}{2}=6$, the value of the 6th data, it is "4".

> Mode
> The most frequent items in a data list

For instance, in the above data list

'3' occurs 2 times

'4' occurs 3 times

'7' occurs 2 times

So, 4 is the mode of the data list above. Sometimes, there are double modes or triple modes in a data list.

7-3 DATA SPREAD

The most common tools to measure the spread of a data set are range, interquartile range, and standard deviation.

> Range: It is the difference between the least value and the greatest value in a data list.

For an example,

Alice's grandfather's blood sugar readings are 135, 138, 152, 142, 143, 138, 133.

Range = the greatest − the least $=152-133=19$

> Quartile range
> Sort a list of data ascending or descending and divide it into four equal parts. The range of each part is called quartile range.

For example:

A sorted data set: 2, 5, 10, 12, 14, 22, 26, 34, 35, 42, 43, 48.

Given 12 data above, divide them into 4 equal parts.

2,	5,	10,	12,	14,	22,	26,	34,	35,	42,	43,	48

Q_1 Q_2 Q_3

★Tip

If $13\times0.25=3.25$, then upgrade to whole number '4'. $Q_1=12$

Q_1:The first 25% of data are below this value

In this case, the position of Q_1 is at $12\times0.25=3$, because '3' is a whole number.

so, the value of Q_1 is equal to the value of the third term adding the value of the 4th term and then divided by 2.

$$Q_1=\frac{10+12}{2}=11$$

Q_2:Half of the data is below this value,

It is also called "Median".

In this case, $12\times0.5=6$

$$\text{median}=\frac{22+26}{2}=24$$

Q_3:The first 75% of data are below it.

In this case, $12\times0.75=9$

$$\therefore\ Q_3=\frac{35+42}{2}=38.5$$

IQR Interquartile range: is the difference between upper and lower quartile data value $\text{IQR}=Q_3-Q_1$

In above case, it is $IQR=38.5-11=27.5$

Standard Deviation

It is a tool to measure the difference between each data value and the mean of the set to observe the dispersion of whole data set.

Standard Deviation (denoted as S or σ),

$$S=\sqrt{\frac{\text{sum of the squares of the difference from the mean}}{\text{number of items in the data set}-1}}$$

$$=\sqrt{\frac{(x_1-\bar{x})^2+(x_2-\bar{x})^2+(x_3-\bar{x})^2+\cdots+(x_n-\bar{x})^2}{n-1}}\ (\bar{x}=\text{mean})$$

Short formula: $S=\sqrt{\dfrac{\sum x^2-\frac{1}{n}(\sum x)^2}{n-1}}$

(Σ means sum) $\sum_{i=1}^{n} x_i^2 = x_1^2+x_2^2+x_3^2+\cdots+x_n^2$

$\sum_{i=1}^{n} x_i = x_1+x_2+x_3+\cdots+x_n$

★Tip

S: represents the standard Deviation of sample data set.

σ: represents the standard Deviation of whole data set.

Proof:

Variance $=S^2=\dfrac{\sum_{i=1}^{n}(x_i-\bar{x})^2}{n-1}=\dfrac{\sum_{i=1}^{n}(x_i^2-2x_i\bar{x}+\bar{x}^2)}{n-1}$ (square of standard deviation)

$$=\frac{\sum_{i=1}^{n}x_i^2-\sum_{i=1}^{n}2x_i\bar{x}+\sum_{i=1}^{n}\bar{x}^2}{n-1}$$

$$=\frac{\sum_{i=1}^{n}x_i^2-2\sum_{i=1}^{n}x_i\bar{x}+\sum_{i=1}^{n}\bar{x}^2}{n-1}$$

$$=\frac{\sum_{i=1}^{n}x_i^2-2\bar{x}\cdot n\bar{x}+\sum_{i=1}^{n}\bar{x}^2}{n-1}\qquad (\sum_{i=1}^{n}x_i=n\bar{x})$$

$$=\frac{\sum_{i=1}^{n}x_i^2-2\cdot n\cdot\bar{x}^2+n\bar{x}^2}{n-1}\qquad (\sum_{i=1}^{n}\bar{x}^2=\bar{x}^2+\bar{x}^2+\cdots+\bar{x}^2=n\bar{x}^2)$$

$$=\frac{\sum_{i=1}^{n}x_i^2-n\bar{x}^2}{n-1}$$

$$=\frac{\sum_{i=1}^{n}x_i^2-\frac{(\sum_{i=1}^{n}x_i)^2}{n}}{n-1}\qquad (\sum_{i=1}^{n}x_i)^2=(n\bar{x})^2=n^2\bar{x}^2$$

$$\therefore \quad S=\sqrt{\frac{\sum x^2-\frac{1}{n}(\sum x)^2}{n-1}}$$

◆Example 1:

The employees' salaries of Katy's company are as follows:

33,000, 35,000, 40,000, 57,000, 45,000, 52,000, 68,000, 72,000, 64,000, 54,000.

$\bar{x}$ (mean)

$$=\frac{33,000+35,000+40,000+57,000+45,000+52,000+68,000+72,000+64,000+54,000}{10}$$

$$=52,000$$

Variance $(S^2)=(33,000-52,000)^2+(35,000-52,000)^2+(40,000-52,000)^2$

$+(57,000-52,000)^2+(45,000-52,000)^2+(52,000-52,000)^2+(68,000-52,000)^2$

$+(72,000-52,000)^2+(64,000-52,000)^2+(54,000-52,000)^2/10-1$

$$=(\frac{361+289+144+25+49+0+256+400+144+4}{9})\times 10^6$$

$$=185.78\times 10^6$$

S(S.D) $=\sqrt{185.78}\times 10^3=13.63\times 10^3 \doteq 13,630$

If we use short formula

$$S=\sqrt{\frac{\sum x^2-\frac{(\sum x)^2}{n}}{n-1}}$$

$$S=\sqrt{\frac{\sum_i^{10} x_i^2-\frac{(\sum_i^{10} x_i)^2}{10}}{10-1}}$$

$$(\sum_{i=1}^{10} x_i)^2=(52,000\times 10)^2=(52\times 10^3\times 10)^2=52^2\times 10^8$$

$$\frac{(\sum_{i=1}^{10} x_i)^2}{n}=\frac{(520)^2\times 10^6}{10}=27,040\times 10^6$$

$$\sum_{i=1}^{10} x_i^2=28,712\times 10^6$$

$$\sum_{i}^{10} x_i^2 - \frac{(\sum_{i}^{10} x_i)^2}{n} = (28,712 - 27,040) \times 10^6 = 1,672 \times 10^6$$

$$S = \sqrt{\frac{1,672}{9} \times 10^6} = 13.63 \times 10^3 = 13,630$$

◆Example 2:

If the lowest temperature of each day of this week are as follows:

20°F, 26°F, 30°F, 28°F, 30°F, 34°F, find its standard deviation.

$$(\sum_{i=1}^{6} x_i)^2 = (20+26+30+28+30+34)^2 = (168)^2 = 28,224$$

$$\sum_{i=1}^{6} x_i^2 = 20^2 + 26^2 + 30^2 + 28^2 + 30^2 + 34^2 = 4,816$$

$$\sigma = \sqrt{\frac{4816 - \frac{28,224}{6}}{6}} = \sqrt{\frac{4,816 - 4,704}{6}} = \sqrt{18.67} = 4.32$$

Standard deviation is 4.32 F°.

7-4 DATA PLOT

A. DOT PLOT

◆Example 1:

Number of marbles	0	1	2	3	4	5	6	7
Number of students	1	3	4	8	6	5	3	2

The table above shows the number of marbles that Ken's schoolmates have.

We may use DOT PLOT to show the data distribution as follows.

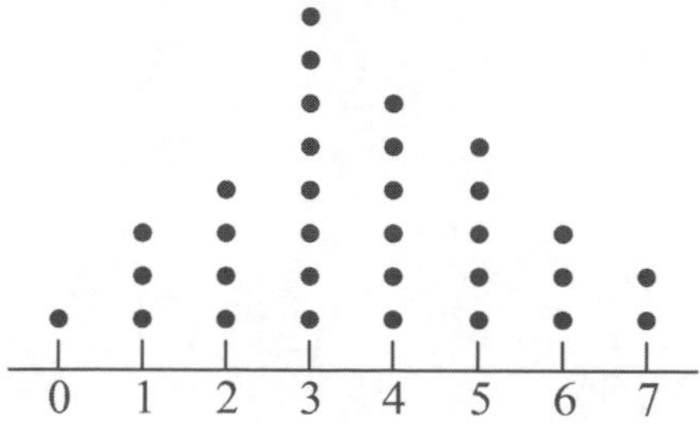

$$\text{Mean} = \frac{1\times0+3\times1+4\times2+8\times3+6\times4+5\times5+3\times6+2\times7}{1+3+4+8+6+5+3+2} = \frac{116}{32} = 3.63$$

Mode = 3

Median:

1. Since there's an even number of data, the median value is the average of the two values in the middle.
2. The middle position $= \frac{32}{2} = 16$

$$\text{Median} = \frac{\text{the value of 16th term} + \text{the value of 17th term}}{2} = \frac{3+4}{2} = 3.5$$

◆Example 2:

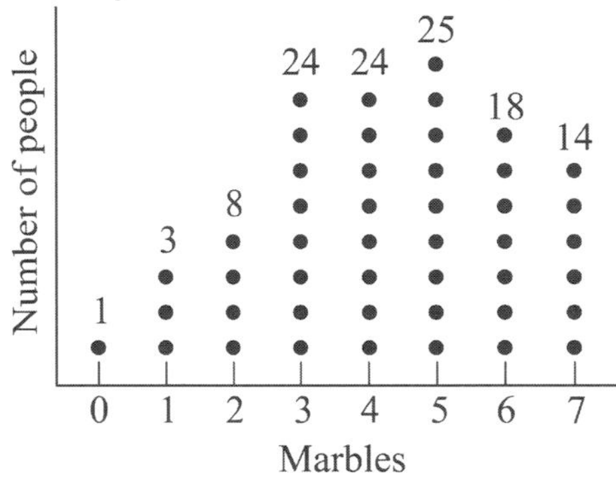

Total number of people $= 1+3+8+24+24+25+18+14 = 117$

Total number of marble $= 1\times0+3\times1+8\times2+24\times3+24\times4+25\times5+18\times6+14\times7 = 518$

Mean $= \frac{518}{117} = 4.43$

The position of median $= \frac{117+1}{2} = 59$ (or, $117\times0.5 = 58.5$, upgrade to 59)

Median $\Rightarrow$ The value of the position of 59th term $\Rightarrow$ it is 4

Range $= 7-0 = 7$

B. Stem-Leaf-Plot

◆Example:

If a data list is:

11, 42, 37, 24, 82, 22, 10, 13, 48, 54, 78, 63, 41, 53, 47, 25, 35

Use stem-leaf-plot shows its distribution

STEM-LEAF-PLOT

8	2
7	8
6	3
5	3, 4
4	1, 2, 7, 8
3	5, 7
2	2, 4, 5
1	0, 1, 3

(Stem is 10's digit, leaf is one's digit)

Answer to Concept Check 7-05

1. $60\times3+61\times2+62\times3+63\times6+64\times4+65\times3+66\times5+67\times6+68\times3+69\times8+70\times4$ $+71\times3+72\times2=3,442$

   mean$(\bar{x})=\dfrac{3442}{52}=66.19$ inches

2. $52\times0.5=26$

   Median $\Rightarrow$ the average of the values of the 26th and 27th term.

   Median $=\dfrac{66+67}{2}=66.5$

3. Mode: 69

   The frequence of 69 is 8 times.

   So, Mode is '69'

4. Range $=72-60=12$ (inches)

Answer to Concept Check 7-06

A sorted data list shown below, draw the stem-leaf-plot of this list, and find the mean, median, mode, and range.

11, 12, 12, 14, 15, 15, 15, 20, 22, 33, 35, 36, 42, 48, 53, 55, 62, 63, 65, 66, 72, 74, 77

Stem	Leaf
7	2, 4, 7
6	2, 3, 5, 6
5	3, 5
4	2, 8
3	3, 5, 6
2	0, 2
1	1, 2, 2, 4, 5, 5, 5

the position of the median:

$$\frac{23+1}{2}=12 \text{ (position)}$$

The value of data at 12th is 36.

Mode: 15

Range: $77-11=66$

$$\text{Mean}=\frac{94+42+104+90+108+256+223}{23}=39.87$$

C. Box-And-Whisker Plot

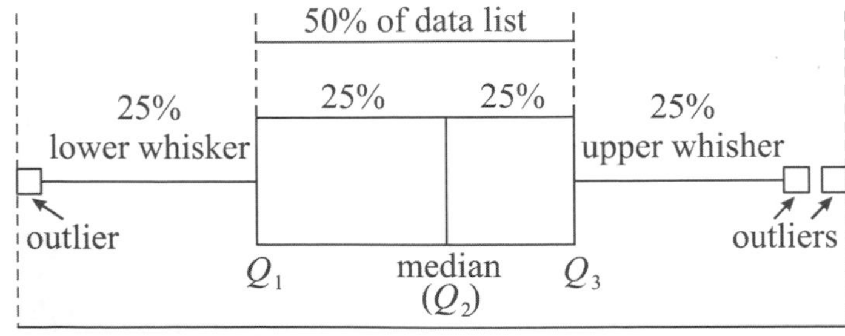

- Q_1 (first quartile): 25 percent of the value of the data in a data set are at or below it.
- Q_2 (second quartile) $\Rightarrow$ median
- Q_3 (third quartile): 25 percent of the value of the data in a data set are at or above it.
- Interquartile range (IQR) $= Q_3 - Q_1$

- Lower outlier: any value $< Q_1 - 1.5 \times \text{IQR}$
- Upper outlier: any value $> Q_3 + 1.5 \times \text{IQR}$

★Tip

How to calculate Q_1, Q_2 and Q_3, there is no universal agreement. The following description is one of the methods. The position of Q_1 is equals to $P_1 = N \times 0.25$ (N is the number of data), if P_1 is a decimal, then the position of Q_1 is P_1 upgrade to whole number, For instance, $18 \times 0.25 = 4.5$, then value of Q_1 is equal to the value of the 5th in the data list.

If $20 \times 0.25 = 5$ (whole number), the value of Q_1 is equal to (value of 5th + value of 6th) ÷ 2 the position of Q_3 is equal to $P_3 = N \times 0.75$ if P_3 is whole number, then the value of Q_3 is equal to $[\text{value of } P_3 + \text{value of } (P_3 + 1)] \div 2$

For instance, $8 \times 0.75 = 6$, the

$$Q_3 = \frac{\text{value of the 6th data} + \text{value of the 7th data}}{2}$$

◆Example:

The owner of BB Company tried to offer an annual bonus to her employees. The bonuses are listed below.

0, 0, 2, 2, 3, 4, 5, 5, 8, 10, 12, 12, 12, 13, 15, 18, 20, 30, unit: $1,000

1. Q_2 (median) $= \frac{18}{2} = 9\,(\text{position})$

   The number of data is even, the position of the median is $\frac{9\text{th} + 10\text{th}}{2}$

   $Q_2\,(\text{median}) \Rightarrow \frac{8+10}{2} = 9\,(\text{value})$

2. $N \times 0.25 = 18 \times 0.25 = 4.5 \Rightarrow$ the 5th position

   so, $Q_1 = 3$

3. $N \times 0.75 = 18 \times 0.75 = 13.5 \Rightarrow$ 14th position

   $Q_3 = 13$

4. Range $= 30 - 0 = 30$
5. $\text{IQR} = Q_3 - Q_1 = 13 - 3 = 10$
6. $\text{IQR} \times 1.5 = 15$, $30 > 15 + 13$

   upper outlier 30.

7. lover outlier: no lower outlier $3 - 1.5 \times 10 = -12 < 0$

The Box-And-Whisker plot

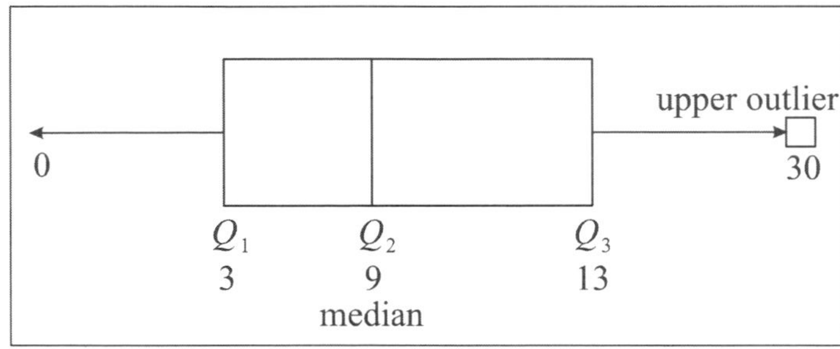

unit: $1,000

Answer to Concept Check 7-07

mean $= (30+46+52+58+60+64+66+70+72+75+75+78+80+80+83+84+85$

$+85+85+85+85+88+88+90+92+95+100)/27 = \dfrac{2,051}{27} = 75.96$

Mode: 85

Range: $100-30=70$

Median (Q_2): $\dfrac{27+1}{2}=14$ at 14th position

Median $=80$

Q_1 (first quartile) The position of Q_1 at $27\times0.25=6.75 \Rightarrow 7$

The value of Q_1 is 66

Q_3 (third quartile) The position of Q_3 at $27\times0.75=20.25 \Rightarrow$ 21st $Q_3=85$

Box-and-whiskers plot

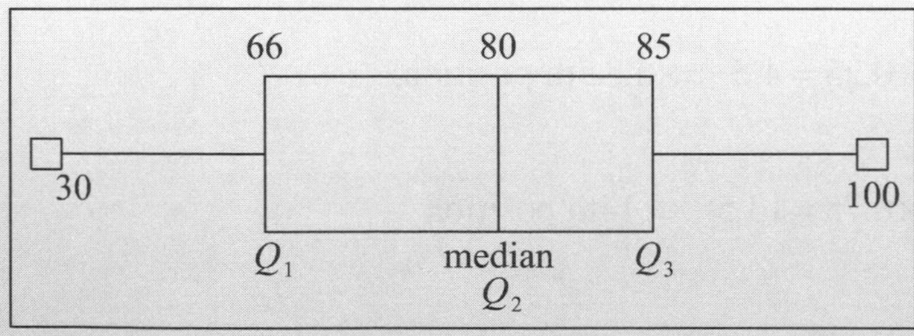

Interquartile range $=Q_3-Q_1=85-66=19$

Lower outliers: (1) $66-30=33$ $33>19\times1.5$ 30 is a lower outlier

(2) $66-46=20<19\times1.5$ not outlier

upper outliers: $100-85=15<28.5$ not outlier

so, there is only one lower outlier:30

D. Scatter-Plot

In an *xy*-plane, the position of a dot is used to represent the relation between two variables, and the graph formed by these dots is called scatter-plot.

◆Example 1:

The following data pairs (x, y), x indicates professional training hours y indicates the sales performance: (1, 2), (2, 2), (2, 4), (2, 5), (3, 3), (3, 4), (3, 5), (4, 3), (4, 6), (5, 4), (5, 6), (5, 8), (6, 5), (6, 7), (6, 8), (7, 9), (7, 10), (8, 7), (8, 8), (8, 10), (9, 12), (10, 10), (10, 11)

For instance (5, 6), represents a sales person who has taken 5 professional training hours and earns $60,000 sales revenue.

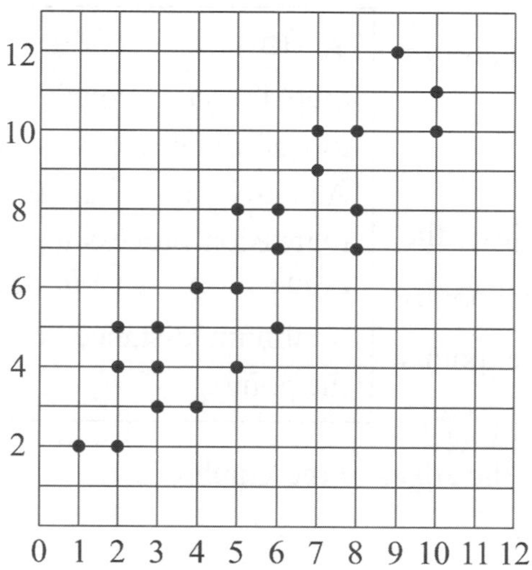

- **Line of best fit**

On a scatter-plot graph, you may draw a straight line which can represent all the characteristics of all the points in the graph. The line of best fit does not necessarily passes through the points on the graph.

1. How to draw a line of best fit

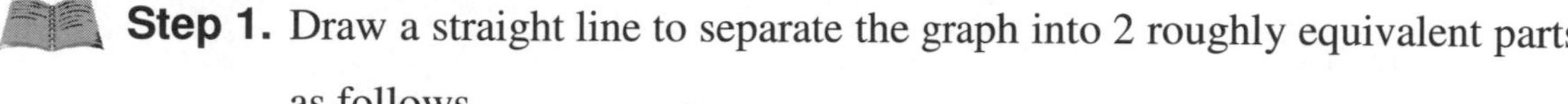

Step 1. Draw a straight line to separate the graph into 2 roughly equivalent parts as follows.

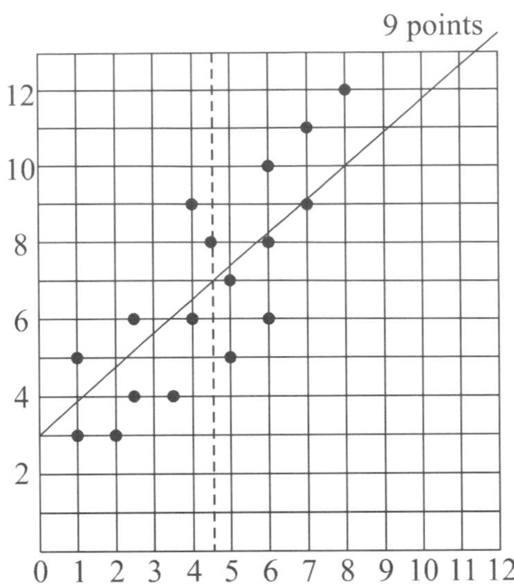

There are 19 points so separate them into 2 parts with each of them having 9 or 10 points.

 **Step 2.** Mark a 'x' as the center (most likely) of each part.

 **Step 3.** Connect 2 'x'.

 **Step 4.** Use a transparent ruler to adjust the corrective line, to make that the points above line are equivalent to the points under the line.

★Tip

Step 4 is just a rough way to find the line of best fit. Actually, it is a linear regression process to find a line of $y = ax + b$ with minimum distance to all the points.

 **Step 5.** Draw a straight line all the way to the edges of the graph.

2. There are two major purposes of the line of best fit.

(1) Rate of change.

In this example, the slope of the line is $\frac{10-3}{8-0} = 0.875$, it is also a unit rate, it means when x ircreases one unit and y incrases 0.875 unit accordingly.

(2) Estimation

The function of the line of best fit is $y = 0.875x + 3$ (It is also called linear regression equation)

so, when $x = 10$, then y should be $(0.875)(10) + 3 = 11.75$

When $x = 12$, then Y could be $(0.875)(12) + 3 = 13.5$

Answer to Concept Check 7-08

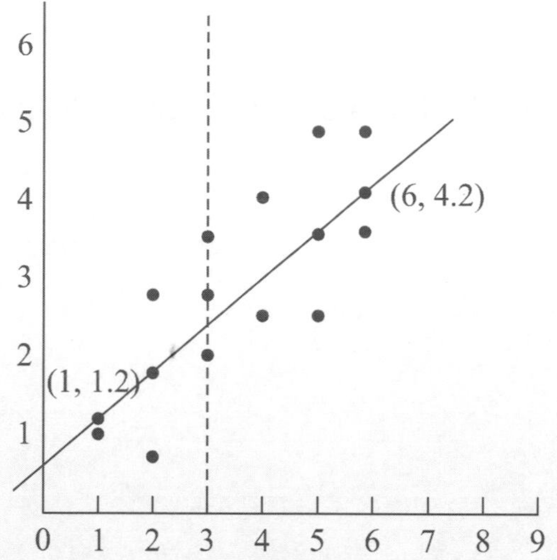

Draw the line

1.(1) Cut 16 points into half

(2) Find the central point of each part then mark an 'x'.

(3) Adjust half of the points above the line and the other half under the line

(4) Draw a line crossing the two points of marked 'x'.

2.Plot two point (6, 4.2) and (1, 1.2)

slope $= \dfrac{4.2-1.2}{6-1} = 0.6$

Let $y = mx + b$

when $4.2 = 0.6 \times (6) + b$

$b = 4.2 - 3.6 = 0.6$

Function: $y = 0.6x + 0.6$

3. No professional training $x = 0$

$y = 0.6x + 0.6$, when $x = 0$, $y = 0.6$

means the player will gain 0.6 point.

4.If he has 7 hours of training.

$y = 0.6 \times 7 + 0.6 = 4.8$

So, he(she) will get 4.8 points.

7-5 DATA SHAPE

A. Left Skewed Distribution

If the number of baseball cards held by students of Jeff's class are represented by the dot plot below.

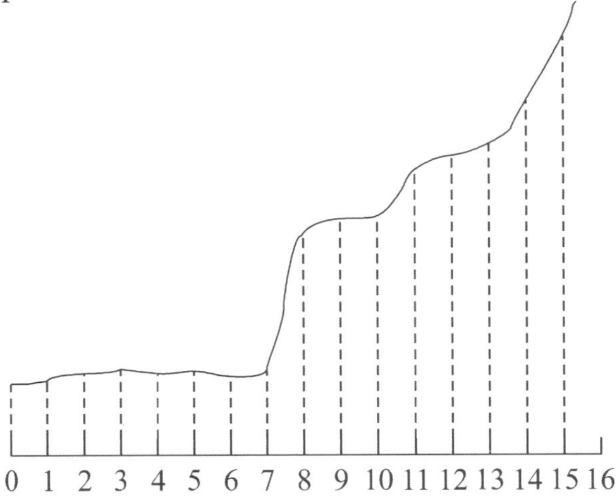

tail on the left, the shape of data distribution is left skewed.

B. Right Skewed Distribution

The annual salaries of Daniel's company is described as the following bar graph.

It is a right skewed distribution

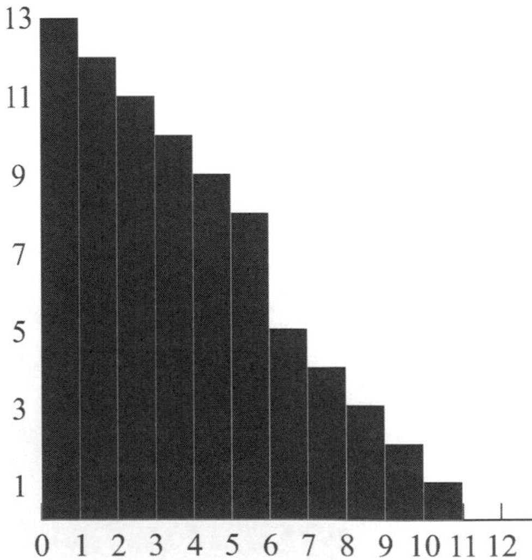

C. Symmetrical Distribution

◆Example:

Scores are 6, 8, 12, 12, 16, 18, 22, 23, 24, 24, 26, 26, 32, 32, 34, 35, 35, 36, 38, 39, 40, 40, 41, 41, 42, 43, 44, 44, 46, 48, 52, 52, 53, 53, 55, 56, 58, 58, 60, 60, 62, 63, 65, 68, 69, 69, 72, 72, 74, 76, 78, 79, 82, 82, 86, 88, 92, 98

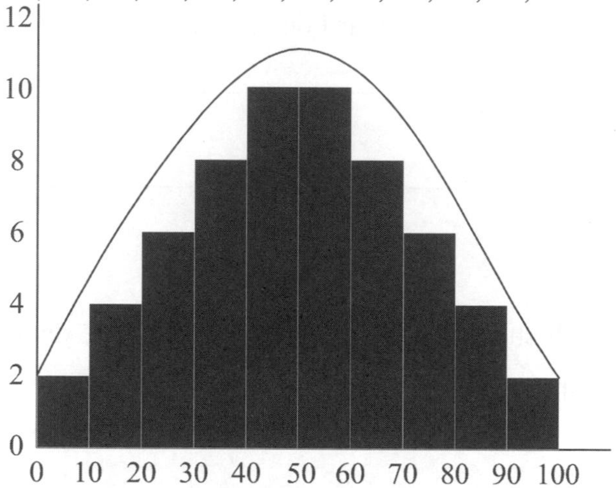

The figure above as a histogram to show the scores of a math exam. The shape of the distribution is symmetrical.

- **Z-score**

In our daily life, many cases of data distribution are represented as a bell-shaped distribution (normal distribution). It is symmetrical structure.

As the figure below,

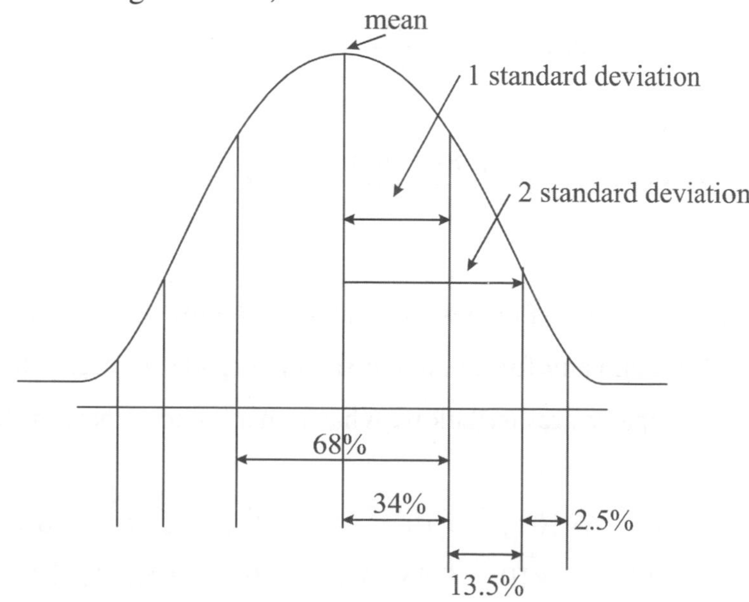

- According to the empirical rule:

  1.Approximately, 68% of the data falls within 1 standard deviation.

  2.Approximately, 95% of the data falls within 2 standard deviation.

  3.Approxinately, 99.75 of the data falls within 3 standard deviation.

- Standardized score (Z-score)

  The distance from the mean in terms of the number of standard deviations.

$$\text{Z-score} = \frac{\text{measurement} - \text{mean}}{\text{standard deviation}} = \frac{x-\mu}{\sigma}$$

positive z-score > mean

negative z-score < mean

- **Percentile:** It is the position of a data set. It shows how many percentage of the data are behind or smeller than the value of this position.

◆Example 1:

The weights of Jens' boxing classmates are 100, 160, 132, 145, 138, 133, 154, 150, 142.

mean = 139.3

$S = (100-139.3)^2 + (160-139.3)^2 + (132-139.3)^2 + (145-139.3)^2$

$+(138-139.3)^2 + (133-139.3)^2 + (154-139.3)^2 + (150-139.3)^2 + (142-139.3)^2 = 2438$

$\sigma = \sqrt{\frac{2438}{9}} = 16.46$

If Jens' weight is 145 pounds then his z-score is $\frac{145-139.3}{16.46} = 0.35$

By z-score table $0.35 \Rightarrow 0.6368 \doteq 64\%$

If Ted's weight is 138 pounds then his z-score is $\frac{138-139.3}{16.46} = -0.08$

By z-score table $-0.08 \Rightarrow 0.46812 \doteq 47\%$

For instance: 90 percentile means the 90% of data whose value are below this value.

Noelle got a offer from ABC-Net company for an annual salary of $60,000. If some of the salaries data of ABC-Net Company are list below, what is her salary's percentile in her company?

5×10^4, 5×10^4, 6×10^4, 6×10^4, 6×10^4, 7×10^4, 7.6×10^4, 7.8×10^4, 8×10^4, 8×10^4, 8.5×10^4, 8.5×10^4, 9×10^4, 9.2×10^4, 9.4×10^4, 10×10^4, 10×10^4, 12×10^4,

14×10^4, 16×10^4, 16×10^4, 20×10^4

$$\text{mean}=\frac{\sum_{i}^{22} x_i}{n}=\frac{209}{22}=9.5\times10^4$$

Variance $\Rightarrow$ Sum of square of difference to mean.

$$S^2\,(\text{Variance}) = (5-9.5)^2+(5-9.5)^2+(6-9.5)^2+(6-9.5)^2+(6-9.5)^2+(7-9.5)^2$$
$$+(7.6-9.5)^2+(7.8-9.5)^2+(8-9.5)^2+(8-9.5)^2+(8.5-9.5)^2+(8.5-9.5)^2+(9-9.5)^2$$
$$+(9.2-9.5)^2+(9.4-9.5)^2+(10-9.5)^2+(10-9.5)^2+(12-9.5)^2+(14-9.5)^2$$
$$+(16-9.5)^2+(16-9.5)^2+(20-9.5)^2=318.59$$

$$\text{Standard Deviation} = \sqrt{\frac{318.59}{22-1}}=3.89$$

Noelle's Salary is \$60,000, the average salary is \$95,000 the difference of Noelle's salary from the center of the normal distribution is $6-9.5=-3.5$ $\quad\sigma=3.89$

z-score is $\frac{-3.5}{3.89}=-0.90$, percentile of -0.90 is 0.18406 $\Rightarrow$ 18%

So, Noelle's salary is at the position of 18% that is Noelle's salary beats 18% of her colleagues. There is another way to find the percentile

$$S^2\,(\text{Variance}) = \frac{\sum_{i=1}^{22} x_i^2-\frac{(\sum x_i)^2}{n}}{n-1}=\frac{2304.1-\frac{(209)^2}{22}}{21}=\frac{318.6}{21}=15.17$$

$$S=\sigma=\sqrt{15.17}=3.89$$

$$\text{Z-score} = \frac{(6-9.5)}{3.89}=-0.90$$

Answer to Concept Check 7-09

1. Peter's classmates' heights are listed as follows

   58, 60, 60, 62, 63, 64, 65, 66, 66, 67, 67, 67, 67, 68, 69, 70, 72, 73, 73, 74.

   If Peter's height is 64 inches, what Percentile is his height in his class ?

   $\bar{x}\,(\text{mean}) = \dfrac{1{,}331}{20} = 66.56$

   $$S^2\,(\text{Variance}) = \frac{\sum_{i=1}^{20} x_i^2 - \frac{(\sum x_i)^2}{n}}{n-1} = \frac{88{,}969 - 88{,}578}{19} = 20.58$$

   $$\sum_{i=1}^{20} x_i^2 = 58^2 + 60^2 + 60^2 + 62^2 + 63^2 + 64^2 + 65^2 + 66^2 \times 2 + 67^2 \times 4 + 68^2$$
   $$+69^2 + 70^2 + 72^2 + 73^2 + 73^2 + 74^2 = 88{,}969$$

   $$(\sum_{i=1}^{20} x_i)^2 = (58 + 60\times2 + 62 + 63 + 64 + 65 + 66\times2 + 67\times4 + 68 + 69 + 70 + 72$$
   $$+73\times2 + 74)^2 / 20 = 88{,}578$$

   $\therefore\ \sigma\,(\text{Standard Deviation}) = \sqrt{20.58} = 4.54$

   $\text{Z-score} = \dfrac{64 - 66.56}{4.54} = -0.56$

   By Z-score table

   Z-score: $-0.56 \Rightarrow 0.28774 \Rightarrow 29\%$ (by z-score table)

2. $\bar{x} = 66.56$

   George's z-score $= \dfrac{70 - 66.56}{4.54} = 0.7577$

   Percentile $\Rightarrow 0.77637 \Rightarrow 78\%$ (By Z-score table)

   The George's percentile in Peter's class is $0.77637 \Rightarrow 78\%$

7-6 CORRELATIONSHIP

Using scatterplot graph to show the relationship between two variables. For instance, the relationship between the learning hours and the test scores, or the education level and the salary. There are three key elements.

1. **Shape:**

   linear or non-linear

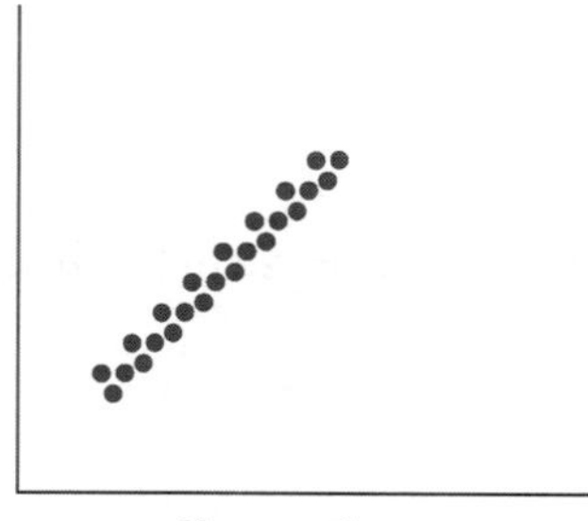

linear shape

non-linear shape

2. **Direction**

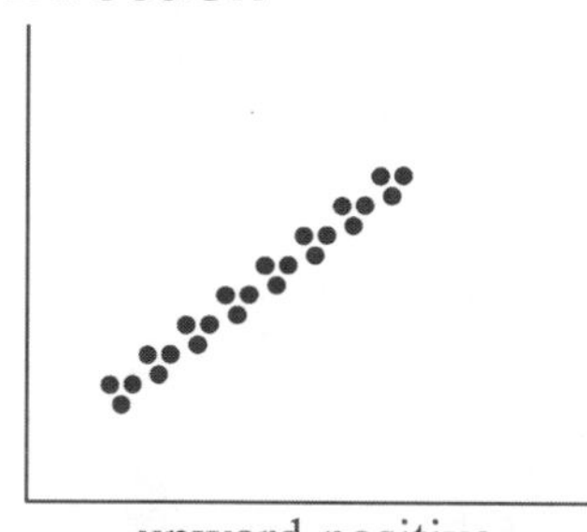

upward-positive

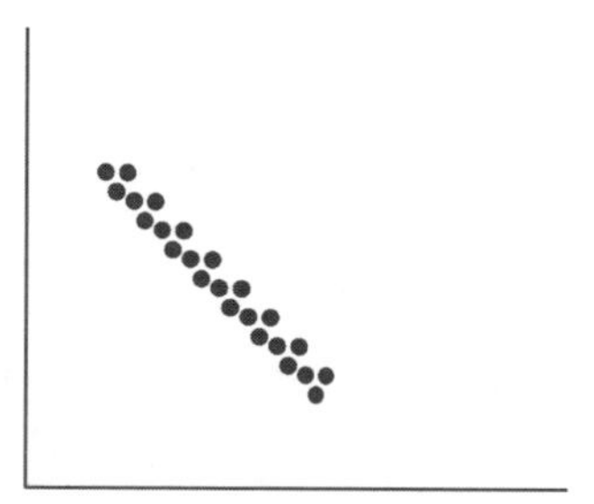

downward-negative

3. **Strengh**

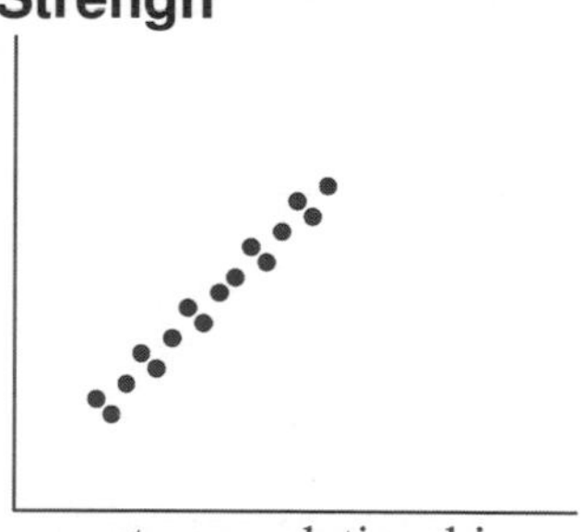

strong relationship

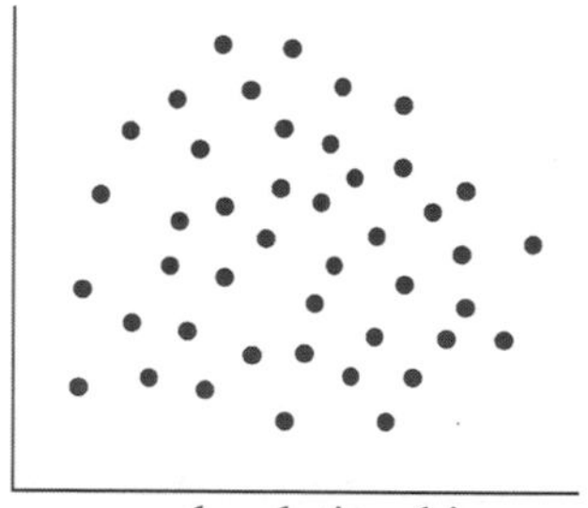

weak relationship

- Correlation Coefficient

  It is an indicator to show how strong the relationship between two variables is.

$$-1 \leq \text{Correlation Coefficient } (r) \leq 1$$

- Positive correlation:

  The direction of the change of the variables are almost the same. That means when one variable becomes larger the other one becomes larger, and vice versa.

- Negative correlation:

  The two variables change in opposite direction.

- Zero correlation

  The change of one variable has nothing to do with that of the other one.

- Complete correlation

  Two variables have very close relationship, almost moving in the same straight line or the same curve.

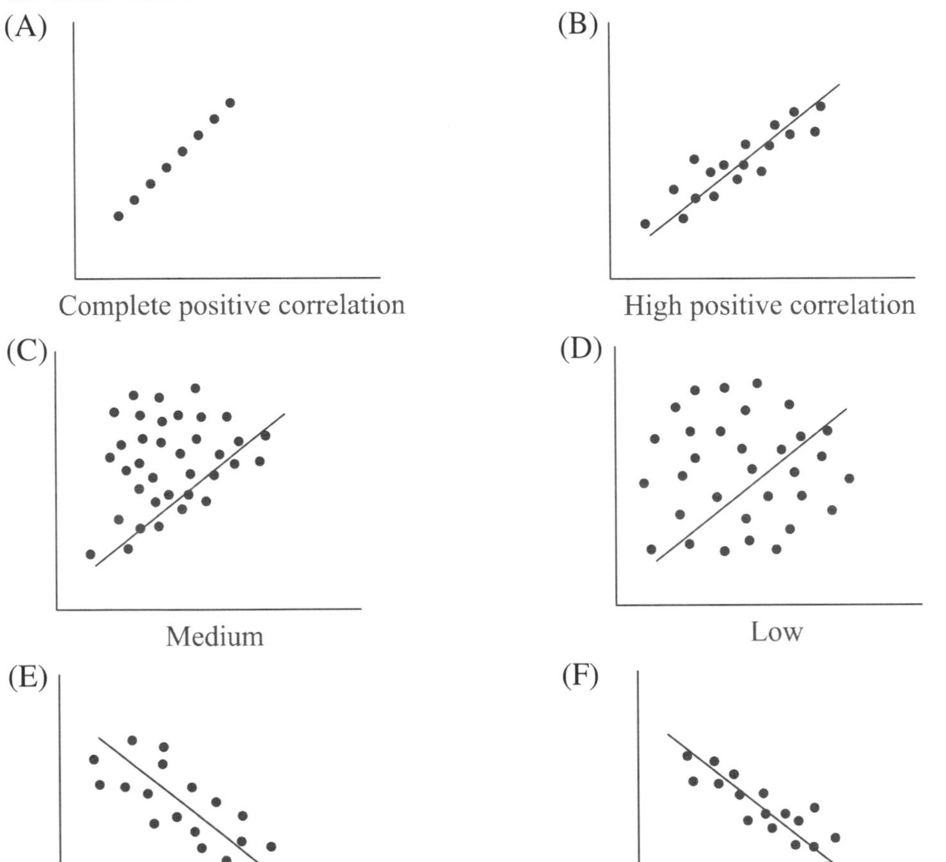

(G)

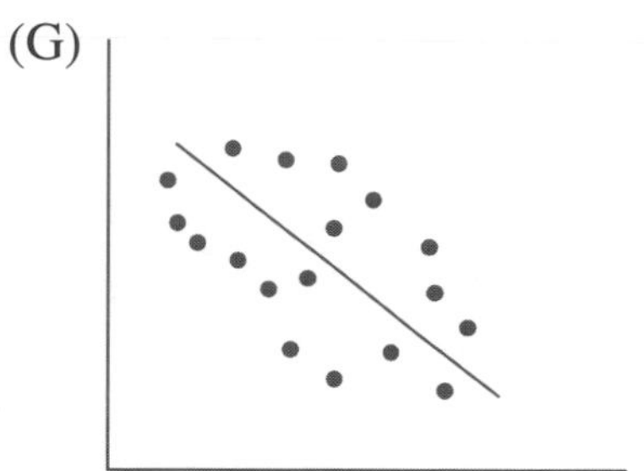

Medium negative correlation

(H)

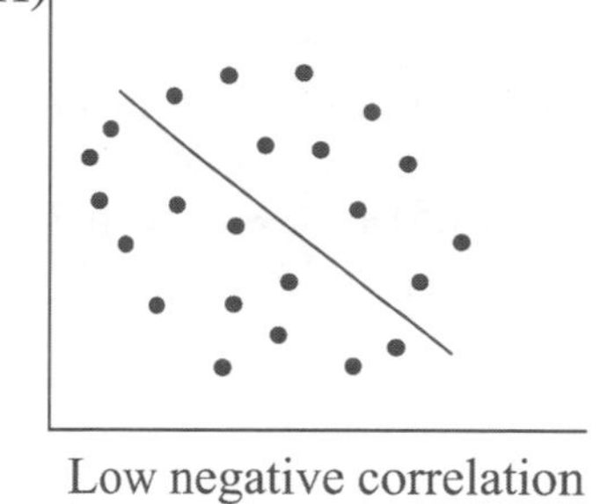

Low negative correlation

(I)

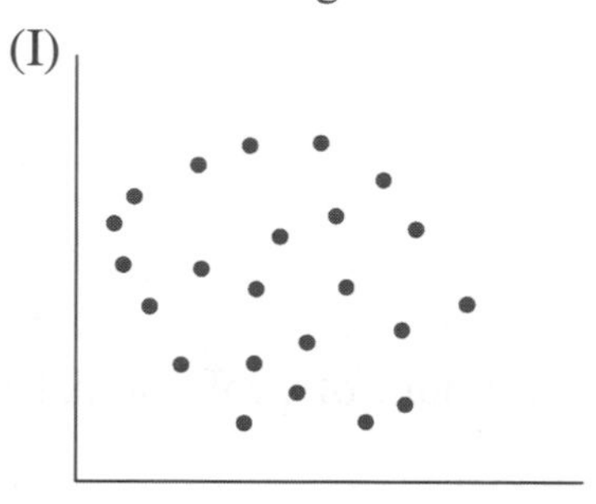

Zero correlation

(J)

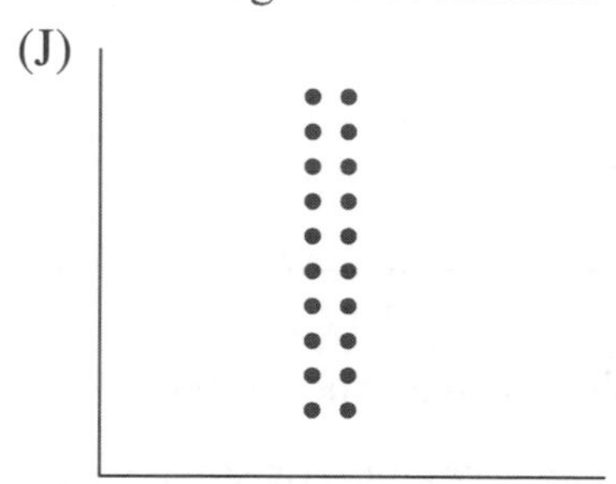

Zero correlation

- $r = 1$ Complete positive relationship
- $r = -1$ Complete negative relationship
- $0.7 \leq |r| < 1$ High relationship
- $0.3 \leq |r| < 0.7$ medium relationship
- $0 \leq |r| < 0.3$ low relationship
- $r = 0$ zero relationship

Formula: $$r = \frac{\sum_{i=1}^{n}(x_i - \bar{x})(y_i - \bar{y})}{\sqrt{\sum_{i=1}^{n}(x_i - \bar{x})^2} \cdot \sqrt{\sum_{i=1}^{n}(y_i - \bar{y})^2}}$$

For easy to memorize $r = \dfrac{S_{xy}}{S_{xx} \cdot S_{yy}}$

$$S_{xy} = \sum_{i=1}^{n}(x_i - \bar{x})(y_i - \bar{y})$$

$$S_{xx} = \sqrt{\sum_{i=1}^{n}(x_i - \bar{x})^2}, \quad S_{yy} = \sqrt{\sum_{i=1}^{n}(y_i - \bar{y})^2}$$

◆Example:

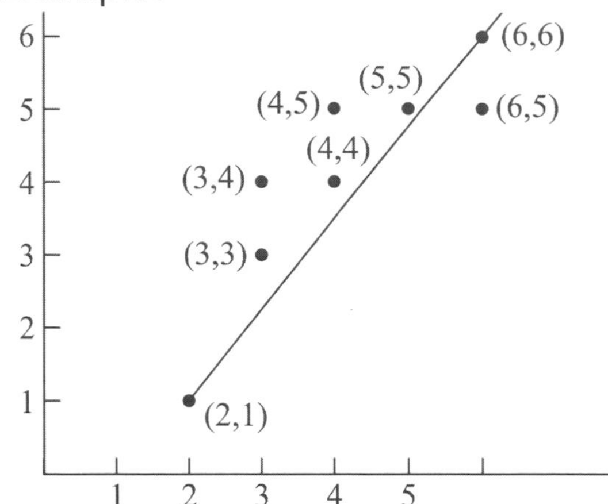

The scatterplot above, shows the relationship between the hours of professional training and the sales performance of some salespeople in the company.

1.Shape: linear

2.Direction: upward (positive)

3.Strength: strong relationship

4.Correlationship coefficients: 0.88

$$\bar{x}(\text{mean}) = \frac{2+3+3+4+4+5+6+6}{8} = 4.125$$

$$\bar{y}(\text{mean}) = \frac{1+3+4+4+5+5+5+6}{8} = 4.125$$

$$\begin{aligned} S_{xy} &= (2-4.125)(1-4.125) + (3-4.125)(3-4.125) + (4-4.125)(3-4.125) \\ &\quad + (4-4.125)(4-4.125) + (4-4.125)(5-4.125) + (5-4.125)(5-4.125) \\ &\quad + (6-4.125)(5-4.125) + (6-4.125)(6-4.125) \\ &= 6.64+1.27+0.14+0.02+(-0.11)+0.77+1.64+3.52 = 13.89 \end{aligned}$$

$$\begin{aligned} S_{xx} &= \sqrt{(2-4.125)^2+(3-4.125)^2+(3-4.125)^2+(4-4.125)^2} \\ &\quad \sqrt{(4-4.125)^2+(5-4.125)^2+(6-4.125)^2+(6-4.125)^2} \\ &= \sqrt{4.52+1.27+1.27+0.02+0.02+3.52+0.77+0.77} = \sqrt{12.16} = 3.49 \end{aligned}$$

$$\begin{aligned} S_{yy} &= \sqrt{(1-4.125)^2+(3-4.125)^2+(4-4.125)^2+(4-4.125)^2} \\ &\quad \sqrt{(5-4.125)^2+(5-4.125)^2+(5-4.125)^2+(6-4.125)^2} \\ &= \sqrt{9.77+1.27+0.02+0.02+0.77+0.77+0.77+3.52} = \sqrt{16.91} = 4.11 \end{aligned}$$

$$r = \frac{S_{xy}}{S_{xx} \cdot S_{yy}} = \frac{13.89}{(3.49)(4.11)} = \frac{13.89}{14.34} = 0.97$$

$0.7 < 0.97 < 1$ High relationship.

Answer to Concept Check 7-10

Shape: linear

Direction: downward

Strength: strong (close to each other)

Correlation: strong

Answer is (B)

7-7 DATA SAMPLING

In the process of data collection, it is often impossible to collect thousands and thousands of data within a limited time and manpower. For instance, censuses in various countries are done once every ten or twenty years, or even not at all. How can we obtain useful and representative data to formulate workable policies? One would need the techniques of sampling methodologies to estimate the over all characteristics and behaviors.

- Group Sampling

Divide a large group into many small groups according to the pre-designed ratio, and then draw the samples from each small group to estimate the characteristics of the overall group.

◆Example:

The principal of Mary's high school wants to know the average math scores of her students. The students can be divided into 9th, 10th, 11th, and 12th grades. Randomly select, equal number of students and calculate their average math score, multiplied by the ratio of each grade and sum it up to get the school's average score in mathematics

9th grades	30%	86
10th grades	25%	84
11th grades	23%	80
12th grades	22%	78

Average score $= 86\times0.3+84\times0.25+80\times0.23+78\times0.22=82.36$

- Cluster Sampling

Divide the whole into many clusters, and each cluster can have the same characteristics as the whole. Randomly select one or several clusters for the census and infer the results of the sample to the whole.

For example, if you want to know the average math score of the students in grade 10, you only need to randomly select a class of grade 10 to infer their average math score to the whole grade 10 ?

- Systematical Sampling

There are couple of steps for systematical sampling:

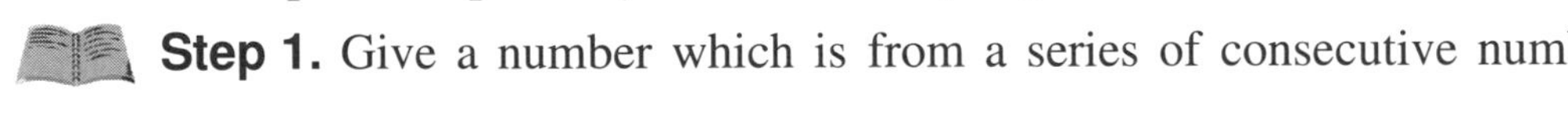

Step 1. Give a number which is from a series of consecutive numbers to each sample.

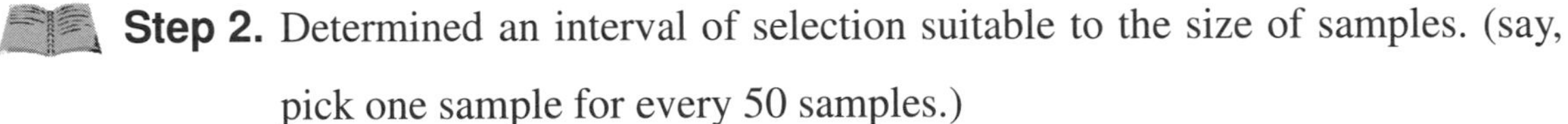

Step 2. Determined an interval of selection suitable to the size of samples. (say, pick one sample for every 50 samples.)

Step 3. Choose a random number to start the sampling.

Step 4. According to the intervals pick up the samples.

◆Example:

Shirley's school will select 10 students from all 148 students in grade 10 to participate in the city's physical fitness test.

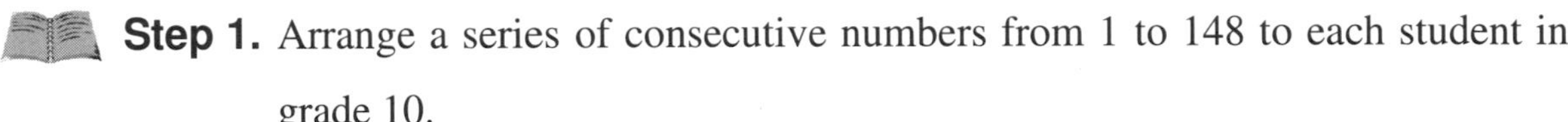

Step 1. Arrange a series of consecutive numbers from 1 to 148 to each student in grade 10.

Step 2. Interval $=\frac{148}{10}=14\,(\text{integer})$

Step 3. Randomly starting number is 5.

Step 4. Pick up 5, $5+14$, $5+14\times2$, $5+14\times3$, $5+14\times4$, $5+14\times5$, $5+14\times6$, $5+14\times7$, $5+14\times8$, $5+14\times9$

so, 5, 19, 33, 47, 61, 75, 89, 103, 117, 131, will be selected

Answer to Concept Check 7-11

The number for each student is listed below 1, 2, 3, [4], 5, 6, 7, 8, 9, 10, 11, 12, 13, 14, 15, 16, 17, 18, 19, 20, 21, 22, 23, 24, 25, 26, 27, 28, 29, 30, 31, 32.

[4]	[14]	[24]	[2]	[12]
	$4+10$	$14+10$	$24+10=34$ $34-32=2$	$2+10$

The number of the 5th student is 12.

No matter which sampling method is used the data obtained must be representative of the characteristics of the entire population, otherwise, the results are meaningless. Randomization is the key to reach the successful data sampling. There are some points should be considered.

- **Right people:**

  The people interviewed must be in the objective group.

- **Right Time:**

  The interview time must be convenient for the people being asked. For instance, it is not a proper time doing telephone survey in working hours, or family visit in dinner time.

- **Right place:**

  The interview should be conducted at the right place. For instance, surveying people's favorite football star at the entrance of a cinema, or surveying people's recent favorite movies at the entrance of the stadium are both inappropriate of them because they are not the right place to do the objective survey.

  Particularly avoid the survey processing in a uncomfortable or scrupulous environment.

- **Right Questions:**

  1. Simple and clear
  2. Avoid using the interview introductory phrases.

  For instance, smoking is harmful to your health. Will you consider to quit smoking ?

Answer to Concept Check 7-12

(A) Not right people

(1)Sampling selection not randomized.

(2) Generally speaking volunteers are interested in learning, their scores can not represent the achievements of whole population. So, the conclusion is failed.

Answer is (A)

7-8 CONFIDENCE LEVEL, CONFIDENCE INTERVAL, and MARGIN OF ERRORS

• Confidence Level:

Confidence Level represents how confident we are that the range of values for a population parameter we obtain from sampling, through statistical calculations and procedures, will include the actual parameter value from the real population. Confidence level is determined prior to surveys are conducted, and is not randomly generated along with sample data. The most commonly used confidence levels are 68%, 90%, 95%, and 99.7%, and among all, 95% is the most common.

It is impractical (if not impossible) to conduct a comprehensive survey for an unknown parameter. For example, if we want to know the proportion of 8-year-old American children who are shorter than 46 inches, it is difficult to do a census survey across the country, and even if we do that, due to lots of variables, the result is hardly representational. Therefore, statisticians create a way of sampling to estimate the result of the actual population.

In statistical analysis, based on Empirical rule, there's a 95% probability that sample data will fall within 2 standard deviations (S.D.). In other words, under the same environment, if a survey is conducted 100 times with different sample groups, 95% of the time we will get the same results. Based on this finding, statisticians can say they have 95% confidence that their sample data result matches the result from the real population parameter. We want to emphasize that Confidence level is bound to sample data but not to real population parameter data. It is a common misconception that Confidence level is the probability that population parameter data fall in the sampling group. It is also a misconception that Confidence level means there's 95% chance that the population parameter data is equal to the sample data result. The reason is that population parameter data are known (although may not be at the present, for example, average height for 8-year-old is 51"), and sample survey results are also known, for example between (46", 50"). So whether the population parameter data fall within the sample result range is a matter of yes or no (0 or 1) and not a probability. In this example, we know for sure that 51" does not fall within the sample result range (46" – 50").

Prior to census survey, nobody knows how many population parameter data will fall in the sampling group. In this example, nobody knows how many 8-year-old children are between 46" to 50" tall.

• Confidence Interval:

Confidence interval is a range of estimates for a parameter. Through sampling, we get our data point, and in order to include the value for the real population, statisticians assign a range for the sampled parameter values, and this range of estimates is called the Confidence Interval (C.I.).

The data we get under confidence level 95% is called 95% confidence internal. The most commonly used confidence intervals are 68%, 90%, 95%, and 99.7%.

The formula for calculation C.I.

$$\text{C.I.} = \hat{p} \pm z\sqrt{\frac{\hat{p}(1-\hat{p})}{n}}$$

$$68\%\ \text{C.I.} = \hat{p} \pm \sqrt{\frac{\hat{p}(1-\hat{p})}{n}}$$

$$95\%\ \text{C.I.} = \hat{p} \pm 2\sqrt{\frac{\hat{p}(1-\hat{p})}{n}}$$

$$99.7\ \text{C.I.} = \hat{p} \pm 3\sqrt{\frac{\hat{p}(1-\hat{p})}{n}}$$

C.I. : Confidence Interval

$\hat{p}$: Sample Proportion

z : z-score

n : Number of Sample

$\frac{\hat{p}(1-\hat{p})}{n}$: Standard Error

Example: A senator candidate wanted to know the result of the election and used a survey service to conduct a sample survey. A total of 892 people answered the survey and out of the 892 people, 512 responded "yes", so the support rate was 0.5740 (512/892). We can show the estimate range in different confidence levels:

$$68\%\ \text{C.I.} = 0.5740 \pm \sqrt{\frac{0.5740(1-0.5740)}{892}} = 0.5740 \pm 0.0166 = [55.74\%, 59.06\%]$$

$$95\%\ \text{C.I.} = 0.5740 \pm 2\sqrt{\frac{0.5740(1-0.5740)}{892}} = 0.5740 \pm 0.0332 = [0.5408, 0.6072]$$

$$99.7\%\ \text{C.I.} = 0.5740 \pm 3\sqrt{\frac{0.5740(1-0.5740)}{892}} = 0.5740 \pm 0.0498 = [0.5242, 0.6238]$$

• Margin of Errors

$$\text{Margin of Error} = \text{z-score} \times \text{standard Error} = z \cdot \sqrt{\frac{\hat{p}(1-\hat{p})}{n}}$$

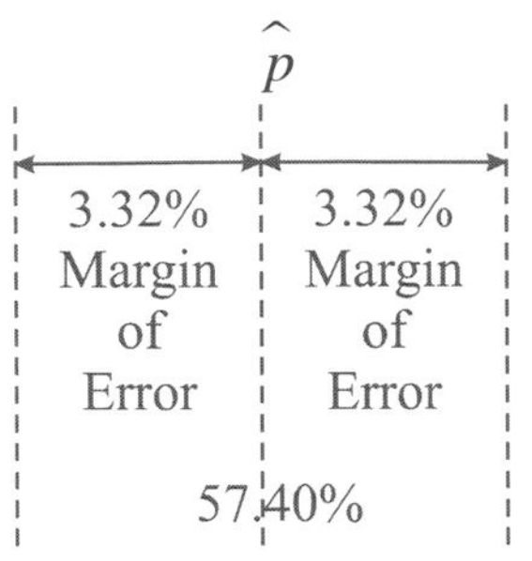

$\hat{p}$

3.32% Margin of Error	3.32% Margin of Error

57.40%

57.40% − 3.32%   57.40% + 3.32%

Margin of Error = ±3.32%

◆Example 1:

A congress candidate commissioned a private survey agency to understand the support rate of k-city voters. If he hoped that the margin of error must be under 2% at 95% confidence level. If the result of confidence interval is [0.68, 0.72], how many voters have they interviewed ?

Sol: Confidence level 95% means within 2 standard deviation

The result of sampling $= \hat{p} = 0.68 + 0.02 = 0.7$

$$e_2 = 2\times\sqrt{\frac{0.7(1-0.7)}{n}}$$

$$0.02 = 2\times\sqrt{\frac{0.21}{n}} \quad 0.01 = \sqrt{\frac{0.21}{n}}$$

$$0.0001 = \frac{0.21}{n} \quad \therefore\ n = \frac{0.21}{0.0001} = 2{,}100$$

interviewed 2,100 people

◆Example 2:

A candidate of congress wants to know how many voters of k-city will support him. He randomly asked 2,800 voters in k-city, and 2,000 of them will vote for him to be a congressman. If at 95% confidence level, what is the confidence interval ?

Sol: $\hat{p} = \frac{2000}{2800} = 0.71$

margin of error of 95% confidence level is $e_2 = 2\sqrt{\frac{\hat{p}(1-\hat{p})}{n}} \Rightarrow 2\sqrt{\frac{0.71(0.29)}{2800}} = 0.017$

$p = \hat{p} \pm 0.017 = 0.71 \pm 0.017$

Confidence interval: [0.693, 0.727]

◆Example 3:

154 students in Mary's school have been surveyed about the location of the field trip, 128 students say 'yes'. At 95% confidence level, what is the confidence interval ?

The result of the sampling $=\dfrac{128}{154} \Rightarrow 0.83$

At 95% confidence level

$$e_2 = 2\sqrt{\frac{0.83(1-0.83)}{154}} = 0.06$$

confidence interval $[0.83-0.06, 0.83+0.06] \Rightarrow [0.77, 0.89]$

Answer to Concept Check 7-13

$$\hat{p} = \frac{800}{1200} = \frac{2}{3} \Rightarrow 0.67$$

95% confidence level, means margin of error in the interval of 2 standard-deviation.

$$e_2 = 2\times\sqrt{\frac{0.67(1-0.67)}{1200}} = 0.027$$

confidence interval $\Rightarrow$ $[0.67-0.027, 0.67+0.027] \Rightarrow [0.64,0.70]$

CHATPER 7
PRACTICE

1. AABB STOCK PRICE TUESDAY

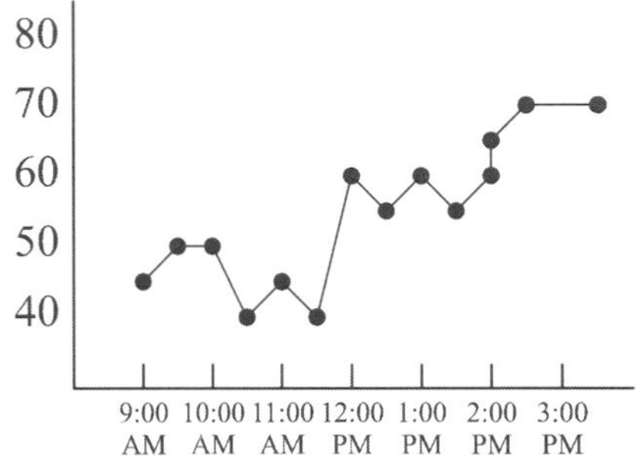

The graph above shows the stock prices at different time period of a certain company on Tuesday. Which period of time fluctuated the most ?

(A) Between 10:00 AM and 11:00 AM
(B) Between 11:00 AM and 12:00 AM
(C) Between 12:00 PM and 1:00 PM
(D) Between 1:00 PM and 2:00 PM

2. 36, 38, 42, 42, 48, 48, 48, 48, 52, 53, 53, 53, 55, 55, 56, 60, 60, 62, 62, 64, 64, 64, 66, 67, 68, 72, 76, 80, 82, 108

The weights of Aden's classmates are listed above.

(1) Find quartile Q_1, Q_2 and Q_3
(2) Find Range
(3) Find interquartile range (IQR)
(4) Find Outlier(s)

3. 170, 163, 164, 159, 156, 174, 144, 166, 142, 150, 146, 170, 174, 172, 168, 148, 168, 153, 158, 170, 172, 152, 164, 159, 165, 177, 172, 165, 165, 174, 178, 172

The height of students in High View High School are shown above.

(1) Show those data on dot plot.
(2) Show those data on stem-leaf-plots

4. A fruit retailer buys a batch of fruits with an average unit price of $2 each, and the standard deviation is $0.6. If each fruit sold at 1.4 times of the purchase price, how much is the standard deviation ?

5. Faith and some of her classmates took a math competition test. There are 40 questions, she got 30 right answers and the number of right answers of her classmates are listed below.

20, 21, 22, 23, 23, 23, 24, 25, 25, 26, 27, 28, 29, 29, 30, 31, 32, 33, 34, 40

If the distribution of the number of the right answers are normal distribution, what is Faith percentile and z-score ?

6. Team A: 2, 3, 4, 4, 4, 5, 7, 8, 8, 9, 9, 12.

Team B: 3, 4, 5, 5, 5, 4, 6, 6, 6, 4, 7, 7.

After a baseball season, the batting scores of the team A, and team B are shown above, which team is more stable ?

7. AVERAGE RAINFALL IN SS-CITY

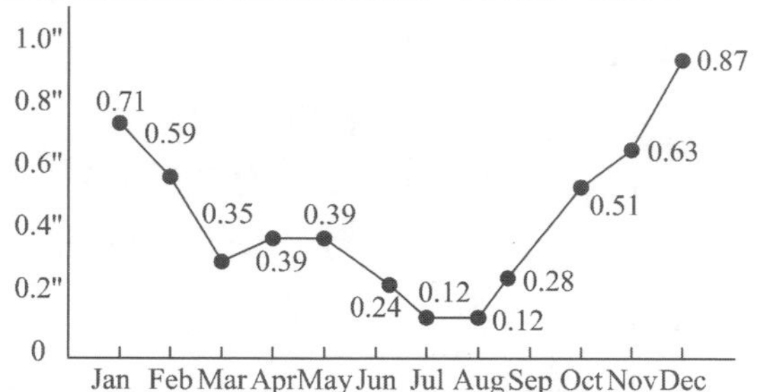

The graph of the average rainfall in SS-City is shown above.

(1) What is the average rainfall ?

(2) What is the range ?

(3) What is the value of the median, and at which month.

(4) What is the values of Q_1, Q_2, and Q_3 ?

(5) What is the IQR ?

(6) Are there any outlier(s) ?

8. The weights of Christopher's schoolmates are listed below 36, 38, 42, 42, 48, 48, 48, 52, 53, 55, 55, 56, 60, 60, 64, 66, 67, 68, 72, 76, 80, 98

(1) Plot the dot-plot and box-and-whisker plot

(2) Find mode

(3) Find range

(4) Find mean

(5) Find median

(6) Find Q_1

(7) Find Q_2

(8) Find Q_3

(9) Find IQR

(10) Find lower outlier(s)

(11) Find upper outlier(s)

9. Samuel swims every day according to the order of first freestyle, then butterfly and breaststroke. He swims faster than breaststroke in freestyle, and swims faster than freestyle in butterfly.

Which of the following graphs could represent the relationship between swimming distance and swimming time ?

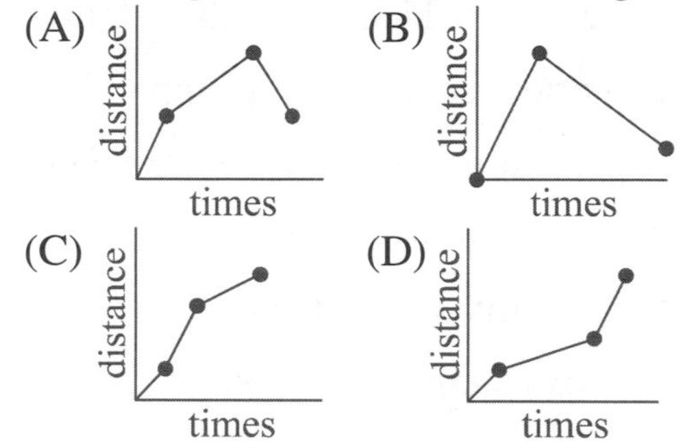

10.

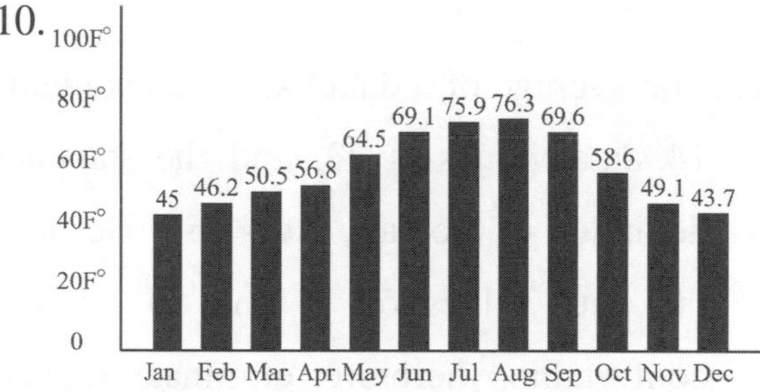

The graph of average high temperature of SS-city are shown above.

(1) What is the average of high temperature in SS-city ?

(2) The median is in which month ?

(3) Range =

(4) $Q_1 =$, $Q_2 =$, $Q_3 =$

(5) IQR =

(6) Outlier(s) =

11. TRANSPORTATION TO SCHOOL

	School bus	walking	Driven by parents	Total
girl	32	51	60	143
boy	41	54	52	147
total	73	105	112	290

A group of 9th-grade students have responded to a survey that asked how they get to school. The survey data were broken down as shown in the table above. Which of the following categories accounts for approximately 14 percent of all the survey respondents ?

(A) Girl walking to school

(B) Girl driven by parents

(C) Boys taking school bus

(D) Boys driven by parents

12. The average of a data list which contains 14 data values is 72, and the standard deviation of the data list is 28. Because the data values of 12 and 14 are not confortable, therefore, eliminate 12 and 14 out from the data list. What is the new standard deviation of the new data list ?

THE RESULT OF THE SURVEY OF CANDIDATES

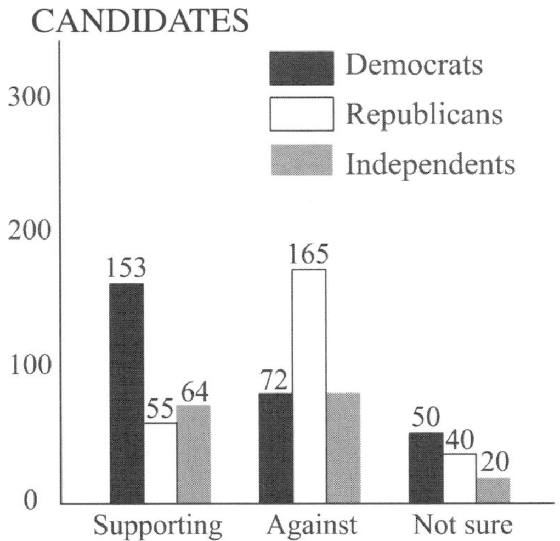

13. A candidate for city councellor wants to know if she will be elected. She surveys 780 registered voters and the result is shown in the graph above. On election day, 86% of the voters who supported her actually voted for her, and 8% of voters who was againsted her then voted for her, and 30% of the voters who are not sure voted for her. If there are 6,200 registered voters in the city, how many votes does she get ? If she is elected, what percentage of voters voted for her ?

14. In a certain county, if the average price of frozen chicken (8.10 ounce) is \$2.10, the standard deviation is \$0.12, and if the average price of one pound of ground beef is \$4.0, the standard deviation is \$0.23.

Amy bought the frozen chicken (8.1 ounce) for $2.29 and one pound of ground beef for $4.29, which one of the following is correct ?

(A) Frozen chicken is relatively expensive comparing to the local price

(B) Ground Beef is relatively expensive comparing to the local price.

(C) They are equally expensive

(D) Not enough information to determine

height	students
60~63	6
64~67	22
68~71	38
72~75	24
76~80	10

15. Otis counts the height of 100 students in his school as shown in the table above, which of the following data is correct?

(A) Average = 69.95
Median = 70
Standard Deviation = 4.26

(B) Average = 70
Median = 73
Standard Deviation = 16.38

(C) Average = 73
Median = 68
Standard deviation = 17.62

(D) Average = 69.95
Median = 68
Standard deviation = 16.38

16. In a state, there are 2,000 high school sophomores who participated in a statewide math competition test. The average score is 53.9 points. The standard deviation is 6 points. How many participants in this test have failed (below 60 points) ?

(A) 1,540

(B) 1,680

(C) 1,720

(D) 1,780

17. A certain survey center conducted a public opinion survey and received a total of 3,200 valid samples, 2,600 samples said "yes". If the margin of error is 1.2% in 95% confidence level, what is the confidence interval of 'yes' ?

(A) [0.69, 0.93]

(B) [0.72, 0.93]

(C) [0.798, 0.822]

(D) [0.72, 0.813]

18. A certain newspaper successfully interviewed 840 people in an opinion survey on whether to build a light-rail in S city. 620 people said "yes", what is the confidence interval at 95% confidence level ?

(A) [0.70, 0.74]

(B) [0.71, 0.74]

(C) [0.70, 0.77]

(D) [0.71, 0.77]

mean	8.5
Standard Deviation	2.3
The first Quartile	7.2
The third Quartile	10.2

19. Brian surveys 400 students in his school about the number of hours using computer per week as shown in the above table.

According to the number in the table above which of the following is incorrect ?

(A) $7.2 \leq$ median ≤ 10.2

(B) About 300 students use computer for more than 10.2 hours

(C) The maximum computer use hours per week is $8.5+2\times2.3=13.1$

(D) About 200 student's computer hours between 7.2 and 10.2 hours

STUDENTS' TRANSPORTATION TIME ANTHORY MIDDLE SCHOOL

Time	Students
0~9	3
10~19	6
20~29	7
30~39	11
40~49	16
50~59	8
60~69	2
70~80	1

20. The table above shows the transportation time between school and home. What is the average transportation time ?

21. A statistician analyzed the annual salary of a company's employees with the years of schooling, and came to the conclusion that "for each additional year of schooling, the annual salary increases by an average of $2,500"

Which of the following information can be directly reached from the above conclusion ?

(A) The "average annual salary" Vs "the average years of schooling".

(B) The slope of the line of best fit of "annual salary" to "years of schooling"

(C) The range between "annual salary" and "years of schooling"

(D) The mode of “annual salary” to the mode of “years of schooling”

22. Two 95% Confidence intervals are of different size, then the larger interval provides which of the following ?

(A) More precise estimate of the population mean

(B) Less precise estimate of the population mean

(C) Larger margin error

(D) Larger standard deviation

23. A public opinion polling center is commissioned by the government to conduct a public opinion survey on one of its policies, and requires that the error be less than 2% at the 95% confidence level, how many samples should be at least ?

CHAPTER 7
ANSWERS and EXPLANATIONS

1. (B)

The stock price from under \$40 goes up to above \$60. It is the most fluctuated period on that day.

2. (1) Q_1 at the position of $0.25\times 30=7.5$

$\doteq 8$

The value $=48$

Q_2 at the position of $0.5\times 30=15$

The value $=\dfrac{56+60}{2}=58$

Q_3 at the position of

$0.75\times 30=22.5=23$

The value $=66$

(2) Range $=108-36=72$

(3) IQR $=66-48=18$

(4) Lower Outlier

$\text{IQR}\times 1.5=18\times 1.5=27$

$48-27=21$

$36>21$

There is no lower outlier

upper outlier

$66+18\times 1.5=93<108$

$(Q_3+\text{IQR}\times 1.5)$

There is a upper outlier $=108$

3. DOT PLOT

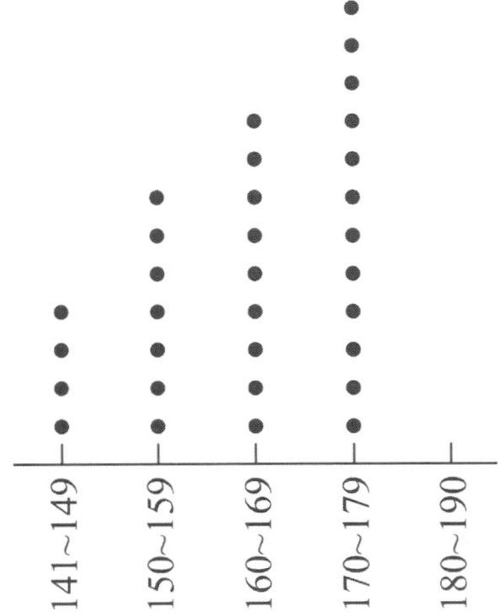

STEM – LEAF – PLOT

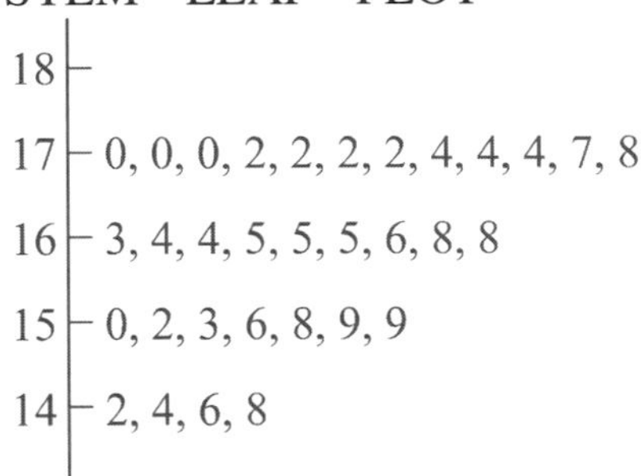

4. Let the average purchase price $=x$

Let the average sell price $=y$

$y=1.4x$,

Since $s^2(x)=\dfrac{\sum x_i^2-\dfrac{(\sum x_i)^2}{n}}{n}$

$s^2(y)=s^2(1.4x)=(1.4)^2 s^2(x)$

$s(y)=(1.4)\cdot s(x)=1.4\times 0.6=0.84$

New Standard Deviation $=\$0.84$

5. $\bar{x} = (20+21+22+23+23+23+24+25$
$+25+26+27+28+29+29+30+31$
$+32+33+34+40)/20$
$= 27.25$

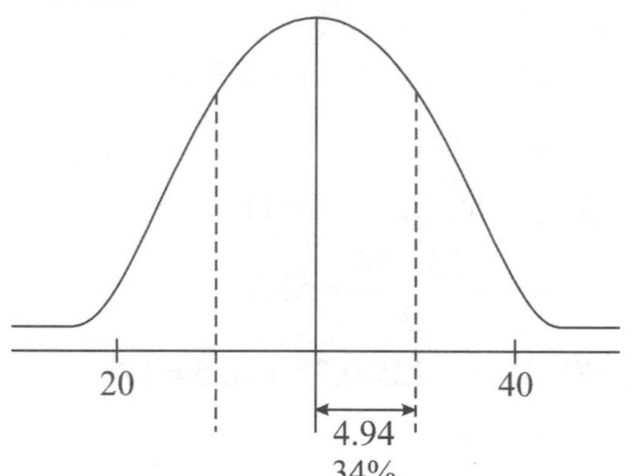

$$S^2 = \frac{\sum_{i=1}^{20} x^2 - \frac{1}{n}(\sum_{i=1}^{20} x)^2}{n}$$

$= \frac{15,339}{20} - 742.56$

$= 24.39$

$\therefore \quad S = \sqrt{24.39} = 4.94$

(one standard deviation)

(1) z-score $= \frac{30-27.25}{4.94} = 0.56$

(2) Percentile $= 71\%$

(z-score table)

6. The standard deviation of team A

$\overline{x_A} = \frac{2+3+4+4+4+5+7+8+8+9+9+12}{12}$

$=6.25$

$S_A^2 = (2-6.25)^2 + (3-6.25)^2 + (4-6.25)^2 \times 3$
$+(5-6.25)^2 + (7-6.25)^2 + (8-6.25)^2 \times 2$
$+(9-6.25)^2 \times 2 + (12-6.25)^2$
$=18.0625+10.5625+15.1875+1.5625$
$+0.5625+6.125+15.125+33.0625$
$=100.25$

S.D. of Team A

$= \sqrt{\frac{S^2}{n}} = \sqrt{\frac{100.25}{12}} = 2.89$

$\overline{x_B} = \frac{3+4+5+5+5+4+6+6+6+4+7+7}{12}$

$=5.17$

$S_B^2 = (3-5.17)^2 + (4-5.17)^2 \times 3$
$+(5-5.17)^2 \times 3 + (6-5.17)^2 \times 3$
$+(7-5.17)^2 \times 2$
$= 4.71+4.11+0.09+2.07+6.70$
$=17.68$

S.D. $= \sqrt{\frac{17.68}{12}} = 1.2137$

Team B is more stable.

7. (1) Average rainfall
$= (0.71+0.59+0.35+0.39+0.39$
$+0.24+0.12+0.12+0.28+0.51+0.63$
$+0.87)/12$
$= 0.43$

(2) Range $= 0.87 - 0.12 = 0.75$

(3) Median is the average of 6th and 7th.

Median $= \frac{0.39+0.39}{2} = 0.39$

Which is on April or May

(4) Q_1, at $12 \times 0.25 = 3$

$Q_1 = \frac{0.24+0.28}{2} = 0.26$

Q_2, at $12 \times 0.5 = 6$

$Q_2 = \frac{0.39+0.39}{2} = 0.39$

Q_3, at $12\times0.75=9$

$$Q_3=\frac{0.59+0.63}{2}=0.61$$

Rain Fall data list

0.12, 0.12, 0.24, 0.28, 0.35, 0.39, 0.39, 0.51, 0.59, 0.63, 0.71, 0.89

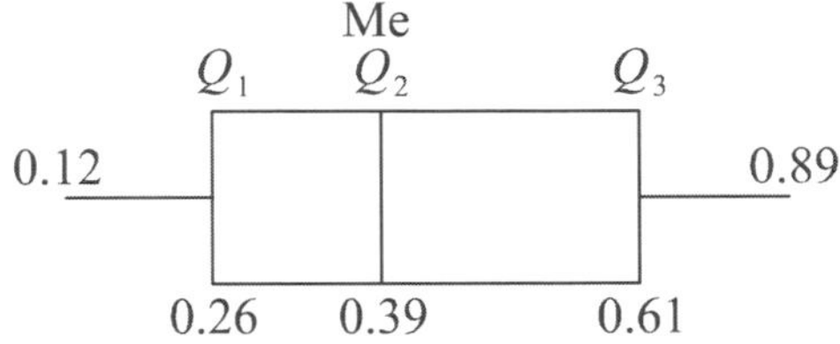

(5) $\text{IQR}=Q_3-Q_1$

$=0.61-0.26=0.35$

(6) Outliers

$0.26-0.35\times1.5=-0.27$

$0.12>-0.27$

There is no lower outlier

$0.61+0.35\times1.5=1.14$

Since $1.14>0.89$

There is no upper outlier

8. (1) A. DOT PLOT

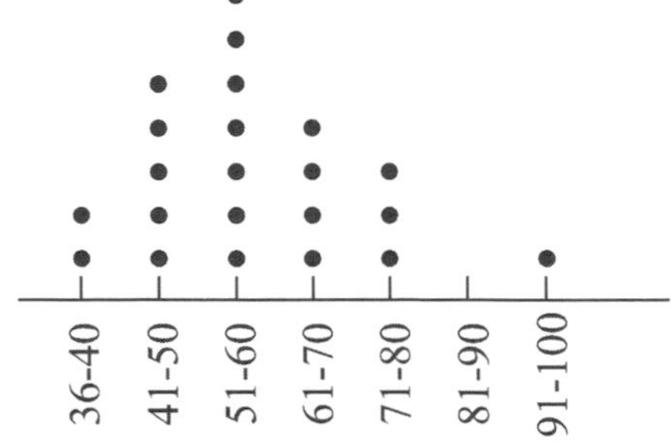

B. BOX-AND-WHISKER PLOT

Me

Q_1 Q_2 $\overline{x}$ Q_3

48 55.5 58.36 67

(2) Mode: 48

(3) Range: $98-36=62$

(4) Mean $=\frac{1284}{22}=58.4$

(5) Median $=55.5$

(6) Q_1 at $22\times0.25=5.5=6$

$Q_1=48$

(7) Q_2 at $22\times0.5=11$

$$Q_2=\frac{55+56}{2}=55.5$$

(8) Q_3 at $22\times0.75=16.5\doteq17$

$Q_3=67$

(9) $\text{IQR}=67-48=19$

(10) Lower outlier

$48-1.5\times19=19.5$

$36>19.5$ no lower outlier

(11) upper outlier

$67+1.5\times19=95.5$

$98>95.5$ one upper outlier

9. The speed of freestyle is slower than butterfly and faster than breaststroke. The graph c fits this nature.

Answer is (C).

10. The list of the data

43.7, 45, 46.2, 49.1, 50.5, 56.8, 58.6, 64.5, 69.1, 69.6, 75.9, 76.3

(1) Mean $=\frac{705.3}{12}=58.78$

(2) Median at $12\times0.5=6$

$$\text{Median value}=\frac{56.8+58.6}{2}=57.7$$

(3) Range $=76.3-43.7=32.6$

(4) Q_1 at $12\times0.25=3$

$$Q_1=\frac{46.2+49.1}{2}=47.65$$

Q_2 at Median $=57.7$

Q_3 at $12\times0.75=9$

$$Q_3=\frac{69.1+69.6}{2}=69.35$$

(5) IQR $=Q_3-Q_1=69.35-47.65$

$=21.7$

(6) Outlier(s)

Since $Q_1-1.5\times\text{IQR}$

$=47.65-21.7\times1.5=15.1$

$43.7>15.1$ no lower outlier

Since $Q_3+1.5\times\text{IQR}$

$=69.35+1.5\times21.7=101.9$

$101.9>76.3$ no upper outlier

11. Total number of survey are 290

so, 14% means $290\times0.14=40.6$

Which matches the number of boys who are taking the school bus

Answer is (C)

12. Since $\sigma=\sqrt{\frac{\sum_{i=1}^{n}x_i^2-\frac{1}{n}(n\bar{x})^2}{n}}=\sqrt{\frac{\sum_{i=1}^{n}x_i^2}{n}-\bar{x}^2}$

$$28=\sqrt{\frac{\sum_{i=1}^{12}x_i^2+12^2+14^2}{14}-72^2}$$

$$784=\frac{\sum_{i=1}^{12}x_i^2}{14}+\frac{12^2+14^2}{14}-72^2$$

$$\therefore\ \sum_{i=1}^{12}x_i^2=(784+72^2)\times14-12^2-14^2$$

$=83212$

Since $\overline{X'}=\frac{14\times72-12-14}{12}=81.83$

Therefore,

$$\sigma'=\sqrt{\frac{\sum_{i=1}^{12}x_i^2}{12}-(81.83)^2}$$

$$=\sqrt{\frac{83,212}{12}-(81.83)^2}=15.43$$

13. Supporting $=153+55+64=272$

Against $=72+165+x=237+x$

Not sure $=50+40+20=110$

$x=780-272-110-237=161$

Against $=237+161=398$

Votes

$=272\times0.86+398\times0.08+110\times0.3$

$=298.76$

$\frac{298.76}{780}=0.383$

$6,200\times0.383=2,375$

She gets 2,375 votes and won with 38.3% of all votes.

14.

	Average price	Bought price	S.D
Chicken	2.10	2.29	0.12
Beef	4.0	4.29	0.23

For chicken,

The difference of chicken

$=2.29-2.10=0.19$

The difference of standard deviation

$=0.19-0.12=0.07$

For Beef

The difference of beef price

$= 4.29 - 4.0 = 0.29$

The difference of standard deviation

$= 0.29 - 0.23 = 0.06$

$0.07 > 0.06$

Chicken is more expensive

Answer is (A)

15. $6 \times \dfrac{60+63}{2} = 369$

$22 \times \dfrac{64+67}{2} = 1,441$

$38 \times \dfrac{68+71}{2} = 2,641$

$24 \times \dfrac{72+75}{2} = 1,764$

$10 \times \dfrac{76+80}{2} = 780$

(1) $A = \dfrac{369 + 1,441 + 2,641 + 1,764 + 780}{100}$

$= \dfrac{6995}{100} = 69.95$

(2) Median should be at the position of the middle, which is 50% $\Rightarrow$ 50th students.

Median should be in the 68~71 interval

(3) Standard deviation

$$S^2 = [(69.95-61.5)^2 \times 6 + (69.95-65.5)^2 \times 22 + (69.95-69.5)^2 \times 38 + (69.95-73.5)^2 \times 24 + (69.95-78)^2 \times 10] \div 100 = 18.15$$

$S = 4.26$

Answer is (A)

16.

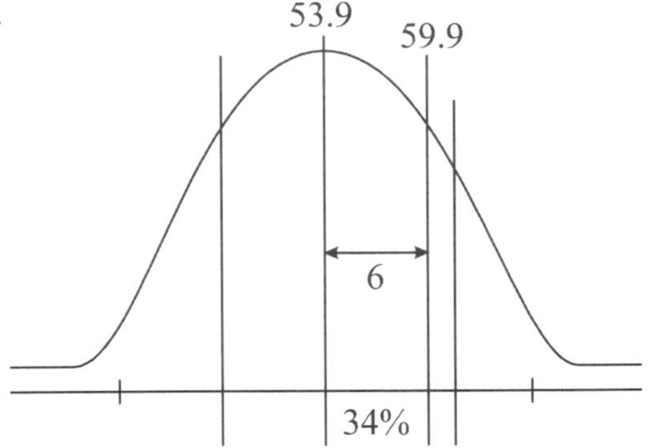

Average = 53.9

1 Standard deviation = 6

The percentage below 60 points is

$50\% + 34\% = 84\%$

$2,000 \times 0.84 = 1,680$

Answer: 1,680 students (B)

17. 'yes' = 2,600 people total number of valid sample = 3,200

The percentage of 'yes' $= \dfrac{2600}{3200} = 0.81$

Confidence interval

$= [0.81 - 0.012, 0.81 + 0.012]$

$= [0.798, 0.822]$

Answer is (C)

18. Since the probability of 'Yes' is $\hat{p} = \dfrac{620}{840} = 0.74$

In confidence level 95% the margin error is in 2 standard deviation

$$e_2 = 2\sqrt{\frac{\hat{p}(1-\hat{p})}{n}} = 2\sqrt{\frac{0.74(1-0.74)}{840}}$$

$= 0.03$

Confidence interval

$= [0.74 - 0.03, 0.74 + 0.03] = [0.71, 0.77]$

Answer is (D)

19.

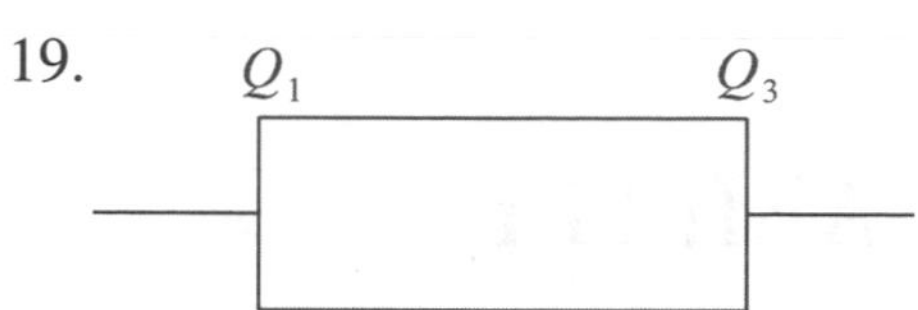

(A) $7.2 \leq \text{median} \leq 10.2$

The median has to be somewhere in the interquartile range, (A) is correct.

(B) $400 \times 0.75 = 300$ is correct

(C) Maximum hours

There is no upper outlier information, so, maximum hours $= 13.1$ is incorrect.

(D) The people in IQR is about 50%, $400 \times 0.5 = 200$ correct

Answer is (C).

20. $3 \times \dfrac{0+9}{2} = 13.5$

$6 \times \dfrac{10+19}{2} = 87$

$7 \times \dfrac{20+29}{2} = 171.5$

$11 \times \dfrac{30+39}{2} = 379.5$

$16 \times \dfrac{40+49}{2} = 712$

$8 \times \dfrac{50+59}{2} = 436$

$2 \times \dfrac{60+69}{2} = 129$

$1 \times \dfrac{70+80}{2} = 75$

$A = (13.5 + 87 + 171.5 + 379.5 + 712 + 436 + 129 + 75) / (3 + 6 + 7 + 11 + 16 + 8 + 2 + 1) = \dfrac{2003.5}{54} = 37$

Answer:

The average time is 37 minutes.

21. The conclusion "For each additional year of schooling, the annually salary increases by an average of \$2,500"

That means it is a linear progressing, which can be modeled y(salary) = 2,500 (year) + original annual salary

$y = mx + b$

So, Answer is (B)

22. (C)

The larger interval means the larger margin error

Answer is (C)

23. Let the number of samples be n, the margin of error at 95% confidence level is $\pm 2\sqrt{\dfrac{\hat{p}(1-\hat{p})}{n}}$

So, $2\sqrt{\dfrac{\hat{p}(1-\hat{p})}{n}} \leq 0.02$

$\Rightarrow \sqrt{\dfrac{\hat{p}(1-\hat{p})}{n}} \leq 0.01$

$\Rightarrow n \geq 10000 \times \hat{p}(1-\hat{p}) \Rightarrow n \geq 10000 \times \dfrac{1}{4}$

Since, when $\hat{p} = 0.5$, $\hat{p}(1-\hat{p})$ gives the maximum value $0.5 \times (1-0.5) = 0.25$

$\therefore\ n \geq 2,500$

CHAPTER 8
COUNTING and PROBABILITY

8-1 COUNTING PRINCIPLE

A. Counting by Addition

When counting items or actions in different groups or events, if the items or actions in each group/event are completely independent from those in other groups/events, the total count is simply the sum of all items/actions.

event A	event B	event C
M possible actions	N possible actions	P possible actions

Total counts/number of actions $= M + N + P$

◆Example:

There are three cabinets, cabinet A has 5 sets of different uniforms, cabinet B has 3 sets of different uniforms, and cabinet C has 2 sets of different uniforms. All the uniforms in each cabinet and among all cabinets are different. The total number of unique uniform choices is $5+3+2=10$.

Cabinet A		Cabinet B		Cabinet C
5 uniforms	+	3 uniforms	+	2 uniforms

Total selections of uniforms $=5+3+2=10$

B. Counting by Multiplication

event A	event B	event C
M ways to do it	N ways to do it	P ways to do it

If each content or action in event A contains all the contents or actions in event B, and similarly, each content or action in event B contains all the contents or actions in event C, then total counts/number of actions $= M \times N \times P$

◆Example:

Using the example above, if there are 5 shirts in cabinet A and 3 pants in cabinet B, and 2 hats in cabinet C, then the total number of different outfits we can create from these items are $5\times3\times2=30$

5 shirts × 3 pants × 2 hats

Answer to Concept Check 8-01

Group 1: There are 3 scenarios where the sum of two dice is below 4, (1, 1), (1, 2), (2, 1).

Group 2: There are 6 scenarios where the points on the two dice are equal (1, 1), (2, 2), (3, 3), (4, 4), (5, 5), (6, 6).

Note (1, 1) in group 1 and group 2, so it counts only once.

Therefore, the number of outcomes $3+6-1=8$

(Because these two scenarios are completely independent from each other, counting with addition rule applies.)

Answer to Concept Check 8-02

Group 1: 5 selections

Group 2: 3 selections

Group 3: 4 selections

Total selections $=5\times3\times4=60$

◆Example 1:

At the same time, rolling the two fair dice.

Event A the sum of 2 face-up points is equal to 5.

Event B the sum of 2 face-up points is a prime number.

Event C both of the face-up points are prime numbers.

Event A The sum of the face-up points is 5. There are, (1, 4), (2, 3), two situations. (4, 1) is the same as (1, 4).

Event B The sum is a prime number. There are, (1, 1), (1, 2), (1, 4), (1, 6), (2, 3), (2, 5), (3, 4), ~~(4, 1)~~, ~~(4, 3)~~, ~~(5, 2)~~, ~~(5, 4)~~, (5, 6); 8 situations

event C Both of the points are prime numbers. There are, (2, 2), (2, 3), (2, 5), (3, 3), (3, 5), (5, 5) 6 pairs.

Since event A, B, and C have nothing to do with one another so, total situations of these 3 events are $2+8+6=16$

◆Example 2:

In a state, the first three digits of an automobile license plate are letters and the last three digits are numbers. How may maximum number of plates can the state issue ?

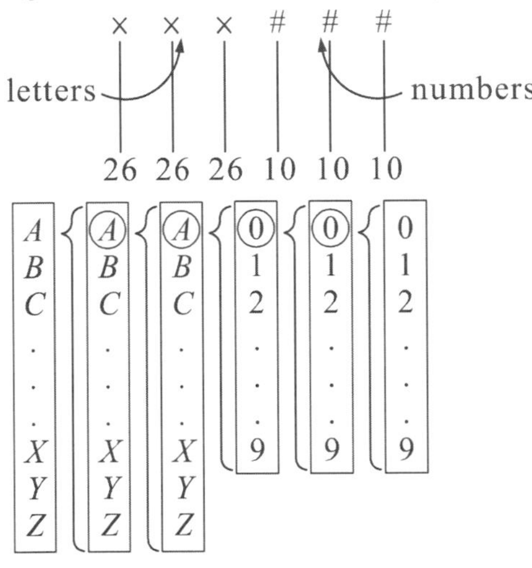

$26 \times 26 \times 26 \times 10 \times 10 \times 10$

Total $=26\times26\times26\times10\times10\times10=17{,}576{,}000$

◆Example 3:

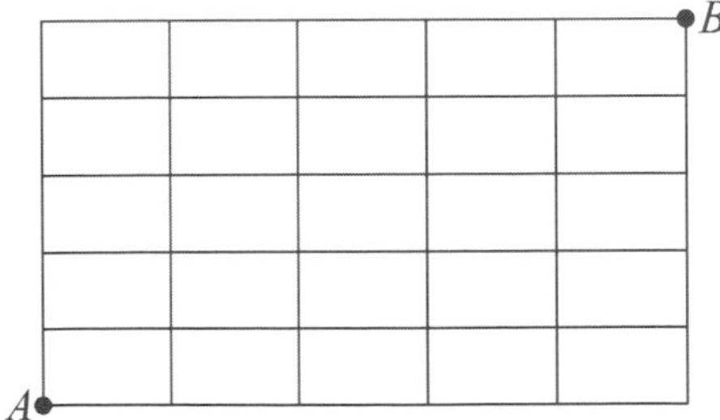

In the grid above, From point A to point B, can only go up or right, how many paths are there to go from A to B ?

Key: The number of paths to get to each intersection point is the sum of all paths that lead to the intersection point. When we get to the last intersection, point B, we get the answer.

	1	6	21	56	126	B
1	1	5 5	15 15	35 35	70 70	126
1	1	4 4	10 10	20 20	35 35	56
1	1	3 3	6 6 10	10 15	15	21
1	1	2 2 3	3 4	4 5	5	6
1	1	1 1 1	1 1	1 1	1	1
A						

Total $=126+126=252$ paths (Addition Rule Applies)

Answer to Concept Check 8-03

5	4	3	2	1	
A	B	C	D	E	F

$A \leftrightarrow B$
$A \leftrightarrow C$ $B \leftrightarrow C$
$A \leftrightarrow D$ $B \leftrightarrow D$ $C \leftrightarrow D$
$A \leftrightarrow E$ $B \leftrightarrow E$ $C \leftrightarrow E$ $D \leftrightarrow E$
$A \leftrightarrow F$ $B \leftrightarrow F$ $C \leftrightarrow F$ $D \leftrightarrow F$ $E \leftrightarrow F$

Each person shaking hands with another person is an independent event. Therefore this problem should be solved by counting with addition.

$5+4+3+2+1=15$ handshakes.

8-2 PROBABILITY (I)

- Probability is a representation of measuring the odds of a specific event.

$$\text{Probability} = \frac{\text{The number of desired outcomes.}}{\text{The number of all possible outcomes.}}$$

For instance, there are three balls in a bag, one red, one blue, and one white. When you pick one ball from the bag, each one has the equal odds to be selected.

$$\text{Probability} = \frac{\text{The number of desired outcomes : 1 red ball}}{\text{The number of all possible outcomes : 3 balls}} = \frac{1}{3}$$

- The probability of event A, denoted as $P(A)$, event B denoted as $P(B)$, ...
- The probability of event A and event B, denoted as $P(A \cap B)$. Read as probability of A and B.

- The probability of event A or event B, denoted as $P(A \cup B)$. Read it as probability of A or B.

◆Example 1:

In a particular state, the automobile license plate is composed of 3 uppercase letters followed by 3 digits. What is the probability of randomly picking one license plate from this state with first 3 letters 'xxx'.

There are 26 uppercase letters that could be used for each letter and 10 digits that can be used for each digit (0~9).

The number of desired outcomes are 'xxx' 000~999 $= 10 \times 10 \times 10 = 1,000$

The number of total possible outcomes $= 26 \times 26 \times 26 \times 10 \times 10 \times 10 = 17,576,000$

$P(\text{plate number are 'xxx'}) = \frac{1000}{17,576,000} = \frac{1}{17,576}$

◆Example 2:

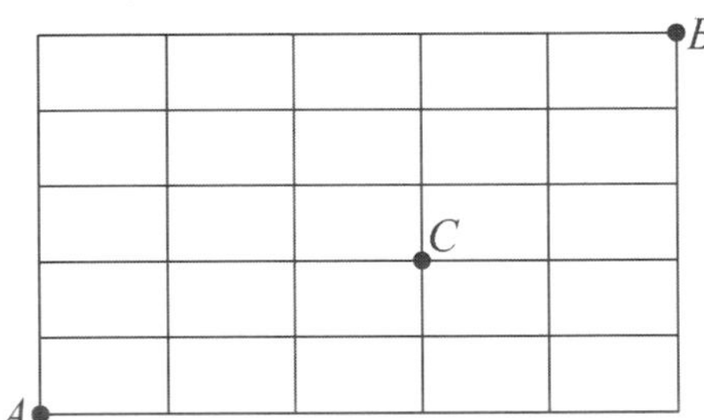

As figure above, from point A to point B, can only go up or right. Point C is blocked for road work. If your path must across point C, it means you fail to reach to point B. What is the probability that you will succeed to arrive at the point B ?

	1	6	21	56	111	B
1	1	5 5	15 15	35 35	55 55	81
1	1	4 4	10 10	20 20	20 20	26
1	1	3 3	6 6	10 10	0 0	6
1	1	2 2	3 3	4 4	5 5	6
1 A	1	1 1	1 1	1 1	1 1	1

Desired outcomes = 192

Total possible outcomes = 252

$\text{P(success from } A \text{ to } B) = \frac{192}{252} = \frac{16}{21}$

◆Example 3:

Roll out a fair dice,

(1) What is the probability of appearing of point 3 ?

(2) What is the probability of appearing of an odd number ?

(3) What is the probability that it shows a number which is greater than 2 and less then 5 ?

(1) There is only one point 3 of a dice.

$$P(3)=\frac{\text{desired outcome}}{\text{total outcomes}}=\frac{1}{6}$$

(2) The odd numbers are 1, 3, 5,

$$P(\text{odd number})=\frac{3}{6}=\frac{1}{2}$$

(3) In $2<N<5$ case, there are 3, 4

$$P(2<N<5)=\frac{2}{6}=\frac{1}{3}$$

◆Example 4:

What is the probability that draw one card from a deck of 52 poker cards is an aces ?

There are 4 cards of 'ACE'.

$$P(\text{ACE})=\frac{4}{52}=\frac{1}{13}$$

Answer to Concept Check 8-04

$$\text{P(white)}=\frac{\text{The number of white balls}}{\text{Total number of balls}}=\frac{4}{3+4+5}=\frac{4}{12}=\frac{1}{3}$$

8-3 PERMUTATION

A. Line Permutation

- P_m^n : m objects are randomly selected from n different objects and line them up in order denoted as P_m^n or ${}_nP_m$

$$P_m^n=\frac{n!}{(n-m)!}$$

* $n!$ (Factorials) $= n \cdot (n-1) \cdot (n-2), \ldots\ldots\ldots 1$

$$P_n^n = \frac{n!}{(n-n)!} = n! \quad (0! \text{ or } 1! \text{ are equal to } 1)$$

◆Example 1:

How many different ways are there to line up 5 students, (A, B, C, D, and E) in a row ?

1	2	3	4	5
5	4	3	2	1

1. For the first place we can arrange any one of A, B, C, D, and E, so then are 5 choices.
2. Once one of them places at the first place, the second place only has 4 choices.
3. The third place only has 3 choices.
4. The fourth place only has 2 choices.
5. The fifth place only has 1 choice.

Because one complete event is fulfilled by completing 5 events. So the total ways by counting of multiplication.

Total arrangement $= P_5^5 = \frac{5!}{(5-5)!} = 5! = 5 \times 4 \times 3 \times 2 \times 1 = 120$

◆Example 2:

How many ways are there to line up 3 out of 5 students (A, B, C, D, and E) in a row ?

☐	☐	☐
5	4	3

1. There are 5 choices for the first place.
2. There are 4 choices for the second place (1 student has been placed already).
3. The third place we have 3 choices.

Total we have $5 \times 4 \times 3 = 60$ placements

$$P_3^5 = \frac{5!}{(5-3)!} = \frac{5!}{2!} = \frac{5 \times 4 \times 3 \times \cancel{2} \times \cancel{1}}{\cancel{2} \times \cancel{1}} = 5 \times 4 \times 3 = 60$$

◆Example 3:

5 girls and 4 boys are lining up for lunch, how many different arrangements are there in each of the scenario below ?

1. Girls must line up together, boys must line up together as well.
2. Girls are required to line up together.

3. Boys must be separated

4. Girls and boys must be separated (no two boys or girls can line up together)

Solutions:

1. Treat girls and boys as 2 groups solve the permutation within each group first, then apply rule of multiplication for the two groups to get the total number of permutation.

(1) (g_1, g_2, g_3, g_4, g_5) (b_1, b_2, b_3, b_4)

2 × 1 = 2!

(2) □ □ □ □ □ five girls

5 × 4 × 3 × 2 × 1 = 5!

□ □ □ □ four boys

4 × 3 × 2 × 1 = 4!

(3) $2! \times 5! \times 4! = 2 \times 120 \times 24 = 5,760$

2. (g_1, g_2, g_3, g_4, g_5) b_1, b_2, b_3, b_4

Since girls need to be lined up together, 5 girls can be arranged as one group. There are 5! arrangements within the girl's group.

The permutation for lining up one girl group and 4 boys is again 5!.

By multiplication $5! \times 5! = 14400$

3. ○ $\boxed{g_1}$ ○ $\boxed{g_2}$ ○ $\boxed{g_3}$ ○ $\boxed{g_4}$ ○ $\boxed{g_5}$ ○

Girls line up $\Rightarrow 5!$

There are six available places to insert 4 boys.

$$P_4^6 = \frac{6!}{(6-4)!} = \frac{6!}{2!} = \frac{6 \cdot 5 \cdot 4 \cdot 3 \cdot \not{2} \cdot \not{1}}{\not{2} \cdot \not{1}} = 360$$

$5! \times 360 = 43,200$

4. $\boxed{g_1}$ ○ $\boxed{g_2}$ ○ $\boxed{g_3}$ ○ $\boxed{g_4}$ ○ $\boxed{g_5}$

There are only 4 places where we can insert the boys, not at the beginning or at the end, since both boys and girls need to be separated.

Total $= 5! \times 4! = 2,880$

Only 4 gaps for 4 boys, so, the arrangements are 4!

Answer to Concept Check 8-05

1	2	3	4	5
(A)	0	(B)	0	(C)

(1) Suppose player A, B, and C are in the best. The coach can arrange any one of A, B, and C in the first, third, and fifth match.

the first match		the third match		the fifth match	
3	×	2	×	1	$=3!$

(2) Arrange 2 player to the second and the forth matches. The selection has P_2^5 ways.

(3) Total arrangements $=3!\times P_2^5=3!\times=\dfrac{5!}{(5-2)!}=3!\times\dfrac{5\cdot 4\cdot \cancel{3!}}{\cancel{3!}}=120$

B. Reverse Counting

When calculating permutation, sometimes there are so many scenarios needed to be counted, so, it is easy only counting the total number of scenarios that are not allowed.

P(Total allowed scenarios) = P(Total scenarios) – P(Total not allowed scenarios)

Where P(Total scenarios) represents permutation calculation.

◆Example 1:

In Sam's class, three of 8 volunteers are to be elected to be the class representative, secretary. and the finance. Jennifer and Jason are among the candidates, however, they cannot be both elected at the same time. How many possible election results are there ?

Desired outcomes $=P_3^8-1\times6\times3!$

- P_3^8: total number of outcomes
- 1: both of Jennifer and Jason are elected
- 6: only 1 out of the 6 students left can be elected
- $3!$: permutation of 3 students

◆Example 2:

Al (A), Bella (B), Caroline (C), Daniel (D), and Edward (E) are lining up to attend a show.

1. How many arrangements are there where A and B lined up with each other, and C and D are lined up with each other ?
2. How many arrangements are there where A and B are lined with each other but C, and D are not ?

3. How many arrangements are there where A is never next to B, and C is never next to D ?

Solution:

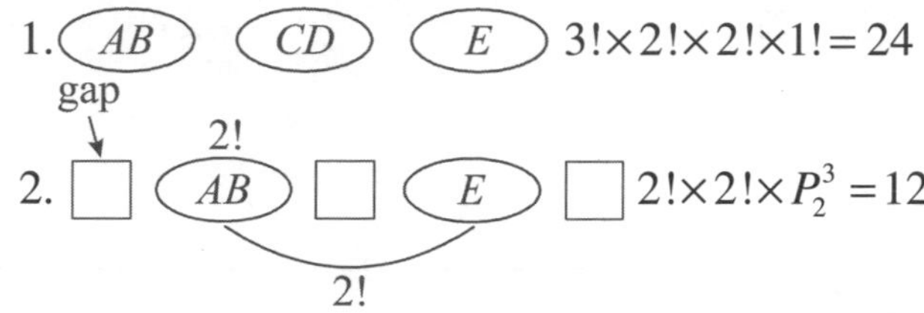

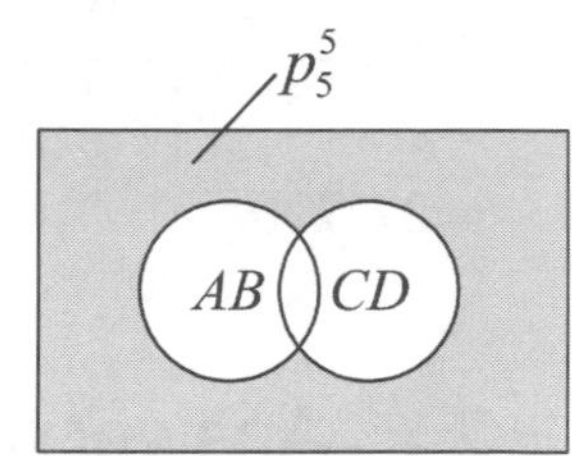

C and D inserted into the gaps.

3. Using reverse counting

$P_5^5 - [P(AB\text{ together}) + P(CD\text{ together})] - P(A, B\text{ together and } C, D\text{ together})$

$P\ (AB\text{ together}) = 4!\times 2!$ (A, B) (C) (D) (E)

$P\ (CD\text{ together}) = 4!\times 2!$ (C, D) (A) (B) (E)

$P\ (AB\text{ together and } CD\text{ together}) = 3!\times 2!\times 2!$ (A, B) (C, D) E

Total ways of arrangement

$= P_5^5 - [4!\times 2! + 4!\times 2! - 3!\times 2!\times 2!]$

$= 5! - [2\times 4!\times 2! - 3!\times 2!\times 2!]$

$= 5! - [96 - 24]$

$= 5! - 72 = 120 - 72 = 48$

C. Permutation of the Same Objects

If we have $\underbrace{a, a, a, \ldots\ldots}_{p}, \underbrace{b, b, b, \ldots\ldots}_{q}, \underbrace{c, c, c, \ldots\ldots}_{r}$

then total of permutation $= \dfrac{n!}{p!q!r!}$

Because the p times of arrangement of 'a' only counted as '1', in other words the total be divided by p time.

Same concept applied to q times and r times as well.

◆Example:

How many ways to line up "mississippi"

m	i	s	p
1	4	4	2

$$P_{11}^{11} = \frac{11!}{1!\cdot 4!\cdot 4!\cdot 2!} = \frac{11\cdot 10\cdot 9\cdot 8\cdot 7\cdot 6\cdot 5\cdot \cancel{4!}}{\cancel{4!}\cdot 4!\cdot 2!} = 34,650$$

How many ways to line up 5 white boxes, 4 blue boxes, and 3 yellow boxes ?

$$P_{12}^{12}=\frac{(5+4+3)!}{5!\cdot 4!\cdot 3!}=\frac{12!}{5!\cdot 4!\cdot 3!}=\frac{12\cdot 11\cdot 10\cdot 9\cdot 8\cdot 7\cdot 6\cdot 5!}{5!\cdot 2!\cdot 3!}=11\times 10\times 9\times 8\times 7\times\frac{1}{2}$$

$=27,720$

Answer to Concept Check 8-06

1	2	3	4	5	6

coffee 3, coke 2, water 1

Total arrangements $=\frac{6!}{3!\cdot 2!\cdot 1!}=\frac{6\cdot 5\cdot 4\cdot 3!}{3!\cdot 2!}=60$

D. Circular Permutation

Circular permutation refers to the number of distinct ways of selecting r objects from n different objects and ordering them in a circular fashion. Circle permutation is denoted as $Q(n, r)$ and is calculated per following equation:

$$Q(n, r)=\frac{P_r^n}{r}$$

* $P(n, r)$ represents the permutation when we select r objects out of n objects and line them up linearly. If the line becomes a circle, we will be over counting by r times (depicted by the picture below). Therefore, we divide $P(n, r)$ by r.

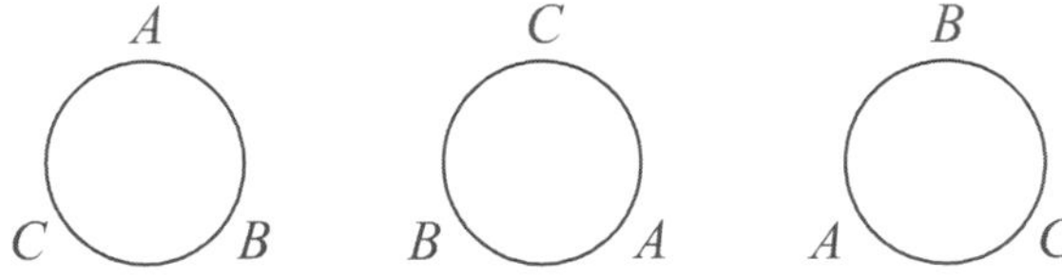

$$Q(n, n)=\frac{P_n^n}{n}=(n-1)!$$

◆Example 1:

select 4 people from 6 people, and arrange these 4 people to be seated on a table.

then we can write the arrangement as $Q(6, 4)=\frac{P_4^6}{4}=\frac{6!}{(6-4)!\cdot 4}\ \frac{6\cdot 5\cdot 4\cdot 3\cdot 2!}{4\cdot 2!}=90$

◆Example 2:

From 6 girls and 4 boys select 6 people to be seated at a rounded table. How many arrangements are there ?

$$Q(10,6)=\frac{P_6^{10}}{6}=\frac{10!}{(10-6)!\cdot 6}=\frac{10\cdot 9\cdot 8\cdot 7\cdot \not{6}\cdot 5\cdot \not{4}!}{\not{4}!\cdot \not{6}}=25,200$$

◆Example 3:

How many ways are there to seat two parents and 4 children at a round table ?

$$Q(6,6)=\frac{P_6^6}{6}=\frac{6!}{6}=(6-1)!=5!=120$$

◆Example 4:

As example 2 above, if the parents must be seated directly facing each other, how many arrangements ?

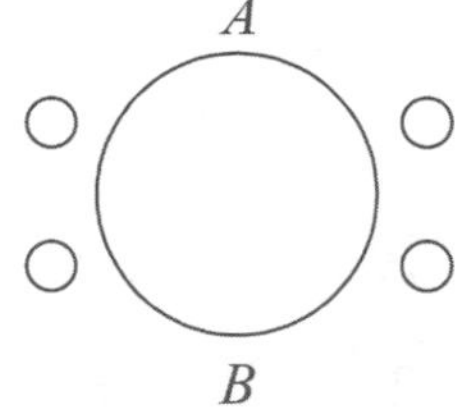

Arrange one parent to be seated in the chair opposite to the other parent.

Seating the 4 children becomes a linear permutation problem.

Therefore $\frac{2!}{2}\cdot 4!=4!=24$

- $4!$ → seating 4 children becomes line permutation
- $\frac{2!}{2}$ → parents sit opposited to each other,

◆Example 5:

Four couples *Aa*, *Bb*, *Cc*, and *Dd* would like to be seated at a round table, where *A*, *B*, *C*, and *D* are male and *a*, *b*, *c*, and d are female.

1. How many ways, if the couple must be seated together ?
2. How many ways, if the same gender must be separated ?
3. All couples must be seated in diagonal positions.
4. A and a are next to each other but *B* and *b* are not ?

1. Treat each couple as a single group object.

$\frac{4!}{4} \times 2^4 = 96$

→ Permutation within each group is 2!, there are 4 groups.

→ 4 couples seated in circular fashion.

2. 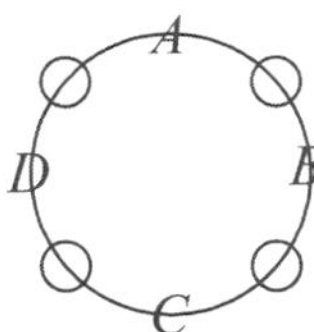

(1)Seated A, B, C, and D first $\Rightarrow$ circular permutation is $\frac{4!}{4}$.

(2)Then seat a, b, c, and $d \Rightarrow$ linear permutation 4!

Total ways $= \frac{4!}{4} \cdot 4! = 3! \cdot 4! = 144$

3. 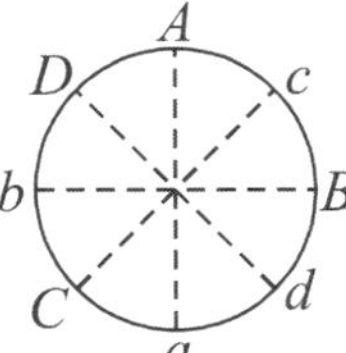

(1)Seat A and a in diagonal position first $\Rightarrow$ circular permutation is $\frac{2!}{2}$

(2)Seat B, C, and D $\Rightarrow$ linear permutation is 3!

(3)Seat (B, b, C, c, D, d switchable) $2 \times 2 \times 2 = 2^3$

Total $\frac{2!}{2} \times 3! \times 2^3 = 6 \times 8 = 48$

4. 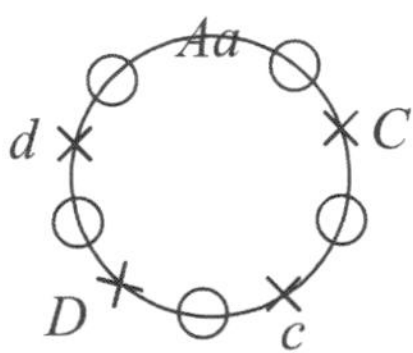

In the graph above, cross represents spaces where C, c, D, d can be seated. Empty circle represents spaces B and b can be inserted.

(1)Seat Aa (as a group), C, c, D, and d first $\Rightarrow$ circular permutation is $\frac{5!}{5} \times 2!$

(2)Since B and b cannot sit next to each other, they can be inserted into the five available seats shown in the graph above P_2^5

Total permutation $=\frac{5!}{5}\times 2!\times P_2^5=4!\times 2!\times\frac{5!}{(5-2)!}=4!\times 2!\times 5\times 4=960$

Answer to Concept Check 8-07

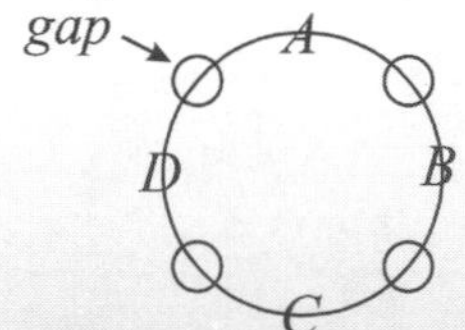

1.Seated *A*, *B*, *C*, and *D* first, $\frac{4!}{4}$

2. Seated *a*, *b*, *c*, and d to the 4 gaps as *a* straight line permutation 4!

Total warp $\frac{4!}{4}\cdot 4!=3!\cdot 4!=144$

8-4 COMBINATION

- The number of ways of selecting *r* desired objects from *n* different objects without any ordering or arrangements is referred to as combination and is denoted as C_r^n.

$$C_r^n=\frac{P_r^n}{r!}$$ (It is the same as P_r^n, but $r!$ ways counted just as "1"way)

For instance, select 3 people from 5 people, to form a group.

it is $C_3^5=\frac{P_3^5}{3!}$ (3! arrangements are just counted as 1)

$$\therefore\ C_3^5=\frac{5!}{(5-3)!\cdot 3!}=\frac{5\cdot 2\cdot \cancel{3}\cdot \cancel{2}\cdot \cancel{1}}{\cancel{2!}\cdot \cancel{3!}}=10$$

- Pascal's rule

$$C_r^n=C_{r-1}^{n-1}+C_r^{n-1}$$

Proof. $C_{r-1}^{n-1}=\frac{(n-1)!}{[(n-1)-(r-1)]!(r-1)!}=\frac{(n-1)!}{(n-r)!(r-1)!}$

$$C_r^{n-1}=\frac{(n-1)!}{[(n-1)-r]!r!}=\frac{(n-1)!}{(n-r-1)!r!}$$

$$C_{r-1}^{n-1}+C_r^{n-1}=\frac{(n-1)!}{(n-r)!(r-1)!}+\frac{(n-1)!}{(n-r-1)!r!}$$

$$=\frac{(n-1)!}{(n-r)\cdot(n-r-1)!(r-1)!}+\frac{(n-1)!}{(n-r-1)!\cdot r\cdot(r-1)!}$$

$$=\frac{(n-1)!\bullet r+(n-1)!(n-r)}{(n-r-1)!(r-1)!(n-r)\bullet r}$$

$$=\frac{(n-1)!(r+n-r)}{(n-r)!r!}=\frac{n\bullet(n-1)!}{(n-r)!r!}$$

$$=\frac{n!}{(n-r)!r!}=C_r^n$$

◆Example 1:

Nine people *A*, *B*, *C*, *D*, *E*, *F*, *G*, *H*, I are forming 3 teams to play basketball game. Each team will have exactly 3 people. *A* and *B* must be on the same team. How many ways can the teams be formed ?

Solution:

Event A: tie up A and B together, then select 1 from the rest 7 people.

Event B: C_3^6, C_3^3, 2 of them only counted as 1.

Total $=C_1^7\bullet C_3^6\bullet C_3^3\bullet\frac{1}{2!}=\frac{7!}{6!\bullet 1!}\bullet\frac{6!}{3!\bullet 3!}\bullet\frac{3!}{0!3!}\times\frac{1}{2}=7\times 20\times 1\times\frac{1}{2}=70$

◆Example 2:

As the example 1 above, if *C* and *D* must not in the same teams, how many ways can 3-people basketball teams be formed ?

1.Total combination $=C_3^9\bullet C_3^6\bullet C_3^3\bullet\frac{1}{3!}=280$ (because team (ABC) is counted for 6 times, actually, it is only one way of selection)

2.*C* and *D* must be in the same team $C_1^7\bullet C_3^6\bullet C_3^3\bullet\frac{1}{2!}=70$

3.*C* and *D* must not in the same team $C_3^9\bullet C_3^6\bullet C_3^3\bullet\frac{1}{3!}-C_1^7C_3^6C_3^3\bullet\frac{1}{2!}=210$

◆Example 3:

The track and field team at David's school has five girls and five boys. Four people must be selected to represent the school to participate in the city's track and field competitions. How many ways to form this team ?

If at least one boy and a girl should be on the team, how many ways are there can the team be formed ?

Solution:

1. C_4^{10}

2. Total combination = combination of selecting 1 boy and 3 girls + combination of selecting 2 boys and 2 girls + combination of selecting 3 boys and 1 girl.

$$C_1^5 \cdot C_3^5 + C_2^5 \cdot C_2^5 + C_3^5 \cdot C_1^5 = 5 \times \frac{5!}{2!\cdot 3!} + \frac{5!}{3!\cdot 2!} \cdot \frac{5!}{3!2!} + \frac{5!}{2!3!} \cdot 5 = 200$$

#of boys #of girls

8-5 PROBABILITY (II)

A. The Properties of Probability

- $p(A)$ represents the probability of event A happening.

  $$P(A) = \frac{\text{Desired outcomes}}{\text{Total possible outcomes}}$$

  $P(A')$ represents the probability of "not" event A.

  $P(A') = 1 - P(A)$

  $P(S) = 1$ (100% chance of happening)

  $P(\phi) = 0$ (100% chance of not happening)

- $P(A \cup B) = P(A) + P(B) - P(A \cap B)$

  The probabilities of event A or even B is equivalent to the probability of event A plus the probability of event B minus the probability of event A and event B.

- $P(A \cap B)$: probabilities of both event A and event B.

  In a probability question, A, B could be any form, such as $P(A' \cup B)$, $P(A' \cup B')$, $P[(A \cup B) \cap B']$, ...

◆Example 1:

If $P(A) = \frac{1}{3}$, $P(B) = \frac{1}{4}$, if event A and event B are mutual exclusive what is the probability of $P(A \cup B)$?

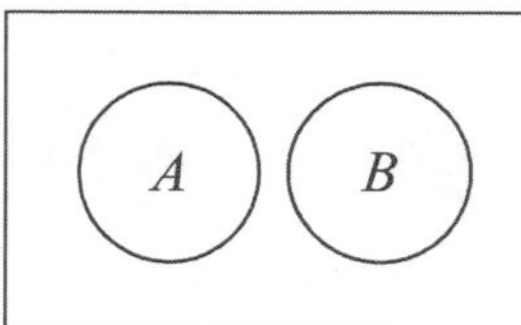

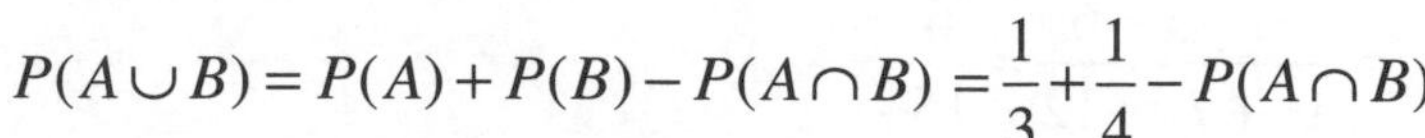

$$P(A \cup B) = P(A) + P(B) - P(A \cap B) = \frac{1}{3} + \frac{1}{4} - P(A \cap B)$$

$\because$ $P(A \cap B) = 0$

$\therefore$ $P(A \cup B) = \frac{1}{3} + \frac{1}{4} - 0 = \frac{4+3}{12} = \frac{7}{12}$

◆Example 2:

If $P(A\cap B)=\frac{1}{5}$, $P(B')=0.6$, $P(A\cup B)=\frac{1}{2}$, what is $P(A)$?

Since $P(A\cup B)=P(A)+P(B)-P(A\cap B)$

$$\frac{1}{2}=P(A)+[1-P(B')]-P(A\cap B)=P(A)+(1-0.6)-\frac{1}{5}=P(A)+0.4-0.2$$

$\therefore\ P(A)=0.5-0.4+0.2=0.3$

◆Example 3:

There are 1 white, 3 yellow, and x red marbles in a bag. If the probability of taking out 2 same color marbles from the bag simultaneously is $\frac{6}{15}$. How many red marbles in the bag ?

Note: Since there is only one white marble, the probability of taking out two white marbles simultaneously is O.

White marbles	yellow marbles	red marbles
1	3	x

The probability of the same color $=\dfrac{C_2^3+C_2^x}{C_2^{4+x}}$

$$=\frac{\frac{3!}{1!2!}+\frac{x!}{(x-2)!2!}}{\frac{(4+x)!}{(4+x-2)!2!}}$$

$$=\frac{3+\frac{x!}{(x-2)!2!}}{\frac{(x+4)(x+3)(x+2)!}{(x+2)!2!}}$$

$$=\frac{3+\frac{(x)(x-1)}{2}}{\frac{(x+4)(x+3)}{2}}=\frac{\frac{6+x^2-x}{2}}{\frac{x^2+7x+12}{2}}$$

$$=\frac{x^2-x+6}{x^2+7x+12}=\frac{6}{15}$$

$15x^2-15x+90=6x^2+42x+72$

$9x^2-57x+18=0$

$(9x-3)(x-6)=0$

$x=\frac{1}{3}$, $x=6$ (integer)

The number of red marbles are 6.

◆Example 4:

When throwing 3 dice simultaneously, what is the probability of rolling the same number on 2 of the dice, and a different number on the 3rd ?

For example, (1, 1, 2), (3, 3, 2), ...

Solution:

1. First, we calculate the number of outcomes where 2 dice have the same number, and the 3rd dice has a different number.

   Select one number for the two dice, and another for the 3rd.

   $C_1^6 \cdot C_1^5$

   C_1^5 → Select 1 from 5

   C_1^6 → Select 1 from 6

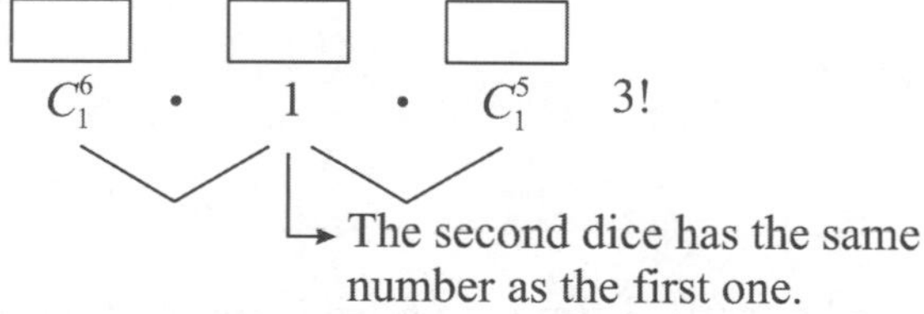

   Total desired outcomes $=C_1^6 \cdot C_1^5 \cdot 3! \cdot \frac{1}{2!}$

2. All possible outcomes $=6 \cdot 6 \cdot 6=6^3$

3. P(two are the same points, one is different) $=C_1^6 \cdot C_1^5 \cdot 3! \cdot \frac{1}{2!} \cdot (\frac{1}{6})^3$

Answer to Concept Check 8-08

The number of situations of passing the quality control inspection

1.take none of bad cups means all are good $=C_0^4 \cdot C_3^6$

2.take one of bad cups $=C_1^4 \cdot C_2^6$

The probability of passing $P(\text{passing})=\dfrac{C_0^4C_3^6+C_1^4 \cdot C_2^6}{C_3^{10}}$

3.The probability of elimination (two cups or more than two cups are bad)

$$=1-\frac{C_0^4C_3^6+C_1^4 \cdot C_2^6}{C_3^{10}}=\frac{C_3^{10}-C_0^4C_3^6-C_1^4C_2^6}{C_3^{10}}=\frac{120-20-60}{120}=\frac{40}{120}=\frac{1}{3}$$

B. Repeating Experiments

If the probability of an event happen is P(event), then never happen denoted as q, $q=1-P$, so in n independent trials:

A. The probability of an event happening in n trials $=P^n$

B. Never happening in n trials $=q^n=(1-P)^n$

C. At least happening once in a trials $=1-q^n$

D. Exactly happening r times in n trials $=C_r^nP^rq^{n-r}$

$\overbrace{ppp\cdots}^{r}\overbrace{qqq\cdots}^{n-r}$, $\overbrace{p\cdot p\cdot p\cdots}^{r}\overbrace{q\cdot q\cdot q\cdots}^{n-r}$, C_r^n

For instance, the probability of an event happened exactly 3 times in 6 trials of an experiments $C_3^6P^3q^3$

An event happening exact 2 times in 8 trials $C_2^8P^2q^6$

$ppq\cdots q, pqpq\cdots q, pqqpq\cdots q,$ $\cdots\cdots \to C_2^8$ (There are only 2 'p' s in 8 trials.)

◆Example 1:

If we continuously throw a fair dice 8 times, what is the probability that the dice rolls a 5 exactly 6 times ?

The probability of showing number 5 is $\dfrac{1}{6}$

$$P = C_6^8(\frac{1}{6})^6 \cdot (\frac{5}{6})^2 = \frac{8!}{(8-6)!6!} \times \frac{1}{6^6} \times \frac{5^2}{6^2} = \frac{8!}{(8-6)!6!} \cdot \frac{25}{6^8} = \frac{8 \cdot 7}{2} \times \frac{25}{6^8}$$

$$= 28 \times 25 \times 6^{-8} = \frac{700}{1,679,616} = 0.0004$$

◆Example 2:

If we continuously toss three coins 5 times, what is the probability that the three coins land on the same side exactly 3 times ?

Three coins landed on the same side

H, H, H or T, T, T

$$P(H) = \frac{1}{2} \times \frac{1}{2} \times \frac{1}{2}$$

$$P(T) = \frac{1}{2} \times \frac{1}{2} \times \frac{1}{2}$$

$$P(\text{landed on the same side}) = \frac{1}{2} \times \frac{1}{2} \times \frac{1}{2} \times 2 = \frac{1}{4}$$

$$\therefore \ P = \frac{1}{4}, \ q = \frac{3}{4} \text{(fail)}$$

Exact 3 times in throwing 5 times

$$P = C_3^5(\frac{1}{4})^3(\frac{3}{4})^2 = \frac{5 \cdot 2 \cdot \cancel{3!}}{\cancel{3!} \cdot \cancel{2!}} \times \frac{1}{4^3} \times \frac{3^2}{4^2} = \frac{10 \times 9}{4^5} = \frac{90}{1024} = \frac{45}{512} = 0.088$$

Answer to Concept Check 8-09

$P(hit) = 0.25$, $P(hit') = 0.75$

$$P(2\ hits) = C_2^6 P(hit)^2 \cdot P(hit')^4 = \frac{6 \cdot 5 \cdot 4!}{4!2!} \times (\frac{1}{4})^2 \times (\frac{3}{4})^4 = 15 \times \frac{1 \times 3^4}{4^6} = \frac{15 \times 81}{4096} \doteq 0.3$$

C. Conditional Probability

- A probability of an event B that will happen in a certain condition of even A Denoted as $P(B \mid A)$.

$$P(B \mid A) = \frac{n(A \cap B)}{n(A)} = \frac{P(A \cap B)}{P(A)}$$

◆Example 1:

If $P(A|B)=0.6$, $P(A\cap B)=0.3$, $P(A')=0.3$ what is the value of $P(A\cup B)$?

$$P(A\cup B)=P(A)+P(B)-P(A\cap B)$$

$$P(A|B)=0.6=\frac{P(A\cap B)}{P(B)}$$

$$P(B)=\frac{P(A\cap B)}{P(A|B)}=\frac{0.3}{0.6}=\frac{1}{2}$$

$$P(A)=1-P(A')=0.7$$

$$P(A\cup B)=P(A)+P(B)-P(A\cap B)=0.7+0.5-0.3=0.9$$

◆Example 2:

	Bachelor	Master	PhD	Total
Male	80	120	50	250
Female	60	110	40	210
Total	140	230	90	460

The table above shows the educational level of employees in a company. What is the probability that one randomly selected female has a PhD degree ?

(1) Method 1, from number of people

Let B represents PhD and A represents female (The condition.)

$$P(\text{female with PhD})=\frac{n(A\cap B)}{n(B)}=\frac{40}{210}=\frac{4}{21}$$

(2) Method 2, from conditional probability P(Female with PhD)

$$P(B|A)=\frac{P(A\cap B)}{P(A)}=\frac{\frac{40}{460}}{\frac{210}{460}}=\frac{4}{21}$$

◆Example 3:

There are 4 white balls and 4 black balls in bag A, and 5 white balls and 3 black balls in bag B. Now rolling a fair dice, if it appears number of 1, 2, 3, or 4 then pick one ball from bag A, if it appears number of 5 or 6, pick one ball from bag B. What is the probability of picking one white ball ?

Solution:

bag A		bag B	
W	B	W	B
4	4	5	3

Let A be the probability of rolling 1, 2, 3, or 4 on the dice

$$P(1,2,3,4)=\frac{4}{6}=\frac{2}{3}$$

Let B be the probability of rolling 5 or 6 on the dice

$$P(5,6)=\frac{2}{6}=\frac{1}{3}$$

White ball in situation A

$$P(W\mid A)=\frac{2}{3}\times\frac{4}{8}=\frac{2}{6}=\frac{1}{3}$$

White ball in situation B

$$P(W\mid B)=\frac{1}{3}\times\frac{5}{8}=\frac{5}{24}$$

They are excluded to each other.

$$P(W)=\frac{1}{3}+\frac{5}{24}=\frac{8+5}{24}=\frac{13}{24}$$

◆Example 4:

In a certain city, 52% of population are male and 48% are female. It is known that 30% of male are smokers and 12% of female are smokers.

1. What is the percentage of smokers in the city ?
2. If a smoker is arbitrarily selected from the city, what is the probability that this person is male ?

(1)smokers $=0.3\times0.52+0.12\times0.48=0.2136$

	smoking	non-smoking	total
male	30%/15.6	70%	52%
Female	12%/5.7	88%	48%
	21.3		

(2) $P(M\mid\text{smoking})=\dfrac{P(\text{male}\cap\text{smoking})}{P(\text{smoking})}=\dfrac{0.156}{0.213}=0.732\ \Rightarrow 73.2\%$

Answer to Concept Check 8-10

Suppose Kim will get \$45 after x games.

The difference between wins and losses are $(12-4)\cdot x \geq 45$, so x must be 6, 7, 8.

The player must have won 6 games, 7 games, or 8 games, then Kim will receive \$45.

$$P(\$45) = C_6^8(\frac{1}{3})^6(\frac{2}{3})^2 + C_7^8(\frac{1}{3})^7(\frac{2}{3})^1 + C_8^8(\frac{1}{3})^8 = \frac{8!}{(8-6)!6!}\times\frac{4}{3^8}+\frac{8!}{7!}\times\frac{2}{3^8}+\frac{8!}{8!}\cdot\frac{1}{3^8}$$

$$\frac{1}{3^8}(112+16+1)=\frac{129}{3^8} \Rightarrow 0.02$$

CHAPTER 8
PRACTICE

1. How many positive four-digit integers have 5 at the thousands' place and 2 at one's place ?
   (A) 68
   (B) 88
   (C) 100
   (D) 112

2. If 8 out of 32 cars in a parking lot are black, what is the probability that we get a car that is not black when selecting a car randomly from the parking lot ?
   (A) $\frac{1}{4}$
   (B) $\frac{1}{2}$
   (C) $\frac{3}{4}$
   (D) $\frac{1}{3}$

3. In a tennis tournament, each player must compete with other players once and only once. It is known that 66 games have been scheduled. How many players are there in this tournament ?

4. A city bridge club holds a friendship match once a year. Eight teams will compete in the match. If the club adopts a round robin system (every team has to play with other team once and only once) and the team with top scores is the winner, how many games will be arranged ?
   If the club adopts the knockout system (the winner of the two teams moves up a level as shown below), how many games will be arranged ?

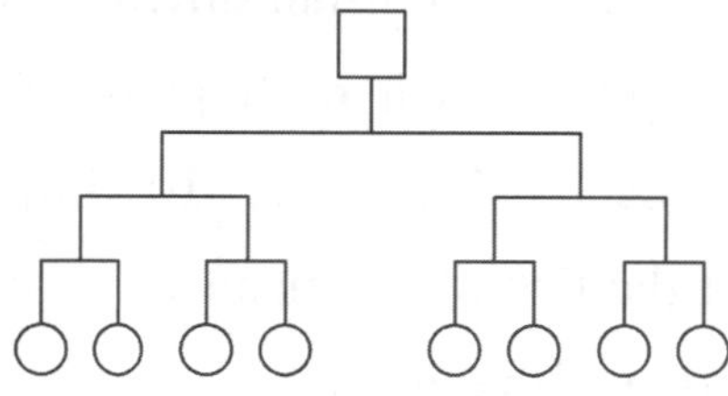

5. In a deck of 52-card poker cards. How many Full-house, (three cards with the same number, and the other 2-with the other same number, such as 88833, AAAQQ, JJJ22, …, and there are four cards for each number.) can be made from a 52-card poker deck ?

6. A light rail in a certain city has 8 carriages. Each carriage must be painted with one flower selected from the national flower, state flower, and city flower (they are distinct). The city would like to paint 2 carriages with the

national flower, 3 carriages with the state flower, and 3 carriages with the city flower. However, the national flower, state flower, and city flower must each appear on the first 3 carriages. How many ways are there for painting all the carriages ?

7. A new Ferris wheel was built in the amusement park near Edward's house. In order to create a landscape at night, the park decided to install different color of the light to each of 8 parts of the Ferris wheel. There are 10 different color of the light, how many ways of the installation of the lights ?

8. In Allison class, 5 boys and 7 girls would like to be selected to form an eight-person cheerleading team. If a cheerleading team is composed of at least 3 boys and 3 girls, how many ways are there to form the team ?

9. A factory produces 16 batteries into a box. Because of the high yield rate, the factory decides that each box will be talking out of 5 batteries for inspection. If there are 2 or more batteries bad, the whole box will be eliminated. What are the chances of the box being obsolete with 5 bad batteries ?

10. $A = \{-2, -1, 0, 3, 4\}$

$B = \{-1, 2, -5, 6, 7\}$

If a is a number that is randomly selected from set A, and b is a number that is randomly selected from set B, what is the probability that $ab < 0$?

(A) $\frac{1}{4}$

(B) $\frac{1}{3}$

(C) $\frac{2}{5}$

(D) $\frac{4}{9}$

11. Brian bought a one-year fire insurance for his house that covers 1 million dollars, and a premium of \$1,000. If the probability of a fire is 0.0005, what is the insurance company expected profit from the premium ?

12. Katy's class has 16 boys and 14 girls. If the teacher randomly chooses 3 students to perform a task, what is the probability that both genders are being selected among the 3 students ?

13. There are 5 white balls and 4 black balls in bag A, and 4 white balls and 6 black

balls in bag B. Now throw a fair die, if the point is 1 or 2, pick a ball from bag A, otherwise, pick a ball from bag B. What is the probability of picking a black ball ?

14. Mr. White is trying to withdraw money from ATM but realized that he forgot his pin. However he does remember that his pin has 4 digits and is composed of number 2, 7, 5, and 5. What is the probability that Mr. White enters the correct pin on the first guess ?

15. What is the probability that exactly 6 of the 10 rolls of a fair die will result in a 2 ?

16. There are 16 boys and 18 girls in Mary's class. Three of them will be elected to participate in a school-wide activity. What are the odds for the elected representatives to include both boys and girls ?

17. A Computer company has production lines in Region A, Region B, and Region C. According to past statistics, There are 5%, 2%, and 2% of the defective computers manufactured in Region A, B, and C respectively.
If the company hopes that no more than $\frac{6}{13}$ of all defective products are manufactured in region A. What is the maximum percentage of computers produced in region A of the total production ? (Rounded to hundredth)

18. What is the probability of getting the sum of 2 dice as a prime number ?

19. There are 3 white marbles and 6 black marbles in a box. Connor and Jens take turns to take one marble from the box. Whoever gets the white marble first wins the game.
   (1) Connor starts the game first. If anyone gets a black marble, it has to be put back into the box. What is the probability of winning for both boys ?
   (2) If anyone gets a black marble, it does not get put back into the box. What are the probability of winning for both boys ?

20. In a place where medical standards are lagging behind. There is a particular infectious disease with two types-A and B. Type A accounts for 60% the cases and can be treated by drug A; the success rate of the treatment is 60%. If failed,

then treated by drug B, the success rate of treatment is 60%. Type B accounts for 40% of the case and can be treated by drug B; the success rate of the treatment is 48%. If failed, then treated by drug A, the success rate of treatment is also 48%. Currently, there is no test to classify whether a patient has type A or B. Doctor prescribes the treatment on a trial-and-error basis. What is the success rate of the second treatment after the first has failed ?

CHAPTER 8
ANSWERS and EXPLANATIONS

1. (C)

× × × ×

↓ 5 (first digit)

↳ 2 (the forth digit)

For the second digit (0~9), and the third digit (0~9), we have $10 \times 10 = 100$

2. (C)

$P(\text{Black}) = \frac{8}{32} = \frac{1}{4}$

$1 - \frac{1}{4} = \frac{3}{4}$

3. 12 players

$C_2^n = 66$, $\frac{n(n-1)(n-2)!}{(n-2)!2!} = \frac{n^2 - n}{2} = 66$

$\therefore\ n^n - n - 132 = 0$

$(n-12)(n+11) = 0$

$\therefore\ n = 12$, or $n = -11$

Number of people can not be negative.

Therefore $n = 12$ is the number of players

4. 28 games, 7 games

(1) A, B, C, D, E, F, G, H

↓ 7 games; B, C, D, E, F, G, H ↓ 6 5 4 3 2 1 0

Total $= 7+6+5+4+3+2+1+0$

$= 28$

(2) For knockout

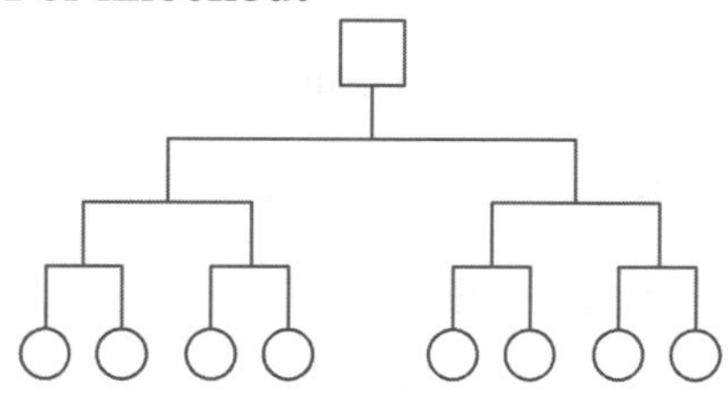

Round 1: 4 games

Round 2 : 2 games

Round 3 (final): 1

Total $= 4+2+1 = 7$ games

5. 3,744

A, K, Q, J, 10, 9, 8, 7, 6, 5, 4, 3, 2

$\Rightarrow\ C_1^{13}$

A, A, A, A $\Rightarrow\ C_3^4$

K, Q, J, 10, 9, 8, 7, 6, 5, 4, 3, 2 $\Rightarrow\ C_1^{12}$

K, K, K, K $\Rightarrow\ C_2^4$

A, A, A, K, K $= C_1^{13} \cdot C_3^4 \cdot C_1^{12} \cdot C_2^4$

$= 13 \cdot 4 \cdot 12 \cdot 6 = 3,744$

6. 180

$\boxed{1} - \boxed{2} - \boxed{3} - \boxed{4} - \boxed{5} - \boxed{6} - \boxed{7} - \boxed{8}$

$3!$ ↙ Three different flowers:

$\frac{5!}{2!2!}$ ↙ same kind of flowers

National flower

State flower

City flower permutation

Total $= 3! \times \frac{5!}{2!2!} = 6 \times 30 = 180$

7. 226,800

8 colours from 10 colours $= C_8^{10} = 45$

put these 8 colours for circular permutation we have $\frac{8!}{8} = 7! = 5,040$

Total $= 45 \times 5,040 = 226,800$

8. 420

$b, b, b, b, b, \quad g, g, g, g, g, g, g,$

$b, b, b, + g, g, g, + x + x$

$C_3^5 \cdot C_5^7 + C_4^5 \cdot C_4^7 + C_5^5 \cdot C_3^7 =$

$\frac{5 \cdot 4 \cdot 3!}{2! \cdot 3!} \times \frac{7 \cdot 6 \cdot 5!}{2! \cdot 5!} + \frac{5 \cdot 4!}{1! \cdot 4!} \times \frac{7 \cdot 6 \cdot 5 \cdot 4!}{3! \cdot 4!}$

$+1 \times \frac{7 \cdot 6 \cdot 5 \cdot 4 \cdot 3!}{4! \cdot 3!}$

$= 210 + 175 + 35 = 420$

9. If there are 2 or 2^+ bad batteries, the whole box of batteries will be obsolete.

$P(2 \text{ or } 2^+ \text{ batteries})$

$= 1 - P(\text{all good}) - P(1 \text{ bad})$

$= 1 - \frac{C_0^5 \cdot C_5^{11} + C_1^5 \cdot C_4^{11}}{C_5^{16}}$

$= 1 - \frac{1 \cdot \frac{11 \cdot 10 \cdot 9 \cdot 8 \cdot 7 \cdot 6!}{6! \cdot 5!} + 5 \cdot \frac{11 \cdot 10 \cdot 9 \cdot 8 \cdot 7!}{7! \cdot 4!}}{\frac{16 \cdot 15 \cdot 14 \cdot 13 \cdot 12 \cdot 11!}{11! \cdot 5!}}$

$= 1 - \frac{462 + 1650}{4368} = \frac{2112}{4368} = 0.4835$

$P = 48.35\%$

10. $A = \{-2, -1, 0, 3, 4\}$

$B = \{-1, 2, -5, 6, 7\}$

$P(ab < 0)$

when "-2" selected $ab < 0$, 3 cases

when "-1" selected $ab < 0$, 3 cases

when "0" selected $ab < 0$, none

when "3" selected $ab < 0$, 2 cases

when "4" selected $ab < 0$, 2 cases

Total $= 3 + 3 + 0 + 2 + 2 = 10$

Total of cases are $5 \times 5 = 25$

$P(ab < 0) = \frac{10}{25} = \frac{2}{5}$

11. $1,000 - 1,000,000 \times 0.0005 = 500$

12. Total outcomes $= C_3^{30}$

$= \frac{30 \cdot 29 \cdot 28 \cdot 27!}{(30-3)! \cdot 3!} = 4060$

2 boys and 1 girl $= C_2^{16} \times C_1^{14} = 1,680$

2 girls and 1 boy $= C_2^{14} \times C_1^{16} = 1456$

Desired outcomes

$= 1,680 + 1,456 = 3,136$

$P(\text{boys, girls}) = \frac{3136}{4060} = 77.2\%$

The other way,

3 boys $= C_3^{16}$

3 girls $= C_3^{14}$

$P(\text{only boys}) = \frac{C_3^{16}}{C_3^{30}} = \frac{560}{4060}$

$P(\text{only girls}) = \frac{C_3^{14}}{C_3^{30}} \frac{364}{4060}$

$1 - \frac{924}{4060} = \frac{3136}{4060} = 77.2\%$

13. bag A bag B

W B W B

5 4 4 6

The probability of picking one ball from bag A is $\frac{2}{6}=\frac{1}{3}$

The probability of picking one ball from bag B is $\frac{4}{6}=\frac{2}{3}$

So, the probability of picking one black ball P(a black ball from bag A)$=\frac{1}{3}\times\frac{4}{9}$

P(a black ball from bag B)$=\frac{2}{3}\times\frac{6}{10}$

P(a black)$=\frac{1}{3}\times\frac{4}{9}+\frac{2}{3}\times\frac{6}{10} \quad =\frac{4}{27}+\frac{12}{30}$

$=0.15+0.4=0.55$

14. Total outcome is the permutation of 2, 7, 5, 5 $=\frac{4!}{2!}=12$ o, $P=\frac{1}{12}$

15. The probability of rolling a fair dice resulting in a 2 is $\frac{1}{6}$, and not resulting in a 2 is $\frac{5}{6}$.

So, the probability of exactly 6 of the 10 rolls is $C_6^{10}(\frac{1}{6})^6(\frac{5}{6})^4$

$=\frac{10\cdot9\cdot8\cdot7\cdot6!}{4!6!}(\frac{1}{6})^6(\frac{5}{6})^4=210\times\frac{5^4}{6^{10}}$

$=\frac{210\times625}{60466176}=\frac{131250}{60466176}=0.22\%$

16. 0.77

P(both boys and girls)

$=\frac{C_3^{34}-C_3^{16}-C_3^{18}}{C_3^{34}}=\frac{5984-560-816}{5,984}$

$=0.77$

or, $\frac{C_1^{16}\cdot C_2^{18}+C_2^{16}\cdot C_1^{18}}{C_3^{34}}=0.77$

17. Let the production volume of area A, B, and C be x, y, and z respectively.

Total Volume
- $x\%$ — A < good / Bad (5%)
- $y\%$ — B < good / Bad (2%)
- $z\%$ — C < good / Bad (2%)

Total defective $=0.05x+0.02y+0.02z$

Total volume $=x+y+z$

The defective products of region $A\leq\frac{6}{13}$

$\frac{5x}{5x+2y+2z}\leq\frac{6}{13}$

$65x\leq30x+12y+12z$

$35x\leq12(y+z) \qquad 35x\leq12(1-x)$

$35x+12x\leq12$, $47x\leq12$, $\therefore\ x\leq\frac{12}{47}$

$\Rightarrow 0.2553 \Rightarrow 25.53\%$

Answer is 25.53%

18. $\frac{2}{9}$

There are 8 outcomes (1, 1), (1, 2), (1, 4), (1, 6), (2, 3), (2, 5), (3, 4), (5, 6)

the sum is a prime number.

$P(\text{prime})=\frac{8}{36}=\frac{2}{9}$

19.(1)

	first try	second	third	fourth
Connor	$\frac{1}{3}$	$\frac{2}{3}\bullet\frac{2}{3}\bullet\frac{1}{3}$	$(\frac{2}{3})^4\bullet\frac{1}{3}$	$(\frac{2}{3})^6\bullet\frac{1}{3}$
Jens	$\frac{2}{3}\bullet\frac{1}{3}$	$(\frac{2}{3})^2\bullet\frac{2}{3}\bullet\frac{1}{3}$	$(\frac{2}{3})^4\bullet(\frac{2}{3})\bullet\frac{1}{3}$	$(\frac{2}{3})^7\bullet\frac{1}{3}$

The winning rate for Connor

$$P(\text{Connor})=\frac{1}{3}+(\frac{2}{3})^2\bullet\frac{1}{3}+(\frac{2}{3})^4\bullet\frac{1}{3}+(\frac{2}{3})^6\bullet\frac{1}{3}+\cdots=\frac{1}{3}[1+(\frac{2}{3})^2+(\frac{2}{3})^4+(\frac{2}{3})^6+\cdots]$$

$$=\frac{1}{3}[\frac{1}{1-(\frac{2}{3})^2}]=\frac{1}{3}\bullet\frac{1}{\frac{5}{9}}=\frac{3}{5}$$

$$P(\text{Jens})=1-\frac{3}{5}=\frac{2}{5}$$

(2) white

3

Connor

Round 1: $\frac{3}{9}=\frac{1}{3}$

Round 2: $\frac{2}{3}\times\frac{5}{8}\times\frac{3}{7}=\frac{5}{28}$

Round 3: $\frac{2}{3}\times\frac{5}{8}\times\frac{4}{7}\times\frac{3}{6}\times\frac{3}{5}=\frac{1}{14}$

Round 4: $\frac{2}{3}\times\frac{5}{8}\times\frac{4}{7}\times\frac{3}{6}\times\frac{2}{5}\times\frac{1}{4}\times\frac{3}{3}=\frac{1}{84}$

Black

6

Jens

$(1-\frac{1}{3})\times\frac{3}{8}=\frac{2}{3}\times\frac{3}{8}=\frac{1}{4}$

$\frac{2}{3}\times\frac{5}{8}\times\frac{4}{7}\times\frac{3}{6}=\frac{5}{42}$

$\frac{2}{3}\times\frac{5}{8}\times\frac{4}{7}\times\frac{3}{6}\times\frac{2}{5}\times\frac{3}{4}=\frac{1}{28}$

So, P(Connor)

$$=\frac{1}{3}+\frac{2}{3}\times\frac{5}{8}\times\frac{3}{7}+\frac{\not 2}{\not 3}\times\frac{\not 5}{\not 8}\times\frac{\not 4}{7}\times\frac{\not 3}{\not 6}\times\frac{\not 3}{\not 5}$$

$$+\frac{\not 2}{3}\times\frac{\not 5}{\not 8}\times\frac{\not 4}{7}\times\frac{\not 3}{\not 6}\times\frac{\not 2}{5}\times\frac{1}{\not 4}\times 1$$

$$=\frac{1}{3}+\frac{5}{28}+\frac{1}{14}+\frac{1}{28\times 3}$$

$$=\frac{28+15+6+1}{28\times 3}=\frac{50}{84}=\frac{25}{42}$$

$$P(\text{Jens})=1-\frac{25}{42}=\frac{17}{42}$$

20.

Disease
- A 60%
  - 60% success
  - 40% fail
    - 60% success
    - 40% fail
- B 40%
  - 48% success
  - 52% fail
    - 48% success
    - 52% fail

P(first times failure)

$=0.6\times 0.4+0.4\times 0.52$

P(second time suceess after first failure)

$$=\frac{0.6\times 0.4\times 0.6+0.4\times 0.52\times 0.48}{0.6\times 0.4+0.4\times 0.52}$$

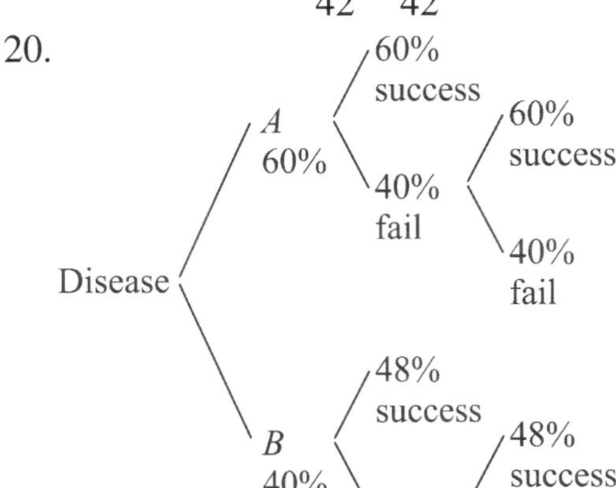

$$=\frac{0.24384}{0.448}=0.54$$

CHAPTER 9
PLANE and SOLID GEOMETRY

9-1 GEOMETRICAL TERMINOLOGY

Terms	Figures	Properties
Point	•A, •B, ······	Only represents the "position" no sense of size
Straight line	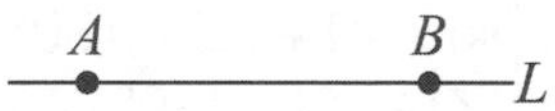 A B L	A straight line passes two points and extends infinitely in both sides denoted as $\overleftrightarrow{AB}$ or $\overleftrightarrow{L}$
Line segment	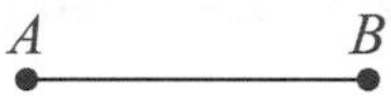 A B	a straight line starts from point A and ends at point B is denoted as $\overline{AB}$
Ray	 A B	starts at point A and passes though point B and extends infinitely in one direction denoted $\overrightarrow{AB}$
Angle	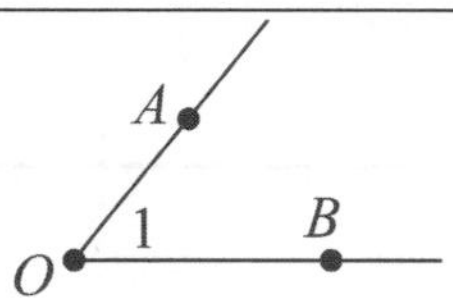 A O 1 B	point 0 is the vertex $\overrightarrow{OA}$, $\overrightarrow{OB}$ are sides denoted $\angle AOB$ or $\angle 1$, $\angle 2$, $\cdots$ $\angle\alpha$, $\angle\beta$
Acute angle	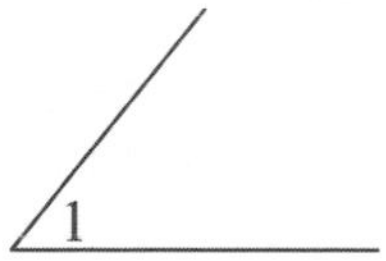 1	$\angle 1 < 90°$
obtuse angle	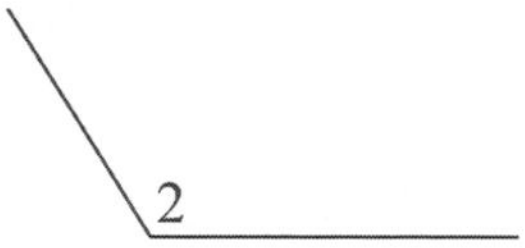 2	$\angle 2 > 90°$
Right angle	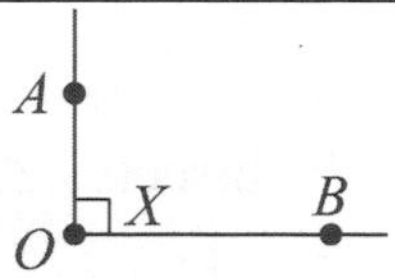 A O X B	$\angle X = 90°$ $\overline{OA} \perp \overline{OB}$ $\overline{OA}$ is perpendicular to $\overline{OB}$
Complementary angle	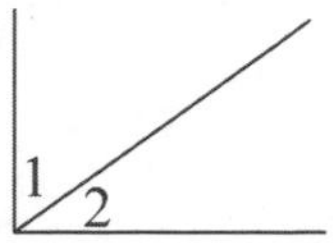 1 2	If $\angle 1 + \angle 2 = 90°$ $\angle 1$ and $\angle 2$ are complementary angles.
Straight angle	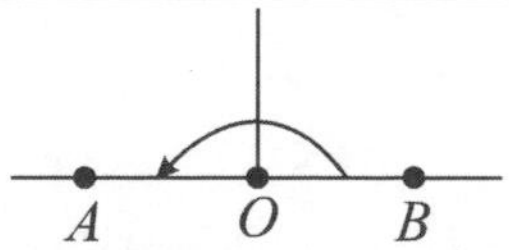 A O B	$\angle AOB = 180°$

Terms	Figures	Properties
Supplementary angles	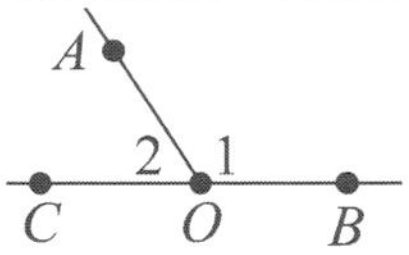	$\angle 1+\angle 2=180°$ $\angle 1$ and $\angle 2$ are supplementary
Vertical angles	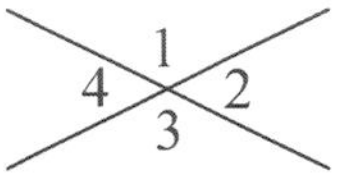	$\angle 1$ and $\angle 3$, $\angle 2$ and $\angle 4$ are opposite to each other, called vertical angles. They are congruent.
Adjacent angles	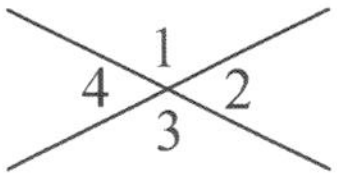	$\angle 1$ and $\angle 4$, $\angle 2$ and $\angle 3$, $\cdots$ are adjacent angles. so, $\angle 1+\angle 2=180°$ $\angle 1+\angle 4=180°$ $\angle 3+\angle 2=180°$ $\angle 3+\angle 4=180°$ $\vdots$
Bisector	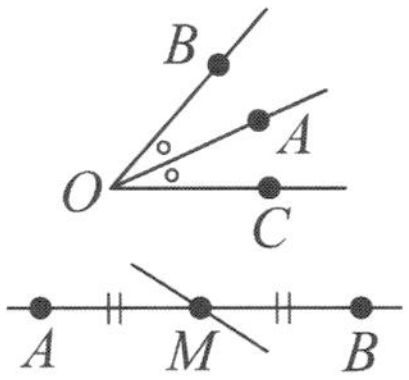	$\overline{OA}$: Bisector of $\angle BOC$, $\angle BOA=\angle AOC$ $\overline{AM}=\overline{BM}$

◆Example 1:

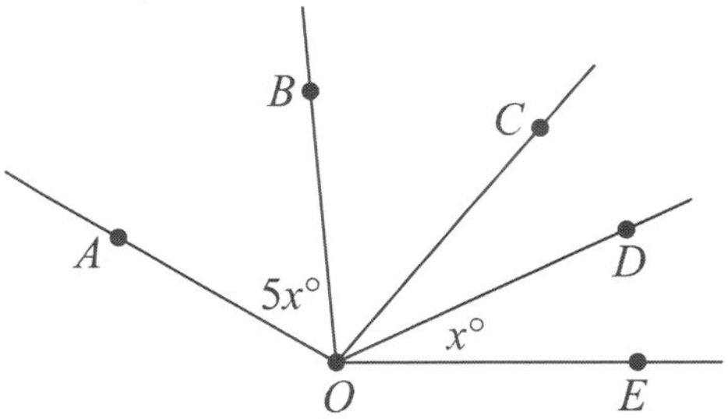

(Note: Figure not drawn to scale)

In the figure above, $m\angle AOE=120°$, $\overline{OB}$ bisects $\angle AOC$, $\overline{OD}$ bisects $\angle COE$, what is the value of x?

Solution: $x+x+5x+5x=12x$

$12x=120 \quad \therefore \quad x=10$

◆Example 2:

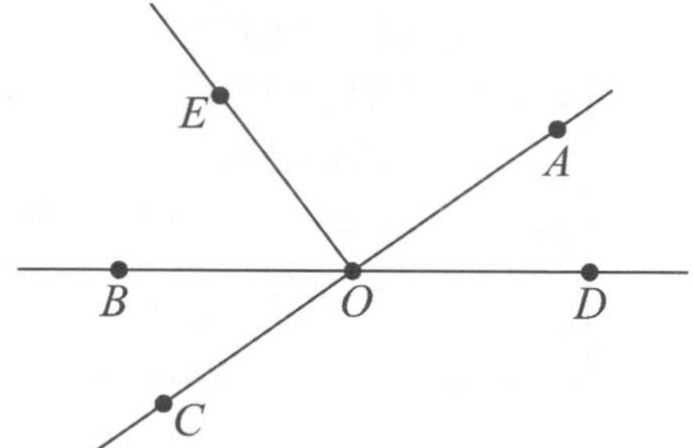

(Note: Figure not drawn to scale)

The figure above, $\overline{EO} \perp \overline{CO}$, $m\angle COD = 138°$, what is the measure of $\angle EOB$?

Solution: $m\angle BOC = 180 - 138 = 42°$

$m\angle EOB = 90 - 42 = 48°$

Answer to Concept Check 9-01

$2(2x-8)+2(3x+3)=180$

$4x-16+6x+6=180$

$10x=190$

$x=19$

9-2 TRIANGLE

A. Basic Properties

Properties	Figure	Explanation
1. The sum of the 3 interior angles $=180°$	A, B, C $m\angle A + m\angle B + m\angle C = 180°$	A, L, 1, 2, B, C 1. Line $L \parallel \overline{BC}$ 2. $\angle 1 = \angle B$, $\angle 2 = \angle C$ 3. $\angle BAC + \angle B + \angle C$ $= \angle 1 + \angle BAC + \angle C = 180°$

Properties	Figure	Explanation
2. The sum of the 3 exterior angles $=360°$	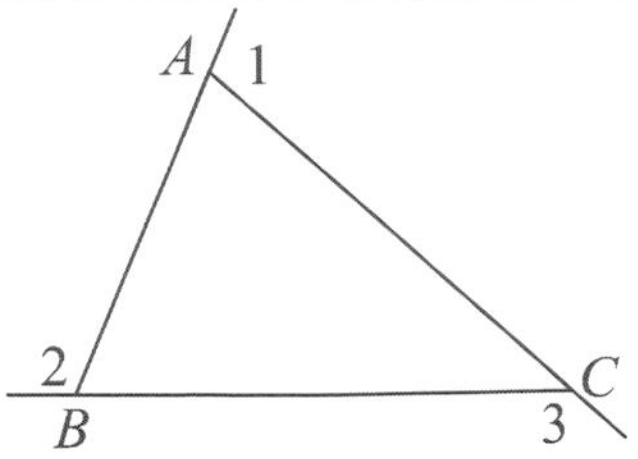 $\angle 1+\angle 2+\angle 3=360$	$\angle 1+\angle BAC=180°$ $\angle 2+\angle ABC=180°$ $\angle 3+\angle ACB=180°$ $\angle 1+\angle 2+\angle 3=3(180)-180$ $=2(180)=360°$
3. The measure of an exterior angle is equal to the sum of the other two interior angles.	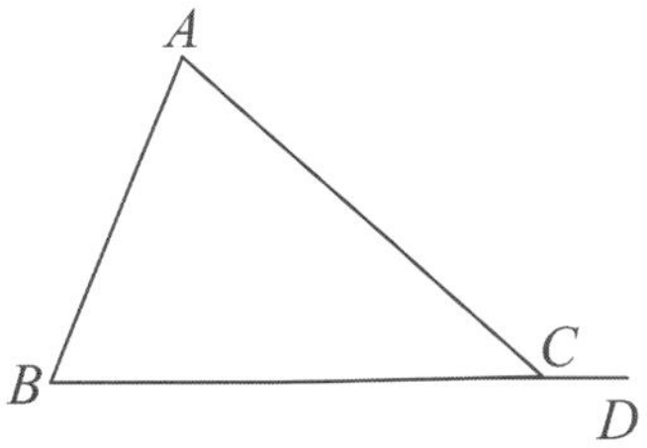 $\angle ACD=\angle A+\angle B$	$\angle A+\angle B+\angle ACB=180°$ $\angle ACD+\angle ACB=180°$ $\therefore\ \angle ACD=\angle A+\angle B$
4. Big sides opposite big angles. vice versa	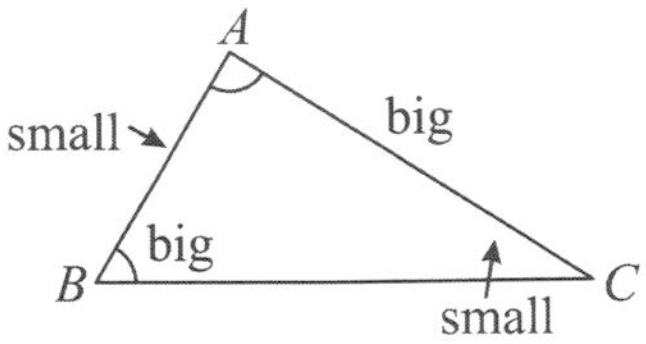	$\angle B>\angle C$ then $\overline{AC}>\overline{AB}$, $\angle A>\angle C$ then $\overline{BC}>\overline{AB}$
5. Equal sides opposite equal angles, vice versa	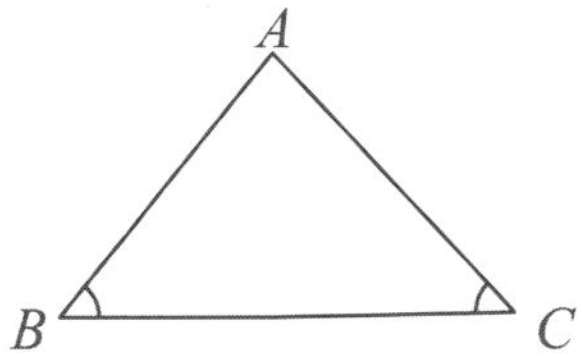	If $\angle B=\angle C$ then $\overline{AC}=\overline{AB}$
6. The sum of lengths of any two sides is greater than the third side	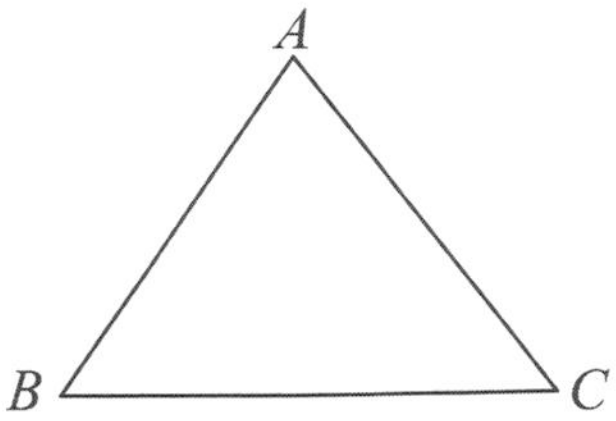	$\overline{AB}+\overline{AC}>\overline{BC}$ or $\overline{BC}+\overline{AC}>\overline{AB}$ or $\overline{AB}+\overline{BC}>\overline{AC}$
7. The difference of any two sides is less than the third side	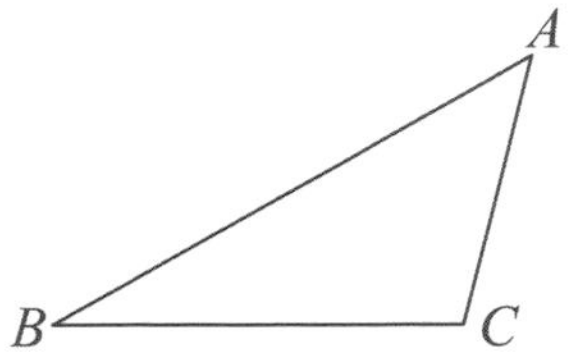	$\overline{AB}-\overline{BC}<\overline{AC}$
8. area $=\frac{1}{2}$ Base$\times$height	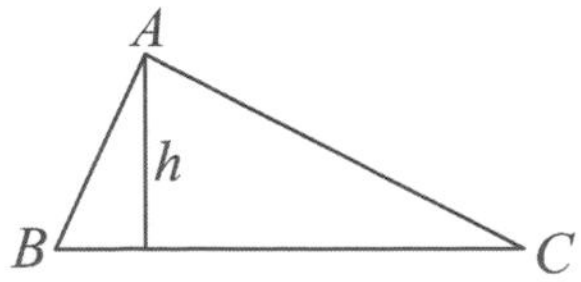	area$=\frac{1}{2}(h\times\overline{BC})$

◆Example 1:

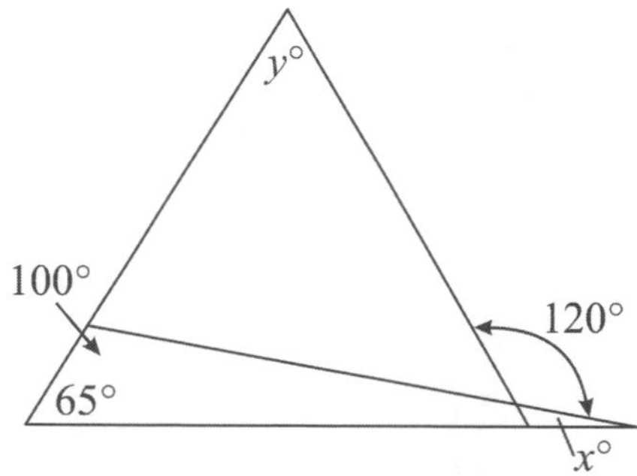

(Note: Figure not drawn to scale)

In the figure above, what are values of x and y ?

Solution: $x+100+65=180$

$x=15$

$120=y+80$

$y=120-80=40$

◆Example 2:

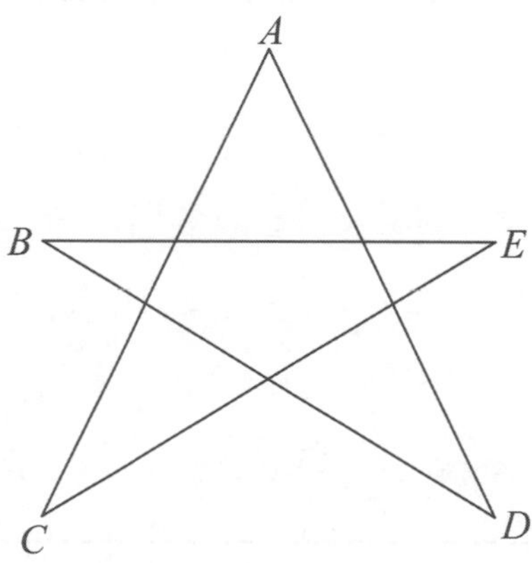

(Note: Figure not drawn to scale)

In the figure above, $m\angle B=32°$, $m\angle C=27°$, $m\angle D=42°$, $m\angle E=35°$, what is the measurement of $\angle A$?

Solution:

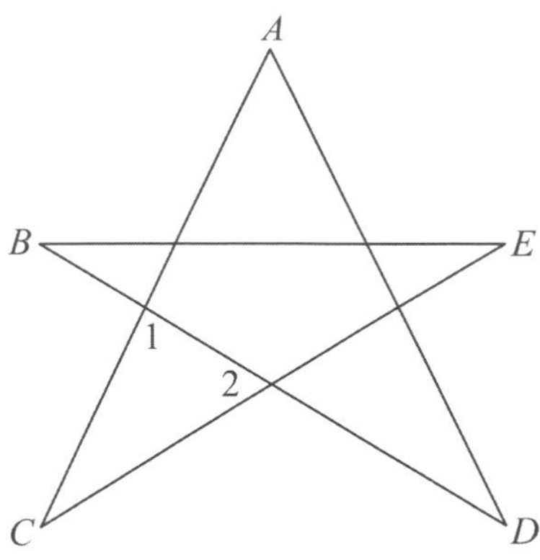

(Note: Figure not drawn to scale)

$\angle 1 = \angle A + \angle D$, $\angle 2 = \angle B + \angle E$

$\because \angle C + \angle 1 + \angle 2 = 180°$

$\therefore \angle 1 + \angle 2 = 180 - \angle C = 180 - 27$

$\therefore \angle A + \angle D + \angle B + \angle E = 180 - 27$

$\therefore \angle A = 180 - 27 - 32 - 42 - 35 = 44°$

Answer to Concept Check 9-02

$2(m\angle B + m\angle C) = m\angle A$

$m\angle A + m\angle B + m\angle C = 180$

$m\angle A = 180 - (m\angle B + m\angle C)$

$2m\angle A = 360 - 2(m\angle B + m\angle C) = 360 - m\angle A$

$3m\angle A = 360°$

$\therefore m\angle A = 120°$

B. Congruent of Triangles

Congruent	Figures	Description
S.S.S.	A, B, C; A', B', C'	$\overline{AB} = \overline{A'B'}$, $\overline{BC} = \overline{B'C'}$, and $\overline{AC} = \overline{A'C'}$ $\triangle ABC \cong \triangle A'B'C'$

Congruent	Figures	Description
S.A.S.	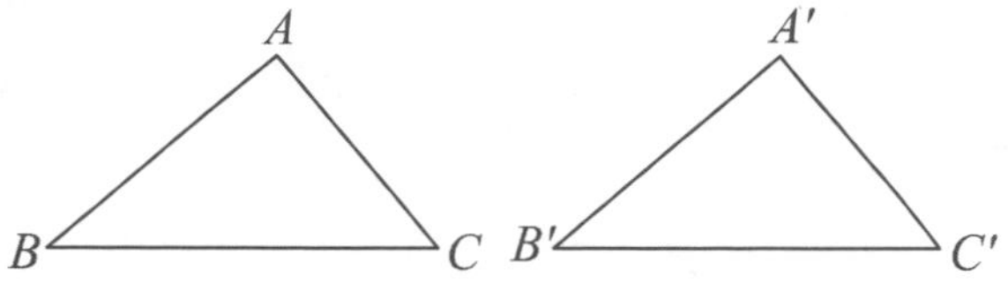	$\overline{AB} = \overline{A'B'}$, $\angle A = \angle A'$, and $\overline{AC} = \overline{A'C'}$ $\triangle ABC \cong \triangle A'B'C'$
A.S.A.	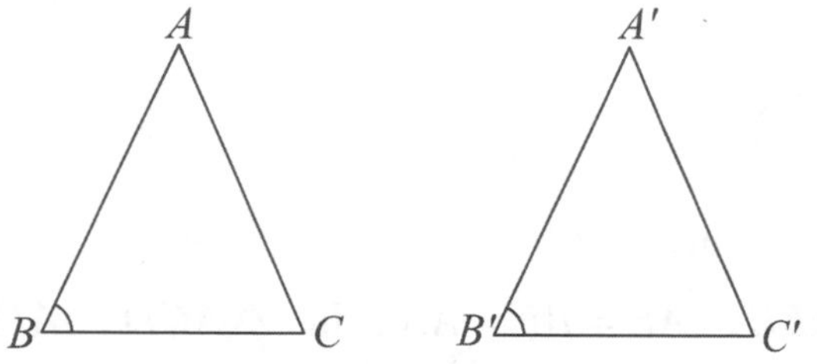	$\angle B = \angle B'$, $\overline{BC} = \overline{B'C'}$, and $\angle C = \angle C'$ $\triangle ABC \cong \triangle A'B'C'$
A.A.S.	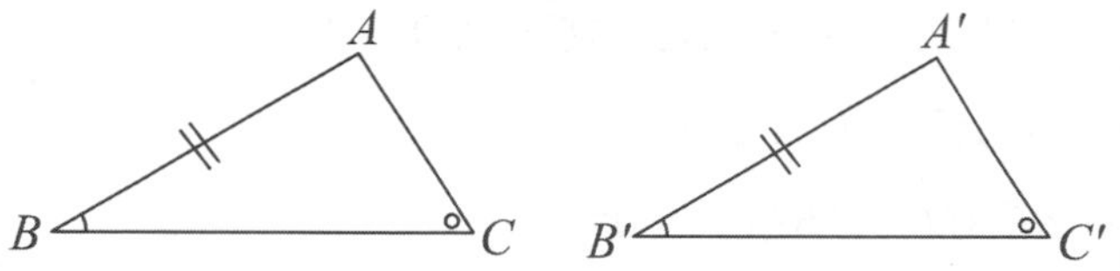	$\angle B = \angle B'$, $\angle C = \angle C'$, and $\overline{AB} = \overline{A'B'}$ or $\overline{AC} = \overline{A'C'}$ or $\overline{BC} = \overline{B'C'}$ $\triangle ABC \cong \triangle A'B'C'$ It is similar to A.S.A.

◆Example 1:

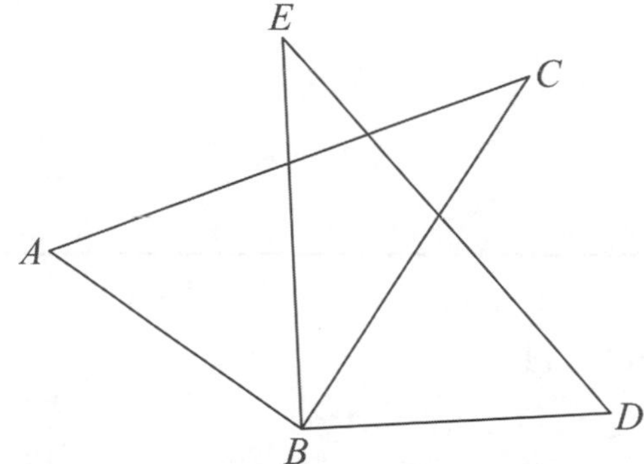

(Note: Figure not drawn to scale)

In the figure above, $\overline{AB} = \overline{BD}$, $\overline{ED} = \overline{BC}$, and $\overline{AC} = \overline{BE}$. If $m\angle ABD = 175°$, $m\angle EBC = 45°$, What is the measure of $\angle ABC$?

Solution: $\because$ $\overline{AB} = \overline{BD}$, $\overline{ED} = \overline{BC}$, and $\overline{AC} = \overline{BE}$, $\triangle ABC \cong \triangle BDE$,

For, $\angle ABE = \frac{1}{2}(175 - 45) = 65$ ($\angle ABC = \angle EBD$)

$\therefore$ $m\angle ABC = 65 + 45 = 110$

◆Example 2:

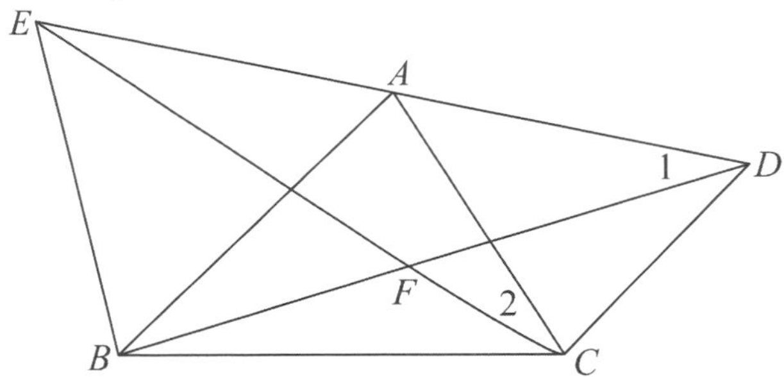

(Note: Figure not drawn to scale)

In the figure above, In $\triangle ABE$, $\overline{AE}=\overline{BE}=\overline{AB}$, In $\triangle ACD$, $\overline{AD}=\overline{DC}=\overline{AC}$, $\overline{BD}$ intersects $\overline{EC}$ at point F. What is the sum of $\angle FBC+\angle FCB$?

Solution: In $\triangle EAC$ and $\triangle BAD$, $\overline{EA}=\overline{AB}$, $\overline{AC}=\overline{AD}$

$\angle EAC=\angle EAB+\angle BAC=60°+\angle BAC$

$\angle BAD=\angle BAC+\angle CAD=\angle BAC+60°$

$\therefore$ $\angle EAC=\angle BAD$, $\triangle EAC\cong\triangle BAD$ (S.A.S.)

$\therefore$ $\angle 1=\angle 2$, and $\angle DFC+\angle 2=\angle CAD+\angle 1$

$\therefore$ $\angle DFC=\angle CAD=60°$

$\angle BFC=180°-60°=120°$

$\angle FBC+\angle FCB=180°-120°=60°$

Answer to Concept Check 9-03

$m\angle BDC=180-m\angle ABC-m\angle BCD=180°-35°-30°=115°$

$m\angle BAE=m\angle BDC-m\angle DFA=m\angle BDC-m\angle EFC=115°-40°=75°$

Answer to Concept Check 9-04

(A) $\overline{AB} = \overline{A'B'}$, $\overline{BC} = \overline{B'C'}$, and $\overline{AC} = \overline{A'C'}$

$\triangle ABC$ and $\triangle A'B'C'$ are overlaid completely so, $\triangle ABC \cong \triangle A'B'C'$ (S.S.S.)

(B) $\overline{AB} = \overline{A'B'}$, $\overline{AC} = \overline{A'C'}$ and $\angle A \cong \angle A'$

Since $\angle A \cong \angle A'$, then $\overline{BC} = \overline{B'C'}$

$\therefore$ $\triangle ABC \cong \triangle A'B'C'$ (S.A.S.)

(C) $\overline{AB} = \overline{A'B'}$, $\overline{AC} = \overline{A'C'}$ and $\angle B \cong \angle B'$

$\angle A$ may not be equivalent to $\angle A'$

$\overline{BC} \neq \overline{B'C'}$, $\triangle ABC$ not congruent to $\triangle A'B'C'$

(D) $\overline{AB} = \overline{A'B'}$, $m\angle B = m\angle B'$, $m\angle C = m\angle C'$, in $\triangle ABC$ and $\triangle A'B'C'$,

then $m\angle A = m\angle A'$, $\overline{BC} = \overline{B'C'}$, $\overline{AC} = \overline{A'C'}$

$\triangle ABC \cong \triangle A'B'C'$ (A.A.S.)

Answer is (C)

Answer to Concept Check 9-05

1. Congruent, S.S.S.
2. Congruent, S.A.S.
3. Congruent, A.S.A.
4. Congruent, A.A.S.
5. Not Congruent, A.A.A.
6. Not Congruent, S.S.A.

For Example:

$\triangle ABC$, and $\triangle ABC'$, $AB = AB$,

$AC = AC'$, $\angle B = \angle B'$

But they are not congruent

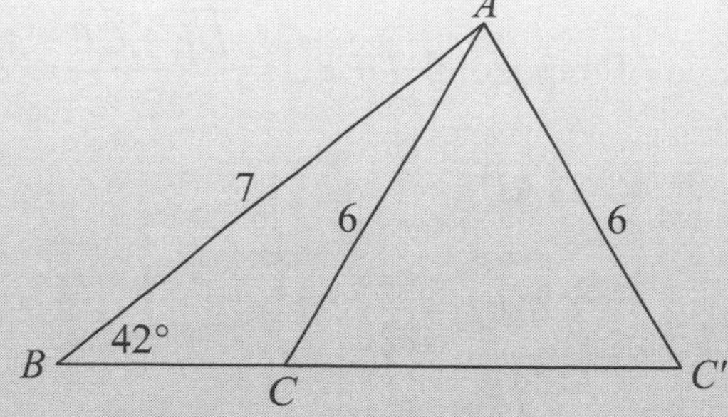

Answer to Concept Check 9-06

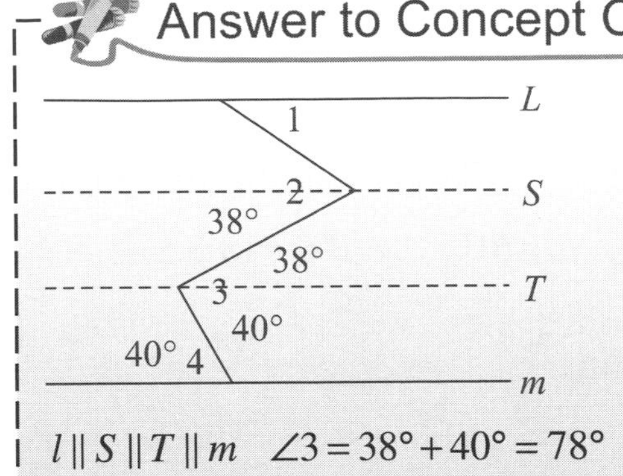

$l \parallel S \parallel T \parallel m \quad \angle 3 = 38° + 40° = 78°$

Answer to Concept Check 9-07

$a = 4$

$b = 13$

$c = 17$

$d = 7$

Answer to Concept Check 9-08

A. $a' = a$, $b' = b$, $\angle ADE = \angle ACB = 90°$ in $\triangle ADE \cong \triangle ACB$ (S.A.S.)

B. $\angle 3 = \angle 4$, $\angle 5 = \angle 1 (\triangle ADE \cong \triangle ACB)$

$\angle 1 + \angle 4 = 90°$, $\angle 1 + \angle 3 = 90°$

$\angle 2 = 180° - \angle 1 - \angle 2 = 90°$

C. Area of trapezoid $DEBC = \dfrac{\overline{DE} + \overline{CB}}{2} \times (AD + AC) = \dfrac{1}{2}(a+b)(a+b)$

Area of $\triangle ADE = \dfrac{1}{2}a'b'$

Area of $\triangle ABE = \dfrac{1}{2}cc'$

Area of $\triangle ACB = \dfrac{1}{2}ab$

$\therefore \ \dfrac{1}{2}(a+b)^2 = \dfrac{1}{2}ab + \dfrac{1}{2}ab + \dfrac{1}{2}c^2$

$\therefore \ a^2 + b^2 = c^2$

C. Special Triangles

• Isosceles Triangles

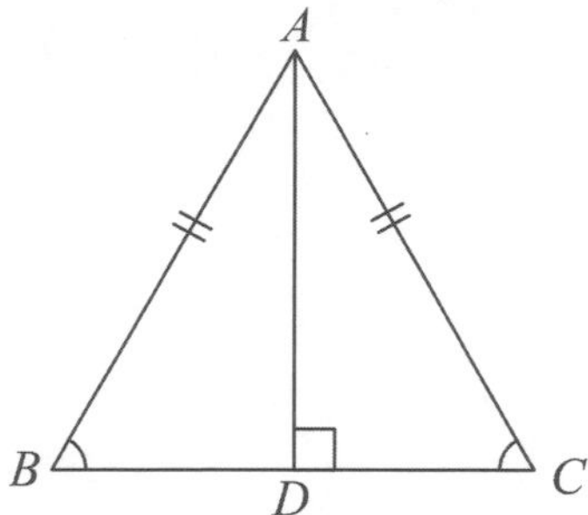

(Note: Figure not drawn to scale)

$\angle B = \angle C$ or $\overline{AB} = \overline{AC}$

$\triangle ABC$ is isosceles triangle

1. If D is the middle point of $\overline{BC}$, then $\overline{AD}$ must be perpendicular to line segment $\overline{BC}$.

   Vice versa.

   Bisector line, middle line, perpendicular line, all are overlaid

   For $\triangle ABD \cong \triangle ADC$ ($\overline{AB} = \overline{AC}$, $\angle B = \angle C$, $\overline{BD} = \overline{DC}$, S.A.S.)

   So, $\angle ADB = \angle ADC$

   $\angle ADB + \angle ADC = 180°$ $\quad \therefore \angle ADB = 90°$

   $\overline{AD} \perp \overline{BC}$, vice versa.

2. 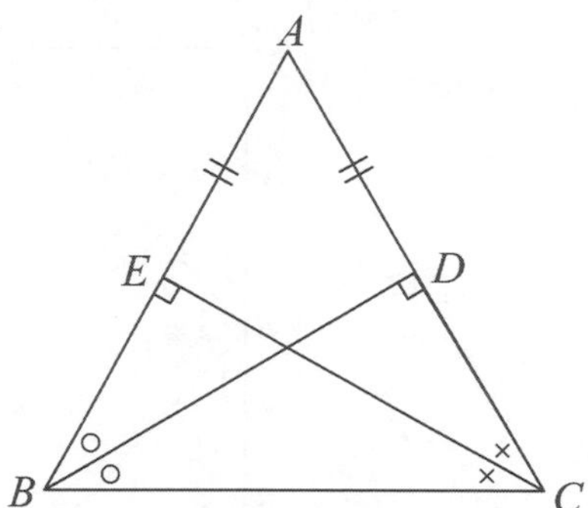

   (Note: Figure not drawn to scale)

   Two perpendicular lines $\overline{BD}$ and $\overline{CE}$ have the same measure as two angles bisecting lines $\overline{BD}$ and $\overline{CE}$ of $\angle B$ and $\angle C$.

   For $\triangle BDC \cong \triangle CBE$, $\overline{BD} = \overline{CE}$, $\angle ABD = \angle DBC$, $\angle ACE = \angle BCE$

   $\therefore$ Perpendicular line = bisecting line

• Equilateral Triangles

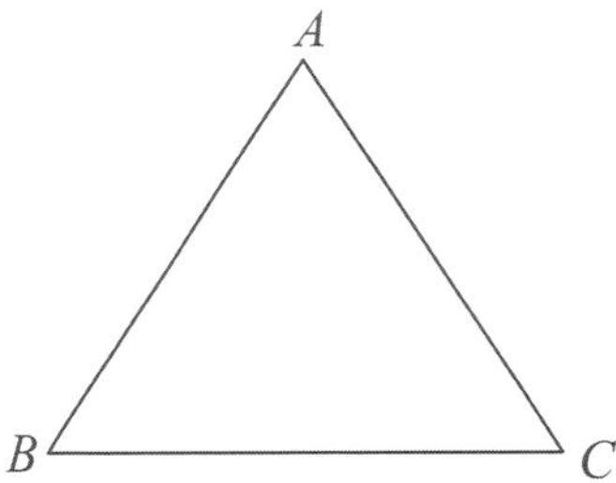

(Note: Figure not drawn to scale)

1. $\angle A = \angle B = \angle C = 60°$
2. $\overline{AB} = \overline{AC} = \overline{BC}$

Has all characteristics of isosceles triangle, and every angle has the measure of $60°$.

• Right Triangles

IF one angle of a triangle is right angle, it's called a right triangle.

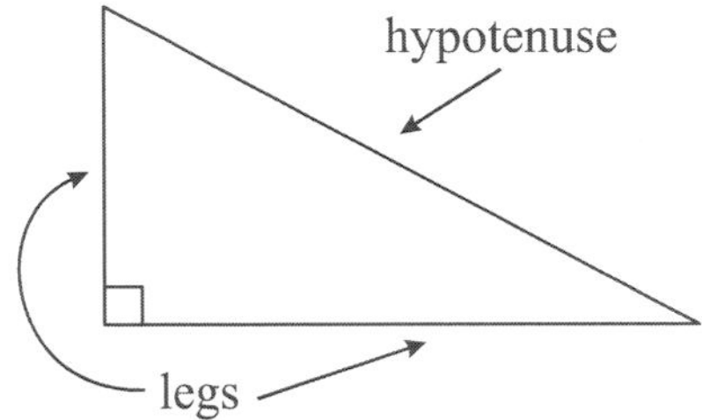

(Note: Figure not drawn to scale)

1. Pythagorean Theorem

$a^2 + b^2 = c^2$

Proof

There are so many ways to prove this theorem, here is a writer's favorite one.

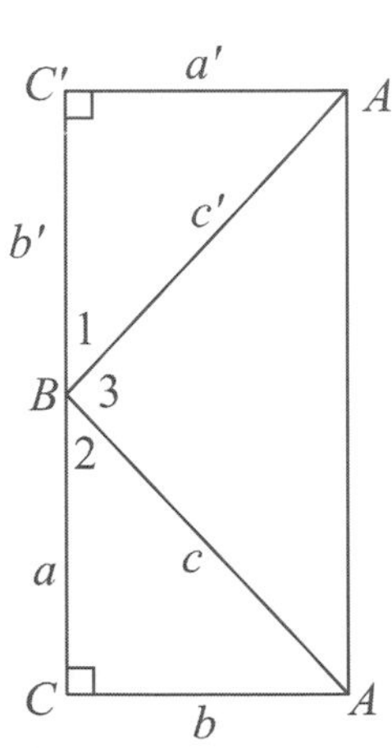

Let $b' = b$, $a' = a$, $\triangle A'BC' \cong \triangle ABC$

$\angle 1 + \angle 2 = 90°$, $\angle 3 = 90°$ $(\angle 2 = \angle C'A'B)$

Trapezoid $ACC'A' = \triangle A'BC' + \triangle ABC + \triangle A'BA$

$(a' + b) \times \frac{1}{2}(a + b') = \frac{1}{2}a'b' + \frac{1}{2}ab + \frac{1}{2}cc'$

$\frac{1}{2}(a+b)(a+b) = \frac{1}{2}ab + \frac{1}{2}ab + \frac{1}{2}c^2$

$(a+b)^2 = 2ab + c^2$

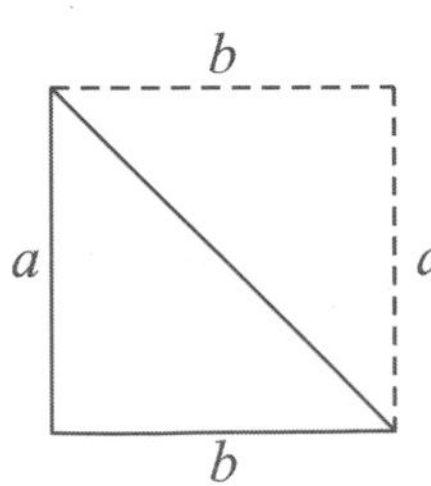

$a^2 + b^2 + \cancel{2ab} = \cancel{2ab} + c^2$

$\boxed{a^2 + b^2 = c^2}$

2. Pythagorean Triplets

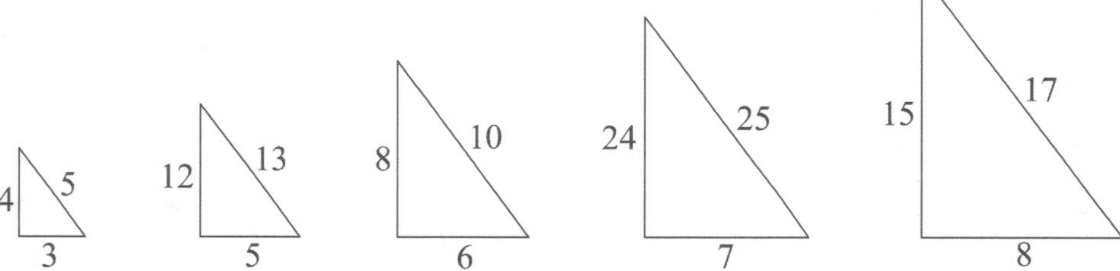

(Note: Figure not drawn to scale)

They all have integer lengths, and any triangle has the same ratio of these triplets has to be a right triangle.

For instance, 30:40:50 is a multiple of 3:4:5, 20:48:52 is a multiple of 5:12:13.

$30:40:50 = 3\times10:4\times10:5\times10 = 3:4:5$

$(3\times10)^2 + (4\times10)^2 = 3^2\times10^2 + 4^2\times10^2 = 10^2(3^2+4^2) = 2500 = (5\times10)^2 = 10^2(5^2)$

$20:48:52 = 4\times5:4\times12:4\times13 = 5:12:13$

$10:24:26 = 2\times5:2\times12:2\times13 = 5:12:13$

3. Special internal angles of a right triangle

- $45°-45°-90°$ triangles

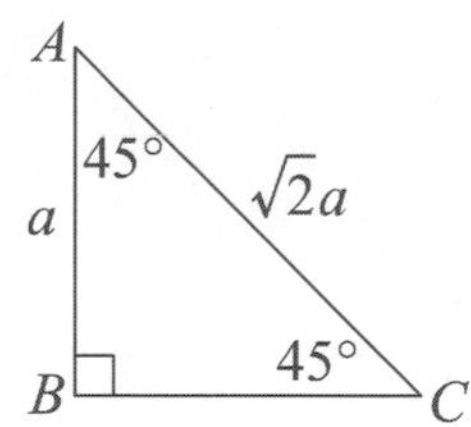

$\overline{AB} = \overline{BC} = a$

$\overline{AC}^2 = \overline{AB}^2 + \overline{BC}^2 = 2\overline{AB}^2 = 2a^2$

$\therefore\ \overline{AC} = \sqrt{2}a$

- $30°-60°-90°$

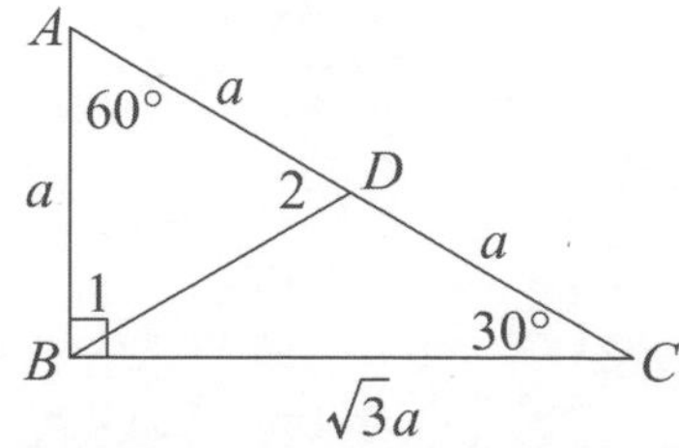

(Note: Figure not drawn to scale)

Let $\overline{AD} = a$, $\angle 1 = \angle 2 = 60°$

$\angle DBC = 30° = \angle C$

$\therefore \ \overline{BD} = \overline{AB} = \overline{CD} = a$

$\therefore \ \overline{AC}^2 = (a+a)^2 = (2a)^2$

$\therefore \ \overline{AC} = 2a = 2\overline{AB}$

$\overline{BC}^2 = (2a)^2 - a^2 = 3a^2$

$\therefore \ \overline{BC} = \sqrt{3}a$

D. Parallel Lines and Transversal Lines

- Parallel Line: Two lines that are never intersecting each other, are parallel to each other. It is Denoted as $\overleftrightarrow{AB} \parallel \overleftrightarrow{CD}$
- Perpendicular line: Two lines intersect each other and formed a right angle. Denoted as $\overleftrightarrow{AB} \perp \overleftrightarrow{CD}$.

9-3 PARALLEL LINES INTERSECTED BY A TRANSVERSAL LINE

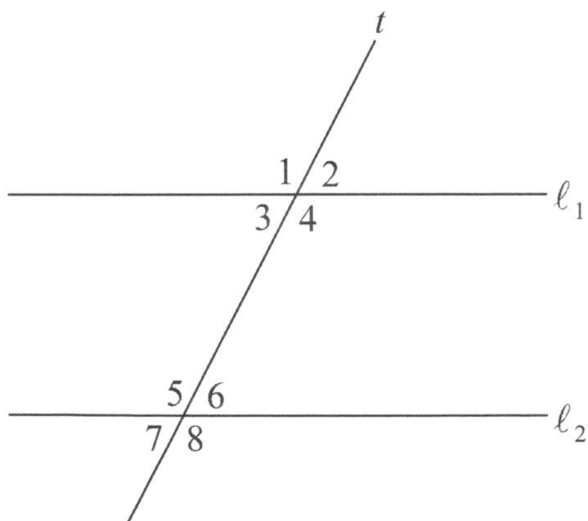

(Note: Figure not drawn to scale)

Big equals big
small equals small
Big plus small equals straight
$\ell_1 \parallel \ell_2$ and intersected by transversal line. There are 8 angles.

- $\angle 1$ and $\angle 5$, $\angle 2$ and $\angle 6$, $\angle 3$ and $\angle 7$, and $\angle 4$, and $\angle 8$ are corresponding angles, they all are congruent.
- $\angle 1$ and $\angle 4$, $\angle 2$ and $\angle 3$, $\angle 5$ and $\angle 8$, $\angle 6$ and $\angle 7$ are vertical angles, they all are congruent.
- $\angle 3$ and $\angle 6$, and $\angle 4$ and $\angle 5$ are alternate interior angles, they all are congruent.
- $\angle 4+\angle 6=180°$, $\angle 3+\angle 5=180°$

◆Example:

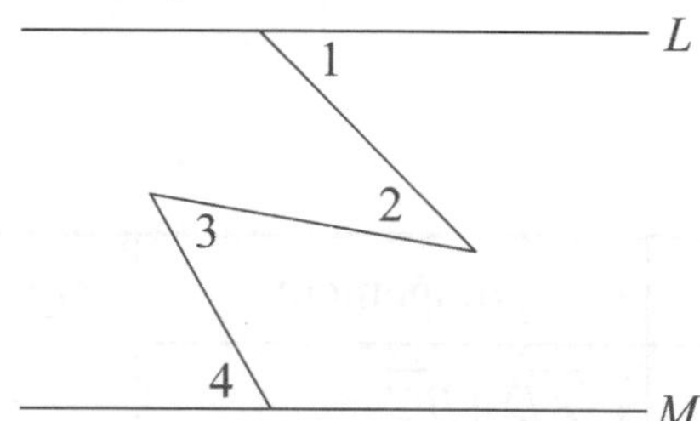

(Note: Figure not drawn to scale)

In the figure above, $\angle 1=50°$, $\angle 4=64°$, $\angle 2+\angle 3=80°$, what is the measure of $\angle 3$?

Solution: Let line *S*, *T* parallel to *L*.

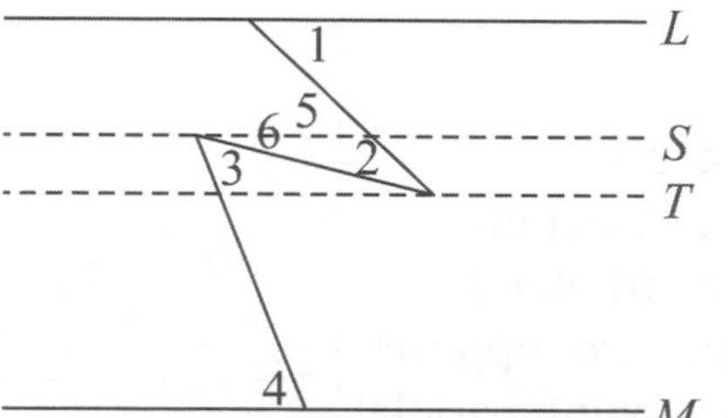

$\angle 5=\angle 1=50°=\angle 2+\angle 6$

$\angle 4=\angle 3+\angle 6=64°$

$\angle 1+\angle 4=50+64=114=\angle 2+\angle 3+2\angle 6$

$2\angle 6=114-\angle 2-\angle 3=34$

$\angle 6=17$

$\angle 3=64-17=47°$

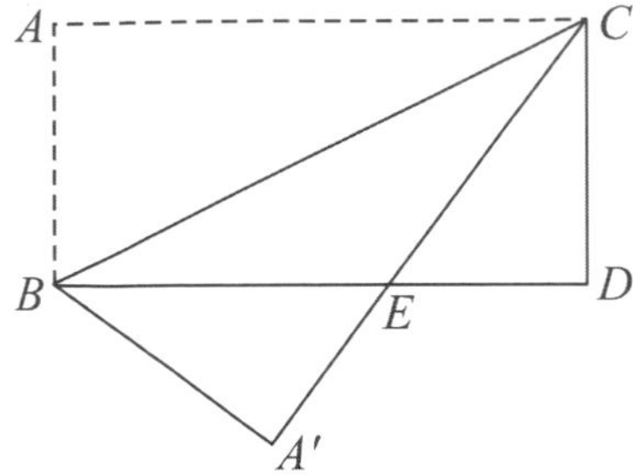

(Note: Figure not drawn to scale)

The figure above shows a rectangle that was folded along the diagonal line. Let the point A landed on point A'.

If $\angle ABC = 72°$, what is the measure of $\angle A'CB$ and $\angle A'ED$?

Solution: $\triangle ABC \cong \triangle A'BC$ ($\overline{AC} = \overline{A'C}$, $\overline{BC} = \overline{BC}$, $\overline{AB} = \overline{A'B}$, SSS), $\angle ACB = \angle A'CB$

$\therefore\ \angle ACB = 90° - 72° = 18°$ $\quad \therefore\ \angle A'CB = 18°$

$\because\ \angle CED = \angle ACB + \angle A'CB = 2 \times 18 = 36°$, ($m\angle ACB = m\angle CBD$)

$\therefore\ \angle A'ED = 180° - 36° = 144°$

9-4 PARALLELOGRAM

Parallelogram	Figures	Properties	Judgement	area
General Parallelogram "Quadrilaterals with two opposite sides being parallel to each other"		1. $\overline{AD} \parallel \overline{BC}$, $\overline{AB} \parallel \overline{CD}$	1. $\overline{AD} \parallel \overline{BC}$, $\overline{AB} \parallel \overline{CD}$ Two opposite sides are parallel	
		2. $\triangle ACD \cong \triangle ABC$, $\triangle ABD \cong \triangle BCD$, $\triangle ABE \cong \triangle CDE$, $\triangle AED \cong \triangle BEC$	S.S.S.	
		3. $\overline{AB} = \overline{CD}$, $\overline{AD} = \overline{BC}$	2. Two opposite sides are equal	
		4. $\angle BAD = \angle BCD$, $\angle ABC = \angle ADC$	3. Two opposite angles are equal	
		5. $\overline{AE} = \overline{EC}$, $\overline{BE} = \overline{ED}$	4. Two diagonals bisect each other	
		6. $\overline{AD} \underset{=}{\parallel} \overline{BC}$ or $AB \underset{=}{\parallel} CD$	5. Two opposite sides are parallel and equal.	

Special Parallelogram 1.Rectangle	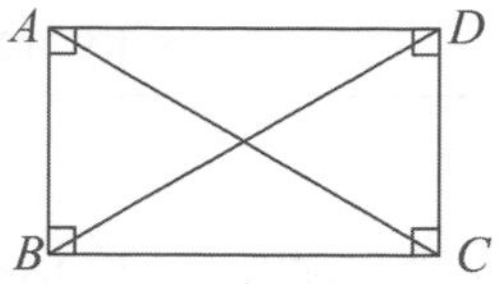	1.interior angles $=90°$ 2. $\overline{AC}=\overline{BD}$	1.parallelogram 2.each angle $=90°$ 3.Diagonals are equal	area $=AB \times BC$
2.Rhombus	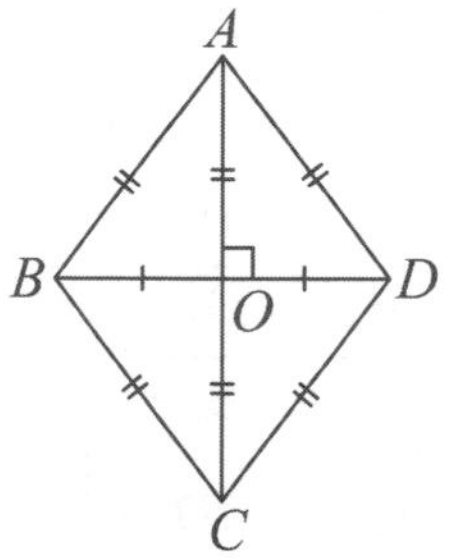	1. $\overline{AB}=\overline{BC}=\overline{CD}=\overline{AD}$ 2. $\overline{AC}\perp\overline{BD}$ 3. $\overline{AO}=\overline{CO}$, $\overline{BO}=\overline{DO}$	1.4 sides are equal 2.diagonals bisect each other and are perpendicular to each other	area $=\frac{1}{2}(BD \times AC)$
3.Square	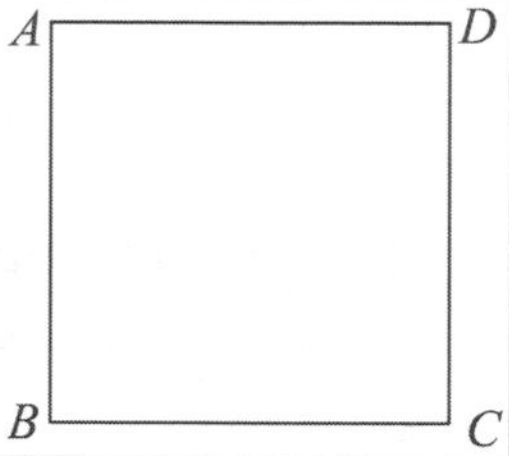	1.all properties of special parallelograms	Rectangle and 4 sides are equal	area $=\overline{CD}^2$

9-5 SPECIAL QUADRILATERALS

	Figure	Characteristics
1.Rhombus	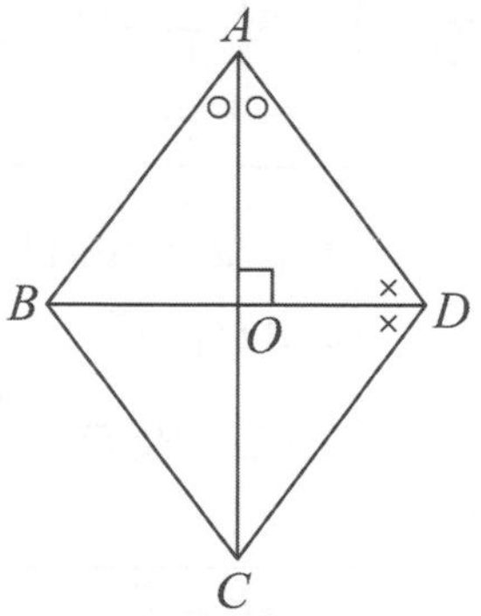	1.parallelogram 2. $\overline{AC}\perp\overline{BD}$ 3. $\overline{AO}=\overline{CO}$, $\overline{BO}=\overline{DO}$ 4. $\angle BAO=\angle DAO$ $\angle BCO=\angle DCO$ $\angle ABO=\angle CBO$ $\angle ADO=\angle CDO$ 5. $\square ABCD=\frac{1}{2}(\overline{AC}\times\overline{BD})$

	Figure	Characteristics
2.Kite	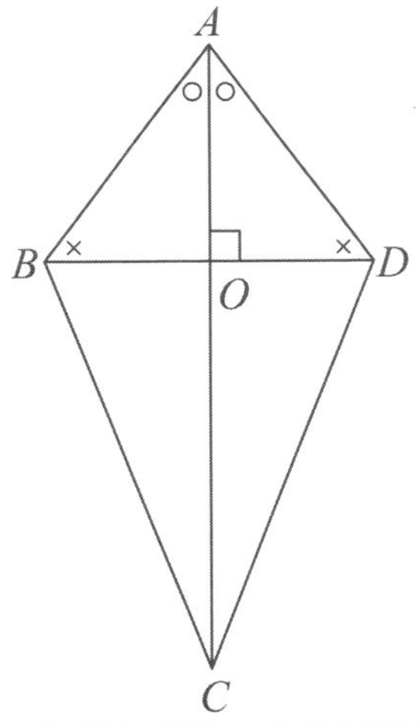 A, B, O, D, C	1."NOT" parallelogram 2. $\overline{BO}=\overline{DO}$ 3. $\angle BAC=\angle DAC$ 4. $\overline{AB}=\overline{AD}$ $\overline{BC}=\overline{DC}$ 5. $\overline{AC}\perp\overline{BD}$ 6. $\square ABCD=\frac{1}{2}\times\overline{AC}\times\overline{BD}$
Rectangle	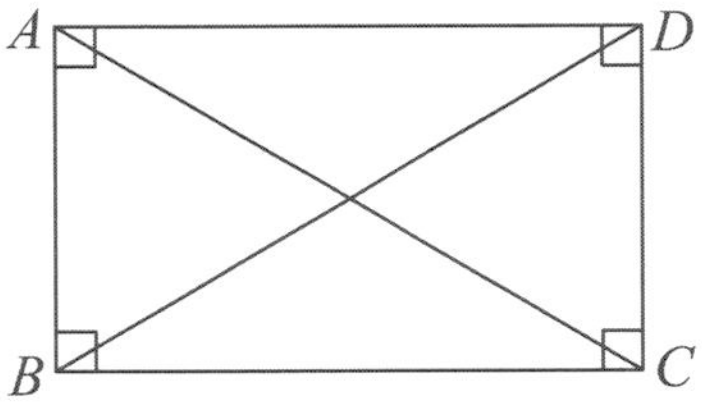 A, D, B, C	1.All Characteristics of parallelogram 2. All internal angles are equal to 90° 3. $\overline{AC}=\overline{BD}$ 4. $\square ABCD=\overline{AB}\times\overline{BC}$
Square	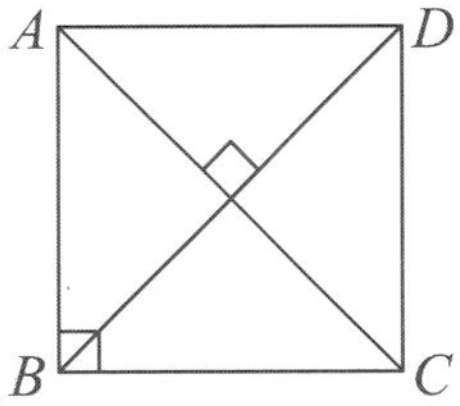 A, D, B, C	1.All characteristics of parallelogram 2. $\overline{AC}=\overline{BD}$ $\overline{AC}\perp BD$ $\square ABCD=\overline{AB}^2$
Trapezoids	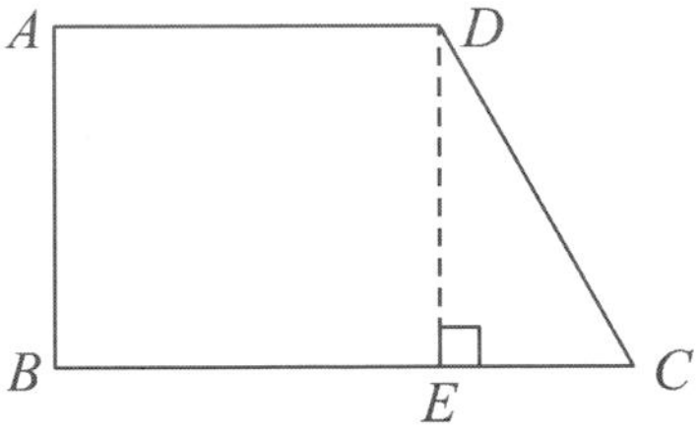 A, D, B, E, C	1.quadrilateral 2. $\overline{AD}\parallel\overline{BC}$ 3. $\square ABCD=\frac{1}{2}[(\overline{AD}+\overline{BC})]\times\overline{DE}$
Isosceles Trapezoids	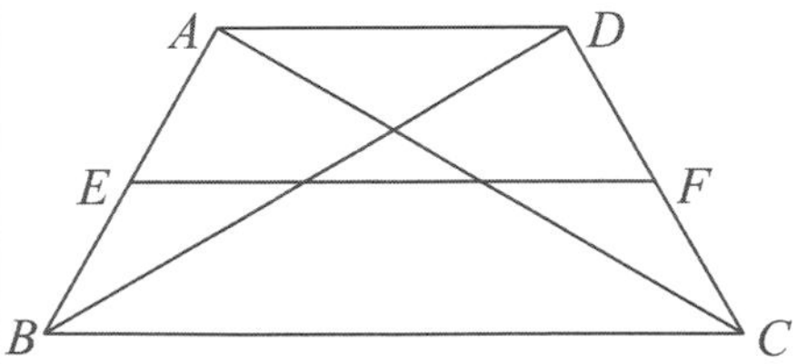 A, D, E, F, B, C	1. $AB=DC$ 2. $AC=DB$ 3.If E and F are the middle points of $\overline{AB}$ and $\overline{CD}$ respectively. then $\overline{EF}=\frac{1}{2}(\overline{AD}+\overline{BC})$

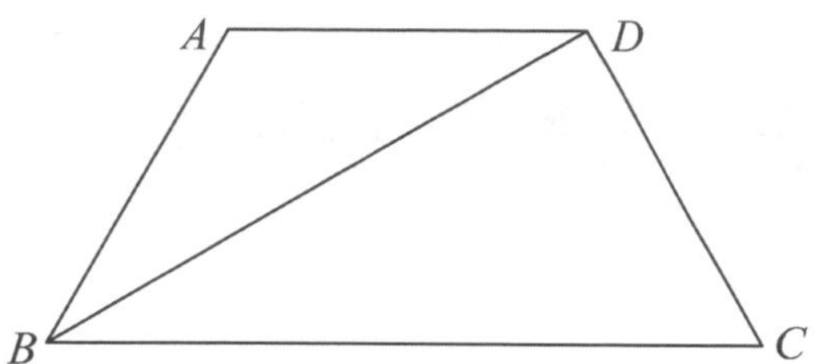

(Note: Figure not drawn to scale)

In the above trapezoid, $\overline{AD} \parallel \overline{BC}$, $\overline{AB} = \overline{CD} = 10$, $\angle C = 60°$, $\angle BDC = 90°$, what is the length of $\overline{BC}$ and area of $\square ABCD$?

Solution: Let $\overline{AE} \perp BC$, $\overline{DF} \perp \overline{BC}$, $\triangle DCF$ is a $30° - 60° - 90°$ triangle, since $AB = DC = 10$, $m\angle DCF = 60°$, so $m\angle CDF = 30°$, since $DC = 10$, $FC = 5$, $DF = \sqrt{10^2 - 5^2} = 5\sqrt{3}$

From $\triangle BDF$, $BD = 2 \times 5\sqrt{3} = 10\sqrt{3}$, $FB = \sqrt{(10\sqrt{3})^2 - (5\sqrt{3})^2} = 15$,

$EF = BF - BE = 15 - 5 = 10$, $BC = 15 + 5 = 20$

area $= \dfrac{1}{2}(10 + 20) \times 5\sqrt{3} = 75\sqrt{3}$

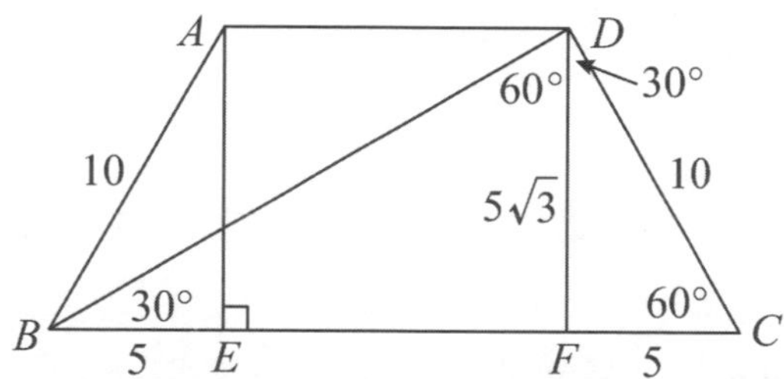

(Note: Figure not drawn to scale)

Answer to Concept Check 9-09

(A) $\overline{AD} = \overline{DC}$ and $\overline{AD} \parallel \overline{BC}$

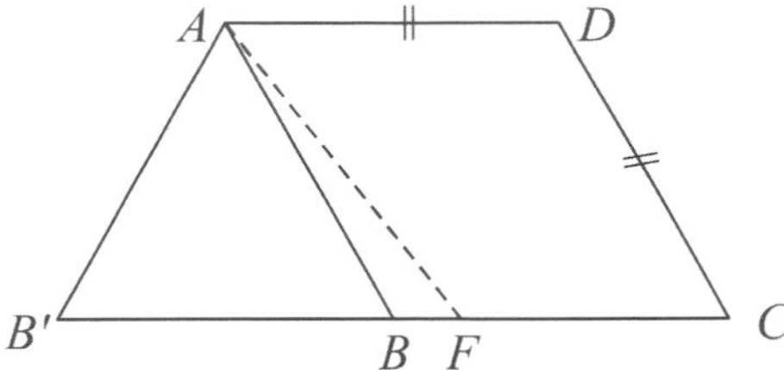

$\square ADCF$ and $\square ABCD$ are not parallelogram

(B) $\overline{AD} = \overline{BC}$ and $\overline{AB} = \overline{DC}$

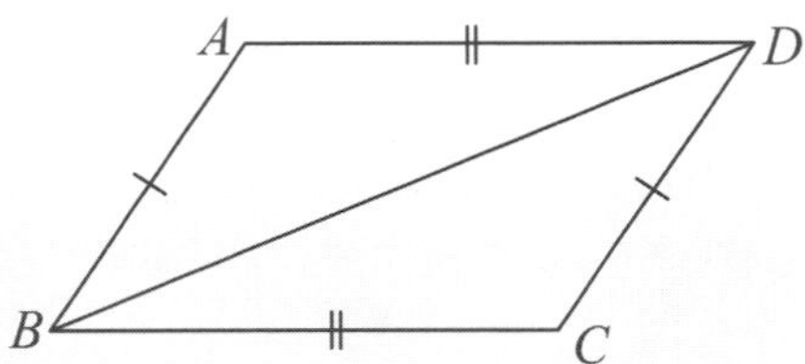

$\triangle ABC \cong \triangle BDC$ (S.S.S.)

$\angle ADB = \angle DBC$ (alternate interior angles)

$\therefore\ \overline{AD} \parallel \overline{BC}$ and $\overline{AB} \parallel \overline{CD}$

It is a parallelogram

(C) $\overline{AD} = \overline{BC}$ and $\overline{AD} \parallel \overline{BC}$

$\angle ADB \cong \angle DBC$ (alternate interior angles)

$\triangle ABD \cong \triangle BDC$ (S.A.S.)

$\therefore\ \overline{AB} = \overline{CD}$

It is a parallelogram

(D) $\overline{AD} \parallel \overline{BC}$ and $\overline{AB} \parallel \overline{DC}$

$\angle DBC = \angle ADB$, $\angle BDC = \angle ABD$, $\overline{BD} = \overline{BD}$

$\therefore\ \triangle ABD \cong \triangle BDC$ (A.S.A), $\therefore\ \overline{AD} = \overline{BC}$ and $\overline{AB} = \overline{CD}$

$\square ABCD$ is a parallelogram.

Answer is (A)

Answer to Concept Check 9-10

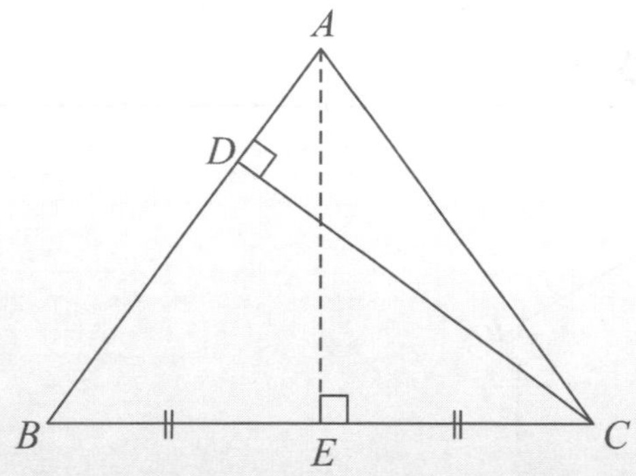

Draw $\overline{AE} \perp \overline{BC}$

$\overline{BE} = \overline{EC} = \frac{32}{2} = 16$ ($AB = AC$, AE bisects $\overline{BC}$)

$\overline{AB} = 20$, $\overline{AE}^2 = \overline{AB}^2 - \overline{BE}^2 = 20^2 - 16^2 = 400 - 256 = 144$

$\therefore$ $\overline{AE} = 12$

$\triangle ABC = \frac{1}{2}(12) \cdot (32) = 192$

$\triangle ABC = \frac{1}{2}(\overline{CD} \times \overline{AB}) = \frac{1}{2}(\overline{CD} \times 20) = 192$

$\therefore$ $\overline{CD} = 19.2$

Answer to Concept Check 9-11

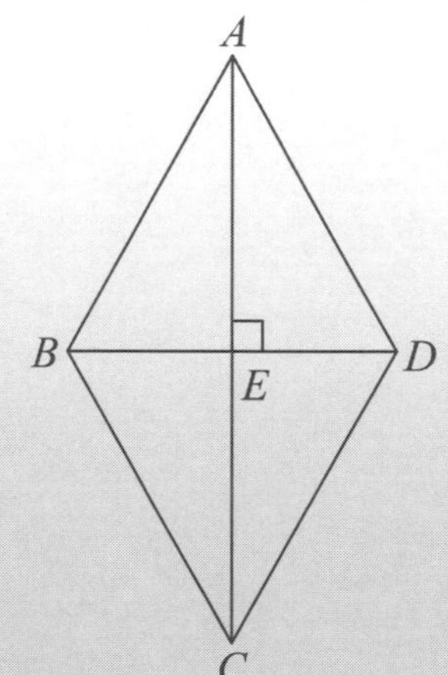

1. connect point A and point C then $\overline{AC} \perp BD$ at E,
   $\overline{BE} = \overline{ED} = \frac{1}{2}\overline{BD} = 3$ (property of the rhombus)
2. $\overline{AE}^2 = \overline{AD}^2 - \overline{ED}^2 = 25 - 9 = 16$
   $\therefore$ $\overline{AE} = 4$
3. The measure of the area of $\square ABCD$ is
   $\frac{1}{2}(\overline{AC} \times \overline{BD}) = \frac{1}{2}(4 \times 2 \times 6) = 24$
   (The property of the rhombus)

9-6 POLYGONS

Sum of the internal angles of an n-sided polygon

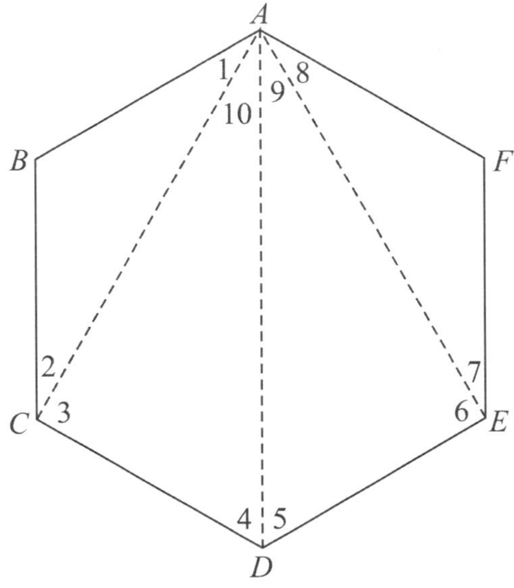

(Note: Figure not drawn to scale)

n-sided polygon can have exactly $(n-2)$ triangles

The sum of the internal angles $(n-2)\times 180°$

The sum of the internal angles of a six-sided polygon $=(6-2)\times 180° = 720°$

Sum of the external angles of n-sided polygon is always equal to 360°

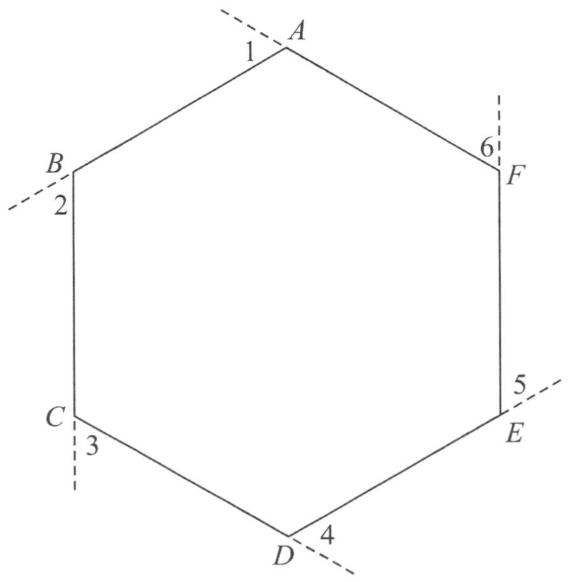

(Note: Figure not drawn to scale)

For instance, the six-sided polygon as above, the sum of external angles of
$\angle 1+\angle 2+\angle 3+\angle 4+\angle 5+\angle 6=(180°-\angle A)+(180°-\angle B)+(180°-\angle C)+(180°-\angle D)$
$+(180°-\angle E)+(180°-\angle F)=6\times 180°-(6-2)\times 180°=2\times 180°=360°$

Regular polygons

- Every internal angle has the some measure in degrees, and every side has the same measure in lengths. For instance, regular five-sided, regular six-sided, and regular eight-sided polygon.

Figure are shown below

Regular five-sided	Regular six-sided	Regular eight-sided
Pentagon	Hexagon	Octagon
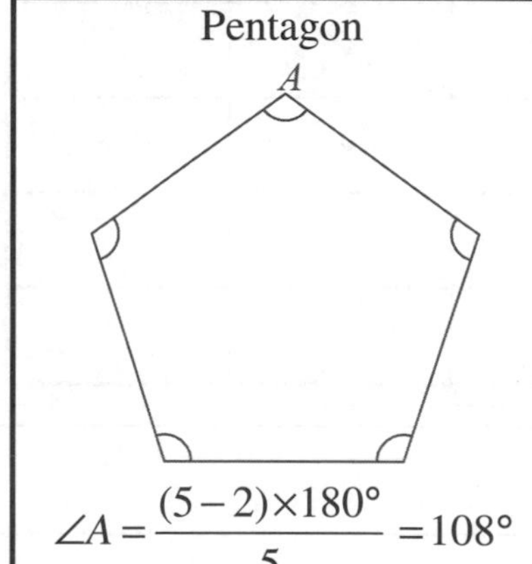	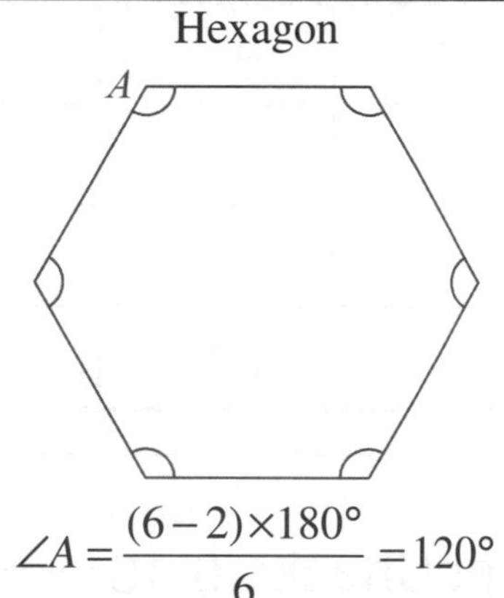	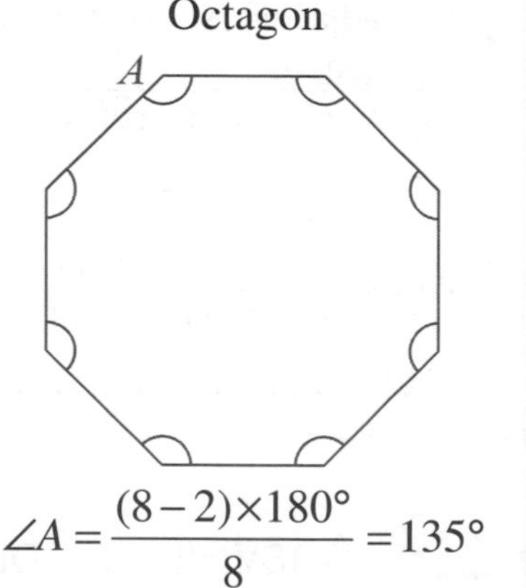
$\angle A=\frac{(5-2)\times 180°}{5}=108°$	$\angle A=\frac{(6-2)\times 180°}{6}=120°$	$\angle A=\frac{(8-2)\times 180°}{8}=135°$

◆Example 1:

What is the measure, in degrees, of an interior angle of a pentagon ?

$\frac{(5-2)\times 180}{5}=108$

◆Example 2:

What is the sum of all exterior angles of a hexagon ?

Sum of exterior angles of n-sided polygon is $360°$

◆Example 3:

What is the sum of all interior angles of a octagon ?

sum $=(8-2)\times 180°=1080°$

Answer to Concept Check 9-12

Name	Parallelogram	Rectangle	Square	Rhombus	kite
Figures					
Two pairs of parallel sides	✓	✓	✓	✓	
Opposite sides are of equal length	✓	✓	✓	✓	
Opposite angles are equal	✓	✓	✓	✓	
4 angles are equal		✓	✓		
4 sides are equal			✓	✓	
Two adjacent sides are of equal length			✓	✓	✓
The diagonals bisect each other	✓	✓	✓	✓	
The diagonals are perpendicular to each other			✓	✓	✓
The diagonals are equal of length		✓	✓		

Answer to Concept Check 9-13

1.

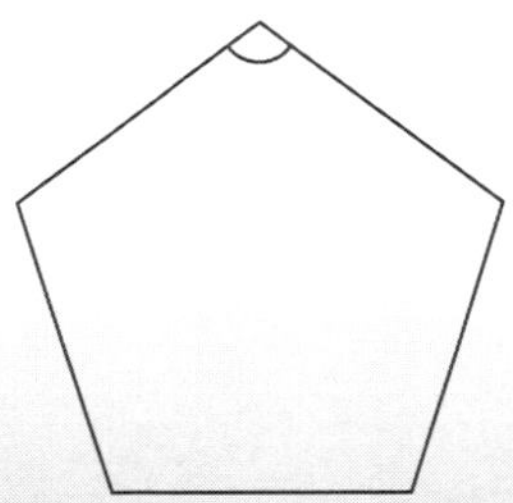

$(5-2)\times 180 = 540°$

2. 360° sum of all external angles is 360°

3. $\frac{(5-2)\times 180}{5} = 108°$

4. $(6-2)\times 180° = 720°$

5. $\frac{720°}{6} = 120°$

9-7 SIMILAR TRIANGLES

- Corresponding angles have the same measures
- Corresponding sides and heights are proportional

Properties	Figures	Description
1. A.A.	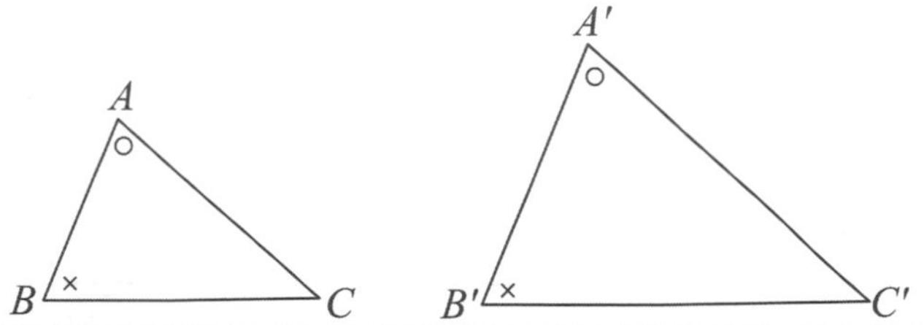 A, B, C; A′, B′, C′	$\angle A = \angle A'$, $\angle B = \angle B'$ Similar Triangles
2. S.A.S.	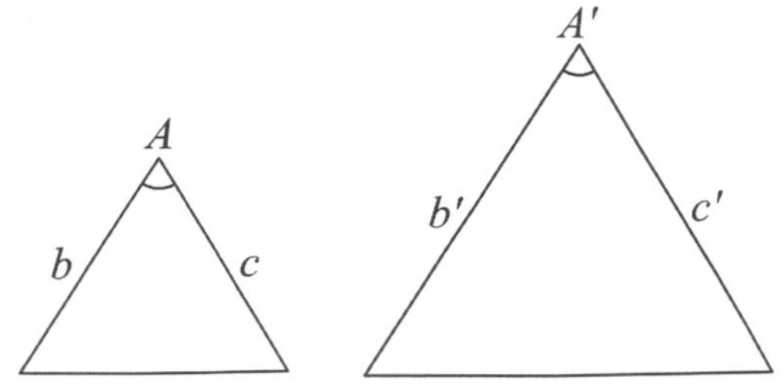 A, b, c; A′, b′, c′	$\angle A = \angle A'$, $\dfrac{b}{b'} = \dfrac{c}{c'}$
3. S.S.S.	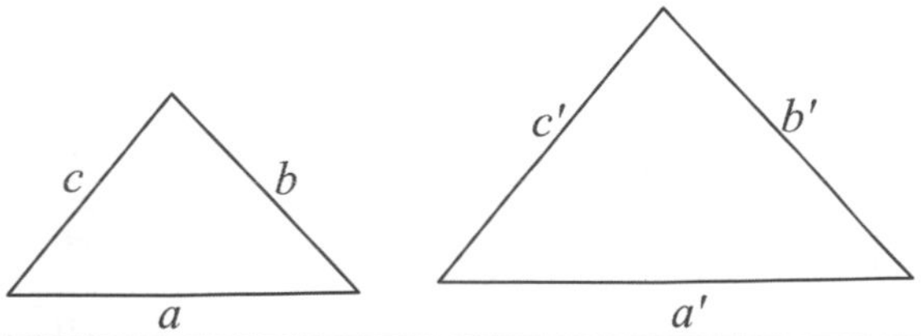 c, b, a; c′, b′, a′	$\dfrac{a}{a'} = \dfrac{b}{b'} = \dfrac{c}{c'}$

• More Properties of Similar Triangles

	Properties	Figures	Descriptions
1.	All corresponding sides and all corresponding lines are proportional	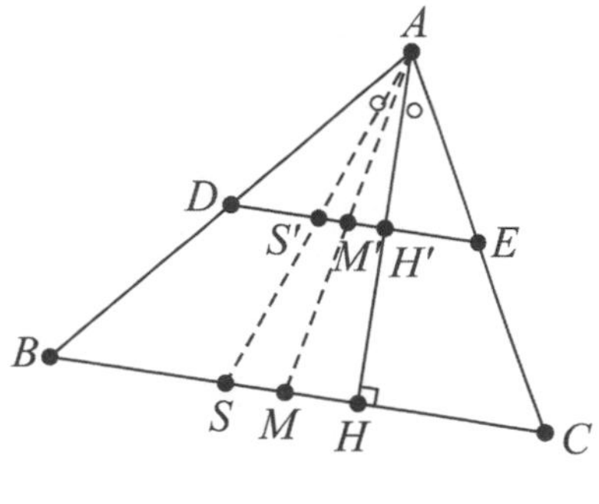 A, D, E, S′, M′, H′, B, S, M, H, C	1. $\dfrac{\overline{AD}}{\overline{AB}} = \dfrac{\overline{AE}}{\overline{AC}} = \dfrac{\overline{AS'}}{\overline{AS}} = \dfrac{\overline{AM'}}{\overline{AM}} = \dfrac{\overline{AH'}}{\overline{AH}} = \dfrac{\overline{DE}}{\overline{BC}} = \dfrac{\overline{AD}+\overline{AE}+\overline{DE}}{\overline{AB}+\overline{BC}+\overline{AC}}$

	Properties	Figures	Descriptions
2.	$\frac{\text{Area of } \triangle ABC}{\text{Area of } \triangle A'B'C'} = \frac{\text{Side}^2 \text{of } \triangle ABC}{\text{Side}^2 \text{of } \triangle A'B'C'}$		2. $\frac{\triangle ABC}{\triangle ADE} = \frac{\frac{1}{2} AH \cdot BC}{\frac{1}{2} AH' \cdot DE}$ $= \frac{\overline{AH}}{\overline{AH'}} \times \frac{\overline{BC}}{\overline{DE}}$ $= \frac{\overline{AB}}{\overline{AD}} \times \frac{\overline{AB}}{\overline{AD}}$ $= \frac{\overline{AB}^2}{\overline{AD}^2}$
	Connecting line of the middle points. $\triangle ADE \backsim \triangle ABC$	A, D, E, B, C	If $\overline{AD} = \overline{BD}$, and $\overline{AE} = \overline{EC}$, then $\overline{DE} = \frac{1}{2}\overline{BC}$ and $\overline{DE} \parallel \overline{BC}$
	Mother-child right triangles are similar triangles 1. $\overline{BC}^2 = \overline{AC} \times \overline{DC}$ 2. $\overline{BD}^2 = \overline{AD} \times \overline{DC}$	A, D, B, C	$\triangle ABC$ and $\triangle BDC$ $\angle ABC = \angle BDC = 90°$ $\angle C = \angle C$ $\therefore \triangle ABC \sim \triangle BDC$ 1. $\frac{\overline{AC}}{\overline{BC}} = \frac{\overline{BC}}{\overline{DC}}$ $\overline{BC}^2 = \overline{AC} \times \overline{DC}$ 2. $\triangle ABD \sim \triangle BDC$ $\frac{\overline{AD}}{\overline{BD}} = \frac{\overline{BD}}{\overline{DC}}$ $\overline{BD}^2 = \overline{AD} \times \overline{DC}$

◆Example 1:

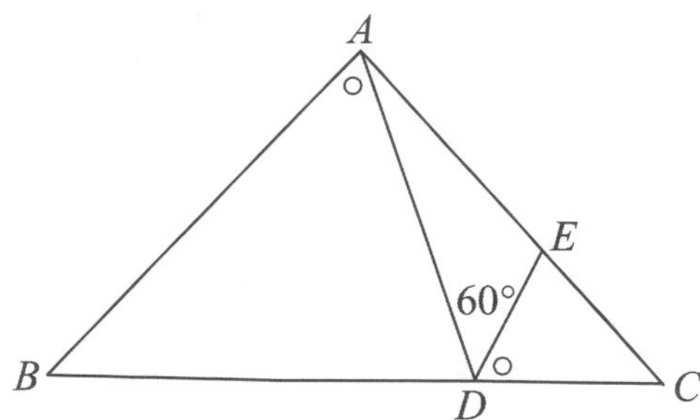

(Note: Figure not drawn to scale)

In the above equilateral triangle *ABC*, if $\angle ADE = 60°$, $BD = 6$ and $CE = 3$, what is the length of $\overline{AB}$?

Solution: $\because$ $\angle EDC + 60° + \angle ADB = 180°$

$\angle BAD + 60° + \angle ADB = 180°$, $\angle EDC = \angle BAD$

$\angle B = \angle C = 60°$, $\triangle ABD \sim \triangle EDC$

$$\frac{\overline{EC}}{\overline{BD}} = \frac{\overline{DC}}{\overline{AB}}$$

$$\frac{3}{6} = \frac{\overline{DC}}{\overline{AB}} = \frac{\overline{DC}}{\overline{BC}} \quad (AB = BC)$$

$$\frac{\overline{DC}}{\overline{BC}} = \frac{1}{2}, \quad DC = \frac{1}{2} \times 6 = 3$$

$$AB = BC = 2 \times 3 = 6$$

◆Example 2:

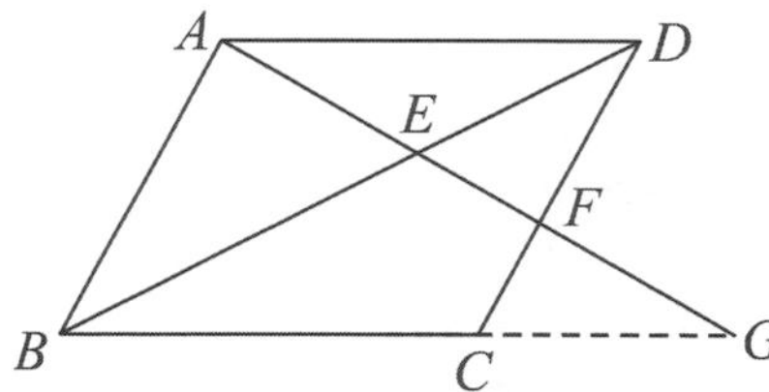

(Note: Figure not drawn to scale)

In parallelogram *ABCD* above, extend $\overline{BC}$ to point *G*, let $CG = \frac{1}{3}\overline{BC}$, $\overline{AG}$ intersects $\overline{BD}$ at E and $\overline{CD}$ at *F*, what is the value of $\overline{AE} : \overline{EF}$?

Solution: $\triangle CFG \sim \triangle AFD$, $(\angle AFD + \angle CFG,\ \angle DAG - \angle BGA)\dfrac{\overline{CG}}{\overline{AD}} = \dfrac{1}{3} = \dfrac{\overline{CF}}{\overline{DF}}$

$$\frac{\overline{FD}}{\overline{CD}} = \frac{3}{4}, \quad \triangle ABE \sim \triangle EFD$$

$$\frac{\overline{EF}}{\overline{AE}}=\frac{\overline{DF}}{\overline{CD}}=\frac{3}{4}(\overline{AB}=\overline{CD})$$

$$\frac{\overline{AE}}{\overline{EF}}=\frac{4}{3}$$

◆Example 3:

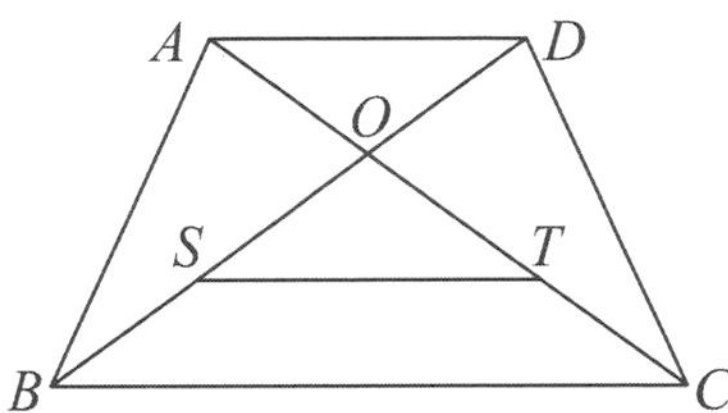

(Note: Figure not drawn to scale)

The trapezoid *ABCD* shown above, $\overline{AC}$ intersects $\overline{BD}$ at *O*, $\overline{ST}\parallel\overline{BC}$, $\overline{DS}:\overline{BS}=3:2$, $AD=14$, $BC=30$, what is the length of $\overline{ST}$?

Solution:

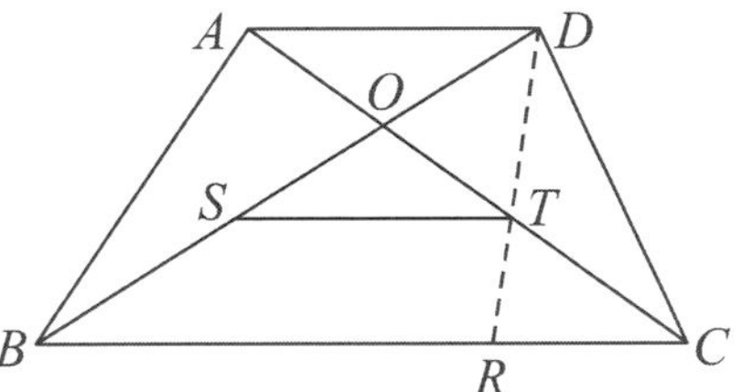

(Note: Figure not drawn to scale)

Draw a line from *D* passes through point *T* and intersects $\overline{BC}$ at *R*.

$\overline{AD}:\overline{RC}=\overline{AT}:\overline{TC}=\overline{DS}:\overline{SB}=3:2$ $\left(\frac{\overline{AT}}{\overline{TC}}=\frac{\overline{DT}}{\overline{TR}}=\frac{\overline{DS}}{\overline{SB}}\right)$

$\therefore\ 14:\overline{RC}=3:2$ $\quad\therefore\ \overline{RC}=\frac{28}{3}$

$\overline{BR}=30-\frac{28}{3}=\frac{62}{3}$, $\overline{ST}:\overline{BR}=\overline{DS}:\overline{BD}=3:(3+2)=3:5$

$\therefore\ \overline{ST}=\frac{3}{5}\times\frac{62}{3}=\frac{62}{5}$

Answer to Concept Check 9-14

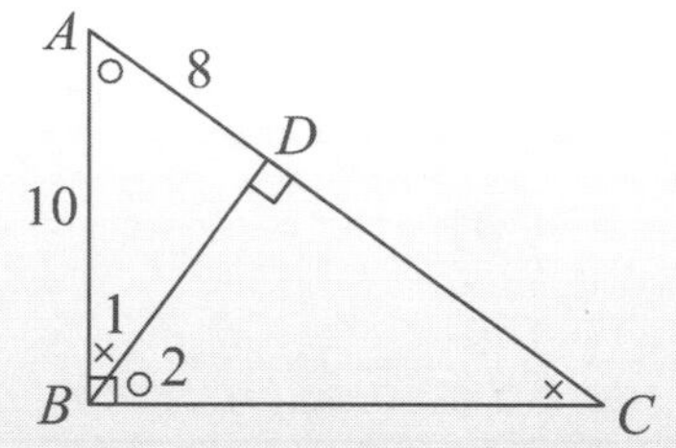

$BD^2 = AB^2 - AD^2 = 10^2 - 8^2 = 36$,

$BD = 6$

$\triangle ABD \sim \triangle BDC$

$\dfrac{AD}{BD} = \dfrac{AB}{BC}$, $\dfrac{8}{6} = \dfrac{10}{BC}$

$\therefore\ BC = \dfrac{10 \times 6}{8} = \dfrac{15}{2}$

9-8 CIRCLE

Term	Figures	Description
Circle		It is a shape which is formed by all the points that have the equivalent distances from a fixed point (called center), denoted as $\odot$.
Radius, Diameter		The equivalent distance is called radius; extend radius to the edge of the circle is called "Diameter", which is twice the length of radius.
Circumference		The perimeter of a circle. $c = \pi \cdot D = 2 \cdot \pi \cdot R$ D the diameter, R is the radius, and π is Circumference/Diameter, which is an infinite decimal 3.14159⋯

Term	Figures	Description
Area	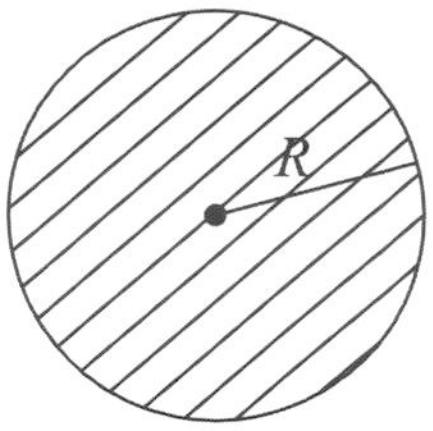 R	$A = \pi R^2$
Sector	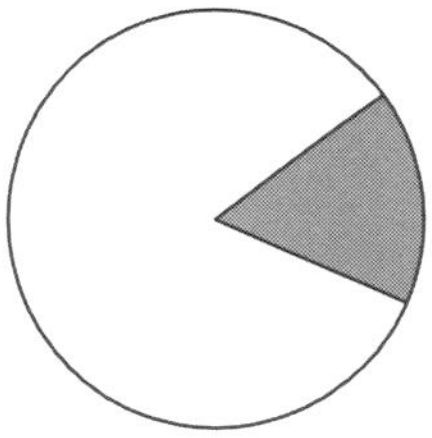	A portion of a circle's area which is formed by two radius.
Central Angle	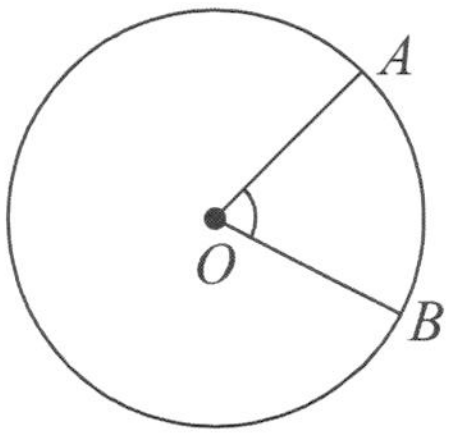 A O B	The angle between two radii. $\angle AOB$
Arc		Arc is a portions of circle's circumference formed by any two points on the edge of the circle.
	Formula: $\dfrac{m\angle AOB}{360°} = \dfrac{m\overset{\frown}{AB}}{2\pi R}$ $= \dfrac{\text{area of sector AOB}}{\pi R^2}$	denoted $\overset{\frown}{AB}$
Inscribed angle	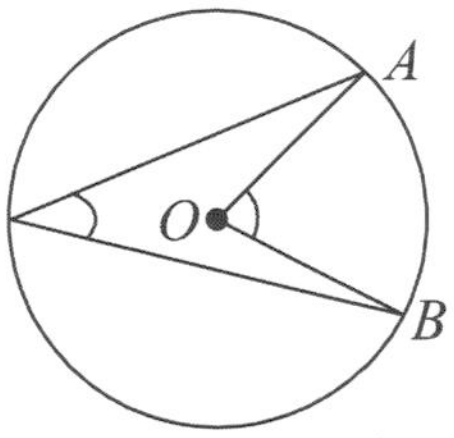 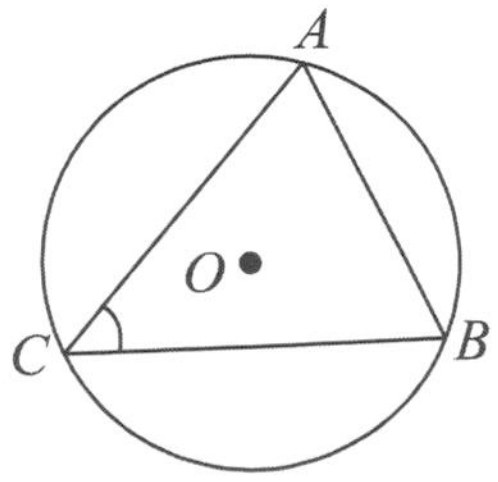 A O B A O C B	$m\angle AOB = m\overset{\frown}{AB}$ $m\angle ACB = \dfrac{1}{2} m\overset{\frown}{AB}$ $\angle C + \angle A + \angle B$ $= 180° = \dfrac{1}{2}$ circumference $\therefore\ \angle C = \dfrac{1}{2} m\overset{\frown}{AB}$

Term	Figures	Description
Chord	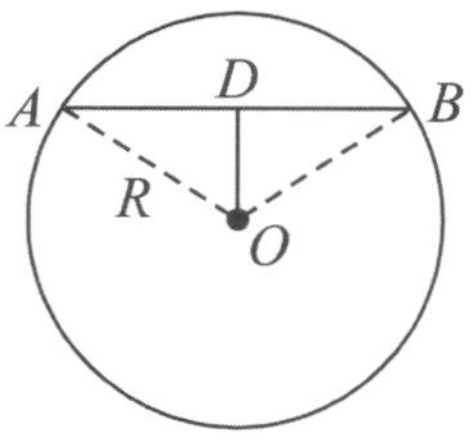	A line connecting 2 points that lie on the circumference of the circle. • $\overline{OD} \perp \overline{AB}$, $\overline{AD} = \overline{BD}$ $R^2 = \overline{OD}^2 + (\frac{1}{2}\overline{AB})^2$
Tangent Chord Angle	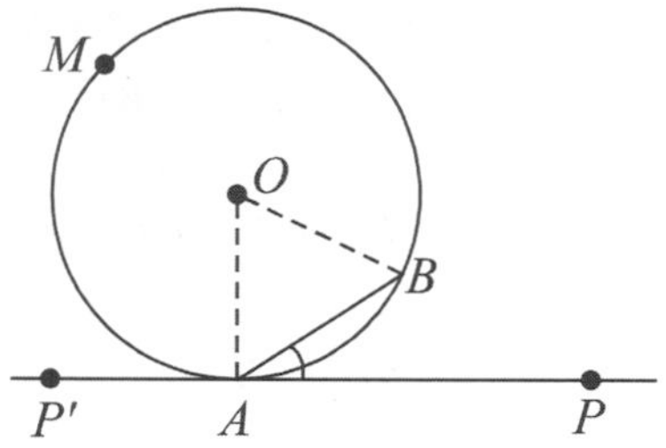	A line $\overleftrightarrow{PA}$ tangent O at one point A 1. $m\angle OAB = m\angle OBA$ ($OA = OB = R$) $\triangle OAB$ is an isosceles triangle) 2. $m\angle AOB = 180° - m\angle OAB - m\angle OBA$ $= 180° - 2m\angle OAB$ 3. $m\angle PAB = 90° - m\angle OAB$ $2m\angle PAB = 180° - 2m\angle OAB$ $= m\angle AOB$ 4. $m\angle PAB = \frac{1}{2} m\angle AOB = \frac{1}{2}\overset{\frown}{AB}$
	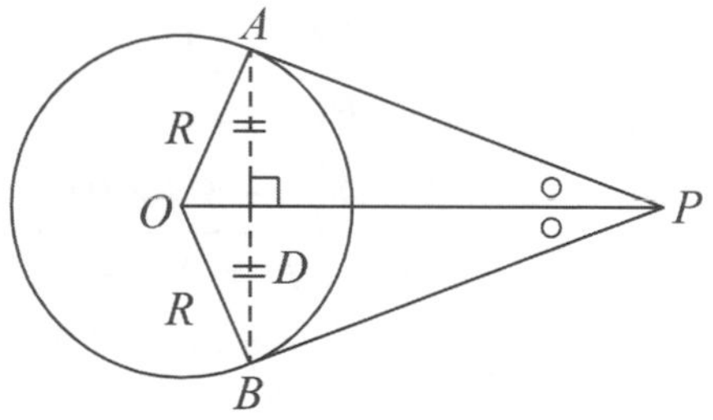	1. $\overline{PA} = \overline{PB}$ 2. $\angle APO = \angle BPO$ 3. $OA = OB = R$ 4. $\overline{AB} \perp \overline{PO}$ 5. $\overline{AD} = \overline{BD}$ 6. $m\angle AOB + m\angle APB = 180°$ Reason: 1. $\triangle AOP \sim \triangle POB$ 2. $\overline{AO} = \overline{BO}$, and $\overline{AP} = \overline{BP}$ $\square AOBP$ is a kite diagonals are perpendicular to each other.
chord-chord angles	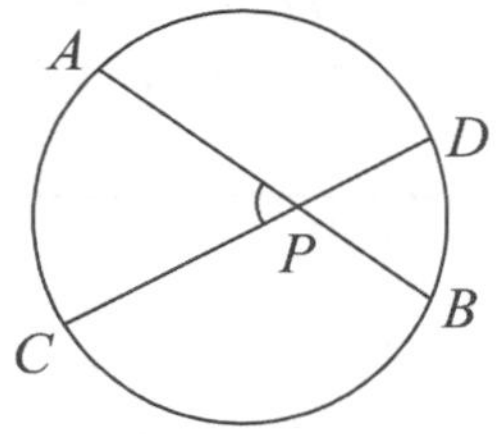	$m\angle APC = \frac{1}{2}(m\overset{\frown}{AC} + m\overset{\frown}{BD})$

Term	Figures	Description
secant-secant angles	A B P D C	$m\angle P = \frac{1}{2}(m\widehat{AC} - m\widehat{BD})$
chord-chord power theorem	A D E B C	$\frac{AE}{DE} = \frac{CE}{BE}$ $AE \times BE = CE \times DE$ $(\triangle AEC \sim \triangle BED)$
secant-secant power theorem	A B P O D C	$PB \times PA = PD \times PC$ $(\triangle PBD \sim \triangle PAC)$ $\frac{\overline{PB}}{\overline{PC}} = \frac{\overline{PD}}{\overline{PA}}$ $m\angle BDP = m\angle BAC$, $m\angle P = m\angle P$
secant-tangent power theorem	C O A B P	$\angle P = \angle P$ $m\angle PCB = m\angle ACB$ $\therefore\ \triangle PBC \sim \triangle APC$ $\frac{\overline{PB}}{\overline{PC}} = \frac{\overline{PC}}{\overline{PA}}$ $\therefore\ \overline{PC}^2 = \overline{PA} \times \overline{PB}$
Inscribed circle for quadrilateral	E A D F H O B G C	$AB + CD = AD + BC$ $(\triangle AEO \cong \triangle AOF)$ $AE = AF$ $FB = BG$ $AB = AF + FB = AE + BG$ $CD = DH + HC = ED + GC$ $AB + CD = AE + ED + BG + GC$ $= AD + BC$
Circumscribed circle for quadrilateral	A B O D C	$m\angle ABC + m\angle ADC = 180°$

Answer to Concept Check 9-15

The circumference $=2\pi \cdot 3 = 6\pi$

$\frac{6\pi}{6}=\frac{2\pi}{x}$ $\quad x=\frac{2\pi}{6\pi}\times 6=2$, $\angle AOB = 2$ radians

$m\angle ACB = \frac{1}{2}m\angle AOB = \frac{2}{2}=1$ radian

Answer to Concept Check 9-16

$m\angle AOB = 2m\angle PAB$

$\therefore\ m\angle PAB = 80 \div 2 = 40°$

Answer to Concept Check 9-17

The circumference of the circle $=2\pi r = 2\pi \cdot (4) = 8\pi$

$\therefore\ \widehat{AC}+\widehat{BD}=120°$

$\frac{120}{360}=\frac{x}{8\pi}$ $\quad \therefore\ x=\frac{8}{3}\pi$

Answer to Concept Check 9-18

$\because\ \overline{PC}^2 = \overline{PA}\times\overline{PB}$

$\because\ \overline{PA}=\frac{\overline{PC}^2}{\overline{PB}}=\frac{16}{3}$

◆Example 1:

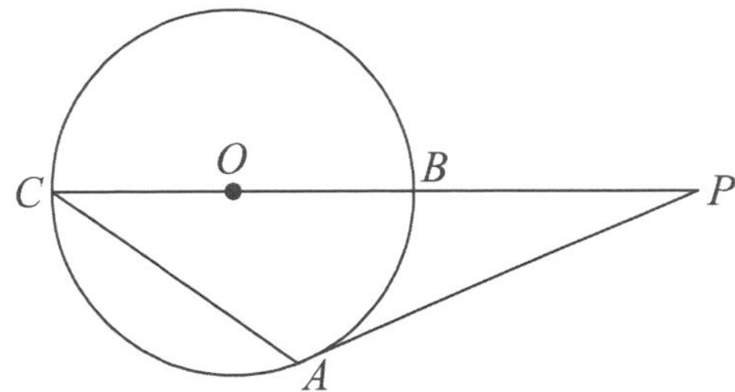

(Note: Figure not drawn to scale)

In the figure above, $\overrightarrow{PA}$ is tangent to circle $\odot$ at point A, if $\angle C = 30°$, and $\overline{PA} = 6\sqrt{3}$ what is the length of $\overline{PB}$?

Analysis:

1.Asking $PB =$?

2.If we know OP then we know PB ($OP = PB + \text{radius}$)

3.From figure above there is no any valuable information.

4.$m\angle C = 30°$, it is a special angle

5.The hidden information, $OC = OA$, and if we connect $\overline{OA}$, then $\angle OAC$ is a special angle which is $30°$

6.$\triangle OAP$ is a right triangle (hidden $\overline{OA} \perp \overline{AP}$)

7.If we know $\overline{AP}$, maybe we will know $\overline{OP}$.

8.Start connecting OA.

Solution: Connecting $\overline{OA}$

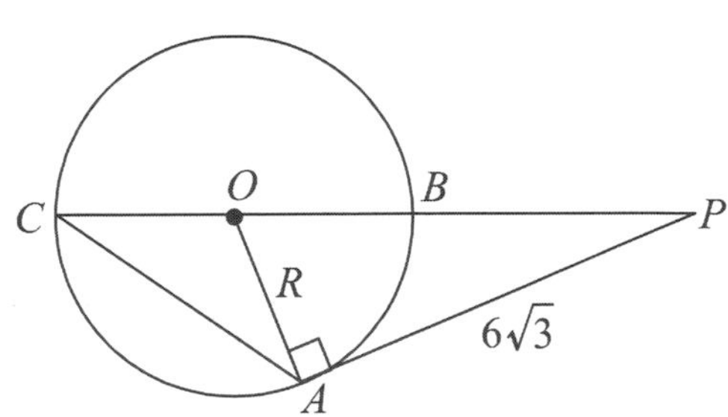

$\angle C = \angle CAO = 30°$

$\angle COA = 180 - 30 - 30 = 120°$

$\angle AOP = 180 - 120 = 60°$

$\therefore\ \angle OPA = 180 - 60 - 90 = 30°$

It is a ($30° - 60° - 90°$)

$OA = R,\ OP = 2R \quad (2R)^2 = R^2 + (6\sqrt{3})^2 \quad 3R^2 = 36 \cdot 3 \quad R^2 = 36 \quad \therefore\ R = 6$

$\therefore\ \overline{OP} = 2R = 2 \times 6 = 12,\ \overline{BP} = 12 - 6 = 6$

◆Example 2:

In the figure below, the diameter of the big semicircle and the diameter of the small semi-circle are overlaid and the radius of big semicircle is twice the radius of small circle. $\overline{AB}$ is tangent to the small-semi circle at point *F*. If the length of $\overline{AB}$ is 20, what is the measure of the shaded area ?

Analysis:

The information needed:

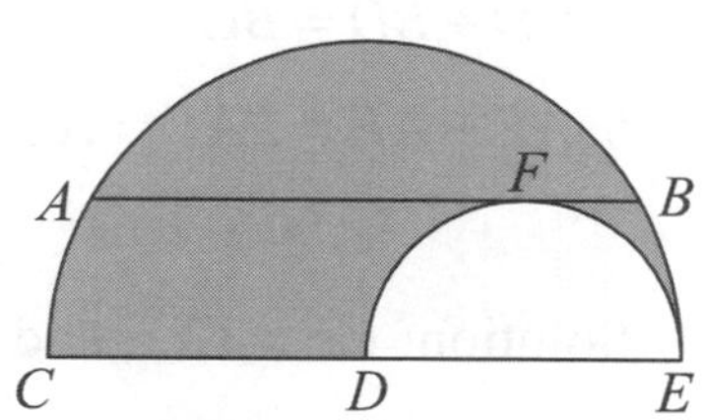

1. The area of bigger circle and the area of smaller circle
2.The radii of two circles.
3.Given $AB = 20$
4.Hidden information; the radius is perpendicular to the chord.
5.Find the radii of both circles.

Solution:

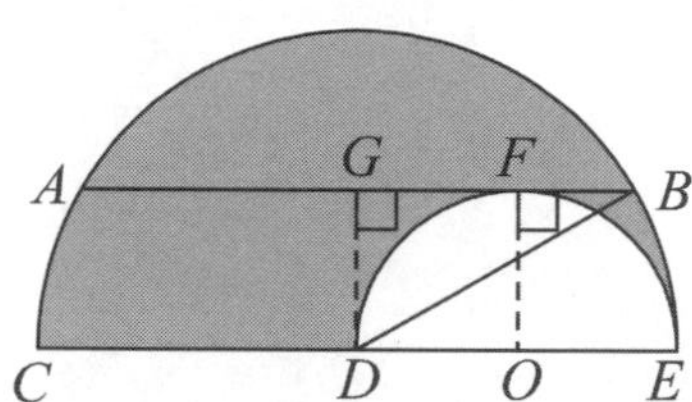

(Note: Figure not drawn to scale)

Set point *D* as the center of big circle connecting $\overline{DB}$, and make $\overline{DG} \perp \overline{AB}$ at point G. The area of big semi-circle $= \frac{1}{2}\pi(\overline{DB})^2$.

The area of small semi-circle $= \frac{1}{2}\pi(\overline{DG})^2$ (For $\overline{DG}$ = the radius of smell circle)

measure of the shaded area $= \frac{1}{2}\pi(\overline{DB}^2 - \overline{DG}^2) = \frac{1}{2}\pi\overline{GB}^2 = \frac{1}{2}(\pi)(10)^2 = 50\pi$

◆Example 3:

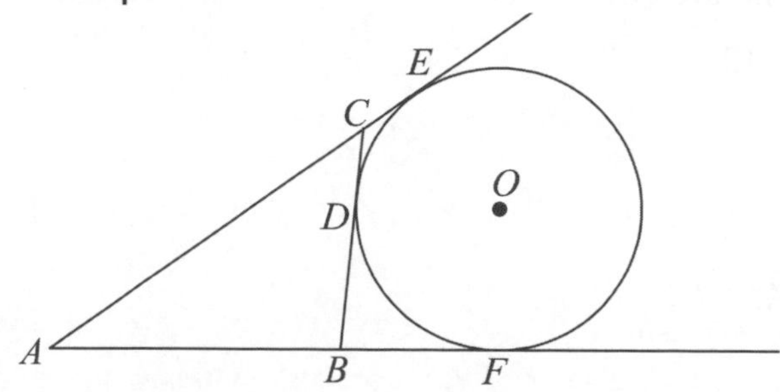

(Note: Figure not drawn to scale)

In the figure above, $\overline{AE}$, $\overline{BC}$, and $\overline{AF}$ are tangent to circle $\odot$ at point *E*, *D*, and *F* respectively. If $AC = 20$, $AB = 18$, and $BC = 12$, what is the length of $\overline{AE}$?

Analysis:

1.Given $AC = 20$, $AB = 18$, $BC = 12$

2. $AE = AC + CE = AC + CD$

3.Hidden information: $CE = DC$, $BD = BF$, $AE = AF$

4. $CD + BD = BC$

5. $AE + AF = 2AE = AC + CD + AB + BD$ got it

Solution: $\overline{CE} = \overline{CD}$, and $\overline{BD} = \overline{BF}$, $\overline{AE} = \overline{AF}$

The perimeter of $\triangle ABC = \overline{AB} + \overline{BC} + \overline{AC} = \overline{AB} + \overline{BF} + \overline{AC} + \overline{CE}$

$= \overline{AE} + \overline{AF} = 2\overline{AE}$

$\therefore\ 20 + 18 + 12 = 2AE \quad \therefore\ AE = 25$

◆Example 4:

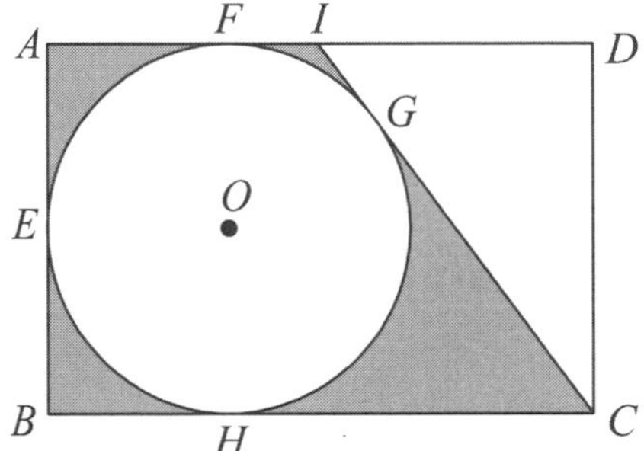

In the figure above, $\square ABCD$ is a rectangle and the four sides of trapezoid *ABCI* are tangent to $\odot\ O$ at *E*, *F*, *G*, *H*, respectively. If $AB = 12$, $CI = 13$ what is the measure of the shaded area ?

Analysis:

1.We have to know the area of $\square ABCI$ and area of $\odot\ O$.

2.Given $AB = 12$, $CI = 13$, we know CD=12

3. $DI^2 = CI^2 - CD^2 = 13^2 - 12^2 = 25$

4.Hidden information: $FO = R = \frac{1}{2}AB = 6$

5.We know area of $\triangle CDI$ and rectangle *ABCD*.

6. The problem could be solved.

Solution:

$DI^2 = CI^2 - CD^2 = 13^2 - 12^2 = 25 \quad \therefore \ DI = 5$

$\because \ AB + CI = AE + BE + CG + GI = AI + BC = AI + AI + DI$

$= 2AI + DI = 12 + 13$

$2AI + 5 = 25 \quad \therefore \ AI = 10, \ BC = 10 + 5 = 15$

$\therefore$ measure of shaded area $= 12\times15 - \pi \cdot (36) - \dfrac{1}{2}(12)(5) = 180 - 30 - 36\pi$

$= 150 - 36\pi = 36.96$

◆Example 5:

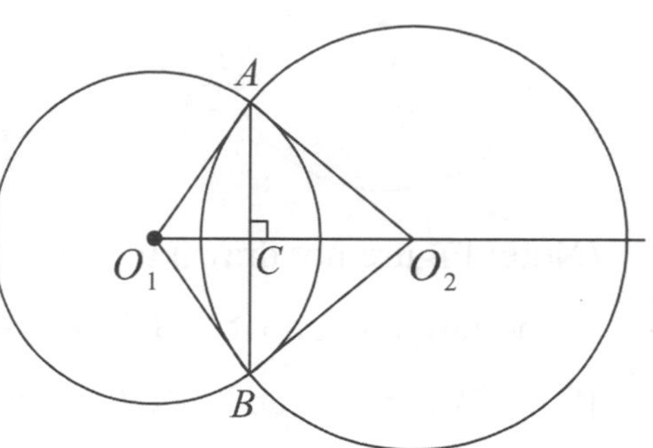

In the figure on the right, O_1 intersects O_2 at point A and point B respectively, if the radius of O_1 is 5, and the radius of O_2 is 10, and if the length of $\overline{AB}$ is 8, what is the value of $\overline{O_1O_2}$?

Analysis:

1.Find the length of $\overline{O_1O_2}$.

2.Given $O_1A = 5$, $O_2A = 10$, $AB = 8$.

3.$\triangle AO_1O_2$ is not a right triangle, can not apply Pythagorean's law.

4.Hidden information $\overline{AB} \perp \overline{O_1O_2}$

5. $\overline{AB}$ intersects with $\overline{O_1O_2}$ at C then $\overline{AB} \perp \overline{O_1O_2}$ and $AC = BC = 4$ for $\overline{O_1A} = O_1B = 5$

6.$\triangle ACO_1$ is a right triangle.

7. We now have all the information to solve the problem.

Solution:

$\because \ \overline{AO_2} = \overline{BO_2} = 10$, and $\overline{AO_1} = \overline{BO_1} = 5$

$\square AO_1BO_2$ is a kite, $\overline{AB} \perp \overline{O_1O_2}$, and $\overline{AC} = \overline{BC}$

$\therefore \ \overline{O_1C}^2 = \overline{O_1A}^2 - \overline{AC}^2 = 5^2 - 4^2 = 9 \quad \therefore \ \overline{O_1C} = 3$

$\overline{O_2C}^2 = \overline{O_2A}^2 - \overline{AC}^2 = 10^2 - 4^2 = 84 \quad \overline{O_2C} = 2\sqrt{21}$

$\therefore \ O_1O_2 = 3 + 2\sqrt{21}$

◆Example 6:

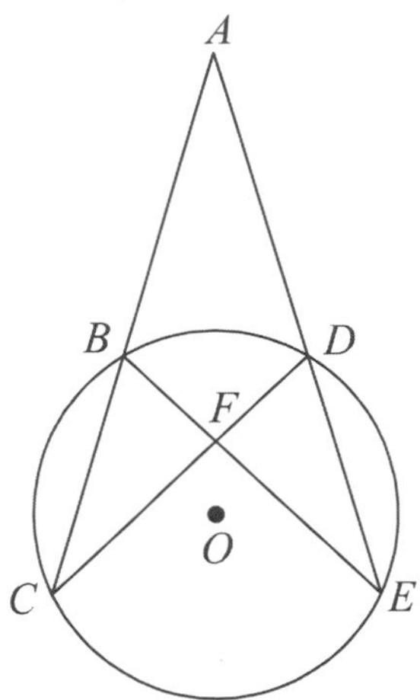

(Note: Figure not drawn to scale)

In the circle O above, $\overline{BE}$ intersects $\overline{CD}$ at point F and $\overline{EA}$ intersects $\overline{BA}$ at point A. If $\angle CAE = 50°$ and $\angle CFE = 80°$, what is the ratio of $\overset{\frown}{CE} : \overset{\frown}{BD}$?

Analysis:

1. $\overset{\frown}{CE} : \overset{\frown}{BD}$ we should find the measurement first.

2. Given $\angle A = 50°$, $\angle CFE = 80°$, $\angle A$ is secant-secant angle $\angle CFE$ is chord-chord angle $\angle A = \frac{1}{2}(\overset{\frown}{CE} - \overset{\frown}{BD}) = 50$, $\angle CFE = \frac{1}{2}(\overset{\frown}{CE} + \overset{\frown}{BD}) = 80$

3. Got $\overset{\frown}{CE}$ and $\overset{\frown}{BD}$, problem can be solved.

Solution:

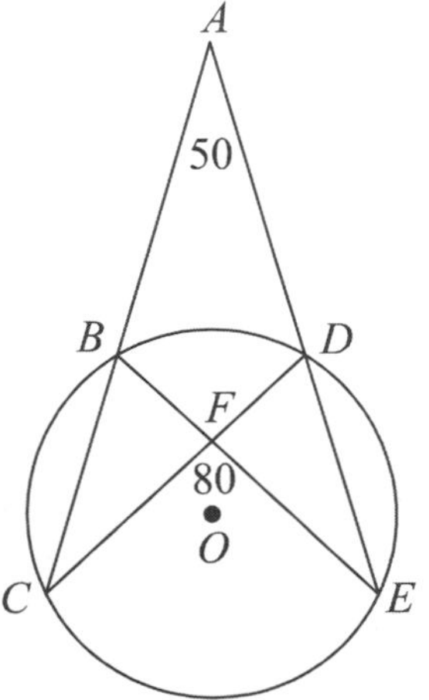

(Note: Figure not drawn to scale)

$\angle CFE = \frac{1}{2}(\overset{\frown}{CE} + \overset{\frown}{BD}) = 80$, $\angle A = \frac{1}{2}(\overset{\frown}{CE} - \overset{\frown}{BD}) = 50$, $80 + 50 = \frac{1}{2}(2\overset{\frown}{CE}) = \overset{\frown}{CE}$,

$80 - 50 = \overset{\frown}{BD}$, $\overset{\frown}{CE} = 130$, $\overset{\frown}{BD} = 30$, $\frac{\overset{\frown}{CE}}{\overset{\frown}{BD}} = \frac{130}{30} = \frac{13}{3}$

9-9 PRISMS

Terms	Figure	Vertices (2n)	Edges (3n)	Faces $(n+2)$
Triangular prism	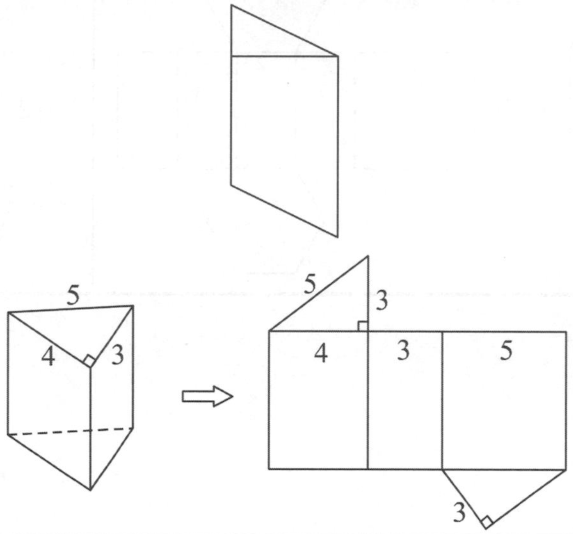 5, 4, 3 ⇨ 5, 3, 4, 3, 5, 3	6	9	5
Rectangular solid	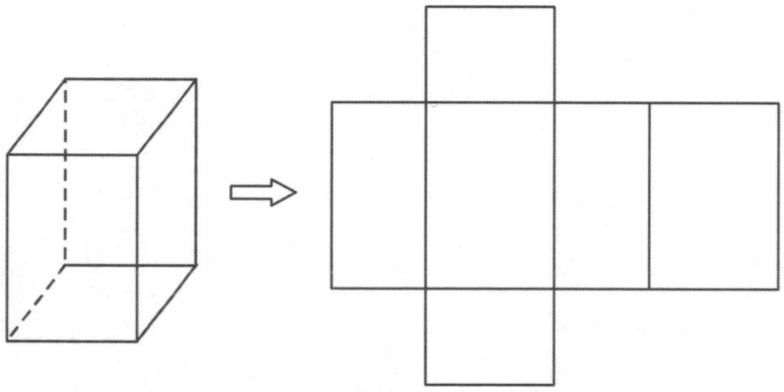 ⇨	8	12	6
Cube	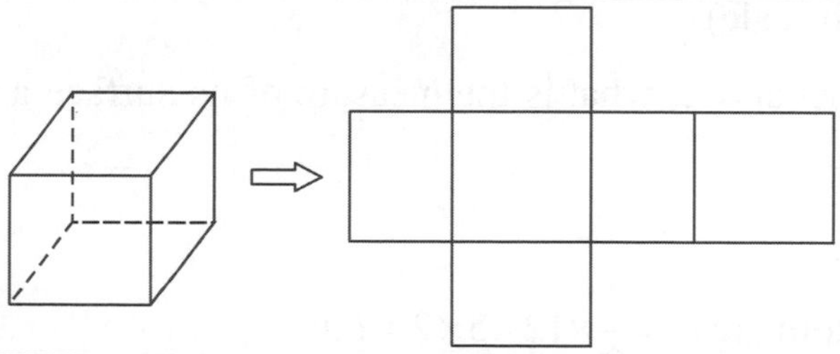 ⇨	8	12	6
Pentahedron	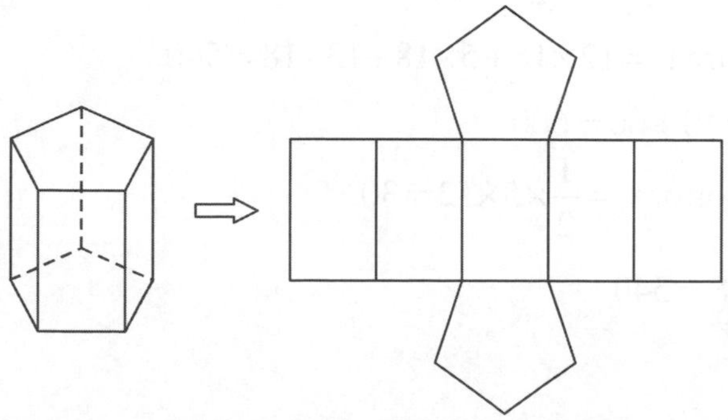 ⇨	10	15	7

Terms	Figure	Vertices (2n)	Edges (3n)	Faces ($n+2$)
Hexahedron		12	18	8

◆Example 1:

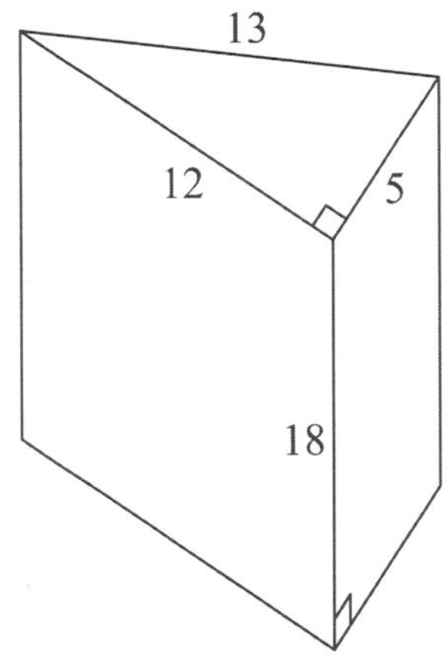

(Note: Figure not drawn to scale)

The triangular prism shown above, what is the measure of its surface area, and what is the measure of its volume ?

Solution:

The top and bottom area $=\frac{1}{2}\times12\times5\times2=60$

The rectangular area $=12\times18+5\times18+13\times18=540$

Surface area $=540+60=600$

The area of the bottom $=\frac{1}{2}\times5\times12=30$

Volume $=30\times18=540$

◆Example 2:

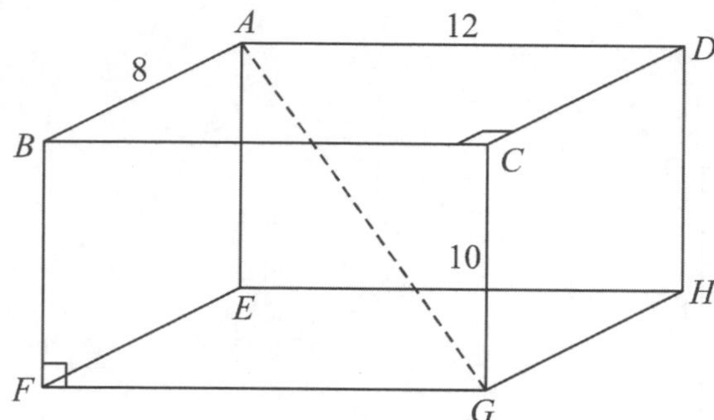

(Note: Figure not drawn to scale)

A rectangular solid is shown above, what is the length of the diagonal $\overline{AG}$?

($AB=8$, $AD=12$, $CG=10$)

Solution:

$$\overline{AH}^2 = 12^2 + 10^2$$

$$\overline{AG}^2 = \overline{AH}^2 + \overline{GH}^2 = 12^2 + 10^2 + 8^2 = 144 + 100 + 64 = 308$$

$\therefore\ AG = 17.55$

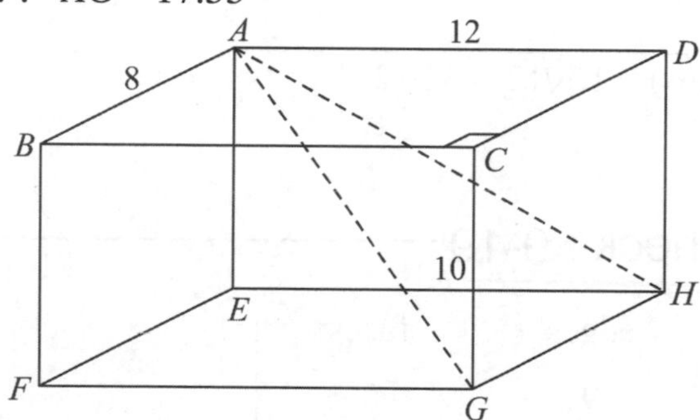

(Note: Figure not drawn to scale)

◆Example 3:

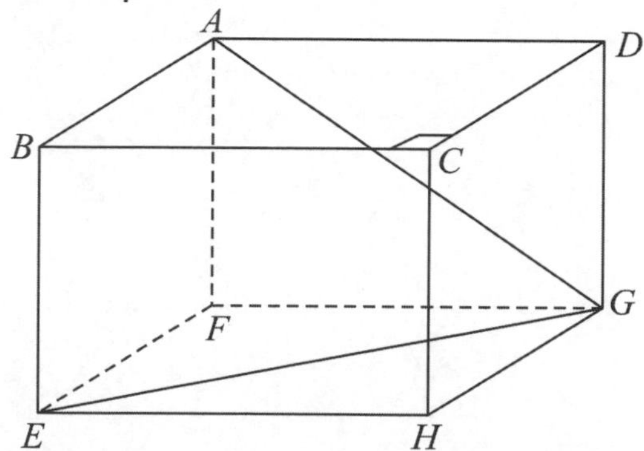

(Note: Figure not drawn to scale)

The cube shown above, the length of the side of the cube is 8, what is the measure of the area of $\triangle AGE$?

Solution:

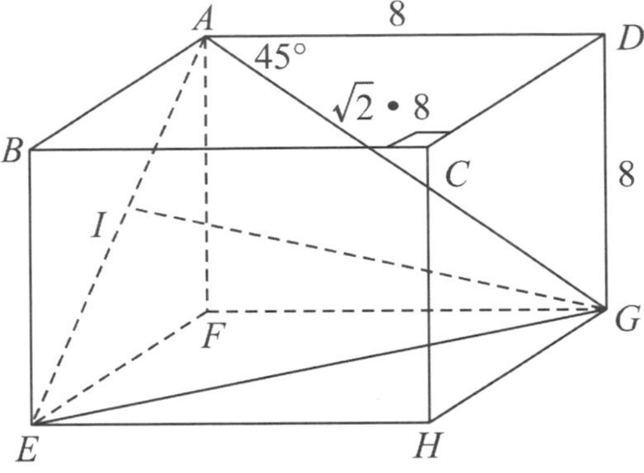

(Note: Figure not drawn to scale)

$\triangle AGE$ is an equilateral triangle

$\overline{AG} = \overline{EG} = \overline{AE} = \sqrt{2} \cdot 8$

It $\overline{GI}$ perpendicular to $\overline{AE}$, $\overline{GI} = \dfrac{\sqrt{3}}{2} \cdot (8\sqrt{2}) = 4\sqrt{6}$

$\therefore\ \triangle AGE = \dfrac{1}{2} \cdot (8\sqrt{2})(4\sqrt{6}) = 16\sqrt{12} = 32\sqrt{2}$

Answer to Concept Check 9-19

	vertices	edges	faces
1.Triangular Prism	6	9	5
2.Pentahedron	10	15	7
3.Hexaheddron	12	18	8

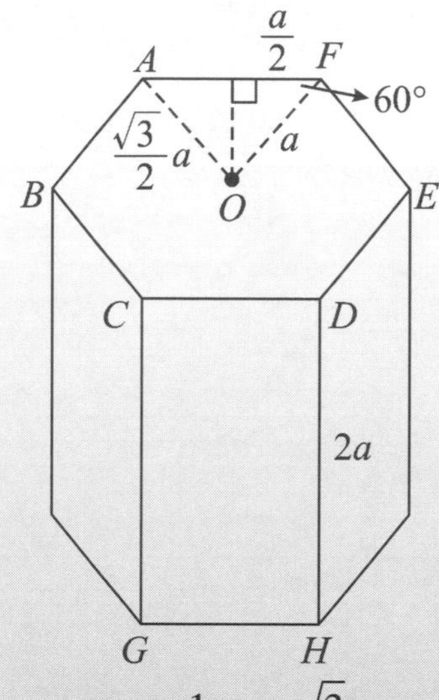

$\triangle AOF = \frac{1}{2}(\frac{a}{2} \times \frac{\sqrt{3}}{2}a) \times 2 = \frac{\sqrt{3}}{4}a^2$, Since Hexagon area $= 6 \times \triangle AOF$

Hexagon $ABCDEF = 6 \times \frac{\sqrt{3}}{4}a^2 = \frac{3}{2}\sqrt{3}a^2$

$\square CDHG = a \times 2a = 2a^2$

Area of hexahedron $= 2 \times \frac{3\sqrt{3}}{2}a^2 + 12a^2 = 3(\sqrt{3}+4)a^2$

Volume $= B \times h = \frac{3}{2}\sqrt{3}a^2 \times 2a = 3\sqrt{3}a^3$

9-10 PYRAMID

A 3-dimension shape with a polygon base and several triangular lateral faces slant up to meet at a single point (apex). The figure shows a 4-sided base pyramid.

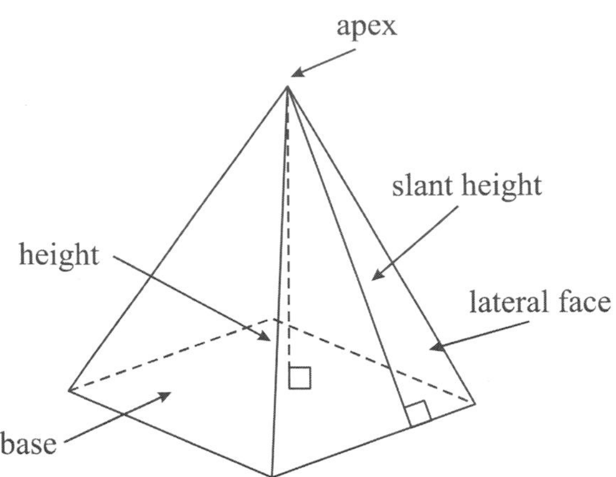

(Note: Figure not drawn to scale)

- Surface area = Base area + all lateral face area
- Volume $= \frac{1}{3}$(Base area $\times$ height)

- N-sided base Pyramids.

Pyramid	Figures	Vertices	Edges	Faces
3-sided base		4	6	4
4-sided base		5	8	5
5-sided base		6	10	6

6-sided base		7	12	7

◆Example 1:

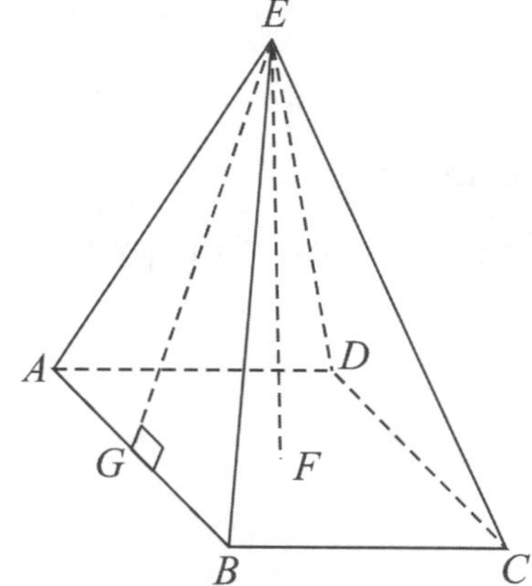

(Note: Figure not drawn to scale)

The square-based Pyramid shown above, if the side of the base is 10, and the side of the isosceles $\triangle AEB$ is 13, and the height of the Pyramid $EF=8$. What is the area and volume of the pyramid ?

Solution:

area of the base $=10\times10$

surface area of the triangle $\triangle ABE$

$\because\ \overline{EG}^2=13^2-5^2=12^2 \quad \overline{EG}=12$

$\triangle AEB=\frac{1}{2}\times12\times10=60$

surface area of pyramid $=4\times60+10\times10=340$

Volume of pyramid $=\frac{1}{3}(100\times8)=\frac{800}{3}$

◆Example 2:

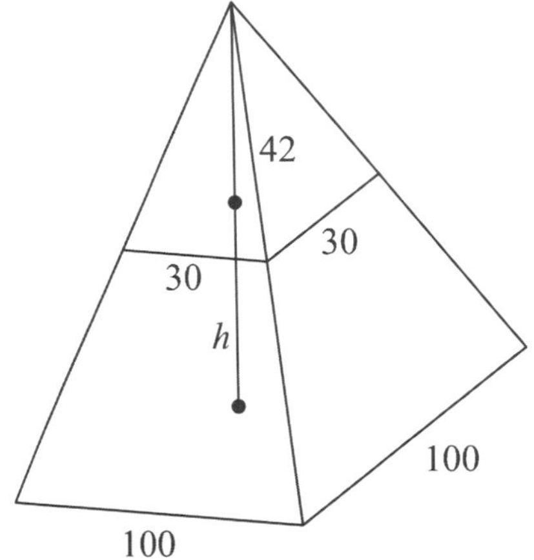

(Note: Figure not drawn to scale)

The length of the side of a square based Pyramid is 100 shown above.

The vertical distance from the cross section (base length = 30) to the apex of the pyramid is 42. What is the height of the pyramid ?

Solution: $\frac{30}{100}=\frac{42}{h}$ $\therefore$ $\frac{100\times 42}{30}=140$

9-11 CYLINDERS

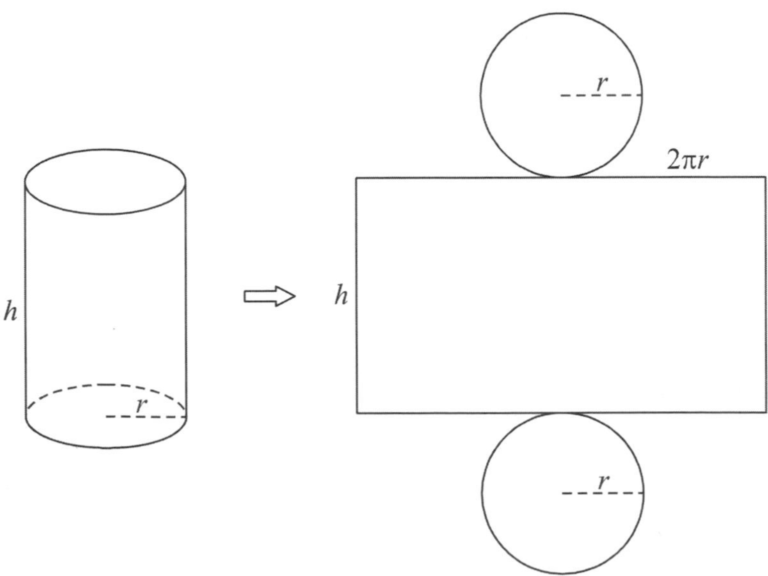

(Note: Figure not drawn to scale)

Area $=2\times\pi r^2+2\pi r\times h$

Volume $=\pi r^2\times h$

◆Example:

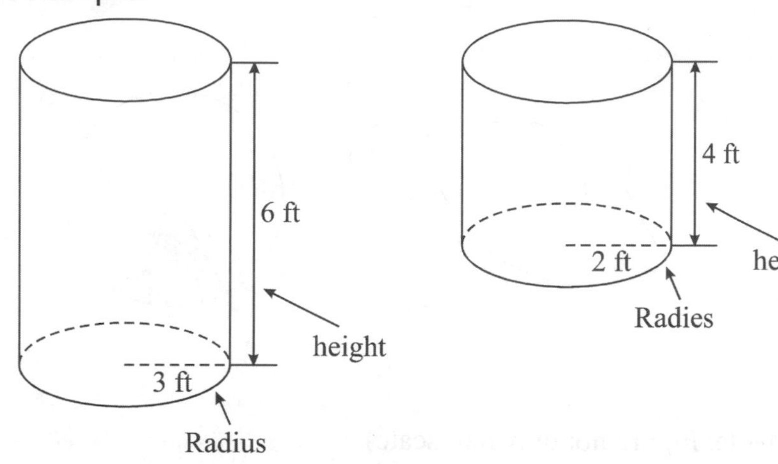

(Note: Figure not drawn to scale)

Two cylinders are shown above, what are the differences of areas and volumes between these 2 cylinders ?

Solution: (1) Area of bigger one $=(2)\pi(3)^2+2\pi(3)\times 6=54\pi$

Area of smaller one $=(2)(\pi)(2)^2+2\pi(2)\cdot(4)=24\pi$

$54\pi-24\pi=30\pi$

(2) Volume of bigger one $=\pi(3)^2\cdot 6=54\pi$

Volume of smeller one $=\pi(2)^2\cdot 4=16\pi$

$54\pi-16\pi=38\pi$

9-12 CONE

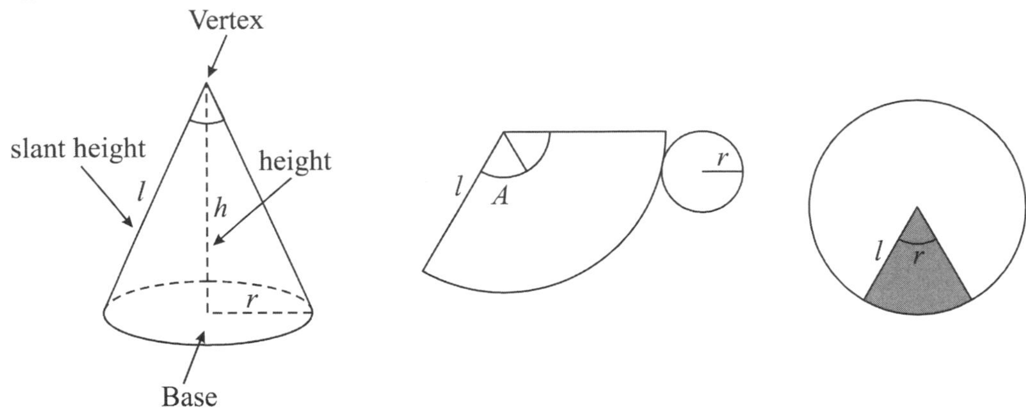

(Note: Figure not drawn to scale)

Volume $=\frac{1}{3}(\pi r)^2(h)$

Surface $=\pi r^2+\pi r\ell$

$\frac{2\pi r}{2\pi\ell}=\frac{x}{\pi\ell^2}$ $\quad x=\frac{2\pi r\cdot\pi\ell^2}{2\pi\ell}=\pi r\ell$ (x:lateral surface area)

$\therefore$ lateral surface area $=\frac{r}{\ell}\times\pi\ell^2=\pi r\ell$

◆Example:

If $h:H=1:3$, and $r:R=1:3$ (Shown as figure below)

What is the ratio of the volume of bigger one to the smaller one?

Solution:

The volume of bigger one: $\frac{1}{3}\pi R^2\cdot H$

The volume of smaller one: $\frac{1}{3}\pi r^2\cdot h$

$$\frac{\text{Volume of bigger cone}}{\text{Volume of smaller cone}}=\frac{\frac{1}{3}\pi R^2\cdot H}{\frac{1}{3}\pi r^2\cdot h}=\frac{\frac{1}{3}\pi(3r)^2\cdot(3h)}{\frac{1}{3}\pi r^2\cdot h}=\frac{27}{1}$$

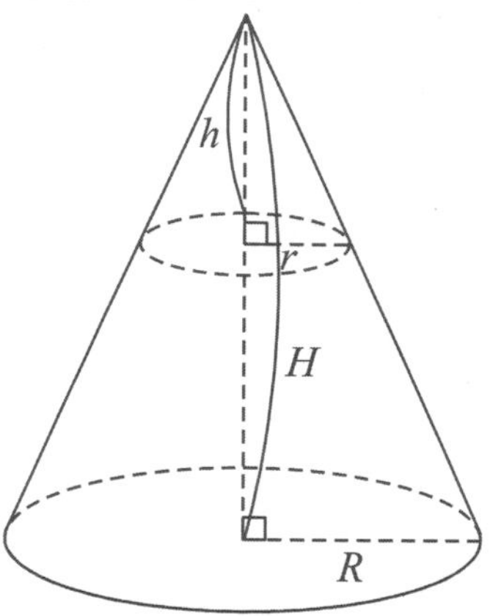

(Note: Figure not drawn to scale)

Answer to Concept Check 9-21

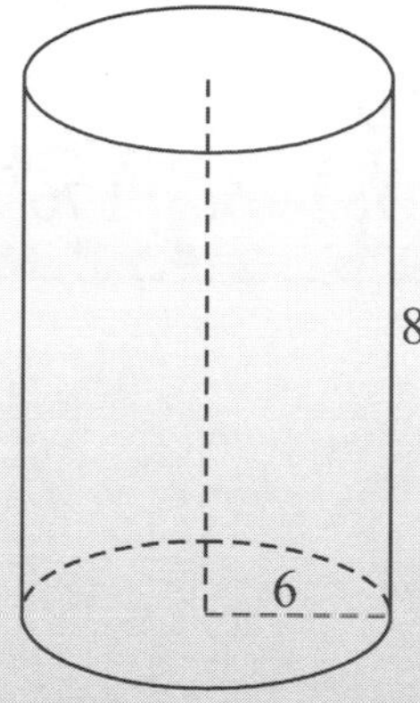

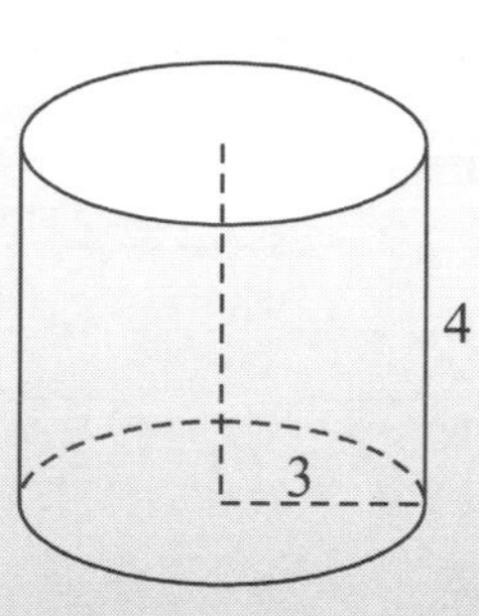

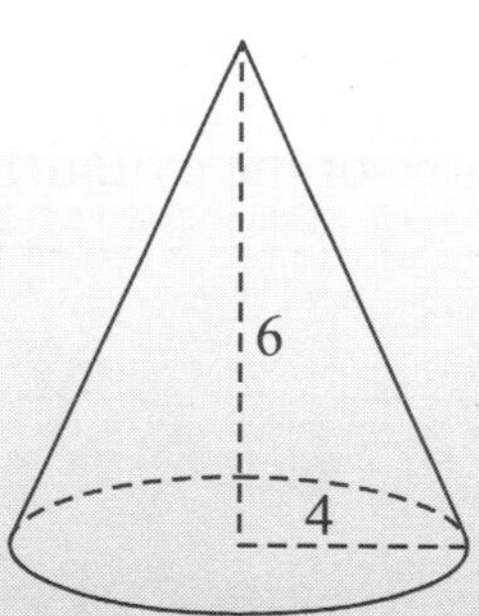

Two cylinders are shown above, what is the ratio of their volumes ?

Solution: The volume of the bigger cylinder $=\pi(6)^2\times 8=288\pi$

The smaller cylinder $=\pi(3)^2\times 4=36\pi$

The volume of the cone $=\dfrac{1}{3}\pi(4)^2\times 6=32\pi$

The ratio $=288\pi:36\pi:32\pi=4\times 72\pi:4\times 9\pi:4\times 8\pi=72:9:8$

Answer to Concept Check 9-22

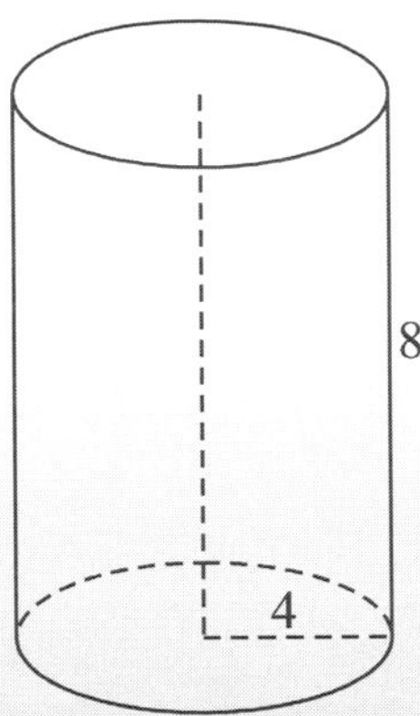

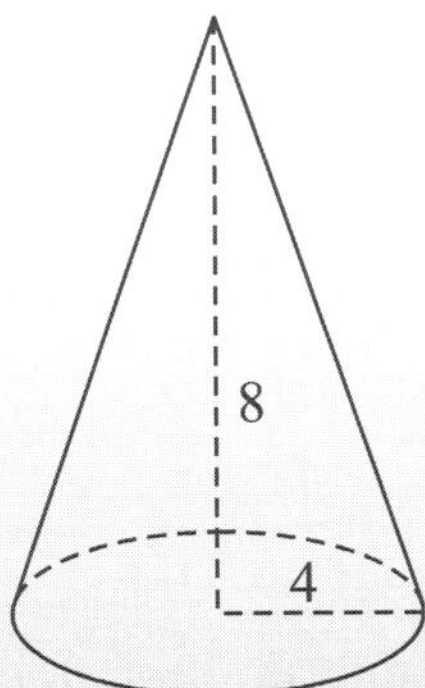

According to the figures of cylinder and cone above, what is ratio of their surface areas ?

Solution: The area of the cylinder: $2\pi(4)^2 + 2\pi(4) \cdot 8 = 96\pi$

The area of the cone:

The slant height $= \sqrt{8^2 + 4^2} = 8.94$

Area $= \pi(4)^2 + \pi r\ell = 16\pi + (4)(8.94)\pi = 16\pi + 35.76\pi = 51.76\pi$

The area of the cylinder to the area of the cone $= 96\pi : 51.76\pi = 96 : 51.76$

CHAPTER 9
PRACTICE

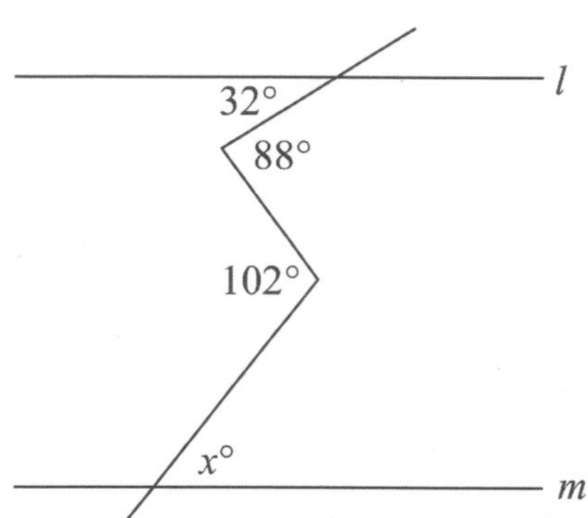

1. In the figure above, $\ell \parallel m$, what is the value of x ?

(A) 32

(B) 36

(C) 42

(D) 46

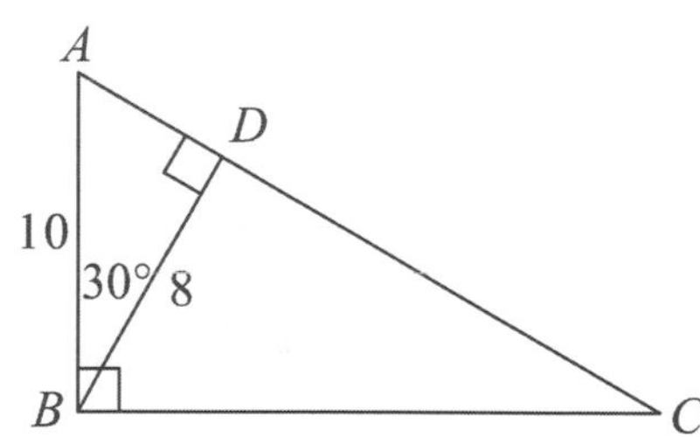

2. In the figure above, $\overline{BD} \perp \overline{AC}$, $m\angle ABD = 30°$, the lengths of $\overline{AB}$ and $\overline{BD}$ are 10 and 8 respecttively, what is the length of $\overline{AC}$?

(A) $\frac{32}{3}$

(B) $\frac{50}{3}$

(C) $\frac{64}{3}$

(D) $\frac{100}{3}$

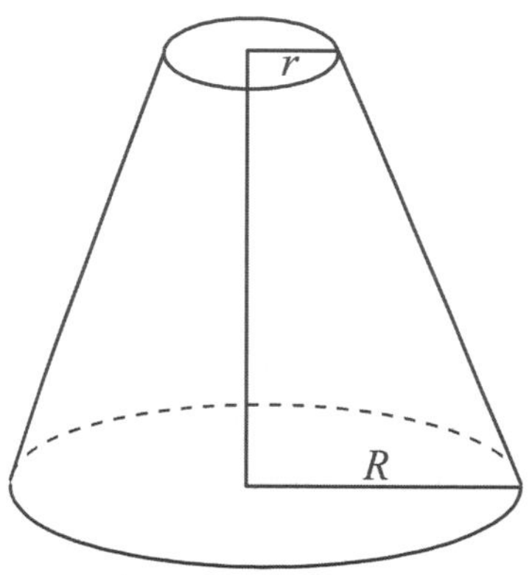

3. In the figure above is a part of a cone. The length of big circle $R=6$, and the length of small circle $r=2$. What is the valume of the part of the cone (As figure above) ?

(A) 104π

(B) 86π

(C) 72π

(D) 64π

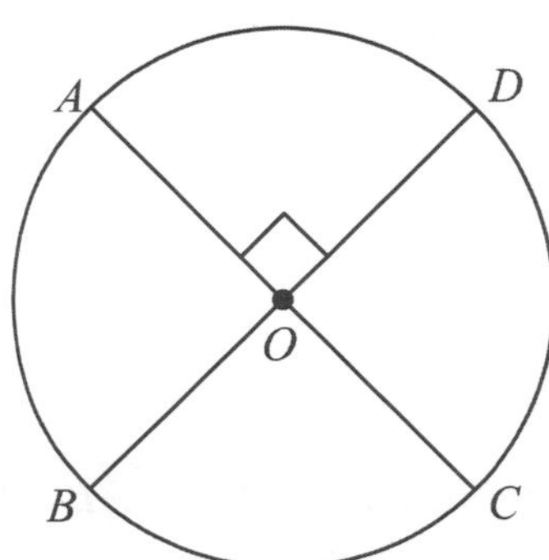

4. In the figure above, $\overline{AC}$ and $\overline{BD}$ are the diameters of circle 0 and $\overline{AC} \perp \overline{BD}$, if the area of circle 0 is 36π, what is the length of the minor arc $\overset{\frown}{AD}$?

(A) 6π

(B) 5π

(C) 4π

(D) 3π

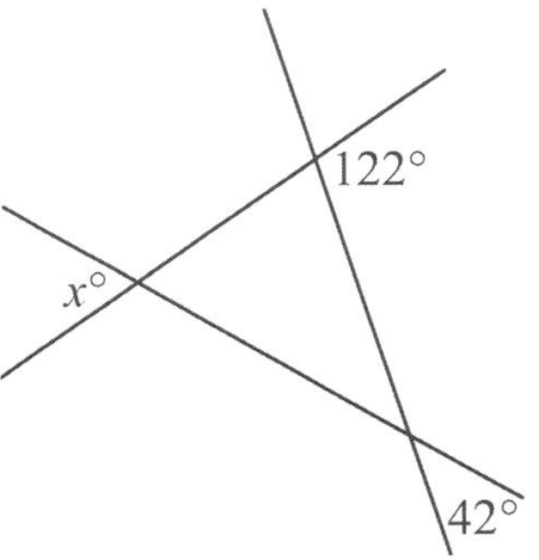

5. In the figure above, what is the measure of x ?

(A) 80°

(B) 60°

(C) 40°

(D) 20°

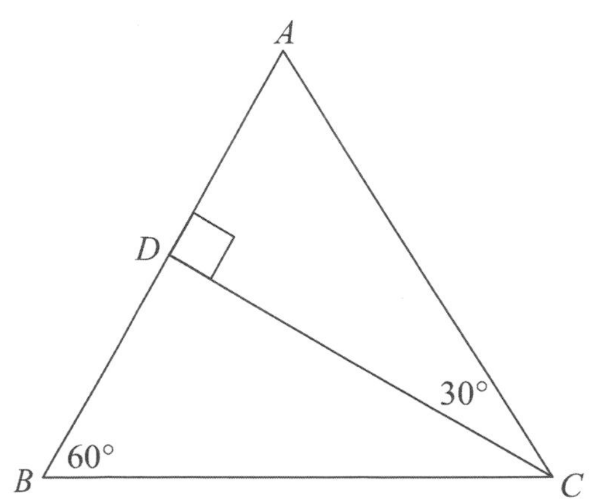

6. In the figure above $\overline{CD} \perp \overline{AB}$, if $AB=8$, what is the length of $\overline{CD}$?

(A) $2\sqrt{2}$

(B) $3\sqrt{3}$

(C) $4\sqrt{3}$

(D) $5\sqrt{2}$

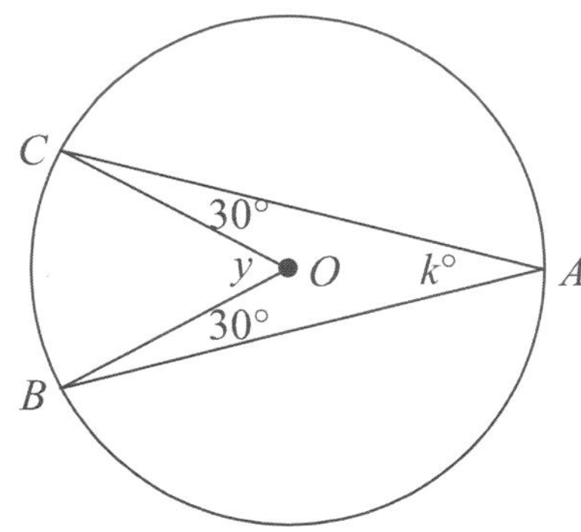

7. In the figure of circle 0 above, $m\angle ACO$ and $\angle ABO$ are both equal to 30°, what is the measure of k.

(A) 30°

(B) 40°

(C) 50°

(D) 60°

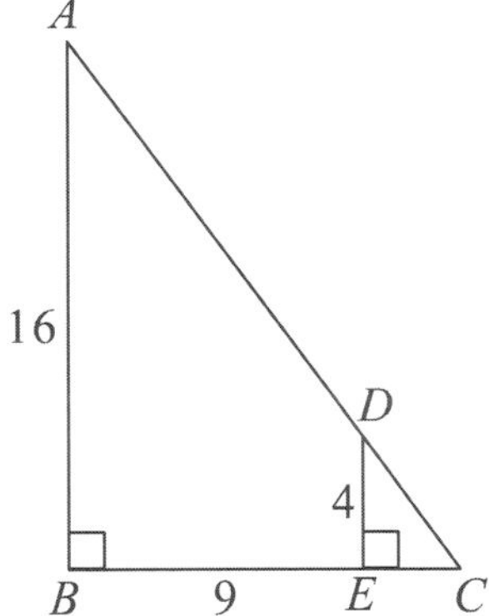

8. In the figure above, $\triangle ABC \sim \triangle DEC$, and $m\angle ABC = m\angle DEC = 90°$, if $AB=16$, $DE=4$, and $BE=9$, what is the length of $\overline{AC}$?

(A) 10

(B) 16

(C) 20

(D) 26

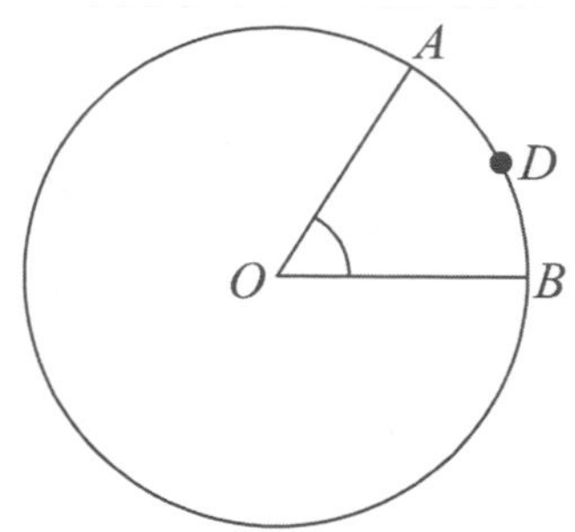

9. In the figure above, the area of circle 0 is 16π, and if $m\angle AOB = \frac{\pi}{5}$, what is length of the are $\overset{\frown}{ADB}$?

(A) $\frac{5}{4}\pi$

(B) $\frac{4}{5}\pi$

(C) $\frac{4}{6}\pi$

(D) $\frac{6}{4}\pi$

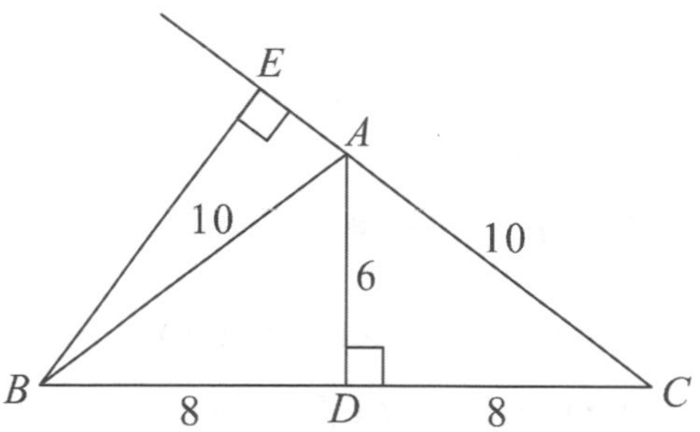

10. In an isosceles triangle ABC above, $AB = AC = 10$. $BC = 16$, $\overline{AD} \perp \overline{BC}$ and $\overline{BE} \perp \overline{CE}$, what is the length of $\overline{AE}$?

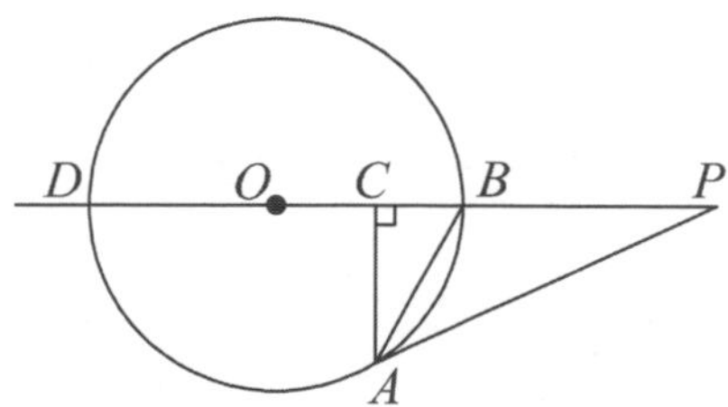

11. In the circle 0 above, $\overline{PA}$ is tangent to circle O at point A, $\overline{AC} \perp \overline{PD}$, if $\angle P = 32°$, what is the measure of $\angle CAB$?

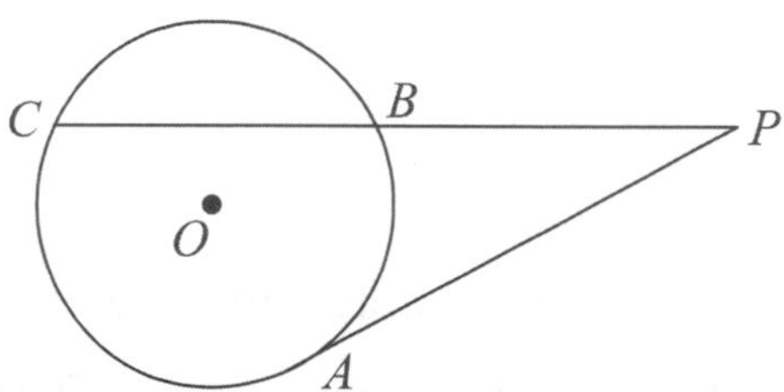

12. In the circle 0 above, $\overline{PA}$ is tangent to 0 at A, $\overline{PC}$ intersects 0 at B and C. If $\overline{PA} = 3x + 2$, $\overline{PC} = 9x - 2$, and $\overline{BC} = 12$, what is the length of $\overline{PA}$?

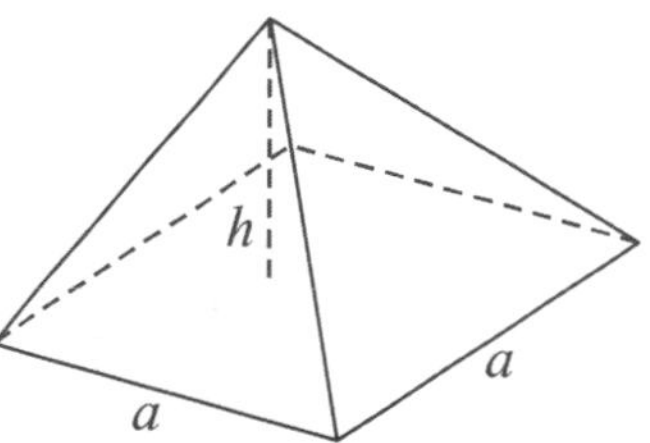

13. A pyramid with a square base shown as above, if $a = 230$ m, $h = 139$ m, total weight $= 5,900,000$ ton, then how many tons per cubic meter ?

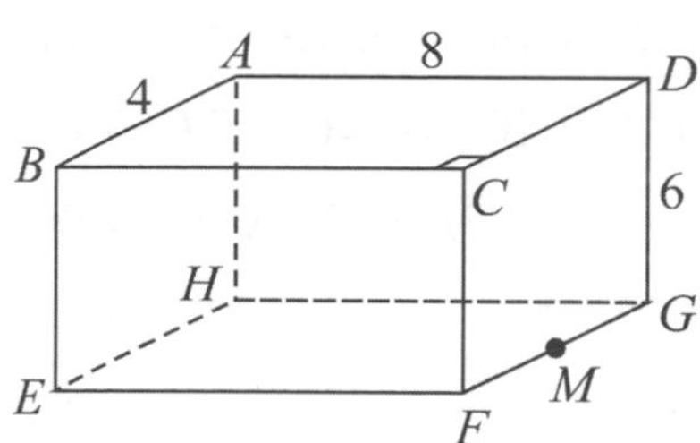

14. In the rectangular solid $ABCDEFGH$ above, M is the midpoint of $\overline{FG}$, connected $\overline{AM}$, $\overline{EM}$ and $\overline{AE}$, what

is the perimeter of the $\triangle AME$ (not shown) ?

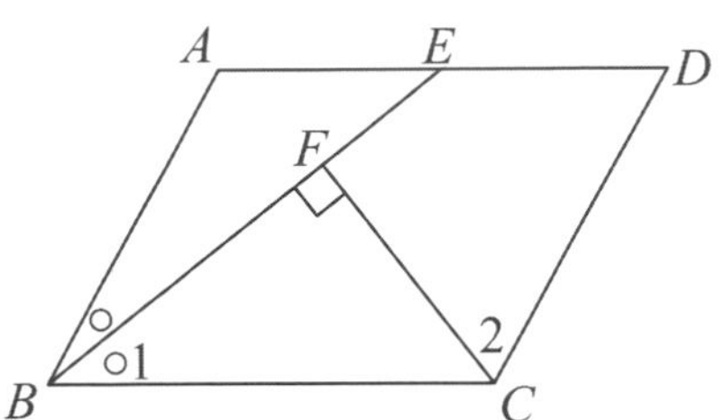

Figure not drawn to scale

15. In the figure above, in $\square ABCD$ parallelogram, $\overline{BE}$ bisects $\angle ABC$, $\overline{CF} \perp \overline{BE}$,
if $\angle A = 120°$, what is the measure of $\angle FED$?

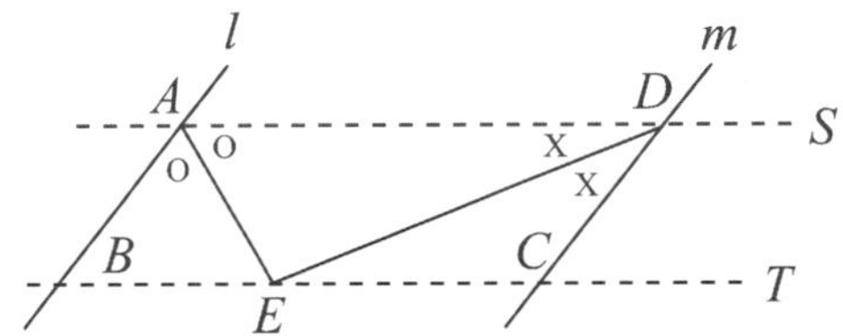

16. In the figure above, $\ell \parallel m$, line s intersect ℓ and m at A and D respectively. $\overline{AE}$ bisects $\angle BAD$, $\overline{DE}$ bisects $\angle ADC$, point E is the intersected point of $\overline{AE}$ and $\overline{DE}$. Line T is parallel to line s and passes point E. If AE=6, DE=14, and the area of $\triangle ABE$=22, what is the area of $\triangle DEC$?

(A) 42

(B) 30

(C) 28

(D) 20

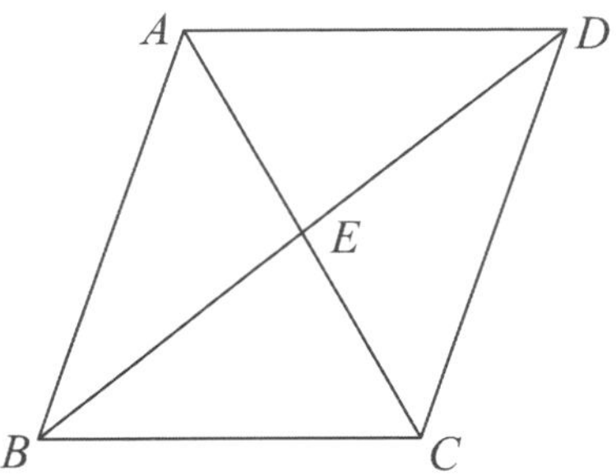

Not drawn to scale

17. In the parallelogram above, the perimeter of $\triangle ABE$ is 8 units longer than the Perimeter of $\triangle BEC$. The perimeter of parallelogram $ABCD$ is 88 units, what is the length of segment $\overline{AD}$?

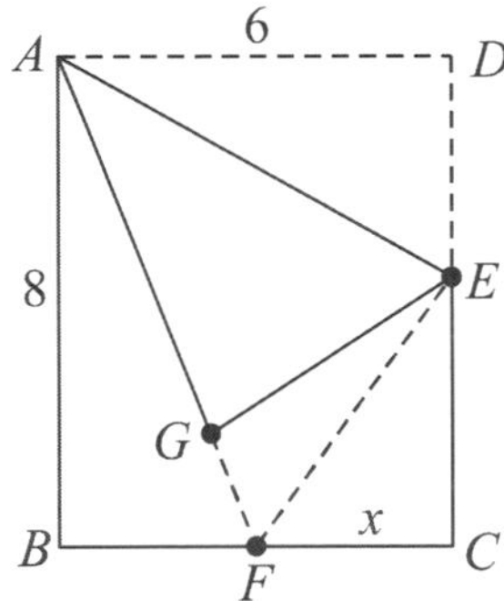

Note: Figure not drawn to scale

18. In rectangle $ABCD$ above, $AD = 6$, $AB = 8$, E is midpoint of segment $\overline{DC}$, F lies on segment $\overline{BC}$. If fold $\triangle ADE$ along segment $\overline{AE}$ and let point D overlay point G. What is the length of $\overline{FC}$?

CHAPTER 9
ANSWERS and EXPLANATIONS

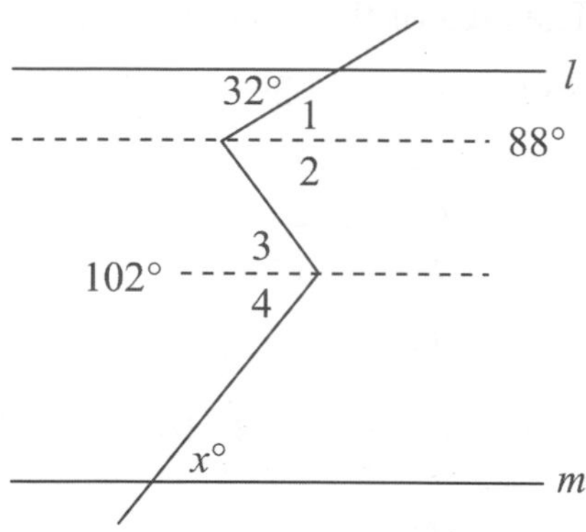

1. Analysis:

The value of x cannot be directly derived from the existing data, but the value of x can be obtained from the relationship between the relative angles of parallel lines.

Operation:

(1) Draw 2 auxiliary parallel lines as the figure above.

(2) $m\angle 1=32°$, $m\angle 2=88-32=56°$

(3) $m\angle 3=56°$, $m\angle 4=102-56=46°$

(4) $x=m\angle 4=46°$

Answer is (D)

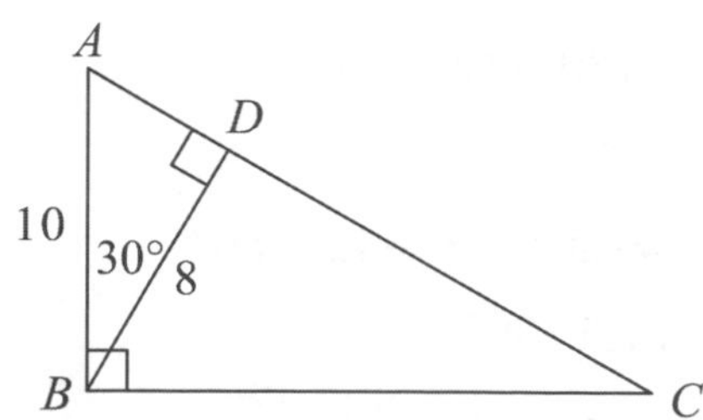

2. Analysis:

(1) $AC=AD+DC$

(2) $m\angle ABD=30°$, $m\angle ADB=rt\angle=90°$

$m\angle BAD=90-30=60°$

$rt\triangle ABD$ is a right triangle and follows the rules of Pythagorian's Triplets.

So, the length of $\overline{AD}$ is 6

(3) $rt\triangle ABC$ and $rt\triangle BDC$ are mother-child right triangles.

Operation:

(1) $AD=6$

(2) $BD^2=AD\times DC$

($\triangle ABD\sim\triangle BDC, \frac{BD}{CD}=\frac{AD}{BD}$)

(3) $8^2=6\times DC$, $DC=\frac{8^2}{6}=\frac{32}{3}$

(4) $AC=6+\frac{32}{3}=\frac{18+32}{3}=\frac{50}{3}$

Answer is (B)

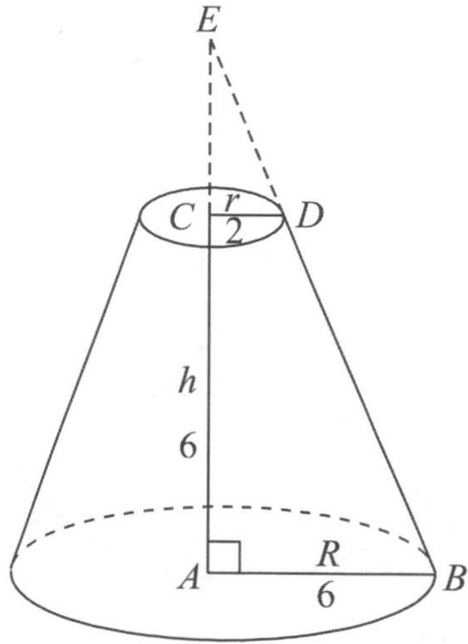

3. Analysis:

(1) Deplot some reference points make it

readable, and extend $\overline{BD}$ and $\overline{AC}$ to intersect at point E. It is a complete cone.

(2) $rt\triangle AEB \sim rt\triangle CDE$

$\dfrac{\overline{AE}}{\overline{EC}}=\dfrac{R}{r}=\dfrac{6}{2}=\dfrac{3}{1}$

Operation:

(1) $\dfrac{EC}{AE}=\dfrac{1}{3}$, $\therefore$ $6+EC=3EC$

$\therefore$ $EC=\dfrac{6}{2}=3$

(2) Volume of the cone

$=\dfrac{1}{3}\pi R^2 \cdot H=\dfrac{1}{3}\pi(6)^2 \cdot (6+3)$

$=(3)(6)^2\pi$

Volume of the small cone

$=\dfrac{1}{3}\pi(2)^2 \cdot (3)=4\pi$

(3) Volume of the part of the cone

$=108\pi-4\pi=104\pi$

Answer is (A)

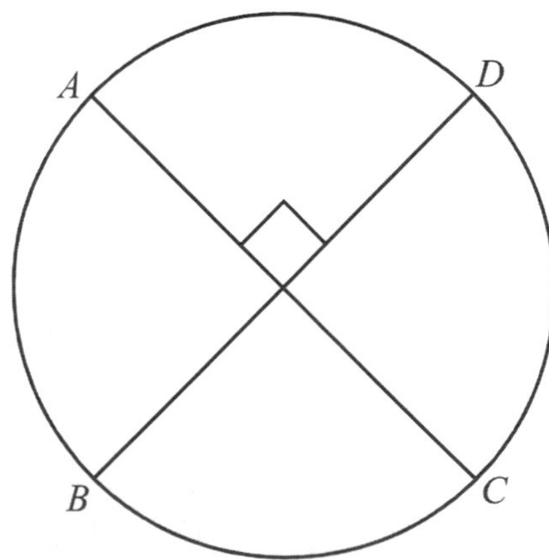

4. Analysis:

The length of the arc is directly proportional to the measure of its central angle.

Operation:

(1) The area of the circle is 36π,

$36\pi=\pi r^2$, $r^2=36$, $r=6$

(2) $\dfrac{\overset{\frown}{AD}}{2\pi r}=\dfrac{\overset{\frown}{AD}}{2\pi\times 6}=\dfrac{\frac{\pi}{2}}{2\pi}$

$\therefore$ The length of

$\overset{\frown}{AD}=2\pi r\times\dfrac{1}{4}=2\cdot\pi\cdot(6)\cdot\dfrac{1}{4}=3\pi$

Answer is (D)

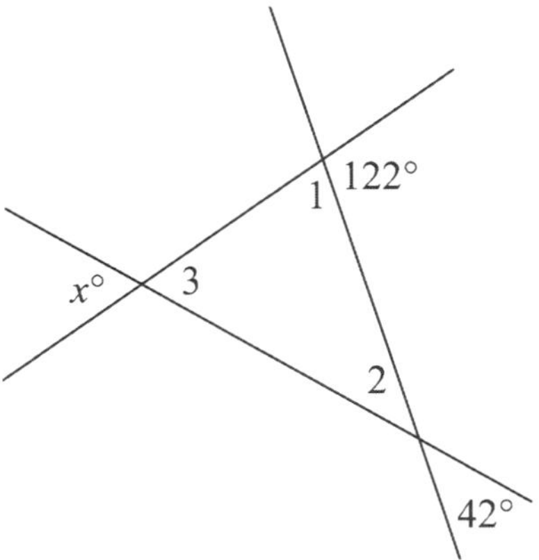

5. Analysis:

This problem can be solved by the vertex angle, the plane angle and the sum of the inner angles of the triangle.

Operation:

$m\angle 1=180°-122°=58°$

$m\angle 2=42°$

$m\angle 3=180°-58°-42°=80°$

$x°=m\angle 3=80°$

Answer is (A)

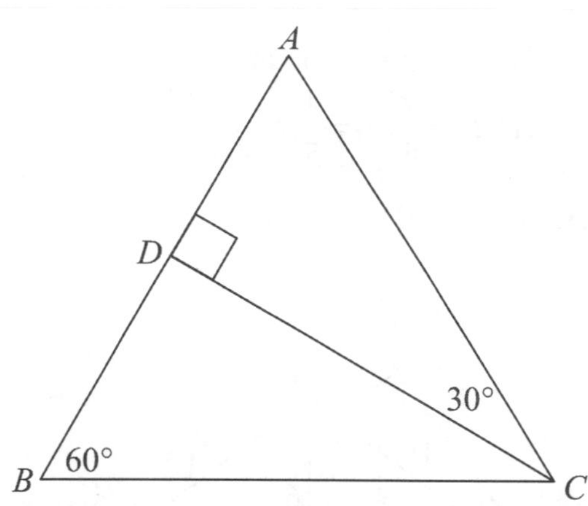

6. Analysis:

(1) $m\angle B=60°$, $m\angle A=90°-30°=60°$

$\triangle ABC$ is an equilateral triangle

(2) $\overline{CD}\perp\overline{AB}$, then $AD=BD$

($\triangle ADC\cong\triangle BDC$)

(3) $AD=\frac{8}{2}=4$, $AC=8$

(4) $CD^2=AC^2-AD^2$

Operation:

$CD^2=8^2-4^2=64-16=48$

$\therefore$ $CD=4\sqrt{3}$

Answer is (C)

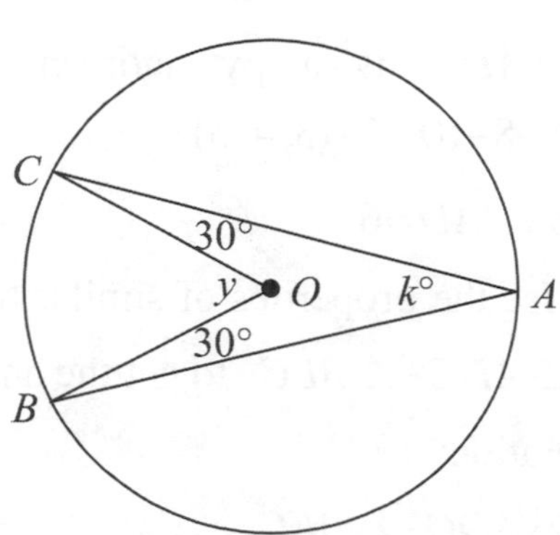

7. In the circle ⊙ above, what is the measure of k ?

Analysis:

(1) $m\angle CAB=k°=\frac{1}{2}m\angle COB$, it doesn't help.

(2) From the sum of internal angles of a quadrilateral, we can find $m\angle COB$, then $k°=\frac{1}{2}m\angle COB$.

Operation:

(1) Denote the center angle $\angle COB$ as $\angle y$

(2) $360°=k+2\times30+(360-y)$ (The sum of internal angles of a quadrilateral is 360°)

(3) $k=360-60-(360-y)$

$=360-60-360+y$

$=-60+2k$

(The measure of central angle is twice the measure of the inscribed angle)

$k=60$

Answer is (D)

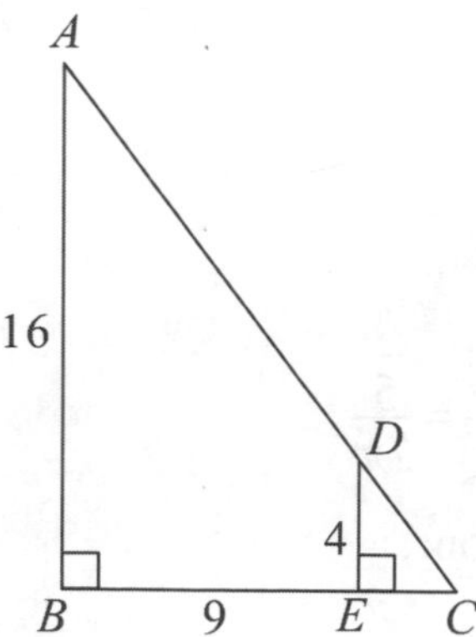

8. Analysis:

(1) If we know the length of $\overline{EC}$ then we know the length of $\overline{AC}$, since $AC^2=AB^2+BC^2$

(2) $BC=BE+EC=9+EC$

(3) Since $\triangle ABC \sim \triangle EDC$

$$\frac{EC}{EC+9}=\frac{DE}{AB}=\frac{4}{16}=\frac{1}{4}$$

Operation:

(1) Let $EC=x$

$BC=9+x$

(2) Since $\triangle ABC \sim \triangle EDC$

$\therefore \frac{x}{x+9}=\frac{1}{4}$, $4x=x+9$

$3x=9$, $x=3$

(3) $BC=9+3=12$

$AC^2=AB^2+BC^2=16^2+12^2=400$

$AC=20$

Answer is (C)

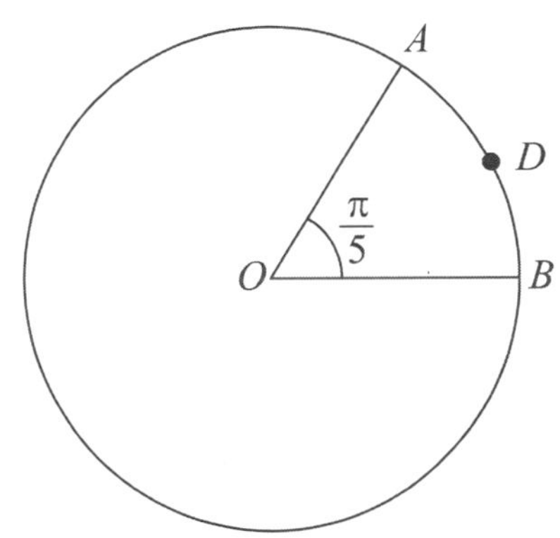

9. Analysis:

(1) $\frac{\overset{\frown}{ADB}}{2\pi r}=\frac{\frac{\pi}{5}}{2\pi}$

Operation:

(1) $\frac{\overset{\frown}{ADB}}{2\pi r}=\frac{\frac{\pi}{5}}{2\pi}$, $\overset{\frown}{ADB}=\frac{2\pi r}{10}=\frac{\pi r}{5}$

(2) given area $=16\pi=\pi r^2$

$\therefore r^2=16$, $r=4$

(3) The length of the arc

$\overset{\frown}{ADB}=\frac{\pi}{5}\times 4=\frac{4}{5}\pi$

Answer is (B)

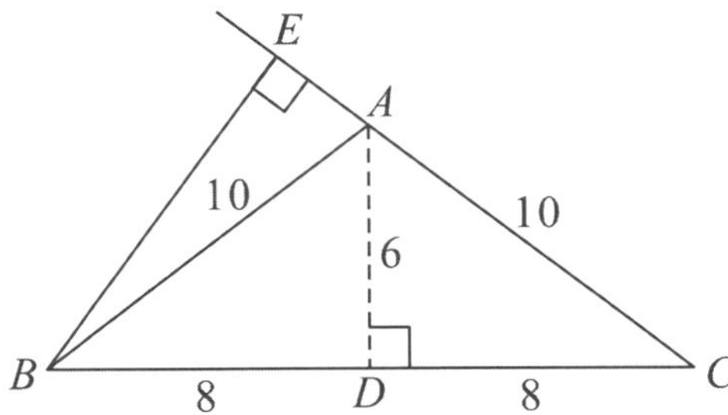

10. (2, 8)

Analysis:

(1) Given $AB=10$ (hidden $AB=AC$)

(2) $AE^2=AB^2-BE^2$

(3) So, we have to find BE

(4) Hidden information of isosceles triangles: Perpendicular line is also the middle line

So, draw an auxiliary perpendicular line $\overline{AD}\perp\overline{BC}$, and $BD=CD=\frac{16}{2}=8$

(5) $\triangle ADC$ is a pythagorean's triplet 6–8–10 2×(3, 4, 5)

So, $AD=6$

(6) Use the properties of similar triangles $\triangle ADC \sim \triangle BEC$ to get the answer

Operation:

(1) $\triangle ADC \sim \triangle BEC$

($\angle C=\angle C, \angle ADC=\angle BEC=90°$)

(2) $\frac{AC}{BC}=\frac{AD}{BE}$, $\frac{10}{16}=\frac{6}{BE}$,

$BE=\frac{16\times 6}{10}=\frac{96}{10}$

(3) $AE^2 = 10^2 - (\frac{96}{10})^2 = 100 - \frac{9216}{100}$

$= \frac{784}{100} = 7.84$

(4) $AE = 2.8$

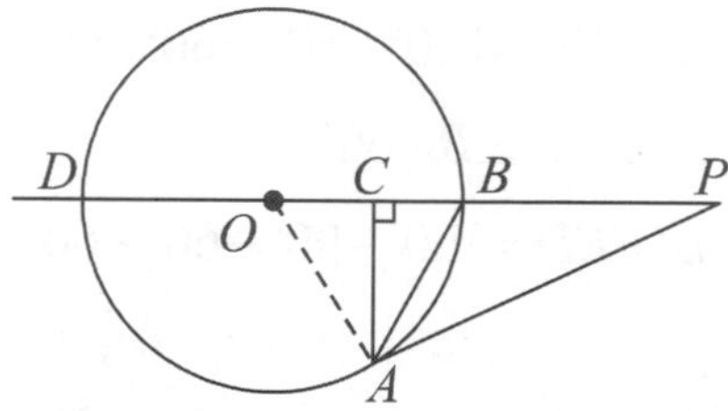

11. Analysis:

(1) $m\angle CAB$, you may get the information from $m\angle BAP$ or $m\angle CBA$ because $m\angle C = 90°$. Obviously, there is no superfacial information, you have to dig deeper to find out the hidden information.

(2) There are some properties of tangent lines, so, connect center with tangent point, get $\overline{OA}$, $m\angle OAP = 90°$

(3) $m\angle POA = 90° - m\angle P = 90° - 32° = 58°$

(4) $m\angle PAB = \frac{1}{2} m\angle POA = \frac{1}{2} \times 58° = 29°$

(Tangent Chord Angle)

(5) In $\triangle CAP$, $m\angle CAP = 90° - 32° = 58°$

(6) $m\angle CAB = 58° - 29° = 29°$

Operation:

(1) connect $\overline{OA}$

(2) $m\angle AOP = 90° - 32° = 58°$

(3) $m\angle PAB = \frac{1}{2} m\angle AOP = \frac{1}{2} \times 58 = 29°$

(4) $m\angle CAB = 90° - 32° - 29° = 29°$

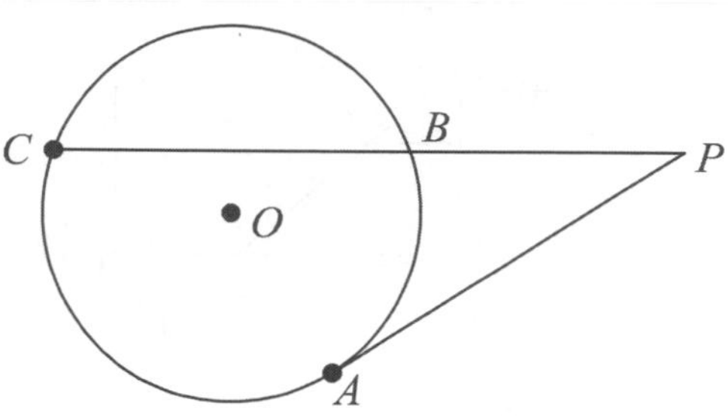

12. Analysis:

(1) $PA^2 = PB \times PC$ (secant-tangent power theorem)

(2) substitute $PA = 3x+2$, $PC = 9x-2$

$(3x+2)^2 = PB \times (9x-2)$

$= (9x-2-BC)(9x-2)$

Operation:

$(3x+2)^2 = PB \times PC = (9x-2-12) \times (9x-2)$

$9x^2 + 12x + 4 = (9x-2)^2 - 12(9x-2)$

$= 81x^2 - 36x + 4 - 108x + 24$

$72x^2 - 156x + 24 = 0$

$6x^2 - 13x + 2 = 0$

$(6x-1)(x-2) = 0$

$x = \frac{1}{6}$ or 2

$PA = 3 \times \frac{1}{6} + 2 = 2\frac{1}{2}$ or $PA = 3 \times 2 + 2 = 8$

13. $V = \frac{1}{3} \times B \times h = \frac{1}{3} \times 230^2 \times 139$

$= 2{,}451{,}000$

$\frac{5{,}900{,}000}{2{,}451{,}000} \doteq 2.4 \text{ ton/m}^3$

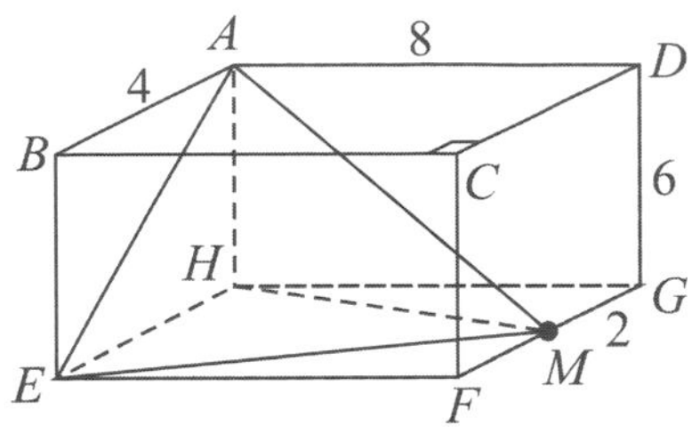

14. Connect $\overline{HM}$

$\overline{HM}^2 = 8^2 + 2^2 = 68$

$\overline{AM}^2 = 6^2 + 68 = 104$

$\therefore \ \overline{AM} = \sqrt{104}$

$\overline{EM}^2 = 8^2 + 2^2 = 68$

$\overline{EM} = \sqrt{68}$

$\overline{AE}^2 = 6^2 + 4^2 = 52$

$\overline{AE} = \sqrt{52}$

the perimeter of

$\triangle AME = \overline{AM} + \overline{EM} + \overline{AE}$

$= \sqrt{104} + \sqrt{68} + \sqrt{52}$

$= 10.20 + 8.25 + 7.21$

$= 25.66$

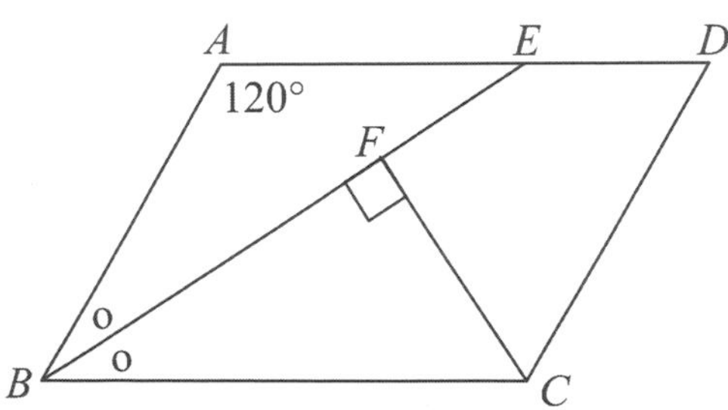

15. Analysis:

(1) Since $\square FEDC$ is a quadrilateral so,

$m\angle EFC + m\angle FCD + m\angle D + mFED$

$= 360°$

(2) Once you can find the measure of other three angles, then you know the measure of $m\angle FED$.

Operation:

(1) $m\angle C = m\angle A = 120°$ (parallelogram)

(2) $m\angle B = 180° - 120° = 60°$

(3) $m\angle FCB = 90° - 30° = 60°$

(4) $m\angle FCD = 120° - 60° = 60°$

(5) $m\angle D = m\angle B = 60°$

(6) $m\angle FED = 360° - 90° - 60° - 60° = 150°$

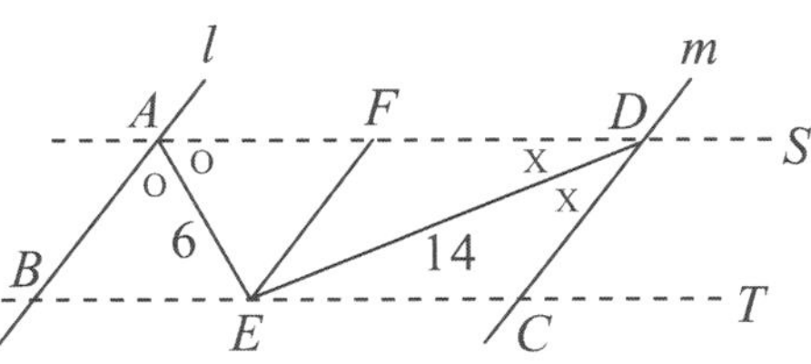

16. Analysis:

(1) The area of $\triangle DEC = \frac{1}{2}(EC \times h)$.

From given information, there is no way to get the lengths of $\overline{EC}$ or h.

(2) We have the lengths of $\overline{AE}$ and $\overline{ED}$.

If they are perpendicular to each other, then we know the area of $\triangle AED$

(3) $m\angle A + m\angle D = 180°$,

$m\angle DAE + m\angle ADE = 90°$,

so, $\overline{AE} \perp \overline{DE}$, we now know the area of $\triangle AED$.

(4) If we know the area of $\square ABCD$ then the area of

$\triangle DEC = \square ABCD - \triangle ABE - \triangle AED$

(5) Draw $\overline{EF} \parallel \overline{CD}$, you will find $\square ABCD = 2\triangle AED$

(6) We reach the answer.

Operation:

(1) Draw $\overline{EF} // \overline{CD}$.

(2) $m\angle FDE = m\angle DEC$,

$m\angle FED = m\angle EDC$, $\overline{ED} = \overline{ED}$,

$\triangle EFD \cong \triangle EDC$

(3) $m\angle FAE = m\angle AEB$,

$m\angle BAE = m\angle AEF$, $\overline{AE} = \overline{AE}$,

$\triangle ABE \cong \triangle AEF$

(4) $m\angle A + m\angle D = 180°$

$m\angle FAE + m\angle FDE = \frac{1}{2} \times 180° = 90°$

$m\angle AED = 180° - (m\angle FAE + m\angle FDE)$

$= 180° - 90° = 90°$

(5) The area of

$\triangle AED = \frac{1}{2}(AE \times ED) = \frac{1}{2}(6 \times 14) = 42$

(6) The area of $\square ABCD = 2 \times 42 = 84$

(7) The area of

$\triangle DEC = \square ABCD - \triangle ABE - \triangle AED$

$= 84 - 22 - 42 = 20$

Answer is (D)

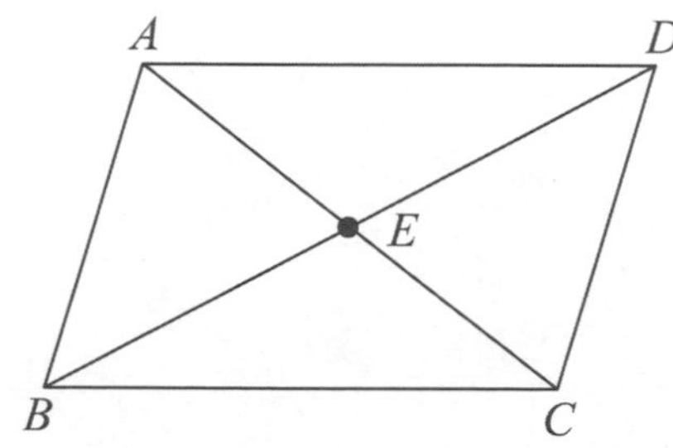

(Note: Figure not drawn to scale)

17. Analysis:

(1) $\square ABCD$ is a parallelogram, if we know the length of $\overline{BC}$ then we know the length of $\overline{AD}$.

(2) The perimeter of $\triangle ABE$ is 8 units longer than the perimeter of $\triangle BEC$ actually, in $\triangle ABE$ and $\triangle BEC$, $AE = EC$ (Bisect each other), $BE = BE$, so, we get $AB = BC + 8$

(3) The perimeter of

$\square ABCD = AD + DC + AB + BC$

$= 2(AB + BC) = 88$

(4) $AB + BC = 44$, $BC + 8 + BC = 44$,

$BC = 18$, that means $AD = 18$

Operation:

(1) $AB + AE + BE - BC + BE + AC = 8$

$AB - BC = 8$, $AB = BC + 8$

(2) $AB + BC + AD + DC = 88$

$2(AB + BC) = 88$, $AB + BC = 44$

(3) $BC = 44 - AB = 44 - (BC + 8) = 36 - BC$

$\therefore$ $BC = 18$, $BC = AD = 18$.

Answer is (C)

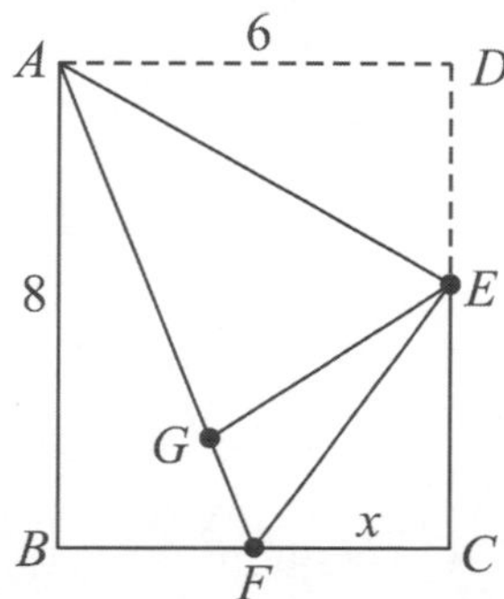

18. Analysis:

(1) Given $\triangle ADE \cong \triangle AGE$

(2) $GE = EC = \frac{1}{2} DC$

(3) $m\angle AGE = m\angle ADE = 90°$

$\therefore$ $m\angle EGF = 90° = m\angle ECF$

($\square ABCD$ is a rectangle)

(4) $\triangle EGF \cong \triangle EFC$ (S.S. rt $\angle$)

(5) $GF = FC = x$

(6) $AF^2 = AB^2 + BF^2$,

$(AG+x)^2 = 8^2 + BF^2$

(7) $(6+x)^2 = 8^2 + (6-x)^2$

(8) We will find x.

Operation:

(1) Let $Fc = x$

(2) $m\angle EGF = 90°$, $GE = EC$, $EF = EF$

$\triangle EGF \cong \triangle EFC$, $GF = x$

(3) $AF^2 - AB^2 = BF^2 = (6-x)^2$

(4) $(6+x)^2 - 8^2 = (6-x)^2$

$36 + 12x + x^2 - 8^2 = 36 - 12x + x^2$

$24x = 64 \quad \therefore \ x = \dfrac{8}{3}$

CHAPTER 10
TRIGONOMETRY

10-1 MEASUREMENT OF ANGLES and ARCS

- Initial side: A designated vector/line segment usually pointing straight right from the origin.
- Terminal side: A vector/line segment after a theta (θ) rotation.
- Theta: The angle difference from the initial side to the terminal side around a point (usually the origin)
- θ is a positive value, if the terminal side rotates counterclockwise.
- θ is a negative value, if the terminal side rotates clockwise.

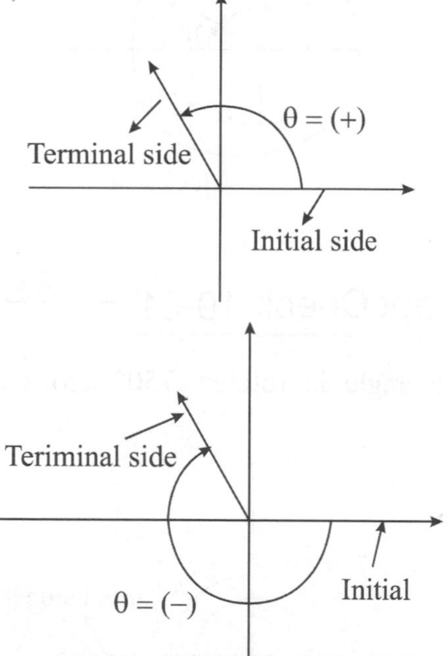

The measurements of two or more angles are the same if they have the same initial side and terminal side

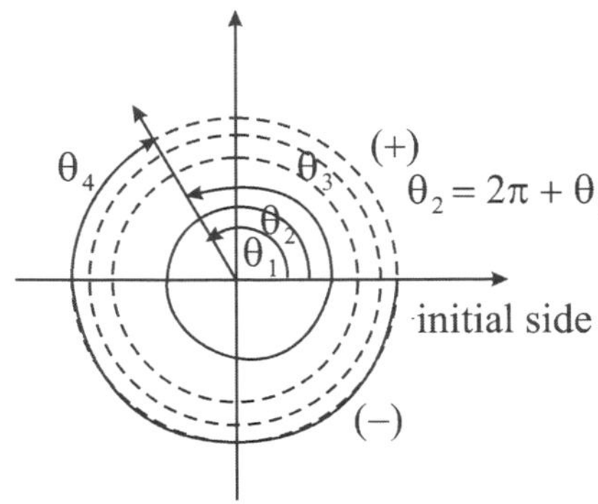

$\theta_1 = \theta_2(2\pi + \theta_1) = \theta_3(4\pi + \theta_1) = \theta_4(6\pi + \theta_1)\cdots$

For instance, $\theta = 30° = -330° = 360° + 30° = 390°$

Same initial and terminal sides

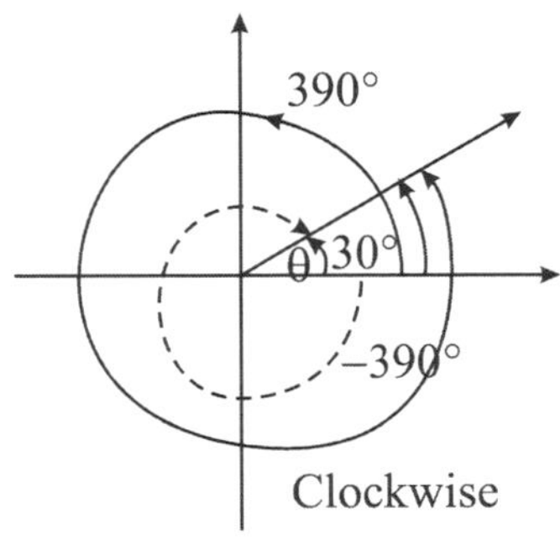

Answer to Concept Check 10-01

$-750°$ terminal side of the angle is rotated $750°$ clockwise. As the figure shows below.

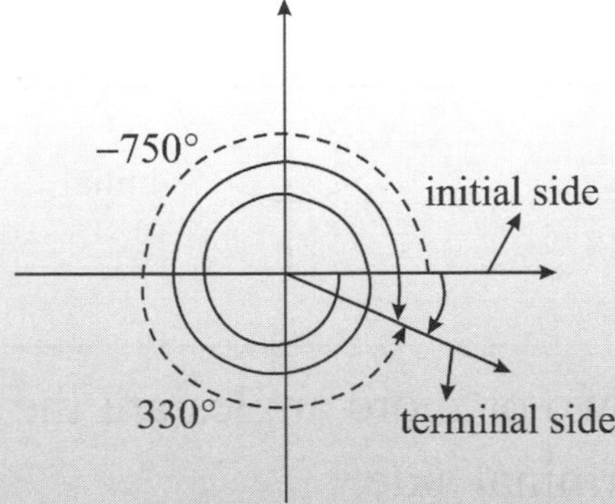

$-750°$ has the same initial rotated side and terminal side as $330°$ (counter-clockwise) so, $-750° = 330°$

Answer is (C)

Radians: an alternate way to measure angles.
Definition: $360°$ is equivalent to $2\pi^R$

$$\frac{360}{\text{degrees}} = \frac{2\pi}{\text{radians}}$$

$$1° = \frac{2\pi}{360} \text{ radians} \quad 1^R = \frac{360}{2\pi} \text{ degrees}$$

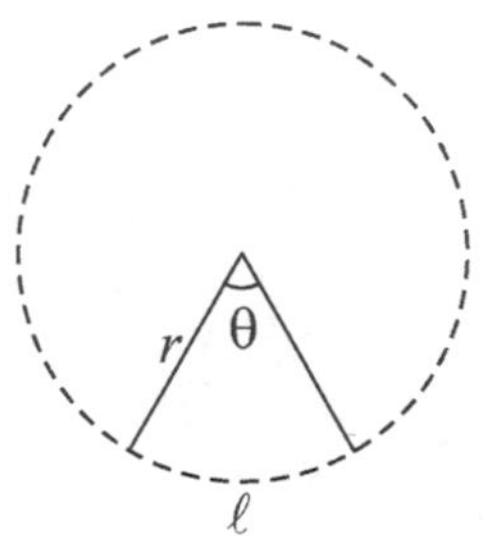

$l = r \cdot \theta$

$A = \frac{1}{2} r^2 \theta$

l: arc length
A: area of the sector
r: radius of the circle
θ: central angle (measured in radians)

Proof: $\frac{l}{2\pi r} = \frac{\theta}{2\pi}$

$$\therefore\ l = \frac{\theta}{2\pi} \cdot 2\pi r = r \cdot \theta$$

$$\frac{\pi r^2}{A} = \frac{2\pi}{\theta} \quad \therefore\ A = \frac{\pi r^2}{2\pi}\theta = \frac{1}{2} r^2 \theta$$

For instance, $r = 4$, $\theta = \frac{\pi}{4}$

then $l = 4 \cdot \frac{\pi}{4} = \pi$

and $A = \frac{1}{2}(4)^2 \cdot \frac{\pi}{4} = 2\pi$

◆Example:

If a sector of a circle has an area of 4π square inches and an arc length of π inches, what is the value of the central angle, in radians, of a sector which has an area of 6π square inches in the same circle ?

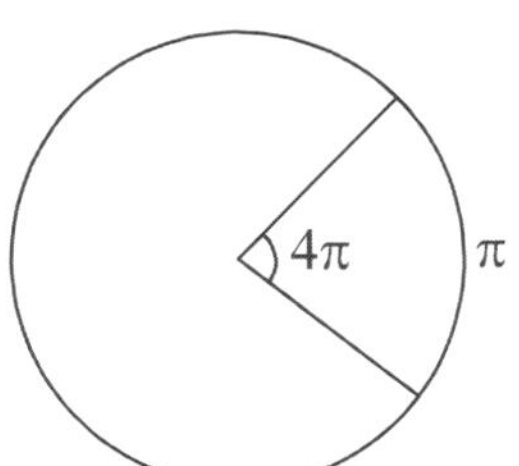

$A=\frac{1}{2}r^2\theta$, $4\pi=\frac{1}{2}r^2\theta$

$l=r\cdot\theta$, $\pi=r\theta$, $\theta=\frac{\pi}{r}$

So, $4\pi=\frac{1}{2}r^2\cdot\frac{\pi}{r}$, $r=8$

Therefore, $6\pi=\frac{1}{2}(8)^2\cdot\theta$, $\theta=\frac{3}{16}\pi$

Answer to Concept Check 10-02

Since $360°=2\pi$ radians

$\frac{15}{2\pi}=\frac{x}{360}$, $x=\frac{15}{2\pi}\times 360=\frac{2700}{\pi}$

Answer is (E)

Answer to Concept Check 10-03

Circumference of the circle is $2\pi\times 6=12\pi$

the arc length of the sector $=3\pi$

$\frac{12\pi}{3\pi}=\frac{2\pi}{x}$ $x=\frac{2\pi\times 3\pi}{12\pi}=\frac{\pi}{2}$ or $l=r\cdot\theta$

$\theta=\frac{l}{r}=\frac{3\pi}{6}=\frac{\pi}{2}$

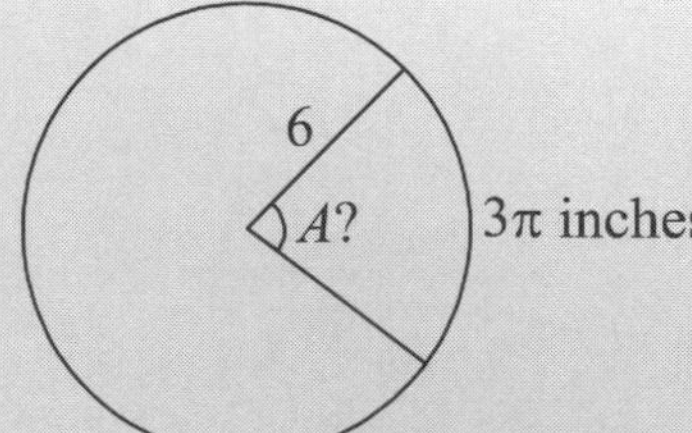

10-2 TRIGONOMETRIC FUNCTIONS

A. Definition:

$P(x, y)$ is a point on xy-plane, r is the distance from the origin. θ is the angle formed by r and x-axis.

Sine of angle θ is the ratio $\frac{y}{r}$, denoted as $\sin\theta=\frac{y}{r}$.

Cosine, is denoted as $\cos\theta=\frac{x}{r}$.

Tangent, is denoted as $\tan\theta=\frac{y}{x}$.

Cotangent, is denoted as $\cot\theta=\frac{x}{y}$.

Secant, is denoted as $\sec\theta=\frac{r}{y}$.

cosecant, is denoted as $\csc\theta=\frac{r}{x}$.

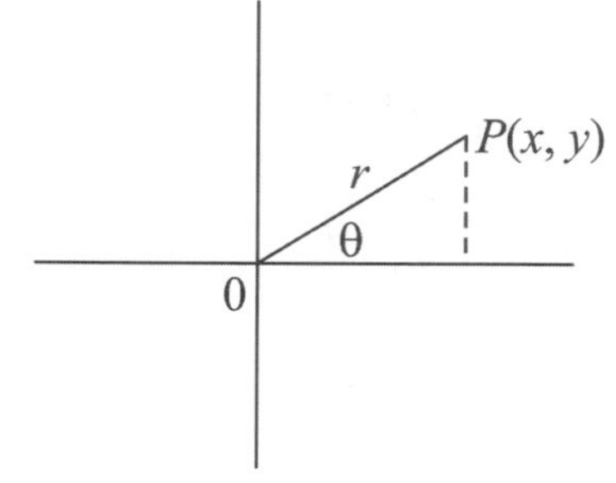

- Soh-Cah-Toa:

Sine: Opposite/Hypotenuse

Cosine: Adjacent/Hypotenuse

Tangent: Opposite/adjacent

Definition

$\sin\theta \xleftrightarrow{\text{cofunction}} \cos\theta$

$\tan\theta \xleftrightarrow{\text{cofunction}} \cot\theta$

$\sec\theta \xleftrightarrow{\text{cofunction}} \csc\theta$

Since $\sin\theta=\frac{y}{r}$, and $\cos(\frac{\pi}{2}-\theta)=\frac{y}{r}$

so, $\sin\theta=\cos(\frac{\pi}{2}-\theta)$

The same true for:

$\tan\theta = \cot(\frac{\pi}{2} - \theta)$

$\sec\theta = \csc(\frac{\pi}{2} - \theta)$

B. Inverse Trigonometric Function

$\sin\theta = \frac{y}{r}$, $\cos\theta = \frac{x}{r}$, $\tan\theta = \frac{y}{x}$

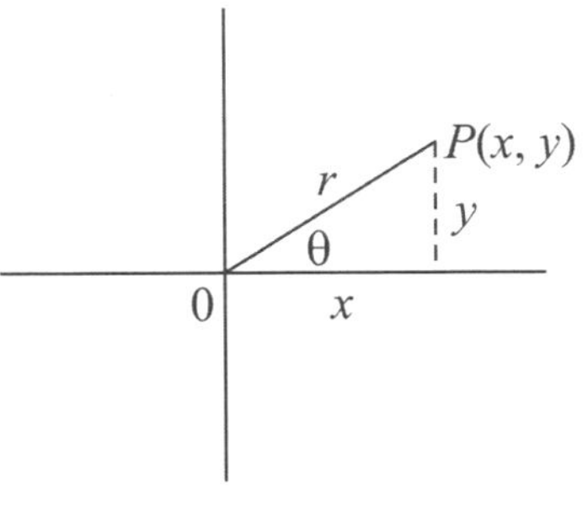

Define: $\sin^{-1}(\frac{y}{r}) = \theta$

$\cos^{-1}(\frac{x}{r}) = \theta$

$\tan^{-1}(\frac{y}{x}) = \theta$

For instance, if $\sin 30° = \frac{1}{2}$, then $\sin^{-1}(\frac{1}{2}) = 30°$

◆Example:

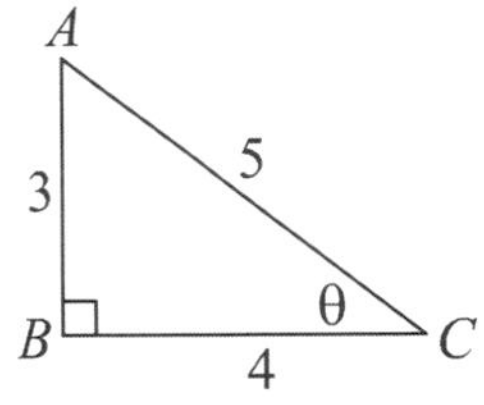

$\sin\theta = \frac{3}{5}$, $\sin^{-1}(\frac{3}{5}) = 30°$

$\cos\theta = \frac{4}{5}$, $\cos^{-1}(\frac{4}{5}) = 30°$

$\tan\theta = \frac{3}{4}$, $\tan^{-1}(\frac{3}{4}) = 30°$

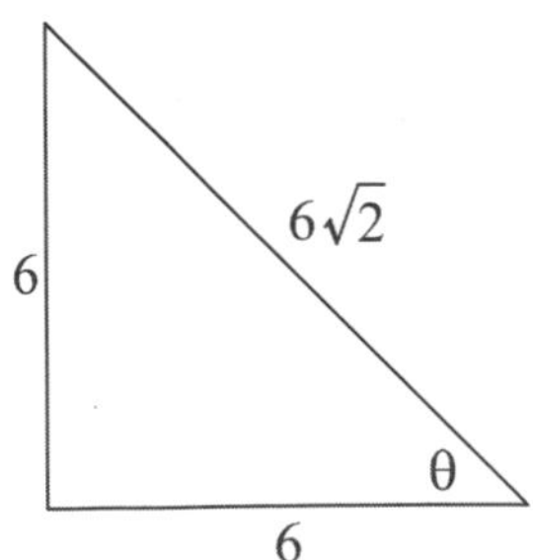

$\tan\theta = \frac{6}{6} = 1$

$\tan^{-1}(1) = 45°$

$$\sin\theta = \frac{6}{6\sqrt{2}} = \frac{1}{\sqrt{2}} = \frac{\sqrt{2}}{2}$$

$$\sin^{-1}(\frac{\sqrt{2}}{2}) = 45°$$

$$\cos^{-1}(\frac{\sqrt{2}}{2}) = 45°$$

C. The Signs Of Trigonometric Functions

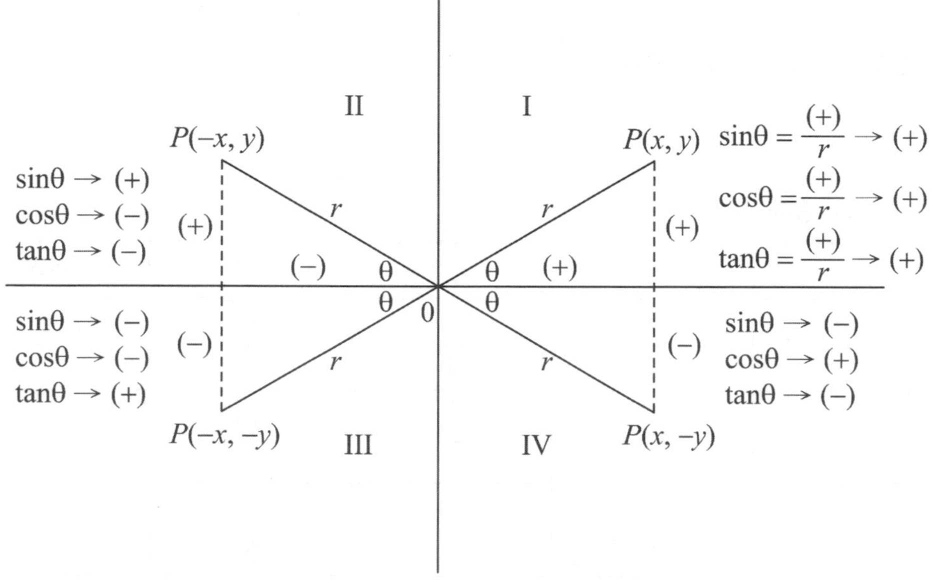

(r is always positive)

10-3 THE REFERENCE ANGLES

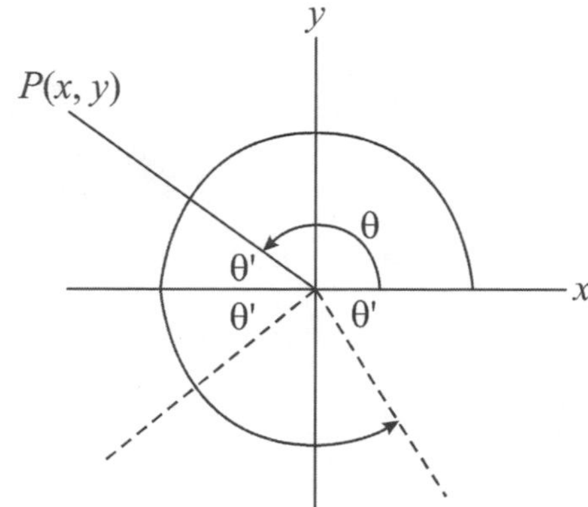

The reference angle θ' is the acute angle formed by terminal side and x-axis.
An absolute value of a trigonometric function of θ is equal to the absolute value of the same trigonometric function of θ'.
The sign of the function of θ' is determined by the quadrant on which the terminal side lies.

◆Example:

$\sin(120°) = \sin(180° - 120°) = \sin(60°) = +\frac{\sqrt{3}}{2}$

(lies on quadrant II)

The value of special angles

Degrees	0°	30°	45°	60°	90°
Radians	0	$\frac{\pi}{6}$	$\frac{\pi}{4}$	$\frac{\pi}{3}$	$\frac{\pi}{2}$
$\sin\theta$	0	$\frac{1}{2}$	$\frac{\sqrt{2}}{2}$	$\frac{\sqrt{3}}{2}$	$\frac{2}{2}=1$
$\cos\theta$	1	$\frac{\sqrt{3}}{2}$	$\frac{\sqrt{2}}{2}$	$\frac{1}{2}$	0
$\tan\theta$	0	$\frac{1}{\sqrt{3}}$	1	$\sqrt{3}$	undefined

Answer to Concept Check 10-04

Asking: $\cos 300° + \cos 120° =$

$\cos 300° = \cos(360° - 60°) = \cos(-60°) = \cos 60° = \dfrac{1}{2}$

since it is in quadrant IV

so, $\cos\theta$ is positive

$\cos(120°) = \cos(180° - 60°)$ (in quadrat II)

$= -\cos 60° = -\dfrac{1}{2}$

$\cos 300° + \cos 120° = \dfrac{1}{2} + (-\dfrac{1}{2}) = 0$

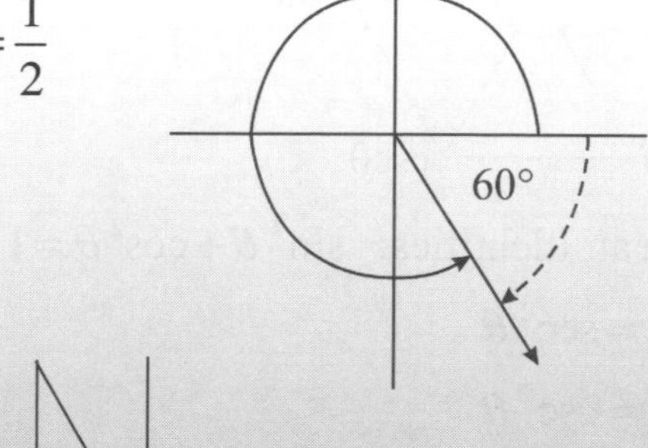

Answer is (C)

Answer to Concept Check 10-05

Asking: $\sin(\frac{\pi}{2} + \theta) + \sin(\frac{\pi}{2} - \theta) + \cos(\frac{\pi}{2} + \theta) + \cos(\frac{\pi}{2} - \theta) + \sin(\pi + \theta) + \cos(\pi - \theta) =$

$\sin(\frac{\pi}{2} + \theta) = -\cos\theta$ (quadrant II)

$\sin(\frac{\pi}{2} - \theta) = \cos\theta$ (quadrant I)

$\cos(\frac{\pi}{2} + \theta) = \sin\theta$ (quadrant II)

$\cos(\frac{\pi}{2} - \theta) = \sin\theta$ (quadrant I)

$\sin(\pi + \theta) = -\sin\theta$ (quadrant III)

$\cos(\pi - \theta) = -\cos\theta$ (quadrant II)

$-\cancel{\cos\theta} + \cancel{\cos\theta} + \sin\theta + \cancel{\sin\theta} - \cancel{\sin\theta} - \cos\theta = \sin\theta - \cos\theta$

Answer is (D)

10-4 TRIGONOMETRIC IDENTITIES

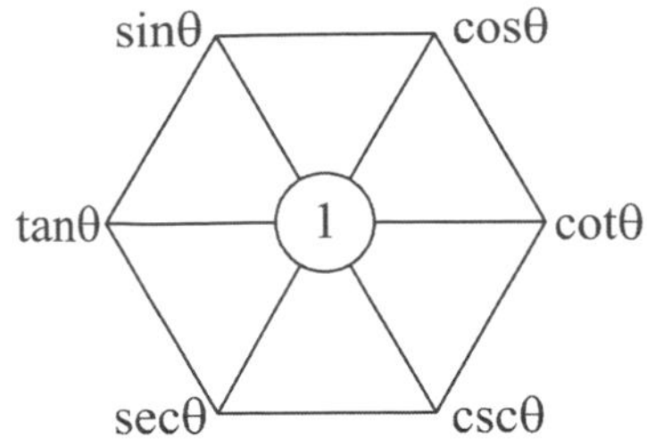

Pythagorean Identities: $\sin^2\theta+\cos^2\theta=1$

$\tan^2\theta+1=\sec^2\theta$

$\cot^2\theta+1=\csc^2\theta$

Reciprocal Identities: $\sin\theta=\dfrac{1}{\csc\theta}$

$\cos\theta=\dfrac{1}{\sec\theta}$

$\tan\theta=\dfrac{1}{\cot\theta}$

Quotient Identities: $\tan\theta=\dfrac{\sin\theta}{\cos\theta}$

$\cot\theta=\dfrac{\cos\theta}{\sin\theta}$

◆Example 1:

If $\cos x=-0.408$, then the value of $(\cos x)(\sin x)(\cot x)$ can be re-written as $(\cos x)(\cancel{\sin x})(\dfrac{\cos x}{\cancel{\sin x}})=\cos^2 x=(-0.408)^2=0.17$

◆Example 2:

$$\frac{1}{\sin\alpha}-(\cos\alpha)(\cot\alpha)=\frac{1}{\sin\alpha}-(\cos\alpha)(\frac{\cos\alpha}{\sin\alpha})=\frac{1}{\sin\alpha}-\frac{\cos^2\alpha}{\sin\alpha}$$

$$=\frac{1-\cos^2\alpha}{\sin\alpha}=\frac{\sin^2\alpha}{\sin\alpha}=\sin\alpha$$

◆Example 3:

if $\sin A-\cos A=\dfrac{1}{5}$, what is the value of $\tan A+\cot A$?

since $(\sin A-\cos A)^2=\sin^2 A+\cos^2 A-2\sin A\cos A=1-2\sin A\cos A=(\dfrac{1}{5})^2$

$$\therefore\ \ 2\sin A\cos A = 1-\frac{1}{25}=\frac{24}{25}$$

$$\sin A\cos A=\frac{12}{25}$$

$$\tan A+\cot A=\frac{\sin A}{\cos A}+\frac{\cos A}{\sin A}=\frac{\sin^2 A+\cos^2 A}{\sin A\cos A}=\frac{1}{\frac{12}{25}}=\frac{25}{12}$$

Answer to Concept Check 10-06

Asking: $\dfrac{\cos 315°+\tan 240°}{(\cot 210°)(\sin 300°)}=$

Since $\cos 315° = \cos(360°-45°) = \cos 45° = \dfrac{\sqrt{2}}{2}$ (quadrant IV)

$\tan 240° = \tan(180°+60°) = \tan 60° = \sqrt{3}$ (quadrant III)

$\cot 210° = \cot(180°+30°) = \cot 30° = \sqrt{3}$ (quadrant III)

$\sin 300° = \sin(360°-60°) = \sin(-60°) = -\dfrac{\sqrt{3}}{2}$ (quadrant IV)

$$\frac{\cos 315°+\tan 240°}{(\cot 210°)(\sin 300°)}=\frac{\frac{\sqrt{2}}{2}+\sqrt{3}}{(\sqrt{3})(-\frac{\sqrt{3}}{2})}=\frac{\sqrt{2}+2\sqrt{3}}{-3}$$

Answer is (C)

Answer to Concept Check 10-07

Asking: $\cot\theta$?

Since $\pi\le\theta\le\frac{3}{2}\pi$, θ is in quadrant III draw a figure as follows

$\csc\theta=-\dfrac{5}{3}$

$\sqrt{5^2-(-3)^2}=\sqrt{25-9}=\sqrt{16}=4$

$\therefore\ \ \cot\theta=\dfrac{-4}{-3}=\dfrac{4}{3}$

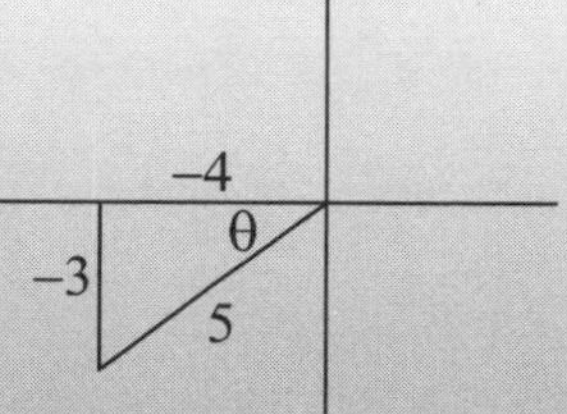

Answer is (A)

Answer to Concept Check 10-08

Asking: $\cos\theta$?

Since $\sin\theta = \frac{3}{5}$, θ must be in quadrant I or quadrant II

and since $\cot\theta < 0$, θ must be in quadrant II or quadrant IV

so, θ must be in quadrant II

Draw a graph below

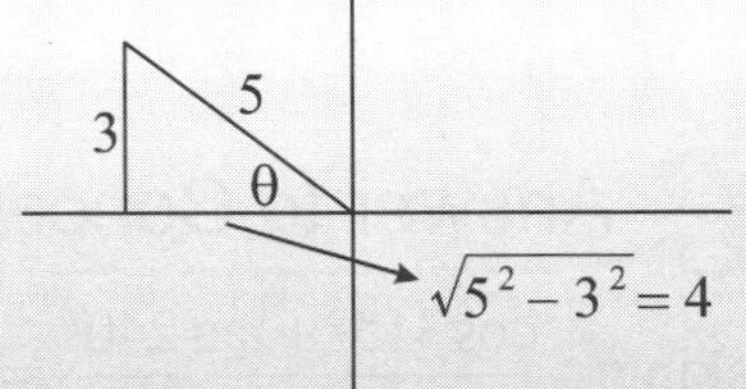

$\cos\theta = -\frac{4}{5}$

Answer is (A)

Answer to Concept Check 10-09

Asking $\dfrac{\frac{1}{\csc x}+\frac{1}{\sec x}}{(\cot x)(\sin x)}$?

$\frac{1}{\csc x} = \sin x$, $\frac{1}{\sec x} = \cos x$, $\cot x = \frac{\cos x}{\sin x}$

$$\frac{\frac{1}{\csc}+\frac{1}{\sec x}}{(\cot x)(\sin x)} = \frac{\sin x + \cos x}{(\frac{\cos x}{\sin x})(\sin x)} = \frac{\sin x + \cos x}{\cos x} = \tan x + 1$$

Answer is (D)

10-5 TRIGONOMETRIC GRAPHS

A. Properties of Trigonometric Functions

Functions	Domain	Range	Normal period
$y=\sin x$	All real numbers	$-1 \le y \le 1$	2π
$y=\cos x$	All real numbers	$-1 \le y \le 1$	2π
$y=\tan x$	All real numbers but $x \ne n\pi + \frac{\pi}{2}$ (n: any integer)	All real numbers	π
$y=\cot x$	All real numbers but $x \ne n\pi$ (n: any integer)	All real numbers	π
$y=\csc x$	All real numbers but $x \ne n\pi$ (n: any integer)	$\lvert \csc x \rvert \ge 1$ $\csc x \le -1$ or $\csc x \ge 1$	2π
$y=\sec x$	All real numbers but $x \ne n\pi + \frac{\pi}{2}$ (n: any integer)	$\lvert \sec x \rvert \ge 1$ $\sec x \le -1$ or $\sec x \ge 1$	2π

B. Normal Graphs of Sine and Cosine

- $y=\sin x$

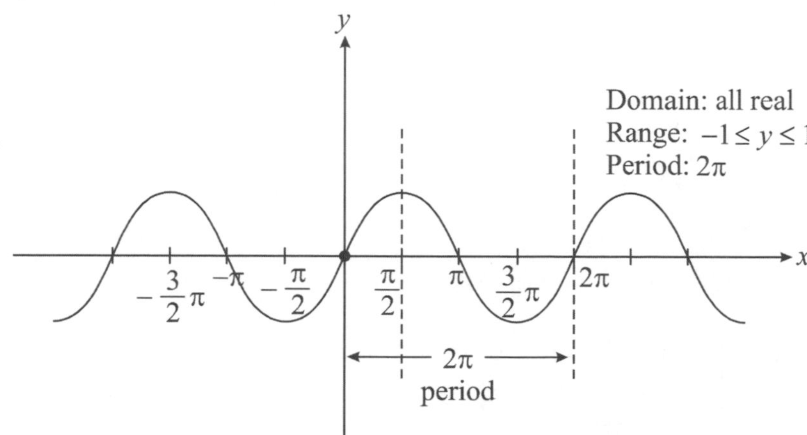

- $y = \cos x$

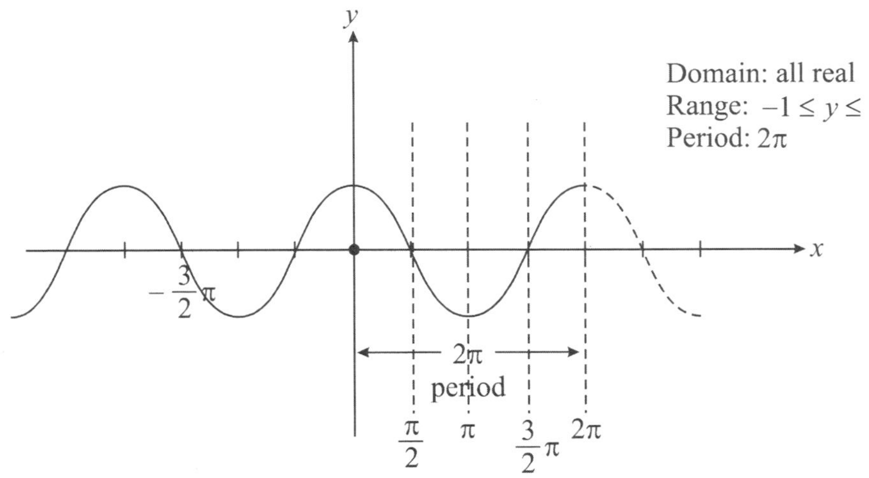

- $y = \csc x$

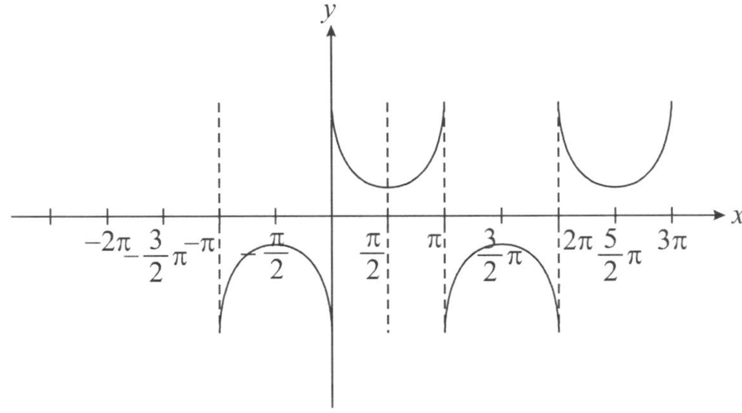

Domain: all real

Except: $n\pi$ (n is an integer)

Range: $y \le -1$ or $y \ge 1$

Period: 2π

- $y = \sec x$

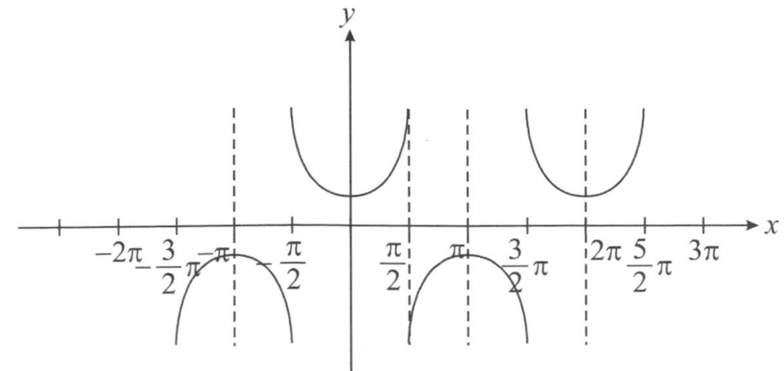

Domain: all real

Except $n\pi+\dfrac{\pi}{2}$ (n is an integer)

Range: $|y|\geq 1$

$y\geq 1$ or $y\leq -1$

Period: 2π

- $y=\tan x$

Domain: All real number

Except $x=n\cdot\pi+\dfrac{\pi}{2}$ (n is an integer)

Range: all real

Period: π

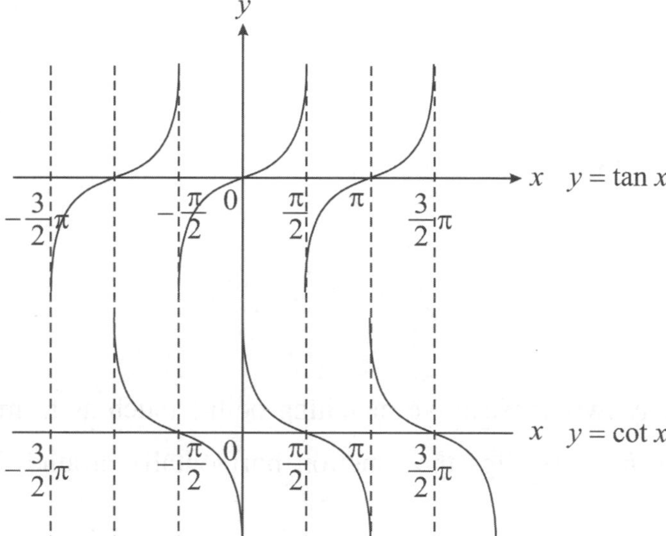

- $y=\cot x$

Domain: all real numbers

Except: $x=n\pi$ (n is an integer)

period: π

10-6 GENERAL FORM OF TRIGONOMETRIC FUNCTION

$$y=a\cdot f(bx+c)+d$$

A. Amplitude

The amplitude is the height from the center line of the function to the peak. A negative value flip the function vertically around the middle line (or commonly known as the center line.)

◆Example:

$y = 2 \cdot f(x+0) + 2 = 2\sin(x+0) + 2$

$|a| = 2, \ b = 1$

$\text{Period} = \dfrac{2\pi}{1} = 2\pi$

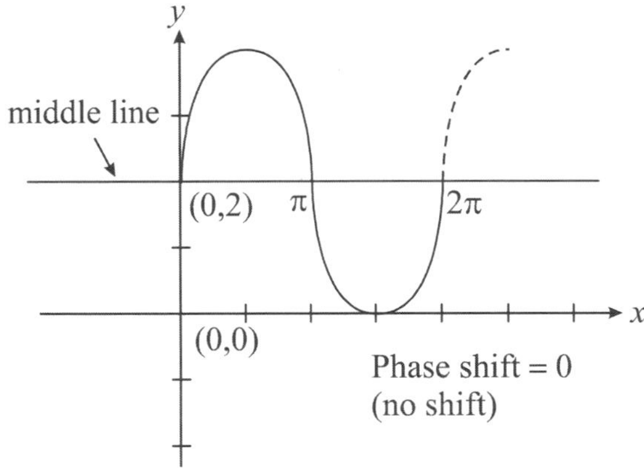

B. Period

The period is the distance between any two consecutive matching points (such as from a peak to the next peak). A negative b value flips the function horizontally around the y-axis.

For example: $\sin bx = \dfrac{2\pi}{b}$

$\cos bx = \dfrac{2\pi}{b}$

$\sec bx = \dfrac{2\pi}{b}$

$\csc bx = \dfrac{2\pi}{b}$

$\tan bx = \dfrac{\pi}{b}$

$\cot bx = \dfrac{\pi}{b}$

C. Phase Shift

x-shift: $-\dfrac{c}{b}$

Positive: shift to left, Negative: shift to right

D. Middle Line Shift

Middle line (y-shift): d

Positive: shift upward

Negative: shift downward.

◆Example 1:

$y = 2\sin(x)$

General form: $y = 2\sin(x+0)+0$

Amplitude $= |2| = 2$

Direction goes up because of positive amplitude.

Period $= \dfrac{2\pi}{1} = 2\pi$

Phase shift $= \dfrac{0}{1} = 0$ (remain in the same starting position on x-axis)

Graph as follows

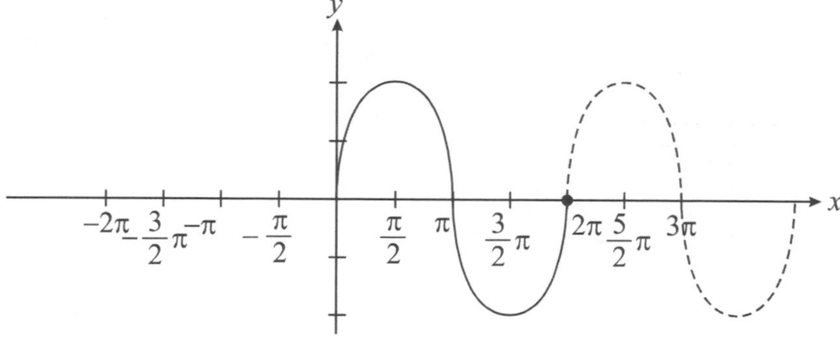

◆Example 2:

$y = 2\sin(2x)$

Amplitude $= |2| = 2$

Direction: goes up

Period $= \dfrac{2\pi}{2} = \pi$

Phase shift $= \frac{0}{2} = 0$

Middle line $= 0$

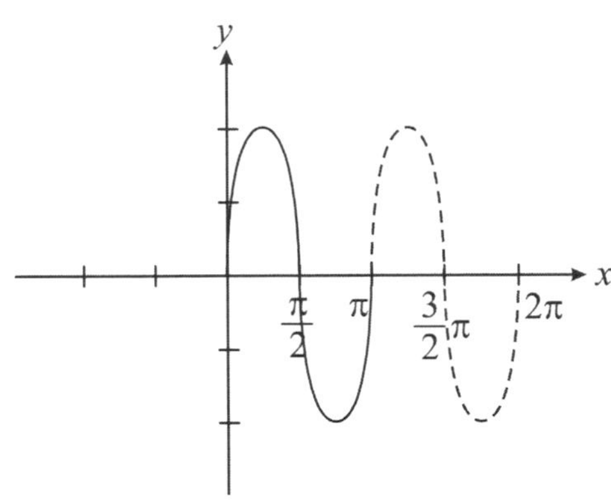

◆Example 3:

$y = \sin(2x - \pi)$

Amplitude $= |1| = 1$

Period $= \frac{2\pi}{2} = \pi$

Phase shift $= -\frac{-\pi}{2} = \frac{\pi}{2}$ (shift to right)

Middle line $= 0$

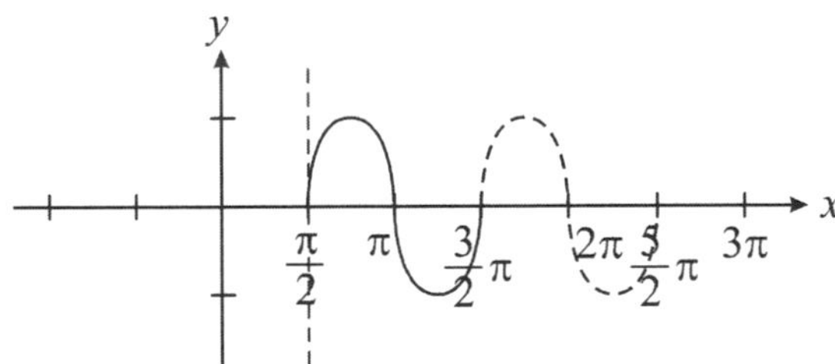

◆Example 4:

$y = \sin(2x - \pi) + 2$

Amplitude $= |1| = 1$

Period $= \frac{2\pi}{2} = \pi$

Phase shift $= -\frac{-\pi}{2} = \frac{\pi}{2}$ (shift to right)

Middle line $= 2$

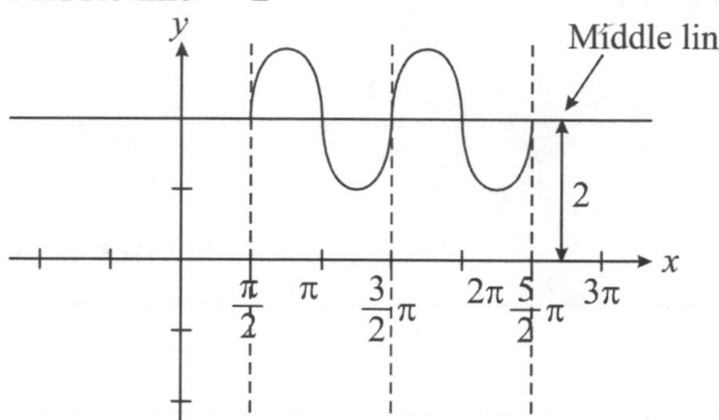

◆Example 5:

$y = -\sin(2x - \pi) + 2$

Amplitude $= |-1| = 1$

Direction: goes down (since it was vertically flipped)

Period $= \dfrac{2\pi}{2} = \pi$

Phase shift $= -\dfrac{-\pi}{2} = \dfrac{\pi}{2}$ (right)

Middle line $= 2$

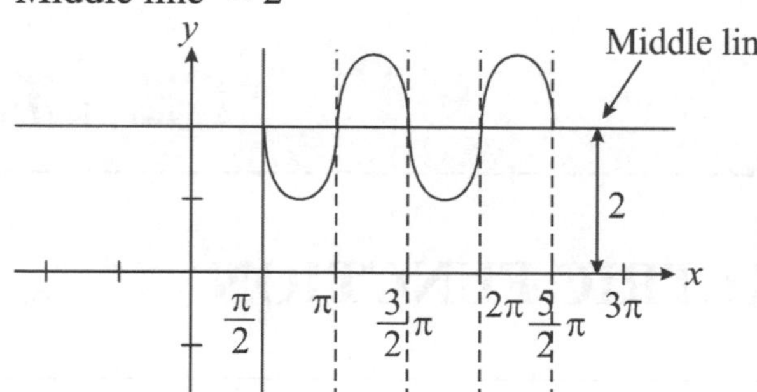

Answer to Concept Check 10-10

Asking: periods?

Period $= \dfrac{2\pi}{2} = \pi$

Answer is (D)

Answer to Concept Check 10-11

Asking: Amplitude ?

Amplitude $=|a|=|\frac{1}{3}|=\frac{1}{3}$

Answer is (A)

Answer to Concept Check 10-12

Asking: Phase shift

Phase shift $=-\frac{c}{b}=-\frac{\frac{\pi}{6}}{2}=-\frac{\pi}{12}$

Answer is (A)

Answer to Concept Check 10-13

Asking: Midline line

Midline line $=d=-2$

Answer is (C)

10-7 INVERSE TRIGONOMETRIC FUNCTION

- The inverse function of $f(x)$ is denoted as $f^{-1}(x)$. The inverse of $f(x)=\sin\theta$, is denoted as $\sin^{-1}\theta$, similarly, the inverse of $\cos\theta$, denoted as $\cos^{-1}\theta$, and the inverse of $\tan\theta$, denoted as $\tan^{-1}\theta$, and so on.

Since $\sin\theta=\frac{1}{2}$, θ could be a lot of values that fit the function $\sin\theta=\frac{1}{2}$, such as, $\theta=30°$, $150°$, $390°$, ⋯

So, The default domain and range are defined by a convention as the table below:

Function	Range	Domain
$\sin^{-1}(x)$	$-\frac{\pi}{2} \le \sin^{-1} x \le \frac{\pi}{2}$	$-1 \le x \le 1$
$\cos^{-1}(x)$	$0 \le \cos^{-1} x \le \pi$	$-1 \le x \le 1$
$\tan^{-1}(x)$	$-\frac{\pi}{2} < \tan^{-1} x < \frac{\pi}{2}$	All real numbers
$\cot^{-1}(x)$	$0 < \cot^{-1} x < \pi$	All real numbers
$\sec^{-1}(x)$	$0 \le \sec^{-1} x \le \pi$ and $\sec^{-1} x \ne \frac{\pi}{2}$	$\lvert x \rvert \ge 1$
$\csc^{-1}(x)$	$-\frac{\pi}{2} \le \csc^{-1} x \le \frac{\pi}{2}$ and $x \ne \frac{\pi}{2}$	$\lvert x \rvert \ge 1$

θ	$\sin^{-1} x$	$\cos^{-1} x$	$\tan^{-1} x$
30°	$x = 0.5$	$x = 0.866$	$x = 0.577$
45°	$x = 0.707$	$x = 0.707$	$x = 1$
60°	$x = 0.866$	$x = 0.5$	$x = 1.732$

◆Example 1:

What is the degree measure of $\sin^{-1} 0.62$?

Enter $\sin(0.62)$ into your calculator in degree mode, and you will get $38°$.

◆Example 2:

Find the value of $\sin(\tan^{-1}\frac{2}{3})$? $[0 \le \theta \le \frac{\pi}{2}]$

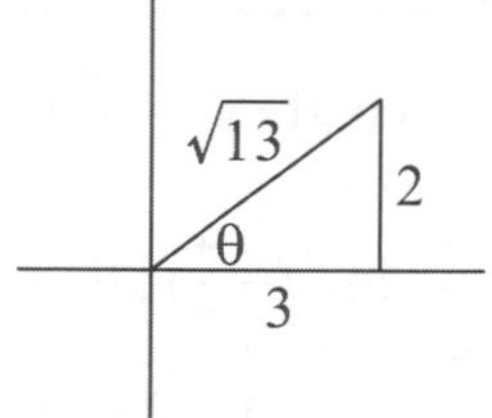

$\sin\theta = \frac{2}{\sqrt{13}} = 0.55$

◆Example 3:

Evaluate $\tan^{-1}(\tan 290°)$

$\tan^{-1}(\tan x) = x$, $\tan^{-1}(\tan 290°) = 290°$

Wait! By convention, θ of tan should be between $-\frac{\pi}{2}$ and $\frac{\pi}{2}$.

$290° - 360° = -70°$ has the same reference angle as $290°$

$-70°$ is the correct answer

◆Example 4:

What is the value of $\cos(\sin^{-1}\frac{5}{13})$?

Let $\theta = \sin^{-1}\frac{5}{13}$, $\sin\theta = \frac{5}{13}$ draw a right triangle

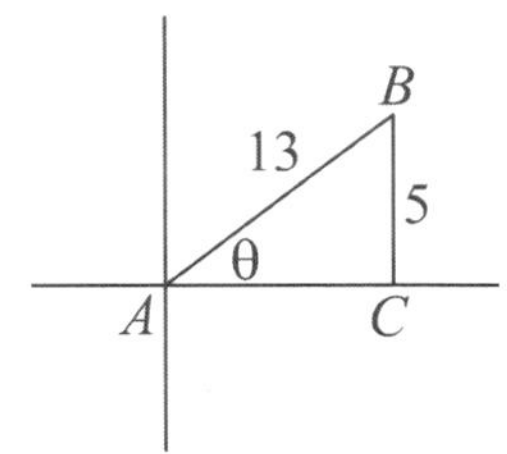

$\overline{AC}^2 = 13^2 - 5^2 = 144$

$\therefore\ \overline{AC} = 12$

$\cos\theta = \frac{\overline{AC}}{\overline{AB}} = \frac{12}{13}$

Answer to Concept Check 10-14

Asking: $\tan^{-1}(\tan(-2))$?

Since $\tan^{-1}(\tan(\theta)) = \theta$,

But θ must be in the interval of $-\frac{\pi}{2} < \theta < \frac{\pi}{2}$ (convention)

so, $\tan^{-1}(\tan(-2)) \neq -2$

because $-2 < -\frac{\pi}{2}$ as shown below

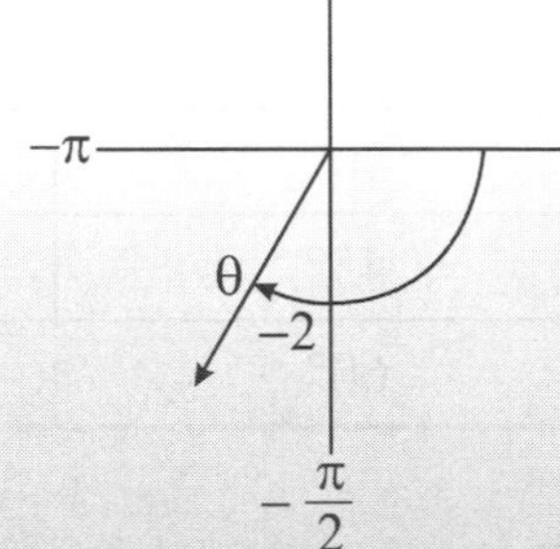

so, the answer is (E)

Answer to Concept Check 10-15

Asking: $\cos(\tan^{-1}(-\frac{4}{3}))$?

Let $\theta = \tan^{-1}(-\frac{4}{3})$, then $\tan\theta = -\frac{4}{3}$

θ must be $-\frac{\pi}{2} < \theta < \frac{\pi}{2}$ (convention)

so, θ is in quadrant IV,

side is $\sqrt{3^2 + 4^2} = 5$

$\cos\theta = \frac{3}{5}$

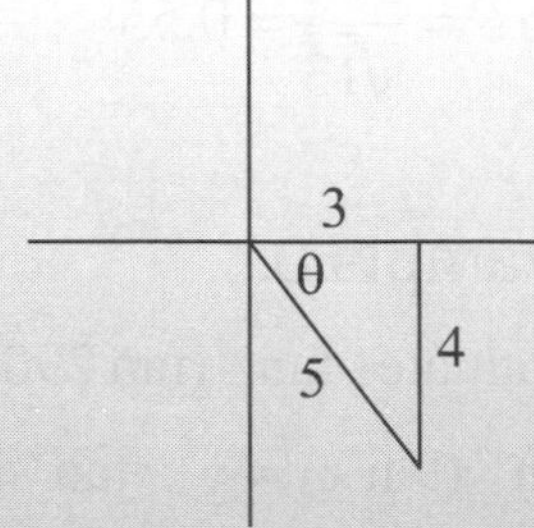

Answer is (D)

Answer to Concept Check 10-16

Asking: $\tan(\sin^{-1}(-\frac{1}{3}))$?

Let $\theta = \sin^{-1}(-\frac{1}{3})$, $\sin\theta = -\frac{1}{3}$

θ must be between $-\frac{\pi}{2}$ and $\frac{\pi}{2}$,

so, θ is in quadrant IV, as shown below

The length of the initial side is $\sqrt{3^2 - 1^2} = \sqrt{8}$

$\tan\theta = \frac{-1}{\sqrt{8}} = -\frac{\sqrt{2}}{4}$

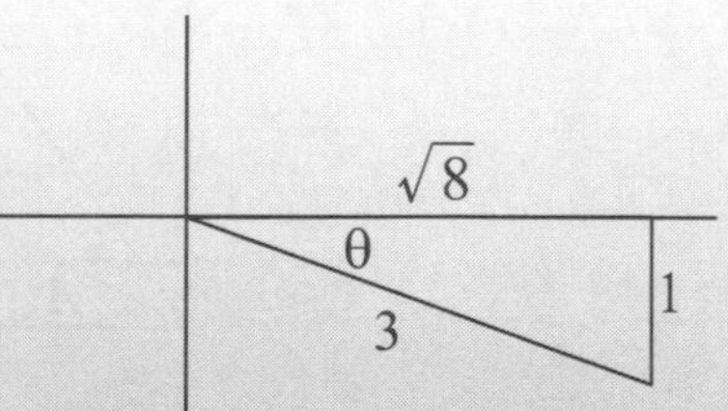

Answer is (D)

Answer to Concept Check 10-17

Asking: $\tan(\tan^{-1}(-2))$?

Let $\theta = \tan^{-1}(-2)$

$\tan\theta = -2$ (It is a value of the function, not the value of an angle)

Answer is (A)

10-8 SUM and DIFFERENCE OF ANGLES (LEVEL 2 ONLY)

$$\sin(A+B) = \sin A\cos B + \cos A\sin B$$
$$\sin(A-B) = \sin A\cos B - \cos A\sin B$$

$$\cos(A+B) = \cos A\cos B - \sin A\sin B$$
$$\cos(A-B) = \cos A\cos B + \sin A\sin B$$

$$\tan(A+B) = \frac{\tan A + \tan B}{1 - \tan A\tan B}$$
$$\tan(A-B) = \frac{\tan A - \tan B}{1 + \tan A\tan B}$$

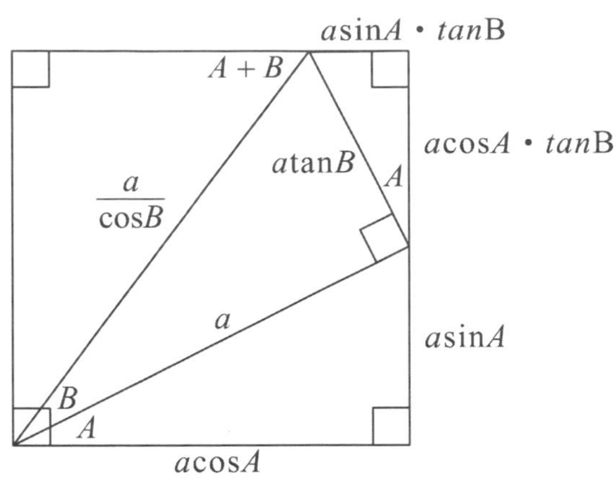

$$\sin(A+B)=\frac{a\sin A+a\cos A\cdot\tan B}{\dfrac{a}{\cos B}}=(\sin A+\cos A\cdot\tan B)\cdot\cos B$$
$$=\sin A\cos B+\cos A\cdot\frac{\sin B}{\cos B}\cdot\cos B$$
$$=\sin A\cos B+\cos A\sin B$$

$$\cos(A+B)=\frac{a\cos A-a\sin A\cdot\tan B}{\dfrac{a}{\cos B}}=\frac{\cos A-\sin A\cdot\dfrac{\sin B}{\cos B}}{\dfrac{1}{\cos B}}$$
$$=\cos A\cos B-\sin A\sin B$$

$$\tan(A+B)=\frac{\sin A\cos B+\cos A\sin B}{\cos A\cos B-\sin A\sin B}=\frac{\dfrac{\sin A}{\cos A}+\dfrac{\sin B}{\cos B}}{1-\dfrac{\sin A\sin B}{\cos A\cos B}}$$
$$=\frac{\tan A+\tan B}{1-\tan A\cdot\tan B}$$

10-9 DOUBLE and HALF OF ANGLES (LEVEL 2 ONLY)

$$\sin 2A = 2\sin A\cos A$$

$\sin 2A = \sin(A+A) = \sin A\cos A + \cos A\sin A = 2\sin A\cos A$

$$\cos 2A = \cos^2 A - \sin^2 A = 1 - 2\sin^2 A = 2\cos^2 A - 1$$

$\cos 2A = \cos(A+A) = \cos A\cos A - \sin A\sin A = \cos^2 A - \sin^2 A$

$\cos 2A = (1-\sin^2 A) - \sin^2 A = 1 - 2\sin^2 A$

$\cos 2A = \cos^2 A - (1-\cos^2 A) = 2\cos^2 A - 1$

$\tan 2A = \dfrac{2\tan A}{1-\tan^2 A}$

$$\tan 2A = \frac{\sin 2A}{\cos 2A} = \frac{2\sin A\cos A}{\cos^2 A - \sin^2 A} = \frac{2\dfrac{\sin A}{\cos A}}{1-\dfrac{\sin^2 A}{\cos^2 A}} = \boxed{\frac{2\tan A}{1-\tan^2 A}}$$

$$\sin\frac{A}{2} = \pm\sqrt{\frac{1-\cos A}{2}}$$

$\because \quad \cos A = \cos^2\dfrac{A}{2} - \sin^2\dfrac{A}{2} = 1 - 2\sin^2\dfrac{A}{2}$

$\therefore \quad 2\sin^2\dfrac{A}{2} = 1 - \cos A$

$\therefore \quad \sin^2\dfrac{A}{2} = \dfrac{1-\cos A}{2}$

$\therefore \quad \sin\dfrac{A}{2} = \pm\sqrt{\dfrac{1-\cos A}{2}}$

$\cos A = 2\cos^2\dfrac{A}{2} - 1$

$2\cos^2\dfrac{A}{2} = 1 + \cos A$

$\cos^2\dfrac{A}{2} = \dfrac{1+\cos A}{2}$

$$\therefore \quad \cos\frac{A}{2}=\pm\sqrt{\frac{1+\cos A}{2}}$$

$$\tan\frac{A}{2}=\frac{\sin\frac{A}{2}}{\cos\frac{A}{2}}=\pm\sqrt{\frac{1-\cos A}{1+\cos A}}$$

10-10 SUM, DIFFERENCE, and PRODUCT OF TRIGONOMETRIC FUNCTIONS (LEVEL 2 ONLY)

$$\sin A+\sin B=2\sin\frac{A+B}{2}\cos\frac{A-B}{2}$$

Let $\alpha=\frac{A+B}{2}$, $\beta=\frac{A-B}{2}$

$\sin(\alpha+\beta)=\sin\alpha\cos\beta+\cos\alpha\sin\beta$

$\sin(\alpha-\beta)=\sin\alpha\cos\beta-\cos\alpha\sin\beta$

$\therefore \quad \sin(\alpha+\beta)+\sin(\alpha-\beta)=2\sin\alpha\cos\beta$

$\sin(\frac{A+B}{2}+\frac{A-B}{2})+\sin(\frac{A+B}{2}-\frac{A-B}{2})=\sin A+\sin B=2\sin\frac{A+B}{2}\cos\frac{A-B}{2}$

$$\sin A-\sin B=2\cos\frac{A+B}{2}\sin\frac{A-B}{2}$$

$\because \quad \sin(\alpha+\beta)-\sin(\alpha-\beta)=2\cos\alpha\sin\beta$

$\sin A-\sin B=2\cos\frac{A+B}{2}\sin\frac{A-B}{2}$

$$\cos A+\cos B=2\cos\frac{A+B}{2}\cos\frac{A-B}{2}$$

Let $\alpha=\frac{A+B}{2}$, $\beta=\frac{A-B}{2}$

$\because \quad \cos(\alpha+\beta)+\cos(\alpha-\beta)=\cos\alpha\cos\beta-\sin\alpha\sin\beta+\cos\alpha\cos\beta+\sin\alpha\sin\beta$

$\therefore\ \cos(\alpha+\beta)=\cos A$

$\cos(\alpha-\beta)=\cos B$

$\therefore\ \cos A+\cos B=2\cos\frac{A+B}{2}\cos\frac{A-B}{2}$

$$\cos A-\cos B=-2\sin\frac{A+B}{2}\sin\frac{A-B}{2}$$

$\cos(\alpha+\beta)-\cos(\alpha-\beta)=\cos\alpha\cos\beta-\sin\alpha\sin\beta-[\cos\alpha\cos\beta+\sin\alpha\sin\beta]$

$=\cancel{\cos\alpha\cos\beta}-\sin\alpha\sin\beta-\cancel{\cos\alpha\cos\beta}-\sin\alpha\sin\beta=-2\sin\alpha\sin\beta$

$\therefore\ \cos A-\cos B=-2\sin\frac{A+B}{2}\sin\frac{A-B}{2}$

$$2\sin\alpha\cos\beta=\sin(\alpha+\beta)+\sin(\alpha-\beta)$$

$\because\ \sin A+\sin B=2\sin\frac{A+B}{2}\cos\frac{A-B}{2}$

Let $\alpha=\frac{A+B}{2}$, $\beta=\frac{A-B}{2}$

$\therefore\ 2\sin\alpha\cos\beta=\sin A+\sin B$

$=\sin(\alpha+\beta)+\sin(\alpha-\beta)$

$\because\ \alpha+\beta=\frac{A+B}{2}+\frac{A-B}{2}=\frac{A}{2}+\frac{A}{2}=A$

$\alpha-\beta=\frac{A+B}{2}-\frac{A-B}{2}=\frac{B}{2}+\frac{B}{2}=B$

★Tip

The proofs ot those formulas are the same.

Firstly, let $\alpha=\frac{A+B}{2}$, $\beta=\frac{A-B}{2}$

then $2\cos\alpha\sin\beta=2\cos\frac{A+B}{2}\sin\frac{A-B}{2}$

$=\sin(\frac{A+B}{2}+\frac{A-B}{2})-\sin(\frac{A+B}{2}-\frac{A-B}{2})$

$=\sin A-\sin B=\sin(\alpha+\beta)-\sin(\alpha-\beta)$

$$2\cos\alpha\sin\beta=\sin(\alpha+\beta)-\sin(\alpha-\beta)$$

$$2\cos\alpha\cos\beta=\cos(\alpha+\beta)+\cos(\alpha-\beta)$$

$$-2\sin\alpha\sin\beta=\cos(\alpha+\beta)-\cos(\alpha-\beta)$$

10-11 TRIGONOMETRIC EQUATIONS

Treat the trigonometric equation just like a general equation.

◆Example:

Solve the equation of $\cos 2x + \cos x = 0 \quad (0 < x < 360°)$

Since $\cos 2x = 2\cos^2 x - 1$

$\therefore \cos 2x + \cos x = 2\cos^2 x - 1 + \cos x = 2\cos^2 x + \cos x - 1 = (2\cos x - 1)(\cos x + 1) = 0$

$\therefore 2\cos x - 1 = 0$ or $\cos x + 1 = 0$

$\therefore \cos x = \frac{1}{2}$ or $\cos x = -1$

When $\cos x = \frac{1}{2}$, x is in quadrant

(I) or (IV), $x = 60°$, or $360° - 60° = 300°$

When $\cos x = -1$, x is in quadrant (II), so, $x = 180°$

So, the solutions are

$x = 60°,\ 180°$, or $300°$

10-12 THE LAW OF SINES

$$\frac{a}{\sin A} = \frac{b}{\sin B} = \frac{c}{\sin C}$$

$\triangle ABC$ is an arbitrary triangle with sides a, b, and c.

Proof:

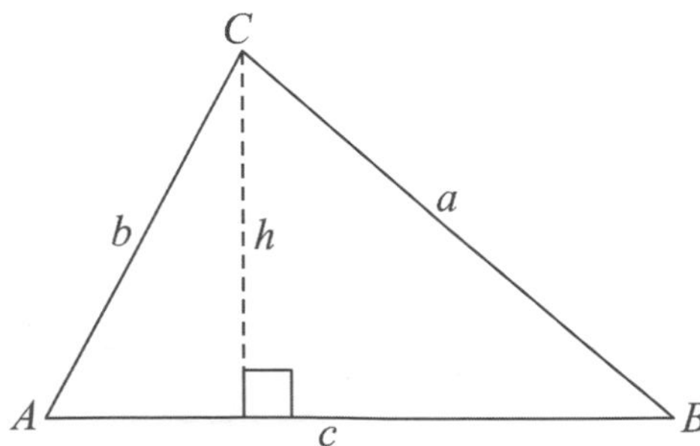

Area of $\triangle ABC = \frac{1}{2} \cdot c \cdot (h) = \frac{1}{2} c \cdot b \sin A$

$= \frac{1}{2} a \cdot c \sin B$

$$=\frac{1}{2}a \cdot b\sin C$$

$\therefore$ Divided by abc

$$\frac{a}{\sin A}=\frac{b}{\sin B}=\frac{c}{\sin C}$$

10-13 THE LAW OF COSINES

$$a^2=b^2+c^2-2bc\cos A$$
$$b^2=a^2+c^2-2ac\cos B$$
$$c^2=a^2+b^2-2ab\cos C$$

$$\cos A=\frac{b^2+c^2-a^2}{2bc}$$
$$\cos B=\frac{a^2+c^2-b^2}{2ac}$$
$$\cos C=\frac{a^2+b^2-c^2}{2ab}$$

Proof:

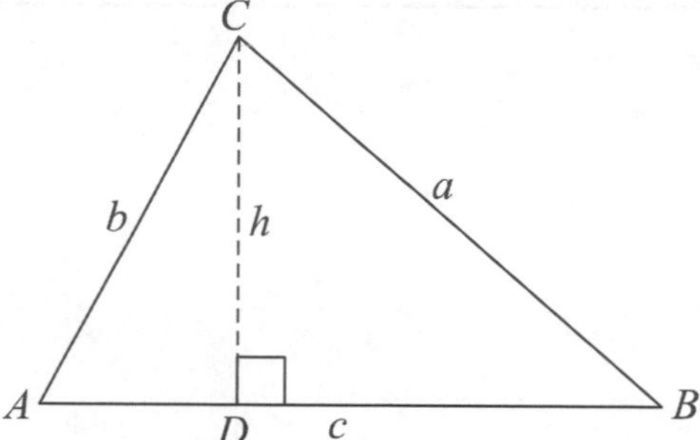

$$\begin{aligned}a^2 &= BD^2+h^2\\ &= BD^2+b^2-AD^2\\ &= (c-AD)^2+b^2-AD^2\\ &= c^2-2c\cdot AD+\cancel{AD^2}+b^2-\cancel{AD^2}\\ &= b^2+c^2-2c\cdot b\cos A\\ &= b^2+c^2-2bc\cos A\end{aligned}$$

Answer to Concept Check 10-18

Asking: $\cos\theta$

Since $\theta < \frac{\pi}{2}$, $\cot(\frac{\pi}{3}) = \tan(\frac{\pi}{2} - \frac{\pi}{3}) = \tan(\theta + \frac{\pi}{6})$

$\therefore\ \theta + \frac{\pi}{6} = \frac{\pi}{2} - \frac{\pi}{3}$ $\quad\therefore\ \theta = \frac{\pi}{6} - \frac{\pi}{6} = 0$

$\cos(0°) = 1$

Answer is (C)

Answer to Concept Check 10-19

Asking: solve the equation

$2\cot x \sin x + 1 = 0$

$2\frac{\cos x}{\sin x}\sin x + 1 = 0$

$2\cos x + 1 = 0$

$\cos x = -\frac{1}{2}$, $\pi < x < \frac{3}{2}\pi$ is in quadrant III

$x = 60 + 180 = 240°$

Answer is (E)

10-14 Heron's Formula

Area of triangle $= \sqrt{s(s-a)(s-b)(s-c)}$

(a, b, and c are the sides of a triangle, $s = \frac{a+b+c}{2}$)

Proof: Area $= \frac{1}{2}(\text{base}) \times (\text{height}) = \frac{1}{2}ab\sin\theta$

Since $\sin\theta = \sqrt{1-\cos^2\theta}$, and $\cos c = \frac{a^2+b^2-c^2}{2ab}$

So, $\sin c = \sqrt{1-(\frac{a^2+b^2-c^2}{2ab})^2} = \sqrt{\frac{4a^2b^2-(a^2+b^2-c^2)^2}{4a^2b^2}} = \frac{\sqrt{(2ab)^2-(a^2+b^2-c^2)^2}}{2ab}$

$= \frac{\sqrt{[(2ab)-(a^2+b^2-c^2)][(2ab)+(a^2+b^2-c^2)]}}{2ab}$

$= \frac{\sqrt{(2ab-a^2-b^2+c^2)(2ab+a^2+b^2-c^2)}}{2ab}$

$$=\frac{\sqrt{[c^2-(a^2+b^2-2ab)][(a^2+b^2+2ab)-c^2]}}{2ab}=\frac{\sqrt{[(c^2-(a-b)^2][(a+b)^2-c^2]}}{2ab}$$

$$=\frac{\sqrt{[c-(a-b)][c+(a-b)][(a+b)-c][(a+b)+c]}}{2ab}$$

$$=\frac{\sqrt{(b+c-a)(a+c-b)(a+b-c)(a+b+c)}}{2ab}$$

$$A=\frac{1}{2}ab\sin c=\frac{1}{4}\sin c=\sqrt{\frac{(b+c-a)(a+c-b)(a+b-c)(a+b+c)}{16}}$$

$$=\sqrt{(\frac{b+c-a}{2})(\frac{a+c-b}{2})(\frac{a+b-c}{2})(\frac{a+b+c}{2})}=\sqrt{s(s-a)(s-b)(s-c)}$$

For Example: If the 3 sides of a triangle are 5, 12, and 13 respectively, what is the area of the triangle ?

$$s=\frac{5+12+13}{2}=15$$

$$A=\sqrt{(15)(10)(3)(2)}=\sqrt{900}=30$$

Answer to Concept Check 10-20

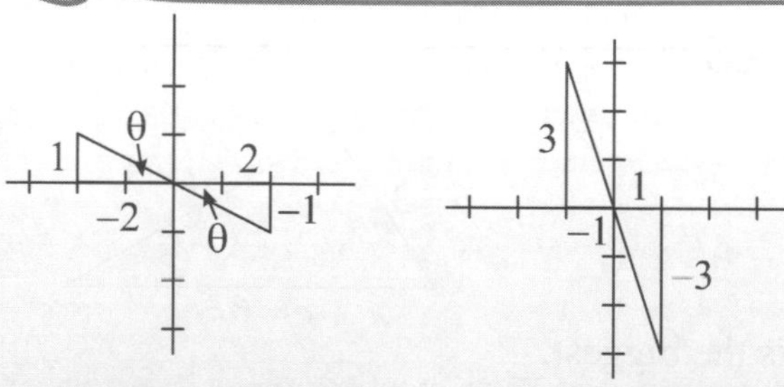

$2\sec^2\theta+7\tan\theta+1=0$

$2(1+\tan^2\theta)+7\tan\theta+1=0$

$2+2\tan^2\theta+7\tan\theta+1=0$

$2\tan^2\theta+7\tan\theta+3=0$

$(2\tan\theta+1)(\tan\theta+3)=0$

$\tan\theta=-\frac{1}{2}\qquad\tan\theta=-3$

There are two solutions in quadrant II and IV respectively, the total number of the solutions are 4

Answer is (D)

Answer to Concept Check 10-21

$$\frac{3}{\sin 45°} = \frac{c}{\sin 30°}, \quad \frac{3}{\frac{\sqrt{2}}{2}} \cdot \sin 30° = c$$

$$\frac{6}{\sqrt{2}} \cdot \frac{1}{2} = c, \quad c = \frac{3}{\sqrt{2}} = \frac{3\sqrt{2}}{2}$$

Answer is (D)

Answer to Concept Check 10-22

$$b^2 = a^2 + c^2 - 2ac\cos B$$

$$9 = 36 + 25 - 2 \cdot (6)(5)\cos B = 61 - 60\cos B$$

$$60\cos B = 52, \quad \cos B = \frac{26}{30} = \frac{13}{15}$$

Answer is (E)

Answer to Concept Check 10-23

$\cos^{-1}(-\frac{1}{15})$

Let $c:a:b = 6:5:3$

then $c = 6k$, $a = 5k$, and $b = 3k$, so, $m\angle C$ is the biggest.

Asking: measure of the largest angle

$b = 3$, $a = 5$, $c = 6$

since $c^2 = a^2 + b^2 - 2ab\cos C$

$(6k)^2 = (5k)^2 + (3k)^2 - 2(5k)(3k)\cos C$

$36k^2 = 25k^2 + 9k^2 - 30k^2\cos C \Rightarrow 36 = 25 + 9 - 30\cos C$

$2 = -30\cos C$

$\therefore \cos C = -\frac{2}{30} = -\frac{1}{15}$

since $m\angle C$ is between $0°$ and $180°$, $\angle C$ is in quadrant II.

$\therefore m\angle C = \cos^{-1}(-\frac{1}{15})$

Answer is $\cos^{-1}(-\frac{1}{15})$

Answer to Concept Check 10-24

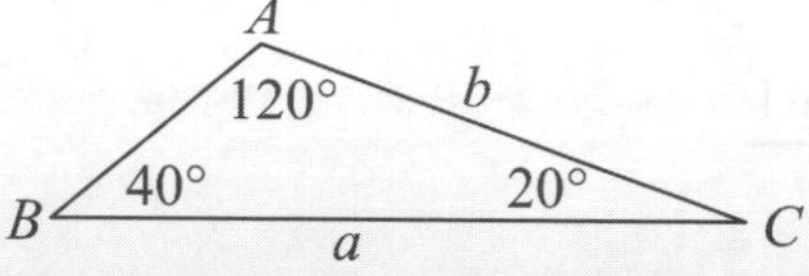

Asking a:b?

$$\frac{b}{\sin B}=\frac{a}{\sin A},\quad A=180°-40°-20°=120°$$

$$\text{so, }\frac{b}{\sin 40°}=\frac{a}{\sin 120°}=\frac{a}{\sin 60°}$$

$$\frac{a}{b}=\frac{\sin 60°}{\sin 40°}=\frac{0.87}{0.64}=1.36$$

Answer is (1.36)

Answer to Concept Check 10-25

By Heron's Formula

$$\text{area}=\sqrt{s(s-a)(s-b)(s-c)}$$

$$s=\frac{a+b+c}{2}=\frac{3+4+5}{2}=6$$

$$\text{area}=\sqrt{6(6-3)(6-4)(6-5)}=\sqrt{6\cdot 3\cdot 2\cdot 1}=6$$

Answer is (6)

CHAPTER 10
PRACTICE

1. Find the period of $f(x)=\dfrac{\sin x}{\sec x}$?

(A) $\dfrac{\pi}{4}$

(B) $\dfrac{\pi}{3}$

(C) $\dfrac{\pi}{2}$

(D) $\dfrac{2\pi}{3}$

(E) π

2. If $y=3-\dfrac{1}{2}\cos^2\dfrac{x}{2}$, where $0<x<2\pi$, what value of x will give a maximum value for y ?

(A) $-\dfrac{\pi}{2}$

(B) $\dfrac{\pi}{2}$

(C) $-\pi$

(D) π

(E) $-\dfrac{3}{2}\pi$

3. prove $\sin 2A=\dfrac{2\tan A}{1+\tan^2 A}$

4. if $\sin 30°=\dfrac{1}{2}$ and $\sin 45°=\dfrac{\sqrt{2}}{2}$, what is the value of $\cos(15°)$?

(A) $\dfrac{\sqrt{2}-1}{2}$

(B) $\dfrac{\sqrt{2}+1}{2}$

(C) $\dfrac{\sqrt{2}}{4}(\sqrt{3}-1)$

(D) $\dfrac{\sqrt{2}}{4}(\sqrt{3}+1)$

(E) $\dfrac{\sqrt{3}}{4}(\sqrt{2}+1)$

5. If $\sin\dfrac{A}{2}=\dfrac{1}{2}$, what is the value of $\cos A$?

(A) $-\dfrac{1}{2}$

(B) $\dfrac{1}{2}$

(C) $\dfrac{\sqrt{2}}{2}$

(D) $-\dfrac{\sqrt{2}}{2}$

(E) 1

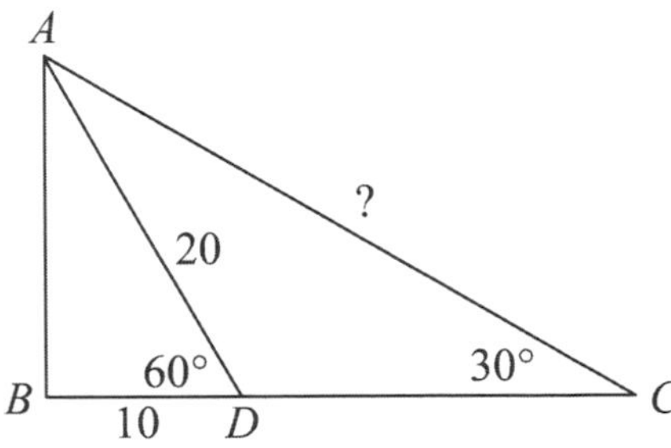

6. In figure above, if $BD=10$, what is the length of $\overline{AC}$?

7. If $\sin x = t$ for all x in the interval $\frac{\pi}{2} < x < \pi$, then $\sin 2x =$

(A) $\frac{t}{2}$

(B) $2t$

(C) $2t\sqrt{1-t^2}$

(D) $2t\sqrt{1+t^2}$

(E) $2t\sqrt{1+t}$

8. What is the value of sin (arc $\cos(-\frac{4}{5})$) ?

$(0 \le \theta \le \pi)$

(A) $\frac{3}{5}$

(B) $-\frac{3}{5}$

(C) $\frac{2}{3}$

(D) $-\frac{2}{3}$

(D) $\frac{1}{3}$

9. If $\cos\theta + 2\sin\theta = 2$, $0° < \theta < 90°$, what is the value of $\sin\theta + \cos\theta$?

(A) 0

(B) 1

(C) $\frac{7}{5}$

(D) $\frac{5}{7}$

(E) $-\frac{5}{7}$

10. Find the value of

$\cos(\sin^{-1}\frac{1}{2} + \cos^{-1}\frac{1}{2})(0 \le \theta \le \frac{\pi}{2})$

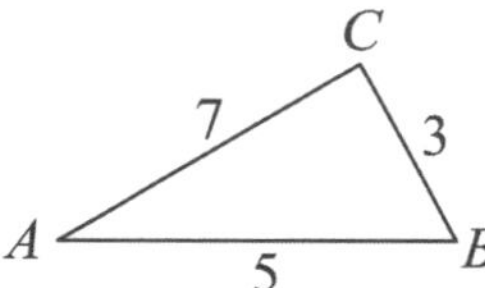

11. In $\triangle ABC$ as figure above, $a = 3$, $b = 7$, and $c = 5$, find the measurement of $\angle C$.

12. As question 11 above, what is the measurement of area of $\triangle ABC$?

13. Which of the following is the period of the graph of $y = -2\sin(3\pi x - \frac{2}{3}\pi) + \frac{1}{2}$?

(A) $\frac{2}{3}$

(B) $\frac{3}{2}$

(C) $\frac{1}{2}$

(D) $\frac{1}{3}$

(E) π

14. If $2\sin x + \cos x = 0$, where $0 \le x \le 2\pi$, which of the following is the value of x ?

(A) −34

(B) −26.6

(C) 27.6

(D) 34

(E) None of them

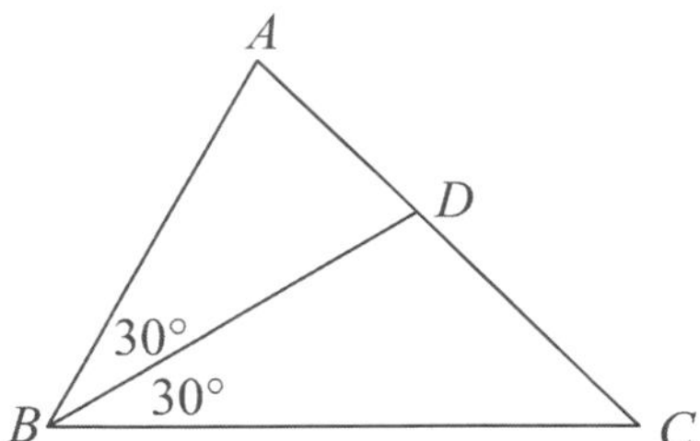

Note: Figure not drawn to scale

15. $\triangle ABC$ shown above, $\angle B = 60°$, bisected by $\overline{BD}$, $AB = 9$, $BD = 3\sqrt{3}$, What is the length of $\overline{BC}$?

16. Find the value of $\cos 120° \sin(-150°) + \cos 300° \sin 240°$

17. Given $\tan\alpha$, $\tan\beta$ are the two roots of $2x^2 - 3x + 1 = 0$, where $0 < \alpha < \beta < \frac{\pi}{2}$, what is the value of $\tan(\alpha - \beta)$?

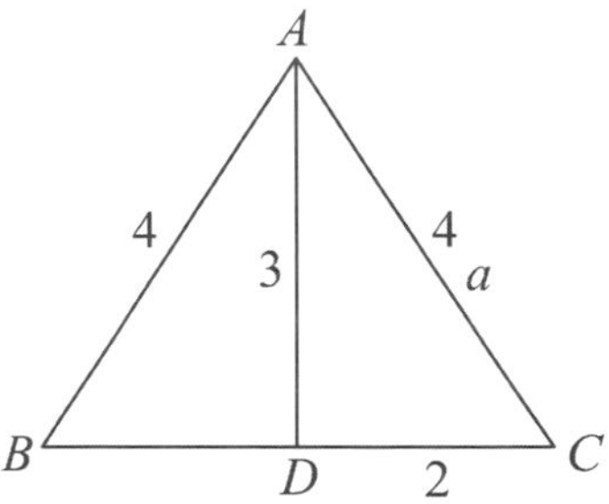

Note: Figure not drawn to scale

18. In the figure above, $AB = 4$, $AD = 3$, and $AC = 4$, $DC = 2$
Find the length of $\overline{BD}$

19. If $y=2\cos^2\theta+2\sin\theta+5$, what is the sum of maximum value and minimum value ?

(A) $\frac{21}{2}$

(B) $\frac{17}{2}$

(C) 7

(D) 5

(E) 3

20. Given $\sin\theta=-\frac{3}{5}$, $\pi<\theta<\frac{3}{2}\pi$, Find $\sin 2\theta$, $\cos 2\theta$, and $\tan\frac{\theta}{2}$.

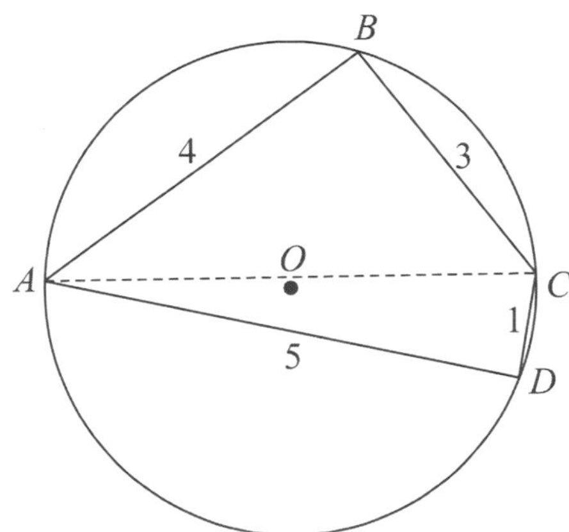

Note: Figure not drown to scale

21. In the figure above, quadrilateral $ABCD$ is inscribed in circle 0. If $AB=4$, $BC=3$, $CD=1$, and $AD=5$, what is the length of $\overline{AC}$? (Rounded to Hundredth)

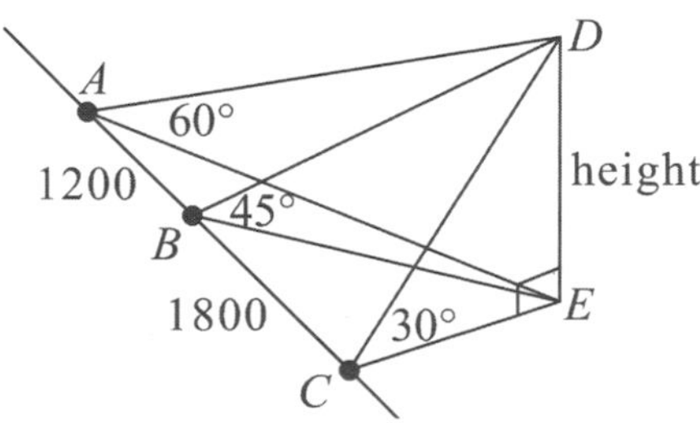

Note: Figure not drown to scale

22. The figure above is a schematic diagram of measuring the height of a mountain. Draw a line on the road to take three points, A, B, and C, $AB=1200$ feet and $\overline{BC}=1800$ feet, the elevation angles of the measured mountain heights are $60°$, $45°$, and $30°$ as shown in the figure respectively. What is the height ($\overline{DE}$) of the mountain ? (Rounded to whole number)

(A) 2,323 ft

(B) 3,412 ft

(C) 4,289 ft

(D) 5,103 ft

(E) 6,324 ft

CHAPTER 10
ANSWERS and EXPLANATIONS

1. (E)

$$y = \frac{\sin x}{\sec x} = \sin x \cos x = \frac{1}{2}\sin 2x$$

$$\text{period} = \frac{2\pi}{2} = \pi$$

2. (D)

$$y = 3 - \frac{1}{2}\cos^2\frac{x}{2} = 3 - \frac{1}{2}\cdot\frac{1+\cos x}{2}$$

$$= 3 - \frac{1}{4}(1+\cos x) = \frac{11}{4} - \frac{1}{4}\cos x$$

(since $\cos^2\frac{x}{2} = \frac{1+\cos x}{2}$)

when $\cos x = -1$, y is the largest and equivalent to 3

so, $x = \pi$

Answer is (D)

3. Since, $\dfrac{\tan A + \tan A}{1+\tan^2 A} = \dfrac{2\dfrac{\sin A}{\cos A}}{\sec^2 A}$

$$= \frac{2\dfrac{\sin A}{\cos A}}{\dfrac{1}{\cos^2 A}} = \frac{2\sin A \cancel{\cos A} \cos A}{\cancel{\cos A}}$$

$$= 2\sin A\cos A$$

$$= \sin 2A$$

4. (D)

Since $\cos(\theta_1 - \theta_2)$

$= \cos\theta_1 \cos\theta_2 + \sin\theta_1 \sin\theta_2$

$\therefore\ \cos(15°) = \cos(45° - 30°)$

$= \cos 45° \cos 30° + \sin 45° \sin 30°$

Since, $\sin 45° = \dfrac{\sqrt{2}}{2}$, $\cos 45° = \dfrac{\sqrt{2}}{2}$

$\sin 30° = \dfrac{1}{2}$, $\cos 30° = \dfrac{\sqrt{3}}{2}$

$\therefore\ \cos 45° \cos 30° + \sin 45° \sin 30°$

$$= \frac{\sqrt{2}}{2}\cdot\frac{\sqrt{3}}{2} + \frac{\sqrt{2}}{2}\cdot\frac{1}{2} = \frac{\sqrt{2}}{2}\left(\frac{\sqrt{3}}{2} + \frac{1}{2}\right)$$

$$= \frac{\sqrt{2}}{4}(\sqrt{3}+1)$$

5. (B)

Since $\sin\dfrac{A}{2} = \pm\sqrt{\dfrac{1-\cos A}{2}}$

$$\therefore\ \pm\sqrt{\frac{1-\cos A}{2}} = \frac{1}{2}$$

$$\therefore\ \frac{1-\cos A}{2} = \frac{1}{4}$$

$1 - \cos A = \dfrac{1}{2}$, $\cos A = 1 - \dfrac{1}{2} = \dfrac{1}{2}$

6. $20\sqrt{3}$

$$20^2 - 10^2 = 300 = AB^2$$

$$\sin C = \frac{AB}{AC} = \frac{10\sqrt{3}}{AC}$$

$$\sin C = \sin 30° = \frac{1}{2} = \frac{10\sqrt{3}}{AC}$$

$$\therefore\ AC = 20\sqrt{3}$$

7. (C)

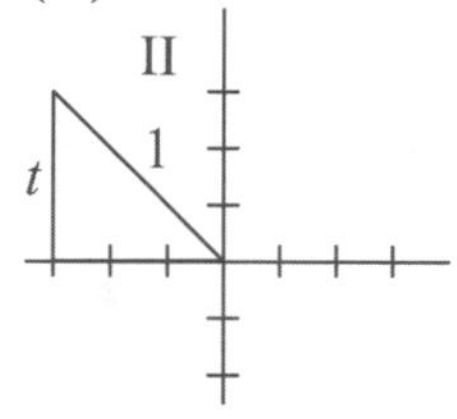

$\sin 2x = 2\sin x \cos x = 2 \cdot t \cdot \sqrt{1-t^2}$

$\sin 2x = 2t\sqrt{1-t^2}$

8. (A)

Let θ=arc $(\cos(-\frac{4}{5}))$, $\cos\theta = -\frac{4}{5}$

$(0 \leq \theta \leq \pi)$

Therefore, θ is in Quadrant II

So, $\sin\theta = \frac{3}{5}$

Answer is (A)

9. (C)

$\cos\theta + 2\sin\theta = 2$, $\cos\theta = 2 - 2\sin\theta$

$(\cos\theta)^2 = (2-2\sin\theta)^2$

$1-\sin^2\theta = 4-8\sin\theta+4\sin^2\theta$

$5\sin^2\theta - 8\sin\theta + 3 = 0$

$\sin\theta = \frac{8\pm\sqrt{64-60}}{10} = \frac{8\pm 2}{10}$

$\sin\theta = 1$ or $\frac{3}{5}$

$\because\ 0° < \theta < 90°\quad \sin\theta \neq 1$, $\cos\theta = \frac{4}{5}$

So, $\sin\theta = \frac{3}{5}$,

Since $\cos\theta = 2 - 2 \cdot \frac{3}{5} = \frac{4}{5}$

$\sin\theta + \cos\theta = \frac{3}{5} + \frac{4}{5} = \frac{7}{5}$

Answer is (C)

10. 0

Set $\theta_1 = \sin^{-1}\frac{1}{2}$, $\sin\theta_1 = \frac{1}{2}$, $\theta_1 = 30°$

Set $\theta_2 = \cos^{-1}\frac{1}{2}$, $\cos\theta_2 = \frac{1}{2}$, $\theta_2 = 60°$

$\therefore\ \cos(\sin^{-1}\frac{1}{2} + \cos^{-1}\frac{1}{2})$

$= \cos(30° + 60°) = \cos 90° = 0$

11. (37.8°)

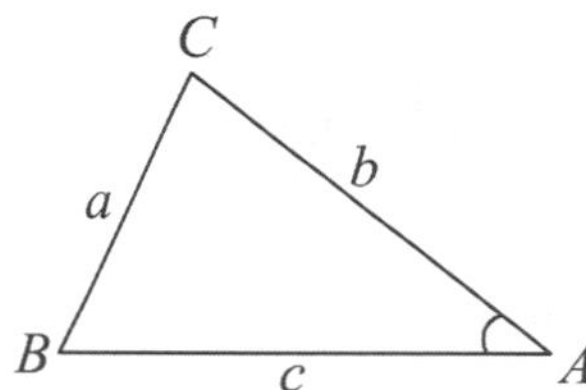

Since $\cos C = \frac{a^2+b^2-c^2}{2ab} = \frac{3^2+7^2-5^2}{(2)(3)(7)}$

$= \frac{9+49-25}{42} = \frac{33}{42} = 0.79$

$m\angle C = \cos^{-1} 0.79$

$\therefore\ m\angle C \doteq 37.8°$

12. (6.5)

Since Area $= \sqrt{s(s-a)(s-b)(s-c)}$,

when $s = \frac{a+b+c}{2}$

$\therefore\ \triangle ABC = \sqrt{7.5(4.5)(0.5)(2.5)} = 6.50$

13. (A)

The period of sin is 2π

so, period $= \frac{2\pi}{3\pi} = \frac{2}{3}$

Answer is (A)

14. (B)

$2\sin x = -\cos x$

$\frac{\sin x}{\cos x} = -\frac{1}{2}$

$\tan x = -\frac{1}{2}, \quad x = \tan^{-1}(-\frac{1}{2}) = -26.57$

Answer is (B)

15. $(\frac{27}{6})$

Since $\triangle ABC = \triangle ABD + \triangle BDC$

$= \frac{1}{2}[AB\sin 30° \times BD + BD\sin 30° \times BC]$

$= \frac{1}{2}[\frac{9}{2} \times 3\sqrt{3} + 3\sqrt{3} \times \frac{1}{2} \times BC]$

$= \frac{27\sqrt{3}}{4} + \frac{3\sqrt{3}}{4}BC$

Since $\triangle ABC = \frac{1}{2} \cdot 9 \cdot BC \cdot \sin 60°$

$= \frac{9\sqrt{3}}{4}BC$

$\therefore \; \frac{9\sqrt{3}}{4}BC = \frac{27\sqrt{3}}{4} + \frac{3\sqrt{3}}{4}BC$

$\therefore \; BC = \frac{27}{6}$

16. $[\frac{1}{4}(1-\sqrt{3})]$

$\cos 120° \sin(-150°) + \cos 300° \sin 240°$

$= -\cos 60°(-\sin 30°) + \cos 60°(-\sin 60°)$

$= \cos 60° \sin 30° - \cos 60° \sin 60°$

$= \cos 60°(\sin 30° - \sin 60°)$

$= \frac{1}{2}(\frac{1}{2} - \frac{\sqrt{3}}{2}) = \frac{1}{4}(1-\sqrt{3})$

17. $-\frac{1}{3}$

since $\tan\alpha + \tan\beta = \frac{3}{2}$,

$\tan\alpha \cdot \tan\beta = \frac{1}{2}$

$(\tan\alpha - \tan\beta)^2 = (\tan\alpha + \tan\beta)^2$

$-4\tan\alpha\tan\beta$

$= (\frac{3}{2})^2 - (4) \cdot (\frac{1}{2}) = \frac{9}{4} - 2 = \frac{1}{4}$

Since $\alpha < \beta$, $\tan\alpha < \tan\beta$

$\tan\alpha - \tan\beta = -\frac{1}{2}$

$\tan(\alpha - \beta) = \frac{\tan\alpha - \tan\beta}{1 + \tan\alpha\tan\beta}$

$= \frac{-\frac{1}{2}}{1 + \frac{1}{2}} = \frac{-\frac{1}{2}}{\frac{3}{2}} = -\frac{1}{3}$

18. $\frac{7}{2}$

Use the theorem of cosine

$c^2 = a^2 + b^2 - 2ab\cos c$

$\therefore \; \cos c = \frac{a^2 + b^2 - c^2}{2ab}$

$= \frac{4^2 + 2^2 - 3^2}{2 \cdot 4 \cdot 2}$

$= \frac{11}{16}$

$\cos c = \frac{4^2 + (2+x)^2 - 4^2}{2 \cdot 4 \cdot (2+x)}$

$= \frac{2+x}{8}$

$\therefore \; \frac{11}{16} = \frac{2+x}{8} \quad \therefore \; 11 = 4 + 2x$

$\therefore \quad x=\frac{7}{2}$

$BD=\frac{7}{2}$

19. (A)

$y=2\cos^2\theta+2\sin\theta+5$

$=2(1-\sin^2\theta)+2\sin\theta+5$

$=2-2\sin^2\theta+2\sin\theta+5$

$=-2(\sin^2\theta-\sin\theta+\frac{1}{4})+\frac{2}{4}+2+5$

$=-2(\sin\theta-\frac{1}{2})^2+7\frac{1}{2}$

maximum value, when $\sin\theta-\frac{1}{2}=0$,

$\sin\theta=\frac{1}{2}$

maximum value of $y=7\frac{1}{2}$

minimum value, when, $(\sin\theta-\frac{1}{2})^2$ is the greatest, $\sin\theta=-1$, $(\sin\theta-\frac{1}{2})^2$ is the greatest, the minimum value of $y=7\frac{1}{2}-2(-\frac{3}{2})^2=7\frac{1}{2}-\frac{18}{4}=\frac{12}{4}$, sum of max and min$=\frac{30}{4}+\frac{12}{4}=\frac{42}{4}=\frac{21}{2}$

Answer is (A)

20. $\sin\theta=-\frac{3}{5}$, $\pi<\theta<\frac{3\pi}{2}$, $\cos\theta=-\frac{4}{5}$,

$\sin 2\theta=2\sin\theta\cos\theta=2\cdot(-\frac{3}{5})(-\frac{4}{5})=\frac{24}{25}$

$\cos 2\theta=2\cos^2\theta-1=2(-\frac{4}{5})^2-1=\frac{32}{25}-1$

$=\frac{7}{25}$

$\tan\frac{a}{2}=\pm\sqrt{\frac{1-\cos\theta}{1+\cos\theta}}=\pm\sqrt{\frac{1+\frac{4}{5}}{1-\frac{4}{5}}}$

$=\pm\sqrt{\frac{\frac{9}{5}}{\frac{1}{5}}}=\pm 3$

$\pi<\theta<\frac{3\pi}{2}$, $\tan\frac{\theta}{2}=3$

21. Let the length of $\overline{AC}$ be x,

$\cos B=\frac{4^2+3^2-x^2}{2\cdot 4\cdot 3}=\frac{25-x^2}{24}$

$\cos D=\frac{5^2+1^2-x^2}{2\cdot 5\cdot 1}=\frac{26-x^2}{10}$

Since $m\angle B+m\angle D=180°$,

$m\angle B=180°-m\angle D$

$\cos B=\cos(180°-m\angle D)=-\cos D$

$\therefore \frac{25-x^2}{24}=-\frac{26-x^2}{10}=\frac{-26+x^2}{10}$

$\Rightarrow 250-10x^2=-624+24x^2$

$x^2=25.71$

$x=5.07$

Answer is $AC=5.07$

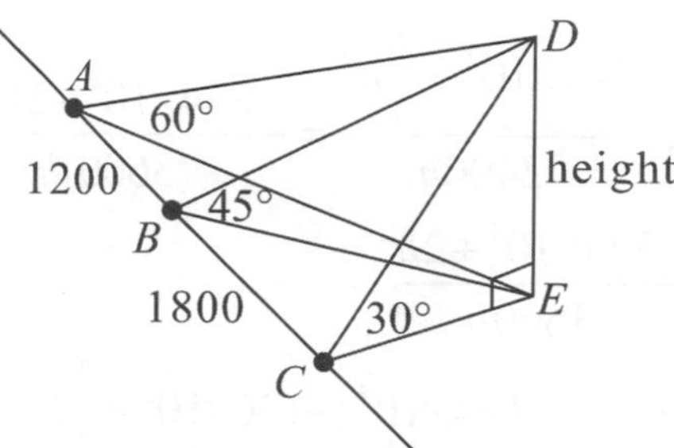

Note: Figure not drawn to scale

22. In $\triangle ADE$, $\tan 60° = \frac{h}{AE}$, $\sqrt{3} = \frac{h}{AE}$,

$AE = \frac{h}{\sqrt{3}}$

in $\triangle CDE$, $\tan 30° = \frac{h}{CE}$

$\frac{1}{\sqrt{3}} = \frac{h}{CE}$, $CE = \sqrt{3}h$

in $\triangle BDE$, $\tan 45° = \frac{h}{BE}$

$1 = \frac{h}{BE}$, $BE = h$

$$\cos(\angle ABE) = \frac{1200^2 + h^2 - (\frac{h}{\sqrt{3}})^2}{(2)(1200)(h)} \ (h = BE)$$

$$= \frac{144\times10^4 + h^2 - \frac{h^2}{3}}{2400h}$$

$$= \frac{144\times10^4 + \frac{2}{3}h^2}{2400h}$$

$$\cos(\angle EBC) = \frac{1800^2 + h^2 - (\sqrt{3}h)^2}{(2)(1800)(h)} \ (h = BE)$$

$$= \frac{324\times10^4 - 2h^2}{(3600)(h)}$$

Since $\cos(\angle EBC)$

$= \cos(180° - m\angle ABE)$

$= -\cos(\angle ABE)$

So, $\frac{144\times10^4 + \frac{2}{3}h^2}{2400h} = -\frac{324\times10^4 - 2h^2}{(3600)h}$

$= \frac{-324\times10^4 + 2h^2}{3600h}$

$(36\times10^2)(144\times10^4) + (36\times10^2)(\frac{2}{3}h^2)$

$= (-324\times10^4)(2400) + (2400)(2h^2)$

$5184\times10^6 + 2400h^2 = -7776\times10^6 + 4800h^2$

$2400h^2 = 12{,}960\times10^6$

$h^2 = 540\times10^4$,

$h = 6\sqrt{15}\times10^2$

Answer is (A)

CHAPTER 11
CONTEXT PROBLEMS

All mathematics problems that require written descriptions are context problems. The purpose of this chapter is to provide students standardized steps and ways of thinking for context problem solving. This chapter also talks about problems that we encounter in our daily lives and how to use mathematical methods to solve them.

11-1 THE STEPS OF SOLVING CONTEXT PROBLEMS

 **Step 1.** Understand the question

- Read the question carefully and thoroughly and make sure you fully understand each sentence and the information provided.
- It is helpful to underline the key-words, numbers, and phrases.
- In order not to forget key information as you read along the problems, transport the descriptions into something that you are most comfortable with. You can use the simplest, most direct, and most logical forms to record the information you are given. For instance, graphs, charts, diagrams, tables, etc; to help you grasp the essence of the problems. Certainly, you may re-read the question to make sure everything recorded is correct and logical.

 **Step 2.** Make sure you understand what is being asked.

 **Step 3.** If you cannot find the answer then try to find the hidden information, or you may infer new information from the given information to help solve the problem.

 **Step 4.** List the mathematical operations and solve them.

 **Step 5.** Check your answer

Substitute your answer to the question and check.

In order not to be tricked by the problem you may plug your answer into the problems, and check to see if it adheres to the essences of the problem, and it is logical and reasonable.

11-2 SOME EXAMPLES OF CONTEXT PROBLEMS

Although the steps to solve problems are fixed. In actual problem-solving, many steps are just a quick operation in the mind, and it is not necessary to write down every step, especially for some easy problems. Whether you follow the method in the book step by step or skip a few steps to jump to the answer directly, it depends on your mathematical literacy.

For instance a question as below: If $2x+4y=8$, what is the value of $4x+8y$?

It is obvious, the answer is 16.

It is not necessarily to follow all steps.

◆Example 1:

Otis mixed up 3 ounces (fluid ounce) of a 30% alcohol solution with 5 ounces (fluid ounce) of a 50% alcohol solution. What was the percent of alcohol in the final solution ?

 **Step 1.** understand the question

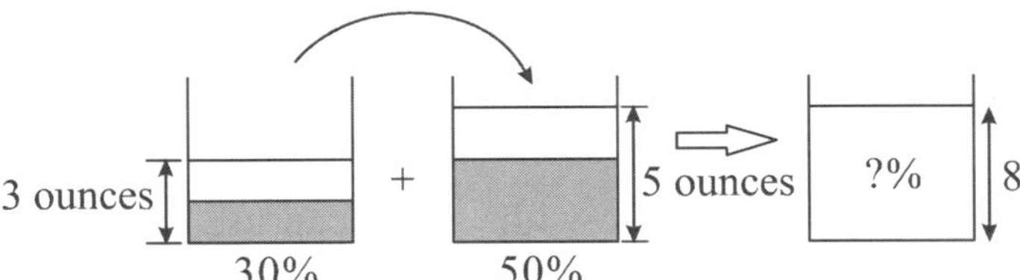

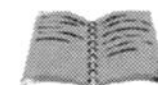 **Step 2.** Analysis

Key: $\dfrac{\text{part1}+\text{part2}}{\text{whole}}\times 100\% = ?\%$

> **★Tip**
> Graphs are better tools to make the contexts easier to understand.

 **Step 3.** Operations

$$\frac{0.3\times 3+0.5\times 5}{3+5}\times 100\% = 42.5\%$$

 **Step 4.** check your answer

$3\times 0.3+5\times 0.5=3.4$

$42.5\%\times 8=3.4$

$\therefore$ left side = right side

◆Example 2:

Allison has 5 ounces of a 50% alcohol solution. How many ounces of 30% alcohol solution are needed to create a 42.5% alcohol solution ?

 **Step 1.** Understand the question

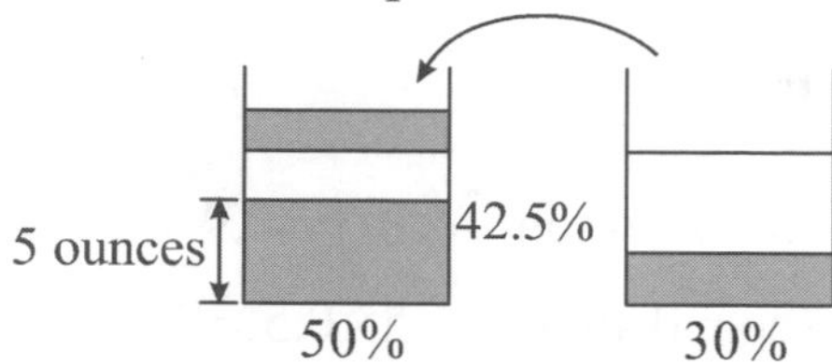

 **Step 2.** Analyze

Key: $\dfrac{\text{part1}+\text{part2}}{\text{whole}}\times 100\% = ?\%$

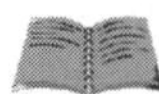 **Step 3.** Operations

Assume x ounces of a 30% alcohol solution are needed.

so, we added $0.3x$ alcohol to the 50% alcohol solution.

$$\frac{0.3x+0.5\times 5}{x+5}\times 100\% = 42.5\%$$

$\therefore\ 0.3x+2.5=0.425x+2.125$

$0.425x-0.3x=2.5-2.125$

$0.125x=0.375$

$\therefore\ x=3$

We need to add 3 ounces of 30% alcohol solution to the 50% alcohol solution.

 **Step 4.** Check the result $\dfrac{0.3\times 3+0.5\times 5}{3+5}\times 100\% = 42.5\%$

The answer is right.

◆Example 3:

Following up with the previous example: If Allison first mixes the 30% (3 ounces) and 50% (5 ounces) solution and then pours some of the third alcohol solution to the mixture alcohol solution. If Allison wants to create the final mixture is 42%, how many ounces of the third alcohol solution 40% should be added in?

 **Step 1.** Understand the question

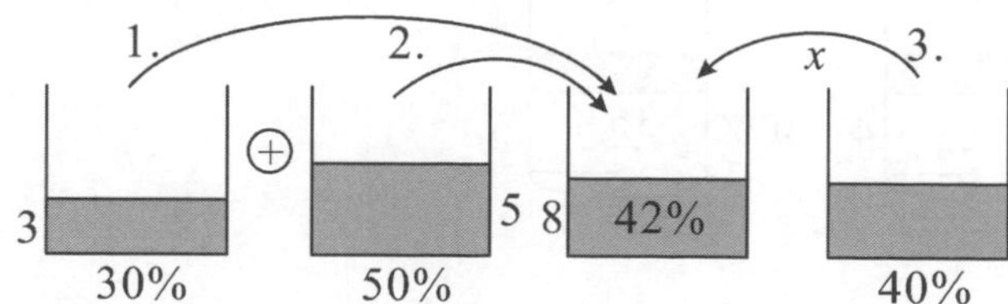

 **Step 2.** Analyzing the problem

Key: $\frac{\text{part1}+\text{part2}}{\text{whole}}\times 100\%$

part 1: Alcohol in the first solution $3\times 0.3=0.9$

part 2: Alcohol in the second solution $5\times 0.5=2.5$

Whole: $3+5=8$

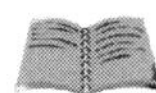 **Step 3.** Alcohol concentration: $\frac{0.9+2.5}{8}\times 100\%=42\%$

$$0.42=\frac{3.4+0.4x}{8+x}$$

$$3.36+0.42x=3.4+0.4x$$

$$0.02x=3.4-3.36=0.04$$

$$x=\frac{0.04}{0.02}=2$$

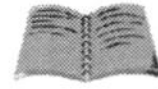 **Step 4.** Check your answer

$$\frac{3.4+0.4\times 2}{8+2}\times 100\%=\frac{4.2}{10}\times 100\%=42\%$$

Therefore 2 ounces is correct.

Even for more complicated problem, the solving steps are the same.

◆Example 4:

Amy has 2 cylindrical cups, cup A and cup B.

Cup A: Contains 40% of alcohol solution and cup B contains 30% of alcohol solution.

The radius of Cup A and Cup B are 2 and 3 inches respectively. The height of these two cups are 4 and 3 inches respectively.

Amy poured all the solution of Cup A into Cup B and made the solution of Cup B 36% concentration. How many inches did the height of Cup B rise ?

Step 1. Understand the problem

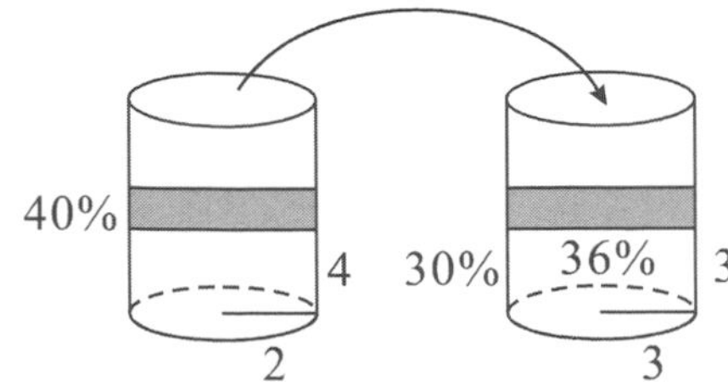

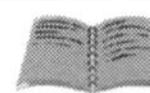 **Step 2.** Analyze the problem

$$\text{Key} = \frac{\text{part1} + \text{part2}}{\text{whole}} \times 100\% = 36\%$$

$$\text{part 1} = \pi(2)^2 \bullet 4 \bullet 40\%$$

$$\text{part 2} = \pi(3)^2 \bullet 3 \bullet 30\%$$

$$\text{whole} = x + \pi(3)^2 \bullet 3$$

$$h = \frac{x}{\pi(3)^2}$$

 **Step 3.** Operations

$$36 = \frac{\pi(2)^2 \bullet 4 \bullet 0.4 + \pi(3)^2 \bullet 3 \bullet 0.3}{x + \pi(3)^2 \bullet 3} \times 100$$

$$36 = \frac{6.4\pi + 8.1\pi}{x + 27\pi} \times 100 = \frac{14.5\pi}{x + 27\pi} \times 100$$

$$\therefore\ 36x + 972\pi = 1450\pi$$

$$\therefore\ 36x = 478\pi$$

$$\therefore\ x = \frac{478 \times \pi}{36}$$

$$\therefore\ h = \frac{\frac{478}{36}\pi}{9\pi} = 1.4753 \text{ (rounded to hundredth)}$$

$$\therefore\ h = 1.48 \text{ (in)}$$

 **Step 4.** Check your answer

$$\frac{6.4\pi + 8.1\pi}{13.28\pi + 27\pi} = \frac{14.5\pi}{40.28\pi} \doteq 0.36 \doteq 36\% \quad \text{(checked)}$$

◆Example 5:

Charles' age is 5 times as old as Brain's, and 6 years later Charles is twice as old as Brain. How old is Brain now ? (easy)

 **Step 1.** Understanding

$$c = 5B \cdots ①$$

$$c + 6 = 2(B + 6) \cdots ②$$

 **Step 2.** Analysis

Key: solve equation

Step 3. Operation

$5B+6=2(B+6)$

$5B+6=2B+12$

$\therefore\ 3B=6,\ B=2$

Brain is 2 years old now.

◆Example 6:

Christine's age is 1 year more than 6 times of Alice's age, and Bonnie's age is 1 year less than twice of Alice's age.

Five years later, Christine's age is 3 times Alice's age. How old is Alice now ? (medium)

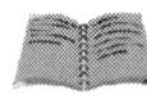 **Step 1.** Understand the question.

$C=6A+1\cdots$①

$B=2A-1\cdots$②

$(C+5)=3(A+5)\cdots$③

 **Step 2.** Analyze the question

Key: solve the equations

 **Step 3.** Operation

$C=6A+1$, substitute to (3)

$6A+1+5=3A+15$

$3A=9,\ \therefore\ A=3,\ C=19$

$B=2A-1,\ \therefore\ B=5$

So, Alice is 3 years old.

◆Example 7:

The distance from Agnes' home to her office is 72 miles. It takes her 1 hour and 20 minutes to drive to work. She takes the same route back home and at an average speed of 55 miles per hour. What is the average speed for the entire route ? (Rounded to hundredth)

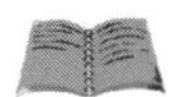 **Step 1.** Understand the question

$H \xrightarrow{\text{72 miles}} O$, time = 1 hr and 20 minutes

average speed = 55 m/hr back home

asking: average speed ?

 **Step 2.** Analyze the question

Key: Average speed $=\dfrac{\text{total distance}}{\text{total time}}$

 **Step 3.** Operations

Total distance $=72\times 2=144$ miles

Total time $=1\dfrac{20}{60}+\dfrac{72}{55}=1.33+1.31=2.64$ hrs

$\therefore$ Average speed $=\dfrac{144}{2.64}=54.55$ miles/hour

◆Example 8:

The distance from Wayne's home to his office is 46 miles. He often rides a bicycle to and back from work. When he gets to work his average speed is 22 miles per hour, because it is uphill on the way back home, his average speed is 18 miles/hour. What is the average speed back and forth from work ?

(A) 22

(B) 23.83

(C) 24

(D) 24.78

(E) 25

 **Step 1.** Understand the question

$H \xrightarrow{\text{46 miles}} O$

Average speed $=22$ miles/hr

$H \longleftarrow O$ average speed $=18$ miles/hr

asking: Average speed of back and forth

 **Step 2.** Analyze the question

Key $=\dfrac{\text{total distance}}{\text{total time}}$

> Caution: Average speed ≠ forward average speed + backward average speed

 **Step 3.** Operations

Time to office $=\dfrac{46}{22}=2.09$ hours

Time to home $=\dfrac{46}{18}=2.56$ hours

Total distance $=46\times 2=92$

Total time $=2.09+2.56=4.65$

$\therefore$ Average speed $=\dfrac{92}{4.65}=19.78$ miles/hr

Caution: Average speed $\neq 20(\dfrac{18+22}{2})$ mile/hr

◆Example 9:

A job can be completed by Chloe alone in 30 days and the same job also can be completed in 15 days by Tina alone. If these guys work together then how many days are needed to complete this job ?

(A) 23

(B) 20

(C) 15

(D) 13

(E) 10

 **Step 1.** Understand the question

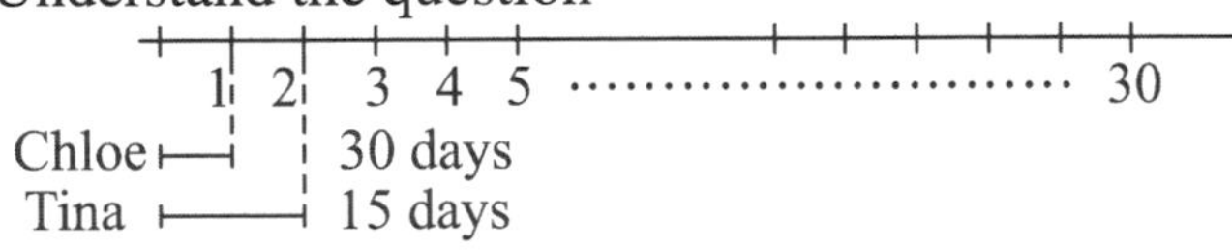

Asking: two people work together how may days to complete the job

 **Step 2.** Analyze the question

Key: Chloe performs $\dfrac{1}{30}$ the job in one day

Tina performs $\dfrac{1}{15}$ of the job in one day

Work together $\dfrac{1}{30}+\dfrac{1}{15}$ of the job in one day

★Tip

The whole job is counted as "1".

Step 3. Operations

Performance in one day $=\dfrac{1}{30}+\dfrac{1}{15}=\dfrac{3}{30}=\dfrac{1}{10}$

So, it is needed $\dfrac{1}{\frac{1}{10}}=10$ days to complete the job.

◆Example 10:

A job can be completed by John in 30 days and the same job can be completed by David in 15 days

If after they work together for 2 days, David can not continue to work, and John must complete the rest of work alone. How many days are needed for John to complete the rest of the job ?

(A) 20

(B) 21

(C) 22

(D) 23

(E) 24

Step 1. Understand the question

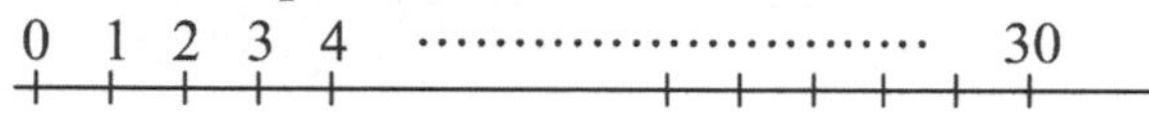

John ⊢⊣ needs 30 days
David ⊢——⊣ needs 15 days

Asking: Work together for two days, and John has to complete the rest of the job by himself.

Step 2. Analyze the question

Key: Work together $\frac{1}{30}+\frac{1}{15}=\frac{3}{30}$

For two days $\frac{3}{30}\times 2=\frac{1}{5}$ has been completed

$\frac{1}{5}$ completed

Step 3. Operations

The rest of work is $1-\frac{1}{5}=\frac{4}{5}$

John has to work alone $=\dfrac{\frac{4}{5}}{\frac{1}{30}}=\frac{30\times 4}{5}=24$ days

◆Example 11:

In Cynthia's class, there are 12 boys and 10 girls. What fraction of students are girls ? (easy)

 **Step 1.** Understand the question

Boys $=12$

Girls $=10$

Asking the fraction of student for Girls.

 **Step 2.** Analyze the question

Key: fraction $=\dfrac{\text{part}}{\text{whole}}$

 **Step 3.** Operation

Fraction (girls) $=\dfrac{10}{12+10}=\dfrac{10}{22}=\dfrac{5}{11}$

◆Example 12:

In a certain bridge club, $\dfrac{2}{3}$ of the members are men, and the ratio of the members who are 45 years or older to the member who are younger than 45 is 3:5.

If $\dfrac{4}{5}$ of women whose age is less than 45 years old. What fraction is male members who are 45 years or older ? (medium)

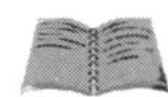 **Step 1.** Understand the question

use 2-dimensional table

	>45	<45	
men	?		$\dfrac{2}{3}$
women		$\dfrac{4}{5}\times\dfrac{1}{3}$	$\dfrac{1}{3}$
	$\dfrac{3}{8}$	$\dfrac{5}{8}$	

3:5 means $\dfrac{3}{8}$ and $\dfrac{5}{8}$

Asking: $\dfrac{\text{men}\geq 45}{\text{total}}=?$

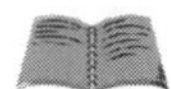 **Step 2.** Analyze the question

Key: fill up the table

 **Step 3.** Operations

Men and less than 45 years old $=1-\frac{3}{8}=\frac{5}{8}$

	>45	<45	
men	$A=\frac{37}{120}$	$B=\frac{43}{120}$	$\frac{2}{3}$
women	C	$\frac{4}{5}\times\frac{1}{3}=\frac{4}{15}$	$\frac{1}{3}$
	$\frac{3}{8}$	$\frac{5}{8}$	

$B=\frac{5}{8}-\frac{4}{15}=\frac{75-32}{120}=\frac{43}{120}$

A could be obtained by $A=\frac{2}{3}-B$ or $A=\frac{3}{8}-C$, either way.

$A=\frac{2}{3}-\frac{43}{120}=\frac{80-43}{120}=\frac{37}{120}$

◆Example 13:

To make a special C-grade noodle according to a special recipe, one needs A-grade and B-grade flour.

A-grade flour is a mixture of high-protein flour and medium-protein flour in a weight ratio of 3:2, B-grade flour is a mixture of low-protein flour and medium-protein flour in a weight ratio of 4:3.

Then you mix A-grade flour and B-grade flour into C-grade flour.

Among them, the weight of the high-protein flour in A-grade flour is exactly the same as the weight of the low-protein flour in the B-grade flour. What is the fraction of medium-protein flour in C-grade flour ?

 **Step 1.** Understand the question

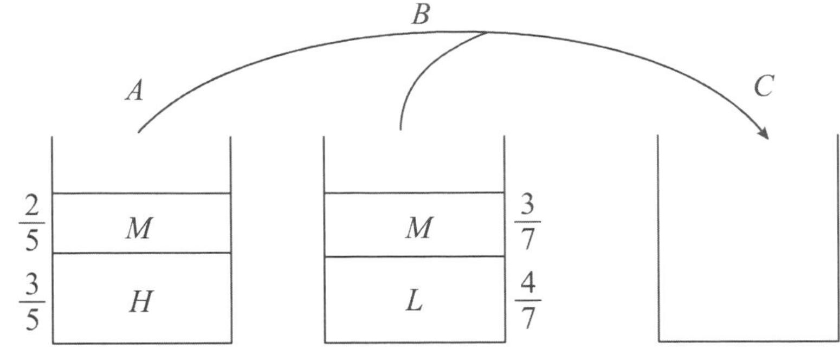

Asking: What is $\frac{M}{C}$ in the condition of $\boxed{H \text{ in } A = L \text{ in } B}$

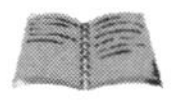 **Step 2.** Analyze the question

Key: Find out the weight of M in C-class flour, then you can find out the fraction of the middle protein flour in C class flour.

Step 3. Operations

Assume the equal quantity of H protein in A and L protein in B flours is k pounds, then, we need $\frac{k}{\frac{3}{5}}$ pounds of A and $\frac{k}{\frac{4}{7}}$ of pounds B

So, $C = \frac{k}{\frac{3}{5}} + \frac{k}{\frac{4}{7}} = \frac{5k}{3} + \frac{7k}{4} = \frac{41}{12}k$

$$M = \underset{\text{in } A}{\underset{\uparrow}{\frac{2}{5}}} \times \frac{k}{\frac{3}{5}} + \underset{\text{in } B}{\underset{\uparrow}{\frac{3}{7}}} \times \frac{k}{\frac{4}{7}} = \frac{2}{5} \times \frac{5k}{3} + \frac{3}{7} \times \frac{7k}{4} = \frac{2}{3}k + \frac{3}{4}k = \frac{17}{12}k$$

$$\therefore \frac{M}{C} = \frac{\frac{17}{12}k}{\frac{41}{12}k} = \frac{17}{41}$$

 **Step 4.** Check your answer

$C = \frac{41}{12}k$

$M = \frac{41}{12}k \times \frac{17}{41} = \frac{17}{12}k$ (Checked)

So, $\frac{M}{C} = \frac{17}{41}$.

11-3 SOME PRACTICAL PROBLEMS IN DAILY LIVES

A. Food

◆Example 1:

Cynthia is a mother bought 3.2 pounds of Fuji apples and 1.8 pounds of honey crisp apples in the supermarket and paid a total of \$14.02. According to the listed price, of honey crisp apples is 3 times the price of Fuji apples plus \$1.28. How much is a pound of honey crisp apples ?

Solution: Let one pound Fuji apples and one pound honey crisp apples cost F and H dollars, respectively.

$3.2F + 1.8H = 14.02$

$H = 3F + 1.28$

$3.2F + 1.8(3F + 1.28) = 14.02$

$3.2F + 5.4F + 2.304 = 14.02$

$F = 1.36,\ H = 3 \times 1.36 + 1.28 = \$5.36$

◆Example 2:

Betty wants to make a cake. According to her recipe, a 10-person serving cake requires $1\frac{2}{3}$ pounds of flour. How many pounds of flour does she need to make a 14-person serving cake ?

Solution: $\frac{10}{14} = \frac{\frac{5}{3}}{x},\ x = \frac{\frac{5}{3} \times 14}{10} = \frac{5 \times 14}{3 \times 10} = \frac{7}{3}$

She needs $2\frac{1}{3}$ pounds of flour

◆Example 3:

There is a seafood stand by the beach. A lobster sells for \$18 and a crab sells for \$12. One day, a total number of 132 crabs and lobsters are sold, for \$1,818. If the number of crabs sold are 54 more than the numbers of lobsters sold. How many lobsters are sold at the stand on that day ?

Solution: $\begin{cases} L+C=132 \\ 18\times L+12C=1,818 \end{cases}$

$18L+12(132-L)=1,818$

$18L-12L=1,818-1,584=234$

$6L=234$

$L=39$

$C=132-39=93$

B. Clothing

◆Example 1:

Company M sells outwear to employees at a 20% discount of the selling price. The original selling price is to mark up 50% of the cost. What is the percentage of the the price that employees pay comparing to the cost ?

Solution: $B=$buying price, $P=$list price

$C=$cost, $B=P\times0.8$, $P=C\times1.5$

$B=0.8\times(C\times1.5)=1.2C$

$\therefore \ \frac{B}{C}=1.2 \Rightarrow 120\%$

◆Example 2:

Faith has 3 hats, 3 scarves, 2 belts, 4 shoes, and 2 over-coats, how many different outfit combination does she have ?

Solution: $3\times3\times2\times4\times2=144$

◆Example 3:

David has 3 pairs of black socks, 3 pairs of white socks, and 4 pairs of red socks. If randomly select one pair of socks, what is the probability that one pair of red socks will be selected?

Solution: Black, white, red

3 3 4

$p(R)=\frac{4}{3+3+4}=\frac{4}{10}=\frac{2}{5}$

C. Housing

◆Example 1:

Samuel and his father decided to repaint their house. Samuel can complete the painting job in 6 days, and his father can complete the same job in 4 days. Two days after they worked together, Samuel's father had to leave the job for some important business. How long would it take for Samuel to complete the remaining work alone ?

Solution: One day can complete $\frac{1}{6}+\frac{1}{4}=\frac{2+3}{12}=\frac{5}{12}$

two days' work $\frac{5}{12}\times 2=\frac{5}{6}$,

remaining $=1-\frac{5}{6}=\frac{1}{6}$

by Samuel alone $=\frac{\frac{1}{6}}{\frac{1}{6}}=1$ day

◆Example 2:

If an ordinary household refrigerator costs $890, the daily electricity consumption is about 1.4 kw/h, and a power-saving refrigerator with the same function costs $1,200, and the daily electricity consumption is about 1.0 kw/h. If household electricity consumption costs 12 cents for every kw. The use limit of the two types of refrigerators is 10 years. How much can you save in 10 years if you buy a power-saving refrigerator in stead of a regular refrigerator ?

Solution: $(1.4-1.0)\times 24\times 365\times 10=35,040$

$0.12\times 35,040=4,205$

$4,205-(1200-890)=3,895$

◆Example 3:

An air conditioning (12,000 BTU) consumes 1.5 kw (kilowatt per hour) of electricity per hour. If it costs 12 cents per kilowatt. In summer, it is used for 12 hours a day for 60 consecutive day. How much is the electricity cost for the air conditioning for the summer ?

Solution: $1.5 \times 12 \times 60 \times 0.12 = 129.6$

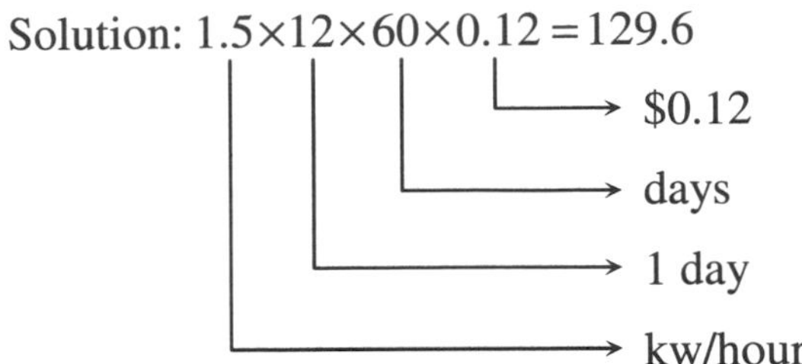

◆Example 4:

Noah owes his parents x dollars, last week he paid $\frac{1}{4}$ of the amount owed by mowing the lawn, and this week he paid $\frac{1}{3}$ of the remaining amount by helping house cleaning plus $40 scholarship. Finally he paid back $40 cash to his parents and paid off all the loan. How much does he owe his parents originally ?

Solution: $\frac{1}{4}x + (1 - \frac{1}{4}) \times \frac{1}{3}x + 40 + 40 = x$

$$\frac{1}{4}x + \frac{1}{4}x + 80 = x$$

$$\frac{1}{2}x = 80$$

$$x = 160$$

D. Transportation

◆Example 1:

The average speed for Shirley to get from her house to work was 0.9 miles per minute and it took her 50 minutes to get to the office today. She took the same route to come back home, but due to traffic jam, it took her 90 minutes to get back to her house. What is her average speed per minute for Shirley to go back and forth to work ?

Solution: Distance $= 0.9 \times 50 = 45$ miles

$$\text{Average speed} = \frac{\text{Distance}}{\text{Time}} = \frac{45 \times 2}{50 + 90} = \frac{90}{140} = 0.64 \text{ miles/minute}$$

◆Example 2:

Amy drove her car from her home to her grandma's house. Her car used 8.2 gallons of gas, and the distance is 210 miles. If 1 gallon costs $4.299 then what is the cost for her car to drive for 1 mile ?

Solution: $\text{cost} = \dfrac{8.2 \times 4.299}{210} = \$0.17$

◆Example 3:

Brennan flies from Seatlle to New York at 10:00 AM (Local Time). The plane is scheduled to stop in Chicago for 1.5 hours and then fly from Chicago to New York. When Brennan arrives New York, it is 8:30 PM (Local Time). How long is the actual flying time from Seattle to New York ? (excluding the layover time in Chicago) (Pacific time is 3 hours later than Eastern time)

Solution: 10:00 AM → 1:00 PM

(Seattle) (New York)

8:30~1:00 ⇒ 7.5 hours

7.5 hours – 1.5 hours layover = 6 hours

◆Example 4:

There are a total of 160 miles from S-city to V-city. One is expected to advance at a rate of 60 miles per hour. However, due to traffic jam, the average rate for the first 1.5 hours is 38 miles per hour. If the speed of the remaining distance can only be up to 60 miles per hour, how many more hours are needed after the expected arrival time ? (Rounded to hundredth)

Solution: For the first 1.5 hours, he has driven $1.5 \times 38 = 57$ miles,

expected arrival time $= \dfrac{160}{60} = 2.67$

$160 - 57 = 103$, $103 \div 60 = 1.72$

actual arrival time $= 1.5 + 1.72 = 3.22$

later than expect $= 3.22 - 2.67 = 0.55$

E. Schooling

◆Example 1:

Every student in Yohan's school must take at least one foreign language course. There are 51 students in Spanish, 39 in Japanese, 43 students in German, 3 students take both Japanese and Spanish, 8 students take both Spanish and German, and 2 students in all three languages classes. How many students are there in total ?

Solution: Total students $=51+39+43-3-8-2\times2=118$

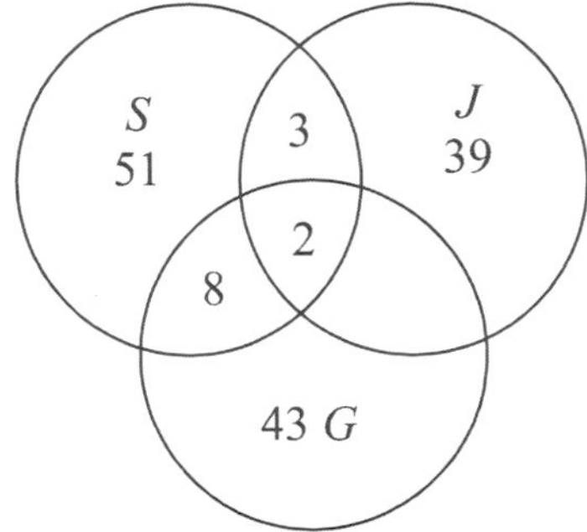

◆Example 2:

The first 5 of Samuel's math scores are 96, 80, 82, 83, and 76. If he wants to get a total average of 85 or more, what is the minimum score for his 6th test ?

Solution: $85\times6=510$

$$96+80+82+83+76+x\geq510$$

$$\therefore\ x\geq93$$

◆Example 3:

In Otis' school, the ratio of boys to girls is 5:4, and the ratio of students who take German course to students who do not take German course is 2:3. If for boys who take German course is one-fourth of the boys who do not take the German course, what is the probability that a student is selected at random from all students is a girl who also takes German course?

Solution: $P=\dfrac{13}{45}$

	boys	girls	Total
Take German	A $\frac{1}{5}\times\frac{5}{9}$	C	$\frac{2}{5}$
Do not take German	B $\frac{4}{5}\times\frac{5}{9}$	D	$\frac{3}{5}$
Total	$\frac{5}{9}$	$\frac{4}{9}$	1

$A=\frac{1}{5}\times\frac{5}{9}=\frac{1}{9}$

$B=\frac{4}{5}\times\frac{5}{9}=\frac{4}{9}$

$C=\frac{2}{5}-\frac{1}{9}=\frac{13}{45}$

$D=\frac{3}{5}-\frac{4}{9}=\frac{7}{45}$

F. Sports

◆Example 1:

In city *W*, there are 6 little league baseball teams to play matches. The rules of the game are that each team has to play a game with the other 5 teams. The four teams with highest points are selected; then the first and the fourth team, the second team and the third team will play knockout. Finally, the last two winning teams to compete for the championship. How many games will be scheduled ?

Solution: A B C D E F

5 4 3 2 1 0

The matches of B has nothing to with the matches of A, so they are all independent.

(1) First they have to schedule $5+4+3+2+1+0=15$ matches.

(2) Knock-out, they have 3 matches $15+3=18$ matches

◆Example 2:

Annie swims 1,000 meters with her dad. Her usual average speed is 46 meter per minute, and her dad's average speed is 56 meters per minute. If Annie and her dad want to arrive at the end point at the same time, how many meters should her dad let Annie swim first ?

Solution: Assume that they will reach the end-point at the same time.

(1)For 1000 meters, Annie needs $\frac{1000}{46}=21.74$ minutes

(2)For 1000 meters, Annie's dad needs $\frac{1000}{56}=17.86$ minutes

(3) $21.74-17.86=3.88$ minutes

(4) $3.88\times 46=178.48$ meters

Annie should swims for 178.48 meters first, then her dad starts to swim.

◆Example 3:

Jens practiced archery with 10 arrows. His score was 2 of 10 points, 5 of 9 points, 1 of 8 points, and 2 of 7 points. How many points does Jens have on average ?

Solution: $\frac{10\times 2+9\times 5+8\times 1+7\times 2}{10}=8.7$

G. Health

◆Example 1:

The formula of body mass index (BMI) BMI $=\frac{W}{(H)^2}$

W is weight represented by kilogram, H is height represented by meters.

(1) If measured by pound for weight and inch for height, the formula would be BMI $=k\cdot\frac{w}{(H)^2}$ what is the value of k ? (rounded to hundredth)

Solution: For 1 kg=2.2 pounds

1 meter =39.37 inches

$1\ \text{BMI}=\frac{2.2}{(39.37)^2}=0.0014\ \text{pound}/(\text{inch})^2$

So, $k=\frac{1}{0.0014}=714.28$

(2) Katy's weight is 132 lbs and height is 68 inches. If the normal BMI, is between 18.5 and 24, is Katy's BMI in the normal range ?

Solution: $\text{BMI}=\frac{132}{(68)^2}\times 714.28=20.39$

She is in standard level.

◆Example 2:

According to statistics, the probability that a woman will develop breast cancer, at age 30, 1 out of 2,212, at 60, 1 out of 23. How many times is the probability of getting breast cancer for a 60-year-old to a 30-year-old woman ?

Solution: $\dfrac{\frac{1}{23}}{\frac{1}{2212}}=\dfrac{2212}{23}=96$

◆Example 3:

There are two types of Vitamin supplements on the market-supplement 1 and 2, that contain vitamin *A*, *B*, and *C*. Supplement 1 contains 8 units of vitamin *A*, 5 units of vitamin *B*, and 3 units of vitamin *C* in each pill. Supplement 2 contains 3 units of vitamin *A*, 5 units of vitamin *B*, and 9 units of vitamin *C* in each pill. Supplement 1 cost $19.20 for a bottle of 360 pills while supplement 2 costs $14.40 for a bottle of 300 pills.
(This example is just for the purpose of explanation, not for the real life.) Shirley needs to take 24 units of vitamin *A*, 25 units of vitamin *B*, and 27 units of vitamin *C* daily. She would like to buy several bottles of supplements 1 and 2 so that she can achieve her daily goal in the most-efficient way. How many bottles of supplement 1 and supplement 2, at minimum, should Shirley buy each time ?

Assume that Shirley needs to take *x* pills of supplement 1 and *y* pills of supplement 2 to get enough vitamin *A*, *B*, and *C* each day, we get following inequalities:

1. x, y must be positive $\quad x>0$ and $y>0$
2. For vitamin *A* $\quad 8x+3y\geq 24$
3. For vitamin *B* $\quad 5x+5y\geq 25$
4. For vitamin *C* $\quad 3x+9y\geq 27$

$$\Rightarrow \begin{cases} x>0, \text{ and } y>0 \cdots\cdots ① \\ 8x+3y\geq 24\cdots\cdots ② \\ x+y\geq 5\cdots\cdots ③ \\ x+3y\geq 9\cdots\cdots ④ \end{cases}$$

From ② and ③ we get $x\geq 1.8$, $y\geq 3.2$

Shirley needs to take at least 1.8 pills of supplement 1 and 3.2 pills of supplement 2 to get enough daily doses of vitamin *A*, *B*, *C*.

From ③ and ④we get $x\geq 3$, $y\geq 2$

Shirley needs to take at least 3 pills of supplement 1 and 2 pills of supplement 2 to get enough daily doses of vitamin A, B, C.

When $x=0$, $y\geq3$, $y\geq5$, and $y\geq8$, so, $y\geq8$

When $y=0$, $x\geq3$, $x\geq5$, and $x\geq9$, so, $x\geq9$

The solution can be represented by the figure below.

In terms of the cost

Supplement 1 costs \$19.20 for a bottle of 360 pills ⇒ each pill costs $\frac{\$19.20}{360}=\$0.053$

Supplement 2 costs \$14.40 for a bottle of 300 pills ⇒ each pill costs $\frac{\$14.40}{300}=\$0.048$

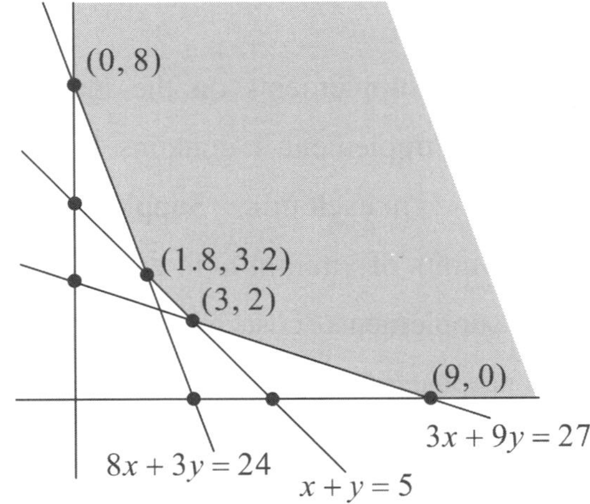

Therefore, as illustrated in below table in order to get enough daily vitamin at the lowest cost. Shirley would pay at the minimum \$0.255 per day with 3 pills of supplement 1 and 2 pills of supplement 2.

intersection (x, y)	cost $\$0.053x+0.048y$
(0, 8)	\$0.384
(1.8, 3.2)	\$0.249
⇒ (2, 4) (Random)	⇒ \$0.298
(3, 2)	**\$0.255**
(9, 0)	\$0.477

Since supplement 1 contains 360 pills per bottle, 3 pills a day would last her 120 days. Supplement 2 contains 300 pills per bottle, 2 pills a day would last her 150 days. The LCM (least common multiple) is 600 days. Therefore, Shirley needs to buy 5 bottles of supplement 1 and 4 bottles of supplement 2 each time that will help her achieve her daily goal at the lowest cost per day at \$0.255.

H. Finance

◆Example 1:

In order to go to university, Ken took out a student loan from a bank. The principle is \$12,500 with APR (annual percentage rate) 6.8%. After graduating from the university he starts to pay back the loan, in 5 years and 60 installments, with monthly of payment \$246. If Ken only wants to pay the interest and pay all principle \$12,500 at the end of 5 years, how much more money does Ken need to pay ?

Solution: Total installment payment $=246\times 60=14,760$

Interest for every month $=12,500\times\frac{0.068}{12}=70.83$

Total payment $=70.83\times 60+12500=16,750$

difference $=16750-14760=1990$

◆Example 2:

Annie invests \$50,000 in a bank with an annual compounded interest rate 3.5%. What is the total amount of principle and interest she will have in 10 years ?

Solution: $T=P(1+r)^{10}=50,000\times(1+0.035)^{10}=50,000\times 1.035^{10}=70,530$

◆Example 3:

Ken bought a house for \$250,000 with the a down payment of \$50,000 and a bank loan \$200,000, the bank loan has an APR (annual percentage rate) of 3.5%, and the monthly payment is \$1,133 for 30 years.

If Ken only wants to pay monthly interest and pays off all the principle at the end of 30 years. How much more does Ken have to pay ?

Solution: $I=(200,000\times\frac{0.035}{12})\times 360=210,000$

The formula for the amortization mortgage is M (fixed monthly pay) $=\frac{pr(1+r)^n}{(1+r)^n-1}$

Where $P=$principle, $r=$interest rate per month $n=$loan period

$$m=\frac{200,000\times\frac{0.035}{12}(1+\frac{0.035}{12})^{360}}{(1+\frac{0.035}{12})^{360}-1}=\frac{583.33\times 2.85328}{2.85328-1}=898.09$$

$$898.09 \times 360 = 323,312.4$$

$$210,000 + 200,000 - 323,312.4 = 86,687.6$$

★Tip

The formula for the amortization mortgage is M (fixed monthly pay) $= \dfrac{pr(1+r)^n}{(1+r)^n - 1}$

M (fixed monthly pay) $= \dfrac{pr(1+r)^n}{(1+r)^n - 1}$

where $P =$ principle,

$r =$ interest rate per month

$n =$ loan period

$A_0 = p$

$A_1 = p + pr - M = p(1+r) - M$

$A_2 = [p(1+r) - M] + [(p(1+r) - M) \cdot r] - M = [p(1+r) - M](1+r) - M$

$\quad = [p(1+r)^2 - M(1+r) - M]$

$A_3 = [p(1+r)^2 - M(1+r) - M] + [p(1+r)^2 - M(1+r) - M]r - M$

$\quad = [p(1+r)^2 - M(1+r) - M](1+r) - M = p(1+r)^3 - M(1+r)^2 - M(1+r) - M$

$\quad \vdots$

$A_n = p(1+r)^n - M(1+r)^{n-1} - M(1+r)^{n-2} \cdots - M$

$\quad = p(1+r)^n - M[(1+r)^{n-1} + (1+r)^{n-2} + (1+r)^{n-3} + \cdots 1]$

$\quad = p(1+r)^n - M(1 + (1+r) + (1+r)^2 + \cdots + (1+r)^{n-1}]$

Since it is a geometric sequence.

$a_1 = 1,\ a_n = (1+r)^{n-1},\ r = 1+r$

$$S_n = \frac{1-(1+r)^n}{1-(1+r)} = \frac{1-(1+r)^n}{-r}$$

$$\therefore\ A_n = p(1+r)^n - M[\frac{1-(1+r)^n}{-r}] = p(1+r)^n - M\frac{(1+r)^n - 1}{r}$$

$A_n = 0$

$$M \cdot \frac{(1+r^n) - 1}{r} = p(1+r)^n$$

$$\therefore\ M = \frac{p(1+r)^n}{\dfrac{(1+r)^n - 1}{r}} = pr \cdot \frac{(1+r)^n}{(1+r)^n - 1}$$

◆Example 4:

Charles' Salary is \$68,000. Assuming the only adjustments to Charles' gross income are 401 k contribution and health insurance premium (For ease of calculation). He contributes 20% of his pre-tax income to his 401 (k) account, and pays \$240 per month for health insurance (a pre-tax deduction after 401 (k) contribution).

Tax	Federal	Federal	Federal	State
Income bracket	\$0~\$9225	\$9226~\$37450	\$37450~90150	gross income
Tax rate	10%	15%	25%	4.5%

(The above table is just for purpose of calculation not for real life)

The table above summarizes his tax liability on his adjusted gross income (remainder after 401 (k)) and insurance payment

He has to pay

(1) 10% on the first \$9,225 of his adjusted gross income.

(2) 15% on adjusted gross income between \$9,225 and 37,450.

(3) 25% on adjusted gross income between \$37,450 and \$90,150.

(4) 4.5% state tax on all of his adjusted gross income.

All taxes are deducted simultaneously. How much does Charles have remaining each month after all deductions have been made ?

Solution:

(1) Charles' adjusted gross income is $\$68,000-(68,000\times 20\%)-(\$240\times 12)=51,520$

(2) He has to pay 10% on the first adjusted gross income \$9,225: $\$9225\times 10\%=\$922.5$

(3) He has to pay 15% on the adjusted gross income between \$9,225 and \$37,450,
$\Rightarrow\ \$28,225\times 15\%=\$4,233.75$

(4) He has to pay 25% on the adjusted gross income between \$37,450 and \$51,520,
$\Rightarrow\ \$14,070\times 25\%=\$3,517.5$

(5) He has to pay 4.5% on all of his adjusted grow income $\Rightarrow\ \$51,520\times 4.5\%=\$2,318.4$

(6) Total tax $=\$922.5+\$4,233.75+\$3,517.5+\$2,318.4=\$10,992.15$

(7) Net income after taxes $=\$51,520-\$10,992.15=\$40,527.85$

(8) Monthly amount (Take home money) $=\dfrac{\$40,527.85}{12}=\$3,377.32$

I. Investment

◆Example 1:

Cash invested in a stock, and one month later, the stock rose by 30% of its original price, but since then, it has started to fall. If Cash got no profit out of this investment, how much percent does the stock price has fallen ?

Solution: Set original price of stock $= P_0$

Let it rose 30% $= P_1$, $P_1 = P_0(1+0.3) = 1.3P_0$

$P_1(1-x\%) = P_0$

$1.3P_0(1-x) = P_0$

$1.3P_0 - 1.3P_0 x = P_0$

$1.3 - 1.3x = 1$, $1.3(1-x) = 1$

$$1 - x = \frac{1}{1.3} = 0.76923$$

$x = 1 - 0.76923 = 0.23077$

$x\% = 0.23077 \times 100\% = 23.08\%$

◆Example 2:

An stock analyst believes that a certain stock will be down 3% each day for 3 continuous days. What is the percentage of the stock price after it's 3-day fall comparing to the original price ?

Solution: Let stock price now $= S$

and let stock price in 3 days before $= S_3$

$S = S_3(1-0.03)^3 = S_3 \times 0.97^3 = 0.91 \cdot S_3$

$$\frac{S}{S_3} = \frac{0.91S_3}{S_3} = 0.91$$

Answer 91%

◆Example 3:

Chloe bought 10 shares of a certain stock for \$78 per share, and sold them for \$104 per share a year later. During the year, the stock paid a dividend ¢35 (35 cents) per share, what rate of return does Chloe earn from this investment ?

Solution: $10\times(104-78)+10\times0.35=260+3.5=263.5$

initial investment $=10\times78=780$

percentage earn $=\dfrac{263.5}{780}=0.3278\doteq 32.78\%$

◆Example 4:

A laptop computer depreciates at an annually rate of 25%, if the selling price is \$500, What is the price after 3 years of depreciation ?

Solution: 1. 1st year's depreciation: $P_1=500\times(1-0.25)$

2. 2nd year's depreciation: $P_2=500\times(1-0.25)\times(1-0.25)=500\times(1-0.25)^2$

3. 3rd year's depreciation: $P_3=500\times(1-0.25)^2\times(1-0.25)=500\times(1-0.25)^3$

$=210.94$

J. Science

◆Example 1:

The distance between the sun and the earth is set by scientists as 1 AU (Astronomical unit), and the light emitted from the surface of the sun takes 8 minutes and 17 seconds to reach the Earth. The speed of the light is 299,792,458 meters per second. How long is one AU in miles (expression in scientific notation) (1 mile = 1.6 km)

Solution: $299{,}792{,}458\times(8\times60+17)=148{,}996{,}851{,}626=148{,}996{,}851$ km

$\dfrac{148{,}996{,}851}{1.6}=93{,}123{,}032$ miles.

$1\text{ AU}=9.3\times10^7$ miles

◆Example 2:

If the half-life of a radioactive substance (The period of the quantity of a radioactive substance is half of the original quantity) is 5,000 years.

How much quantity is left after 25,000 years ?

Solution: $Q=Q_1\times(\frac{1}{2})^{\frac{25{,}000}{5{,}000}}=Q_1(\frac{1}{2})^5=Q_1\bullet\frac{1}{32}$

◆Example 3:

If there are a million bacteria in a tube, and they can grow double in number every 10 minutes. How many bacteria in the tube after 1 hour ?

Solution: $Q = 1{,}000{,}000 \times 2^{\frac{60}{10}} = 1{,}000{,}000 \times 2^6 = 1{,}000{,}000 \times 64 = 64{,}000{,}000$

◆Example 4:

The magnitude of the earthquake is usually expressed by the Richter scale. It is known that every time the Richter scale increases by 1 level, the earthquake intensity increases by approximately 10 times, and the energy release intensity increases approximately by 32 times. Suppose that there are two earthquakes with a difference of 1,000 times in intensity, how many times the differences in energy released ?

Solution: $E(r) = I \cdot 32^r$, $A(r) = M \cdot 10^t$

where r is the Richter scale, M and I are constants.

$\frac{A(r_1)}{A(r_2)} = \frac{M10^{r_1}}{M10^{r_2}} = 10^{(r_1 - r_2)}$, $1{,}000 = 10^3$

$r_1 - r_2 = 3$

$\frac{E(r_1)}{E(r_2)} = \frac{I \cdot 32^{r_1}}{I \cdot 32^{r_2}} = 32^{(r_1 - r_2)}$

$32^{r_1 - r_2} = 32^3 = 32{,}768$

K. Others

◆Example 1:

Noelle's age is 4 times that of her sister Michelle. After 6 years, Noelle's age will be 2 times Michelle's age less 2. How old is Michelle now ?

Solution: $N = 4M$

$N + 6 = 2(M + 6) - 2$

$4M + 6 = 2M + 12 - 2$

$2M = 4$, $M = 2$ Michelle is 2 years old now

◆Example 2:

Mr. Peterson's family and Mr. Johnson's families go together to watch a movie. The total ticket fares for 4 adults and 7 children were \$67.5. If each child's fare is one half of each adult's fare, what is the child's fare ?

Solution: $4A+7C=67.5$

$A=2C$

$4(2C)+7(C)=67.5$

$15C=67.5$

$\therefore\ C=4.5$

◆Example 3:

Luck has triple the number of poker cards as Eva has. If Luck gives Eva 10 of his poker cards, he will have twice the number of Eva's cards minus 9.

How many poker cards more does Luck have than Eva now ?

Solution: $L=3E$

$L-10=2(E+10)-9$

$3E-10=2E+20-9$

$E=21,\ L=63$

$63-21=42$

◆Example 4:

If the sum of p, q and r is 3 times the sum of p minus q and p minus r, what is the value of p in terms of q and r ?

Solution: $p+q+r=3[(p-q)+(p-r)]=3(2p-q-r)=6p-3q-3r$

$$p=\frac{4q+4r}{5}$$

CHAPTER 12
FUNCTION (II)

12-1 ODD and EVEN FUNCTION

A. Even Function

If a function $f(x)=f(-x)$ then that means $f(x)$ is an even function, whether the value of x is positive or negative, the value of y remains the same.

It is symmetric to y-axis

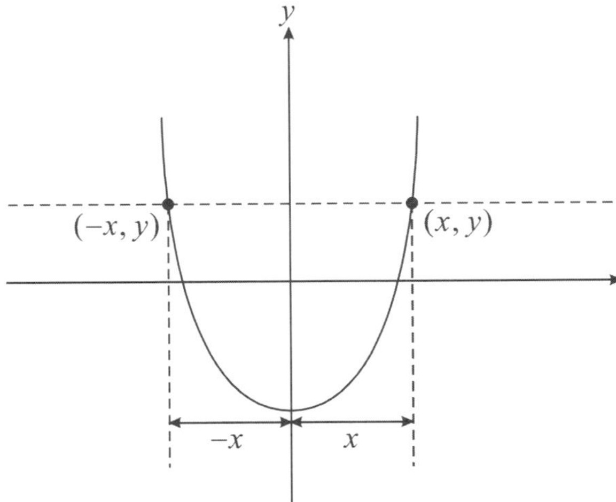

B. Odd Function

If a function $f(-x)=-f(x)$ then that means, when x changes to $-x$, then the value of y changes to negative $(-y)$.

It is symmetric to the origin.

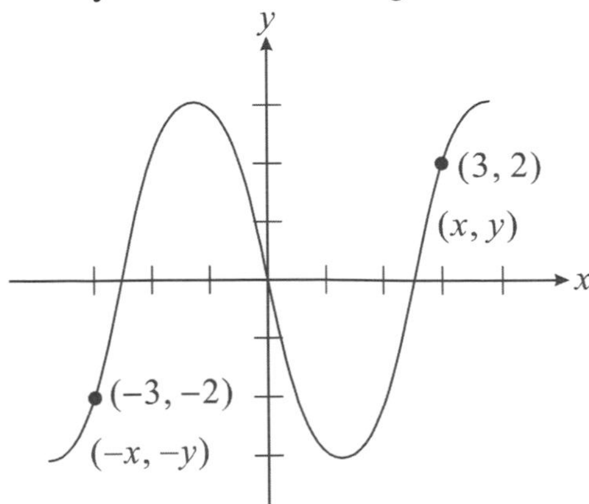

◆Example 1:

$f(x) = x^2 + 2$

$f(-x) = (-x)^2 + 2 = x^2 + 2$

$\therefore\ f(x) = f(-x)$ it is an even function

◆Example 2:

$f(x) = x^3 - 5$

$f(-x) = (-x)^3 - 5 = -x^3 - 5$

$f(x) \neq f(-x)$, it is not an even function

$-f(x) = -x^3 + 5$

$f(x) \neq -f(x)$, it is not an odd function

conclusion $f(x)$ is neither even nor odd function

◆Example 3:

$f(x) = x^4 - 3x^2 - 2$

$f(-x) = (-x)^4 - 3(-x)^2 - 2 = x^4 - 3x^2 - 2$

$f(x) = f(-x)$ it is an even function

◆Example 4:

$f(x) = x^3 + 3x$

$f(-x) = (-x)^3 + 3(-x) = [-x^3 - 3x]$

$-f(x) = -x^3 - 3x$

$f(-x) = -f(x)$ it is an odd function

Answer to Concept Check 12-01

1. $y = x$

   $f(x) = x$, $f(-x) = -x$, $f(-x) = -f(x)$

   It is odd function

2. $x^2 + y^2 = 1$, $y = \pm\sqrt{1-x^2}$

   $f(-x) = \pm\sqrt{1-(-x)^2} = \pm\sqrt{1-x^2}$

   $\therefore\ f(x) = f(-x)$

   It is an even relation

3. $x - 2y + 1 = 0$

   $2y = x + 1$

   $y = \frac{x}{2} + \frac{1}{2}$, $f(x) = \frac{x}{2} + \frac{1}{2}$

   $f(-x) = -\frac{x}{2} + \frac{1}{2}$

   $-f(-x) = \frac{x}{2} - \frac{1}{2}$

   $f(x) \neq -f(-x)$, It is not an odd function,

   $f(x) \neq f(-x)$, It is not an even function.

4. $x - y = 2$, $y = x - 2$, $f(x) = x - 2$, $f(-x) = -x - 2$

   $f(x) \neq f(-x)$, not even $-f(-x) = -(-x-2) = x + 2$

   $f(x) \neq -f(-x)$ not odd

   neither even nor odd

5. $f(x) = x^5 - 1$, $f(-x) = (-x)^5 - 1 = -x^5 - 1$

   $f(x) \neq f(-x)$ not even $-f(-x) = -(-x^5 - 1) = x^5 + 1$

   $f(x) \neq -f(-x)$ not odd

12-2 POLYNOMIAL FUNCTION

Since the linear function $ax + by + c$ and quadratic function $ax^2 + bx + c$ have been already discussed. In this section, we here, only discuss higher-degree polynomial function.

The standard form of a polynomial function $f(x) = a_n x^n + a_{n-1} x^{n-1} + \cdots + a_1 x + a_0$

- N is a non-negative number.
- The coefficients of the x terms are real numbers.
- a_n must not be zero.

The properties of the polynomial functions

- They all are continuous graphs.
- They all are rounded curves.
- If n (the highest exponent) is even, and the coefficient $a_n > 0$, the curve ascends both to the left and right.

  Such as $f(x) = x^4 + 5x^3 - 19x^2 + 83x - 60$

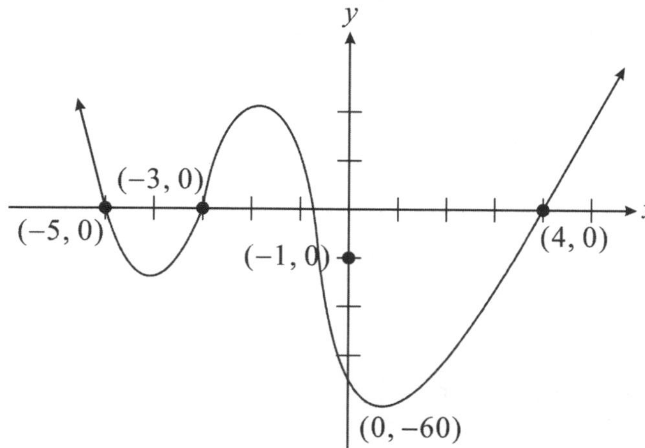

- If n (the highest exponent) is even, and the coefficient $a_n < 0$, the curve descends both to the left and right.

  Such as $f(x) = -x^4 - 5x^3 + 13x^2 + 73x + 60$

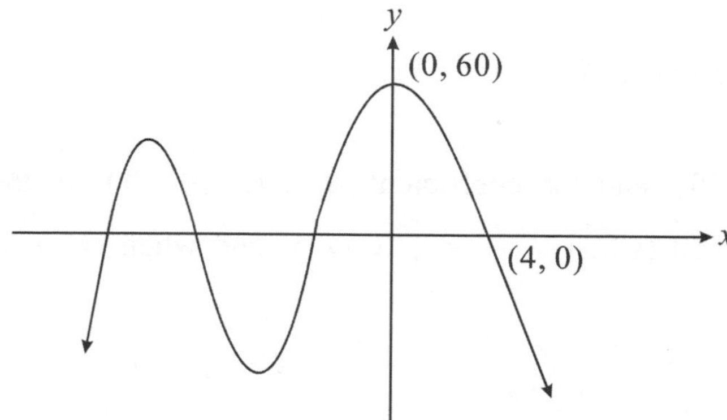

- If n (the highest exponent) is odd and the coefficient $a_n > 0$, the curve falls to the left and assends to the right.

Such as: $f(x)=x^3+x^2-22x-40$

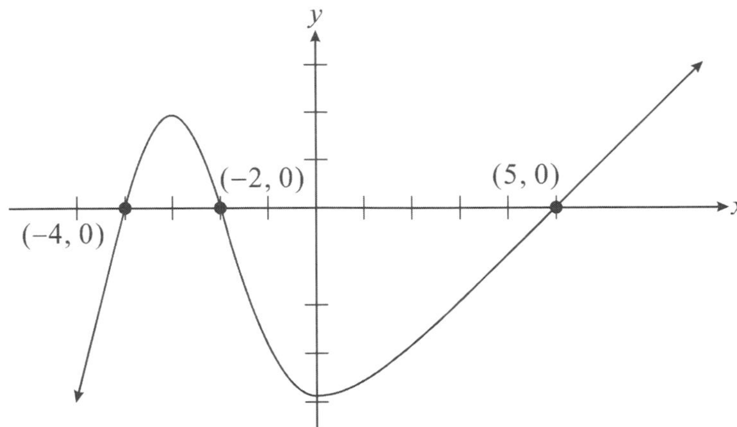

- If n (the highest exponent) is odd and the coefficient $a_n<0$, the curve acends to the left and decends to the right.

  Such as: $f(x)=-x^3-x^2+22x+40$

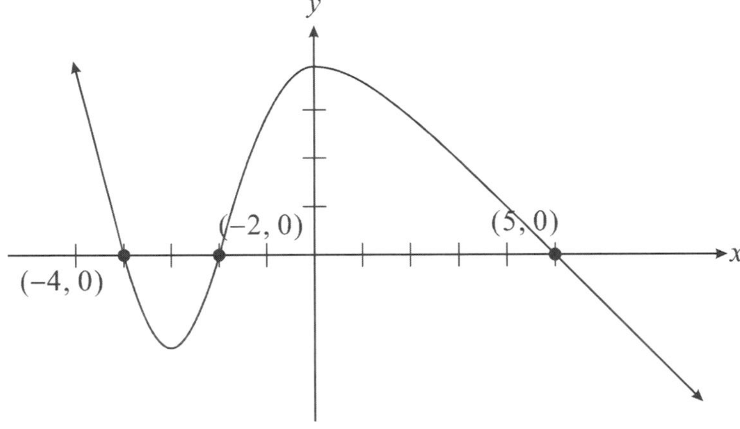

◆Example:

Describe the right and left behavior of the graph of

$f(x)=-x^5+3x^3+1$

The degree of the function is odd ($n=5$), and the coefficient $a_n<0$, $(-1<0)$, so the graph rises to the left and falls to the right (when $x\to-\infty$, $y\to\infty$, and when $x\to\infty$, $y\to-\infty$)

◆Example:

Which of the following graphs could be the graph of

$f(x)=3x^5+2x^4+3x^3-x^2+5x-8$

(A) 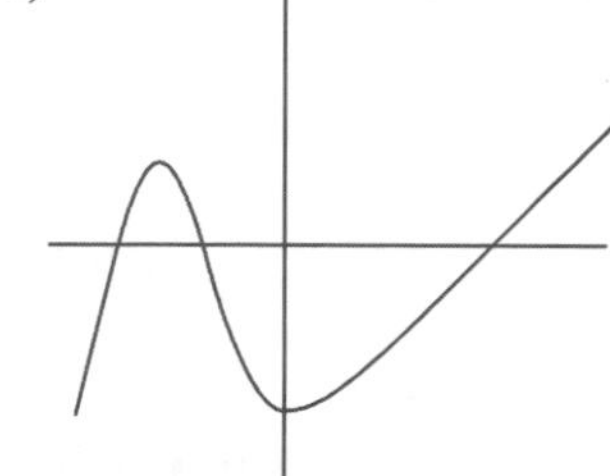

(B) 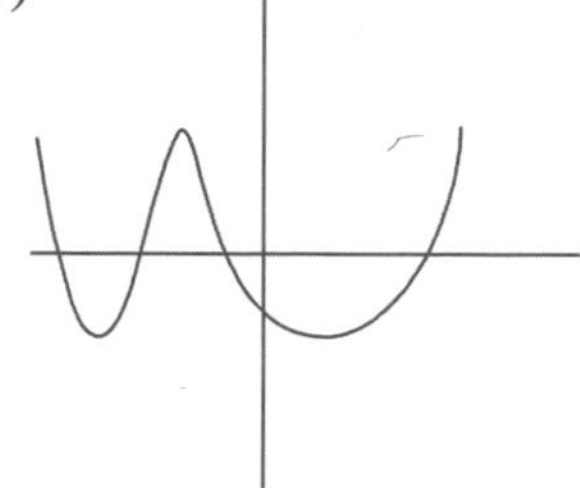

(C) 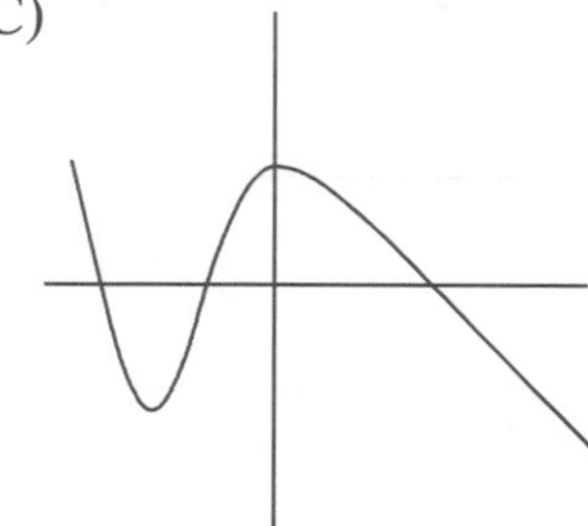

(D) 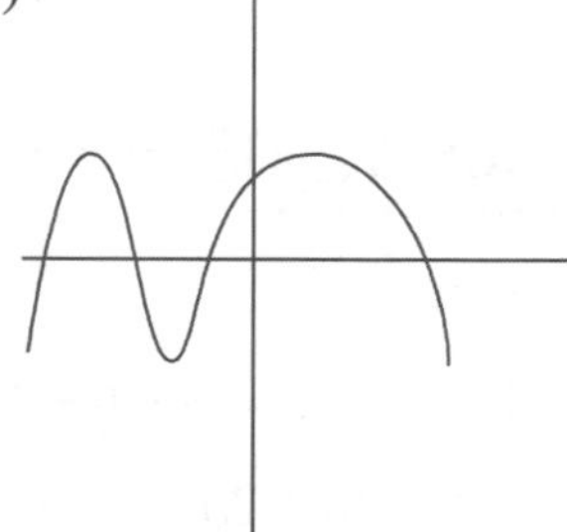

(A) When $x \to -\infty$, $y \to -\infty$ falls to the left

When $x \to \infty$, $y \to \infty$ rises to the right

Answer to Concept Check 12-02

$p(x) = -7x^5 + x^4 - 3x + 1$

The highest order is 5, it is odd number, and negative

So, its curve rises to the left and falls to the right.

The answer to (D)

Answer to Concept Check 12-03

$p(x) = x^3 - 2x^2 + 3x - 6 = x^2(x-2) + 3(x-2) = (x-2)(x^2+3)$

If $p(x)$ can be divided by $x+k$, then

$p(x) = (x+k) \cdot p'(x)$

$\therefore\ k = -2$

Answer is (D)

Answer to Concept Check 12-04

$p(x) = x^{15} + 2x^{14} - x^{13} + 3x^2 - 2x + 3$

divided by $x - 1$

The remainder $= p(1) = 1 + 2 - 1 + 3 - 2 + 3 = 6$

Answer is (C)

Answer to Concept Check 12-05

$p(x) = x^3 + ax^2 + bx - 2 = (x+1)(x-2)q(x)$

$p(-1) = (-1)^3 + a(-1)^2 + b(-1) - 2 = -1 + a - b - 2 = a - b - 3 = 0 \cdots ①$

$p(2) = 2^3 + 2^2 \cdot a + 2b - 2 = 8 + 4a + 2b - 2 = 4a + 2b + 6 = 0 \cdots ②$

$\begin{cases} a - b = 3 \\ 2a + b = -3 \end{cases}$

$\therefore\ a = 0,\ b = -3$

$\therefore\ a + b = 0 - 3 = -3$

Answers in (B)

A. Sum and Product of the Roots

Sum of the Roots $= -\dfrac{a_{n-1}}{a_n}$

Product of the Roots $= (-1)^n \dfrac{a_0}{a_n}$

Where n is the degree of $p(x) = 0$

1. $p(x) = a_n x^n + a_{n-1}x^{n-1} + a_{n-2}x^{n-2} + \cdots + a_0 = 0$

   with integer coefficients

   If $a_1,\ a_2,\ a_3 \cdots a_n$ are the roots of $p(x) = 0$

   Then $(x - a_1)(x - a_2)(x - a_3)\cdots(x - a_n)$

   $= x^n - (a_1 + a_2 + a_3 + \cdots a_n)x^{n-1} + \cdots + (-1)^n(a_1a_2a_3a_4 \cdots a_n)$

So, $a_1 + a_2 + a_3 + \cdots + a_n = -\dfrac{a_{n-1}}{a_n}$

Sum of the roots of $p(x)=0$ is $-\dfrac{a_{n-1}}{a_n}$

2. $(-1)^n(a_1)(a_2)(a_3)\cdots(a_n)$ must equal to $(-1)^n \cdot \dfrac{a_0}{a_n}$

So, product of roots $=(-1)^n \dfrac{a_0}{a_n}$

◆Example 1:

Find the product of all roots of $p(x)=x^5-2x^4-5x^3+5x^2+4x-12=0$

For $p(x)=(x-1)(x+1)(x-2)(x+2)(x-3)$

$\therefore\ a_1=1,\ a_2=-1,\ a_3=2,\ a_4=-2,\ a_5=3$

$a_1 \cdot a_2 \cdot a_3 \cdot a_4 \cdot a_5 = (1)(-1)(2)(-2)(3) = (-1)^5 \cdot \dfrac{-12}{1} = 12$

◆Example 2:

Find the sum of all roots of $p(x)=x^5-3x^4+5x^3+5x^2+4x-12=0$

$\alpha_1+\alpha_2+\alpha_3+\alpha_4+\alpha_5 = -(\dfrac{-3}{1}) = 3$

◆Example 3:

If the roots of $p(x)=2x^2-6x-3=0$ are α and β, what is the value of $\dfrac{1}{\alpha^2}+\dfrac{1}{\beta^2}$?

Since $\alpha+\beta=-(-\dfrac{6}{2})=3$

$\alpha \cdot \beta = (-1)^2 \cdot \dfrac{-3}{2} = -\dfrac{3}{2}$

$$\frac{1}{\alpha^2}+\frac{1}{\beta^2}=\frac{\alpha^2+\beta^2}{(\alpha\beta)^2}=\frac{(\alpha+\beta)^2-2\alpha\beta}{(\alpha\beta)^2}=\frac{3^2-2(-\frac{3}{2})}{(-\frac{3}{2})^2}=\frac{9+3}{\frac{9}{4}}=\frac{12}{\frac{9}{4}}=\frac{48}{9}=\frac{16}{3}$$

Answer to Concept Check 12-06

$a = -(-\frac{3}{1}) = 3$

$(-1)^5 \cdot b = -12$

$b = 12$

$a \times b = 3 \times 12 = 36$

Answers in (B)

Answer to Concept Check 12-07

$\alpha\beta = \frac{3}{2}, \quad \alpha + \beta = -\frac{5}{2}$

$$\frac{1}{\alpha^2} + \frac{1}{\beta^2} = \frac{\alpha^2 + \beta^2}{\alpha^2\beta^2} = \frac{(\alpha+\beta)^2 - 2\alpha\beta}{(\alpha\beta)^2} = \frac{\frac{25}{4} - 2 \cdot \frac{3}{2}}{(\frac{3}{2})^2} = \frac{\frac{25-12}{4}}{\frac{9}{4}} = \frac{13}{9}$$

Answers in (D)

B. The Possible Rational Zeroes of a Polynomial

$\dfrac{\text{factors of constant term}}{\text{factors of leading coefficient}}$

If the polynomial function

$p(x) = a_n x^n + a_{n-1} x^{n-1} + a_{n-2} x^{n-2} + \cdots + a_0$

with integer coefficients,

Then the possible zeros of function $p(x)$ or the possible roots of equation $p(x) = 0$ are $\dfrac{\text{factors of } a_0}{\text{factors of } a_n}$.

The reason is very simple, for the product of all roots is $(-1)^n \dfrac{a_0}{a_n}$, therefore, the root (zero) must be one factor of $\dfrac{a_0}{a_n}$.

◆Example 1:

Find the possible rational zeros of $f(x)=x^5+3x^4-2x^3+x^2-3x+2$

The coefficient of x^5 is 1, The constant is 2

So, $\dfrac{\text{factors of constant}}{\text{factors of leading coefficient}}=\dfrac{\pm1,\pm2}{\pm1}=\pm1,\pm2$

$f(1)=1+3-2+1-3+2=2(\neq0)$

$f(-1)=(-1)^5+3(-1)^4-2(-1)^3+(-1)^2-3(-1)+2=-1+3+2+1+3+2=10(\neq0)$

$f(2)=2^5+3(2)^4-2(2)^3+2^2-3(2)+2=32+48-16+4-6+2=64(\neq0)$

$f(-2)=(-2)^5+3(-2)^4-2(-2)^3+(-2)^2-3(-2)+2$

$=-32+48+16+4+6+2=44(\neq0)$

So, $f(x)$ has no rational zero

◆Example 2:

What are the roots of $p(x)=2x^2+x-1$?

all possible zeros of $P(x)$ are included in $\dfrac{\pm1}{\pm1,\pm2}$

They are $\pm1,\pm\dfrac{1}{2}$

$p(1)=2(1)^2+1-1=2(\neq0)$

$p(-1)=2(-1)^2+(-1)-1=2-1-1=0$ (roots)

$p(\frac{1}{2})=2(\frac{1}{2})^2+\frac{1}{2}-1=\frac{1}{2}+\frac{1}{2}-1=0$ (root)

$p(-\frac{1}{2})=2(-\frac{1}{2})^2+(-\frac{1}{2})-1=\frac{1}{2}-\frac{1}{2}-1=-1(\neq0)$

Therefore the zeros of $p(x)$ are $x=-1$, and $\dfrac{1}{2}$

◆Example 3:

$p(x)=2x^3+9x^2+13x+6=0$

The rational roots of $p(x)=0$

must be some of $\dfrac{\pm1,\pm2,\pm3,\pm6}{\pm1,\pm2}$

$\Rightarrow\ \pm1,\pm2,\pm3,\pm6,\pm\frac{1}{2},\pm\frac{3}{2}$

$p(-1)=2(-1)^3+9(-1)^2+13(-1)+6=-2+9-13+6=0$ (yes)

$p(1)=2(1)^3+9(1)^2+13(1)+6\neq 0$

$p(-2)=2(-2)^3+9(-2)^2+13(-2)+6=-16+36-26+6=0$ (yes)

$p(-\frac{3}{2})=2(-\frac{3}{2})^3+9(-\frac{3}{2})^2+13(-\frac{3}{2})+6$

$=-\frac{27}{4}+\frac{81}{4}-\frac{39}{2}+\frac{12}{2}=\frac{54}{4}-\frac{78}{4}+\frac{24}{4}=0$ (yes)

$$\begin{array}{r|l} 2x^3+9x^2+13x+6 & x+2 \\ \underline{2x^3+4x^2} & 2x^2+5x+3 \\ 5x^2+13x & \\ \underline{5x^2+10x} & \\ 3x+6 & \\ \underline{3x+6} & \\ 0 & \end{array}$$

$p(x)=(x+2)(2x^2+5x+3)=(x+2)(2x+3)(x+1)$

$x=-2,\ -\frac{3}{2},\ -1$

Answer to Concept Check 12-08

If there are zeroes, they must be $\pm 1,\ \pm 2$

$f(-1)=(-1)^5+2(-1)^4+3(-1)^3-2(-1)^2-3(-1)-2=-1+2-3-2+3-2=-3\neq 0$

$f(1)=(1)^5+2(1)^4+3(1)^3-2(1)^2-3(1)-2=1+2+3-2-3-2=-1\neq 0$

$f(-2)=(-2)^5+2(-2)^4+3(-2)^3-2(-2)^2-3(-2)-2$

$=-32+32-24-8+6-2$

$=-28\ (\neq 0)$

$f(2)=2^5+2(2^4)+3(2)^3-2(2)^2-3(2)-2$

$=32+32+24-8-6-2$

$=64+24-16$

$=72\ (\neq 0)$

so, there is no rational zero for the polynomial.

Answer is None

Answer to Concept Check 12-09

$\alpha\beta\gamma = (-1)^n a_0 = -(-15) = 15$

$\alpha + \beta + \gamma = -7$

$\dfrac{\alpha\beta\gamma}{\alpha+\beta+\gamma} = \dfrac{15}{-7} = -\dfrac{15}{7}$

Answer is $-\dfrac{15}{7}$

C. The Location of Roots Theorem

It $f(x)$ is a continuous function, and $f(a) \cdot f(b) < 0$, then $f(x)$ has odd number of real roots between $f(a)$ and $f(b)$, on the other hand, if $f(a) \cdot f(b) > 0$, $f(x)$ has even number of real roots between $f(a)$ and $f(b)$. ("0" is an even number)

Such as,

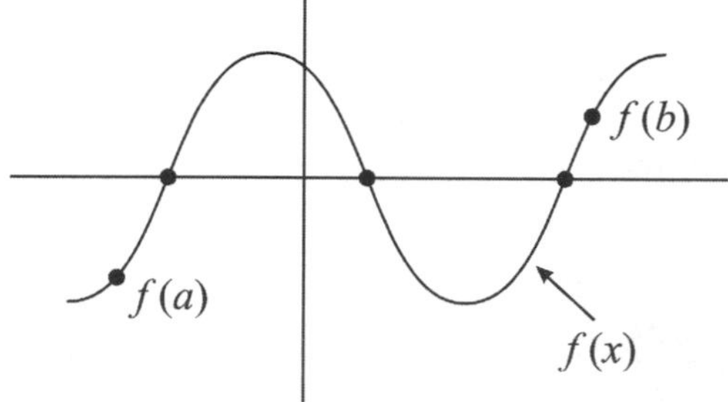

$f(a) \cdot f(b) < 0$, the function of $f(x)$ has to pass the *x*-axis odd number of times that mean 1, 3, 5, ⋯, In this case, there are 3 roots.

(A)

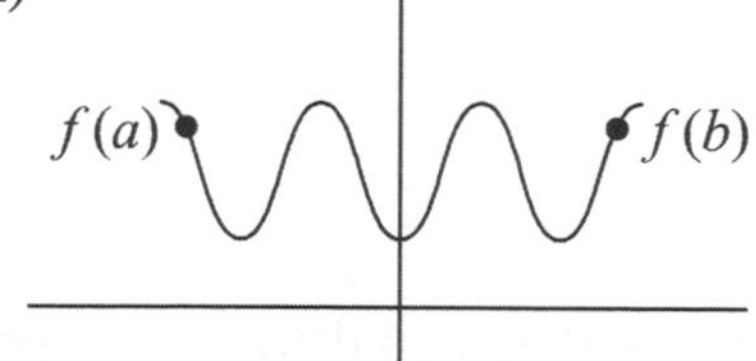
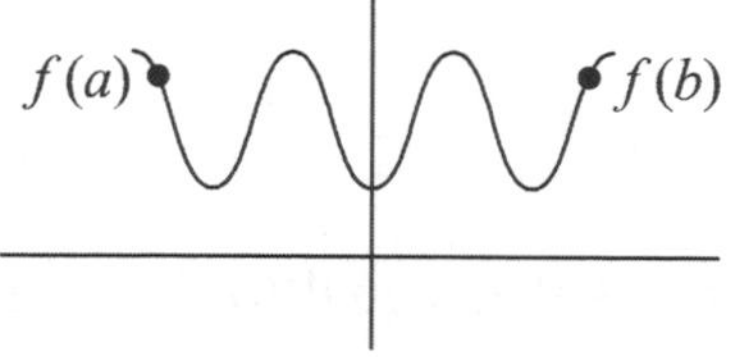

There are no real root

(B)

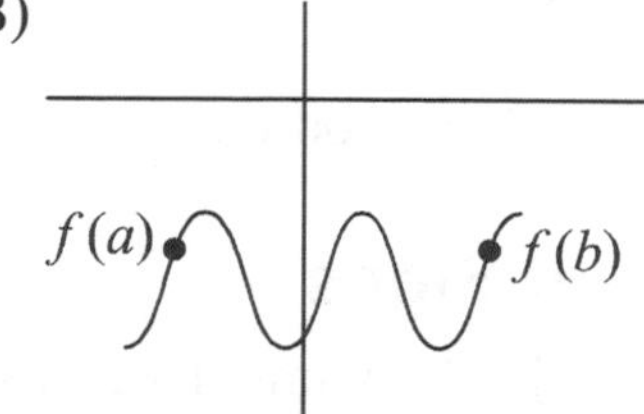

There are no real root

(C)

$f(a) \cdot f(b) > 0$

$f(a)$ $f(b)$

There are 4 real roots

(D)

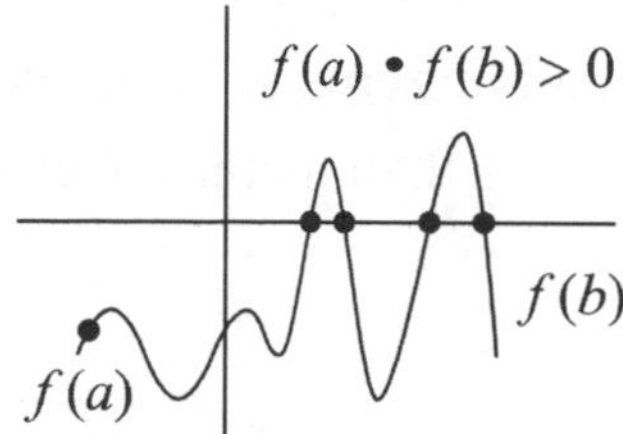

There are 4 real roots

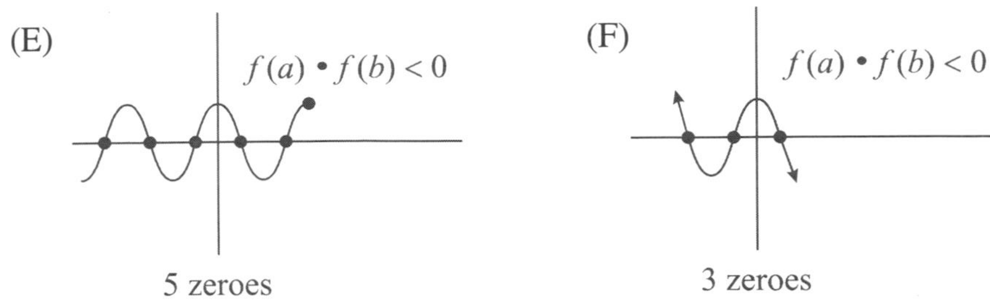

◆Example:

$f(x) = x^3 + 2x^2 - x - 2$ is a continuous function, how many real roots are there between $x = -3$ and $x = 2$?

Since $f(-3) = -27 + 18 + 3 - 2 = -8$

$f(2) = 8 + 8 - 2 - 2 = 12$

$\therefore \; f(-3) \cdot f(2) = (-8)(12) = -96 < 0$

So, there are 3 or 1 real zeros between $x = -3$ and $x = 2$.

D. Conjugate Roots

If $f(x)$ have complex roots, they must appear in pairs

Such as, $f(x) = x^3 + 2x^2 + 2x + 1 = (x+1)(x^2 + x + 1)$

For $x^2 + x + 1 = 0$, $x = \dfrac{-1 \pm \sqrt{1^2 - 4 \cdot (1)(1)}}{2} = -\dfrac{1}{2} \pm \dfrac{\sqrt{3}i}{2}$

There are 2 complex roots

$-\dfrac{1}{2} + \dfrac{\sqrt{3}}{2}i$ and $-\dfrac{1}{2} - \dfrac{\sqrt{3}}{2}i$

they are conjugate roots of each other.

E. Descartes' Rule of Signs

- In a polynomial $p(x)$, the number of positive zeros is either equal to the number of variations in coefficient sign of $p(x)$ or less than that number by an even number.
- The number of negative zeros is either equal to the number of variations in coefficient sign of $p(-x)$ or less than that number by an even number.

◆Eample:

How many possible positive zeros and negative zeros of $p(x) = 3x^3 - 2x^2 + 5x - 8$?

1.In $p(x)$ it has 3 variations in sign. so, it possibly has 3 positive zero or 1 positive zero.

2.In $p(-x)=3(-x)^3-2(-x)^2+5(-x)-8=-3x^3-2x^2-5x-8$ it has zero variation in sign Therefore, it has no negative zero, conclusion: $p(x)$ has 3 positive zeros or 1 positive zero and no negative zero.

◆Eample:

$P(x)=2x^4+3x^3-x^2+x+1$, there are 2 variations in sign.

0 1 1

So, there are 2 positive zeroes or none.

$P(-x)=2(-x)^4+3(-x)^3-(-x)^2+(-x)+1=2x^4-3x^3-x^2-x+1$

1 0 0 1

There are 2 variations in sign, so, there are 2 negative zeroes or none.

Answer to Concept Check 12-10

$x^3-2x^2+2x-1=0$

x^3-2x^2+2x-1

$=x^3-x^2-x^2+x+x-1$

$=x^2(x-1)-x(x-1)+x-1$

$=(x-1)(x^2-x+1)=0$

$x=1$, or $x^2-x+1=0$

$x=\dfrac{1\pm\sqrt{1-4}}{2}=\dfrac{1\pm\sqrt{3}i}{2}$

real root: $x=1$

conjugate roots $x=\dfrac{1\pm\sqrt{3}i}{2}$

Answer to Concept Check 12-11

$5x^5-4x^4+3x^3+2x^2-x+1=0$

positive zeros.

It has 4 signs changed.

It could have 4, 2, 0, positive roots

For negative zeros

$p(-x)=5(-x)^5-4(-x)^4+3(-x)^3+2(-x)^2-(-x)+1$

$=-5x^5-4x^4-3x^3+2x^2+x+1$

It has only one change of signs

so, It could have just 1 negative zero.

Conclusion:

Polynomial $p(x)$ could have 4, 2, 0 positive roots

and may have 1 negative root.

F. Polynomial Inequality

◆Example:

 **Step 1.** Rewrite the $p(x)$ to factors' form

such as $p(x)=2x^3+3x^2-3x-2=(x-1)(x+2)(2x+1)$

 **Step 2.** Graphing

If $p(x)=(x-1)(x+2)(2x+1)>0$

We have $x=1$, $x=-2$, $x=-\frac{1}{2}$ 3 zeros

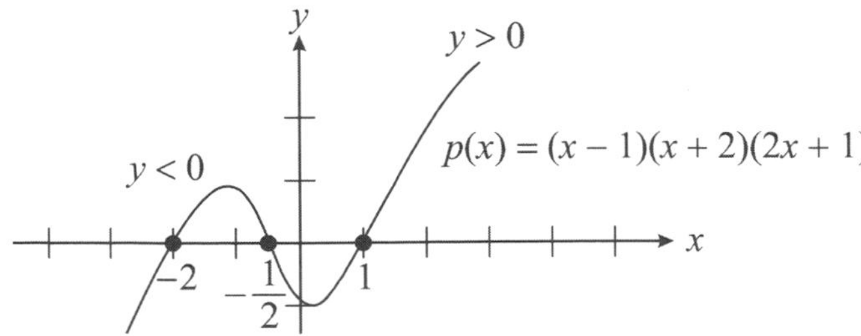

when (1) $x<-2$ $p(x) \Rightarrow (-)\times(-)\times(-)<0$

(2) $-2<x<-\frac{1}{2}$ $p(x) \Rightarrow (-)\times(+)\times(-)>0$

(3) $-\frac{1}{2}<x<1$ $p(x) \Rightarrow (-)\times(+)\times(+)<0$

$(4)\, 1 < x \quad p(x) \Rightarrow (+)\times(+)\times(+) > 0$

So, $-2 < x < -\frac{1}{2}$ or $x > 1$, it makes $P(x) > 0$

◆Example:

Step 1. $p(x) = x^3 + 2x^2 - 5x - 6 > 0 \Rightarrow (x-2)(x+1)(x+3)$

Step 2. Graphing

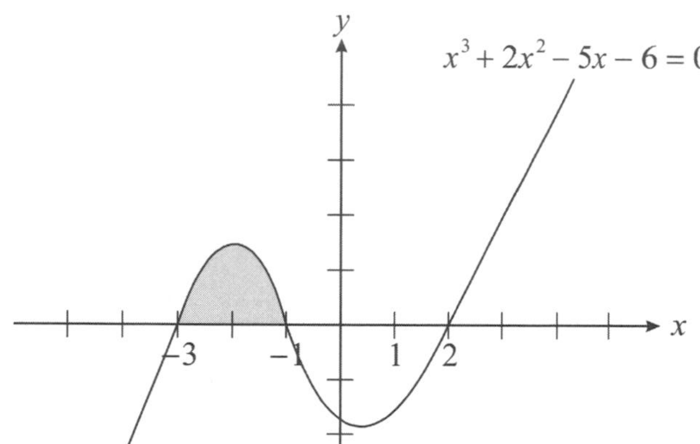

(1) When $x < -3 \quad p(x) \Rightarrow (-)\times(-)\times(-) \Rightarrow < 0$

(2) When $-3 < x < -1 \quad p(x) \Rightarrow (-)\times(-)\times(+) \Rightarrow > 0$

(3) When $-1 < x < 2 \quad p(x) \Rightarrow (-)\times(+)\times(+) \Rightarrow < 0$

(4) When $2 < x \quad p(x) \Rightarrow (+)\times(+)\times(+) \Rightarrow > 0$

(5) When $-3 < x < -1$ or $2 < x$, then $p(x) > 0$

Answer to Concept Check 12-12

$p(x) = 2x^3 + 9x^2 + 13x + 6 = 2x^3 + 2x^2 + 7x^2 + 7x + 6x + 6$

$= 2x^2(x+1) + 7x(x+1) + 6(x+1) = (x+1)(2x^2 + 7x + 6) = (x+1)(2x+3)(x+2) > 0$

When $x < -2$, $p(x) \Rightarrow (-)(-)(-) \Rightarrow (-)$

When $-2 < x < -\frac{3}{2} \quad p(x) \Rightarrow (-)(-)(+) \Rightarrow (+)$

When $-\frac{3}{2} < x < -1 \quad p(x) \Rightarrow (-)(+)(+) \Rightarrow (-)$

When $-1 < x \quad p(x) \Rightarrow (+)(+)(+) \Rightarrow (+)$

When $-2 < x < -\frac{3}{2}$, and $-1 < x$, $p(x) > 0$

$\{-2 < x < -\frac{3}{2}\} \cup \{-1 < x\}$

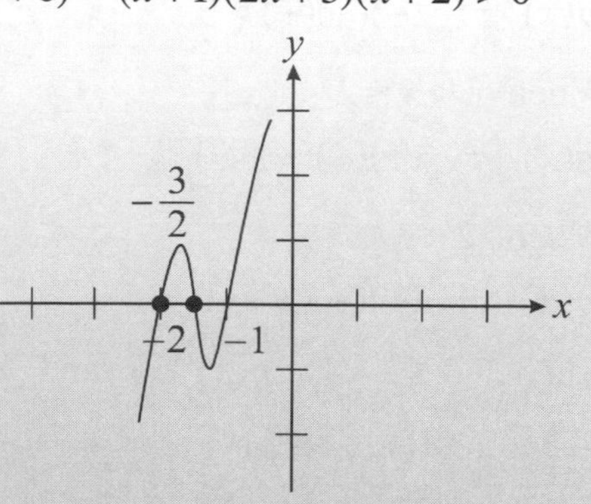

Answer to Concept Check 12-13

$(x-1)^2 \dfrac{(x-2)(x+3)}{x-1} \geq 0$

$\because\ (x-1)^2 \Rightarrow$ positive

$(x-1)(x-2)(x+3) \geq 0$

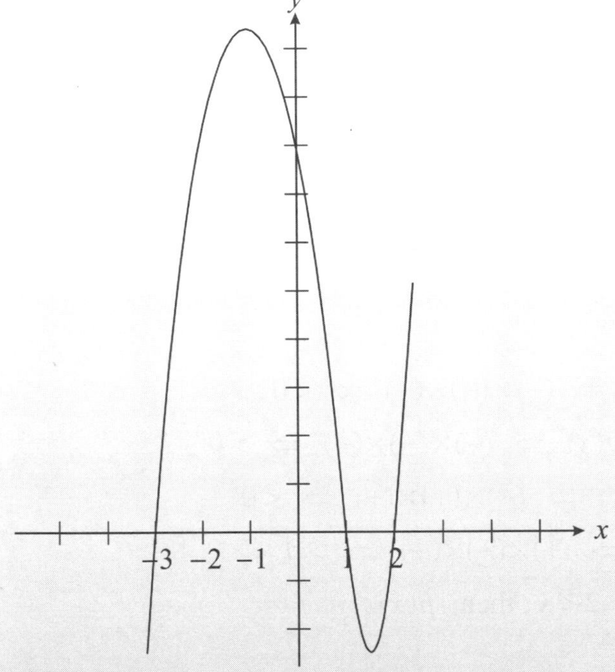

When $x \neq 1 \quad x \leq -3$

$p(x) \Rightarrow (-)(-)(-) \leq 0$

When $-3 \leq x < 1$

$p(x) \Rightarrow (-)(-)(+) > 0$

When $1 < x \leq 2$

$p(x) \Rightarrow (+)(-)(+) < 0$

When $2 < x$

$p(x) \Rightarrow (+)(+)(+) > 0$

when $-3 \leq x < 1$, $p(x) \geq 0$ and $2 < x$, $p(x) \geq 0$

$\{-3 \leq x < 1\} \cup \{2 \leq x\}$

Answer to Concept Check 12-14

$$\frac{(x+1)(x-2)(x-1)^2(x-3)^2}{(x-1)(x-3)} \geq 0$$

$$(x+1)(x-2)(x-1)(x-3) \geq 0$$

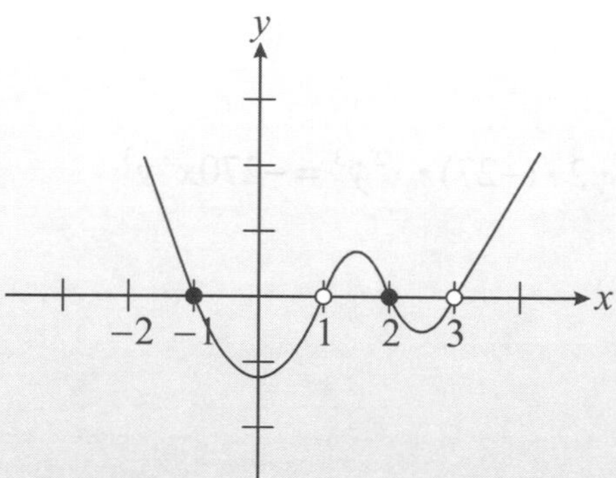

1. When $x \leq -1$

   $p(x) \Rightarrow (-)(-)(-)(-) \geq 0$

2. When $-1 \leq x < 1$

   $p(x) \Rightarrow (+)(-)(-)(-),\ p(x) \leq 0$

3. When $1 < x \leq 2$

   $p(x) \Rightarrow (+)(-)(+)(-),\ p(x) \geq 0$

4. When $2 \leq x < 3$

   $p(x) \Rightarrow (+)(+)(+)(-),\ p(x) < 0$

5. When $3 < x$

   $p(x) \Rightarrow (+)(+)(+)(+),\ p(x) > 0$

   $\{x \leq -1\} \cup \{1 < x \leq 2\} \cup \{3 < x\}$

G. The Binomial Theorem

- $(x+y)^n = (x+y)(x+y)(x+y)\cdots(x+y)$ (where n is a positive integer)

  $= nC_0 x^n (y)^0 + nC_1 x^{n-1}(y)^1 + nC_2 x^{n-2} y^2 + \cdots + nC_{r-1} x^{n-r+1} y^{r-1} + \cdots + nC_n x^0 y^n$

- The rth term of the expansion is $nC_{r-1} x^{n-r+1} y^{r-1}$

  For example, the 5th term is $nC_{5-1} x^{n-5+1} y^{5-1} = nC_4 x^{n-4} y^4$

- For $(x+y)^2 = (x+y)(x+y) = x^2 + 2xy + y^2$

  $= 2C_0 x^2 y^0 + 2C_1 x^1 y^1 + 2C_2 x^0 y^2$

  $= x^2 + 2xy + y^2\ (x^0 = 1,\ y^0 = 1)$

For $(x+y)^3 = 3C_0x^3y^0 + 3C_1x^2y^1 + 3C_2x^1y^2 + 3C_3x^0y^3 = x^3 + 3x^2y + 3xy^2 + y^3$

For $(x+y)^n = nC_0x^ny^0 + nC_1x^{n-1}y^1 + nC_2x^{n-2}y^2 + \cdots + nC_nx^0y^n$

◆Example 1:

What is the fourth term of $(x-3y)^5$?

The fourth term means $5C_{4-1}x^{5-4+1}y^{4-1}$

$$5C_3x^{n-4+1}y^3 = 5C_3x^{5-4+1}(-3y)^3 = \frac{5\cdot4\cdot3!}{(5-3)!3!}x^2(-3y)^3 = 5\cdot2\cdot(-27)\cdot x^2y^3 = -270x^2y^3$$

◆Example 2:

Find the 4th term of $(x-y)^6$.

The fourth term:

$$6C_{4-1}x^{6-4+1}(-y)^{4-1} = 6C_3x^3(-y)^3 = 6C_3x^3(-y)^3 = \frac{6\cdot5\cdot4\cdot3!}{3!3!}x^3(-y)^3 = -20x^3y^3$$

◆Example 3:

What is the value of the fifth term of $(x+\frac{1}{x})^8$?

The 5th term

$$8C_{5-1}x^{8-5+1}(\frac{1}{x})^{5-1} = 8C_4x^4(\frac{1}{x})^4 = \frac{8!}{(8-4)!4!} = \frac{8\cdot7\cdot6\cdot5\cdot4!}{4!\cdot4!} = \frac{8\cdot7\cdot6\cdot5}{4\cdot3\cdot2\cdot1} = 70$$

◆Example 4:

Find the constant term of $(x-\frac{1}{x})^{12}$.

Assume r[th] term is a constant

$$12C_{r-1}(x)^{n-r+1}(\frac{1}{x})^{r-1} = 12C_{r-1}x^{12-r+1}\cdot x^{-(r-1)} = 12C_{r-1}x^{12-r+1-r+1} = 12C_{r-1}x^{12-2r+2}$$

If r[th] is constant, $12-2r+2=0$

$14-2r=0 \quad 2r=14 \quad r=7$

7th term is

$$12C_6x^{12-7+1}(\frac{1}{x})^6 = 12C_6x^{6-6} = 12C_6 = \frac{12!}{(12-6)!6!} = \frac{12!}{6!6!} = \frac{12\cdot11\cdot10\cdot9\cdot8\cdot7\cdot6!}{6!\cdot6!}$$

$$= \frac{\overset{2}{\cancel{12}}\cdot11\cdot\overset{2}{\cancel{10}}\cdot\overset{3}{\cancel{9}}\cdot\cancel{8}\cdot7}{\cancel{6}\cdot\cancel{5}\cdot\cancel{4}\cdot\cancel{3}\cdot\cancel{2}\cdot1} = 11\times2\times3\times2\times7 = 924$$

Answer to Concept Check 12-15

$(2x-y)^5$

The fourth term is

$$5C_{4-1}(2x)^{5-4+1}(-y)^{4-1}=5C_3(2x)^2(-y)^3=-5C_3 2^2\cdot x^2\cdot y^3$$

$$=-4\cdot 5C_3x^2y^3=-4\cdot\frac{5!}{(5-3)!3!}x^2y^3=-4\cdot\frac{5\times4\times3!}{2!\cdot3!}x^2y^3$$

$$=(-4)\cdot(5)\cdot(2)\cdot x^2y^3=-40x^2y^3$$

Answer to Concept Check 12-16

$(x-\frac{1}{x})^{12}$

For $x^0=x^{n-r+1}\cdot(\frac{1}{x})^{r-1}=x^{n-r+1-(r-1)}=x^{n+2-2r}$

$n=12,\ 12+2-2r=0$

$r=7$

The term is the 7th term

The term is $12C_{7-1}x^{12-7+1}\cdot(\frac{1}{x})^{7-1}\cdot(-1)^{7-1}=12C_6(-1)^6x^6(\frac{1}{x})^6=12C_6(-1)^6=12C_6$

$$=\frac{12!}{(12-6)!6!}=\frac{12\times11\times10\times9\times8\times7\times6!}{6!\cdot6!}$$

$$=\frac{12\times11\times10\times9\times8\times7}{6\times5\times4\times3\times2\times1}=924$$

Answer to Concept Check 12-17

$f(x)=x^3+2x^2+2x+1$

$f(-2)=-8+8-4+1=-3<0$

$f(2)=8+8+4+1=21>0$

$f(-2)\cdot f(2)=-63<0$

There must be odd number of roots between them.

one or three.

12-3 EXPONENTIAL and LOGARITHMIC FUNCTION

A. Exponential Function

Standard From: a^x

- $a > 0$

Case 1: $0 < a < 1$

x becomes bigger; a^x becomes smaller,

a^x is a decreasing function

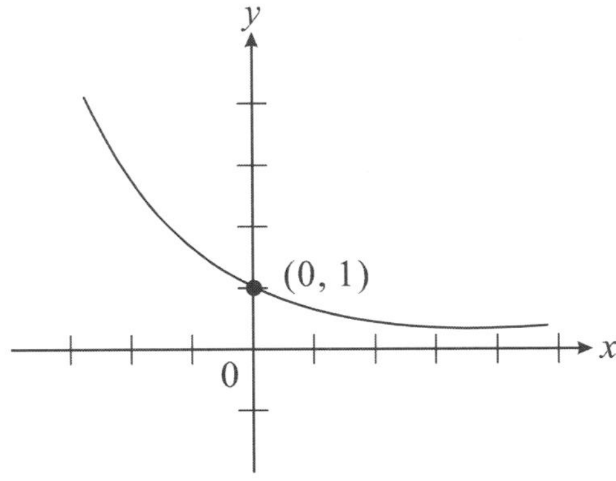

Case 2: $a > 1$

x becomes bigger; a^x becomes bigger,

a^x is an increasing function

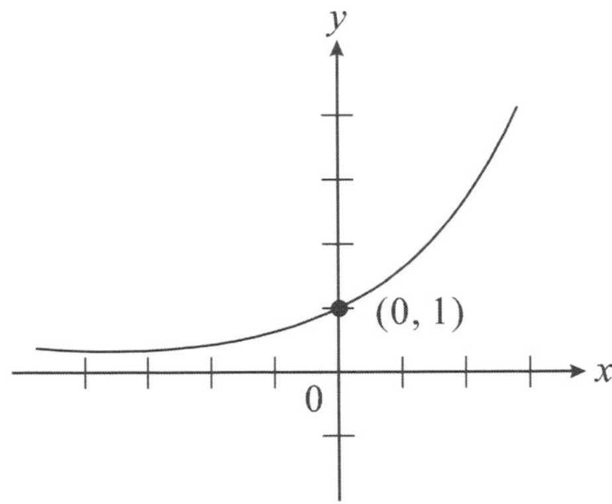

- Both cases the graph passes through the point (0, 1), and x axis is the asymptote.

◆Example:

Which of the following graphs is a possible graph of $f(x)=a^x$?

(A)

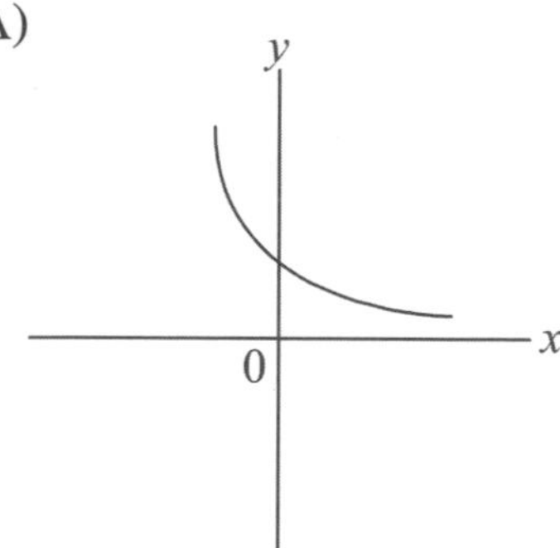

(B)

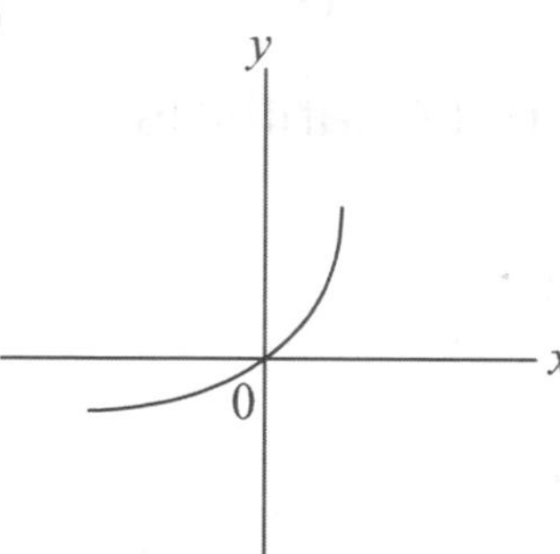

(C)

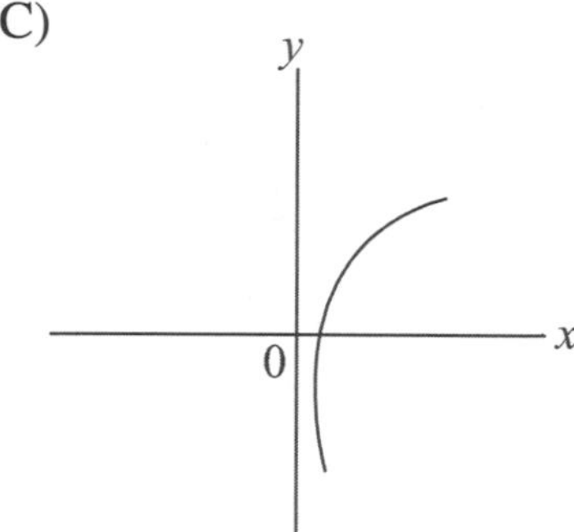

(D)

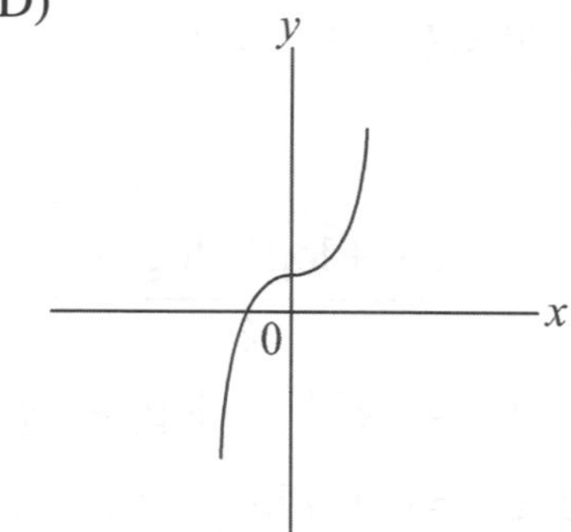

(A) The answer is *A*, because a^x is always greater then 0, $0<a<1$, $f(x)$ is decreasing function.

(B) The graph of *B* is not a graph of a^x, because it passes origin, and $a^x>0$.

(C) $y=a^x$, never <0, so the graph of *C* is not $y=a^x$

(D) a^x is never less than zero.

◆Example:

Although the terms of exponential functions are different, they can be the same function. For examples: $f(x)=64(8^{-x})$ and $g(x)=(\frac{1}{8})^{x-2}$ are the same.

Because: $f(x)=64(8^{-x})=8^2(8^{-x})=8^{2-x}$

$g(x)=(\frac{1}{8})^{x-2}=(8^{-1})^{x-2}=8^{2-x}$

$f(x)=g(x)$

They are the same.

B. Logarithmic Function

If $a^x = y$, define $\log_a^y = x \begin{cases} a \neq 1 \\ a > 0 \end{cases}$

- The properties of Logarithms

1. $\log_a^a = 1$

   $a^1 = a$, $\log_a^a = 1$

2. $\log_a^1 = 0$

   $a^0 = 1$, $\log_a^1 = 0$

3. $a^{\log_a^x} = x$, Let $\log_a^x = y$ $\quad a^y = x \quad a^{\log_a^x} = x$

   Let $\log_b^k = y \quad b^y = k \quad b^{\log_b^k} = k$

4. $\log_a^{b^k} = k \cdot \log_a^b$

   $$\log_a^{b^k} = \log_a^{\overbrace{b \cdot b \cdot b \cdot \cdots}^{k}} = \underbrace{\log_a^b + \log_a^b + \log_a^b + \cdots}_{k}$$

   For example: $\log_a^{b^3} = \log_a^{b \cdot b \cdot b} = \log_a^b + \log_a^b + \log_a^b = 3\log_a^b$

   $\log_2^{x^3} = \log_2^x + \log_2^x + \log_2^x = 3\log_2^x$

5. $\log_a^{xy} = \log_a^x + \log_a^y$

   Since $a^{\log_a^x} = x$, $a^{\log_a^y} = y$

   $\log_a^{xy} = \log_a^{(a\log_a^x)\cdot(a\log_a^y)} = (\log_a^x + \log_a^y) \cdot \log_a^a = \log_a^x + \log_a^y$

6. $\log_a^{\frac{x}{y}} = \log_a^x - \log_a^y$

   $\log_a^{\frac{x}{y}} = \log_a \dfrac{a^{\log_a^x}}{a^{\log_a^y}} = \log_a a^{\log_a^x - \log_a^y} = (\log_a^x - \log_a^y)\log_a^a = \log_a^x - \log_a^y$

7. $\log_a^x = \dfrac{\log_b^x}{\log_b^a}$

   For $\log_b^x = \log_b^{a^{\log_a^x}} = \log_a^x \cdot \log_b^a$

   $\therefore \ \log_a^x = \dfrac{\log_b^x}{\log_b^a}$

8. If $\log_b^x = \log_b^y$, then $x = y$

9. $\log_{10}^x \Rightarrow \log^x$ (Common logarithmic 10 omitted)

10. $\log_e^x \Rightarrow \ln x = \log e^x$ (natual logarithmic e omitted)

11. $\log_a^x = \log_{a^n}^{x^n}$

Since $\log_{a^n}^{x^n} = \dfrac{\log_a^{x^n}}{\log_a^{a^n}} - \dfrac{n \cdot \log_a^x}{n \cdot \log_a^a} = \dfrac{n \cdot \log_a^x}{n \cdot 1} = \log_a^x$

- Graphs of Logarithmic Functions

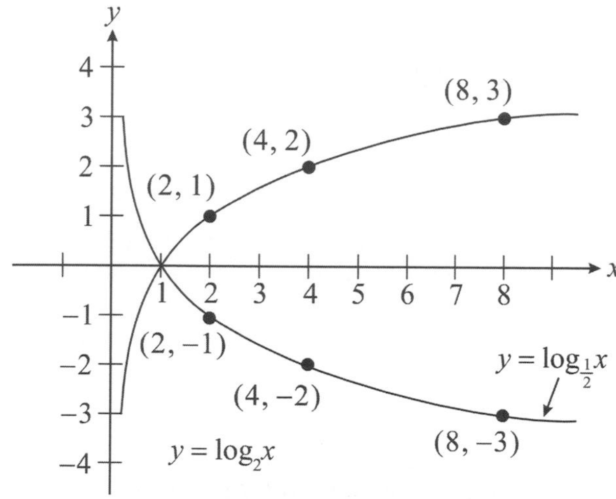

The reflection of $y = \log_2^x$ about the x-axis is the graph of $y = \log_{\frac{1}{2}}^x$.

◆Example 1:

If $(\log_2^{(x+3)}) + (\log_2^{(x-2)}) = \log_2^{6x}$, what is the value of x ?

$(x+3)(x-2) = 6x$

$x^2 + x - 6 = 6x$

$x^2 - 5x - 6 = 0$

$(x-6)(x+1) = 0$

$x = 6$ or $x = -1$

Since $x > 0$, so $x = 6$

◆Example 2:

Sean has invested \$2,000 in an account at an interest rate of 2%, compounded quarterly. The agreement is $A = De^{rt}$, where t represents the number of quarters. How long does it take for the balance to triple ?

$6,000 = 2000e^{0.02t}$

$\dfrac{6000}{2000} = e^{0.02t}$

$3 = e^{0.02t}$

$0.02t = \ln^3, \quad t = \dfrac{\ln^3}{0.02} = 55$

◆Example 3:

$$3 \cdot \log^{\frac{3}{2}} - \log^{\frac{7}{3}} + 3\log^2 + \frac{1}{2}\log^4 = \log^{(\frac{3}{2})^3} - \log^{\frac{7}{3}} + \log^{2^3} + \log^{(4)^{\frac{1}{2}}}$$

$$= \log\frac{27}{8} - \log\frac{7}{3} + \log 8 + \log 2$$

$$= \log\frac{\frac{27}{8}\cdot 8 \cdot 2}{\frac{7}{3}} = \log\frac{54}{\frac{7}{3}} = \log\frac{162}{7} = 1.36$$

◆Example 4:

If $\log_3^2 = a$, and $\log_2^3 = b$, what is the value of $\log_4^6$ represented by a and b ?

$$\log_4^6 = \frac{\log_2^6}{\log_2^4} = \frac{\log_2^{2\times 3}}{\log_2^{2^2}} = \frac{\log_2^2 + \log_2^3}{2\log_2^2} = \frac{1+b}{2}$$

◆Example 5:

Shirley hopes her salary will be raised to at least double the current salary at the beginning of the 13th year. What is the fixed rate of raise each year ?

If Shirley's current annual salary is \$60,000, and if her fixed annual raise is only 1.2%, how much will be her annual salary at the beginning of 13th year ?

A. Let S be current salary

$S(1+r\%)^{12} \geq 2S$

$(1+r\%)^{12} \geq 2$

$12\log^{(1+r\%)} \geq \log^2, \quad \log^{(1+r\%)} \geq \dfrac{0.3}{12} \Rightarrow 0.025$

$1+r\% \geq 10^{0.025} \Rightarrow 1+r\% \geq 1+0.0593$

$\therefore \ r=6$, raising 6% each year within 12 years.

B. $60,000\times(1+0.012)^{12} = 60,000\times 1.15389 = 69,233$

Shirley will receive 9,233 more at the beginning of the 13th year.

Example 6:

If $\log a$, $\log b$ are the two roots of equation $3x^2-8x+2=0$, what is the value of $\log_a b+\log_b a$?

$$\log a+\log b=\frac{8}{3},\quad \log a \cdot \log b=\frac{2}{3}$$

$$\log_a b+\log_b a=\frac{\log b}{\log a}+\frac{\log a}{\log b}=\frac{\log b^2+\log a^2}{(\log a)(\log b)}=\frac{(\log a+\log b)^2-2\log a\log b}{(\log a)(\log b)}$$

$$=\frac{(\frac{8}{3})^2-2\cdot\frac{2}{3}}{\frac{2}{3}}=\frac{\frac{64}{9}-\frac{4}{3}}{\frac{2}{3}}=\frac{\frac{64-12}{9}}{\frac{2}{3}}=\frac{\frac{52}{9}}{\frac{2}{3}}=\frac{26}{3}$$

Answer to Concept Check 12-18

1. $2^{\log_2^x} = x$
2. $\log_2^{2^3} = 3$
3. $\log_2^{3^2} = 2\log_2^3$
4. $\log_8^{2\times3} = \log_8^2+\log_8^3$
5. $\log_2^{\frac{2}{3}} = \log_2^2-\log_2^3 = 1-\log_2^3$
6. $\ln^e = \log_e^e = 1$
7. $\dfrac{\log_2^5}{\log_2^3} = \log_3^5$
8. $\log_{2^3}^{x^3} = \log_2^x$
9. $\log^{(2x+1)}+\log^{(x-2)} = \log^7$

   $(2x+1)(x-2)=7$

   $2x^2-4x+x-2=7$

   $2x^2-3x-9=0$

   $(2x+3)(x-3)=0$

   $x=-\dfrac{3}{2}$ or $x=3$, When $x=-\dfrac{3}{2}$, $2x+1<0$

   $\therefore\ x=3$

Answer to Concept Check 12-19

$\ln(x+1)+\ln(2x-3)=\ln^{4x}$

$\ln^{(x+1)(2x-3)}=\ln^{4x}$

$(x+1)(2x-3)=4x$

$2x^2-3x+2x-3=4x$

$2x^2-5x-3=0$

$(2x+3)(x-3)=0$

$x=-\dfrac{3}{2}$ or $x=3$

$x>0$, so, $x=3$

Answer to Concept Check 12-20

$\log^{(2x+1)}+\log^{(x-2)}=\log^{7}$

$(2x+1)(x-2)=7$

$2x^2-4x+x-2=7$

$2x^2-3x-9=0$

$(2x+3)(x-3)=0$

$x=-\dfrac{3}{2}$ or $x=3$

$x>0 \quad \therefore \quad x=3$

12-4 RATIONAL FUNCTIONS

A. Standard Form

$$F(x)=\frac{P(x)}{Q(x)}[Q(x)\neq 0]$$

$P(x)$ and $Q(x)$ are both polynomial functions, and $Q(x)\neq 0$

For instance, $F(x)=\dfrac{x^3+3x^2-x+2}{x+3}$

B. Asymptotes

In a rational function of $f(x)$, when x approaches a certain value, say 'a', yielding the function $f(x)$ approaching to $+\infty$ or $-\infty$. This line is called asymptote.

- Vertical Asymptote:

In $f(x)$, when x approaches to a certain value 'a', yielding $f(x)$ approaching to ∞ or $-\infty$, then $x=a$ is called vertical asymptote.

- Horizontal Asymptote:

In $f(x)$, when x approaches to ∞ or $-\infty$, yields $f(x)$ approaches to a certain value 'b', $y=b$ is called horizontal asymptote.

For instance, $f(x)=\dfrac{1}{x+5}$, when x approaches to '-5' either from left side or right side, $f(x)$ gets to $+\infty$ or $-\infty$ then $x=-5$ is a vertical asymptote. On the other hand, when x approaches $+\infty$ or $-\infty$, $f(x)$ approaches to zero, then $y=0$ is a horizontal asymptote.

The graph of $f(x)=\dfrac{1}{x+5}$, shown below

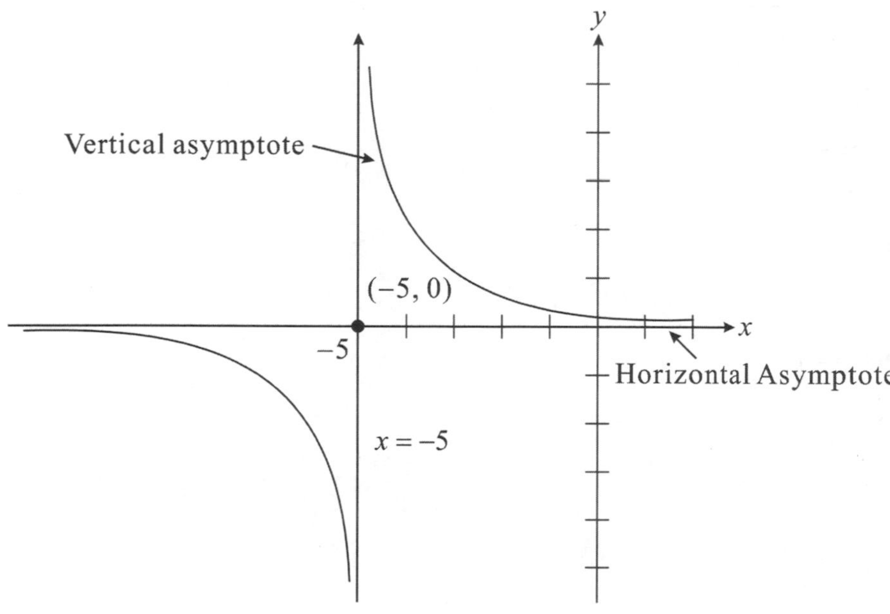

Example 1:

Find the asymptotes of $f(x)=\dfrac{3x}{x+1}$.

When $x=-1$, $x+1=0$

Vertical asymptotes is $x=-1$

Since $\lim\limits_{x\to\infty} f(x)=\lim\limits_{x\to\infty}\dfrac{3x}{x+1}=\lim\limits_{x\to\infty}\dfrac{3}{1+\dfrac{1}{x}}=3$

So, $y=3$ is the horizontal asymptote

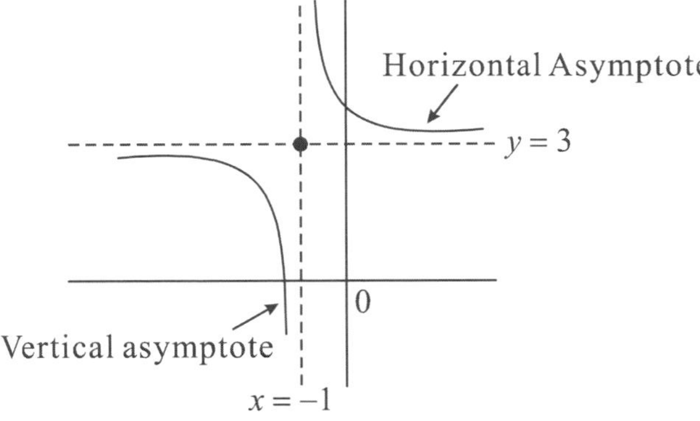

Example 2:

Find the Asymptote(s) of $f(x)=\dfrac{x^2+x-6}{x^2+4x+3}$

$$\frac{x^2+x-6}{x^2+4x+3}=\frac{\cancel{(x+3)}(x-2)}{\cancel{(x+3)}(x+1)}=\frac{x-2}{x+1}$$

When $x=-1$, $x+1=0$

Vertical Asymptote $x=-1$

$$\lim_{x\to\infty}\frac{x-2}{x+1}=\lim_{x\to\infty}\frac{1-\dfrac{2}{x}}{1+\dfrac{1}{x}}=1$$

Horizontal asymptote $y=1$

The hole is at $x=-3$

$$y=\frac{-3-2}{1-3}=\frac{5}{2}$$

(−3, 2.5)

y = 1

(−1, −1)

x = −1

Example 3:

If the vertical asymptote of the rational function $f(x)=\frac{x^2+x+a}{x^2-3x+2}$ is $x=1$, what is the value of a ?

$\frac{x^2+x+a}{x^2-3x+2}=\frac{x^2+x+a}{(x-1)(x-2)}$, if the vertical asymptote is $x=1$, then x^2+x+a must have the factor of $x-1$. In other words, $1^2+1+a=0 \quad \therefore \ a=-2$

C. Holes

If $P(x)$ and $Q(x)$ in $F(x)=\frac{P(x)}{Q(x)}$ have the same factor or factors that can be removed from the function then the graphs of $f(x)$ and $F(x)$ are the same. Only difference is that simplified graph of $f(x)$ has holes.

For instance, $F(x)=\frac{x^2-4}{x-2}$

Since $x^2-4=(x-2)(x+2)$, so, $x-2$

Can be removed, $F(x)=\frac{(x-2)(x+2)}{x-2}=x+2$

So, $f(x)=x+2$ and $F(x)=\frac{x^2-4}{x-2}$

have the same graph, but $x=2$ still undefined $f(x)=x+2$ has the hole at (2, 4)

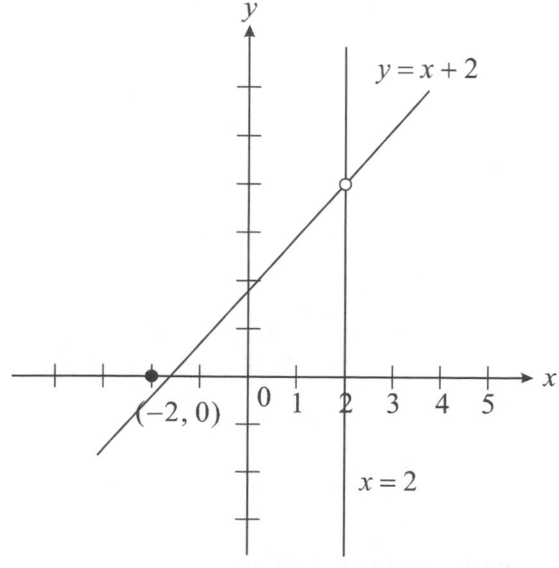

★Tip:
Most of rational function are discontinuous.

◆Example:

$$F(x)=\frac{x^4+4x^3-x^2-16x-12}{x+2}$$

By division

$$\begin{array}{r|l}
x^4+4x^3-x^2-16x-12 & x+2 \\ \cline{2-2}
x^4+2x^3 \qquad\qquad\qquad & x^3+2x^2-5x-6 \\ \hline
2x^3-x^2-16x-12 & \\
2x^3+4x^2 \qquad\qquad & \\ \hline
-5x^2-16x-12 & \\
-5x^2-10x \qquad & \\ \hline
-6x-12 & \\
-6x-12 & \\ \hline
0 &
\end{array}$$

$f(x)=x^3+2x^2-5x-6$

$f(1)=1+2-5-6\neq 0$

$f(-1)=-1+2+5-6=0$ $(x+1)$ is a factor

$f(2)=8+8-10-6=0$ $(x-2)$ is a factor

$f(3)=27+18-15-6\neq 0$

$f(-3)=-27+18+15-6=0$ $(x+3)$ is a factor

So, $F(x)=\dfrac{(x+1)(x-2)\cancel{(x+2)}(x+3)}{\cancel{x+2}}=x^3+2x^2-5x-6$

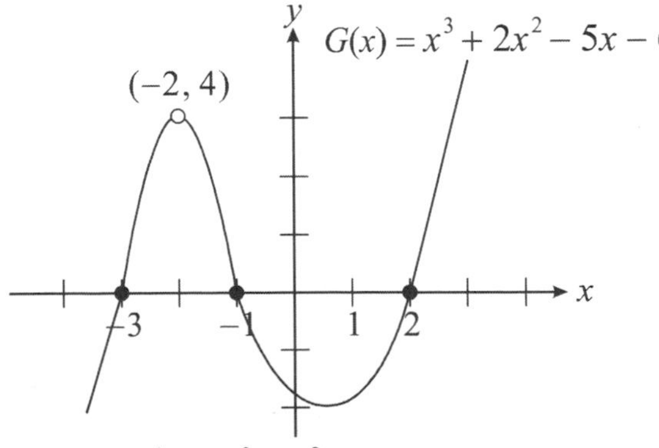

$$F(x)=\frac{x^4+4x^3-x^2-16x-12}{x+2}$$

$G(x)=x^3+2x^2-5x-6$

Both of $F(x)$ and $G(x)$ have the same graph, with a hole at $(-2, 4)$

◆Example:

$$F(x)=\frac{x^4-5x^3+5x^2+5x-6}{x-2}$$

$$F(x)=g(x)=\frac{(x-2)(x^3-3x^2-x+3)}{x-2}=x^3-3x^2-x+3$$

3 has the factors of ±1, ±3

$g(1)=1-3-1+3=0$

$g(-1)=-1-3+1+3=0$

$g(3)=27-27-3+3=0$

$g(-3)=-27-27+3+3\neq0$

$g(x)=x^3-3x^2-x+3=(x+1)(x-1)(x-3)$

$g(2)=8-12-2+3=-3$

There is a hole at $(2,-3)$

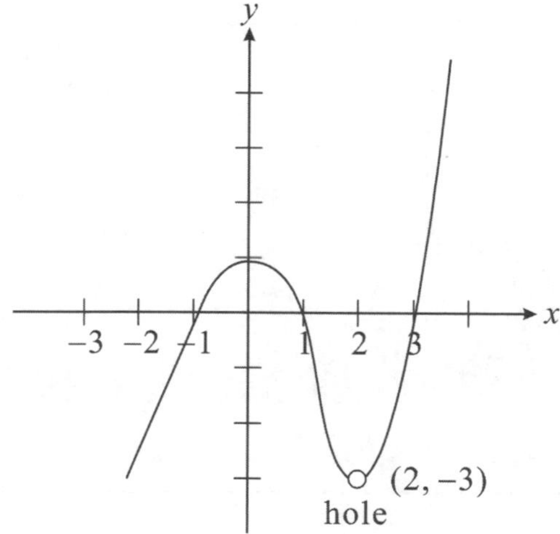

Answer to Concept Check 12-21

$$f(x)=\frac{x^4+x^3-7x^2-x+6}{x+3}$$

$$\begin{array}{r|l}
x^4+\ x^3-7x^2-\ x+6 & x+3 \\
\underline{x^4+3x^3\qquad\qquad\qquad} & x^3-2x^2-x+2 \\
-2x^3-7x^2-\ x+6 & \\
\underline{-2x^3-6x^2\qquad\quad} & \\
-x^2-\ x+6 & \\
\underline{-x^2-3x\quad} & \\
2x+6 & \\
\underline{2x+6} & \\
0 &
\end{array}$$

$g(x)=x^3-2x^2-x+2$

2 has the factors of $\pm1,\ \pm2$

$g(1)=1-2-1+2=0$

$g(-1)=-1-2+1+2=0$

$g(2)=8-8-2+2=0$

The graph of $g(x)$ and $f(x)$ is the same except holes.

$g(x)$ is an odd function

$\because\ g(-3)=-27-18+3+2=-40$

a hole at $(-3,-40)$

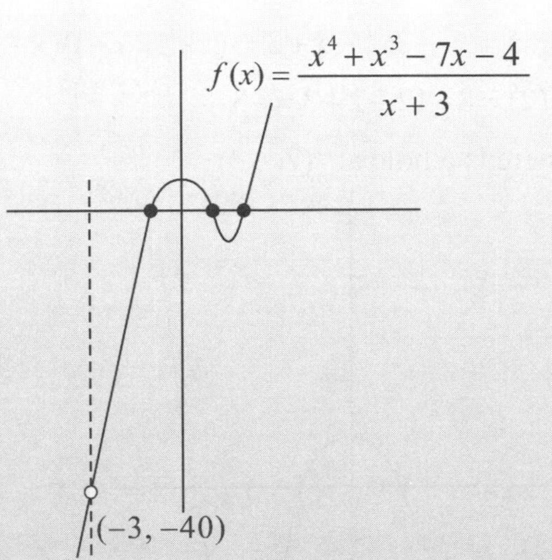

D. Rational inequality

 **Step 1.** Eliminate the denominator by multiplying the denominator of both sides.

 **Step 2.** Use the polynomial graphs to find out the inequality.

◆Example:

Solve $\dfrac{(x-2)(x+3)}{x+1}\geq 0\ \ (x\neq -1)$

 **Step 1.** Since $(x+1)^2$ is positive multiply $(x+1)^2$ to both sides of the function

yield $(x+1)^2\cdot\dfrac{(x-2)(x+3)}{x+1}\geq 0\cdot(x+1)^2$

$\Rightarrow 0$

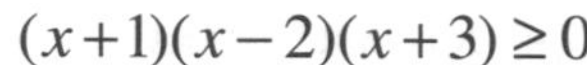

$(x+1)(x-2)(x+3) \geq 0$

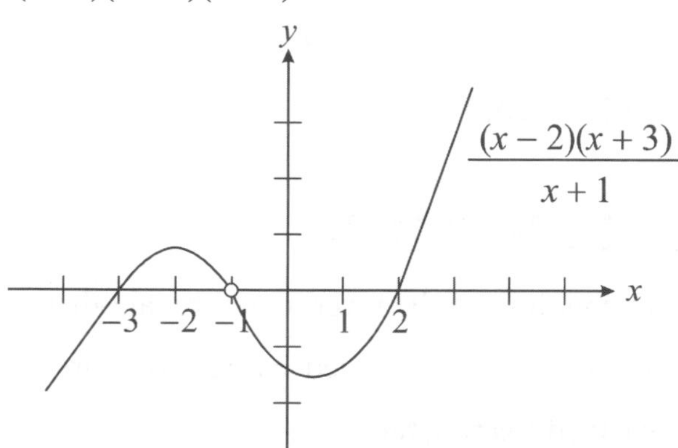

When $x \leq -3$ $\quad p(x) \Rightarrow (-)\times(-)\times(-) < 0$

When $-3 \leq x < -1$ $\quad p(x) \Rightarrow (-)\times(-)\times(+) \geq 0$

When $-1 < x \leq 2$ $\quad p(x) \Rightarrow (+)\times(-)\times(+) \leq 0$

When $2 \leq x$ $\quad p(x) \Rightarrow (+)\times(+)\times(+) \geq 0$

◆Example:

Solve $\dfrac{(x-1)(x-3)}{(x+1)(x-2)} \leq 0$

 **Step 1.** Multiply $(x+1)^2(x-2)^2$ to both sides.

$$\frac{(x-1)(x-3)}{(x+1)(x-2)} \cdot (x+1)^2(x-2)^2 \leq 0$$

$$(x+1)(x-1)(x-2)(x-3) \leq 0$$

 **Step 2.** Graphing

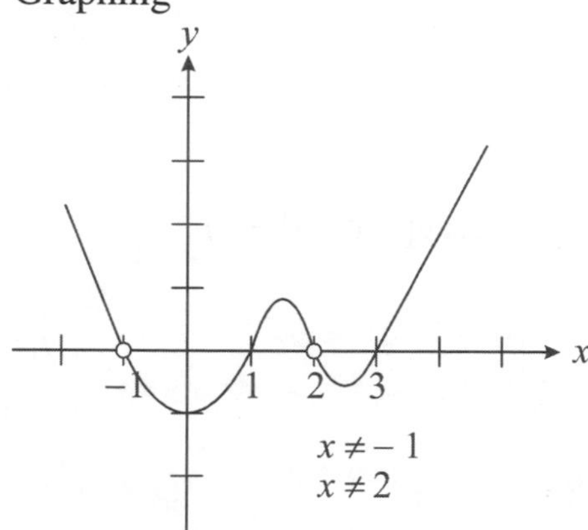

(1) When $x < -1$ $\quad x \neq -1$ $\quad p(x) \Rightarrow (-)\times(-)\times(-)\times(-) > 0$

(2) When $-1 < x \leq 1$ $\quad p(x) \Rightarrow (+)\times(-)\times(-)\times(-) \leq 0$

(3) When $1 \leq x < 2$ $\quad p(x) \Rightarrow (+)\times(+)\times(-)\times(-) \geq 0$

(4) When $2 < x \leq 3$ $\quad p(x) \Rightarrow (+)\times(+)\times(+)\times(-) \leq 0$

(5) When $3 < x$ $\quad p(x) \Rightarrow (+)\times(+)\times(+)\times(+) > 0$

$p(x) \leq 0 \quad -1 < x \leq 1$ and $2 < x \leq 3$

$\{-1 < x \leq 1\} \cup \{2 < x \leq 3\}$

12-5 PARAMETRIC FUNCTIONS

Sometimes, the variables of an expression can be represented by another variable to make the expression much easier to manipulate. The changed expressions are so called parametric functions. The third variable is called 'parameter'.

For instance: $y = \dfrac{3x-7}{4}$

Let $x = 4w+1$

then $4y = 3x-7 = 3(4w+1)-7 = 12w+3-7 = 12w-4$

$\therefore \ y = 3w-1$

$$\begin{cases} x = 4w+1 \\ y = 3w-1 \end{cases}$$

points position (x, y) is now $(4w+1, 3w-1)$ according to the change of variable w.

◆Example:

Sketch the graph of the parametric functions $\begin{cases} x = 2t^2 \\ y = 5t^2 - 1 \end{cases}$

Since $x = 2t^2 \quad \therefore \ t^2 = \dfrac{x}{2}$

$y = 5(\dfrac{x}{2}) - 1 = \dfrac{5}{2}x - 1$

It is a straight line with slope $= \dfrac{5}{2}$ and y-intercept at $(0, -1)$

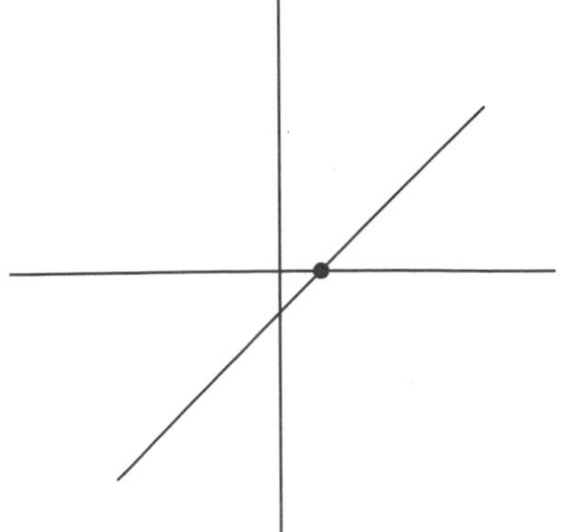

Since $x = 2t^2$, So, $x \geq 0$

The graph is a ray from point $(0, -1)$ along the slope $\frac{5}{2}$ going up.

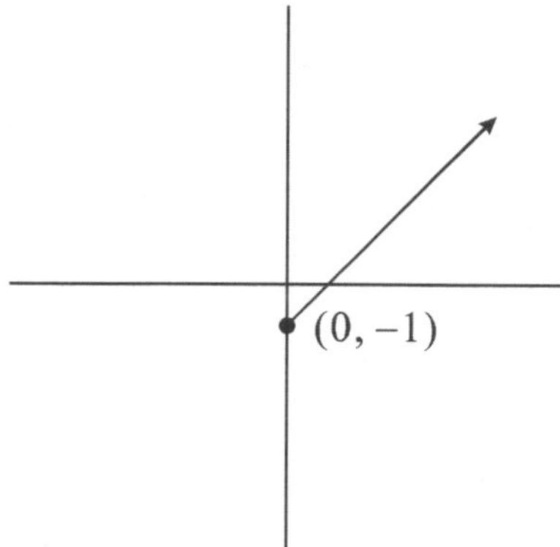

Answer to Concept Check 12-22

$\sqrt{p} = x+3$, $p = (x+3)^2$

$y = p-3$, $y = (x+3)^2 - 3$

It is a parabola

Answer to Concept Check 12-23

$x = \sin\theta$, $y = \cos\theta$

$x^2 = \sin^2\theta$, $y^2 = \cos^2\theta$

$\therefore\ \sin^2\theta + \cos^2\theta = x^2 + y^2 = 1$

It is a circle.

Answer to Concept Check 12-24

$$\begin{cases} x = 2w \\ y = 4w^2 - 8w \end{cases}$$

$x = 2w \quad \therefore \quad w = \dfrac{x}{2}$ substitute in $y = 4w^2 - 8w$, $y = 4(\dfrac{x}{2})^2 - 8(\dfrac{x}{2}) = x^2 - 4x$

$\therefore \quad y = x^2 - 4x + 4 - 4 = (x^2 - 4x + 4) - 4 = (x-2)^2 - 4$

It is a parabola, with vertex at $(2, -4)$

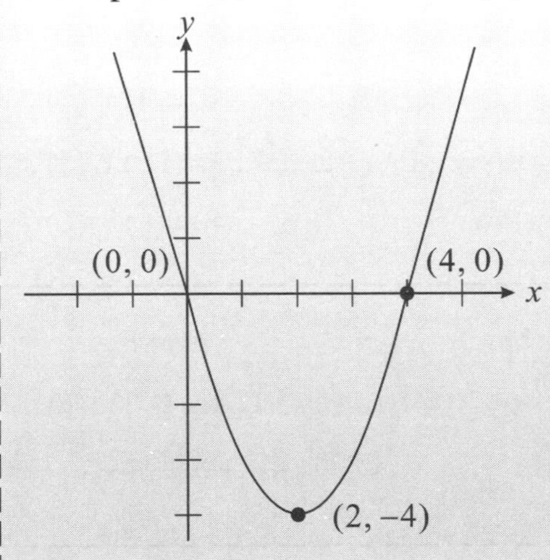

12-6 PIECEWISE FUNCTIONS

A piecewise-defined function is formed by several pieces of functions.

◆Example:

Find the domain and range of $f(x) = \begin{cases} (x+2)^2 - 3 & x < 0 \\ x + 1 & x > 0 \end{cases}$

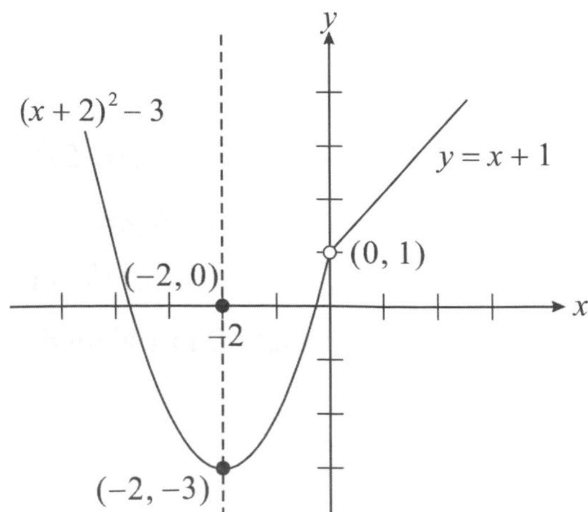

Domain: $x = \{\text{all real}, x \neq 0\}$

Range: $y \geq -3$

◆Example:

What are the domain and range of the piecewise function of

$$f(x) = \begin{cases} (x+3)^2 - 3 \leq 0 \\ 2x-1 \quad x \geq 0 \end{cases}$$

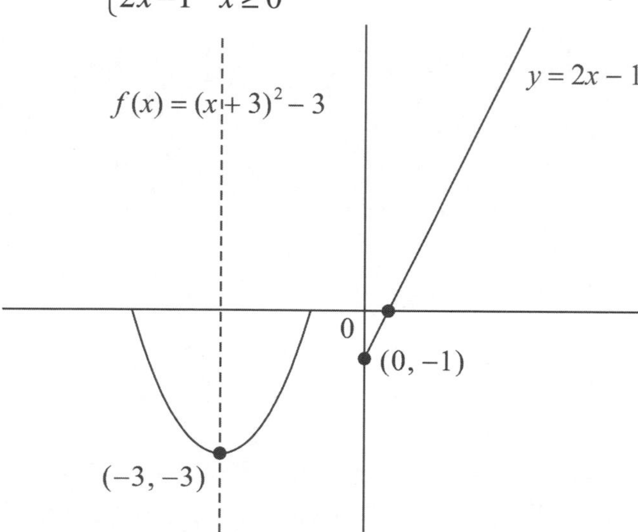

$f(x)=0=(x+3)^2-3$, $(x+3)^2=3$, $x+3=\pm\sqrt{3}$

$x=-3\pm\sqrt{3}$

Domain: $\{-4.73, -1.27\} \cup \{0, \infty\}$

Range: $y \geq -3$

Answer to Concept Check 12-25

$f(x)$ is a piecewise function formed by a parabola $y=(x-2)^2-2$ and a straight line $y=2x+1$.

x-intercept $0=(x-2)^2-2$, $(x-2)^2=2$, $x-2=\pm\sqrt{2}$, $x=2\pm\sqrt{2}$

y-intercept $y=(-2)^2-2=2$

The graph shown below.

Domain: all real numbers

Range: all real number

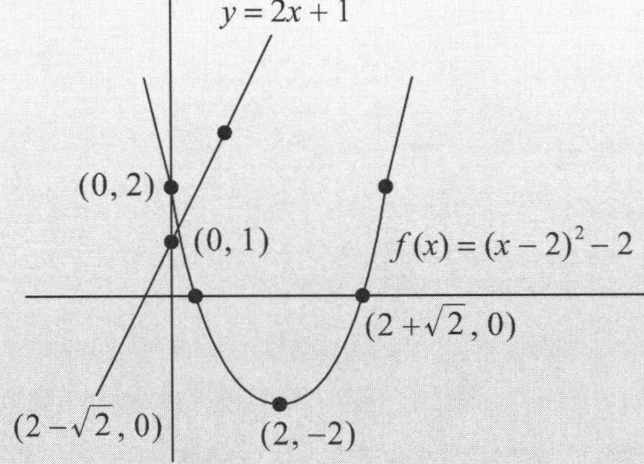

CHAPTER 12
PRACTICE

1. If 1 and $1-\sqrt{2}i$ are the two roots of equation $ax^3+2x^2+bx-1=0$, where a, b are real numbers and $a \neq 0$ what is the value of $a-b$?
(A) $-\frac{4}{3}$
(B) $-\frac{3}{4}$
(C) $\frac{4}{3}$
(D) $\frac{3}{4}$

2. If $f(\frac{x}{2})=6x$, what is the value of $f(2x)$?
(A) $12x$
(B) $24x$
(C) $36x$
(D) $48x$

3. If $f(x)=\frac{x^2-5x+6}{x^2-9}$, what are the asymptotes ?
(A) $x=-3$ and $y=1$
(B) $x=2$ and $y=1$
(C) $x=-2$, and $y=1$
(D) $x=2$, $y=1$, and $x=-3$

4. If $f(x)=3x+2$ and $f(g(2))=4$, what is $g(x)$?
(A) $2x$
(B) $3x$
(C) $\frac{x}{2}$
(D) $\frac{x}{3}$

5. A line has parametric equation $x=2t-1$, and $y=-2+3t$, where t is the parameter, what is the y-intercept of the line ?
(A) $-\frac{1}{2}$
(B) $\frac{1}{2}$
(C) $-\frac{1}{3}$
(D) $\frac{1}{3}$

6. What is the domain of the function defined by $f(x)=\sqrt{3x^2-1}$?
(A) $x \geq \frac{\sqrt{3}}{3}$
(B) $x \geq 0$
(C) All real numbers
(D) $x \geq \frac{\sqrt{3}}{3}$ or $x \leq \frac{\sqrt{3}}{3}$

7. Graph the solution of $\frac{(x-2)(x+1)}{x-1} \geq 0$

8. Solve $(x-3)(x-2)(x+1)>0$ by graph.

9. $\begin{cases} |x-3|-y+3=0 \\ x-3y+1=0 \end{cases}$

solve the system equation above.

10. Which of the following is(are) odd function(s) ?

I. $f(x)=x^4+3x^3+2x^2-3x+1$

II. $f(x)=2x^3+3x+1$

III. $f(x)=x^3-x$

11. If the coordinates of points lie on a graph can be represented by $x=a\cos\theta$, $y=b\sin\theta$, then, what is its shape ?

12. What is the nature of the graph of $P(x) =-x^7+6x^6+3x^5-2x^3+5x+1$

(A) The graph falls to the both sides.

(B) The graph falls to the left and rises to the right.

(C) The graph falls to the right and rises to the left.

(D) The graph rises to the left side and falls to the right side.

13. Find the constant term of $(x+\frac{1}{x})^{10}$

14. Solve $\log_3^{(x-2)}+\log_3^{(x-1)}=\log_3^{(6x-16)}$

15. Solve $x^3+x^2-4x-4\geq 0$ and graph.

16. How long does it take to raise Caroline's salary to double in the circumstance of the annual raise at a fixed rate 2% each year ?

17. How many zeroes for $f(x)=x^5-2x^4+2x^3-4x^2+3x+3$?

18. How many real roots between -2 and 2 for a continuous function $f(x)=x^3-2x^2+x-3$?

(A) only 1

(B) 1 or 3

(C) 2 or 3

(D) 2 or 4

19. How many possible positive zeroes and negative zeroes for $P(x)=2x^5+x^4+2x^2+7x+3$

20. A scientific experiment has only three outcomes A, B, and C. The probabilities of their occurrence are $P(A)$, $P(B)$, and $P(C)$. IF $P(A)=\log_4 k$, $P(B)=\log_{16} k$, and $P(C)=\log_{64} k$, where k is a positive real number, what is the value of $P(B)$?

CHAPTER 12
ANSWERS and EXPLANATIONS

1. If 1, and $1-\sqrt{2}i$ are two roots of the equation $ax^3+2x^2+bx-1=0$, then

$(1-\sqrt{2}i)(1+\sqrt{2}i)(1)=(-1)^3\cdot\frac{-1}{a}=\frac{1}{a}$

$[1^2-(\sqrt{2}i)^2](1)=3=\frac{1}{a}$

$\therefore\ a=\frac{1}{3}$

Since $(1-\sqrt{2}i)(1+\sqrt{2}i)+(1-\sqrt{2}i)(1)$ $+(1+\sqrt{2}i)(1)=\frac{b}{a}$

$3+1-\sqrt{2}i+1+\sqrt{2}i=\frac{b}{a}$

$5=\frac{b}{a}\quad\therefore\ b=5a=\frac{5}{3}$

So, $a-b=\frac{1}{3}-\frac{5}{3}=-\frac{4}{3}$

Answer is (A)

2. Let $x'=\frac{x}{2}$, then $x=2x'$

$f(x')=6(2x')=12x'$

Change x' to x

$f(x)=12x,\ f(2x)=12(2x)=24x$

Answer is (B)

3. $f(x)=\frac{x^2-5x+6}{x^2-9}=\frac{(x-2)(x-3)}{(x+3)(x-3)}$

So, $x=-3$ is an vertical asymptote

Is there any horizontal asymptote ?

Lets take limit of $f(x)$

$f(x)=\frac{(x-2)(x-3)}{(x+3)(x-3)}=\frac{x-2}{x+3}=g(x)$

So, $\lim\limits_{x\to\infty}\frac{x-2}{x+3}$

$=\lim\limits_{x\to\infty}\frac{1-\frac{2}{x}}{1+\frac{3}{x}}=\frac{1-0}{1+0}=1$

So, the horizontal asymptote is $y=1$

Answer is (A)

4. Since $f(x)=3x+2$

$f(g(2))=3\cdot g(2)+2=4$

$\therefore\ 3\cdot g(2)=2,\ g(2)=\frac{2}{3}$

So, $g(x)=\frac{x}{3}$

Answer is (D)

5. $x=2t-1,\ t=\frac{x}{2}+\frac{1}{2}$

$y=-2+3t=-2+3(\frac{x}{2}+\frac{1}{2})$

$=-2+\frac{3}{2}x+\frac{3}{2}$

$=\frac{3}{2}x-\frac{1}{2}$

y-intercept is $y=-\frac{1}{2}$

Answer is (A)

6. $3x^2-1\geq 0$

$3x^2\geq 1\quad x^2\geq\frac{1}{3}$

$x\geq\sqrt{\frac{1}{3}}$ or $x\leq-\sqrt{\frac{1}{3}}$

$x \geq \frac{\sqrt{3}}{3}$ or $x \leq -\frac{\sqrt{3}}{3}$

Answer is (D)

7. $f(x)=\frac{(x-2)(x+1)}{x-1} \geq 0$

Multiply $(x-1)^2$ to both sides

Since $(x-1)^2$ is positive, the inequality sign doesn't change

$(x-1)^2 \times \frac{(x-2)(x+1)}{(x-1)} \geq 0$

$(x-1)(x-2)(x+1) \geq 0$

When $x \leq -1$ $(-)(-)(-) \Rightarrow -$

When $-1 \leq x < 1$ $(-)(-)(+) \Rightarrow +$

When $1 < x \leq 2$ $(+)(-)(+) \Rightarrow -$

When $2 \leq x$ $(+)(+)(+) \Rightarrow +$

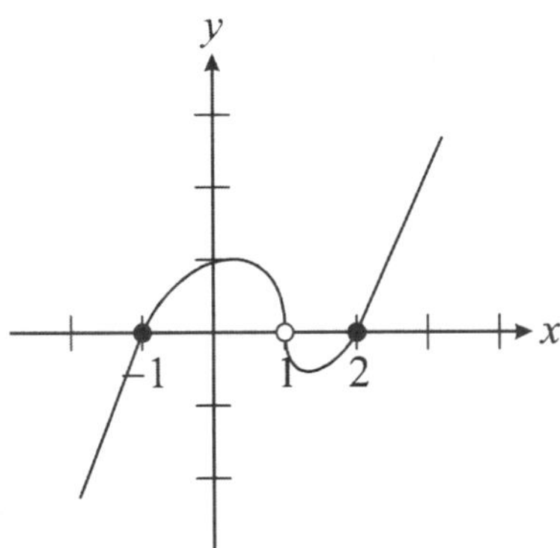

Note: There is a hole at $x=1$

8. Graph the equation

$y=(x-3)(x-2)(x+1)>0$

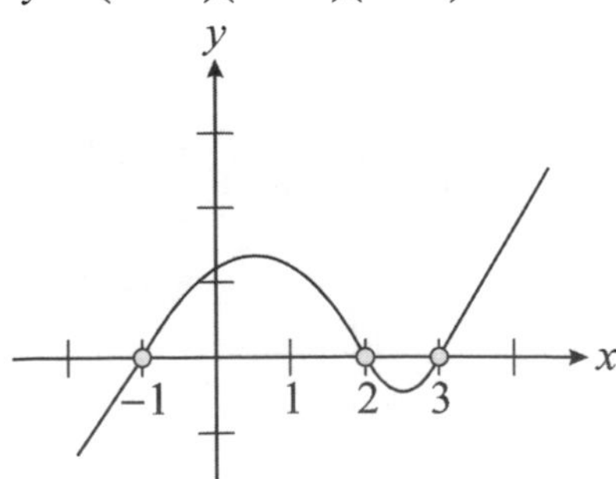

There are 3 holes

When $x<-1$, y:negative

When $-1<x<2$, y: positive

When $2<x<3$, y:negative

When $x>3$, y: positive

9.(1) $\begin{cases} x-3-y+3=0 \cdots ① \\ x-3y+1=0 \cdots ② \end{cases}$

$x-y=0$

$x-3y=-1$

$2y=1$, $y=\frac{1}{2}$, $x=\frac{1}{2}$

(2) $\begin{cases} x-3+y-3=0 \\ x-3y+1=0 \end{cases}$

$x+y=6$

$x-3y=-1$

$4y=7$, $y=\frac{7}{4}$

$x=6-\frac{7}{4}=\frac{17}{4}$

solution $\begin{cases} x=\frac{17}{4} \\ y=\frac{7}{4} \end{cases}$

10. I. $f(-x)$

$=(-x)^4+3(-x)^3+2(-x)^2-3(-x)+1$

$=x^4-3x^3+2x^2+3x+1$

$-f(x)=-x^4-3x^3-2x^2+3x-1$

$f(x) \neq -f(-x)$

II. $f(x)=2x^3+3x+1$

$f(-x)=2(-x)^3+3(-x)+1$

$=-2x^3-3x+1$

$$-f(-x)=2x^3+3x-1$$

$$f(x)\neq -f(-x)$$

III. $f(-x)=(-x)^3-(-x)$

$$=-x^3+x$$

$$=-(x^3-x)$$

$$=-f(x)$$

$$f(-x)=-f(x)=-y$$

$\therefore$ $f(x)$ is an odd function.

11. $\cos\theta=\dfrac{x}{a}$, $\sin\theta=\dfrac{y}{b}$

Since $\cos^2\theta+\sin^2\theta=1$

$\therefore$ $\dfrac{x^2}{a^2}+\dfrac{y^2}{b^2}=1$

It is an ellipse

12. The highest order of the function is odd, so it is up and down. But it is negative, so, it rises to the left and falls to the right. Answer is (D)

13. Assume the rth term is a constant term, then the exponent of x must be zero. The form at the rth term is $nC_{r-1}x^{n-r+1}(\dfrac{1}{x})^{r-1}$,

$n=10$

$10C_{r-1}x^{10-r+1}\cdot(x^{-r+1})=10C_{r-1}x^{10-r+1-r+1}$

$\therefore$ $10-r+1-r+1=10-2r+2=0$

$\therefore$ $2r=12$, $r=6$

$10C_{6-1}x^{10-6+1}(\dfrac{1}{x})^{6-1}=10C_5x^5\cdot(\dfrac{1}{x})^5$

$=\dfrac{10\cdot9\cdot8\cdot7\cdot6\cdot5!}{(10-5)!(5!)}x^{5-5}$

$=\dfrac{10\cdot9\cdot8\cdot7\cdot6}{5\cdot4\cdot3\cdot2\cdot1}$

$=3\cdot2\cdot7\cdot6=252$

14. $\log_3(x-2)+\log_3(x-1)$

$=\log_3(x-2)(x-1)=\log_3 x^2-3x+2$

$=\log_3(6x-16)$

$\therefore$ $x^2-3x+2=6x-16$

$x^2-9x+18=0$, $(x-3)(x-6)=0$

$x=3$, or $x=6$

15. $x^3+x^2-4x-4\geq0$

$x^2(x+1)-4(x+1)=(x+1)(x^2-4)$

$=(x+1)(x-2)(x+2)\geq0$

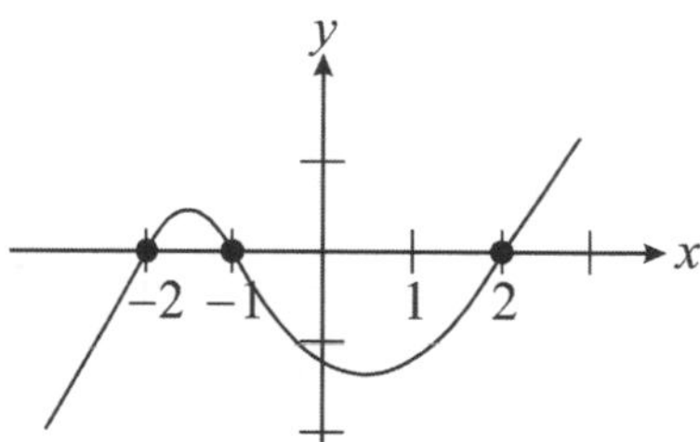

When x	$f(x)$
$x\leq-2$	$f(x)\leq0$
$-2\leq x\leq-1$	$f(x)\geq0$
$-1\leq x\leq2$	$f(x)\leq0$
$2\leq x$	$f(x)\geq0$

So, the graph shown above.

16. Assume Caroline's annual salary is s, then $S(1+r\%)^n\geq2S$

($r\rightarrow$rate, $n\rightarrow$years)

$1.02^n\geq2$

$n\log^{1.02}\geq\log^{2}$

$n\geq\dfrac{\log^{2}}{\log^{1.02}}=35$

About 35 year later.

17. $f(x)=x^5-2x^4+2x^3-4x^2+3x+3$

The zeroes must be the factor(s) of 3,

$3\Rightarrow\ \pm1,\pm3$

(1) $f(1)=1-2+2-4+3+3=3\neq0$

(2) $f(-1)=-1-2-2-4-3+3=-9\neq0$

(3) $f(3)=(3)^5-2(3)^4+2(3)^3-4(3)^2+3(3)$

$+3$

$=243-162+54-36+9+3=111\neq0$

(4) $f(-3)=(-3)^5-2(-3)^4+2(-3)^3$

$-4(-3)^2+3(-3)+3$

$=-243-162-54-36-9+3=-501\neq0$

There is no zero for this function.

In other words, there is no real root for equation $x^5-2x^4+2x^3-4x^2+3x$ $+3=0$

18. How many real roots between -2 and 2 checked by $f(2)\cdot f(-2)<0$

$f(-2)=(-2)^3-2(-2)^2+2(-2)-3$

$=-8-8-4-3=-23$

$f(2)=(2)^3-2(2)^2+2(2)-3$

$=8-8+4-3=1$

$f(-2)\cdot f(2)=(-23)(1)=-23<0$

There is 1 or 3 real zeroes.

19. $P(x)=2x^5+x^4+2x^2+7x+3$

$=2x^5+x^4+0x^3+2x^2+7x+3$

No change of sign, so, no positive zero

$P(-x)=2(-x)^5+(-x)^4+2(-x)^2+7(-x)+3$

$=-2x^5+x^4+0+2x^2-7x+3$

There are 3 changes of signs, so, it has 3 negative zeroes or just one negative zero. (It is $x=-\dfrac{1}{2}$)

20. $P(A)=\log_4 k$

$P(A)+P(B)+P(C)$

$=\log_4 k+\log_{16} k+\log_{64} k=1$

$=\log_4 k+\dfrac{\log_4 k}{\log_4 16}+\dfrac{\log_4 k}{\log_4 64}$

$=\log_4 k+\log_4 k\cdot\dfrac{1}{2}+\log_4 k\cdot\dfrac{1}{3}$

$=(1+\dfrac{1}{2}+\dfrac{1}{3})\log_4 k$

$=(\dfrac{6+3+2}{6})\log_4 k=\dfrac{11}{6}\log_4 k$,

$\therefore\ \log_4 k=\dfrac{6}{11}$

$P(B)=\dfrac{1}{2}\cdot\dfrac{6}{11}=\dfrac{6}{22}=\dfrac{3}{11}$

CHAPTER 13
VECTOR

13-1 VECTOR IN A PLANE

A vector is a quantity with magnitude and direction that is commonly represented as an ordered pair of real numbers in a plane.

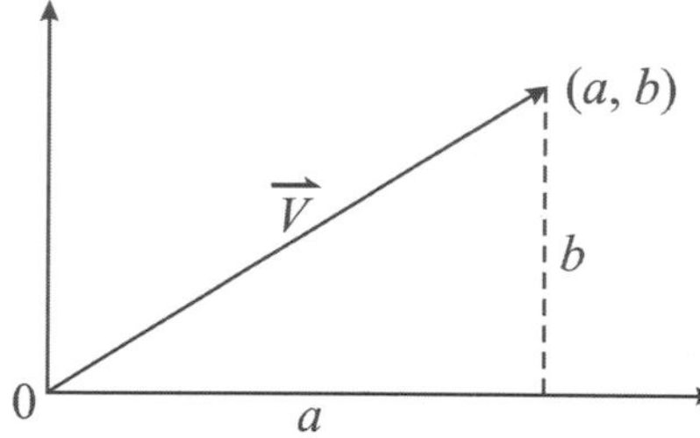

Above figure shows vector V which can be represented in one of the two ways below.

1. $\overrightarrow{V} = (a, b)$
2. $\overrightarrow{V} = (x\overrightarrow{i} + y\overrightarrow{j})$

   where $\overrightarrow{i}$ is the unit vector on x-axis

   where $\overrightarrow{j}$ is the unit vector on y-axis

- The magnitude (norm) of a vector is denoted by $|\overrightarrow{V}|$

  $|\overrightarrow{V}| = \sqrt{a^2 + b^2}$

- The unit vector of a vector is denoted by $\overrightarrow{u} = \dfrac{\overrightarrow{V}}{|\overrightarrow{V}|} = \dfrac{\overrightarrow{V}}{\sqrt{a^2 + b^2}}$

- The starting point of a vector can be any point on a plane that is not necessarily from the origin.

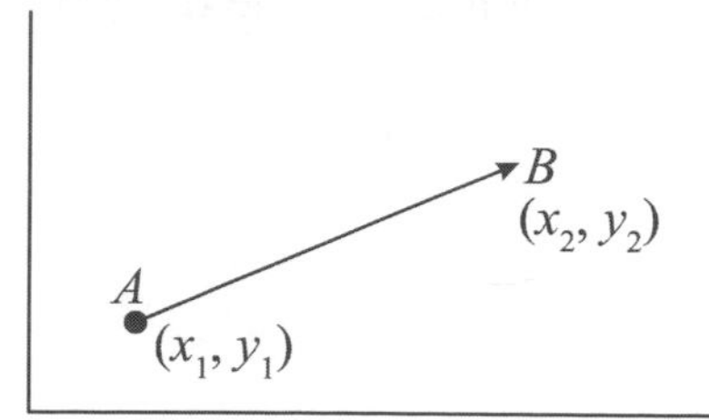

$\overrightarrow{AB} = (x_2 - x_1, y_2 - y_1)$

The norm of $\overrightarrow{AB}$ is $\sqrt{(x_2-x_1)^2+(y_2-y_1)^2}$

The unit vector $\overrightarrow{u}$ of $\overrightarrow{AB}=\dfrac{\overrightarrow{AB}}{|\overrightarrow{AB}|}=\dfrac{\overrightarrow{AB}}{\sqrt{(x_2-x_1)^2+(y_2-y_1)^2}}$

13-2 THE PROPERTIES OF VECTORS

A. Positive/Negative

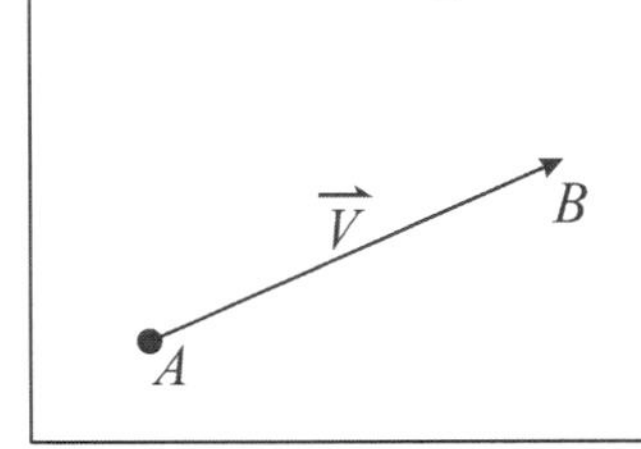

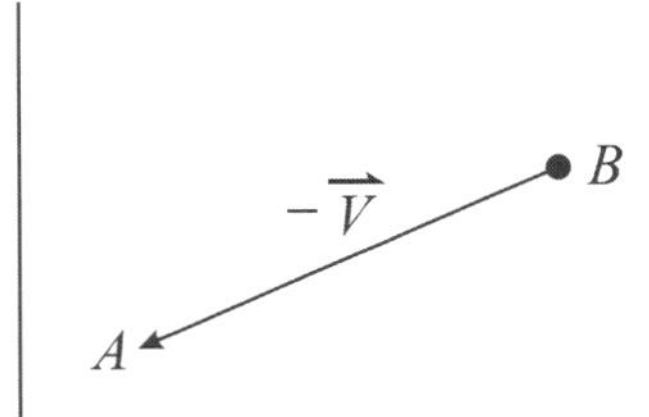

$\overrightarrow{V}=\overrightarrow{AB}$, $-\overrightarrow{V}=\overrightarrow{BA}$. If V is positive $\Rightarrow$ $-V$ is negative.

B. The Resultant of Vectors

Resultant vector is the result of adding two or more individual vectors that give the combined effect of the vectors involved.

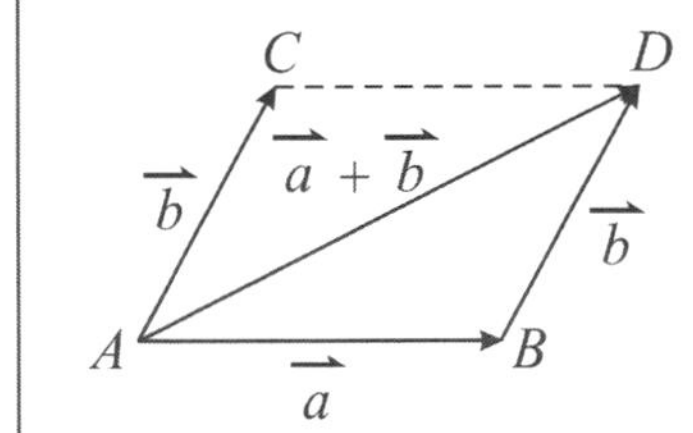

In the graph above, $\overrightarrow{AB}=\overrightarrow{a}$, $\overrightarrow{BD}=\overrightarrow{b}$, the resultant vector of $\overrightarrow{a}+\overrightarrow{b}$ is $\overrightarrow{AD}$

Note that when magnitude and direction are the same for 2 vectors, they are considered the same vector.

Therefore, one can use segments $\overline{AC}$ and $\overline{AB}$ to sketch a parallelogram, the diagonal is the resultant vector of $\overrightarrow{AB}+\overrightarrow{AC}$

Triangle Method

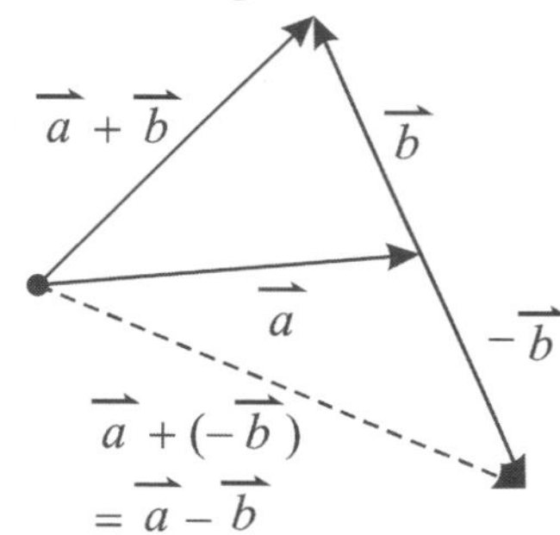

Parallelogram Method

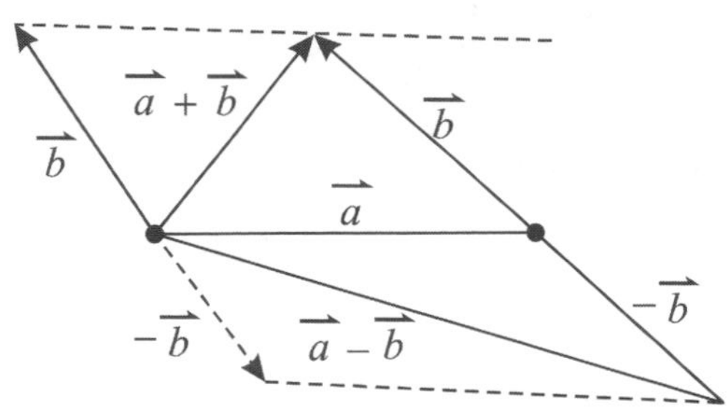

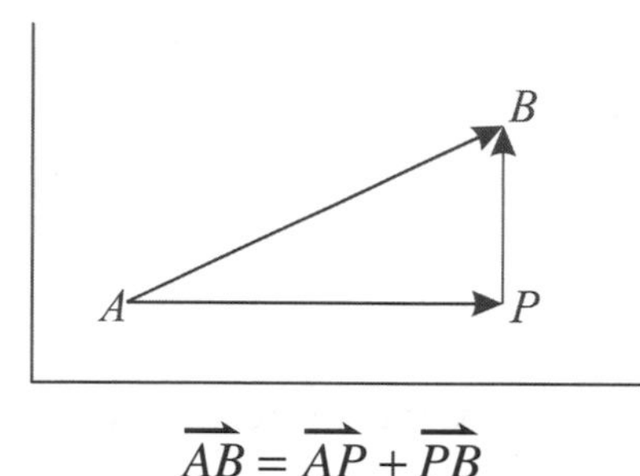

$\overrightarrow{AB}=\overrightarrow{AP}+\overrightarrow{PB}$

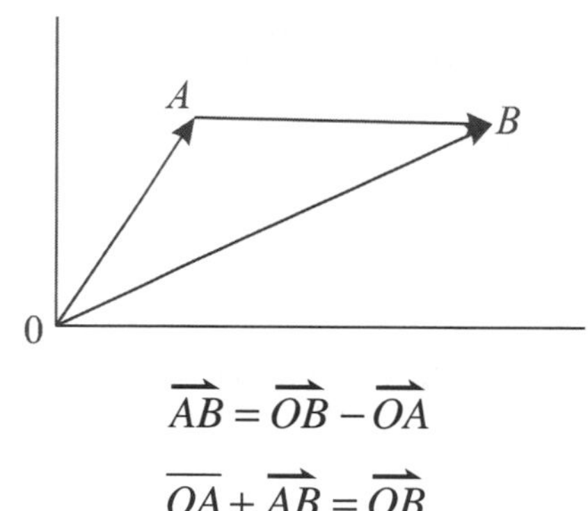

$\overrightarrow{AB}=\overrightarrow{OB}-\overrightarrow{OA}$

$\overline{OA}+\overrightarrow{AB}=\overrightarrow{OB}$

13-3 THE OPERATION OF VECTORS

Let $\overrightarrow{a}=(a_1,a_2)$, $\overrightarrow{b}=(b_1,b_2)$

A. Equal

$\overrightarrow{a}=\overrightarrow{b}$ (vector a equals vector b) $\Rightarrow$ $a_1=b_1$, $a_2=b_2$

B. Parallel

$\overrightarrow{a}\,//\,\overrightarrow{b}$ (parallel to each other) $\Rightarrow$ $\overrightarrow{a}=t\,\overrightarrow{b}$ (where t is a constant) t times the magnitude the direction is the same.

C. Addition/Subtraction

Let $\overrightarrow{a}=(a_1,a_2)$, $\overrightarrow{b}=(b_1,b_2)$

$\overrightarrow{a}+\overrightarrow{b}=(a_1+b_1,a_2+b_2)$

$\overrightarrow{a}-\overrightarrow{b}=(a_1-b_1,a_2-b_2)$

$-\overrightarrow{a}=(-a_1,-a_2)$

$k(\overrightarrow{a})=(ka_1,ka_2)$

$\vec{a}+(\vec{b}+\vec{c})=(\vec{a}+\vec{b})+\vec{c}$

$k(\vec{a}+\vec{b})=k\vec{a}+k\vec{b}$

D. Multiplication

$\vec{a}\cdot\vec{b}$ is defined as $\vec{a}\cdot\vec{b}=(a_1b_1+a_2b_2)$

The result of dot product is a scalar quantity not a vector

$\vec{a}\cdot\vec{b}=\vec{b}\cdot\vec{a}$

$\vec{a}\cdot(\vec{b}+\vec{c})=\vec{a}\cdot\vec{b}+\vec{a}\cdot\vec{c}$

$(k\vec{a})\cdot\vec{b}=k(\vec{a}\cdot\vec{b})$

◆Example 1:

Points $A(3, 2)$ and $B(-5, 4)$ are on the same plane, what is the representation of $\overrightarrow{AB}$ and its norm ?

Solution: $\overrightarrow{AB}=(-5-3, 4-2)=(-8, 2)$

$|\overrightarrow{AB}|=\sqrt{(-5-3)^2+(4-2)^2}=\sqrt{(-8)^2+2^2}=2\sqrt{17}$

◆Example 2:

If $\overrightarrow{AB}=(4, 2)$, and $A(1, 3)$, what are the coordinates of point B ?

Solution: Let $B(b_1, b_2)$

$\overrightarrow{AB}=(b_1-1, b_2-3)=(4, 2)$

$b_1-1=4 \Rightarrow b_1=5$

$b_2-3=2,\ b_2=5$

$B(b_1, b_2)=B(5, 5)$

◆Example 3:

Given $\vec{a}=(1, 4)$ and $\vec{b}=(3, -2)$

Find the resultant of $3\vec{a}+5\vec{b}$

Solution: $3\vec{a}=3(1, 4)=(3, 12)$, $5\vec{b}=5(3, -2)=(15, -10)$

$3\vec{a}+5\vec{b}=[3+15, 12+(-10)]=(18, 2)$

◆Example 4:

As the example above, find the unit vector of $\vec{a}$.

Solution: $\vec{u} = \dfrac{\vec{a}}{|\vec{a}|} = \dfrac{(1, 4)}{\sqrt{17}}$

Answer to Concept Check 13-01

$2\vec{a} = 2(2, 3) = (4, 6)$

$3\vec{b} = 3(4, 5) = (12, 15)$

$2\vec{a} + 3\vec{b} = (4, 6) + (12, 15) = (16, 21)$

Answer is (16, 21)

Answer to Concept Check 13-02

$\therefore$ If $A = (a_1, a_2)$, $B = (b_1, b_2)$

$\overrightarrow{AB} = (b_1 - a_1, b_2 - a_2)$

$\therefore \overrightarrow{AB} = (b_1 - 2, b_2 - 4) = (6, 3)$

$\therefore b_1 - 2 = 6,\ b_2 - 4 = 3$

$\therefore b_1 = 8,\ b_2 = 7$

The coordinates of point *B* are (8, 7)

Answer to Concept Check 13-03

$|\vec{a}| = \sqrt{3^2 + 2^2} = \sqrt{9 + 4} = \sqrt{13}$

$\vec{u} = \dfrac{\vec{a}}{|\vec{a}|} = \dfrac{(3, 2)}{\sqrt{13}}$

Answer is $\dfrac{(3, 2)}{\sqrt{13}}$

• More about dot product

The definition of 'dot product' or 'inner product' of 2 vectors can be defined by

$\vec{a} \cdot \vec{b} = |\vec{a}||\vec{b}|\cos\theta$, or $\vec{a} \cdot \vec{b} = a_1b_1 + a_2b_2$

(Where θ is the angle between $\vec{a}$ and $\vec{b}$)

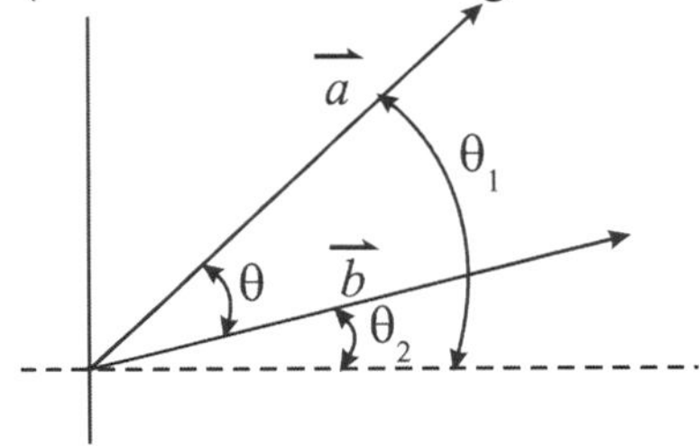

If $\vec{a} = (a_1, a_2)$, $\vec{b} = (b_1, b_2)$

$$\vec{a} \cdot \vec{b} = |\vec{a}||\vec{b}|\cos\theta = |\vec{a}||\vec{b}|\cos(\theta_1 - \theta_2)$$

$$= |\vec{a}||\vec{b}|(\cos\theta_1\cos\theta_2 + \sin\theta_1\sin\theta_2)$$

$$= |\vec{a}|\cos\theta_1|\vec{b}|\cos\theta_2 + |\vec{a}|\sin\theta_1|\vec{b}|\sin\theta_2$$

$$= a_1b_1 + a_2b_2$$

$|\vec{a}|\cos\theta_1 = a_1$, $|\vec{b}|\cos\theta_2 = b_1$, $|\vec{a}|\sin\theta_1 = a_2$, $|\vec{b}|\sin\theta_2 = b_2$

$\vec{a} \perp \vec{b} = \vec{a} \cdot \vec{b} = 0\ (\theta = 90°)$

$\vec{a} \cdot \vec{a} = |\vec{a}||\vec{a}|\cos 0° = |\vec{a}|^2$

$$|\vec{a} + \vec{b}|^2 = (\vec{a} + \vec{b}) \cdot (\vec{a} + \vec{b}) = \vec{a} \cdot \vec{a} + \vec{a} \cdot \vec{b} + \vec{a} \cdot \vec{b} + \vec{b} \cdot \vec{b}$$

$$= |\vec{a}|^2 + |\vec{b}|^2 + 2\vec{a} \cdot \vec{b}$$

13-4 ANGLE BETWEEN TWO VECTORS

If $\vec{a}$ and $\vec{b}$ are nonzero vectors,

$\vec{a} \cdot \vec{b} = |\vec{a}||\vec{b}|\cos\theta$

$$\therefore \cos\theta = \frac{\vec{a} \cdot \vec{b}}{|\vec{a}||\vec{b}|}$$

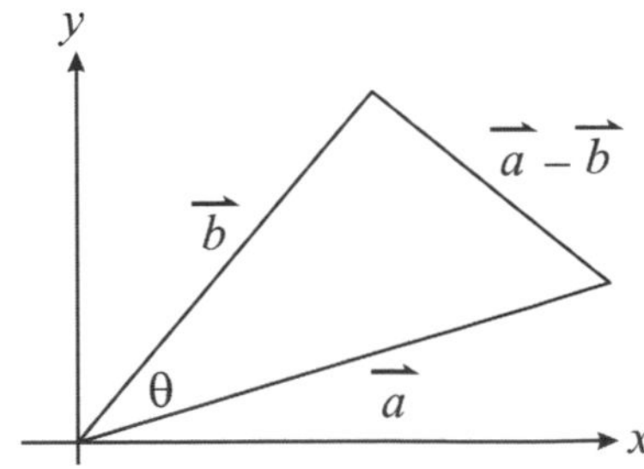

Note, if $\vec{a} \cdot \vec{b} = 0 = \cos\theta$ $\theta = 90°$ (Vector a and b are non-zero vector)

that means vector $\vec{a}$ is perpendicular to vector $\vec{b}$.

◆Examples:

Let $\vec{a}=(1,4)$ and $\vec{b}=(3,-2)$

1.Find the resultant of $\vec{a}+\vec{b}$.

The resultant of $\vec{a}+\vec{b}$ is $[1+3,4+(-2)]=(4,2)$

2.Find the norm (magnitude) of $\vec{b}$

The norm of $\vec{b}$ is $|\vec{b}|=\sqrt{(3)^2+(-2)^2}=\sqrt{9+4}=\sqrt{13}$

3.Are $\vec{a}$ and $\vec{b}$ perpendicular ?

$\vec{a}\cdot\vec{b}=1\cdot 3+4(-2)=3-8=-5$

Since $\vec{a}\cdot\vec{b}\neq 0$

$\vec{a}$ and $\vec{b}$ are not perpendicular.

4.Find the norm of $\vec{a}+\vec{b}$.

$\vec{a}+\vec{b}=(4,2)$ the norm is $\sqrt{4^2+2^2}=\sqrt{20}$

5.Find the resultant of $3\vec{a}+5\vec{b}$.

$3\vec{a}=3(1,4)=(3,12)$, $5\vec{b}=5(3,-2)=(15,-10)$

$3\vec{a}+5\vec{b}=[3+15,12+(-10)]=(18,2)$

6.Find the unit vector of $\vec{a}$.

$\vec{a}=(1,4)$, the unit vector is $\dfrac{\vec{a}}{|\vec{a}|}=\dfrac{(1,4)}{\sqrt{1^2+4^2}}=\dfrac{(1,4)}{\sqrt{17}}$

7. $\vec{a}\perp\vec{b}\Rightarrow\vec{a}\cdot\vec{b}=0$

($\vec{a}\cdot\vec{b}=|\vec{a}||\vec{b}|\cos\theta=0$, $\cos\theta=0$, $\theta=90°$)

8. $\vec{a}\cdot\vec{a}=|\vec{a}|^2$ ($|\vec{a}|\cdot|\vec{a}|\cos 0°=|\vec{a}|^2$)

9.$|3\vec{a}-2\vec{b}|^2=9|\vec{a}|^2+4|\vec{b}|^2-12\vec{a}\cdot\vec{b}=9\times17+4\times13-12\times(-5)=265$

$3\vec{a}-2\vec{b}=3(1,4)-2(3,-2)=(3,12)-(6,-4)=(-3,16)$

$|3\vec{a}-2\vec{b}|^2=(-3)^2+16^2=265$

13-5 VECTOR PROJECTION

The Projection of $\vec{a}$ onto $\vec{b}$ is $\dfrac{\vec{a}\cdot\vec{b}}{|\vec{b}|}\cdot\dfrac{\vec{b}}{|\vec{b}|}=\dfrac{\vec{a}\cdot\vec{b}}{|\vec{b}|^2}\vec{b}$

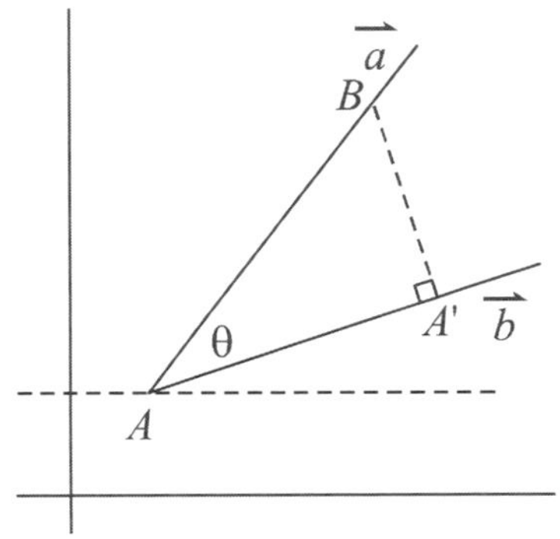

> **★Tip**
> The "scale projection" is a vector's length projected onto another vector (no direction). The vector projection is a scale projection with direction added to it.

- The scalar projection of $\overrightarrow{AB}$ onto $\vec{b}$ is AA'

$\cos\theta=\dfrac{AA'}{AB}$, $AA'=AB\cos\theta=|\vec{a}|\cos\theta$

$\because\ \vec{a}\cdot\vec{b}=|\vec{a}||\vec{b}|\cos\theta=|\vec{b}|AA'$

$\therefore\ AA'=\dfrac{\vec{a}\cdot\vec{b}}{|\vec{b}|}$

The unit vector of $\vec{b}$ is $\dfrac{\vec{b}}{|\vec{b}|}$

The vector projection for $\vec{a}$ onto $\vec{b}$ is $(AA')\vec{u}=(\dfrac{\vec{a}\cdot\vec{b}}{|\vec{b}|})\dfrac{\vec{b}}{|\vec{b}|}=\dfrac{\vec{a}\cdot\vec{b}}{|\vec{b}|^2}\vec{b}$

◆Example 1:

If $|\vec{a}|=3$, $|\vec{b}|=8$, the angle between $\vec{a}$ and $\vec{b}$ is 30°, what is the value of $|\vec{a}-2\vec{b}|$?

Solution: $\because\ \vec{a}\cdot\vec{b}=|\vec{a}||\vec{b}|\cos\theta=3\times8\times\cos30°=24\times\dfrac{\sqrt{3}}{2}=12\sqrt{3}$

$$
\begin{aligned}
|\vec{a}-2\vec{b}|^2&=(\vec{a}-2\vec{b})\cdot(\vec{a}-2\vec{b})\\
&=\vec{a}\cdot\vec{a}-2\vec{a}\cdot\vec{b}-2\vec{a}\cdot\vec{b}+4\vec{b}\cdot\vec{b}\\
&=\vec{a}\cdot\vec{a}-4\vec{a}\cdot\vec{b}+4\vec{b}\cdot\vec{b}\\
&=|\vec{a}|^2-4\vec{a}\cdot\vec{b}+4|\vec{b}|^2
\end{aligned}
$$

$$= 9 - 4 \times 12\sqrt{3} + 256$$
$$= 265 - 48\sqrt{3} = 265 - 83.14 = 181.86$$
$$|\vec{a} - 2\vec{b}| = \sqrt{181.86} = 13.49$$

◆Example 2:

Calculate the vector projection of $\vec{a} = (3, -2)$ on to vector $\vec{b} = (-2, 3)$

Solution: projection $\vec{p} = \dfrac{\vec{a} \cdot \vec{b}}{|\vec{b}|^2}\vec{b} = \dfrac{3 \cdot (-2) + (-2) \cdot (3)}{(-2)^2 + 3^2}(-2, 3)$

$$= \frac{-12}{13}(-2, 3) = (\frac{24}{13}, -\frac{36}{13})$$

◆Example 3:

Calculate the scalar projection of $\vec{a} = (1, 5)$ on vector $\vec{b} = (2, 3)$

$$\frac{\vec{a} \cdot \vec{b}}{|\vec{b}|} = \frac{(1)(2) + (5)(3)}{\sqrt{2^2 + 3^2}} = \frac{2 + 15}{\sqrt{4 + 9}} = \frac{17}{\sqrt{13}} = \frac{17\sqrt{13}}{13}$$

Answer to Concept Check 13-04

$|\vec{a}| = 2$, $|\vec{b}| = 3$

$\because\ \vec{a} \cdot \vec{b} = |\vec{a}||\vec{b}|\cos 60° = 2 \times 3 \times \dfrac{1}{2} = 3$

$$|\vec{a} + 3\vec{b}|^2 = (\vec{a} + 3\vec{b}) \cdot (\vec{a} + 3\vec{b})$$
$$= \vec{a} \cdot \vec{a} + 3\vec{a} \cdot \vec{b} + 3\vec{a} \cdot \vec{b} + 9\vec{b} \cdot \vec{b}$$
$$= |\vec{a}|^2 + 6\vec{a} \cdot \vec{b} + 9|\vec{b}|^2$$
$$= 4 + 6 \times 3 + 81$$
$$= 103$$

$\therefore\ |\vec{a} + 3\vec{b}| = \sqrt{103} = 10.15$

Answer is 10.15

◆Example 4:

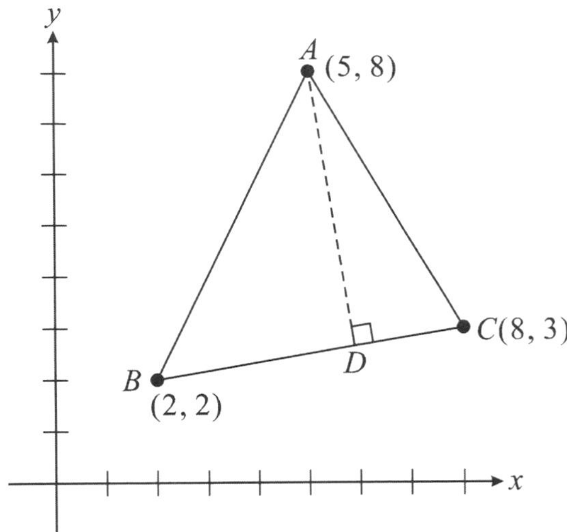

$\triangle ABC$ shown on above,

Find its normal projection from $\overrightarrow{BA}$ on the vector $\overrightarrow{BC}$

$$\overrightarrow{BD}=\frac{\overrightarrow{BA}\cdot\overrightarrow{BC}}{|\overrightarrow{BC}|^2}\overrightarrow{BC}=\frac{[(5-2),(8-2)]\cdot[(8-2),(3-2)]}{|\overrightarrow{BC}|^2}\overrightarrow{BC}=\frac{(3,6)\cdot(6,1)}{6^2+1^2}\overrightarrow{BC}$$

$$=\frac{18+6}{37}(6,1)=\frac{24}{37}(6,1)$$

$$\frac{\overrightarrow{BA}\cdot\overrightarrow{BC}}{|\overrightarrow{BC}|}=\frac{24}{\sqrt{37}}$$

The length of normal projection $=\dfrac{24\sqrt{37}}{37}$

Answer to Concept Check 13-05

$\overrightarrow{AB}=[(3)-(-2),4-1]=(5,3)$

$\overrightarrow{AC}=[(2)-(-2),0-1]=(4,-1)$

$$\overrightarrow{AD}=\frac{\overrightarrow{AB}\cdot\overrightarrow{AC}}{|\overrightarrow{AC}|^2}\overrightarrow{AC}=\frac{(5,3)\cdot(4,-1)}{|\sqrt{4^2+(-1)^2}|^2}(4,-1)=\frac{20-3}{|\sqrt{17}|^2}(4,-1)=\frac{17}{17}(4,-1)=(4,-1)$$

The length of $\overrightarrow{AD}=\sqrt{4^2+(-1)^2}=\sqrt{17}$

Answer is $\sqrt{17}$

13-6 THE AREA OF TRIANGLES

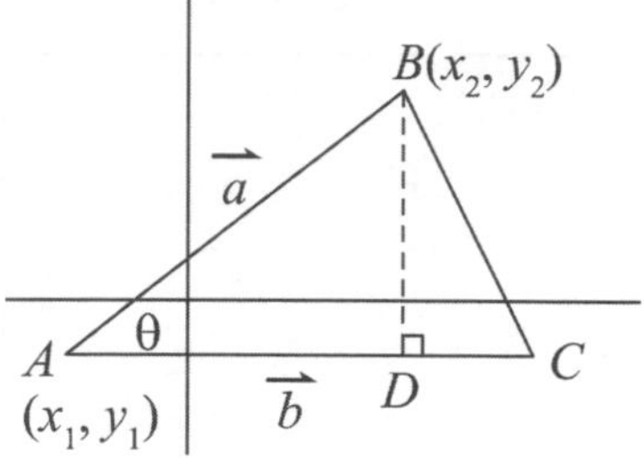

$$\triangle ABC = \frac{1}{2}(\overline{BD} \times \overline{AC})$$

$$\overline{BD} = |\vec{a}|\sin\theta$$

$$\overline{AC} = |\vec{b}|$$

$$\triangle ABC = \frac{1}{2}|\vec{a}||\vec{b}|\sin\theta = \frac{1}{2}|\vec{a}||\vec{b}|\cdot(\sqrt{1-\cos^2\theta}) = \frac{1}{2}\sqrt{|\vec{a}|^2|\vec{b}|^2 - |\vec{a}|^2|\vec{b}|^2\cos^2\theta}$$

$$= \frac{1}{2}\sqrt{|\vec{a}|^2|\vec{b}|^2 - (\vec{a}\cdot\vec{b})^2}$$

◆Example:

What is the area of triangle ABC shown on the right ?

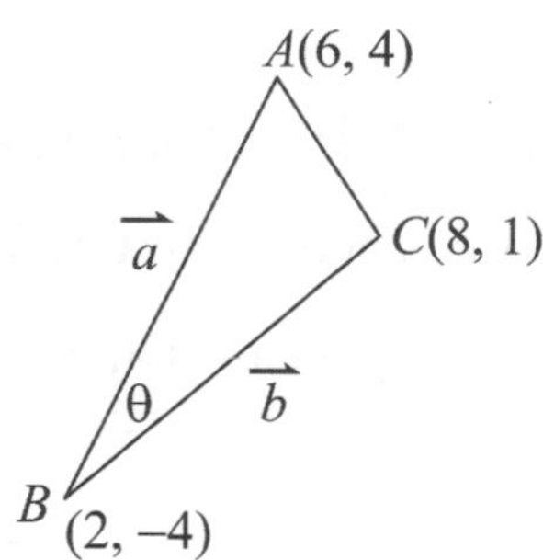

Solution: $\vec{a} = [6-2, 4-(-4)] = (4, 8)$

$\vec{b} = [8-2, 1-(-4)] = (6, 5)$

$\vec{a}\cdot\vec{b} = (4\times6+8\times5) = (24+40) = 64$

$|\vec{a}|^2 = 4^2 + 8^2 = 80$

$|\vec{b}|^2 = 6^2 + 5^2 = 61$

$$\triangle ABC = \frac{1}{2}\sqrt{(80)\times(61)-64^2} = \frac{1}{2}\sqrt{4880-4096} = \frac{1}{2}\sqrt{784} = \frac{1}{2}\cdot 28 = 14$$

Answer to Concept Check 13-06

$$\triangle ABC = \frac{1}{2}|\overrightarrow{AB}||\overrightarrow{AC}|\sin\theta$$

$$= \frac{1}{2}|\overrightarrow{AB}||\overrightarrow{AC}| \cdot \sqrt{1-\cos^2\theta}$$

$$= \frac{1}{2}\sqrt{|\overrightarrow{AB}|^2|\overrightarrow{AC}|^2 - (\overrightarrow{AB}\cdot\overrightarrow{AC})^2}$$

Since $\overrightarrow{AB} = [6-(-3), 5-2] = (9, 3)$

$\overrightarrow{AC} = [(8-(-3), (-3)-2] = (11, -5)$

$\overrightarrow{AB}\cdot\overrightarrow{AC} = 99-15 = 84$

$|\overrightarrow{AB}|^2 = 81+9 = 90$, $|\overrightarrow{AC}|^2 = 121+25 = 146$

$\triangle ABC = \frac{1}{2}\sqrt{(90)(146)-(84)^2} = \frac{1}{2}\sqrt{13140-7056} = \frac{1}{2}\sqrt{6084} = 39$

Answer is 39

13-7 VECTORS IN THREE-DIMENSIONAL SPACE

All rules and properties of two-dimensional vectors also apply to three-dimensional vectors.

A three dimensional vector is denoted by $\overrightarrow{a} = (a_1, a_2, a_3)$ or $\overrightarrow{a} = x\overrightarrow{i} + y\overrightarrow{j} + z\overrightarrow{k}$ (base form)

Where $\overrightarrow{i}$, $\overrightarrow{j}$, and $\overrightarrow{k}$ are unit vectors in x, y, and z axes respectively.

Let $\overrightarrow{a} = (a_1, a_2, a_3) = (a_x, a_y, a_z)$, $\overrightarrow{b} = (b_1, b_2, b_3) = (b_x, b_y, b_z)$

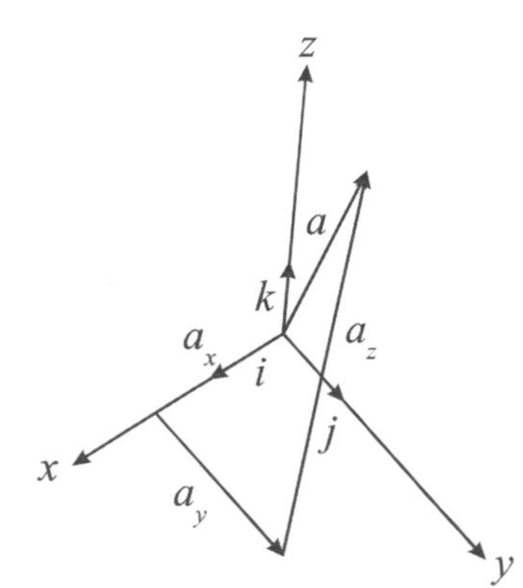

1. Distance from A to B

   $AB = \sqrt{(a_1-b_1)^2 + (a_2-b_2)^2 + (a_3-b_3)^2}$

2. The magnitude of $\overrightarrow{a}$

   $|\overrightarrow{a}| = \sqrt{a_1^2 + a_2^2 + a_3^2}$

3. The direction angles are α, β, γ

   then $\cos\alpha = \dfrac{a_1}{\sqrt{a_1^2 + a_2^2 + a_3^2}}$

$$\cos\beta = \frac{a_2}{\sqrt{a_1^2 + a_2^2 + a_3^2}}$$

$$\cos\gamma = \frac{a_3}{\sqrt{a_1^2 + a_2^2 + a_3^2}}$$

$$\cos^2\alpha + \cos^2\beta + \cos^2\gamma = 1$$

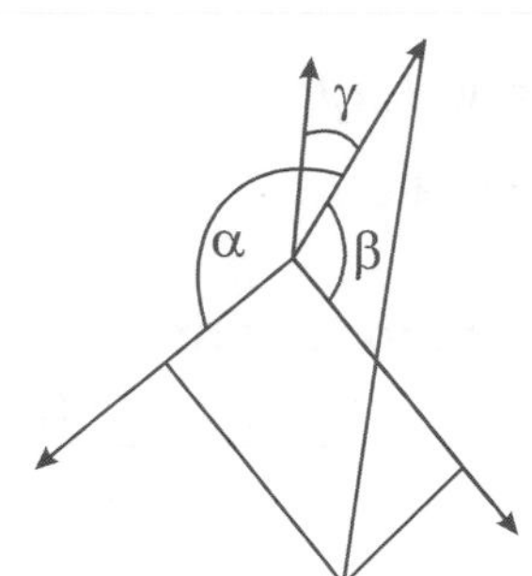

13-8 LINES IN THREE-DIMENSIONAL SPACE

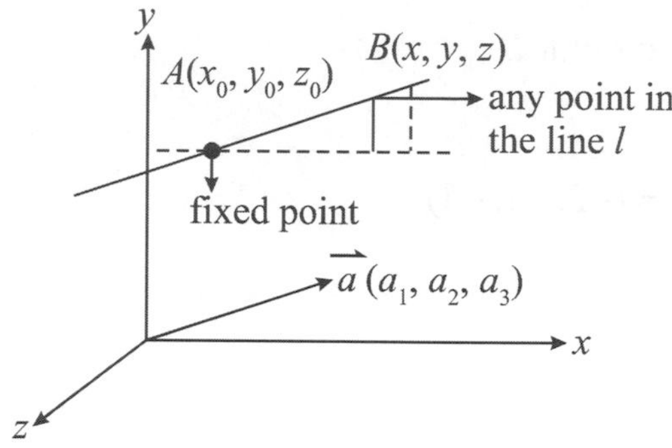

$A(x_0, y_0, z_0)$ is a fixed point on line l, $B(x, y, z)$ any arbitrary point on line l. (the unit vector in the same direction)

1. $\vec{a}$ is the direction vector of line ℓ
2. The x component of $|\vec{a}|$ is a_1, y component of $|\vec{a}|$ is a_2, and the z component of $|\vec{a}|$ in a_3.
3. $\frac{x - x_0}{a_1}$ is equivalent to $\frac{y - y_0}{a_2}$ and $\frac{z - z_0}{a_3}$

- The equations of straight line in 3-dimension are:

  Since p_0, p_1 and p are collinear. Their direction numbers are proportional.

$$\frac{x - x_0}{x_1 - x_0} = \frac{y - y_0}{y_1 - y_0} = \frac{z - z_0}{z_1 - z_0} = t$$

$$x = x_0 + (x_1 - x_0)t = x_0 + a_1 t$$

$$y = y_0 + (y_1 - y_0)t = y_0 + a_2 t$$

$$z = z_0 + (z_1 - z_0)t = z_0 + a_3 t$$

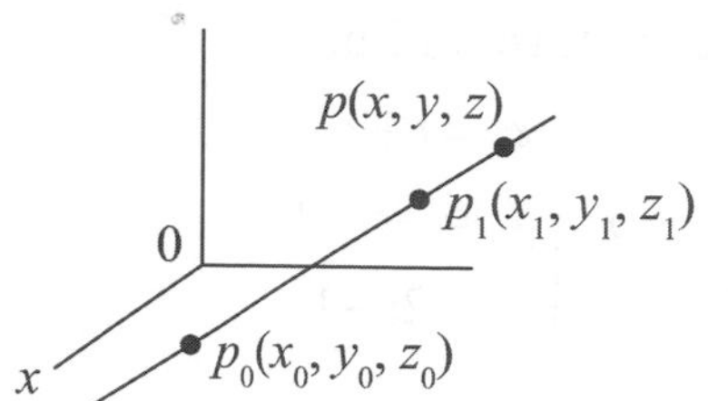

1. parametric form

$$x = x_0 + a_1 t$$

$y = y_0 + a_2 t$

$z = z_0 + a_3 t$

2. Proportional form

$$\frac{x - x_0}{a_1} = \frac{y - y_0}{a_2} = \frac{z - z_0}{a_3}$$

◆Example 1:

Find the equation of a straight line passing point A and point B.

Point $A(-1, 2, 3)$ and point $B(-3, 1, 2)$

1. Parametric form: For $\overrightarrow{AB} = (-3-(-1), 1-2, 2-3) = (-2, -1, -1)$

The equation is $x = x_0 + a_1 t = -1 - 2t$

$y = y_0 + a_2 t = 2 - t$

$z = z_0 + a_3 t = 3 - t$

2. Proportional form

$$\frac{x+1}{-2} = \frac{y-(2)}{-1} = \frac{z-(3)}{-1}$$

◆Example 2:

Find the equation of line ℓ_1 which passes through point $(1, 2, 1)$ and is perpendicular to line $l_2 : \frac{x-1}{1} = \frac{y-2}{-1} = \frac{z+1}{2}$.

Solution:

1. Find the coordinates of point B

(intersection point ⇒ perpendicular point)

The parametric form of line l_2:

$$\begin{cases} x = t+1 \\ y = -t+2 \\ z = 2t-1 \end{cases} \text{ so } \begin{cases} x_0 = t+1 \\ y_0 = -t+2 \\ z_0 = 2t-1 \end{cases}$$

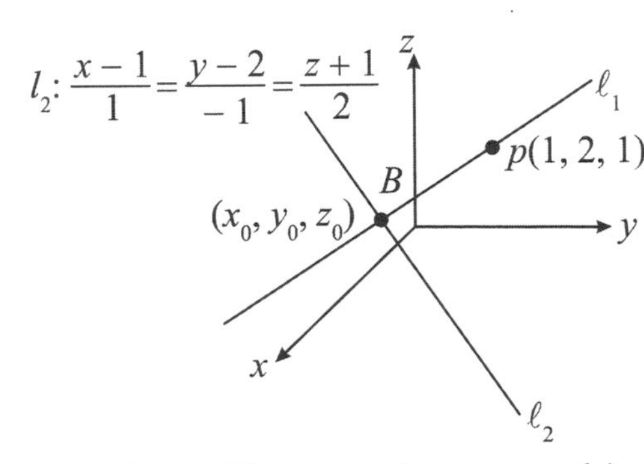

(Note: Figure not drawn to scale)

2. Find $\overrightarrow{PB}$

$\overrightarrow{PB} = (x_0 - x_1, y_0 - y_1, z_0 - z_1) \Rightarrow (t+1-1, -t+2-2, 2t-1-1) = (t, -t, 2t-2)$

3. Find t

Since $\overrightarrow{PB} \perp \overrightarrow{l_2}$, $\overrightarrow{PB} \cdot \overrightarrow{l_2} = 0 \Rightarrow (t \cdot 1 + (-t) \cdot (-1) + (2t-2) \cdot 2)$

$\Rightarrow\ (t+t+4t-4)=0$

$\therefore\ t=\dfrac{2}{3}$

The coordinates of point B are $\Rightarrow\ (\dfrac{2}{3}+1, -\dfrac{2}{3}+2, \dfrac{4}{3}-1)$

4. Find the equation of $\overrightarrow{\ell_1}$

Since the vector number of $\overrightarrow{\ell_1}$ are $(t, -t, 2t-2)\ \Rightarrow\ (\dfrac{2}{3}, -\dfrac{2}{3}, -\dfrac{2}{3})$

The equation of $l_1 : \dfrac{x-1}{\frac{2}{3}} = \dfrac{y-2}{\frac{-2}{3}} = \dfrac{z-1}{-\frac{2}{3}}$

Answer to Concept Check 13-07

$D=\sqrt{(-1-3)^2+(3+1)^2+(2-2)^2}$

$=\sqrt{16+16+0}$

$=\sqrt{32}$

$=4\sqrt{2}$

Answer to Concept Check 13-08

$A(2,1,3)$, $B(3,-2,1)$

$\overrightarrow{AB}=(3-2,-2-1,1-3)=(1,-3,-2)$ (direction of the vector)

A. The equation of line ℓ in parametric form

$x=2+t$

$y=1+(-3)t=1-3t$

$z=3+(-2)t=3-2t$

B. The equation of line ℓ in proportional form $\dfrac{x-2}{1}=\dfrac{y-1}{-3}=\dfrac{z-3}{-2}$

13-9 THE DISTANCE FROM A POINT TO A LINE

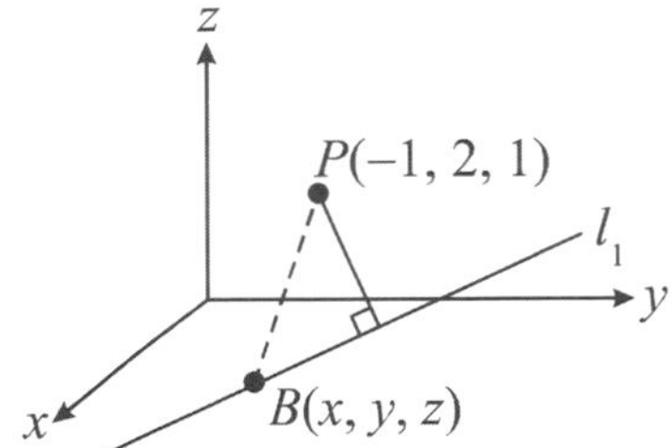

Note: Figure not drawn to scale

There are two methods to find the shortest distance from point p to line l.

Method 1. Let point B (x, y, z) be an arbitrary point on line l.

Set point $P(-1, 2, 1)$ in space, and the function of line $\ell_1 : \dfrac{x-1}{2} = \dfrac{y+2}{1} = \dfrac{z-1}{-1}$

Since point B lies on ℓ_1, so its coordinates must be $x = 2t+1$, $y = t-2$, $z = -t+1$

$$\therefore\ |\overrightarrow{PB}|^2 = (2t+1+1)^2 + [(t-2)-2]^2 + [-t+1-1]^2$$
$$= (2t+2)^2 + (t-4)^2 + (-t)^2$$
$$= 4t^2 + 8t + 4 + t^2 - 8t + 16 + t^2$$
$$= 6t^2 + 20$$

$PB = \sqrt{6t^2 + 20}$

when $t = 0$, it is the shortert distance $PB = \sqrt{20}$

Method 2. Set point $P(-1, 2, 1)$ and line $\ell_1 : \dfrac{x-1}{2} = \dfrac{y+2}{1} = \dfrac{z-1}{-1}$ point B is any point that lies on the line ℓ_1

So, the coordinates of point $B(x, y, z)$ must be $x = 2t+1$, $y = t-2$, $z = -t+1$

$(x = x_0 + a_1 t,\ y = y_0 + a_2 t,\ z = z_0 + a_3 t)$

$\overrightarrow{PB} = [(2t+1)+1, (t-2)-2, (-t+1)-1] = (2t+2, t-4, -t)$

The vector direction of ℓ_1, $\overrightarrow{a}(2, 1, -1)$

If point B is the perpendicular point then

$$\overrightarrow{PB} \cdot \overrightarrow{a} = 0 = [(2t+2) \cdot 2 + (t-4) \cdot 1 + (-t) \cdot (-1)]$$
$$= (4t + 4 + t - 4 + t) = 6t$$
$$= 0$$

$t = 0$

So, the coordinates of point $B:(1,-2,1)$

$\overline{PB}^2=[1-(-1)]^2+[(-2)-2]^2+(1-1)^2=4+16=20$

$\therefore\ PB=\sqrt{20}$

13-10 THE DISTANCE BETWEEN TWO PARALLEL LINES

Below examples show how to calculate the distance between two parallel lines utilizing parametric and proportional forms of line equation.

◆Example:

Line 1: $\dfrac{x+1}{2}=\dfrac{y+2}{1}=\dfrac{z-1}{1}$

Line 2: $\dfrac{x+2}{2}=\dfrac{y-1}{1}=\dfrac{z+1}{1}$

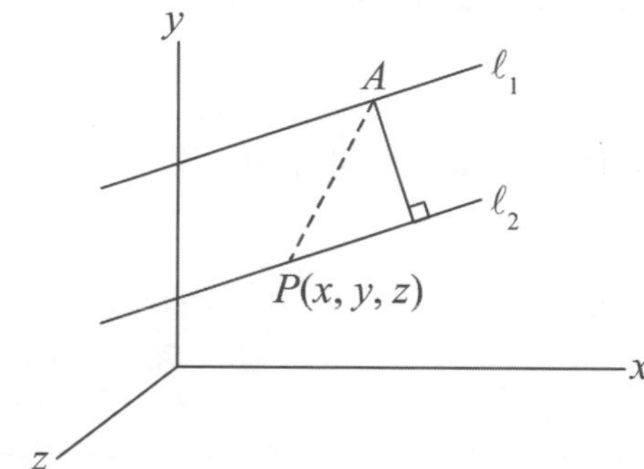

1. We find that point $A(3, 0, 3)$ lies on line 1, because $A(3, 0, 3)$ satisfies the equation of line 1.

   $\dfrac{3+1}{2}=\dfrac{0+2}{1}=\dfrac{3-1}{1}$

2. Let $p(x, y, z)$ be an arbitrary point on l_2, then it satisfies the equation of l_2 and p can be represented by

   $x=2t-2$

   $y=t+1$

   $z=t-1$

   $\therefore\ \overrightarrow{AP}=(2t-2-3, t+1-0, t-1-3)=(2t-5, t+1, t-4)$

   $\overrightarrow{AP}\cdot\overrightarrow{\ell_2}=(4t-10+t+1+t-4)=6t-13=0,\ t=\dfrac{13}{6}$

When $t=\frac{13}{6}$, $\overrightarrow{AP}\perp\overrightarrow{\ell_2}$

The perpendicular point at $(2t-2,t+1,t-1)=(\frac{13}{6}\times2-2,\frac{13}{6}+1,\frac{13}{6}-1)=(\frac{7}{3},\frac{19}{6},\frac{7}{6})$

The shortest distance is

$$D=\sqrt{(3-\frac{7}{3})^2+(-\frac{19}{6})^2+(3-\frac{7}{6})^2}=\sqrt{(\frac{2}{3})^2+(\frac{19}{6})^2+(\frac{11}{6})^2}=3.72$$

Answer to Concept Check 13-09

An arbitrary point B is on the line, then its coordinates are:

$x=2t-1$

$y=-2t+1$

$z=t-2$

$\therefore\ \overrightarrow{PB}=[(2t-1)-1,(-2t+1)-1,(t-2)-2]=(2t-2,-2t,t-4)$

The direction vector of line ℓ is $(2,-2,1)$

$\overrightarrow{PB}\cdot\overrightarrow{\ell}=(2t-2)\times2+(-2t)\times(-2)+(t-4)\times1=4t-4+4t+t-4=9t-8$

$\because\ \overrightarrow{PB}\perp\overrightarrow{\ell}\quad\therefore\ 9t-8=0,\ t=\frac{8}{9}$

So, the perpendicular point at $x=2\times\frac{8}{9}-1=\frac{7}{9}$

$y=-2\times\frac{8}{9}+1=\frac{-7}{9}$

$z=\frac{8}{9}-2=-\frac{10}{9}$

The shortest distance between point $P(1,1,2)$ and point B is

$$D=\sqrt{(1-\frac{7}{9})^2+(1+\frac{7}{9})^2+(2+\frac{10}{9})^2}=\sqrt{\frac{2^2}{9^2}+\frac{16^2}{9^2}+\frac{28^2}{9^2}}=\frac{32.37}{9}=3.60$$

Answer is 3.60

13-11 THE EQUATION OF A PLANE

• The normal vector of a plane

The normal vector $\vec{n}$, or just "normal", to a plane E is a vector that is perpendicular to its surface at a given point.

• The standard form of the equation of a plane

E: $ax+by+cz+d=0$, where a, b, c are the value of normal vector $\vec{n}(a,b,c)$, and d is a constant.

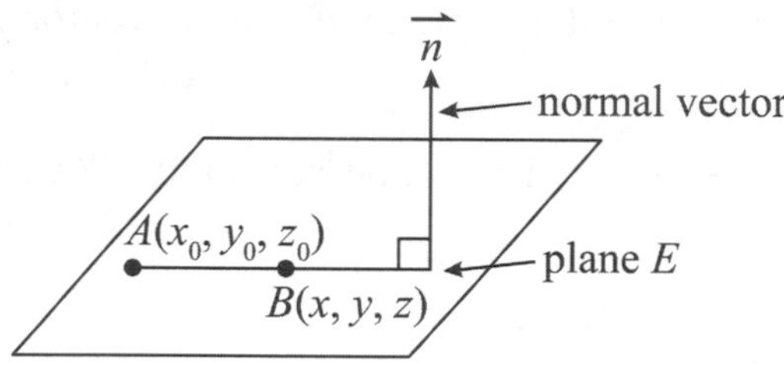

1. Set a fixed point $A(x_0, y_0, z_0)$ and an arbitrary point $B(x, y, z)$ on plane E. $\vec{n}$ is the normal vector of plane E.

   then $\overrightarrow{AB}=(x-x_0,\ y-y_0,\ z-z_0)$

2. Since $\vec{n}$ is normal vector, $\vec{n}\cdot\overrightarrow{AB}=0$

$$\begin{aligned}\vec{n}\cdot\overrightarrow{AB}&=[a(x-x_0)+b(y-y_0)+c(z-z_0)]\\&=ax-ax_0+by-by_0+cz-cz_0\\&=ax+by+cz-(ax_0+by_0+cz_0)\\&=ax+by+cz+d=0\end{aligned}$$

(Since $A(x_0, y_0, z_0)$ lies on plane E, so $ax_0+by_0+cz_0=-d$)

For instance, if the normal vector of plane E is $\vec{n}=(2,3,1)$ and a point $(1,-1,2)$ is on plane E, then the equation of plane E is

E: $2x+3y+z-(2-3+2)=2x+3y+z-1=0$

Answer to Concept Check 13-10

E: $x+2y+3z-(1+(-2)+6)=x+2y+3z-5=0$

13-12 THE NORMAL PROJECTION FROM A POINT TO A PLANE

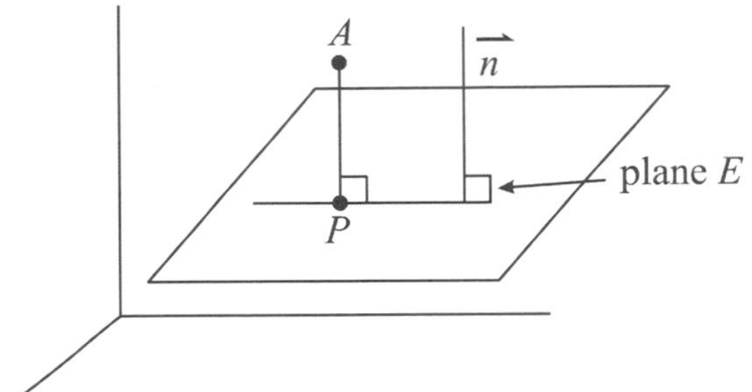

The normal projection of point A on plane E, is a point on plane E where vector AP is perpendicular to plane E.

For instance, set point $A(2, 3, 4)$ and plane E: $x + y - 2z - 1 = 0$, and set point $P(x_1, y_1, z_1)$ as the normal projection of point A.

then $\overrightarrow{AP} = (x_1 - 2, y_1 - 3, z_1 - 4)$, $\overrightarrow{n} = (1, 1, -2)$, since $\overrightarrow{AP} \parallel \overrightarrow{n}$.

Therefore, $\dfrac{x_1 - 2}{1} = \dfrac{y_1 - 3}{1} = \dfrac{z_1 - 4}{-2} = t$

$\therefore\ x_1 = t + 2,\ y_1 = t + 3,\ z_1 = -2t + 4$

Since $P(x_1, y_1, z_1)$ is on the plane E: $x + y - 2z = 1$

$(t + 2) + (t + 3) - 2(-2t + 4) - 1 = 0$

$\therefore\ 6t - 4 = 0 \quad t = \dfrac{4}{6} = \dfrac{2}{3}$

Substitute $x_1 = \dfrac{2}{3} + 2 = \dfrac{8}{3}$, $y_1 = \dfrac{2}{3} + 3 = \dfrac{11}{3}$, $z_1 = (-2)(\dfrac{2}{3}) + 4 = -\dfrac{4}{3} + 4 = \dfrac{8}{3}$

$P(\dfrac{8}{3}, \dfrac{11}{3}, \dfrac{8}{3})$

Answer to Concept Check 13-11

Let point $P(x_1, y_1, z_1)$ be the normal projection of point $A(2,-1,2)$ on plane. Based on standard equation of plane E, its normal vector is, $\overrightarrow{n}=(2,1,-1)$

$\overrightarrow{AP}=(x_1-2,\ y_1+1,\ z_1-2)$

Since $\overrightarrow{AP} \parallel \overrightarrow{n}$, $\dfrac{x_1-2}{2}=\dfrac{y_1+1}{1}=\dfrac{z_1-2}{-1}=t$

$x_1=2t+2,\ \ y_1=t-1,\ \ z_1=-t+2$

Since point P is a point on the plane E

$$\begin{aligned}2x+y-z+2&=2(2t+2)+(t-1)-(-t+2)+2\\&=4t+4+t-1+t-2+2\\&=6t+3=0\end{aligned}$$

$6t=-3$

$\therefore\ t=-\dfrac{1}{2}$

Plug t back into $x_1=2t+2,\ \ y_1=t-1,\ \ z_1=-t+2$

$x_1=2(-\dfrac{1}{2})+2=1$

$y_1=(-\dfrac{1}{2})-1=-\dfrac{3}{2}$

$z_1=-(-\dfrac{1}{2})+2=\dfrac{5}{2}$

The coordinates of point $P:(1,-\dfrac{3}{2},\dfrac{5}{2})$

13-13 THE DISTANCE FROM A POINT TO A PLANE

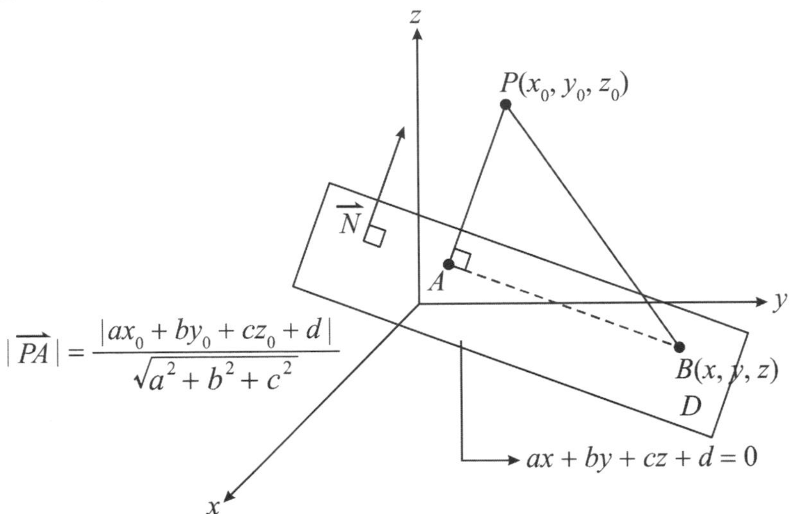

1. In the graph above the equation of plane D is $ax+by+cz+d=0$
2. Set an outside point $P(x_0, y_0, z_0)$
3. The normal vector of plane D is $\vec{N}=(a, b, c)$
4. $B(x, y, z)$ represents an arbitrary point on the plane D.

Then $\overrightarrow{BP}=(x_0-x, y_0-y, z_0-z)$

The distance from point P to plane D is the length of the projection of $\overrightarrow{BP}$ on the normal vector $\vec{N}$ of plane D.

Distance $= |\text{projn}\,\overrightarrow{BP}|$ (norm of the projection)

$$= \frac{\overrightarrow{BP} \cdot \vec{N}}{|\vec{N}|} = \frac{|(x_0-x)\cdot a+(y_0-y)\cdot b+(z_0-z)\cdot c|}{\sqrt{a^2+b^2+c^2}}$$

$$= \frac{|ax_0+by_0+cz_0-ax-by-zc|}{\sqrt{a^2+b^2+c^2}} \quad (ax+by+zc=-d)$$

$$= \frac{|ax_0+by_0+cz_0+d|}{\sqrt{a^2+b^2+c^2}}$$

Answer to Concept Check 13-12

$$D=\frac{|ax_0+by_0+cz_0+d|}{\sqrt{a^2+b^2+c^2}}$$

$a=1,\ b=2,\ c=-1,\ d=3$

$x_0=1,\ y_0=3,\ z_0=-2$

$$\therefore\ D=\frac{|1+6+2+3|}{\sqrt{1^2+2^2+(-1)^2}}=\frac{12}{\sqrt{6}}=4.90$$

Answer is 4.90

13-14 THE DISTANCE BETWEEN TWO PARALLEL PLANES

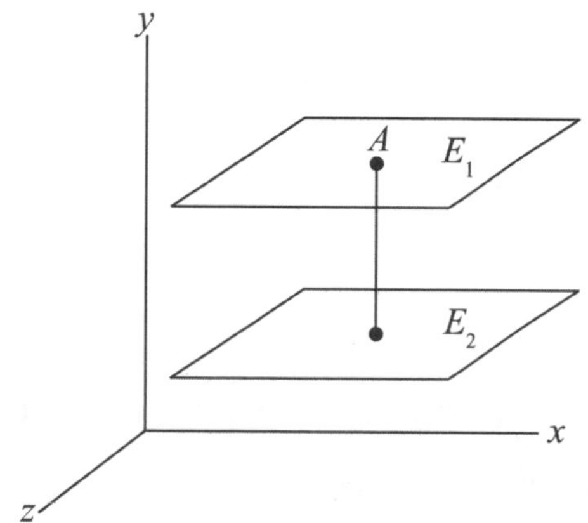

In the graph above, $E_1 \parallel E_2$,

$E_1: ax+by+cz+d_1=0$

$E_2: ax+by+cz+d_2=0$

Point $A(x_0, y_0, z_0)$ is a point on plane E_1

The distance from point A to plane E_2 is the distance from plane E_1 to plane E_2.

$$D=\frac{|ax_0+by_0+cz_0+d_2|}{\sqrt{a^2+b^2+c^2}}$$

point A is also on the plane E_1, $ax_0+by_0+cz_0=-d_1$

$$\therefore\ D=\frac{|d_2-d_1|}{\sqrt{a^2+b^2+c^2}}$$

Answer to Concept Check 13-13

$E_1 = x-2y+z+2=0$, $E_2 = 2x-4y+2z+8=0$; $E_2 = x-2y+z+4=0$

$a=1$, $b=-2$, $c=1$, $d_1=2$, $d_2=4$

$$D=\frac{|4-2|}{\sqrt{1^2+(-2)^2+1^2}}=\frac{2}{\sqrt{6}}=\frac{2\sqrt{6}}{6}=\frac{\sqrt{6}}{3}$$

Answer is $\frac{\sqrt{6}}{3}$

13-15 THE ANGLES BETWEEN TWO PLANES

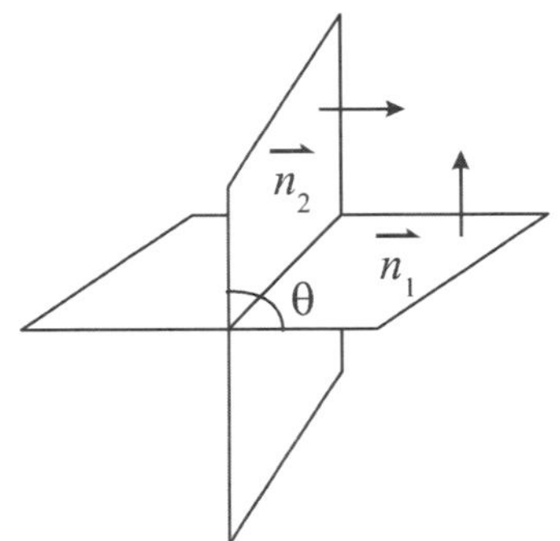

The angle between two planes E_1 and E_2 is the same as the angle between two normal vectors of plane E_1 and plane E_2 respectively.

$E_1 : a_1x+b_1y+c_1z+d_1=0$, $\overrightarrow{n_1}=(a_1, b_1, c_1)$

$E_2 : a_2x+b_2y+c_2z+d_2=0$, $\overrightarrow{n_2}=(a_2, b_2, c_2)$

$$\cos\theta=\pm\frac{\overrightarrow{n_1}\cdot\overrightarrow{n_2}}{|\overrightarrow{n_1}||\overrightarrow{n_2}|}=\pm\frac{a_1a_2+b_1b_2+c_1c_2}{\sqrt{a_1^2+b_1^2+c_1^2}\cdot\sqrt{a_2^2+b_2^2+c_2^2}}$$

For instance, given $E_1 : x+y+z-3=0$, $E_2 = 2x-y+3z-1=0$

If θ is the angle between plane E_1 and plane E_2, what is the value of $\cos\theta$?

$$\cos\theta=\pm\frac{2-1+3}{\sqrt{1+1+1}\cdot\sqrt{2^2+(-1)^2+3^2}}=\pm\frac{4}{(\sqrt{3})(\sqrt{14})}=\pm 0.62$$

Answer to Concept Check 13-14

$E_1 : x + y - z + 3 = 0$, $E_2 = 2x + y - z + 1 = 0$

$\overrightarrow{n_1} = (1, 1, -1)$, $\overrightarrow{n_2} = (2, 1, -1)$

$$\cos\theta = \pm\frac{\overrightarrow{n_1} \cdot \overrightarrow{n_2}}{|\overrightarrow{n_1}||\overrightarrow{n_2}|} = \pm\frac{(2+1+1)}{\sqrt{1^2+1^2+(-1)^2} \cdot \sqrt{2^2+1^2+(-1)^2}} = \pm\frac{4}{\sqrt{3} \cdot \sqrt{6}} = \pm\frac{4}{\sqrt{18}} = \pm 0.94$$

Answer is ± 0.94

CHAPTER 13 PRACTICE

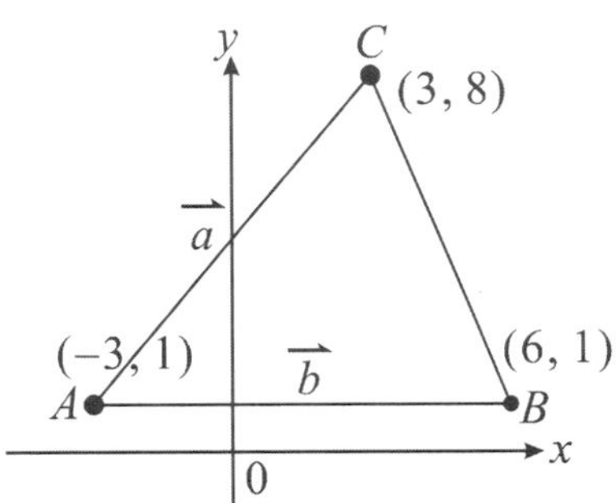

Note: Figure not drawn to scale

1. What is the area of $\triangle ABC$ shown above ? (use method of vector)

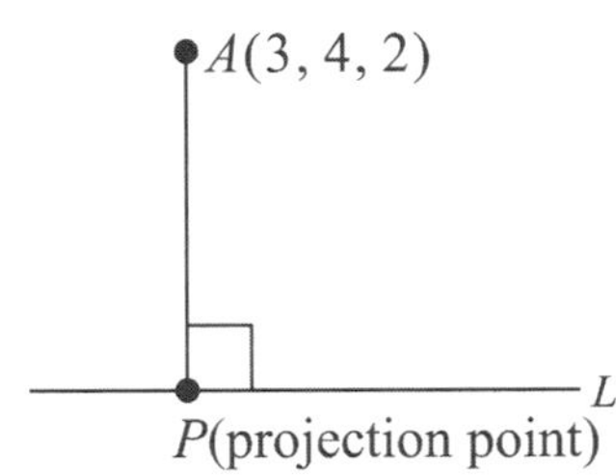

Note: Figure not drawn to scale.

2. In the figure above, line L is the intersection line of plane $x-y+z=2$ and plane $x+y-z=2$, point P is the projection point of point A on line L. What are the coordinates of point P ?

3. In a three-dimensional space, which of the following statements is correct ?
   (A) Through a point outside of a given line, there is exactly one plane perpendicular to this line
   (B) Through a point outside of a given plane, there is exactly one plane parallel to this plane
   (C) Through a point outside of a given plane, there is exactly one straight line parallel to this plane.
   (D) Through a point outside of a given plane, there is exactly one plane perpendicular to this plane

4. Given $|\vec{b}|=12$, and the direction angle of $\vec{b}$ is $\frac{\pi}{3}$, what is $\vec{b}$?
   (A) $(6, 6\sqrt{3})$
   (B) $(6, 6\sqrt{2})$
   (C) $(8, 8\sqrt{2})$
   (D) $(8, 8\sqrt{3})$

5. In $\triangle ABC$, $\overrightarrow{AB}=(10,-6)$, $\overrightarrow{BC}=(-4,12)$, what is the length of the perimeter of $\triangle ABC$?
   (A) 24.42
   (B) 28.64
   (C) 32.80
   (D) 38.48

6. What is the shortest distance from $p(1,2,1)$ to plane $E: 2x-y+z=-1$?
   (A) $\frac{\sqrt{6}}{3}$

(B) $\frac{\sqrt{3}}{4}$

(C) $\frac{3}{8}$

(D) $\frac{2}{7}$

7. If line L: $\frac{x+2}{-1}=\frac{y-1}{1}=\frac{z+a}{-b}$ lies on plane E: $x+2y+z-1=0$, what is the value of $a+b$?

(A) -1

(B) 0

(C) 1

(D) 2

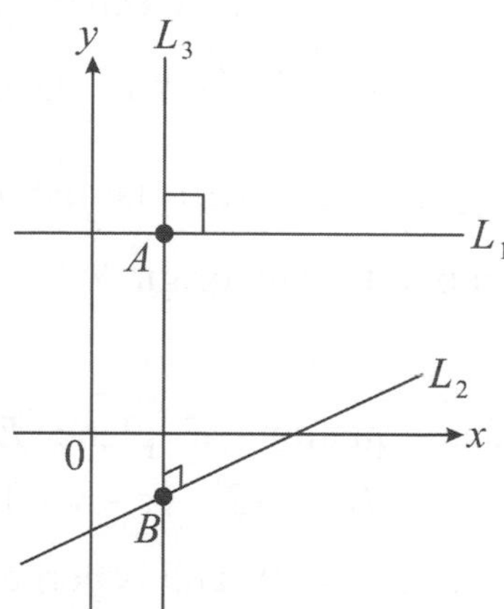

Note: Figure not drawn to scale

8. In the figure above, line L_3 is the common perpendicular line to both of $L_1:\frac{x}{-2}=\frac{y+1}{1}=\frac{z-1}{2}$ and $L_2:\frac{x-1}{2}=\frac{y}{2}=\frac{z-2}{1}$, and intersect with L_1 and L_2 at point A and point B respectively, what is the length of $\overline{AB}$? and what is the equation of $\overleftrightarrow{AB}$?

9. Given $\overrightarrow{a}=(2,3)$, $\overrightarrow{b}=(1,-2)$, what is the unit vector of $3\overrightarrow{a}-2\overrightarrow{b}$?

(A) $\frac{(4,13)}{13.6}$

(B) $\frac{(13,4)}{13.6}$

(C) $\frac{(-4,3)}{13.6}$

(D) $\frac{(4,-3)}{13.6}$

10. If $|\overrightarrow{a}|=2$, $|\overrightarrow{b}|=3$, and the angle between $\overrightarrow{a}$ and $\overrightarrow{b}$ is $30°$, find $|2\overrightarrow{a}-3\overrightarrow{b}|^2$

(A) 5.89

(B) 11

(C) 34.72

(D) 121

11. Which of following is the normal vector projection of $\overrightarrow{a}=(2,2)$ on to vector $\overrightarrow{b}(2,1)$?

(A) $\frac{2}{3}(2,1)$

(B) $\frac{2}{5}(2,1)$

(C) $\frac{4}{5}(2,1)$

(D) $\frac{2}{6}(2,1)$

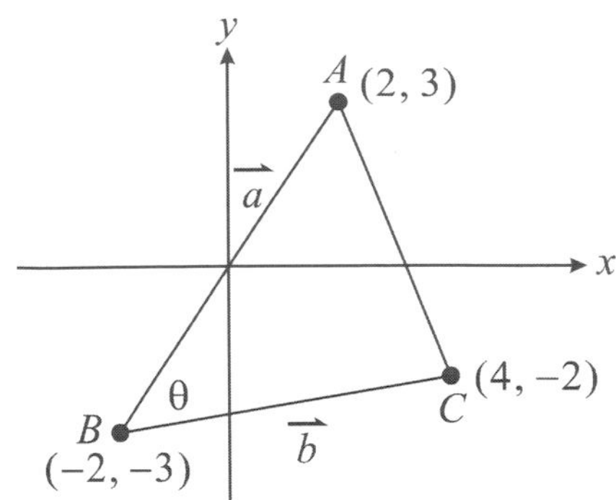

12. In the triangle shown above, what is the area of $\triangle ABC$? (use the concept of the vector to solve it)

13. What is the shortest distance from point $(1, 1, 2)$ to the line $\frac{x-1}{1}=\frac{y-1}{2}=\frac{z+1}{1}$?

14. Point $A(2, 1, 3)$ is on plane E whose normal vector is $\vec{n}(2, 2, 2)$. What is the equation of plane E ?

15. Find the coordinates of the normal projection point of point $A(1, 2, 3)$ on to plane $E: x-y+2z+1=0$

16. What is the equation of a straight line which passes point $A(2, 1, 3)$ and point $B(3, -1, 2)$?

17. Find the shortest distance between

$L_1: \frac{x-1}{2}=\frac{y+1}{1}=\frac{z-1}{1}$

$L_2: \frac{x+2}{2}=\frac{y-2}{1}=\frac{z+1}{1}$

Where $L_1 \parallel L_2$.

(A) 3.21

(B) 4.22

(C) 5.78

(D) 6.31

18. There are two planes:

$E_1: x-y+z=2$

$E_2: 2x+y-z=1$

If θ is the angle between plane E_1 and E_2 what is the value of $\cos\theta$?

19. If point A is the intersection point of $L: \frac{x-1}{2}=\frac{y+1}{-1}=\frac{z-2}{1}$ and plane $E: 2x-y+z=1$, what is the distance between point A and origin ?

20. Find the equation of plane E which contains $L_1: x-2y-z-3=0$ and $L_2: 2x+y+2z-1=0$ and is perpendicular to the plane $x+3y-z=2$

21. Connect point $A(1, 0, -1)$ and $B(-1, 1, 1)$, if $\overline{AB}$ intersects line $L: \frac{x}{1}=\frac{y-2}{1}=\frac{z+1}{-1}$ at point P, what is the ratio of $\overline{AP}:\overline{BP}$?

CHAPTER 13
ANSWERS and EXPLANATIONS

1. Since $\triangle=\frac{1}{2}\sqrt{|\vec{a}|^2|\vec{b}|^2-(\vec{a}\cdot\vec{b})^2}$

$\vec{a}=(3+3,8-1)=(6,7)$,

$|\vec{a}|^2=6^2+7^2=85$

$\vec{b}=(6+3,1-1)=(9,0)$,

$|\vec{b}|^2=9^2=81$

$\vec{a}\cdot\vec{b}=6\times9+7\times0=54$,

$(\vec{a}\cdot\vec{b})^2=2916$

$\triangle=\frac{1}{2}\sqrt{85\times81-2916}=\frac{63}{2}=31.5$

2. Since line L is on both planes, so, solve the system equations of

$\begin{cases}x-y+z=2\cdots\cdots①\\x+y-z=2\cdots\cdots②\end{cases}$

①+② $2x=4$, $x=2$

$\begin{cases}2-y+z=2\\2+y-z=2\end{cases}$, $y=z$

Let $y=z=t$

The equation of line L is $\begin{cases}x=2\\y=t\\z=t\end{cases}$

Since point P lies on line L, the coordinates of point P is $(2,t,t)$

$\overrightarrow{AP}=(2-3,t-4,t-2)=(-1,t-4,t-2)$

$|\overrightarrow{AP}|=\sqrt{(-1)^2+(t-4)^2+(t-2)^2}$

$=\sqrt{1+t^2-8t+16+t^2-4t+4}$

$=\sqrt{2t^2-12t+21}$

$=\sqrt{2(t^2-6t+9)-18+21}$

$=\sqrt{2(t-3)^2+3}$

When $t=3$, $|\overrightarrow{AP}|$ is the smallest

So, the coordinates of point P is (2, 3, 3)

3. (A) incorrect (lots of planes)

(B) correct

(C) incorrect (lots of lines)

(D) incorrect (lots of planes)

4. $\vec{b}=(12\cos\frac{\pi}{3},12\sin\frac{\pi}{3})$

$=(12\times\frac{1}{2},12\times\frac{\sqrt{3}}{2})$

$=(6,6\sqrt{3})$

Answer is (A)

5. $|\overrightarrow{AB}|=\sqrt{10^2+(-6)^2}=\sqrt{136}$

$|\overrightarrow{BC}|=\sqrt{(-4)^2+(12)^2}=\sqrt{160}$

$\overrightarrow{AC}=\overrightarrow{AB}+\overrightarrow{BC}=(10-4,-6+12)$

$=(6,6)$

$\therefore\ |\overrightarrow{AC}|=\sqrt{6^2+6^2}=6\sqrt{2}$

The length of perimeter

$=\sqrt{136}+\sqrt{160}+6\sqrt{2}=32.80$

Answer is (C)

6. (A)

Since the shortest distance from a point

to the plane $D=\dfrac{|ax_0+by_0+cz_0+d|}{\sqrt{a^2+b^2+c^2}}$

$x_0=1$, $y_0=2$, and $z_0=1$

$a=2$, $b=-1$, $c=1$ and $d=1$

Therefore

$D:\dfrac{|2+(-2)+1+1|}{\sqrt{2^2+(-1)^2+(1)^2}}=\dfrac{2}{\sqrt{6}}=\dfrac{\sqrt{6}}{3}$

Answer is (A)

7. Since line L is on plane E, the vector direction of line L is perpendicular to the normal vector of plane E.

$(-1,1,-b)\cdot(1,2,1)=-1+2-b=0$,

$b=1$

$x=-t-2$, $y=t+1$, $z=-bt-a$

Since all points on line L lie on plane E, substitute point $(-t-2,t+1,-bt-a)$ into

$(-t-2)+2(t+1)+(-bt-a)-1=0$

$-t-2+2t+2-t-a-1=0$

$-a-1=0 \qquad a=-1$

$a+b=(-1)+1=0$

Answer is (B)

8. Since $L_1:\begin{cases}x=-2t+0\\y=t-1\\z=2t+1\end{cases}$ $L_2:\begin{cases}x=2k+1\\y=2k\\z=k+2\end{cases}$

Set the coordinates of point A $(-2t,t-1,2t+1)$ and point B $(2k+1,2k,k+2)$ respectively.

Therefore, $\overrightarrow{AB}=[(2k+1)-(-2t),(2k)-(t-1),(k+2)-(2t+1)]$

$=[2k+1+2t,2k-t+1,k+2-2t-1]$

$=(2k+2t+1,2k-t+1,k-2t+1)$

Since $\overrightarrow{AB}\perp L_1$, and $\overrightarrow{AB}\perp L_2$

So, $(-2)\cdot(2k+2t+1)+(2k-t+1)$
$+2(k-2t+1)=0$

$-4k-4t-2+2k-t+1+2k-4t+2=0$

$-9t+1=0$, $t=\dfrac{1}{9}$

and $(2)(2k+2t+1)+2(2k-t+1)$
$+(k-2t+1)=0$

$4k+4t+2+4k-2t+2+k-2t+1=0$

$9k+5=0$, $k=-\dfrac{5}{9}$

point $A(-2\times\dfrac{1}{9},\dfrac{1}{9}-1,2\times\dfrac{1}{9}+1)$

$\Rightarrow$ point $A(-\dfrac{2}{9},-\dfrac{8}{9},\dfrac{11}{9})$

point $B(2\times\dfrac{-5}{9}+1,2\times(-\dfrac{5}{9}),-\dfrac{5}{9}+2)$

$\Rightarrow$ point $B(-\dfrac{1}{9},\dfrac{-10}{9},\dfrac{13}{9})$

$\overrightarrow{AB}=[(-\dfrac{1}{9})-(-\dfrac{2}{9}),(-\dfrac{10}{9})-(-\dfrac{8}{9}),\dfrac{13}{9}-\dfrac{11}{9}]$

$=(\dfrac{1}{9},-\dfrac{2}{9},\dfrac{2}{9})$

$\therefore\ |\overrightarrow{AB}|=\sqrt{(\dfrac{1}{9})^2+(-\dfrac{2}{9})^2+(\dfrac{2}{9})^2}=\dfrac{1}{3}$

9. $\vec{a}=(2,3)$, $3\vec{a}=(3\times2,3\times3)=(6,9)$

$\vec{b}=(1,-2)$,

$2\vec{b}=[2\times1,2\times(-2)]=(2,-4)$

$3\vec{a}-2\vec{b}=(6-2,9-(-4)=(4,13)$

Norm $=|3\vec{a}-2\vec{b}|=\sqrt{4^2+13^2}$

$= \sqrt{16+169} = \sqrt{185}$

The unit vector $= \dfrac{(4,13)}{\sqrt{185}} = \dfrac{(4,13)}{13.6}$

Answer is (A)

10. Since $|2\vec{a}-3\vec{b}|^2$

$= (2\vec{a}-3\vec{b})\cdot(2\vec{a}-3\vec{b})$

$= (2\vec{a})(2\vec{a}) - 6(\vec{a})(\vec{b}) - 6(\vec{a})(\vec{b})$

$+9(\vec{b})(\vec{b})$

$= 4|\vec{a}|^2 - 12(\vec{a})(\vec{b}) + 9|\vec{b}|^2$

$= 4\times 2^2 + 9\times 3^2 - 12(\vec{a})(\vec{b})$

$= 16+81-12(\vec{a})(\vec{b})$

Since $(\vec{a})(\vec{b}) = |\vec{a}|\cdot|\vec{b}|\cos 30°$

$= |\vec{a}||\vec{b}|\cdot\dfrac{\sqrt{3}}{2}$

$= (2)(3)(\dfrac{\sqrt{3}}{2}) = 3\sqrt{3}$

So, $|2\vec{a}-3\vec{b}|^2 = 97-12(3\sqrt{3}) = 34.65$

$|2\vec{a}-3\vec{b}| = \sqrt{34.65} = 5.89$

11. Since $\vec{p} = \dfrac{\vec{a}\cdot\vec{b}}{|\vec{b}|^2}\vec{b}$

$\vec{a} = (1,2)$, $\vec{b} = (2,1)$

$\vec{p} = \dfrac{(1,2)\cdot(2,1)}{|\sqrt{2^2+1^2}|^2}(2,1)$

$= \dfrac{(1\times 2+2\times 1)}{5}(2,1) = \dfrac{4}{5}(2,1)$

Answer is (C)

12. Let $\vec{BA} = \vec{a}$, and $\vec{BC} = \vec{b}$

$\triangle ABC = \dfrac{1}{2}|\vec{a}||\vec{b}|\sin\theta$

$= \dfrac{1}{2}|\vec{a}||\vec{b}|\sqrt{1-\cos^2\theta}$

$= \dfrac{1}{2}\sqrt{|\vec{a}|^2|\vec{b}|^2 - |\vec{a}|^2|\vec{b}|^2\cos^2\theta}$

$= \dfrac{1}{2}\sqrt{|\vec{a}|^2|\vec{b}|^2 - (\vec{a}\cdot\vec{b})^2}$

Since

$\vec{BA} = \vec{a} = (2-(-2), 3-(-3)) = (4,6)$

$\vec{BC} = \vec{b} = (4-(-2), -2-(-3)) = (6,1)$

$|\vec{a}|^2 = (\sqrt{4^2+6^2})^2 = 52$,

$|\vec{b}|^2 = (\sqrt{6^2+1^2}) = 37$

$(\vec{a}\cdot\vec{b})^2 = (4\times 6+6\times 1)^2 = 900$

$\therefore\ \triangle ABC = \dfrac{1}{2}\sqrt{52\times 37-900}$

$= \dfrac{1}{2}\times 32 = 16$

13. Let $B(x, y, z)$ be an arbitrary point on line $l: \dfrac{x-1}{1} = \dfrac{y-1}{2} = \dfrac{z+1}{1}$

So, $x = t+1$

$y = 2t+1$

$z = t-1$

Then,

$\vec{PB} = (t+1-1, 2t+1-1, t-1-2)$

$= (t, 2t, t-3)$

Since $\vec{PB} \perp \vec{l_1}$, then $|\vec{PB}|$ is the shortest distance

$\vec{PB}\cdot\vec{l_1} = (t\cdot 1 + 2t\cdot 2 + 1\cdot(t-3))$

$= t+4t+t-3 = 6t-3 = 0$

$\therefore\ 6t-3=0,\ 6t=3,\ t=\frac{1}{2}$

$x=\frac{1}{2}+1=\frac{3}{2},\ y=2\times\frac{1}{2}+1=2,$

$z=\frac{1}{2}-1=-\frac{1}{2}$

The coordinates of normal projection point at $(\frac{3}{2}, 2, -\frac{1}{2})$

$|\overrightarrow{PB}|^2=(1-\frac{3}{2})^2+(1-2)^2+(2-(-\frac{1}{2}))^2$

$=(-\frac{1}{2})^2+(-1)^2+\frac{25}{4}=\frac{15}{2},$

$|\overrightarrow{PB}|=2.74$

The shortest distance is 2.74

14. Point $A(2,1,3)$ is on plane E. Set an arbitrary point $P(x, y, z)$ on plane E. Then the vector of

$\overrightarrow{AP}=(x-2,\ y-1,\ z-3)$

Since the normal vector $\vec{n}$ is perpendicular to $\overrightarrow{AP}$, so, $\overrightarrow{AP}\cdot\vec{n}=0$

$\overrightarrow{AP}\cdot\vec{n}=(x-2,\ y-1,\ z-3)\cdot(2, 2, 2)$

$=[2(x-2)+2(y-1)+2(z-3)]$

$=2x-4+2y-2+2z-6$

$=2x+2y+2z-12=0$

Plane $E: 2x+2y+2z-12=0$

$E: x+y+z-6=0$

15. Set an arbitrary point $P(x_1,\ y_1,\ z_1)$ lie on the plane $E: x-y+2z+1=0$

$\overrightarrow{PA}=(1-x_1, 2-y_1, 3-z_1)$. If $|\overrightarrow{PA}|$ is the shortest distance from point A to plane E, then point P is the perpendicular point.

According to the equation of plane E the normal vector $\vec{n}$ of plane E is $\vec{n}=(1,-1,2)$

Since $\overrightarrow{PA} \parallel$ place E,

Therefore, $\frac{1-x_1}{1}=\frac{2-y_1}{-1}=\frac{2-z_1}{2}=t$

(the equation of line $\overrightarrow{AP}$)

$1-x_1=t,\ 2-y_1=-t,\ 2-z_1=2t$

$\therefore\ x_1=1-t,\ y_1=2+t,\ z_1=2-2t$

Since point P lies on plane E

$(1-t)-(2+t)+2(2-2t)+1=0$

$1-t-2-t+4-4t+1=0$

$-6t=-4$

$t=\frac{4}{6}=\frac{2}{3}$

So, the coordinates of normal projection point are $x_1=1-\frac{4}{6}=1-\frac{2}{3}=\frac{1}{3}$

$y_1=2+t=2+\frac{2}{3}=\frac{8}{3},$

$z_1=2-2t=2-2\cdot\frac{2}{3}=\frac{2}{3}$

$P(\frac{1}{3},\frac{8}{3},\frac{2}{3})$

16. $A(2,1,3),\ B(3,-1,2)$

$\overrightarrow{AB}=(3-2,-1-1,2-3)=(1,-2,-1)$

Parametric form of a straight line:

$x=x_0+at$

$y=y_0+bt$

$z = z_0 + ct$

$x_0 = 2,\ y_0 = 1,\ z_0 = 3$

$a = 1,\ b = -2,\ c = -1$

Therefore the equation of the straight line $l: \begin{cases} x = 2+t \\ y = 1-2t \\ z = 3-t \end{cases}$

17. (B)

(1) Set point $p(5,1,3)$ on the line L_1.

Because $(5,1,3)$ satisfies the straight line equation: $\frac{5-1}{2} = \frac{1+1}{1} = \frac{3-1}{1}$

(2) Set an arbitrary point $B(x, y, z)$ on the line L_2.

$x = 2t-2$

$y = t+2$

$z = t-1$

(3) $\overrightarrow{AB} = (2t-2-5, t+2-1, t-1-3)$

$= (2t-7, t+1, t-4)$

(4) Because $\overrightarrow{AB} \perp L_2$

Therefore $\overrightarrow{AB} \cdot \overrightarrow{L_2} = 0$

Since the direction vector is $(2,1,1)$

$\overrightarrow{AB} \cdot \overrightarrow{L_2}$

$= [(2t-7)\times 2 + (t+1)\times 1 + (t-4)\times 1]$

$= 6t - 17 = 0$

$\therefore\ t = \frac{17}{6}$

The coordinates of the intersection point is

$x = 2 \cdot \frac{17}{6} - 2 = \frac{11}{3}$

$y = \frac{17}{6} + 2 = \frac{29}{6}$

$z = \frac{17}{6} - 1 = \frac{11}{6}$

$D = \sqrt{(5-\frac{11}{3})^2 + (1-\frac{29}{6})^2 + (3-\frac{11}{6})^2}$

$= \sqrt{(\frac{4}{3})^2 + (-\frac{23}{6})^2 + (\frac{7}{6})^2} = 4.22$

Answer is (B)

18. The normal vector of E_1 is

$\overrightarrow{n_1} = (1, -1, 1)$

The normal vector of E_2 is

$\overrightarrow{n_2} = (2, 1, -1)$

Since

$\cos\theta = \pm \frac{\overrightarrow{n_1} \cdot \overrightarrow{n_2}}{|\overrightarrow{n_1}||\overrightarrow{n_2}|}$

$= \pm \frac{(1,-1,1)\cdot(2,1,-1)}{\sqrt{1^2+(-1)^2+1^2} \cdot \sqrt{2^2+1^2+(-1)^2}}$

$= \pm \frac{(1\times 2 + (-1)\times 1 + (1)\times(-1))}{\sqrt{3}\cdot\sqrt{6}}$

$= \pm \frac{2-1-1}{3\sqrt{2}} = \frac{0}{3\sqrt{2}} = 0$

19. Parametric form of line L

$x = 2t+1$

$y = -t-1$

$z = t+2$

The coordinates of intersection point is the same for both lines and plane E.

$E: 2x - y + z = 1$

$2(2t+1) - (-t-1) + (t+2) = 1$

$4t+2+t+1+t+2-1=0$

$6t-4=0, \ t=\dfrac{2}{3}$

Substitute parametric form of straight line

$x=2\cdot\dfrac{2}{3}+1=\dfrac{4}{3}+1=\dfrac{7}{3}$

$y=-\dfrac{2}{3}-1=-\dfrac{5}{3}$

$z=\dfrac{2}{3}+2=\dfrac{8}{3}$

The distance from the origin

$$D=\sqrt{(0-\frac{7}{3})^2+(0+\frac{5}{3})^2+(0-\frac{8}{3})^2}$$

$$=\sqrt{\frac{49}{9}+\frac{25}{9}+\frac{64}{9}}$$

$=3.92$

20. Since plane E contains $L_1: x-2y-z-3=0$ and $L_2: 2x+y+2z-1=0$.

Therefore, set the equation of plane as

$(x-2y-z-3)+t(2x+y+2z-1)=0$

$(1+2t)x+(t-2)y+(2t-1)z-(t+3)=0$

Since plane $E \perp$ plane E_1

So, $n_E \cdot n_{E_1}=[(1+2t+3(t-2)-(2t-1)]$

$=0$

$1+2t+3t-6-2t+1=3t-4=0$

$\therefore \ t=\dfrac{4}{3}$

Plane $E: (1+\dfrac{8}{3})x-\dfrac{2}{3}y+\dfrac{5}{3}z-\dfrac{13}{3}=0$

$\dfrac{11}{3}x-\dfrac{2}{3}y+\dfrac{5}{3}z-\dfrac{13}{3}=0$

$11x-2y+5z-13=0$

21. Analysis:

Draw a figure below.

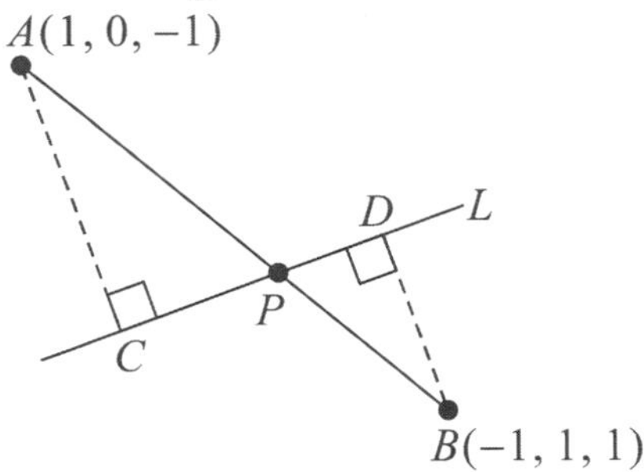

$\dfrac{x}{1}=\dfrac{y-2}{1}=\dfrac{z+1}{-1}$

make $\overline{AC}\perp L, \ \overline{BD}\perp L$

then $\triangle ACP \sim \triangle BDP$,

then $\overline{AP}:\overline{BP}=\overline{AC}:\overline{BD}$

Operation:

Since both point C and point D lie on the line $\dfrac{x}{1}=\dfrac{y-2}{1}=\dfrac{z+1}{-1}$, then set the coordinates as $C(t,t+2,-t-1)$, $D(t,t+2,-t-1)$

$\overrightarrow{AC}=(t-1,t+2,-t-1+1)$,

$\overrightarrow{BD}=(t+1,t+2-1,-t-1-1)$

Since $\overrightarrow{AC}\perp\overrightarrow{L}$,

$\overrightarrow{AC}\cdot\overrightarrow{L}=0=(t-1+t+2+t)=3t+1$,

$t=-\dfrac{1}{3}$

Point $C=(-\dfrac{1}{3},\dfrac{5}{3},\dfrac{-2}{3})$

$$AC=\sqrt{(1+\frac{1}{3})^2+(-\frac{5}{3})^2+(-\frac{1}{3})^2}=2.16$$

$\overrightarrow{BD}\perp L$,

$\overrightarrow{BD}\cdot L=(t+1+t+1+t+2)=3t+4=0$

$\therefore \ t=-\dfrac{4}{3}$

$$D = (-\frac{4}{3}, -\frac{4}{3}+2, -\frac{4}{3}-1)$$

$$= (-\frac{4}{3}, \frac{2}{3}, -\frac{7}{3})$$

$$BD = \sqrt{(-\frac{4}{3}+1)^2 + (\frac{2}{3}-1)^2 + (-\frac{7}{3}-1)^2}$$

$$= \sqrt{(-\frac{1}{3})^2 + (-\frac{1}{3})^2 + (-\frac{10}{3})^2}$$

$$= 3.37$$

Since $\frac{AP}{BP} = \frac{AC}{BD} = \frac{2.16}{3.37} = 0.64$

CHAPTER 14
MATRIX, SERIES, and LIMIT

14-1 MATRIX

A. Terminologies and Properties

- Matrix–a rectangular array of numbers, symbols, or expressions.
- Dimension (order)–number of rows by number of columns

$$\text{bracket}\begin{bmatrix} a_{11} & a_{12} & a_{13} & \cdots & a_{1n} \\ a_{21} & a_{22} & a_{23} & \cdots & a_{2n} \\ \vdots & \vdots & \vdots & \vdots & \vdots \\ a_{n1} & a_{n2} & a_{n3} & \cdots & a_{nn} \end{bmatrix}\begin{matrix} \text{row } 1 \\ \text{row } 2 \\ \vdots \\ \text{row } n \end{matrix}$$

column 1 column 2 column n

a_{ij} $\Rightarrow$ the element at the position row i and column j.

a_{23} $\Rightarrow$ the element at the position of row 2 and column 3.

- Square Matrix: the number of rows equals to the number of columns.
- Zero matrix: all elements are zeroes

  for instance,

$$O_{3\times2}=\begin{bmatrix} 0 & 0 \\ 0 & 0 \\ 0 & 0 \end{bmatrix},\quad O_{3\times3}=\begin{bmatrix} 0 & 0 & 0 \\ 0 & 0 & 0 \\ 0 & 0 & 0 \end{bmatrix}$$

- Identity matrix

  Denoted: I_n represents $n\times n$ square matrix, all diagonal elements are equal to 1, and all other elements are equal to 0.

  $n\times n$ square matrix A

  $AI_n = I_nA = A$

$$I_2=\begin{bmatrix} 1 & 0 \\ 0 & 1 \end{bmatrix},\quad I_3=\begin{bmatrix} 1 & 0 & 0 \\ 0 & 1 & 0 \\ 0 & 0 & 1 \end{bmatrix} \text{ diagonal}$$

• Equal matrices: every element in matrix A is equal to every corresponding element in matrix B.

For instance, $A=\begin{bmatrix} a & c \\ b & d \end{bmatrix}$, $B=\begin{bmatrix} 1 & 3 \\ 2 & 4 \end{bmatrix}$ if $A=B$, then $a=1$, $b=2$, $c=3$, and $d=4$

• Representation $A=[\ ]$ or $[a_{ij}]_{m\times n}$ A (capital letter),

For instance, $A=\begin{bmatrix} 3 & 1 \\ 2 & 0 \end{bmatrix}$ or $[a_{ij}]_{2\times2}$ or just $\begin{bmatrix} 3 & 1 \\ 2 & 0 \end{bmatrix}$

For instance, $(a_{ij})3\times3=\begin{bmatrix} 1 & 2 & 3 \\ 2 & 1 & 2 \\ 3 & 2 & 1 \end{bmatrix}$, $(a_{ij})2\times2=\begin{bmatrix} 2 & -1 \\ 1 & 3 \end{bmatrix}$

◆Examples

$$A=\begin{bmatrix} 3 & 0 & 2 \\ 1 & 2 & 3 \end{bmatrix}=[a_{ij}]_{2\times3}$$

$$B=\begin{bmatrix} a & b \\ c & d \\ e & f \end{bmatrix}=[b_{ij}]_{3\times2}$$

$$C=\begin{bmatrix} a_1 \\ a_2 \\ a_3 \\ a_4 \end{bmatrix}$$ column matrix

$D=[b_1, b_2, b_3, b_4]$ row matrix

• The value of a matrix

Determinant: It is a real number to represent the value of a matrix, defined as $\det(x)=\begin{vmatrix} a & c \\ b & d \end{vmatrix}=ad-bc$

For order 3 $\det(x)=\begin{vmatrix} a_1 & b_1 & c_1 \\ a_2 & b_2 & c_2 \\ a_3 & b_3 & c_3 \end{vmatrix}$

Defined as:

Extend 2 columns 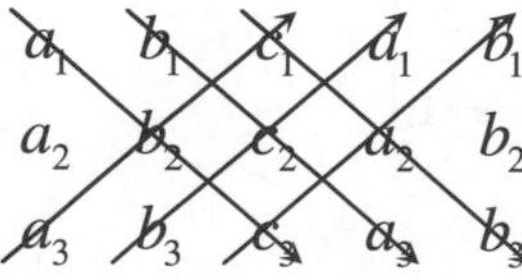

Then solve for $((a_1 \cdot b_2 \cdot c_3 + b_1 \cdot c_2 \cdot a_3 + c_1 \cdot a_2 \cdot b_3) - (a_3 \cdot b_2 \cdot c_1 + b_3 \cdot c_2 \cdot a_1 + c_3 \cdot a_2 \cdot b_1)$

For instance $A = \begin{bmatrix} 1 & 0 & -3 \\ 2 & -1 & 1 \\ 3 & 2 & -2 \end{bmatrix}$

The determinant of $A \Rightarrow \det A = -21$

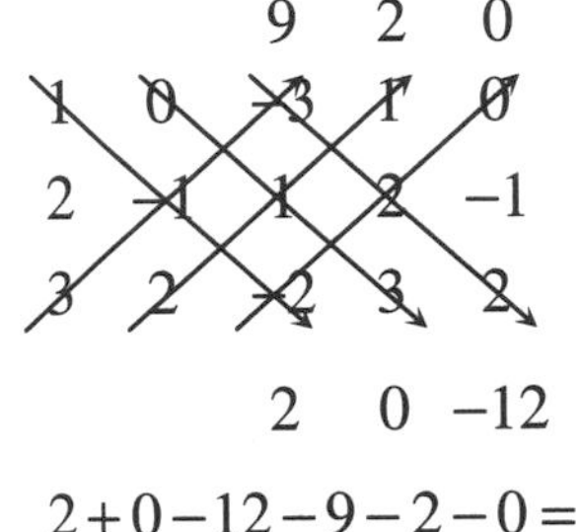

$2+0-12-9-2-0=-21$

B. Properties of Matrices:

(1) Transposing corresponding row and column in a matrix the value of the matrix remains unchanged.

Proof: Set $A=\begin{bmatrix} a & b \\ c & d \end{bmatrix}$, transposing row 1 and column 1, $B=\begin{bmatrix} a & c \\ b & d \end{bmatrix}$

$\det(A)=ad-cb$, $\det(B)=ad-bc$

$\det(A)=\det(B)$

For example: $\begin{bmatrix} 2 & -1 \\ 1 & 3 \end{bmatrix} \Rightarrow \begin{bmatrix} 2 & 1 \\ -1 & 3 \end{bmatrix}$

The determinant of $\begin{bmatrix} 2 & -1 \\ 1 & 3 \end{bmatrix}=6+1=7$

The determinant of $\begin{bmatrix} 2 & 1 \\ -1 & 3 \end{bmatrix}=6+1=7$

(2) Transposing rows or columns, the sign of the value changed.

Proof: Set $\det(A)=\begin{bmatrix} a & b \\ c & d \end{bmatrix}=ad-bc$

$B=\begin{bmatrix} c & d \\ a & b \end{bmatrix}=cb-ad=-(ad-cb)=-\det(A)$

For example: $A=\begin{bmatrix} 2 & -1 \\ 1 & 3 \end{bmatrix}$ changes to $B=\begin{bmatrix} 1 & 3 \\ 2 & -1 \end{bmatrix}$

$\det(B)=-1-6=-7=-\det(A)$

For example: $A=\begin{bmatrix} 2 & -1 \\ 1 & 3 \end{bmatrix}$, $B=\begin{bmatrix} -1 & 2 \\ 3 & 1 \end{bmatrix}$

$\det(A)=6+1=7$, $\det(B)=-1-6=-7$, $\det(A)=-\det(B)$

(3) The elements in any two rows or columns are proportional then the value of the matrix is equal to zero.

Proof: If $A=\begin{bmatrix} a & b \\ ka & kb \end{bmatrix}$, then $\det(A)=kab-kab=0$

For Example: If $A=\begin{bmatrix} 2 & -1 \\ 4 & -2 \end{bmatrix}$, then $\det(A)=-4+4=0$

(4) Any row or column is zero, then the value of the matrix is zero.

Proof: $A=\begin{bmatrix} 0 & 0 \\ c & d \end{bmatrix}$, $\det(A)=0+0=0$ or $B=\begin{bmatrix} 0 & b \\ 0 & c \end{bmatrix}$, $\det(B)=0+0=0$

For Example: $A=\begin{bmatrix} 0 & 0 \\ 2 & 3 \end{bmatrix}=\det A=0+0=0$

(5) The properties of the determinant.

$$\begin{vmatrix} 4+4 & 2+1 \\ 1 & 2 \end{vmatrix}=16-3=13$$

$$\begin{vmatrix} 4 & 2 \\ 1 & 2 \end{vmatrix}+\begin{vmatrix} 4 & 1 \\ 1 & 2 \end{vmatrix}=(8-2)+(8-1)=6+7=13$$

$$\begin{vmatrix} ka & kb \\ kc & kd \end{vmatrix}=k^2\begin{vmatrix} a & b \\ c & d \end{vmatrix}$$

$$3\begin{vmatrix} 1 & -3 \\ -2 & 4 \end{vmatrix}=\begin{vmatrix} 3 & -9 \\ -2 & 4 \end{vmatrix}, \quad \begin{vmatrix} 3\times1 & 3\times(-3) \\ 3\times(-2) & 3\times4 \end{vmatrix}=3^2\begin{vmatrix} 1 & -3 \\ -2 & 4 \end{vmatrix}$$

$\det A=4-6=-2$

$$\begin{vmatrix} 3 & -9 \\ -2 & 4 \end{vmatrix}=12-18=-6$$

$$3\begin{vmatrix} 1 & -3 \\ -2 & 4 \end{vmatrix}=3\times(4-6)=3\times(-2)=-6$$

(6) Any row (column) times k, then added to other row (column) the value of matrix remain unchanged

Proof: $A=\begin{bmatrix} a & b \\ c & d \end{bmatrix}\times k \to =\begin{bmatrix} a & b \\ ak+c & bk+d \end{bmatrix}=B$, $\det(A)=ad-bc$

$\det(B)=a(bk+d)-b(ak+c)-b=\cancel{abk}+ad-\cancel{abk}-bc=ad-bc=\det(A)$

Example: $A=\begin{bmatrix} 2 & -1 \\ 1 & 3 \end{bmatrix}$

$$\begin{bmatrix} 2 & -1 \\ 1 & 3 \end{bmatrix} \times\left(-\frac{1}{2}\right) = \begin{bmatrix} 2 & -1 \\ 0 & 3\frac{1}{2} \end{bmatrix} = B$$

$\det(A)=6+1=7$, $\det(B)=2\times\frac{7}{2}+0=7$

$\det(A)=\det(B)$

(7) If $A=\begin{vmatrix} a & b \\ kc & kd \end{vmatrix} \Rightarrow k\begin{vmatrix} a & b \\ c & d \end{vmatrix}$

$\det A=akd-bkc=k(ad-bc)$

Example: $A=\begin{bmatrix} 2 & -1 \\ 3 & 9 \end{bmatrix}$,

$$\det(A)=\begin{vmatrix} 2 & -1 \\ 3 & 9 \end{vmatrix}=3\begin{vmatrix} 2 & -1 \\ 1 & 3 \end{vmatrix}=3(2\times3-(-1))=3(6+1)=3(7)=21$$

(8) Matrices can be down graded

For instance, $A=\begin{bmatrix} a_{11} & a_{12} & a_{13} \\ a_{21} & a_{22} & a_{23} \\ a_{31} & a_{32} & a_{33} \end{bmatrix}$

Down grade according to row 1

$$A=(-1)^{i+j}\cdot a_{11}\cdot\begin{bmatrix} a_{22} & a_{23} \\ a_{32} & a_{33} \end{bmatrix}+(-1)^{i+j}\cdot a_{12}\begin{bmatrix} a_{21} & a_{23} \\ a_{31} & a_{33} \end{bmatrix}+(-1)^{i+j}\cdot a_{13}\begin{bmatrix} a_{21} & a_{22} \\ a_{31} & a_{32} \end{bmatrix}$$

$$\text{If } A=\begin{bmatrix} 1 & -1 & 2 \\ 3 & 0 & 1 \\ 1 & -2 & -1 \end{bmatrix}=(-1)^{1+1}\cdot(1)\cdot\begin{bmatrix} 0 & 1 \\ -2 & -1 \end{bmatrix}+(-1)^{1+2}\cdot(-1)\cdot\begin{bmatrix} 3 & 1 \\ 1 & -1 \end{bmatrix}$$

$$+(-1)^{1+3}\cdot(2)\cdot\begin{bmatrix} 3 & 0 \\ 1 & -2 \end{bmatrix}=\begin{bmatrix} 0 & 1 \\ -2 & -1 \end{bmatrix}+\begin{bmatrix} 3 & 1 \\ 1 & -1 \end{bmatrix}+2\begin{bmatrix} 3 & 0 \\ 1 & -2 \end{bmatrix}$$

$\det A=(0+2)+(-3-1)+(2)(-6-0)=+2-4-12=-14$

$$\det A'=3\begin{matrix} 1 & -1 & 2 & 1 & -1 \\ 0 & 1 & 3 & 0 \\ 1 & -2 & -1 & 1 & -2 \end{matrix}$$

$0+(-1)+(-12)-0-(-2)-3=-14$

$\therefore\ \det A=\det A'$

Answer to Concept Check 14-01

If we transpose row1 and row2, the matrix will change sign.

That means $\det(A) = -\det(B)$

Answer is (C)

Answer to Concept Check 14-02

Row 1 transposing with column 1, the matrix will remain unchanged

So, $\det A = \det B$

Answer is (A)

C. Operations

Adding and Subtracting of matrices.

(1) Apply all rules of numbering operations.

(2) The two matrices must be in the same order.

(3) By adding and subtracting corresponding elements.

For instance, if $A = \begin{bmatrix} 2 & 1 \\ 3 & -2 \end{bmatrix}$, $B = \begin{bmatrix} 1 & 2 \\ 0 & 3 \end{bmatrix}$

$$A + B = \begin{bmatrix} 2+1 & 1+2 \\ 3+0 & -2+3 \end{bmatrix} = \begin{bmatrix} 3 & 3 \\ 3 & 1 \end{bmatrix}$$

$$A - B = \begin{bmatrix} 2-1 & 1-2 \\ 3-0 & -2-3 \end{bmatrix} = \begin{bmatrix} 1 & -1 \\ 3 & -5 \end{bmatrix}$$

- Scalar of matrix is multiplied by a constant, then each element in the matrixs multiplied by this constant.

  For instance,

$$2 \times A = \begin{bmatrix} 1\times2 & 2\times2 & 3\times2 \\ -1\times2 & 2\times2 & 1\times2 \\ 0\times2 & 1\times2 & 2\times2 \end{bmatrix} = \begin{bmatrix} 2 & 4 & 6 \\ -2 & 4 & 2 \\ 0 & 2 & 4 \end{bmatrix}$$

- Multiplication of matrices

  (1) $A \times B$, if $A = a_{ij}$ and $B = b_{mn}$

  j (columns of A) must be equivalent to m (rows of B)

For instance, $A=\begin{bmatrix}2 & 3\\ 1 & -1\\ -2 & 1\end{bmatrix}$, $B=\begin{bmatrix}1 & 3 & -1\\ 2 & 3 & -2\end{bmatrix}$

$A=a_{3\times2}$, $B=b_{2\times3}$, $j=m \Rightarrow 2=2$

so, $A\times B$ works

if $A=\begin{bmatrix}1 & 3\\ 2 & 3\end{bmatrix}$, $B=\begin{bmatrix}2 & 3\\ 1 & -1\\ -2 & 1\end{bmatrix}$

columns of $A_{2\times2}$ is 2, rows of $B_{3\times2}$ is 3

so, $2\neq3$,

$A\times B$ undefined

(2) $A\times B$ defined by each element in a row of A multiplied by each element in corresponding column of B then summing up as a new element of new matrix.

For instance, $A=\begin{bmatrix}3 & -1 & 3\\ 2 & 2 & 1\\ 1 & 1 & -2\end{bmatrix}$, $B=\begin{bmatrix}2 & 1\\ 3 & 2\\ -1 & 3\end{bmatrix}$

$A=(a_{ij})_{3\times3}$, $B=(b_{mn})_{3\times2}$

$$A\times B=\begin{bmatrix}3 & -1 & 3\\ 2 & 2 & 1\\ 1 & 1 & -2\end{bmatrix}\times\begin{bmatrix}2 & 1\\ 3 & 2\\ 1 & 3\end{bmatrix}=\begin{bmatrix}3\times2+(-1)\times3+3\times1 & 3\times1+(-1)\times2+3\times3\\ 2\times2+2\times3+1\times1 & 2\times1+2\times2+1\times3\\ 1\times2+1\times3+(-2)\times1 & 1\times1+1\times2+(-2)\times3\end{bmatrix}$$

$$=\begin{bmatrix}6 & 10\\ 11 & 9\\ 3 & -3\end{bmatrix}$$

The new matrix will be a_{in}, in this case it is $a_{3\times2}$

◆Example 1:

If $A=\begin{bmatrix}2 & 2 & 3\\ 3 & 0 & -1\end{bmatrix}$, $B=\begin{bmatrix}1 & -2 & 3\\ 0 & 1 & 2\end{bmatrix}$

What is (1) $A+B$, (2) $-2B$, (3) $3A-2B$

(1) $A+B=\begin{bmatrix}2+1 & 2-2 & 3+3\\ 3+0 & 0+1 & -1+2\end{bmatrix}=\begin{bmatrix}3 & 0 & 6\\ 3 & 1 & 1\end{bmatrix}$

(2) $-2B$ $\quad (-2)\begin{bmatrix} 1 & -2 & 3 \\ 0 & 1 & 2 \end{bmatrix} = \begin{bmatrix} (-2)\times 1 & (-2)\times(-2) & (-2)\times 3 \\ (-2)\times 0 & (-2)\times 1 & (-2)\times 2 \end{bmatrix}$

$$= \begin{bmatrix} -2 & 4 & -6 \\ 0 & -2 & -4 \end{bmatrix}$$

(3) $3A - 2B$

$$3A = 3\begin{bmatrix} 2 & 2 & 3 \\ 3 & 0 & -1 \end{bmatrix} = \begin{bmatrix} 3\times 2 & 3\times 2 & 3\times 3 \\ 3\times 3 & 3\times 0 & 3\times(-1) \end{bmatrix} = \begin{bmatrix} 6 & 6 & 9 \\ 9 & 0 & -3 \end{bmatrix}$$

$$2B = 2\begin{bmatrix} 1 & -2 & 3 \\ 0 & 1 & 2 \end{bmatrix} = \begin{bmatrix} 2 & -4 & 6 \\ 0 & 2 & 4 \end{bmatrix}$$

$$3A - 2B = \begin{bmatrix} 6-2 & 6-(-4) & 9-6 \\ 9-0 & 0-2 & (-3)-4 \end{bmatrix} = \begin{bmatrix} 4 & 10 & 3 \\ 9 & -2 & -7 \end{bmatrix}$$

- Multiplication of matrices all rules of numbering multiplication applies.

Except:

(1) $AB \neq BA$

For instance, if $A = \begin{bmatrix} 1 & 3 \\ 2 & 4 \end{bmatrix}$, $B = \begin{bmatrix} 1 & -1 \\ 0 & 2 \end{bmatrix}$

$A \times B = \begin{bmatrix} 1 & 5 \\ 2 & 6 \end{bmatrix}$ $\quad B \times A = \begin{bmatrix} -1 & -1 \\ 4 & 8 \end{bmatrix}$

$AB \neq BA$

$A \times B \neq B \times A$

$$\begin{bmatrix} 3 & 2 \\ -1 & 4 \end{bmatrix}\begin{bmatrix} 0 & -3 \\ 2 & -1 \end{bmatrix} \neq \begin{bmatrix} 0 & -3 \\ 2 & -1 \end{bmatrix}\begin{bmatrix} 3 & 2 \\ -1 & 4 \end{bmatrix}$$

(2) $AP = AQ$ if $\det(A) \neq 0$

$P = Q$ if $\det(A) = 0$

$P \neq Q$

$AB = AC$ if $\det(A)=0$, $B \neq C$

$$\begin{bmatrix} 2 & 0 \\ 0 & 0 \end{bmatrix}\begin{bmatrix} 4 & 0 \\ 1 & 2 \end{bmatrix} = \begin{bmatrix} 2 & 0 \\ 0 & 0 \end{bmatrix}\begin{bmatrix} 4 & 0 \\ 5 & 1 \end{bmatrix} = \begin{bmatrix} 8 & 0 \\ 0 & 0 \end{bmatrix}$$

$$\begin{bmatrix} 4 & 0 \\ 1 & 2 \end{bmatrix} \neq \begin{bmatrix} 4 & 0 \\ 5 & 1 \end{bmatrix}$$

◆Example 1:

If $A=\begin{bmatrix}2 & -1\\ 1 & 2\end{bmatrix}$, $B=\begin{bmatrix}3 & 1\\ 2 & 0\end{bmatrix}$, and $C=\begin{bmatrix}3 & 1\\ -1 & -2\end{bmatrix}$

What is the result of $AB-AC$?

since $AB=\begin{bmatrix}2 & -1\\ 1 & 2\end{bmatrix}\times\begin{bmatrix}3 & 1\\ 2 & 0\end{bmatrix}=\begin{bmatrix}6-2 & 2+0\\ 3+4 & 1+0\end{bmatrix}=\begin{bmatrix}4 & 2\\ 7 & 1\end{bmatrix}$

$AC=\begin{bmatrix}2 & -1\\ 1 & 2\end{bmatrix}\times\begin{bmatrix}3 & 1\\ -1 & -2\end{bmatrix}=\begin{bmatrix}6+1 & 2+2\\ 3-2 & 1-4\end{bmatrix}=\begin{bmatrix}7 & 4\\ 1 & -3\end{bmatrix}$

$AB-AC=\begin{bmatrix}4-7 & 2-4\\ 7-1 & 1-(-3)\end{bmatrix}=\begin{bmatrix}-3 & -2\\ 6 & 4\end{bmatrix}$

◆Example 2:

If $A=\begin{bmatrix}1 & 2 & -1\\ 2 & 0 & 1\end{bmatrix}$, $B=\begin{bmatrix}1 & -2 & 1\\ 0 & -1 & 2\\ -1 & 1 & 1\end{bmatrix}$, $C=\begin{bmatrix}4 & 1 & -2\\ 1 & 0 & -1\end{bmatrix}$,

Find the result of $AB-3C$.

$\because\ AB=\begin{bmatrix}1 & 2 & -1\\ 2 & 0 & 1\end{bmatrix}\begin{bmatrix}1 & -2 & 1\\ 0 & -1 & 2\\ -1 & 1 & 1\end{bmatrix}=\begin{bmatrix}1+0+1 & -2-2-1 & 1+4-1\\ 2+0-1 & -4+0+1 & 2+0+1\end{bmatrix}$

$=\begin{bmatrix}2 & -5 & 4\\ 1 & -3 & 3\end{bmatrix}$

$\because\ 3C=3\begin{bmatrix}4 & 1 & -2\\ 1 & 0 & -1\end{bmatrix}=\begin{bmatrix}12 & 3 & -6\\ 3 & 0 & -3\end{bmatrix}$

so, $AB-3C=\begin{bmatrix}2 & -5 & 4\\ 1 & -3 & 3\end{bmatrix}-\begin{bmatrix}12 & 3 & -6\\ 3 & 0 & -3\end{bmatrix}=\begin{bmatrix}-10 & -8 & 10\\ -2 & -3 & 6\end{bmatrix}$

Answer to Concept Check 14-03

$A=\begin{bmatrix}-1 & 2\\ 0 & 3\end{bmatrix}$, $B=\begin{bmatrix}-1 & 3\\ 2 & 1\end{bmatrix}$

$3A=\begin{bmatrix}-3 & 6\\ 0 & 9\end{bmatrix}$, $2B=\begin{bmatrix}-2 & 6\\ 4 & 2\end{bmatrix}$

$3A+2B=\begin{bmatrix}-3 & 6\\ 0 & 9\end{bmatrix}+\begin{bmatrix}-2 & 6\\ 4 & 2\end{bmatrix}=\begin{bmatrix}-3-2 & 6+6\\ 0+4 & 9+2\end{bmatrix}=\begin{bmatrix}-5 & 12\\ 4 & 11\end{bmatrix}$

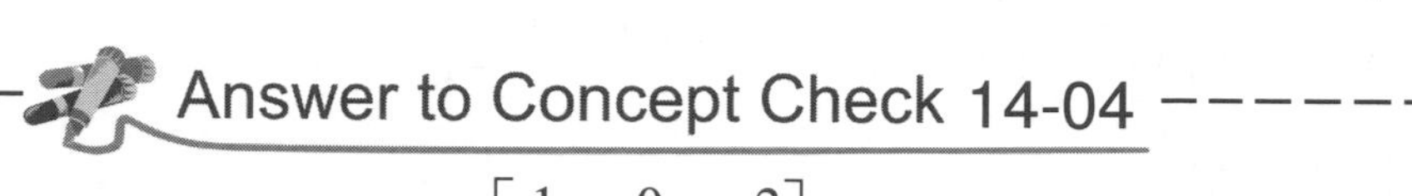

Answer to Concept Check 14-04

$$A=\begin{bmatrix}2&1&2\\1&0&2\end{bmatrix},\quad B=\begin{bmatrix}1&0&-2\\-2&1&1\\1&-1&0\end{bmatrix},\quad C=\begin{bmatrix}2&3&-2\\1&0&1\end{bmatrix}$$

$$AB=\begin{bmatrix}2-2+2&0+1-2&-4+1+0\\1+0+2&0+0-2&-2+0+0\end{bmatrix}=\begin{bmatrix}2&-1&-3\\3&-2&-2\end{bmatrix}$$

$$CB=\begin{bmatrix}2-6-2&0+3+2&-4+3+0\\1+0+1&0+0-1&-2+0+0\end{bmatrix}=\begin{bmatrix}-6&5&-1\\2&-1&-2\end{bmatrix}$$

$$3AB-2CB=\begin{bmatrix}6&-3&-9\\9&-6&-6\end{bmatrix}-\begin{bmatrix}-12&10&-2\\4&-2&-4\end{bmatrix}=\begin{bmatrix}18&-13&-7\\5&-4&-2\end{bmatrix}$$

D. Inverse matrix

If $AB=I$ I is identity (square matrix)

$$I_2=\begin{bmatrix}1&0\\0&1\end{bmatrix},\quad I_3=\begin{bmatrix}1&0&0\\0&1&0\\0&0&1\end{bmatrix},\quad I_4=\begin{bmatrix}1&0&0&0\\0&1&0&0\\0&0&1&0\\0&0&0&1\end{bmatrix}$$

then B is the inverse matrix of A, or A is the inverse matrix of B.

B is denoted as A^{-1}

If $A=\begin{bmatrix}a&b\\c&d\end{bmatrix}$

then $A^{-1}=\dfrac{1}{\det(A)}\begin{bmatrix}d&-b\\-c&a\end{bmatrix}$ a, d exchange positions, and b, c change signs

Proof: $AA^{-1}=\begin{bmatrix}a&b\\c&d\end{bmatrix}\cdot\dfrac{1}{\det(A)}\begin{bmatrix}d&-b\\-c&a\end{bmatrix}=\dfrac{1}{\det(A)}\begin{bmatrix}a&b\\c&d\end{bmatrix}\begin{bmatrix}d&-b\\-c&a\end{bmatrix}$

$$=\frac{1}{\det(A)}\begin{bmatrix}ad-bc&-ab+ab\\cd-cd&-bc+ad\end{bmatrix}=\frac{1}{\det(A)}\begin{bmatrix}ad-bc&0\\0&-bc+ad\end{bmatrix}=\frac{ad-bc}{ad-bc}\begin{bmatrix}1&0\\0&1\end{bmatrix}=I$$

◆Example:

If $AC=\begin{bmatrix}-1&0\\-3&2\end{bmatrix}$, $AB=\begin{bmatrix}-2&-1\\4&1\end{bmatrix}$, $B+C=\begin{bmatrix}-3&-1\\-1&1\end{bmatrix}$, Find A.

Since $AC+AB=A(C+B)=A(B+C)$

$$\begin{bmatrix}-1 & 0\\-3 & 2\end{bmatrix}+\begin{bmatrix}-2 & -1\\4 & 1\end{bmatrix}=\begin{bmatrix}-1-2 & -1\\-3+4 & 2+1\end{bmatrix}=\begin{bmatrix}-3 & -1\\1 & 3\end{bmatrix}=A(B+C)=A\begin{bmatrix}-3 & -1\\-1 & 1\end{bmatrix}$$

both sides multiplied by $\begin{bmatrix}-3 & -1\\-1 & 1\end{bmatrix}^{-1}$

$$\begin{bmatrix}-3 & -1\\1 & 3\end{bmatrix}\times\begin{bmatrix}-3 & -1\\-1 & 1\end{bmatrix}^{-1}=A\begin{bmatrix}-3 & -1\\-1 & 1\end{bmatrix}\bullet\begin{bmatrix}-3 & -1\\-1 & 1\end{bmatrix}^{-1}$$

Since, $\begin{bmatrix}-3 & -1\\-1 & 1\end{bmatrix}^{-1}=\dfrac{1}{-3-1}\begin{bmatrix}1 & 1\\1 & -3\end{bmatrix}=\dfrac{1}{-4}\begin{bmatrix}1 & 1\\1 & -3\end{bmatrix}$

$$\therefore\ A=\begin{bmatrix}-3 & -1\\1 & 3\end{bmatrix}\times\frac{1}{-4}\begin{bmatrix}1 & 1\\1 & -3\end{bmatrix}=-\frac{1}{4}\begin{bmatrix}-3-1 & -3+3\\1+3 & 1-9\end{bmatrix}$$

$$=-\frac{1}{4}\begin{bmatrix}-4 & 0\\4 & -8\end{bmatrix}=\begin{bmatrix}1 & 0\\-1 & 2\end{bmatrix}$$

Answer to Concept Check 14-05

Does not make sense because $B_{3\times3}\times C_{2\times3}$ the operation is undefined.

Answer to Concept Check 14-06

$A=\begin{bmatrix}2 & -1\\1 & 0\end{bmatrix}$ $A^{-1}=\dfrac{1}{\det A}\begin{bmatrix}0 & 1\\-1 & 2\end{bmatrix}=\dfrac{1}{2\times0-(-1)}\begin{bmatrix}0 & 1\\-1 & 2\end{bmatrix}=\begin{bmatrix}0 & 1\\-1 & 2\end{bmatrix}$

Verify $\begin{bmatrix}2 & -1\\1 & 0\end{bmatrix}\times\begin{bmatrix}0 & 1\\-1 & 2\end{bmatrix}=\begin{bmatrix}1 & 0\\0 & 1\end{bmatrix}$.

Answer to Concept Check 14-07

Since $AB+AC=A(B+C)$

$\therefore\ (AB+AC)(B+C)^{-1}=A(B+C)(B+C)^{-1}=AI=A$

Since $AB+AC=\begin{bmatrix}-1 & 2\\ -1 & -2\end{bmatrix}+\begin{bmatrix}4 & -1\\ -2 & -1\end{bmatrix}=\begin{bmatrix}-1+4 & 2-1\\ -1-2 & -2-1\end{bmatrix}=\begin{bmatrix}3 & 1\\ -3 & -3\end{bmatrix}$

Since $B+C=\begin{bmatrix}0 & -1\\ 3 & 3\end{bmatrix}$

$$\begin{bmatrix}3 & 1\\ -3 & -3\end{bmatrix}\cdot\begin{bmatrix}0 & -1\\ 3 & 3\end{bmatrix}^{-1}=A$$

$$\begin{bmatrix}0 & -1\\ 3 & 3\end{bmatrix}^{-1}=\frac{1}{0-(-3)}\begin{bmatrix}3 & 1\\ -3 & 0\end{bmatrix}=\frac{1}{3}\begin{bmatrix}3 & 1\\ -3 & 0\end{bmatrix}=\begin{bmatrix}1 & \frac{1}{3}\\ -1 & 0\end{bmatrix}$$

$$\begin{bmatrix}3 & 1\\ -3 & -3\end{bmatrix}\begin{bmatrix}1 & \frac{1}{3}\\ -1 & 0\end{bmatrix}=\begin{bmatrix}3-1 & 1+0\\ -3+3 & -1+0\end{bmatrix}=\begin{bmatrix}2 & 1\\ 0 & -1\end{bmatrix}$$

$$\therefore\ A=\begin{bmatrix}2 & 1\\ 0 & -1\end{bmatrix}$$

E. Solving Linear System Equations By Matrices

• Method 1: Using Inverse Matrix

(1) Two variables system equations

$$\begin{cases}a_1x+b_1y=c_1\\ a_2x+b_2y=c_2\end{cases}$$

Matrix form

$$\begin{bmatrix}a_1 & b_1\\ a_2 & b_2\end{bmatrix}\begin{bmatrix}x\\ y\end{bmatrix}=\begin{bmatrix}c_1\\ c_2\end{bmatrix}$$

From property of matrix

If $AV=B$, then $A^{-1}AV=A^{-1}B$ $[\det(A)\neq 0]$

It means $V=A^{-1}B \Rightarrow \begin{bmatrix}x\\ y\end{bmatrix}=A^{-1}B$

For instance, solve $\begin{cases} 2x-y=4 \\ x+2y=7 \end{cases}$

$$\begin{bmatrix} 2 & -1 \\ 1 & 2 \end{bmatrix}\begin{bmatrix} x \\ y \end{bmatrix}=\begin{bmatrix} 4 \\ 7 \end{bmatrix}$$

$$A^{-1}=\begin{bmatrix} 2 & -1 \\ 1 & 2 \end{bmatrix}^{-1}=\frac{1}{4+1}\begin{bmatrix} 2 & 1 \\ -1 & 2 \end{bmatrix}=\frac{1}{5}\begin{bmatrix} 2 & 1 \\ -1 & 2 \end{bmatrix}$$

$$\therefore \begin{bmatrix} x \\ y \end{bmatrix}=\frac{1}{5}\begin{bmatrix} 2 & 1 \\ -1 & 2 \end{bmatrix}\times\begin{bmatrix} 4 \\ 7 \end{bmatrix}=\frac{1}{5}\begin{bmatrix} 8+7 \\ -4+14 \end{bmatrix}=\frac{1}{5}\begin{bmatrix} 15 \\ 10 \end{bmatrix}=\begin{bmatrix} 3 \\ 2 \end{bmatrix}$$

$x=3$, $y=2$

(2) Three variables system equations

- Solving by inverse matrix

  Since $[A|I]\Rightarrow[I|A^{-1}]$

  Then $AV=B$, $A^{-1}AV=A^{-1}B$

  The solution matrix $\begin{bmatrix} x \\ y \\ z \end{bmatrix}=A^{-1}B$

  Example: solve $\begin{cases} x+y=3 \\ 2x+3y+z=11 \\ 2x-y+2z=6 \end{cases}$

  The coefficients of the equation can be written as $\underset{A}{\begin{bmatrix} 1 & 1 & 0 \\ 2 & 3 & 1 \\ 2 & -1 & 2 \end{bmatrix}}\underset{V}{\begin{bmatrix} x \\ y \\ z \end{bmatrix}}=\underset{B}{\begin{bmatrix} 3 \\ 11 \\ 6 \end{bmatrix}}$

  So, $\begin{bmatrix} 1 & 1 & 0 \\ 2 & 3 & 1 \\ 2 & -1 & 2 \end{bmatrix}^{-1}\begin{bmatrix} 1 & 1 & 0 \\ 2 & 3 & 1 \\ 2 & -1 & 2 \end{bmatrix}\begin{bmatrix} x \\ y \\ z \end{bmatrix}=\begin{bmatrix} 1 & 1 & 0 \\ 2 & 3 & 1 \\ 2 & -1 & 2 \end{bmatrix}^{-1}\begin{bmatrix} 3 \\ 11 \\ 6 \end{bmatrix}$

$$\begin{bmatrix} 1 & 0 & 0 \\ 0 & 1 & 0 \\ 0 & 0 & 1 \end{bmatrix}\begin{bmatrix} x \\ y \\ z \end{bmatrix}=\begin{bmatrix} 1 & 1 & 0 \\ 2 & 3 & 1 \\ 2 & -1 & 2 \end{bmatrix}^{-1}\begin{bmatrix} 3 \\ 11 \\ 6 \end{bmatrix}$$

Step 1: Form an extended matrix with identity matrix to find the inverse matrix

Example: $[A|I]=\left[\begin{array}{ccc|ccc} 1 & 1 & 0 & 1 & 0 & 0 \\ 2 & 3 & 1 & 0 & 1 & 0 \\ 2 & -1 & 2 & 0 & 0 & 1 \end{array}\right]$

Step 2: Apply row/column operation to change the extended matrix $[A|I] \to [I|A^{-1}]$

①fix row 1 to eliminate a_{21}, a_{31}.

②eliminate a_{12}, a_{13}, and so on.

③make the elements diagonal equal to "1".

$$\left|\begin{array}{ccc|ccc} 1 & 1 & 0 & 1 & 0 & 0 \\ 2 & 3 & 1 & 0 & 1 & 0 \\ 2 & -1 & 2 & 0 & 0 & 1 \end{array}\right| \times(-2) \to \left|\begin{array}{ccc|ccc} 1 & 1 & 0 & 1 & 0 & 0 \\ 0 & 1 & 1 & -2 & 1 & 0 \\ 0 & -3 & 2 & -2 & 0 & 1 \end{array}\right| \begin{array}{l} \times(-1) \\ \times(3) \end{array}$$

$$\to \left|\begin{array}{ccc|ccc} 1 & 0 & -1 & 3 & -1 & 0 \\ 0 & 1 & 1 & -2 & 1 & 0 \\ 0 & 0 & 5 & -8 & 3 & 1 \end{array}\right| \times(\frac{1}{5}) \to \left[\begin{array}{ccc|ccc} 1 & 0 & 0 & \frac{7}{5} & -\frac{2}{5} & \frac{1}{5} \\ 0 & 1 & 1 & -2 & 1 & 0 \\ 0 & 0 & 5 & -8 & 3 & 1 \end{array}\right] \times(-\frac{1}{5})$$

$$\to \left[\begin{array}{ccc|ccc} 1 & 0 & 0 & \frac{7}{5} & -\frac{2}{5} & \frac{1}{5} \\ 0 & 1 & 0 & -\frac{2}{5} & \frac{2}{5} & -\frac{1}{5} \\ 0 & 0 & 5 & -8 & 3 & 1 \end{array}\right] \times(\frac{1}{5}) \to \left[\begin{array}{ccc|ccc} 1 & 0 & 0 & \frac{7}{5} & -\frac{2}{5} & \frac{1}{5} \\ 0 & 1 & 0 & -\frac{2}{5} & \frac{2}{5} & -\frac{1}{5} \\ 0 & 0 & 1 & -\frac{8}{5} & \frac{3}{5} & \frac{1}{5} \end{array}\right]$$

$$A^{-1} = \begin{bmatrix} \frac{7}{5} & -\frac{2}{5} & \frac{1}{5} \\ -\frac{2}{5} & \frac{2}{5} & -\frac{1}{5} \\ -\frac{8}{5} & \frac{3}{5} & \frac{1}{5} \end{bmatrix} = \frac{1}{5}\begin{bmatrix} 7 & -2 & 1 \\ -2 & 2 & -1 \\ -8 & 3 & 1 \end{bmatrix}$$

$$X = A^{-1} \cdot B = \frac{1}{5}\begin{bmatrix} 7 & -2 & 1 \\ -2 & 2 & -1 \\ -8 & 3 & 1 \end{bmatrix}\begin{bmatrix} 3 \\ 11 \\ 6 \end{bmatrix} = \frac{1}{5}\begin{bmatrix} 21-22+6 \\ -6+22-6 \\ -24+33+6 \end{bmatrix} = \frac{1}{5}\begin{bmatrix} 5 \\ 10 \\ 15 \end{bmatrix} = \begin{bmatrix} 1 \\ 2 \\ 3 \end{bmatrix}$$

$\therefore\ x=1,\ y=2,\ z=3$

- Solving by Gaussian elimination.

Step 1: Form an augmented matrix

$$\left[\begin{array}{ccc|c} 1 & 1 & 0 & 3 \\ 2 & 3 & 1 & 11 \\ 2 & -1 & 2 & 6 \end{array}\right]$$

↑ Constant

Step 2: Perform row operation

$$\left[\begin{array}{ccc|c} 1 & 1 & 0 & 3 \\ 2 & 3 & 1 & 11 \\ 2 & -1 & 2 & 6 \end{array}\right] \times(-2) = \left[\begin{array}{ccc|c} 1 & 1 & 0 & 3 \\ 0 & 1 & 1 & 5 \\ 0 & -3 & 2 & 0 \end{array}\right] \times(-1)$$

$$\rightarrow \left[\begin{array}{ccc|c} 1 & 0 & -1 & -2 \\ 0 & 1 & 1 & 5 \\ 0 & -3 & 2 & 0 \end{array}\right] \times(3) \rightarrow \left[\begin{array}{ccc|c} 1 & 0 & -1 & -2 \\ 0 & 1 & 1 & 5 \\ 0 & 0 & 5 & 15 \end{array}\right] \times(\frac{1}{5})$$

$$\rightarrow \left[\begin{array}{ccc|c} 1 & 0 & 0 & 1 \\ 0 & 1 & 1 & 5 \\ 0 & 0 & 5 & 15 \end{array}\right] (-\frac{1}{5}) \rightarrow \left[\begin{array}{ccc|c} 1 & 0 & 0 & 1 \\ 0 & 1 & 0 & 2 \\ 0 & 0 & 5 & 15 \end{array}\right] \times(\frac{1}{5}) \rightarrow \left[\begin{array}{ccc|c} 1 & 0 & 0 & 1 \\ 0 & 1 & 0 & 2 \\ 0 & 0 & 1 & 3 \end{array}\right]$$

$$\begin{cases} x+y=3 \\ 2x+3y+z=11 \\ 2x-y+2z=6 \end{cases} \rightarrow \begin{cases} x+0+0=1 \\ 0+y+0=2, \quad x=1, \quad y=2, \quad z=3 \\ 0+0+z=3 \end{cases}$$

- Solving by Gauss-Jordan elimination

It simplifies Gaussian-elimination and concentrates on the last row to find the solution of that variable first.

Example: Solve $\begin{cases} x+y=3 \\ 2x+3y+z=11 \\ 2x-y+2z=6 \end{cases}$

Step 1. Form an extended matrix

$$\left[\begin{array}{ccc|c} 1 & 1 & 0 & 3 \\ 2 & 3 & 1 & 11 \\ 2 & -1 & 2 & 6 \end{array}\right] \times(-2) \rightarrow \left[\begin{array}{ccc|c} 1 & 1 & 0 & 3 \\ 0 & 1 & 1 & 5 \\ 0 & -3 & 2 & 0 \end{array}\right] \times 3$$

$$\rightarrow \left[\begin{array}{ccc|c} 1 & 1 & 0 & 3 \\ 0 & 1 & 1 & 5 \\ 0 & 0 & 5 & 15 \end{array}\right] \times(\frac{1}{5}) \rightarrow \left[\begin{array}{ccc|c} 1 & 1 & 0 & 3 \\ 0 & 1 & 1 & 5 \\ \boxed{0} & \boxed{0} & \boxed{1} & \boxed{3} \end{array}\right]$$

$$\begin{bmatrix} 1 & 1 & 0 \\ 0 & 1 & 1 \\ 0 & 0 & 1 \end{bmatrix} \begin{bmatrix} x \\ y \\ z \end{bmatrix} = \begin{bmatrix} 3 \\ 5 \\ 3 \end{bmatrix}$$

$0x+0y+z=3, \quad z=3$

Step 2. Substitute to original equation

$x+y=3 \cdots\cdots ①$

$2x+3y+3=11 \cdots\cdots ②$

$① \times 2-②\quad -y=-2,\ y=2$

$x+2=3,\ x=1$

$x=1,\ y=2,\ z=3$

- Solving by Gramer's Elimination

For two variables system equations

$$\begin{cases} a_1x+b_1y=c_1 \cdots\cdots ① \\ a_2x+b_2y=c_2 \cdots\cdots ② \end{cases}$$

Eliminate y $① \times b_2 - ② \times b_1$

$a_1b_2x+b_1b_2y=c_1b_2 \cdots\cdots ③$

$a_2b_1x+b_1b_2y=c_2b_1 \cdots\cdots ④$

$③-④\ (a_1b_2-a_2b_1)x=c_1b_2-c_2b_1$

$$\therefore\ x=\frac{c_1b_2-c_2b_1}{a_1b_2-a_2b_1}$$

Since $a_1b_2-a_2b_2$ is the determinant of matrix

$$\begin{bmatrix} a_1 & b_1 \\ a_2 & b_2 \end{bmatrix}=a_1b_2-a_2b_1=\Delta \text{ (determinant)}$$

$c_1b_2-c_2b_1$ is the determinant of matrix of $\begin{bmatrix} c_1 & b_1 \\ c_2 & b_2 \end{bmatrix}$

$$\text{So, } x=\frac{\begin{vmatrix} c_1 & b_1 \\ c_2 & b_2 \end{vmatrix}}{D}=\frac{D_x}{D}$$

$$\text{The same reason } Y=\frac{\begin{vmatrix} c_1 & b_1 \\ c_2 & b_2 \end{vmatrix}}{D}=\frac{D_y}{D}$$

For all square Matrix with the determinant $\neq 0$ applied.

For three variables system Equations.

$$\begin{cases} a_1x+b_1y+c_1z=d_1 \\ a_2x+b_2y+c_2z=d_2 \\ a_3x+b_3y+c_3z=d_3 \end{cases}$$

$x=\dfrac{D_x}{D}$, $y=\dfrac{D_y}{D}$, $z=\dfrac{D_z}{D}$

D-determinate of matrix $\begin{bmatrix} a_1 & b_1 & c_1 \\ a_2 & b_2 & c_2 \\ a_3 & b_3 & c_3 \end{bmatrix}$

D_x-determinate of matrix $\begin{bmatrix} d_1 & b_1 & c_1 \\ d_2 & b_2 & c_2 \\ d_3 & b_3 & c_3 \end{bmatrix}$

D_y-determinate of matrix $\begin{bmatrix} a_1 & d_1 & c_1 \\ a_2 & d_2 & c_2 \\ a_3 & d_3 & c_3 \end{bmatrix}$

D_z-determinate of matrix $\begin{bmatrix} a_1 & b_1 & d_1 \\ a_2 & b_2 & d_2 \\ a_3 & b_3 & d_3 \end{bmatrix}$

Example: Solve $\begin{cases} x+y=3 \\ 2x+3y+z=11 \\ 2x-y+2z=6 \end{cases}$

 **Step 1.** Find the determinant

$$A=\begin{bmatrix} 1 & 1 & 0 \\ 2 & 3 & 1 \\ 2 & -1 & 2 \end{bmatrix}$$

$0-1+4=3$

$$\begin{matrix} 1 & 1 & 0 & 1 & 1 \\ 2 & 3 & 1 & 2 & 3 \\ 2 & -1 & 2 & 2 & -1 \end{matrix}$$

$6+2=8$

$D=\det(A)=8-3=5$

- Method 2 (down grade)

$$D=(-1)^2\times\begin{vmatrix} 3 & 1 \\ -1 & 2 \end{vmatrix}+(-1)^{1+2}\cdot\begin{vmatrix} 2 & 1 \\ 2 & 2 \end{vmatrix}+(-1)^{1+3}\cdot(0)\begin{vmatrix} 2 & 3 \\ 2 & -1 \end{vmatrix}$$

$$=[6-(-1)]+(-1)[4-2]+0=7-2=5$$

Change the constant column to column 1

 **Step 2.** $D_1 = \begin{vmatrix} 3 & 1 & 0 \\ 11 & 3 & 1 \\ 6 & -1 & 2 \end{vmatrix}$

$$\begin{array}{ccccc} & & 0 & -3 & +22 \\ 3 & 1 & 0 & 3 & 1 \\ 11 & 3 & 1 & 11 & 3 \\ 6 & -1 & 2 & 6 & -1 \\ & & 18 & 6 & 0 \end{array}$$

$D_1 \rightarrow = 18+6+0-0-(-3)-22=5$

Change the Constant Column to Column 2

$$\begin{array}{ccccc} & & 0 & 6 & 12 \\ 1 & 3 & 0 & 1 & 3 \\ 2 & 11 & 1 & 2 & 11 \\ 2 & 6 & 2 & 2 & 6 \\ & & 22 & 6 & 0 \end{array}$$

$D_2 \rightarrow \quad 22+6+0-0-6-12=10$

Change the constant Column to column 3

$$\begin{array}{ccccc} & & 18 & -11 & 12 \\ 1 & 1 & 3 & 1 & 1 \\ 2 & 3 & 11 & 2 & 3 \\ 2 & -1 & 6 & 2 & -1 \\ & & 18 & 22 & -6 \end{array}$$

$D_3 \rightarrow \quad 18+22-6-18+11-12=15$

So, $x=\dfrac{D_1}{D}=\dfrac{5}{5}=1$

$y=\dfrac{D_2}{D}=\dfrac{10}{5}=2$

$z=\dfrac{D_3}{D}=\dfrac{15}{5}=3$

◆Example:

$$\begin{cases} x+y+z=6 \cdots ① \\ 2x+y-z=1 \cdots ② \\ 2x-y+z=3 \cdots ③ \end{cases}$$

$$A=\begin{bmatrix} 1 & 1 & 1 \\ 2 & 1 & -1 \\ 2 & -1 & 1 \end{bmatrix},\quad x=\begin{bmatrix} x \\ y \\ z \end{bmatrix},\quad B=\begin{bmatrix} 6 \\ 1 \\ 3 \end{bmatrix}$$

Step 1. Find determinant of $A \Rightarrow \det(A)$

$$\begin{array}{ccccc} & & 2 & 1 & 2 \\ 1 & 1 & 1 & 1 & 1 \\ 2 & 1 & -1 & 2 & 1 \\ 2 & -1 & 1 & 2 & -1 \\ & & 1 & -2 & -2 \end{array}$$

$$\det(A) = 1 + (-2) + (-2) - (2) - (1) - 2 = -8$$

- Using down grade

$$A = (-1)^2 \begin{vmatrix} 1 & -1 \\ -1 & 1 \end{vmatrix} + (-1)^{1+2} \begin{vmatrix} 2 & -1 \\ 2 & 1 \end{vmatrix} + (-1)^{1+3} \begin{vmatrix} 2 & 1 \\ 2 & -1 \end{vmatrix}$$

$$\det(A) = (1-1) + (-1)[2-(-2)] + (-2-2) = -4 - 4 = -8$$

Step 2. $x = \dfrac{D_x}{D}, \quad y = \dfrac{D_y}{D}, \quad z = \dfrac{D_z}{D}$

Let B in column 1 $D_x =$

$$\begin{array}{ccccc} & & 3 & 6 & 1 \\ 6 & 1 & 1 & 6 & 1 \\ 1 & 1 & -1 & 1 & 1 \\ 3 & -1 & 1 & 3 & -1 \\ & & 6 & -3 & -1 \end{array}$$

$$D_x = 6 + (-3) + (-1) - 3 - 6 - 1 = -8$$

Let B in column 2 $D_y =$

$$\begin{array}{ccccc} & & 2 & -3 & 12 \\ 1 & 6 & 1 & 1 & 6 \\ 2 & 1 & -1 & 2 & 1 \\ 2 & 3 & 1 & 2 & 3 \\ & & 1 & -12 & 6 \end{array}$$

$$D_y = (1) + (-12) + (6) - (2) - (-3) - (12) = -16$$

Let B in column 3 $D_z =$

$$\begin{array}{ccccc} & & 12 & -1 & 6 \\ 1 & 1 & 6 & 1 & 1 \\ 2 & 1 & 1 & 2 & 1 \\ 2 & -1 & 3 & 2 & -1 \\ & & 3 & 2 & -12 \end{array}$$

$$D_z = (3) + (2) + (-12) - (12) - (-1) - (6) = -24$$

$$x = \frac{D_x}{D} = \frac{-8}{-8} = 1$$

$$y = \frac{D_y}{D} = \frac{-16}{-8} = 2$$

$$z = \frac{D_z}{D} = \frac{-24}{-8} = 3$$

Answer to Concept Check 14-08

Step 1: Find the determinant

$$\begin{bmatrix} 1 & 1 & 1 \\ 2 & -1 & 3 \\ 1 & -3 & -1 \end{bmatrix}\begin{bmatrix} x \\ y \\ z \end{bmatrix}=\begin{bmatrix} 2 \\ 9 \\ 2 \end{bmatrix}$$

method 1:

$$\begin{array}{ccccc} & & -1 & -9 & -2 \\ 1 & 1 & 1 & 1 & 1 \\ 2 & -1 & 3 & 2 & -1 \\ 1 & -3 & -1 & 1 & -3 \\ & & 1 & 3 & -6 \end{array}$$

$$\det = 1+3-6-(-1)-(-9)-(-2) = 4-6+1+9+2=10$$

method 2:

$$\begin{bmatrix} 1 & 1 & 1 \\ 2 & -1 & 3 \\ 1 & -3 & -1 \end{bmatrix} = (-1)^{1+1}\times\begin{vmatrix} -1 & 3 \\ -3 & -1 \end{vmatrix}+(-1)^{3}\begin{vmatrix} 2 & 3 \\ 1 & -1 \end{vmatrix}+(-1)^{1+3}\begin{vmatrix} 2 & -1 \\ 1 & -3 \end{vmatrix}$$

$$=\begin{vmatrix} -1 & 3 \\ -3 & -1 \end{vmatrix}-\begin{vmatrix} 2 & 3 \\ 1 & -1 \end{vmatrix}+\begin{vmatrix} 2 & -1 \\ 1 & -3 \end{vmatrix}$$

$$=1-(-9)-(-2-3)+(-6+1)=10$$

Step 2: $x=\dfrac{D_1}{D}$, $y=\dfrac{D_2}{D}$, $z=\dfrac{D_3}{D}$

$$D_1=\begin{vmatrix} 2 & 1 & 1 \\ 9 & -1 & 3 \\ 2 & -3 & -1 \end{vmatrix} \quad \begin{array}{ccccc} & & -2 & -18 & -9 \\ 2 & 1 & 1 & 2 & 1 \\ 9 & -1 & 3 & 9 & -1 \\ 2 & -3 & -1 & 2 & -3 \\ & & 2 & 6 & -27 \end{array} \quad =2+6-27+2+18+9=10$$

$$D_2=\begin{vmatrix} 1 & 2 & 1 \\ 2 & 9 & 3 \\ 1 & 2 & -1 \end{vmatrix} \quad \begin{array}{ccccc} & & 9 & 6 & -4 \\ 1 & 2 & 1 & 1 & 2 \\ 2 & 9 & 3 & 2 & 9 \\ 1 & 2 & -1 & 1 & 2 \\ & & -9 & 6 & 4 \end{array} \quad =-9+6+4-9-6+4=-10$$

$$D_3 = \begin{vmatrix} 1 & 1 & 2 \\ 2 & -1 & 9 \\ 1 & -3 & 2 \end{vmatrix} \begin{array}{ccccc} & & -2 & -27 & 4 \\ 1 & 1 & 2 & 1 & 1 \\ 2 & -1 & 9 & 2 & -1 \\ 1 & -3 & 2 & 1 & -3 \\ & & -2 & 9 & -12 \end{array} = -2+9-12+2+27-4=20$$

$$x=\frac{10}{10}=1,\quad y=\frac{-10}{10}=-1,\quad z=\frac{20}{10}=2$$

CHAPTER 14
MATRIX
PRACTICE

1. Find the resultant matrix of $A+B$,

where $A=\begin{bmatrix} 2 & -1 & 3 \\ 1 & -2 & -1 \\ 3 & 2 & -2 \end{bmatrix}$,

$B=\begin{bmatrix} 1 & 3 & 2 \\ 2 & 1 & 3 \\ 3 & 2 & 1 \end{bmatrix}$

2. If $A=\begin{bmatrix} 2 & 3 \\ -1 & 1 \end{bmatrix}$, $B=\begin{bmatrix} 1 & 2 & 3 \\ 1 & -2 & 2 \end{bmatrix}$,

and $C=\begin{bmatrix} -1 & 3 & 2 \\ 1 & 2 & -1 \end{bmatrix}$

Find $3A+2B$, and $3B+2C$

3. What is the inverse matrix of matrix

$A=\begin{bmatrix} 1 & 2 \\ 3 & 4 \end{bmatrix}$?

(A) $\begin{bmatrix} 2 & 1 \\ 4 & 3 \end{bmatrix}$

(B) $\begin{bmatrix} -1 & 2 \\ -3 & 4 \end{bmatrix}$

(C) $\begin{bmatrix} 1 & -\frac{1}{2} \\ 3 & -\frac{3}{2} \end{bmatrix}$

(D) $\begin{bmatrix} -2 & 1 \\ \frac{3}{2} & -\frac{1}{2} \end{bmatrix}$

(E) $\begin{bmatrix} \frac{3}{2} & 2 \\ -1 & \frac{1}{2} \end{bmatrix}$

4. What is the determinant of matrix

$A=\begin{bmatrix} 2 & -3 & 1 \\ 1 & 2 & -3 \\ 3 & -1 & 2 \end{bmatrix}$?

(A) 28

(B) 30

(C) 32

(D) 34

(E) 36

5. What is the solutions of the system linear

equations $\begin{cases} x+3y+2z=2 \\ 2x-y+2z=7 \\ 2x+3y+z=1 \end{cases}$?

6. If matrix $A=\begin{bmatrix} 2 & 1 \\ -1 & 3 \end{bmatrix}$,

$B=\begin{bmatrix} 1 & 2 & -1 \\ 3 & 1 & 2 \end{bmatrix}$ and $c=\begin{bmatrix} 2 & 3 & -2 \\ 3 & -1 & 0 \end{bmatrix}$,

what is the result after operation of $A(B+2C)$?

7. Find the order of AB and BA, if the orders for A and B are following.

I. A: 3×2, B: 2×3

II. A: 2×3, B: 3×4

III. A: 2×3, B: 2×3

8. Find determinant of matrix

$$A=\begin{bmatrix}1 & -2 & 2\\ 4 & -1 & 3\\ 2 & 3 & 4\end{bmatrix}$$

(A) 21

(B) 25

(C) 27

(D) 31

(E) 35

9. Solve $\begin{cases}x+2z=1\\ 2x+y+3z=3\\ 2x+3y+4z=8\end{cases}$

10. If $AC=\begin{bmatrix}-2 & 4\\ 0 & -2\end{bmatrix}$, $AD=\begin{bmatrix}-1 & 2\\ -1 & -2\end{bmatrix}$, $BC=\begin{bmatrix}0 & -4\\ -1 & 3\end{bmatrix}$, and $BD=\begin{bmatrix}-2 & -4\\ 0 & 2\end{bmatrix}$

What is the value of $(2A-3B)\ (C-D)$?

11. Solve $\begin{cases}x+y+z=2\\ 2x+y-2z=-3\\ x-z=-1\end{cases}$ by inverse

matrix.

12. Solve $\begin{cases}x+y+z=2\\ 2x+y-2z=-3\\ x-z=-1\end{cases}$

by Gaussian Elimination.

13. Solve $\begin{cases}x+y+z=2\\ 2x+y-2z=-3\\ x-z=-1\end{cases}$

by Gauss-Jordan Elimination.

14. Solve $\begin{cases}x+y+z=2\\ 2x+y-2z=-3\\ x-z=-1\end{cases}$

by Cramer's Rule.

CHAPTER 14
MATRIX
ANSWERS and EXPLANATIONS

1. $A+B=\begin{bmatrix} 2+1 & (-1)+3 & 3+2 \\ 1+2 & (-2)+1 & (-1)+3 \\ 3+3 & 2+2 & (-2)+1 \end{bmatrix}$

$=\begin{bmatrix} 3 & 2 & 5 \\ 3 & -1 & 2 \\ 6 & 4 & -1 \end{bmatrix}$

2. $3A+2B=\begin{bmatrix} 6 & 9 \\ -3 & 3 \end{bmatrix}+\begin{bmatrix} 2 & 4 & 6 \\ 2 & -4 & 4 \end{bmatrix}$

undefined

$3B+2C=\begin{bmatrix} 3 & 6 & 9 \\ 3 & -6 & 6 \end{bmatrix}+\begin{bmatrix} -2 & 6 & 4 \\ 2 & 4 & -2 \end{bmatrix}$

$=\begin{bmatrix} 3-2 & 6+6 & 9+4 \\ 3+2 & (-6)+4 & 6+(-2) \end{bmatrix}$

$=\begin{bmatrix} 1 & 12 & 13 \\ 5 & -2 & 4 \end{bmatrix}$

3. Asking A^{-1} form extended matrix

$\left[\begin{array}{cc|cc} 1 & 2 & 1 & 0 \\ 3 & 4 & 0 & 1 \end{array}\right]$ $\times(-3)$ (row 1 added to row 2)

$=\left[\begin{array}{cc|cc} 1 & 2 & 1 & 0 \\ 0 & -2 & -3 & 1 \end{array}\right]$ $\times(1)$ (row 2 added to row 1)

$=\left[\begin{array}{cc|cc} 1 & 0 & -2 & 1 \\ 0 & -2 & -3 & 1 \end{array}\right]\times(-\frac{1}{2})$

$=\left[\begin{array}{cc|cc} 1 & 0 & -2 & 1 \\ 0 & 1 & \frac{3}{2} & -\frac{1}{2} \end{array}\right]$

so, $A^{-1}=\begin{bmatrix} -2 & 1 \\ \frac{3}{2} & -\frac{1}{2} \end{bmatrix}$

or, $A^{-1}=\dfrac{1}{4-6}\begin{bmatrix} 4 & -2 \\ -3 & 1 \end{bmatrix}=\dfrac{1}{-2}\begin{bmatrix} 4 & -2 \\ -3 & 1 \end{bmatrix}$

$=\begin{bmatrix} -2 & 1 \\ \frac{3}{2} & -\frac{1}{2} \end{bmatrix}$

Answer is (D)

4. Asking $D=$

Method 1: $D=2\begin{vmatrix} 2 & -3 \\ -1 & 2 \end{vmatrix}$

$-(-3)\begin{vmatrix} 1 & -3 \\ 3 & 2 \end{vmatrix}+\begin{vmatrix} 1 & 2 \\ 3 & -1 \end{vmatrix}$

$=2(4-3)+3(2+9)+(-1-6)$

$=28$

Method 2: $D=\begin{matrix} 2 & -3 & 1 & 2 & -3 \\ 1 & 2 & -3 & 1 & 2 \\ 3 & -1 & 2 & 3 & -1 \end{matrix}$

$=8+27-1-6-6+6=28$

Answer is (A)

5. Asking $x=,\ y=,\ z=$

Coefficient of Matrix $A=\begin{bmatrix} 1 & 3 & 2 \\ 2 & -1 & 2 \\ 2 & 3 & 1 \end{bmatrix}$

$X=\begin{bmatrix} x \\ y \\ z \end{bmatrix},\ B=\begin{bmatrix} 2 \\ 7 \\ 1 \end{bmatrix}$

Method 1: $x=\dfrac{D_1}{D},\ y=\dfrac{D_2}{D}$

$z=\dfrac{D_3}{D}$

$$D = \begin{matrix} 1 & 3 & 2 & 1 & 3 \\ 2 & -1 & 2 & 2 & -1 \\ 2 & 3 & 1 & 2 & 3 \end{matrix} = 15$$

$$D_1 = \begin{matrix} 2 & 3 & 2 & 2 & 3 \\ 7 & -1 & 2 & 7 & -1 \\ 1 & 3 & 1 & 1 & 3 \end{matrix}$$

$$= -2 + 6 + 42 - (-2) - 12 - 21 = 15$$

$$D_2 = \begin{matrix} 1 & 2 & 2 & 1 & 2 \\ 2 & 7 & 2 & 2 & 7 \\ 2 & 1 & 1 & 2 & 1 \end{matrix}$$

$$= 7 + 8 + 4 - 28 - 2 - 4 = -15$$

$$D_3 = \begin{matrix} 1 & 3 & 2 & 1 & 3 \\ 2 & -1 & 7 & 2 & -1 \\ 2 & 3 & 1 & 2 & 3 \end{matrix}$$

$$= -1 + 42 + 12 - (-4) - 21 - 6 = 30$$

So, $x = \frac{D_1}{D} = \frac{15}{15} = 1$

$y = \frac{D_2}{D} = \frac{-15}{15} = -1$

$z = \frac{D_3}{D} = \frac{30}{15} = 2$

Method 2:

Since $A = \begin{bmatrix} 1 & 3 & 2 \\ 2 & -1 & 2 \\ 2 & 3 & 1 \end{bmatrix} = \begin{bmatrix} -1 & 2 \\ 3 & 1 \end{bmatrix}$

$-(3)\begin{bmatrix} 2 & 2 \\ 2 & 1 \end{bmatrix} + 2\begin{bmatrix} 2 & -1 \\ 2 & 3 \end{bmatrix}$

$D = |A| = (-1-6) - 3(2-4) + 2(6+2)$

$= (-7) + 6 + 16 = 15$

$D_1 = \begin{bmatrix} 2 & 3 & 2 \\ 7 & -1 & 2 \\ 1 & 3 & 1 \end{bmatrix}$

$= 2\begin{bmatrix} -1 & 2 \\ 3 & 1 \end{bmatrix} - (3)\begin{bmatrix} 7 & 2 \\ 1 & 1 \end{bmatrix} + 2\begin{bmatrix} 7 & -1 \\ 1 & 3 \end{bmatrix}$

$= 2(-1-6) - (3)(7-2) + (2)(21+1)$

$= 2(-7) - 15 + 44$

$= -14 - 15 + 44 = 15$

$D_2 = \begin{bmatrix} 1 & 2 & 2 \\ 2 & 7 & 2 \\ 2 & 1 & 1 \end{bmatrix}$

$= \begin{bmatrix} 7 & 2 \\ 1 & 1 \end{bmatrix} - (2)\begin{bmatrix} 2 & 2 \\ 2 & 1 \end{bmatrix} + 2\begin{bmatrix} 2 & 7 \\ 2 & 1 \end{bmatrix}$

$= (7-2) - (2)(2-4) + (2)(2-14)$

$= 5 + 4 - 24 = -15$

$D_3 = \begin{bmatrix} 1 & 3 & 2 \\ 2 & -1 & 7 \\ 2 & 3 & 1 \end{bmatrix}$

$= (1)\begin{bmatrix} -1 & 7 \\ 3 & 1 \end{bmatrix} - (3)\begin{bmatrix} 2 & 7 \\ 2 & 1 \end{bmatrix} + 2\begin{bmatrix} 2 & -1 \\ 2 & 3 \end{bmatrix}$

$= (-1-21) - (3)(2-14) + 2(6+2)$

$= (-22) + 36 + 16 = 30$

So, $x = \frac{15}{15} = 1$, $y = \frac{-15}{15} = -1$,

$z = \frac{30}{15} = 2$

6. Asking: $A(B+2C)$?

$A(B+2C) = AB + 2AC$

$AB = \begin{bmatrix} 2 & 1 \\ -1 & 3 \end{bmatrix}\begin{bmatrix} 1 & 2 & -1 \\ 3 & 1 & 2 \end{bmatrix}$

$= \begin{bmatrix} 2+3 & 4+1 & -2+2 \\ -1+9 & -2+3 & 1+6 \end{bmatrix}$

$= \begin{bmatrix} 5 & 5 & 0 \\ 8 & 1 & 7 \end{bmatrix}$

$2AC = 2\begin{bmatrix} 2 & 1 \\ -1 & 3 \end{bmatrix}\begin{bmatrix} 2 & 3 & -2 \\ 3 & -1 & 0 \end{bmatrix}$

$= 2\begin{bmatrix} 4+3 & 6-1 & -4+0 \\ -2+9 & (-3)+(-3) & 2+0 \end{bmatrix}$

$$=\begin{bmatrix}14 & 10 & -8\\ 14 & -12 & 4\end{bmatrix}$$

$A(B+2C)$

$$=\begin{bmatrix}5 & 5 & 0\\ 8 & 1 & 7\end{bmatrix}+\begin{bmatrix}14 & 10 & -8\\ 14 & -12 & 4\end{bmatrix}$$

$$=\begin{bmatrix}19 & 15 & -8\\ 22 & -11 & 11\end{bmatrix}$$

7. I. $A\times B \Rightarrow [3\times2][2\times3] \Rightarrow (3\times3)$

$B\times A\,[2\times3][3\times2] \Rightarrow (2\times2)$

II. $A\times B\,[2\times3][3\times4]$

$A_{2\times3}\times B_{3\times4} \Rightarrow 2\times4$

$B\times A \quad B_{3\times4}\times A_{2\times3} \Rightarrow$ undefined

III. $A\times B \quad A_{2\times3}\times B_{2\times3}$

$\Rightarrow$ undefined

$B\times A \quad B_{2\times3}\times A_{2\times3}$ undefined

8. $D=(1)\begin{vmatrix}-1 & 3\\ 3 & 4\end{vmatrix}-(-2)\begin{vmatrix}4 & 3\\ 2 & 4\end{vmatrix}$

$+2\begin{vmatrix}4 & -1\\ 2 & 3\end{vmatrix}$

$=(-4-9)+2(16-6)+2(12+2)$

$=-13+20+28$

$=35$

Answer is (E)

9. Solve system equations

$$\begin{cases}x+2z=1\\ 2x+y+3z=3\\ 2x+3y+4z=8\end{cases}$$

Method 1. use A^{-1}

Step 1. Form Matrix

$$\left[\begin{array}{ccc|ccc}1 & 0 & 2 & 1 & 0 & 0\\ 2 & 1 & 3 & 0 & 1 & 0\\ 2 & 3 & 4 & 0 & 0 & 1\end{array}\right]\times(-2)$$

$$=\left[\begin{array}{ccc|ccc}1 & 0 & 2 & 1 & 0 & 0\\ 0 & 1 & -1 & -2 & 1 & 0\\ 0 & 3 & 0 & -2 & 0 & 1\end{array}\right]\times(-3)$$

$$=\left[\begin{array}{ccc|ccc}1 & 0 & 2 & 1 & 0 & 0\\ 0 & 1 & -1 & -2 & 1 & 0\\ 0 & 0 & 3 & 4 & -3 & 1\end{array}\right]\times(\frac{1}{3})$$

$$=\left[\begin{array}{ccc|ccc}1 & 0 & 2 & 1 & 0 & 0\\ 0 & 1 & 0 & -\frac{2}{3} & 0 & \frac{1}{3}\\ 0 & 0 & 3 & 4 & -3 & 1\end{array}\right]\times(-\frac{2}{3})$$

$$=\left[\begin{array}{ccc|ccc}1 & 0 & 0 & -\frac{5}{3} & 2 & -\frac{2}{3}\\ 0 & 1 & 0 & -\frac{2}{3} & 0 & \frac{1}{3}\\ 0 & 0 & 3 & 4 & -3 & 1\end{array}\right]\times\frac{1}{3}$$

$$=\left[\begin{array}{ccc|ccc}1 & 0 & 0 & -\frac{5}{3} & 2 & -\frac{2}{3}\\ 0 & 1 & 0 & -\frac{2}{3} & 0 & \frac{1}{3}\\ 0 & 0 & 1 & \frac{4}{3} & -1 & \frac{1}{3}\end{array}\right]$$

$$A^{-1}=\begin{bmatrix}-\frac{5}{3} & 2 & -\frac{2}{3}\\ -\frac{2}{3} & 0 & \frac{1}{3}\\ \frac{4}{3} & -1 & \frac{1}{3}\end{bmatrix}$$

Step 2.

Since $X=A^{-1}B$

$$\begin{bmatrix} x \\ y \\ z \end{bmatrix} = \begin{bmatrix} \frac{-5}{3} & 2 & -\frac{2}{3} \\ -\frac{2}{3} & 0 & \frac{1}{3} \\ \frac{4}{3} & -1 & \frac{1}{3} \end{bmatrix} \begin{bmatrix} 1 \\ 3 \\ 8 \end{bmatrix}$$

$$= \frac{1}{3} \begin{bmatrix} -5 & 6 & -2 \\ -2 & 0 & 1 \\ 4 & -3 & 1 \end{bmatrix} \begin{bmatrix} 1 \\ 3 \\ 8 \end{bmatrix}$$

$$= \frac{1}{3} \begin{bmatrix} -5+18-16 \\ -2+0+8 \\ 4-9+8 \end{bmatrix} = \frac{1}{3} \begin{bmatrix} -3 \\ 6 \\ 3 \end{bmatrix} = \begin{bmatrix} -1 \\ 2 \\ 1 \end{bmatrix}$$

$x=-1$, $y=2$, $z=1$

Method 2, use Determinants

Step 1. Find determinant

$$\begin{bmatrix} 1 & 0 & 2 \\ 2 & 1 & 3 \\ 2 & 3 & 4 \end{bmatrix}$$

$$= (1)\begin{bmatrix} 1 & 3 \\ 3 & 4 \end{bmatrix} - (0)\begin{bmatrix} 2 & 3 \\ 2 & 4 \end{bmatrix} + (2)\begin{bmatrix} 2 & 1 \\ 2 & 3 \end{bmatrix}$$

$$D = (1\times4-3\times3)+2(2\times3-2\times1)$$

$$= -5+8=+3$$

$$D_1 = \begin{vmatrix} 1 & 0 & 2 \\ 3 & 1 & 3 \\ 8 & 3 & 4 \end{vmatrix}$$

$$= (1)\begin{vmatrix} 1 & 3 \\ 3 & 4 \end{vmatrix} + 0\begin{vmatrix} 3 & 3 \\ 8 & 4 \end{vmatrix} + (2)\begin{vmatrix} 3 & 1 \\ 8 & 3 \end{vmatrix}$$

$$\therefore\ D_1 = (4-9)+(2)(9-8)$$

$$= -5+2=-3$$

$$D_2 = \begin{vmatrix} 1 & 1 & 2 \\ 2 & 3 & 3 \\ 2 & 8 & 4 \end{vmatrix}$$

$$= (1)\begin{vmatrix} 3 & 3 \\ 8 & 4 \end{vmatrix} + (-1)^{1-2}(1)\begin{vmatrix} 2 & 3 \\ 2 & 4 \end{vmatrix}$$

$$+(2)\begin{vmatrix} 2 & 3 \\ 2 & 8 \end{vmatrix}$$

$$D_2 = (12-24)-(8-6)+(2)(16-6)$$

$$= -12-2+20=6$$

$$D_3 = \begin{vmatrix} 1 & 0 & 1 \\ 2 & 1 & 3 \\ 2 & 3 & 8 \end{vmatrix}$$

$$= (1)\begin{vmatrix} 1 & 3 \\ 3 & 8 \end{vmatrix} - 0 + (1)\begin{vmatrix} 2 & 1 \\ 2 & 3 \end{vmatrix}$$

$$\therefore\ D_3 = (1)(8-9)+(1)(6-2)$$

$$= -1+4=3$$

$$x = \frac{D_1}{D} = \frac{-3}{3} = -1$$

$$y = \frac{D_2}{D} = \frac{6}{3} = 2$$

$$z = \frac{D_3}{D} = \frac{3}{3} = 1$$

10. $AC - AD = A(C-D)$

$$= \begin{bmatrix} -2 & 4 \\ 0 & -2 \end{bmatrix} - \begin{bmatrix} -1 & 2 \\ -1 & -2 \end{bmatrix}$$

$$= \begin{bmatrix} -1 & 2 \\ 1 & 0 \end{bmatrix}$$

$$2A(C-D) = 2\begin{bmatrix} -1 & 2 \\ 1 & 0 \end{bmatrix} = \begin{bmatrix} -2 & 4 \\ 2 & 0 \end{bmatrix}$$

$$BC - BD = \begin{bmatrix} 0 & -4 \\ -1 & 3 \end{bmatrix} - \begin{bmatrix} -2 & -4 \\ 0 & 2 \end{bmatrix}$$

$$= \begin{bmatrix} 2 & 0 \\ -1 & 1 \end{bmatrix}$$

$$\therefore\ 3B(C-D) = 3\begin{bmatrix} 2 & 0 \\ -1 & 1 \end{bmatrix} = \begin{bmatrix} 6 & 0 \\ -3 & 3 \end{bmatrix}$$

$$\therefore\ (2A-3B)(C-D)$$

$$= 2A(C-D) - 3B(C-D)$$

$$=\begin{bmatrix}-2 & 4\\ 2 & 0\end{bmatrix}-\begin{bmatrix}6 & 0\\ -3 & 3\end{bmatrix}$$

$$=\begin{bmatrix}-2-6 & 4-0\\ 2-(-3) & 0-3\end{bmatrix}$$

$$=\begin{bmatrix}-8 & 4\\ 5 & -3\end{bmatrix}$$

11. Solve $\begin{cases}x+y+z=2\\ 2x+y-2z=-3\\ x-z=-1\end{cases}$

By inverse matrix.

Let $A=\begin{bmatrix}1 & 1 & 1\\ 2 & 1 & -2\\ 1 & 0 & -1\end{bmatrix}$, $V=\begin{bmatrix}x\\ y\\ z\end{bmatrix}$,

$$B=\begin{bmatrix}2\\ -3\\ -1\end{bmatrix}$$

Form an extended matrix with identity matrix.

$$[A|I]=\left[\begin{array}{ccc|ccc}1 & 1 & 1 & 1 & 0 & 0\\ 2 & 1 & -2 & 0 & 1 & 0\\ 1 & 0 & -1 & 0 & 0 & 1\end{array}\right]\times(-2),\times(-1)$$

$$=\left[\begin{array}{ccc|ccc}1 & 1 & 1 & 1 & 0 & 0\\ 0 & -1 & -4 & -2 & 1 & 0\\ 0 & -1 & -2 & -1 & 0 & 1\end{array}\right]\times(-1)$$

$$=\left[\begin{array}{ccc|ccc}1 & 1 & 1 & 1 & 0 & 0\\ 0 & 1 & 4 & 2 & -1 & 0\\ 0 & -1 & -2 & -1 & 0 & 1\end{array}\right]\times(1)$$

$$=\left[\begin{array}{ccc|ccc}1 & 1 & 1 & 1 & 0 & 0\\ 0 & 1 & 4 & 2 & -1 & 0\\ 0 & 0 & 2 & 1 & -1 & 1\end{array}\right]\times(-1)$$

$$=\left[\begin{array}{ccc|ccc}1 & 0 & -3 & -1 & 1 & 0\\ 0 & 1 & 4 & 2 & -1 & 0\\ 0 & 0 & 2 & 1 & -1 & 1\end{array}\right]\times(-2)$$

$$=\left[\begin{array}{ccc|ccc}1 & 0 & -3 & -1 & 1 & 0\\ 0 & 1 & 0 & 0 & 1 & -2\\ 0 & 0 & 2 & 1 & -1 & 1\end{array}\right]\times(\frac{1}{2})$$

$$=\left[\begin{array}{ccc|ccc}1 & 0 & -3 & -1 & 1 & 0\\ 0 & 1 & 0 & 0 & 1 & -2\\ 0 & 0 & 1 & \frac{1}{2} & -\frac{1}{2} & \frac{1}{2}\end{array}\right]\times(3)$$

$$=\left[\begin{array}{ccc|ccc}1 & 0 & 0 & \frac{1}{2} & -\frac{1}{2} & \frac{3}{2}\\ 0 & 1 & 0 & 0 & 1 & -2\\ 0 & 0 & 1 & \frac{1}{2} & -\frac{1}{2} & \frac{1}{2}\end{array}\right]$$

$$=\frac{1}{2}\left[\begin{array}{ccc|ccc}1 & 0 & 0 & 1 & -1 & 3\\ 0 & 1 & 0 & 0 & 2 & -4\\ 0 & 0 & 1 & 1 & -1 & 1\end{array}\right]$$

$$\therefore\ A^{-1}=\frac{1}{2}\begin{bmatrix}1 & -1 & 3\\ 0 & 2 & -4\\ 1 & -1 & 1\end{bmatrix}$$

Since $AV=B$, $A^{-1}AV=A^{-1}B$,

$V=A^{-1}B$

$$\begin{bmatrix}x\\ y\\ z\end{bmatrix}=\frac{1}{2}\begin{bmatrix}1 & -1 & 3\\ 0 & 2 & -4\\ 1 & -1 & 1\end{bmatrix}\begin{bmatrix}2\\ -3\\ -1\end{bmatrix}=\frac{1}{2}\begin{bmatrix}2+3-3\\ 0-6+4\\ 2+3-1\end{bmatrix}$$

$$=\frac{1}{2}\begin{bmatrix}2\\ -2\\ 4\end{bmatrix}=\begin{bmatrix}1\\ -1\\ 2\end{bmatrix}$$

$\therefore\ x=1,\ y=-1,\ z=2$

12. Solve $\begin{cases} x+y+z=2 \\ 2x+y-2z=-3 \\ x-z=-1 \end{cases}$

Using Gaussian Elimination.

(1) Augmented matrix for the above 3 equations will be:

$$\left[\begin{array}{ccc|c} 1 & 1 & 1 & 2 \\ 2 & 1 & -2 & -3 \\ 1 & 0 & -1 & -1 \end{array}\right]$$

(2) Perform elimination operation.

$$\left[\begin{array}{ccc|c} 1 & 1 & 1 & 2 \\ 2 & 1 & -2 & -3 \\ 1 & 0 & -1 & -1 \end{array}\right] \times(-2), (-1)$$

$$=\left[\begin{array}{ccc|c} 1 & 1 & 1 & 2 \\ 0 & -1 & -4 & -7 \\ 0 & -1 & -2 & -3 \end{array}\right] \times(-1)$$

$$=\left[\begin{array}{ccc|c} 1 & 1 & 1 & 2 \\ 0 & 1 & 4 & 7 \\ 0 & -1 & -2 & -3 \end{array}\right] \times(-1)$$

$$=\left[\begin{array}{ccc|c} 1 & 0 & -3 & -5 \\ 0 & 1 & 4 & 7 \\ 0 & -1 & -2 & -3 \end{array}\right] \times(1)$$

$$=\left[\begin{array}{ccc|c} 1 & 0 & -3 & -5 \\ 0 & 1 & 4 & 7 \\ 0 & 0 & 2 & 4 \end{array}\right] \times(-2)$$

$$=\left[\begin{array}{ccc|c} 1 & 0 & -3 & -5 \\ 0 & 1 & 0 & -1 \\ 0 & 0 & 2 & 4 \end{array}\right] \times\frac{1}{2}$$

$$=\left[\begin{array}{ccc|c} 1 & 0 & -3 & -5 \\ 0 & 1 & 0 & -1 \\ 0 & 0 & 1 & 2 \end{array}\right] \times(3)$$

$$=\left[\begin{array}{ccc|c} 1 & 0 & 0 & 1 \\ 0 & 1 & 0 & -1 \\ 0 & 0 & 1 & 2 \end{array}\right]$$

$x+0y+0z=1$

$0x+y+0z=-1$

$0x+0y+z=2$

$x=1,\quad y=-1,\quad z=2$

13. Solve $\begin{cases} x+y+z=2 \cdots\cdots ① \\ 2x+y-2z=-3 \cdots\cdots ② \\ x-z=-1 \cdots\cdots ③ \end{cases}$

by Gaussian-Jordan Elimination.

(1) Form an augmented matrix

$$\left[\begin{array}{ccc|c} 1 & 1 & 1 & 2 \\ 2 & 1 & -2 & -3 \\ 1 & 0 & -1 & -1 \end{array}\right]$$

(2) Perform elimination operations

$$\left[\begin{array}{ccc|c} 1 & 1 & 1 & 2 \\ 2 & 1 & -2 & -3 \\ 1 & 0 & -1 & -1 \end{array}\right] \times(-2), \times(-1)$$

$$=\left[\begin{array}{ccc|c} 1 & 1 & 1 & 2 \\ 0 & -1 & -4 & -7 \\ 0 & -1 & -2 & -3 \end{array}\right] \times(-1)$$

$$=\left[\begin{array}{ccc|c} 1 & 1 & 1 & 2 \\ 0 & -1 & -4 & -7 \\ 0 & 0 & 2 & 4 \end{array}\right] \begin{array}{l} \\ \times(-1) \\ \times(\frac{1}{2}) \end{array}$$

$$=\left[\begin{array}{ccc|c} 1 & 1 & 1 & 2 \\ 0 & 1 & 4 & 7 \\ 0 & 0 & 1 & 2 \end{array}\right]$$

Once get the pattern then stop.

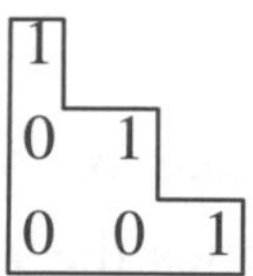

Because, you get

$z=0\times2+0\times7+1\times2=2$

When $z=2$, substitute back to (3)

$x-z=-1,\ x-2=-1,\ x=1$

Substitute to (1) $x+y+z=2$

$1+y+2=2$

$y=2-3=-1$

Sol: (1, −1, 2)

14. Solve $\begin{cases} x+y+z=2 \\ 2x+y-2z=-3 \\ x-z=-1 \end{cases}$ by Cramer's Rule

Step 1. Find determinate of A det(A)

$\det(A)=-1-2+0-1+2=-2$

Step 2. D_1: Let the constant column be column 1.

D_2: Let the constant column be column 2.

D_3: Let the constant column be column 3.

$$D_1:\ \begin{array}{ccccc} 2 & 1 & 1 & 2 & 1 \\ -3 & 1 & -2 & -3 & 1 \\ -1 & 0 & -1 & -1 & 0 \end{array}$$

$\det(D_1)=-2+2+0+1+0-3=-2$

$$D_2:\ \begin{array}{ccccc} 1 & 2 & 1 & 1 & 2 \\ 2 & -3 & -2 & 2 & -3 \\ 1 & -1 & -1 & 1 & -1 \end{array}$$

$\det(D_2)=3-4-2+3-2+4=2$

$$D_3:\ \begin{array}{ccccc} 1 & 1 & 2 & 1 & 1 \\ 2 & 1 & -3 & 2 & 1 \\ 1 & 0 & -1 & 1 & 0 \end{array}$$

$\det(D_3)=-1-3+0-2-0+2=-4$

$$x=\frac{D_1}{D}=\frac{-2}{-2}=1$$

$$y=\frac{D_2}{D}=\frac{2}{-2}=-1$$

$$z=\frac{D_3}{D}=\frac{-4}{-2}=2$$

Sol: (1, −1, 2)

14-2 LIMIT

The value of the limit of a function $f(x)$ could be going to a fixed value of α as x approaches to a certain value of a. It is denoted as $\lim_{x \to a} f(x) = \alpha$. Sometimes the α does not exist.

A. The Properties of the Limits

1. If $f(x)$ is a constant, them $\lim f(x) = \lim C = C$.

2. If $f(x)$ is a defined function, the value of the limit is calculated by substitution.

For Example: $f(x)=x^2+x+1$ is a defined function, then $\lim_{x\to 2} f(x) = \lim_{x\to 2} x^2+x+1$

$=2^2+2+1=4+2+1=7$

3. If $f(x)$ is an undefined function, then use the methods of limit evaluation to evaluate the value of the limit.

For Example: $f(x)=\dfrac{x^2-4}{x-2}$, when $x=2$, the function of $f(x)$ is undefined.

$\lim_{x\to 2}\dfrac{x^2-4}{x-2}=\lim_{x\to 2}\dfrac{(x-2)(x+2)}{x-2}=\lim_{x\to 2}(x+2)=2+2=4$

4. If a rational function $\dfrac{p(x)}{q(x)}$, when $\lim_{x\to a}\dfrac{p(a)}{q(a)}=k$, where $q(a)=0$.

In this case, $p(a)$ must equal to zero, in other words. $p(x)$ must have the factor of $q(x)$.

For Example: $\lim_{x\to 2}\dfrac{x^2+2x+a}{x-2}=b$, when $x=2$, then $p(2)=2^2+4+a=0$. It also means x^2+2x+a has the factor of $x-2$

B. Operation Rule

1. Addition/subtraction rules

$\lim[f(x)\pm\lim g(x)]=\lim f(x)\pm\lim g(x)$

For Example:

$\lim_{x\to 2}[(x^2+1)+(x-1)]=\lim_{x\to 2}(x^2+1)+\lim_{x\to 2}(x-1)=(2^2+1)+(2-1)=5+1=6$

2. Production rule

$\lim_{x\to a}[f(x)\cdot g(x)]=\lim_{x\to a} f(x)\cdot\lim_{x\to a} g(x)$

For Example: $\lim_{x\to 2}[(x^2+1)\cdot(x-1)]=\lim_{x\to 2}(x^2+1)\cdot\lim_{x\to 2}(x-1)=(4+1)\cdot(2-1)=5$

3. Quotient rules

$$\lim_{x\to a}\frac{P(x)}{q(x)}=\frac{\lim_{x\to a}p(x)}{\lim_{x\to a}q(x)}=\frac{p(a)}{q(a)}$$

For Example: $\lim_{x\to 2}\frac{x^2+1}{x-1}=\frac{\lim_{x\to 2}x^2+1}{\lim_{x\to 2}x-1}=\frac{2^2+1}{2-1}=5$

4. Power rule

$$\lim_{x\to a}[f(x)]^n=[\lim_{x\to a}f(x)]^n$$

For Example: $\lim_{x\to 2}(x^2+1)^3=[\lim_{x\to 2}(x^2+1)]^3=(2^2+1)^3=5^3=125$

5. Root rule

$$\lim_{x\to a}\sqrt[n]{f(x)}=\sqrt[n]{\lim_{x\to a}f(x)}$$

(Where n is a positive integer)

For Example:

$$\lim_{x\to 2}\sqrt{x^4-2x^2+1}=\sqrt{\lim_{x\to 2}(x^4-2x^2+1)}=\sqrt{2^4-2(2)^2+1}=\sqrt{16-8+1}=\sqrt{9}=3$$

C. Evaluation of the Limits

1. Direct substitution

For Example: $\lim_{x\to 1}3x^3+2x^2+x+1$

Since when x approaches 1, $f(x)=3x^3+2x^2+x+1$ is a defined function, so, its value can be evaluated by direct substitution so that, $\lim_{x\to 1}3x^3+2x^2+x+1=3(1)^3+2(1)^2+1+1=7$

For Example: $\lim_{x\to -8}\frac{\sqrt{1-x}+3}{3-\sqrt[3]{x}}=\frac{\sqrt{1-(-8)}+3}{3-\sqrt[3]{-8}}=\frac{3+3}{3-(-2)}=\frac{6}{5}$

2. Elimination of common factors

For Example: $\lim_{x\to 0}\frac{(1-x)(1-2x)(1-3x)-1}{x}=\lim_{x\to 0}\frac{(1-2x-x+2x^2)(1-3x)-1}{x}$

$=\lim_{x\to 0}\frac{-6x^3+11x^2-6x}{x}=\lim_{x\to 0}\frac{(-6x^2+11x-6)\not{x}}{\not{x}}=\lim_{x\to 0}(-6x^2+11x-6)=-6$

3. Sometimes it does not exist.

(1) when “a” approaches “o”

Example: $\lim_{x\to 0}f(x)=\lim_{x\to 0}\frac{1}{x}$

The limit does not exist.

(2) When "a" approaches "∞" infinitely large.

Example: $\lim_{n\to\infty}(\frac{4}{3})^n$

It is divergent so the value of limit does not exist.

But, if it is convergent, the limit exist.

Example: $\lim_{n\to\infty}(\frac{1}{3})^n$

When n approaches '∞', so $\lim_{n\to\infty}(\frac{1}{3})^n = 0$

the same as $\lim_{n\to\infty}(\frac{5}{6})^n = 0$

◆Example 1:

$$\lim_{x\to 1}(x^3 - 2x^2 + 2x + 3) = 1 - 2 \cdot 1^2 + 2 + 3 = 4$$

◆Example 2:

$$\lim_{x\to 2}(\sqrt{x+3} - \sqrt{x+2}) = \sqrt{2+3} - \sqrt{2+2} = \sqrt{5} - \sqrt{4} = \sqrt{5} - 2$$

• Some Examples

◆Example 1:

$$\lim_{x\to 1}\frac{2x-1}{x+5} = \frac{2(1)-1}{1+5} = \frac{1}{6}\ (x \neq -5)$$

◆Example 2:

$$\lim_{x\to 2}\frac{x^2+4x-12}{x-2} = \lim_{x\to 2}\frac{\cancel{(x-2)}(x+6)}{\cancel{x-2}} = \lim_{x\to 2} x+6 = 2+6 = 8$$

◆Example 3:

$$\lim_{x\to 4}\frac{(x^2+2x-24)}{\sqrt{x}-2}$$

$$= \lim_{x\to 4}\frac{(x+6)(x-4)(\sqrt{x}+2)}{(\sqrt{x}-2)(\sqrt{x}+2)}$$

$$= \lim_{x\to 4}\frac{(x+6)(\sqrt{x}+2)\cancel{(x-4)}}{\cancel{x-4}}$$

$$= \lim_{x \to 4} (x+6)(\sqrt{x}+2)$$

$$= (4+6)(2+2) = 40$$

◆Example 4:

$$\lim_{n \to \infty} \frac{2^n + 3^n}{3^n} = \lim_{n \to \infty} (\frac{2}{3})^n + 1 = 0 + 1 = 1$$

◆Example 5:

$$\lim_{n \to \infty} \frac{1^n + 3^n + 5^n}{7^n} = \lim_{n \to \infty} (\frac{1}{7})^n + \lim_{n \to \infty} (\frac{3}{7})^n + \lim_{n \to \infty} (\frac{5}{7})^n = 0 + 0 + 0 = 0$$

◆Example 6:

$$\lim_{n \to \infty} \frac{2^n - 1}{2^n - 2^{n-1}} = \lim_{n \to \infty} \frac{2^n - 1}{2^{n-1}(2-1)} = \lim_{n \to \infty} \frac{2^n - 1}{2^{n-1}}$$

$$= \lim_{n \to \infty} \frac{2^n - 1}{\frac{2^n}{2}} = \lim_{n \to \infty} \frac{1 - \frac{1}{2^n}}{\frac{1}{2}} \quad \text{(Divided by } 2^n\text{)}$$

$$= \frac{1-0}{\frac{1}{2}} = 2$$

◆Example 7:

If $\lim_{x \to -1} \frac{a\sqrt{x+2} - b}{x+1} = 2$, what are the values of a and b ?

Since $\lim_{x \to p} \frac{f(x)}{g(x)} = k$, k is a constant, then $g(p) \to 0$, there $f(p)$ must approach to zero.

Since $g(-1) = (-1) + 1 = 0$

So, $f(-1) = a\sqrt{-1+2} - b = 0$, $a - b = 0$

$\therefore\ a = b$

$$\lim_{x \to -1} \frac{a\sqrt{x+2} - a}{x+1} = \lim_{x \to -1} \frac{a(\sqrt{x+2} - 1)(\sqrt{x+2} + 1)}{(x+1)(\sqrt{x+2} + 1)}$$

$$= \lim_{x \to -1} \frac{a[(x+2) - 1]}{(x+1)(\sqrt{x+2} + 1)} = \lim_{x \to -1} \frac{a(x+1)}{(x+1)(\sqrt{x+2} + 1)}$$

$$= \lim_{x \to -1} \frac{a}{\sqrt{x+2}+1} = \frac{a}{1+1} = \frac{a}{2}$$

$\because \ \frac{a}{2} = 2$, $a = 4$, $b = 4$

Answer to Concept Check 14-09

$$\lim_{x \to 3} \frac{2x^3+2x^2+x+1}{x^2+x} = \lim_{x \to 3} \frac{2x^2(x+1)+x+1}{x(x+1)} = \lim_{x \to 3} \frac{(2x^2+1)\cancel{(x+1)}}{x\cancel{(x+1)}}$$

$$= \lim_{x \to 3} \frac{2x^2+1}{x} = \frac{2(3)^2+1}{3} = \frac{19}{3}$$

Answer to Concept Check 14-10

$$\lim_{x \to 0} \frac{x^2+x}{x} = \lim_{x \to 0} \frac{x(x+1)}{x} = \lim_{x \to 0}(x+1) = 1$$

Answer to Concept Check 14-11

$$\lim_{x \to 1} \frac{\sqrt{x+1}+\sqrt{x}}{\sqrt{x^2+1}+1} = \lim_{x \to 1} \frac{(\sqrt{x+1}+\sqrt{x})(\sqrt{x^2+1}-1)}{(\sqrt{x^2+1}+1)(\sqrt{x^2+1}-1)}$$

$$= \lim_{x \to 1} \frac{\sqrt{(x+1)(x^2+1)}+\sqrt{x(x^2+1)}-\sqrt{x+1}-\sqrt{x}}{(x^2+1)-1}$$

$$= \lim_{x \to 1} \frac{\sqrt{x^3+x^2+x+1}+\sqrt{x^3+x}-\sqrt{x+1}-\sqrt{x}}{x^2}$$

$$= \sqrt{4}+\sqrt{2}-\sqrt{2}-1 = 1$$

Answer to Concept Check 14-12

$$\lim_{x\to1}\frac{a\sqrt{x+3}+b}{x-1}$$

A. The value of limit does not exist but $\lim_{x\to1}\frac{a\sqrt{x+3}+b}{x-1}=2$, so, $P(1)=a\sqrt{1+3}+b$ must be equivalent to 'o'.

B. $\therefore\ a\sqrt{1+3}+b=0$ (When $x=1$)

$2a+b=0 \quad b=-2a$

$$\lim_{x\to1}\frac{a\sqrt{x+3}-2a}{x-1}=\lim_{x\to1}\frac{a(\sqrt{x+3}-2)}{x-1}=\lim_{x\to1}\frac{a(\sqrt{x+3}-2)(\sqrt{x+3}+2)}{(x-1)(\sqrt{x+3}+2)}$$

$$=\lim_{x\to1}\frac{a[\sqrt{(x+3)^2}-4]}{(x-1)(\sqrt{x+3}+2)}=a\lim_{x\to1}\frac{x+3-4}{(x-1)(\sqrt{x+3}+2)}$$

$$=a\lim_{x\to1}\frac{x-1}{(x-1)(\sqrt{x+3}+2)}=a\lim_{x\to1}\frac{1}{\sqrt{x+3}+2}=a\cdot\frac{1}{2+2}=\frac{1}{4}a=2$$

$\therefore\ a=8$

Natural Exponential Base: e (Euler's Number)
Defined as

$$e=\lim_{n\to\infty}(1+\frac{1}{n})^n \text{ or } \lim_{n\to0}(1+n)^{\frac{1}{n}}=2.718281828\cdots\cdots \text{ (irrational number)}$$

$$e=\lim_{n\to\infty}(1+\frac{1}{n})^n=\lim_{n\to\infty}[1+nC_1\frac{1}{n}+nC_2(\frac{1}{n})^2+\cdots+nC_{n-1}(\frac{1}{n})^{n-1}+nC_n(\frac{1}{n})^n]$$

$$=\lim_{n\to\infty}[1+\frac{n}{n}+\frac{n!}{(n-2)!2!}(\frac{1}{n})^2+\frac{n!}{(n-3)!3!}(\frac{1}{n})^3+\cdots\cdots+\frac{n!}{(n-n+1)!(n-1)!}(\frac{1}{n})^{n-1}+(\frac{1}{n})^n]$$

$$=\lim_{n\to\infty}1+\lim_{n\to\infty}\frac{n}{n}+\lim_{n\to\infty}\frac{n(n-1)}{2!}(\frac{1}{n})^2+\lim_{n\to\infty}\frac{n(n-1)(n-2)}{3!}(\frac{1}{n})^3+\cdots\cdots+\lim_{n\to\infty}(\frac{1}{1})^{n-1}+\lim_{n\to\infty}(\frac{1}{n})^n$$

$$=2+\lim_{n\to\infty}\frac{1-\frac{1}{n}}{2}+\lim_{n\to\infty}\frac{1}{6}\cdot\frac{1-\frac{3}{n}+\frac{2}{n^2}}{1}+\lim_{n\to\infty}\frac{1}{24}\cdot\frac{1-\frac{6}{n}+\frac{11}{n^2}-\frac{6}{n^3}}{1}+\cdots$$

$$=2+\frac{1}{2}+\frac{1}{6}+\frac{1}{24}+\frac{1}{120}+\frac{1}{720}+\cdots\cdots=2.718281828\cdots\cdots$$

e also can be defined by $e=\frac{1}{0!}+\frac{1}{1!}+\frac{1}{2!}+\frac{1}{3!}+\frac{1}{4!}+\frac{1}{5!}+\frac{1}{6!}+\cdots\cdots$

$=1+1+\frac{1}{2}+\frac{1}{6}+\frac{1}{24}+\frac{1}{120}+\frac{1}{720}+\cdots\cdots$ (Where ! means factorial) $=2.718281828$

CHAPTER 14
LIMIT
PRACTICE

1. $\lim\limits_{x\to1}\dfrac{x-1}{\sqrt{x+3}-2}=$

2. $\lim\limits_{x\to-2}[\dfrac{1}{x^2-x-2}-\dfrac{1}{x^2-5x+6}]=$

3. $\lim\limits_{x\to3}\dfrac{x^2-2x-3}{x-3}=$

4. $\lim\limits_{x\to1}\dfrac{2x^2+x-3}{\sqrt{x}-1}=$

5. $\lim\limits_{x\to0}\dfrac{\sqrt{x+9}-3}{x}=$

6. If $\lim\limits_{x\to1}\dfrac{ax+b}{\sqrt{2x+1}-\sqrt{x+2}}=4\sqrt{3}$, what is the value of a ?

7. Find $\lim\limits_{x\to3}(x^2-3x+5)$

8. Find $\lim\limits_{x\to3}\dfrac{x-3}{x^2-9}$

9. Find $\lim\limits_{x\to3}\dfrac{\sqrt{x}-3}{x-9}$

10. Find $\lim\limits_{x\to\infty}\dfrac{3x+2}{2x+3}$

11. Find $\lim\limits_{x\to0}\dfrac{(x+5)^3-125}{x}$

12. $\lim\limits_{x\to2}\sqrt[3]{\dfrac{x^3-2x+2}{x+4}}=$

13. Find $\lim\limits_{x\to2}\dfrac{\sqrt{x-1}-\sqrt{2x-3}}{\sqrt{x-2}-\sqrt{2x-4}}$, (Where $x>2$)

14. If $f(x)=2x^2-3x+1$, find $\lim\limits_{t\to0}\dfrac{f(1-t)+f(1)}{t}$

15. If $\lim\limits_{x\to2}\dfrac{a\sqrt{x+2}-b}{x-2}=2$, what is the value of $a+b$?

16. $\lim\limits_{n\to\infty}\dfrac{2^{n-1}+4\cdot3^{n+1}-5\cdot4^{n-1}}{5\cdot2^{n+1}-3\cdot3^{n-1}+4\cdot4^{n+1}}=$

CHAPTER 14
LIMIT
ANSWERS and EXPLANATIONS

1. $\lim_{x\to1}\frac{x-1}{\sqrt{x+3}-2}$

$=\lim_{x\to1}\frac{(x-1)(\sqrt{x+3}+2)}{(\sqrt{x+3}-2)(\sqrt{x+3}+2)}$

$=\lim_{x\to1}\frac{(x-1)(\sqrt{x+3}+2)}{(x+3)-4}$

$=\lim_{x\to1}\frac{(x-1)(\sqrt{x+3}+2)}{(x-1)}$

$=\lim_{x\to1}\sqrt{x+3}+2$

$=\sqrt{1+3}+2$

$=4$

2. $\lim_{x\to-2}[\frac{1}{x^2-x-2}-\frac{1}{x^2-5x+6}]$

$=\lim_{x\to-2}[\frac{1}{(x-2)(x+1)}-\frac{1}{(x-2)(x-3)}]$

$=\lim_{x\to-2}\frac{-4}{(x-2)(x+1)(x-3)}$

$=\frac{-4}{(-2-2)(-2+1)(-2-3)}$

$=\frac{-4}{-20}=\frac{1}{5}$

3. $\lim_{x\to3}\frac{x^2-2x-3}{x-3}$

$=\lim_{x\to3}\frac{(x-3)(x+1)}{x-3}$

$=\lim_{x\to3}(x+1)$

$=3+1=4$

4. $\lim_{x\to1}\frac{2x^2+x-3}{\sqrt{x}-1}$

$=\lim_{x\to1}\frac{(2x^2+x-3)(\sqrt{x}+1)}{(\sqrt{x}-1)(\sqrt{x}+1)}$

$=\lim_{x\to1}\frac{(x-1)(2x+3)(\sqrt{x}+1)}{x-1}$

$=\lim_{x\to1}(2x+3)(\sqrt{x}+1)$

$=(2+3)(1+1)$

$=10$

5. $\lim_{x\to0}\frac{\sqrt{x+9}-3}{x}$

$=\lim_{x\to0}\frac{(\sqrt{x+9}-3)(\sqrt{x+9}+3)}{(x)(\sqrt{x+9}+3)}$

$=\lim_{x\to0}\frac{(x+9-9)}{x(\sqrt{x+9}+3)}$

$=\lim_{x\to0}\frac{1}{1(\sqrt{x+9}+3)}$

$=\lim_{x\to0}\frac{1}{\sqrt{x+9}+3}$

$=\frac{1}{3+3}=\frac{1}{6}$

6. If $\lim_{x\to1}\frac{ax+b}{\sqrt{2x+1}-\sqrt{x+2}}=4\sqrt{3}$

Because when x approaches to 1, then $\sqrt{2x+1}-\sqrt{x+2}\to0$

But the value of $\lim_{x\to1}\frac{ax+b}{\sqrt{2x+1}-\sqrt{x+2}}$ $=4\sqrt{3}$, that mean $f(1)=a(1)+b=0$

when $x \to 1$

$\therefore\ b = -a$

$$\lim_{x\to1}\frac{ax+b}{\sqrt{2x+1}-\sqrt{x+2}}$$

$$=\lim_{x\to1}\frac{a(x-1)}{\sqrt{2x+1}-\sqrt{x+2}}$$

$$=\lim_{x\to1}\frac{a(x-1)(\sqrt{2x+1}+\sqrt{x+2})}{(\sqrt{2x+1}-\sqrt{x+2})(\sqrt{2x+1}+\sqrt{x+2})}$$

$$=\lim_{x\to1}\frac{a\cdot(x-1)(\sqrt{2x+1}+\sqrt{x+2})}{(2x+1)-(x+2)}$$

$$=\lim_{x\to1}a(\sqrt{2x+1}+\sqrt{x+2})$$

$$=a(\sqrt{3}+\sqrt{3})=2a\sqrt{3}$$

$$2a\sqrt{3}=4\sqrt{3}$$

$\therefore\ a=2$

7. $\lim_{x\to3}(x^2-3x+5)$

$$=\lim_{x\to3}x^2-\lim_{x\to3}3x+\lim_{x\to3}5$$

$$=9-(\lim_{x\to3}3)\cdot(\lim_{x\to3}x)+\lim_{x\to3}5$$

$$=9-3\cdot3+5=5$$

8. $\lim_{x\to3}\frac{x-3}{x^2-9}=\lim_{x\to3}\frac{\cancel{(x-3)}}{\cancel{(x-3)}(x+3)}$

$$=\lim_{x\to3}\frac{1}{x+3}=\frac{1}{3+3}=\frac{1}{6}$$

9. Asking $\lim_{x\to3}\frac{\sqrt{x}-3}{x-9}$?

$$\lim_{x\to3}\frac{\sqrt{x}-3}{(\sqrt{x}-3)(\sqrt{x}+3)}$$

$$=\lim_{x\to3}\frac{1}{\sqrt{x}+3}=\frac{1}{\sqrt{3}+3}$$

$$=\lim_{x\to3}\frac{\sqrt{3}-3}{(\sqrt{3}+3)(\sqrt{3}-3)}$$

$$=\frac{\sqrt{3}-3}{3-9}=\frac{\sqrt{3}-3}{-6}=\frac{3-\sqrt{3}}{6}$$

or, $\lim_{x\to3}\frac{\sqrt{3}-3}{x-9}=\frac{\sqrt{3}-3}{-6}=\frac{3-\sqrt{3}}{6}$

10. Asking: $\lim_{x\to\infty}\frac{3x+2}{2x+3}$

$$\lim_{x\to\infty}\frac{3x+2}{2x+3}=\lim_{x\to\infty}\frac{3+\frac{2}{x}}{2+\frac{3}{x}}$$

$$=\frac{\lim_{x\to\infty}3+\lim_{x\to\infty}\frac{2}{x}}{\lim_{x\to\infty}2+\lim_{x\to\infty}\frac{3}{x}}$$

$$=\frac{3+0}{2+0}=\frac{3}{2}$$

11. Asking $\lim_{x\to0}\frac{(x+5)^3-125}{x}$

$$=\lim_{x\to0}\frac{x^3+3x^2\cdot(5)+3x\cdot(5)^2+(5)^3-125}{x}$$

$$=\lim_{x\to0}\frac{x^3+15x^2+75x+125-125}{x}$$

$$=\lim_{x\to0}\frac{x(x^2+15x+75)}{x}$$

$$=\lim_{x\to0}x^2+15x+75$$

$$=\lim_{x\to0}x^2+\lim_{x\to0}15x+\lim_{x\to0}75$$

$$=0+0+75$$

$$=75$$

12. $\lim_{x\to2}\sqrt[3]{\frac{x^3-2x+2}{x+4}}=\sqrt[3]{\frac{8-4+2}{2+4}}=\sqrt[3]{\frac{6}{6}}=1$

13. $\lim_{x\to 2}\frac{\sqrt{x-1}-\sqrt{2x-3}}{\sqrt{x-2}-\sqrt{2x-4}}\,(x>2)$

$=\lim_{x\to 2}\frac{\sqrt{x-1}-\sqrt{2x-3}}{\sqrt{x-2}-\sqrt{2x-4}}\cdot\frac{\sqrt{x-2}+\sqrt{2x-4}}{\sqrt{x-2}+\sqrt{2x-4}}$

$\cdot\frac{\sqrt{x-1}+\sqrt{2x-3}}{\sqrt{x-1}+\sqrt{2x-3}}$

$=\lim_{x\to 2}\frac{(x-1)-(2x-3)}{(x-2)-(2x-4)}\cdot\frac{\sqrt{x-2}+\sqrt{2x-4}}{\sqrt{x-1}+\sqrt{2x-3}}$

$=\lim_{x\to 2}\frac{(2-x)}{(2-x)}\cdot\frac{\sqrt{x-2}+\sqrt{2x-4}}{\sqrt{x-1}+\sqrt{2x-3}}$

$=\lim_{x\to 2}\frac{\sqrt{x-2}+\sqrt{2x-4}}{\sqrt{x-1}+\sqrt{2x-3}}$

$=\frac{\sqrt{0}+\sqrt{0}}{\sqrt{2-1}+\sqrt{4-3}}=\frac{0}{2}=0$

14. $f(x)=2x^2-3x+1$

$\lim_{t\to 0}\frac{f(1-t)+f(1)}{t}$

$=\lim_{t\to 0}\frac{2(1-t)^2-3(1-t)+1+2-3+1}{t}$

$=\lim_{t\to 0}\frac{2(1-2t+t^2)+3t-3+1+0}{t}$

$=\lim_{t\to 0}\frac{2-4t+2t^2+3t-2+0}{t}$

$=\lim_{t\to 0}\frac{2t^2-t+0}{t}$

$=\lim_{t\to 0}2t-1$

$=2(0)-1=-1$

15. Since $\lim_{x\to 2}\frac{a\sqrt{x+2}-b}{x-2}=2$, when $x\to 2$, the denominator is $2-2=0$. Because the limit exist, so, the numerator $f(2)$ must equal to zero.

$a\sqrt{2+2}-b=0,\ 2a=b,$

$\lim_{x\to 2}\frac{a\sqrt{x+2}-b}{x-2}$

$=\lim_{x\to 2}\frac{a(\sqrt{x+2}-2)}{x-2}$

$=a\lim_{x\to 2}\frac{\sqrt{x+2}-2}{x-2}$

$=a\lim_{x\to 2}\frac{(\sqrt{x+2}-2)(\sqrt{x+2}+2)}{(x-2)(\sqrt{x+2}+2)}$

$=a\lim_{x\to 2}\frac{(x+2)-4}{(x-2)(\sqrt{x+2}+2)}$

$=a\lim_{x\to 2}\frac{\cancel{x-2}}{(\cancel{x-2})(\sqrt{x+2}+2)}$

$=\lim_{x\to 2}\frac{a}{\sqrt{x+2}+2}=\frac{a}{2+2}=2$

$\therefore\ a=8,\ b=2a=16$

$a+b=8+16=24$

16. $\lim_{n\to\infty}\frac{2^{n-1}+4\cdot 3^{n+1}-5\cdot 4^{n-1}}{5\cdot 2^{n+1}-3\cdot 3^{n-1}+4\cdot 4^{n+1}}$

$=\lim_{n\to\infty}\frac{(\frac{2^n}{2})+4\cdot 3\cdot 3^n-(\frac{5}{4})(4^n)}{10\cdot 2^n-3^n+16\cdot 4^n}$

$=\lim_{n\to\infty}\frac{4^n[\frac{1}{2}(\frac{2}{4})^n+12(\frac{3}{4})^n-\frac{5}{4}]}{4^n[10\cdot(\frac{2}{4})^n-(\frac{3}{4})^n+16]}$

$=\lim_{n\to\infty}\frac{\frac{1}{2}(\frac{1}{2})^n-12(\frac{3}{4})^n-\frac{5}{4}}{10\cdot(\frac{2}{4})^n-(\frac{3}{4})^n+16}$

$=\frac{0-0-\frac{5}{4}}{0-0+16}=-\frac{5}{64}$

14-3 SERIES

- Series is a summation of a sequence of numbers, $S_n = a_1 + a_2 + a_3 + \cdots + a_n$, denoted as $\sum_{i=1}^{n} a_i$, read as 'Sigma', a_i means the 'ith' term.

A. The Properties of Series

1. $\sum_{i=1}^{n} ka_i = k\sum_{i=1}^{n} a_i$
2. $\sum_{i=1}^{n} (a_i \pm b_i) = \sum_{i=1}^{n} a_i \pm \sum_{i=1}^{n} b_i$
3. $\sum_{i=1}^{n} c = n \cdot c$
4. $\sum_{i=1}^{n} i = \frac{n(n+1)}{2}$

◆Example 1:

$$\sum_{n=1}^{\infty} \frac{2^n + 3^n}{6^n} = \sum_{n=1}^{\infty} (\frac{2}{6})^n + \sum_{n=1}^{\infty} (\frac{3}{6})^n = \sum_{n=1}^{\infty} (\frac{1}{3})^n + \sum_{n=1}^{\infty} (\frac{1}{2})^n$$

$$= (\frac{1}{3})^1 + (\frac{1}{3})^2 + (\frac{1}{3})^3 + \cdots + (\frac{1}{3})^n + (\frac{1}{2})^1 + (\frac{1}{2})^2 + (\frac{1}{2})^3 + \cdots + (\frac{1}{2})^n$$

$$= \frac{\frac{1}{3}}{1-\frac{1}{3}} + \frac{\frac{1}{2}}{1-\frac{1}{2}} \quad (S_n = \frac{a}{1-r}) = \frac{\frac{1}{3}}{\frac{2}{3}} + \frac{\frac{1}{2}}{\frac{1}{2}} = 1\frac{1}{2}$$

◆Example 2:

$$0.\dot{2} = 0.22222222\cdots\cdots$$

$$= 0.2 + 0.02 + 0.002 + 0.0002 + \cdots\cdots$$

$$= \frac{2}{10} + \frac{2}{10^2} + \frac{2}{10^3} + \frac{2}{10^4} + \cdots\cdots$$

$$= \frac{2}{10}[1 + \frac{1}{10} + \frac{1}{10^2} + \frac{1}{10^3} + \cdots\cdots$$

$$= \frac{1}{5}\sum_{n=0}^{\infty} \frac{1}{10^n} = \frac{1}{5} \cdot \frac{1}{1-\frac{1}{10}} = \frac{\frac{1}{5}}{\frac{9}{10}} = \frac{2}{9}$$

◆Example 3:

$$S_n = \frac{5}{3^2} + \frac{9}{3^3} + \frac{17}{3^4} + \frac{33}{3^5} + \cdots\cdots$$

$$S_n = \sum_{n=1}^{\infty} \frac{2^{n+1}+1}{3^{n+1}} = \sum_{n=1}^{\infty} \frac{2^{n+1}}{3^{n+1}} + \sum_{n=1}^{\infty} \frac{1}{3^{n+1}} = \sum_{n=2}^{\infty} (\frac{2}{3})^n + \sum_{n=2}^{\infty} (\frac{1}{3})^n$$

$$= \frac{(\frac{2}{3})^2}{1-\frac{2}{3}} + \frac{(\frac{1}{3})^2}{1-\frac{1}{3}} = \frac{\frac{4}{9}}{\frac{1}{2}} + \frac{\frac{1}{9}}{\frac{2}{3}} = \frac{4}{3} + \frac{1}{6} = \frac{3}{2}$$

Or, $\sum_{n=2}^{\infty} (\frac{2}{3})^n + \sum_{n=2}^{\infty} (\frac{1}{3})^n = \sum_{n=1}^{\infty} (\frac{2}{3})^n - \frac{2}{3} + \sum_{n=1}^{\infty} (\frac{1}{3})^n - \frac{1}{3} = \frac{\frac{2}{3}}{1-\frac{2}{3}} - \frac{2}{3} + \frac{\frac{1}{3}}{1-\frac{1}{3}} - \frac{1}{3}$

$$= 2 - \frac{2}{3} + \frac{1}{2} - \frac{1}{3} = \frac{3}{2}$$

B. Divergent/Convergent Sequence

Divergent Sequence:

$\sum_{i=1}^{n} a_i = a_1 + a_2 + a_3 + \cdots a_n$, when n reaches to infinity (∞), then the sum of $\sum_{i=1}^{n} a_i$ reaches to infinity.

The sequence is called divergent sequence.

• Convergent Sequence

On the contrary, when n approaches to '∞', $\sum_{i=1}^{n} a_i = \alpha$ (a fixed number)

It is called convergent sequence.

Answer to Concept Check 14-13

$$\sum_{i=1}^{100} n = 1 + 2 + 3 + \cdots + 20 + \cdots + 100$$

It is are arithmetic sequence

$a_1 = 1$, $d = 1$, $a_{100} = 1 + 1 \times (100 - 1) = 100$

$$\therefore \ S_n = \frac{1+100}{2} \times 100 = 5050$$

Answer to Concept Check 14-14

$$\sum_{n=1}^{\infty} 2^n = 2^1 + 2^2 + 2^3 + 2^4 + \cdots\cdots$$

$a_1 = 2,\ \ r = 2$

It is a Geometrical sequence and the sum goes to infinity.

so,

$S_n = \sum_{n=1}^{\infty} 2^n$ goes to infinity

Answer to Concept Check 14-15

$$\sum_{n=1}^{\infty} \left(\frac{1}{2}\right)^n = \left(\frac{1}{2}\right)^1 + \left(\frac{1}{2}\right)^2 + \left(\frac{1}{2}\right)^3 + \cdots\cdots$$

$a_1 = \frac{1}{2},\ \ r = \frac{1}{2},$

$$S_n = \sum_{n=1}^{\infty} \left(\frac{1}{2}\right)^n = \frac{\frac{1}{2}}{1-\frac{1}{2}} = \frac{\frac{1}{2}}{\frac{1}{2}} = 1$$

Answer to Concept Check 14-16

$$\sum_{n=1}^{\infty} \frac{4^n - 2^n - 1^n}{8^n} = \sum_{n=1}^{\infty} \left(\frac{4}{8}\right)^n - \sum_{n=1}^{\infty} \left(\frac{2}{8}\right)^n - \sum_{n=1}^{\infty} \left(\frac{1}{8}\right)^n$$

$$= \sum_{n=1}^{\infty} \left(\frac{1}{2}\right)^n - \sum_{n=1}^{\infty} \left(\frac{1}{4}\right)^n - \sum_{n=1}^{\infty} \left(\frac{1}{8}\right)^n = \frac{\frac{1}{2}}{1-\frac{1}{2}} - \frac{\frac{1}{4}}{1-\frac{1}{4}} - \frac{\frac{1}{8}}{1-\frac{1}{8}}$$

$$= \frac{\frac{1}{2}}{\frac{1}{2}} - \frac{\frac{1}{4}}{\frac{3}{4}} - \frac{\frac{1}{8}}{\frac{7}{8}} = 1 - \frac{1}{3} - \frac{1}{7} = \frac{21-7-3}{21} = \frac{11}{21}$$

CHAPTER 14
SERIES
PRACTICE

1. $0.\dot{5} =$

2. Find $S_n = \frac{8}{4^2} + \frac{26}{4^3} + \frac{80}{4^4} + \frac{24^2}{4^5} + \cdots\cdots$

3. $S_n = (\frac{3}{2} - \frac{2}{3}) + (\frac{3}{2^2} - \frac{2}{3^2}) + (\frac{3}{2^3} - \frac{2}{3^3})$
$+ \cdots + (\frac{3}{2^n} - \frac{2}{3^n})$

4. $\sum_{n=1}^{\infty} \frac{2^n - 2}{3^{n+2}} =$

5. $\sum_{n=1}^{\infty} \frac{2^{n+1} + 3^{n+1} + 4^n}{5^{n-1}} =$

6. Find $S_n = 1 - \frac{1}{5} + \frac{1}{5^2} - \frac{1}{5^3} + \frac{1}{5^4} - \frac{1}{5^5} \cdots\cdots$

7. What is the sum of the series
$\frac{1}{3} - \frac{2}{5} + \frac{1}{3^2} - \frac{2}{5^2} + \frac{1}{3^3} - \frac{2}{5^3} + \cdots$
$+ (\frac{1}{3^n} - \frac{2}{5^n})$?

8. $\sum_{n=1}^{\infty} (\frac{n+1}{n} - \frac{n+2}{n+1}) =$

9. $\sum_{n=1}^{\infty} \frac{2^{n-1} + 3^n + 4^{n+1}}{5^{n-1}} =$

10. $S_n = (\frac{3}{2} - \frac{4}{3}) + (\frac{4}{3} - \frac{5}{4}) + (\frac{5}{4} - \frac{6}{5}) + \cdots\cdots$
$+ (\frac{n+2}{n+1} - \frac{n+3}{n+2})$

CHAPTER 14
SERIES
ANSWERS and EXPLANATIONS

1. $0.\dot{5} = 0.555555\cdots\cdots$

$$= \frac{5}{10} + \frac{5}{10^2} + \frac{5}{10^3} + \frac{5}{10^4} + \cdots\cdots$$

$$= \frac{1}{2}(1 + \frac{1}{10} + \frac{1}{10^2} + \frac{1}{10^3} + \cdots\cdots)$$

$$= \frac{1}{2}(\frac{1}{1 - \frac{1}{10}})$$

$$= \frac{1}{2} \times \frac{10}{9} = \frac{10}{18} = \frac{5}{9}$$

2. $S_n = \frac{8}{4^2} + \frac{26}{4^3} + \frac{80}{4^4} + \frac{242}{4^5} + \cdots\cdots$

$$= \sum_{n=1}^{\infty} \frac{3^{n+1} - 1}{4^{n+1}} = \sum_{n=1}^{\infty} \frac{3^{n+1}}{4^{n+1}} - \sum_{n=1}^{\infty} \frac{1}{4^{n+1}}$$

$$= \sum_{n=2}^{\infty} \frac{3^n}{4^n} - \sum_{n=2}^{\infty} \frac{1}{4^n} = \sum_{n=2}^{\infty} (\frac{3}{4})^b - \sum_{n=2}^{\infty} (\frac{1}{4})^n$$

$$= \frac{(\frac{3}{4})^2}{1 - \frac{3}{4}} - \frac{(\frac{1}{4})^2}{1 - \frac{1}{4}} = \frac{\frac{9}{16}}{\frac{1}{4}} - \frac{\frac{1}{16}}{\frac{3}{4}}$$

$$= \frac{9}{4} - \frac{1}{12} = \frac{26}{12} = 3$$

3. $Sn = \frac{3}{2}[1 + \frac{1}{2} + \frac{1}{2^2} + \cdots]$

$$- \frac{2}{3}[1 + \frac{1}{3} + \frac{1}{3^2} + \cdots]$$

$$= \frac{3}{2} \bullet \frac{1}{1 - \frac{1}{2}} - \frac{2}{3} \bullet \frac{1}{1 - \frac{1}{3}}$$

$$= \frac{3}{2} \times 2 - \frac{2}{3} \times \frac{3}{2}$$

$$= 3 - 1 = 2$$

4. $\sum_{n=1}^{\infty} \frac{2^n - 2}{3^{n+2}}$

$$= \sum_{n=1}^{\infty} \frac{2^n}{3^{n+2}} - \sum_{n=1}^{\infty} \frac{2}{3^{n+2}}$$

$$= \sum_{n=1}^{\infty} \frac{2^n}{3^n \bullet 3^2} - \sum_{n=1}^{\infty} \frac{2}{3^n \bullet 3^2}$$

$$= \frac{1}{9} \sum_{n=1}^{\infty} (\frac{2}{3})^n - \frac{2}{9} \sum_{n=1}^{\infty} (\frac{1}{3})^n$$

$$= \frac{1}{9} \bullet \frac{\frac{2}{3}}{1 - \frac{2}{3}} - \frac{2}{9} \bullet \frac{\frac{1}{3}}{1 - \frac{1}{3}}$$

$$= \frac{1}{9} \bullet \frac{\frac{2}{3}}{\frac{1}{3}} - \frac{2}{9} \bullet \frac{\frac{1}{3}}{\frac{2}{3}}$$

$$= \frac{2}{9} - \frac{2}{9} \times \frac{1}{2} = \frac{1}{9}$$

5. $\sum_{n=1}^{\infty} \frac{2^{n+1} + 3^{n+1} + 4^n}{5^{n-1}}$

$$= \sum_{n=1}^{\infty} \frac{2^{n+1}}{5^{n-1}} + \sum_{n=1}^{\infty} \frac{3^{n+1}}{5^{n-1}} + \sum_{n=1}^{\infty} \frac{4^n}{5^{n-1}}$$

$$= \sum_{n=1}^{\infty} \frac{2^{n-1} \bullet 2^2}{5^{n-1}} + \sum_{n=1}^{\infty} \frac{3^{n-1} \bullet 3^2}{5^{n-1}} + \sum_{n=1}^{\infty} \frac{4^{n-1} \bullet 4}{5^{n-1}}$$

$$= 4\sum_{n=1}^{\infty} (\frac{2}{5})^{n-1} + 9\sum_{n=1}^{\infty} (\frac{3}{5})^{n-1} + 4\sum_{n=1}^{\infty} (\frac{4}{5})^{n-1}$$

$$= 4 \bullet \frac{1}{1 - \frac{2}{5}} + 9 \bullet \frac{1}{1 - \frac{3}{5}} + 4 \bullet \frac{1}{1 - \frac{4}{5}}$$

$$=\frac{4}{\frac{3}{5}}+\frac{9}{\frac{2}{5}}+\frac{4}{\frac{1}{5}}=\frac{20}{3}+\frac{45}{2}+\frac{20}{1}$$

$$=\frac{40+135+120}{6}=\frac{295}{6}$$

6. $S_n=1-\frac{1}{5}+\frac{1}{5^2}-\frac{1}{5^3}+\frac{1}{5^4}-\frac{1}{5^5}\cdots$

$$=(1+\frac{1}{5^2}+\frac{1}{5^4}\cdots)-\frac{1}{5}(1+\frac{1}{5^2}+\frac{1}{5^4}\cdots)$$

$$=(1-\frac{1}{5})(1+\frac{1}{5^2}+\frac{1}{5^4}+\cdots)$$

$$=\frac{4}{5}(\frac{1}{1-\frac{1}{5^2}})=\frac{4}{5}\bullet\frac{25}{24}=\frac{20}{24}=\frac{5}{6}$$

7. $\frac{1}{3}-\frac{2}{5}+\frac{1}{3^2}-\frac{2}{5^2}+\frac{1}{3^3}-\frac{1}{5^3}+\cdots+(\frac{1}{3^n}-\frac{2}{5^n})$

$$=\frac{1}{3}(1+\frac{1}{3}+\frac{1}{3^2}+\cdots+\frac{1}{3^n})$$

$$-\frac{2}{5}(1+\frac{1}{5}+\frac{1}{5^2}+\cdots+\frac{1}{5^n})$$

$$=\frac{1}{3}(\frac{1}{1-\frac{1}{3}})-\frac{2}{5}(\frac{1}{1-\frac{1}{5}})$$

$$=\frac{1}{3}\bullet\frac{3}{2}-\frac{2}{5}\bullet\frac{5}{4}$$

$$=\frac{1}{2}-\frac{2}{4}=0$$

8. $\sum_{n=1}^{\infty}(\frac{n+1}{n}-\frac{n+2}{n+1})$

$$=\sum_{n=1}^{\infty}\frac{(n+1)^2-n\bullet(n+2)}{n(n+1)}$$

$$=\sum_{n=1}^{\infty}\frac{n^2+2n+1-n^2-2n}{n^2+n}$$

$$=\sum_{n=1}^{\infty}\frac{1}{n^2+n}=\sum_{n=1}^{\infty}\frac{1}{n(n+1)}$$

$$=\frac{1}{1\bullet 2}+\frac{1}{2\bullet 3}+\frac{1}{3\bullet 4}+\cdots+\frac{1}{(n-1)\bullet n}$$

$$=(\frac{1}{1}-\frac{1}{2})+(\frac{1}{2}-\frac{1}{3})+(\frac{1}{3}-\frac{1}{4})+\cdots$$

$$+(\frac{1}{n-1}-\frac{1}{n})=1-\frac{1}{n}=1\quad(\text{when } n\to\infty)$$

9. $\sum_{n=1}^{\infty}\frac{2^{n-1}+3^n+4^{n+1}}{5^{n-1}}$

$$=\sum_{n=1}^{\infty}\frac{2^{n-1}}{5^{n-1}}+\sum_{n=1}^{\infty}\frac{3^n}{5^{n-1}}+\sum_{n=1}^{\infty}\frac{4^{n+1}}{5^{n-1}}$$

$$=\sum_{n=1}^{\infty}(\frac{2}{5})^{n-1}+3\sum_{n=1}^{\infty}(\frac{3}{5})^{n-1}+\sum_{n=1}^{\infty}4^2(\frac{4}{5})^{n-1}$$

$$=\frac{1}{1-\frac{2}{5}}+3\frac{1}{1-\frac{3}{5}}+4^2\frac{1}{1-\frac{4}{5}}$$

$$=\frac{5}{3}+\frac{15}{2}+16\times 5$$

$$=\frac{5}{3}+\frac{15}{2}+80$$

$$=1.67+7.5+80=89.17$$

10. $S_n=(\frac{3}{2}-\frac{4}{3})+(\frac{4}{3}-\frac{5}{4})+(\frac{5}{4}-\frac{6}{5})+\cdots\cdots$

$$+(\frac{n+2}{n+1}-\frac{n+3}{n+2})$$

$$S_n=\frac{3}{2}-\frac{n+3}{n+2},$$

$$\lim_{n\to\infty}S_n=\lim_{n\to\infty}[\frac{3}{2}-\frac{n+3}{n+2}]$$

$$=\lim_{n\to\infty}\frac{3}{2}-\lim_{n\to\infty}\frac{n+3}{n+2}=\frac{3}{2}-\lim_{n\to\infty}\frac{1+\frac{3}{n}}{1+\frac{2}{n}}$$

$$=\frac{3}{2}-1=\frac{1}{2}$$

CHAPTER 15
MODEL TEST 1
LEVEL 1

DIRECTION:

Fill in the correct or the best answer in the proper position on the answer sheet. You may use any available space for scratch work.

NOTES:

1. Calculator is permitted.
2. All figures are not drawn to scale.
3. All numbers are real, unless stated otherwise.
4. Some reference information:

(1) Right Circular Cone

$V = \frac{1}{3}\pi r^2 \cdot h$, $A(\text{lateral surface}) = \frac{1}{2}c\ell = \pi r \cdot \ell$

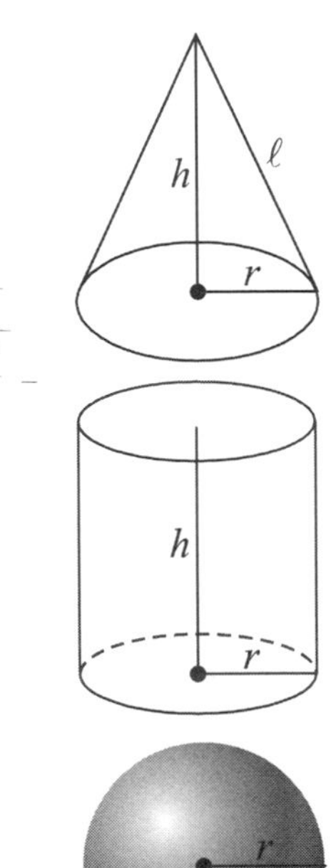

(2) Cylinder

$V = \pi r^2 \cdot h$, $A(\text{surface area}) = 2\pi r \cdot h$

(3) Sphere

$V = \frac{4}{3}\pi r^3 \qquad A = 4\pi r^2$

(4) Pyramid

$V = \frac{1}{3}Bh$ (B: area of base)

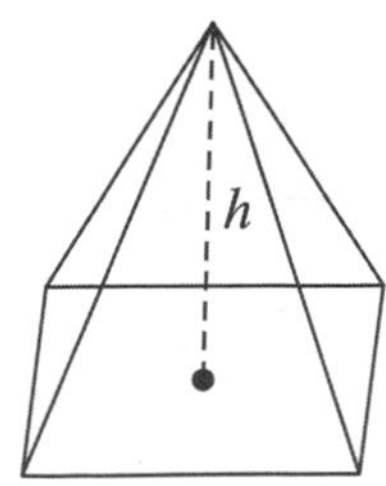

(5) Special Right Triangle

$30° - 60° - 90°$

$45° - 45° - 90°$

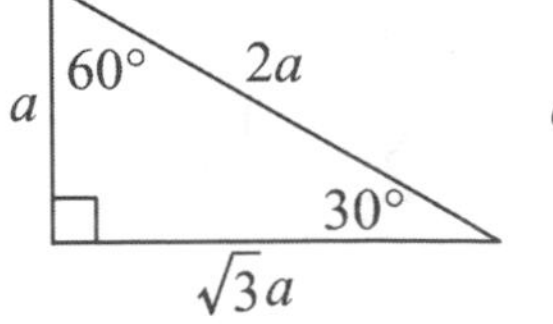

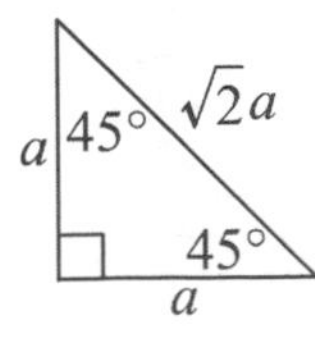

MODEL TEST 1
ANSWER SHEET

Question Number		Your Answers	Answer Key	Question Number		Your Answers	Answer Key
1	(A)(B)(C)(D)	_______	_______	26	(A)(B)(C)(D)	_______	_______
2	(A)(B)(C)(D)	_______	_______	27	(A)(B)(C)(D)	_______	_______
3	(A)(B)(C)(D)	_______	_______	28	(A)(B)(C)(D)	_______	_______
4	(A)(B)(C)(D)	_______	_______	29	(A)(B)(C)(D)	_______	_______
5	(A)(B)(C)(D)	_______	_______	30	(A)(B)(C)(D)	_______	_______
6	(A)(B)(C)(D)	_______	_______	31	(A)(B)(C)(D)	_______	_______
7	(A)(B)(C)(D)	_______	_______	32	(A)(B)(C)(D)	_______	_______
8	(A)(B)(C)(D)	_______	_______	33	(A)(B)(C)(D)	_______	_______
9	(A)(B)(C)(D)	_______	_______	34	(A)(B)(C)(D)	_______	_______
10	(A)(B)(C)(D)	_______	_______	35	(A)(B)(C)(D)	_______	_______
11	(A)(B)(C)(D)	_______	_______	36	(A)(B)(C)(D)	_______	_______
12	(A)(B)(C)(D)	_______	_______	37	(A)(B)(C)(D)	_______	_______
13	(A)(B)(C)(D)	_______	_______	38	(A)(B)(C)(D)	_______	_______
14	(A)(B)(C)(D)	_______	_______	39	(A)(B)(C)(D)	_______	_______
15	(A)(B)(C)(D)	_______	_______	40	(A)(B)(C)(D)	_______	_______
16	(A)(B)(C)(D)	_______	_______	41	(A)(B)(C)(D)	_______	_______
17	(A)(B)(C)(D)	_______	_______	42	(A)(B)(C)(D)	_______	_______
18	(A)(B)(C)(D)	_______	_______	43	(A)(B)(C)(D)	_______	_______
19	(A)(B)(C)(D)	_______	_______	44	(A)(B)(C)(D)	_______	_______
20	(A)(B)(C)(D)	_______	_______	45	(A)(B)(C)(D)	_______	_______
21	(A)(B)(C)(D)	_______	_______	46	(A)(B)(C)(D)	_______	_______
22	(A)(B)(C)(D)	_______	_______	47	(A)(B)(C)(D)	_______	_______
23	(A)(B)(C)(D)	_______	_______	48	(A)(B)(C)(D)	_______	_______
24	(A)(B)(C)(D)	_______	_______	49	(A)(B)(C)(D)	_______	_______
25	(A)(B)(C)(D)	_______	_______	50	(A)(B)(C)(D)	_______	_______

CHAPTER 15
MODEL TEST 1
LEVEL 1

1. If $3x-1=8$, what is the value of $6x+2$?
   (A) 16
   (B) 18
   (C) 20
   (D) 22

2. Which of the following is true ?
   (A) $(\sqrt{3})^2=\pm 3$
   (B) $(\sqrt{-7})^2=7$
   (C) $\sqrt{(-7)^2}=7$
   (D) $(\sqrt{81})^{\frac{1}{2}}=\frac{1}{9}$

3. $g(2x)=6x-2$, which of the following is equivalent to $g(x)$?
   (A) $2x-2$
   (B) $2x+2$
   (C) $3x+2$
   (D) $3x-2$

4. A ball is thrown upward from the roof of a building. The model of the movement is $h(t) = -t^2+14t+31$ what is the meaning of the constant $+31$?
   (A) The highest height the ball can reach.
   (B) The time needed to touch the ground.
   (C) The coordinates of the vertex.
   (D) The height of the building.

5. George and his father decided to repaint their house. George can complete the painting job alone for 6 days, and his father can complete the same job in 4 days. Two days after the two worked together, George's father had to leave the job for some importart business. How long would it take for the remaining work to be completed by George alone?
   (A) 1
   (B) 2
   (C) 3
   (D) 4

6. Samantha wants to lose her weight. She plans that her weight will drop to 212 pounds in 4 months, and 206 pounds in 12 months.
   If she can persist in implementing her plan, how many pounds will be her weight after two years?
   (A) 174
   (B) 188
   (C) 197
   (D) 220

7. There are two lines in the xy-plane, one line has the slope of -2, and passes the point (2, 1). The other line pass point $(-1, 3)$ and (5, 1). If they intersect at point (a, b), what is the value of $a-b$?

(A) $\frac{11}{3}$

(B) $\frac{11}{5}$

(C) $-\frac{32}{5}$

(D) $\frac{32}{5}$

8. There are 6 different varieties of apples with prices ranging from \$1.45 per pound to \$5.15 per pound. If x represents the price of apple per pound. Which of the following inequalities best represents their range ?

(A) $|x-3.7|\le 1.45$

(B) $|x-1.45|\le 3.7$

(C) $|x-1.85|\le 3.7$

(D) $|x-3.3|\le 1.85$

9. Which of the following complex numbers is equivalent to $\frac{2-3i}{6+2i}$?

(Note: $i=\sqrt{-1}$)

(A) $6-3i$

(B) $6+3i$

(C) $\frac{3}{20}-\frac{11}{20}i$

(D) $\frac{3}{20}+\frac{4}{20}i$

10. The distance between the sun and the earth is set by scientists as 1 AU (1 AU = astronomical unit), and the light emitted from the surface of the sun takes 8 minutes and 17 seconds to reach the earth. The speed of the light is 299,792,458 meters per second how long is one AU in miles? (expression in scientic notation) (1 mile = 1.6 km)

(A) 9.3×10^{7}

(B) 1.5×10^{8}

(C) 2.8×10^{9}

(D) 1.3×10^{10}

11. Agnes owes her parents x dollars. Last month she paid $\frac{1}{6}$ of the amount owed. This month she paid them $\frac{1}{6}$ of the remaining amount plus \$20.00. In terms of x, how much money does she still owe?

(A) $\frac{x-20}{6}$

(B) $\frac{5}{6}x-20$

(C) $\frac{5x-20}{6}$

(D) $\frac{25}{36}x-20$

12. If $5^p=400$, what is the value of $5^{\frac{p}{2}+1}$?

(A) 50

(B) 100

(C) 200

(D) 250

13. Jocob bought a simple lunch for his family. He bought a total of two burgers and 3 sandwiches and 5 drinks plus 8% tax for a total of $30.51.
He only remembered that each burger was $1 more than the price of a sandwich, and a drink was just the half the price of a sandwich. How much was a sandwich?

(A) 2.75

(B) 3.15

(C) 3.5

(D) 4.00

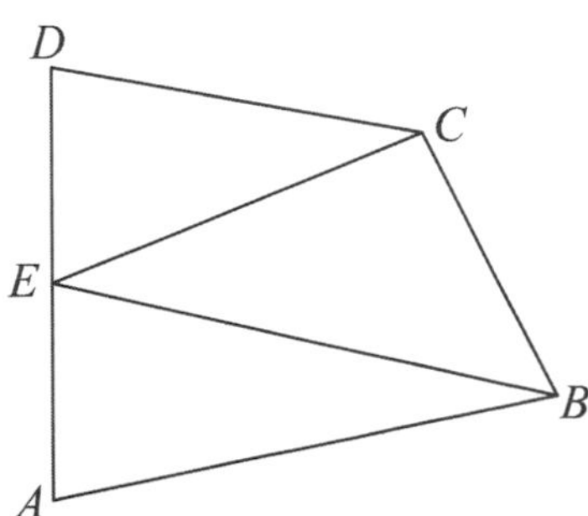

14. The quadrilateral $ABCD$ shown above, $\overline{CE} \parallel \overline{AB}$, $\overline{DC} \parallel \overline{BE}$, and
$\overline{DE} : \overline{AE} = 3:4$,
if the area of $\triangle CDE = 54$, what is the value of the area of $\triangle ABE$?

(A) 96

(B) 112

(C) 128

(D) 144

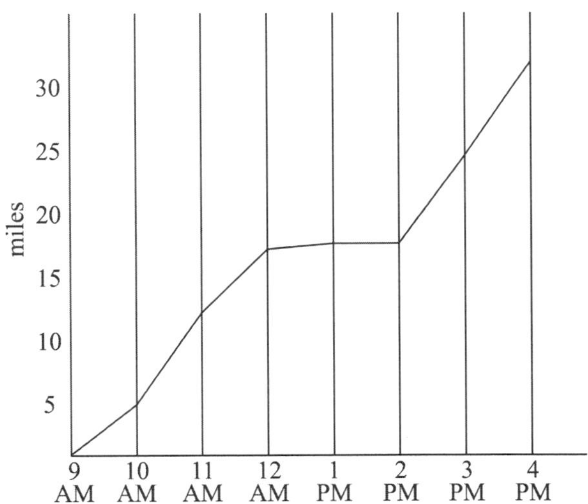

15. The above graph shows the time when Sean's family went hiking in a famous park. What period of time lime did they finish lunch?

(A) 9 AM ~ 11 AM

(B) 11 AM ~ 12 AM

(C) 12 AM ~ 2 PM

(D) 2 PM ~ 3 PM

16. What is the range of $f(x)=x^2+4x+16$?

(A) $\{y \mid y \le 12\}$

(B) $\{y \mid 0 \le y \le 12\}$

(C) $\{y \mid y \ge 12\}$

(D) $\{y \mid 0 \ge y \ge -12\}$

17. It the function $f(x)=\frac{1}{2}x-4$, what is the value of $f^{-1}(f(f^{-1}(4)))$?

(A) 4

(B) 8

(C) 16

(D) 32

18. 36, 38, 42, 42, 48, 50, 50, 50, 52, 53, 53, 54, 54, 54, 55, 60, 60, 62, 64, 64, 64, 66, 67, 68, 70, 72, 72, 74, 76, 80, 82, 108.

The weights of James' schoolmates are listed above. What is interquartile range (IQR)?

(A) 12

(B) 18

(C) 24

(D) 30

19. Samuel has triple as many poker cards as Eva has. If Samuel gives Eva 10 of his poker cards, he will have twice the number of Eva's cards minus 9. How many poker cards did Samuel originally have ?

(A) 48

(B) 55

(C) 63

(D) 72

20.

	fresh-man	sopho-more	Junior	Senior
with glasses	152	162	185	201
without glasses	208	174	136	106
total	360	336	321	307

The above table shows whether wear glasses of M High School students.

If a student was chosen at random, what is the probability that a freshman without glasses will be chosen?

(A) 13.2%

(B) 15.7%

(C) 16.8%

(D) 18.3%

21. The volume of an object is inversely proportional to pressure and directly proportional to temperature. If the pressure increases by 20% and the temperature decreases by 10% how may percentage will the volume increase or decrease?

(A) Decrease 33%

(B) Increase 33%

(C) Decrease 25%

(D)Increase 25%

22. In order to find out the error rate of COVID-19 rapid antigen test kit, 1,000 people were recruited for a study. Among these 1,000 people, 240 who

confirmed positive cases. The result from these antigen test was positive is 210 individuals, but only 188 individuals have really contracted the virus. What is the error rate of this particular rapid antigen test kit ?

(A) 3%

(B) 7.4%

(C) 10.8%

(D) 26.5%

23. In a city, there are 2,400 High School students who participated in a statewide math competition test. If the average score is 58.2 points and the standard deviation is 5 points. Approximately, how many participants in this test have failed (below 60 points) ? (The area $= 0.1406$, when $z - \text{score} = 0.36$)

(A) 1,000

(B) 1,500

(C) 2,000

(D) 2,500

24. What is the measure of $-\frac{3}{5}\pi$ radian in degrees?

(A) $-100°$

(B) $-108°$

(C) $100°$

(D) $108°$

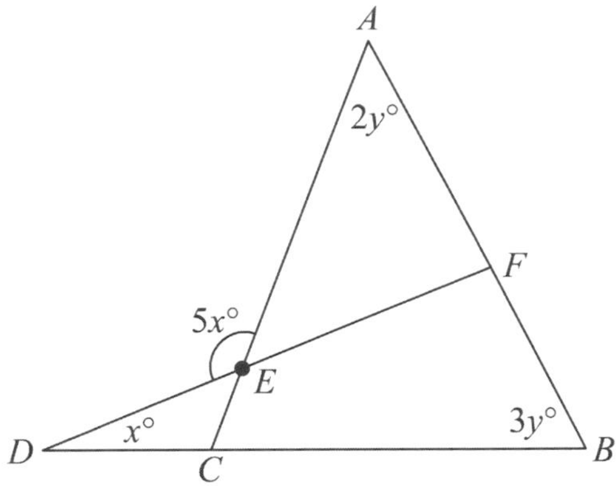

Note: Figure not drawn to scale

25. In the Figure above, what is the value of x in terms of y ?

(A) $\frac{2}{3}y$

(B) $\frac{3}{2}y$

(C) $\frac{4}{5}y$

(D) $\frac{5}{4}y$

26. If $a^5 = 5{,}555$ and $\frac{a^4}{b} = 55$, what is the value of ab ?

(A) 101

(B) 275

(C) 1100

(D) 2020

27. The ratio of x to y is $P:4$, and the ratio of y to z is $6:q$. If the ratio of x to z is $3:2$, what is the ratio of p to q ?

(A) 3 : 5

(B) 5 : 3

(C) 1 : 1

(D) 1 : 3

28. A radioactive substance has an initial mass of 200 grams, and the mass is reduced by 30% every 50 years. Which equation of the following models could be used to calculate the mass, y, remains after x years ?

(A) $y = 200(0.3)^{\frac{x}{50}}$

(B) $y = 200(0.7)^{\frac{x}{50}}$

(C) $y = 200(0.3)^{50 \cdot x}$

(D) $y = 200(0.7)^{50 \cdot x}$

29. If p is the average of $3x+7$, g is the average of $5x+4$, and r is the average of $4x+13$, what is the average of p, q, and r in terms of x?

(A) $x+2$

(B) $2x+4$

(C) $x+3$

(D) $3x+3$

30. The graph of $y=(2x-2)(3-x)$ in the xy-plane is a parabola, which of the following statements is(are) correct ?

(I) The maximum value of y is $y=2$

(II) The vertex is at $(2, 2)$

(III) The graph's line of symmetry is $x=4$

(A) I only

(B) II only

(C) I and II

(D) I and III

31. If the graph of a function $f(x)$ passes through the point $(6, 4)$, and $g(x)$ is defined as $g(x) = -f(x+1)-6$, which of the following points does the graph of $g(x)$ pass ?

(A) (5, 2)

(B) (5, 6)

(C) (5, − 2)

(D) (5, −10)

32. If $\cos(-130°) = t$, what is the value of $\tan(230°)$ in terms of t ?

33. Set $A = \{a+3, a+4\}$, $B = \{b-2, b+2\}$ if $A \cup B = \{0, 4, 6, 7\}$, what is the value of $a+b$?

(A) 5

(B) 6

(C) 7

(D) 8

34. Cynthia and her classmates, total of 32 people, would like to rent beach bikes to cruise on the beach. There are two types of beach bikes available in the store (for 2 riders and for 3 riders).

Cynthia and her classmates can rent total of 14 beach bikes. Because most people in her group prefer 2-rider bikes, what is the maximum number of 2-rider bikes can they rent ?

(A) 6

(B) 8

(C) 10

(D) 12

35. $N = H[1+0.2(1-\frac{H}{M})]$

Rewrite the expression above, make *H* in terms of *N*, and *M*?

36. As the function of question 35 above, if *N* represents the heads of pig next year, and *M* is the maximum heads of pig in any year, and *H* is the heads of pig, this year. If the maximum heads *M* is 2,000, and the head next year are 200, what is the value of *H* (this year)?

(A) 170

(B) 290

(C) 1800

(D) 4200

37. A certain landscaping company charges Wyne for $20 per worker per hour and $S per square inch for their work, plus garbage handling fee $120.

Which of following best to model this estimation for f squre inches of land?

(A) $C = 20 \cdot p \cdot h + f \cdot s + 120$

(B) $C = 20 \cdot p \cdot h + 120 + \frac{f \cdot s}{p}$

(C) $C = 20 \cdot p \cdot h \cdot f \cdot s$

(D) $C = ph + f \cdot s + 120$

38.

	take bus	not take bus
boys	400	290
girls	370	220

The above table shows the students of Jayden's school and whether they take school buses or not.

What is the probability that randomly select one from girls who does not take school bus to school?

(A) 17%

(B) 27%

(C) 37%

(D) 47%

39. If $k < -1$, which of the following graphs could be the equation of $2x + y = k(x - y)$?

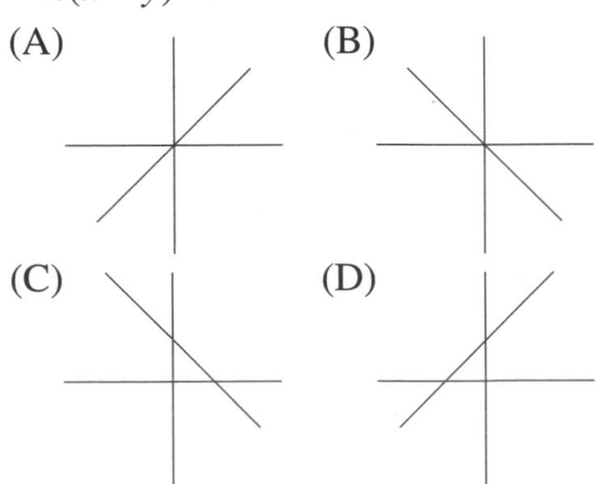

40. What is the value of $\cos(\sin^{-1}\frac{5}{13})$?

41. The temperature in a certain area can be represented by the model $f(t)=-t^2+12t+17$. What is the highest temperature in this area ?

(A) 48

(B) 53

(C) 59

(D) 64

42. A forestist develop the equation $y=1,200x+42,000$ to estimate the number of trees that will re-grow after x years of a forest fire in this small area. What does the number of 42,000 represent in this model?

(A) The estimated yearly increase in the number of trees.

(B) The estimated yearly decrease in the number of trees.

(C) The estimated number of the trees that survived from this forest fire.

(D) The estimated number of trees after x years.

43. A parabola was shifted right by 5 units and down for 4 units, finally flipped about x-axis, the function change to $f(x)=x^2-12x+42$, what is the original function ?

(A) $f(x)=(x-1)^2+2$

(B) $f(x)=(x-1)^2-2$

(C) $f(x)=(x-5)^2+4$

(D) $f(x)=(x+5)^2+4$

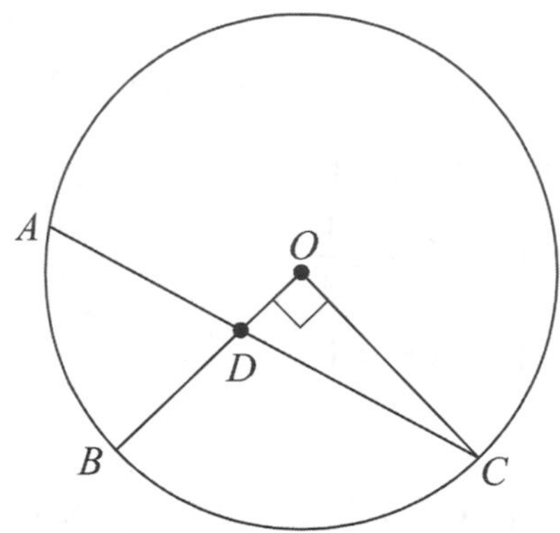

Note: Figure not drawn to scale

44. In the figure above the circle with center O and BD:OD = 3:1, what is the value of $\cos\angle DCO$

(A) 0.33

(B) 0.65

(C) 0.83

(D) 0.97

45. Which of the following could be the distance of x-intercept and y-intercept of the parabola $f(x)=(x+2)^2-9$?

(A) 4.6

(B) 5.1

(C) 5.9

(D) 6.3

46. Where $i=\sqrt{-1}$, $(2+3i)(3-2i)=$

(A) $12+5i$

(B) $5+12i$

(C) $6+5i$

(D) $5+6i$

47. What is the range of $f(x)=\dfrac{x}{x^2+x+1}$?

(A) $f(x)$ can be any real numbers

(B) $-1< f(x)\le 1$

(C) $-1\le f(x)\le \dfrac{1}{3}$

(D) $-1\le f(x)<1$

48. The price of a certain product of *ABC*-Toy company is \$38 each, and is sold for 220 units per month. Based on the company's experience, a reduction of \$1 per unit will increase sales by 10 units. How many dollars does the company need to reduce the price per unit to reach the maximum revenue ?

(A) \$4

(B) \$6

(C) \$8

(D) \$10

49. Sam's classmates heights are listed as follows: 58, 60, 60, 62, 63, 64, 65, 66, 67, 67, 67, 67, 68, 69, 70, 72, 73, 73, 74.
If Sam's height is 70, and the standard deviation (S.D.) is 4.54, what is Sam's estimated percentile in his class ?

(A) 62.34%

(B) 69.21%

(C) 72.35%

(D) 77.3%

50. $\left(\dfrac{1}{\csc\theta}-\dfrac{\tan\theta}{\sin\theta}\right)(\cos\theta)=$

(A) $\sin\theta$

(B) $\cos\theta$

(C) $\sin\theta\cos\theta-1$

(D) $\sin\theta\cos\theta+1$

MODEL TEST 1
ANSWER KEY

1. (C)
2. (C)
3. (D)
4. (D)
5. (A)
6. (C)
7. (C)
8. (D)
9. (C)
10. (A)
11. (D)
12. (B)
13. (C)
14. (A)
15. (C)
16. (C)
17. (C)
18. (B)
19. (C)
20. (B)
21. (C)
22. (D)
23. (B)
24. (B)
25. (D)
26. (A)
27. (C)
28. (B)
29. (B)
30. (C)
31. (D)
32. (C)
33. (A)
34. (C)
35. $H = \dfrac{1.2M \pm \sqrt{(1.2M)^2 - 0.8MN}}{0.4}$
36. (A)
37. (A)
38. (C)
39. (A)
40. (0.92)
41. (B)
42. (C)
43. (B)
44. (D)
45. (B)
46. (A)
47. (C)
48. (C)
49. (D)
50. (C)

MODEL TEST 1
ANSWERS and EXPLANATIONS

1. (C)

$3x-1=8$, $3x=9$

$\therefore\ 6x=18 \quad 6x+2=18+2=20$

2. (C)

(A) $(\sqrt{3})^2=3$

(B) Since the input of the function of square root is a imaginary number, so, the operation must follow the operation rule of imaginary number.
$(\sqrt{-7})^2=(\sqrt{7})^2i^2=-7$

(C) Since $\sqrt{(-7)^2}=\sqrt{(-7)(-7)}$
$=\sqrt{49}=7$

(D) Since $(\sqrt{81})^{\frac{1}{2}}=\dfrac{1}{(\sqrt{81})^2}=\dfrac{1}{81}$

Answer is (C)

3. (D)

$g(2x)=6x-2$

Let $x'=2x$, $x=\dfrac{x'}{2}$

$g(x')=6(\dfrac{x'}{2})-2=3x'-2$

$\therefore\ g(x)=3x-2$

4. (D)

$h(t)=-t^2+14t+31$

when $t=0$, $h(0)=+31$

It means the initial position of the ball is at $+31$.

So, it is the height of the building.

5. (A)

George and his father work togother

One day can complete $\dfrac{1}{6}+\dfrac{1}{4}=\dfrac{2+3}{12}$
$=\dfrac{5}{12}$

two days' work $\dfrac{5}{12}\times 2=\dfrac{5}{6}$,

remaining work $=1-\dfrac{5}{6}=\dfrac{1}{6}$

by George alone $=\dfrac{\frac{1}{6}}{\frac{1}{6}}=1$ day

6. (C)

Month	Weight
4	212
12	206
24	?
$12-4=8$	$212-206=6$

She will drop her weight 6 pounds in 8 months, so, she will drop $\dfrac{6}{8}$ pounds in one month.

For 2 years, drop

$24\times\dfrac{6}{8}=18$ (pounds)

$206-9=197$ (1 year from 206)

Answer is (C)

7. (C)

$\ell_1: y=-2x+b$

Since passes (2, 1)

$1=-2(2)+b,\ b=5$

$\ell_1: y=-2x+5$

$\ell_2: y=mx+b$

passes $(-1,3)$, and $(5,1)$

$3=-m+b \cdots ①$

$1=5m+b \cdots ②$

① − ② $2=-6m,\ m=-\frac{1}{3}$

substitute in ①

$3=-(-\frac{1}{3})+b=\frac{1}{3}+b$

$b=3-\frac{1}{3}=2\frac{2}{3}=\frac{8}{3}$

$\ell_2: y_2=-\frac{1}{3}x+\frac{8}{3}$

Find Their intersection point

$y_1=y_2$

(y-coordinate of ℓ_1 is the same as y-coordinate of ℓ_2)

$-2x+5=-\frac{1}{3}x+\frac{8}{3}$

$\therefore\ -\frac{5}{3}x=-\frac{15}{3}+\frac{8}{3}=-\frac{7}{3}$

$\therefore\ x=\frac{7}{5}$

Substitute $y_1=2x+5$

$\therefore\ y_1=\frac{14}{5}+5=\frac{39}{5}$

$(a,b)=(\frac{7}{5},\frac{39}{5})$

$a-b=\frac{7}{5}-\frac{39}{5}=-\frac{32}{5}$

Answer is (C)

8. (D)

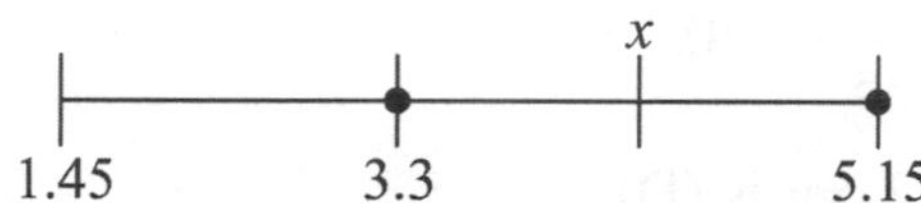

$5.15-1.45=3.7,\ 3.7\div 2=1.85$

$1.85+1.45=3.3$

all prices can be represented by

$|x-3.3|\le 1.85$

Answer is (D)

9. (C)

$$\frac{2-3i}{6+2i}=\frac{(2-3i)(6-2i)}{(6+2i)(6-2i)}$$
$$=\frac{12-18i-4i+6i^2}{6^2-(2i)^2}$$
$$=\frac{12-22i-6}{36-4i^2}$$
$$=\frac{6-22i}{40}$$
$$=\frac{3-11i}{20}$$

10. (A)

$299,792,458\times(8\times 60+17)$

$=148,996,851,626=148,996,851$ km

$\frac{148,996,852}{1.6}=93,123,033$ miles.

Scientific notation $=9.3123033\times 10^7$

or 9.3×10^7 miles

11. (D)

$x-[\frac{x}{6}+\frac{5}{6}x\times\frac{1}{6}+20]$

$=x-(\frac{6x}{36}+\frac{5x}{36}+20)$

$=x-\frac{11x}{36}-20$

$=\frac{25}{36}x-20$

Answer is (D)

12. (B)

$5^p=400$, $5^{\frac{p}{2}}=20$,

$5^{\frac{p}{2}+1}=5^{\frac{p}{2}}\times5=20\times5=100$

Answer is (B)

13. (C)

Let the price of a sandwich be s, then

$30.51=[2(s+1)+3s+(\frac{s}{2})\times5]\times1.08$

$=(2s+2+3s+\frac{5}{2}s)\times1.08$

$=(7.5s+2)\times1.08$

$=8.1s+2.16$

$\therefore\ 28.35=8.1s$

$s=3.5$

Answer is (C)

14. (A)

$\overline{CD}\parallel\overline{BE}$, $\angle D=\angle AEB$, $\overline{EC}\parallel\overline{AB}$,

$\angle DEC=\angle A$, $\triangle DEC\sim\triangle ABE$,

$\frac{\triangle DEC}{\triangle ABE}=\frac{3^2}{4^2}=\frac{9}{16}$

$\therefore\ \triangle ABE=\frac{\triangle DEC\times16}{9}=\frac{54\times16}{9}$

$=96$

Answer is (A)

15. (C)

From 12AM to 2PM there is no miles increased that means her family are in the situation of rest obvionsly, they finish their lunch at the period of 12AM to 2PM.

Answer is (C)

16. (C)

$x^2+4x+16=x^2+4x+4-4+16$

$=(x+2)^2+12$

Vertex at $(-2,12)$

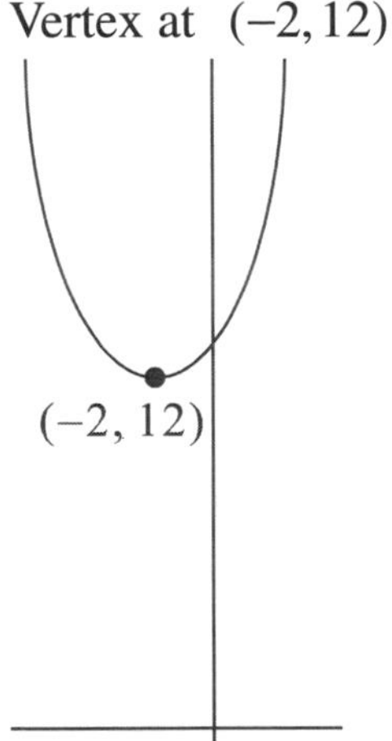

The minimum of y is 12 (concave up)
So, the range of $f(x)$ is $\{y\mid y\geq12\}$

Answer is (C)

17. (C)

$f(x)=\frac{1}{2}x-4$, Let $y=\frac{1}{2}x-4$,

$2y = x - 8$, $x = 2y + 8$, exchange x, y.

So, $f^{-1}(x) = 2x + 8$, $y = 2x + 8$

$f^{-1}(f(f^{-1}(4))) = f^{-1}(f(16)) = 16$

Answer is (C)

18. (B)

$IQR = Q_3 - Q_1$

Q_1 $32 \times 0.25 = 8$

the value of $Q_1 = \dfrac{50+52}{2} = 51$

Q_3 at $32 \times 0.75 = 24$

The value of $Q_3 = \dfrac{68+70}{2} = 69$

$IQR = 69 - 51 = 18$

Answer is (B)

19. (C)

$S = 3E$

$S - 10 = 2(E + 10) - 9$

$3E - 10 = 2E + 20 - 9$

$E = 21$, $S = 63$

Answer is (C)

20. (B)

Freshman without wear glasses $= 208$

Total students $= 1{,}324$

$\dfrac{208}{1324} = 0.157 \Rightarrow 15.7\%$

Answer is (B)

21. (C)

$V_1 = \dfrac{T}{P}$, $V_2 = \dfrac{0.9T}{1.2P} = 0.75\dfrac{T}{P} = 0.75V_1$

$\dfrac{V_2 - V_1}{V_1} = \dfrac{(0.75-1)V_1}{V_1} = -0.25$

Answer is (C)

22. (D)

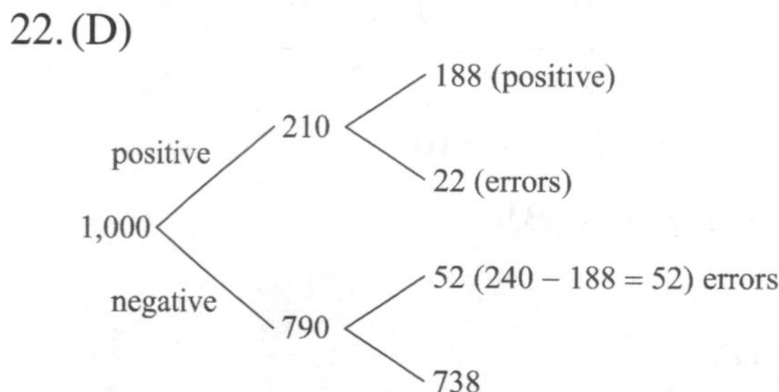

Total errors $= 22 + 52 = 74$

error rate $= \dfrac{74}{1000} = 7.4\%$

Answer is (B)

23. (B)

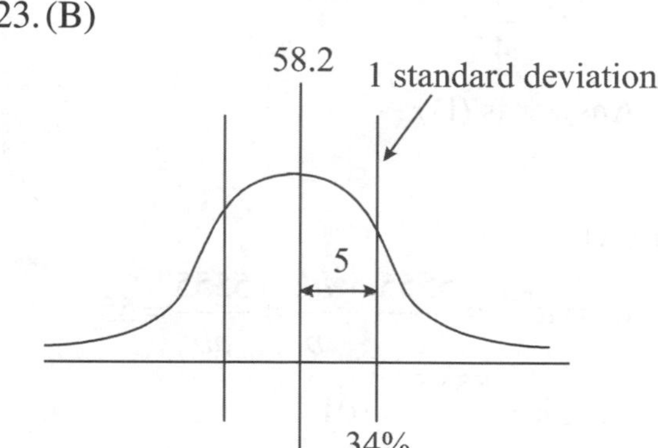

Average $= 58.2$

1 standard deviation $= 5$

$\dfrac{60 - 58.2}{5} = z\text{-score} = 0.36$

From z-score table z-score=0.36

the area =0.1406

0.50+0.1406=0.6406

2400×0.6406=1537.4≐1538

Answer is (B)

24. (B)

$$\frac{-\frac{3}{5}\pi}{2\pi}=\frac{x}{360}$$

$$x=\frac{360\times(-\frac{3}{5})}{2}=\frac{72\times(-3)}{2}$$

$$=36\times(-3)=-108$$

Answer is (B)

25. (D)

Since in $\triangle ABC$,

$3y°+2y°+m\angle ACB=180°$

$\Rightarrow\ 3y°+2y°+x°+m\angle DEC=180°$

$\Rightarrow\ 3y°+2y°+x°+(180°-5x°)=180°$

$\Rightarrow\ 5y+x+180-5x=180\ \Rightarrow\ 5y=4x$

$\therefore\ x=\frac{5}{4}y$

Answer is (D)

26. (A)

$$a^4=a^{5-1}=\frac{5555}{a},\ \frac{a^4}{b}=\frac{5555}{ab}=55$$

$$\therefore\ ab=\frac{5555}{55}=101$$

Answer is (A)

27. (C)

$$\frac{x}{z}=\frac{3}{2},\ x=\frac{3}{2}z;\ \frac{y}{z}=\frac{6}{q},\ y=\frac{6z}{q}$$

$$\therefore\ \frac{x}{y}=\frac{p}{4}=\frac{\frac{3}{2}z}{\frac{6}{q}z}=\frac{3q}{12}=\frac{q}{4}$$

$$\therefore\ \frac{p}{4}=\frac{q}{4},\ \frac{p}{q}=\frac{4}{4}=1$$

Answer is (C)

28. (B)

mass remains $=1-0.3=0.7$

After 50 years, $y=200(0.7)^{\frac{x}{50}}$

Answer is (B)

29. (B)

$$p=\frac{3x+7}{2},\ q=\frac{5x+4}{2},\ r=\frac{4x+13}{2}$$

$$\frac{p+q+r}{3}=\frac{\frac{3x+7}{2}+\frac{5x+4}{2}+\frac{4x+13}{2}}{3}$$

$$=\frac{1}{6}(3x+7+5x+4+4x+13)$$

$$=\frac{1}{6}(12x+24)=2x+4$$

Answer is (B)

30. (C)

$$y=6x-2x^2-6+2x=-2x^2+8x-6$$

$$=-2(x^2-4x+4)+8-6$$

$$=-2(x-2)^2+2$$

Answer is (C)

31. (D)

function	point	shift
$f(x+1)$	$[(6-1),4)]$ $\rightarrow(5,4)$	left 1 unit
$-f(x+1)$	$(5,-4)$	flipped on x-axis
$-f(x+1)$ -6	$(5,-4)$ $\rightarrow(5,-4-6)$ $\rightarrow(5,-10)$	downwards 6 units

Answer is (D)

32. (C)

$\cos(-130°) = \cos(230°) = -\cos(50°) = t$

$\therefore$ $\tan(230°) = \tan(50°)$ (In Quadrant III)

$\tan(230°) = \tan(50°) = \dfrac{-\sqrt{1-t^2}}{t}$

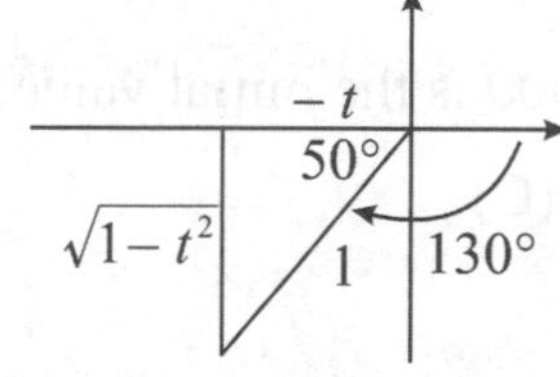

33. (A)

Since $A = \{a+3, a+4\}$, the relation between $(a+3)$ and $(a+4)$ is $(a+4) -(a+3) = 1$

In set $A \cup B = \{0, 4, 6, 7\}$ only elements 6 and 7 fit the condition, so, $a+3=6$, $a+4=7$, $a=3$. Since in $\{b-2, b+2\}$, only $b-2=0$, $b+2=4$, fit the condition, so, $b=2$, $a+b=3+2=5$

Answer is (A)

34. (C)

Solution: Let x be the maximum number of 2-rider bikes, then $2x+(14-x)\times 3 = 32$

$\therefore$ $x = 10$

Answer is (C)

35. $N = H[1+0.2(1-\dfrac{H}{M})]$

$= H[1+0.2(\dfrac{M-H}{M})]$

$= H[1+0.2 \cdot \dfrac{M}{M} - 0.2\dfrac{H}{M}]$

$= H(1+0.2-0.2\dfrac{H}{M})$

$= H(1.2-0.2\dfrac{H}{M})$

$= 1.2H - 0.2\dfrac{H^2}{M}$

$MN = 1.2MH - 0.2H^2$

$0.2H^2 - 1.2MH + MN = 0$

$H = \dfrac{1.2M \pm \sqrt{(1.2M)^2 - 4(0.2)MN}}{2\times 0.2}$

36. (A)

$N = H[1+0.2(1-\dfrac{H}{M})]$

$200 = H[1+0.2(1-\dfrac{H}{2000})]$

$\because$ $H = \dfrac{1.2\times M \pm \sqrt{(1.2M)^2 - 0.8MN}}{0.4}$

$= \dfrac{1.2\times 2000 \pm \sqrt{2400^2 - (0.8)(2000)(200)}}{0.4}$

$= 170$, or $4733 > 2000$ (false)

Answer is (A)

37. (A)

Total manpower costs

$20 \cdot p \cdot h$

Total square inches costs $f \cdot s$

garbage handling \$120

So, total cost $= 20 \cdot p \cdot h + f \cdot s + 120$

Answer is (A)

38. (C)

Girls not taking bus = 220

Total girls $= 370 + 220 = 590$

$P(\text{girl not taking bus}) = \dfrac{220}{590} = 0.37$

$\Rightarrow 37\%$

Answer is (C)

39. (A)

$2x + y = k(x - y)$

$y + ky = kx - 2x$

$(1+k)y = (k-2)x$

$y = \dfrac{k-2}{1+k}x$

Since $k < -1$, $\dfrac{k-2}{1+k}$ is positive

So, the graph passes origin with positive slope

Answer is (A)

40. 0.92

Let $\theta = \sin^{-1}\dfrac{5}{13}$

$\cos\theta = \dfrac{12}{13} = 0.92$

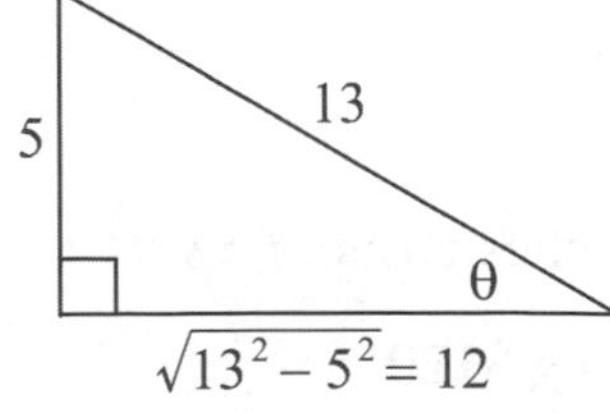

41. (B)

$f(t) = -(t^2 - 12t + 36) + 36 + 17$

$= -(t-6)^2 + 53$

Therefore, the highest temperature is 53 °F in this area.

Answer is (B)

42. (C)

The constant 42,000 is the initial value,

So, the answer is (C)

43. (B)

(1) Rewrite the function for finding the new vertex.

$f(x) = x^2 - 12x + 42$

$= x^2 - 12x + 6^2 - 6^2 + 42$

$= (x-6)^2 + 6$

new vertex at $(6, 6)$

(2) flip about x-axis, vertex change to $(6, -6)$

(3) Shift up for 4 units, vertex change to $(6, -6+4) \Rightarrow (6, -2)$

(4) Shift left by 5 units, vertex change to $(6-5, -2) \Rightarrow (1, -2)$

(5) The original function

$f(x) = (x-1)^2 - 2$

Answer is (B)

44. (D)

Since $\overline{BO} = \overline{CO}$ (Radius)

Since $BD = 3DO$, $BO = 4DO$

$\cos\angle DCO = \dfrac{OC}{CD} = \dfrac{BO}{CD}$

Since $\overline{CD}^2 = \overline{OC}^2 + \overline{DO}^2$

$$= \overline{CO}^2 + (\frac{1}{4}\overline{BO})^2$$
$$= \overline{CO}^2 + \frac{1}{16}\overline{BO}^2$$
$$= \overline{CO}^2 + \frac{1}{16}\overline{CO}^2$$
$$= (\frac{17}{16})\overline{CO}^2$$
$$\therefore\ CD = \frac{\sqrt{17}}{4}CO$$
$$\cos\angle DCO = \frac{CO}{CD} = \frac{4}{\sqrt{17}} = 0.97$$

45. (B)

x-intercept:

When $y=0$

$(x+2)^2-9=0$

$x^2+4x+4-9=0$

$x^2+4x-5=0$

$(x+5)(x-1)=0$

$x=-5,\ x=1$

x-intercept at $(-5,0)$, $(1,0)$

When $x=0$, $y=4-9=-5$

y-intercept at $(0,-5)$

Distance

$=\sqrt{(-5)^2+(0+5)^2}=\sqrt{25+25}=5\sqrt{2}$

Distance

$=\sqrt{(1-0)^2+(0-(-5))^2}=\sqrt{1^2+25}$

$=\sqrt{26}\doteq 5.1$

Answer is (B)

46. (A)

$(2+3i)(3-2i)=6+9i-4i-6i^2$

$=6+5i-6(-1)$

$=6+5i+6$

$=12+5i$

Answer is (A)

47. (C)

Let $f(x)=y=\frac{x}{x^2+x+1}$

$yx^2+yx+y=x,\quad yx^2+(y-1)x+y=0$

Since x is real number, so

$\triangle=b^2-4ac\ge 0\ \Rightarrow\ (y-1)^2-4y^2\ge 0$

$y^2-2y+1-4y^2\ge 0$

$\Rightarrow\ -3y^2-2y+1\ge 0$

$\Rightarrow\ 3y^2+2y-1\le 0$

$(3y-1)(y+1)\le 0$

(1) $3y-1\ge 0$ and $y+1\le 0$ (unreasonable)

$\Rightarrow\ y\ge\frac{1}{3}$ and $y\le -1$

(2) $3y-1\le 0$ and $y+1\ge 0$

$\Rightarrow\ -1\le y\le\frac{1}{3}$

Answer is (C)

48. (C)

1. Original price is \$38 for each product, so total revenue $R=220\times 38=8,360$
2. If decrease $\$x$ for each product, revenue will be

$R=(38-x)(220+10x)$

$=8360-220x+380x-10x^2$

$=-10(x^2-16x+8^2)$

$+10\times 8^2+8360$

$=-10(x-8)^2+9000$

It is a parabola and concave downward, so maximum revenue is at vertex (8, 9000)

3. Decrease \$8 for each product will get the maximum revenue 9,000

   $9000-8360=640$

4. Reduce \$8 for each product will get the maximum revenue.

Answer is (C)

49. (D)

$\bar{x}=66.58$

Sam's z-score$=\dfrac{70-66.58}{4.54}=0.75$

The area in z-score table $0.75 \Rightarrow 0.2734$

sam's percentile in his class is

$0.5+0.2734=0.7734 \Rightarrow 77.3\%$

Answer is (D)

50. (C)

$$\left(\frac{1}{\csc\theta}-\frac{\tan\theta}{\sin\theta}\right)(\cos\theta)$$

$$=\left(\frac{1}{\frac{1}{\sin\theta}}-\frac{\frac{\sin\theta}{\cos\theta}}{\sin\theta}\right)(\cos\theta)$$

$$=\left(\sin\theta-\frac{1}{\cos\theta}\right)\cos\theta$$

$$=\sin\theta\cos\theta-1$$

Answer is (C)

CHAPTER 15
MODEL TEST 2
LEVEL 1

DIRECTION:

Fill in the correct or the best answer in the proper position on the answer sheet. You may use any available space for scratch work.

NOTES:

1. Calculator is permitted.
2. All figures are not drawn to scale.
3. All numbers are real, unless stated otherwise.
4. Some reference information:

(1) Right Circular Cone

$$V=\frac{1}{3}\pi r^2\cdot h,\ A(\text{lateral surface})=\frac{1}{2}c\ell=\pi r\cdot\ell$$

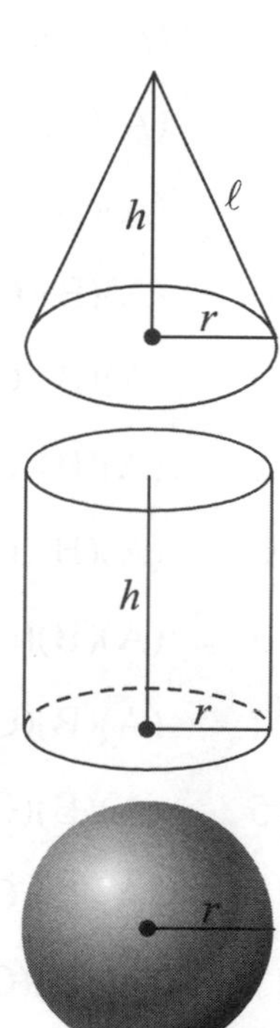

(2) Cylinder

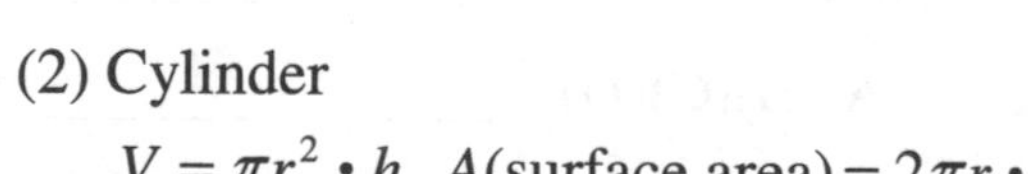

$$V=\pi r^2\cdot h,\ A(\text{surface area})=2\pi r\cdot h$$

(3) Sphere

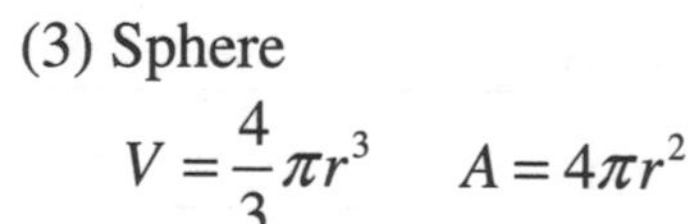

$$V=\frac{4}{3}\pi r^3 \qquad A=4\pi r^2$$

(4) Pyramid

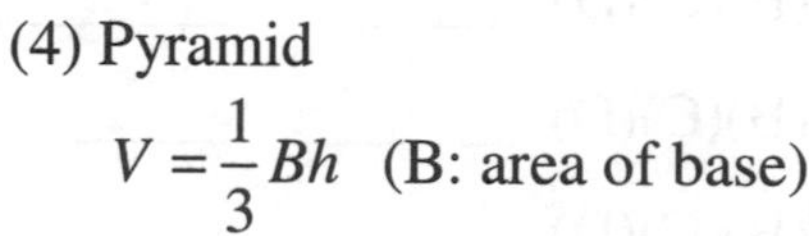

$$V=\frac{1}{3}Bh \quad (\text{B: area of base})$$

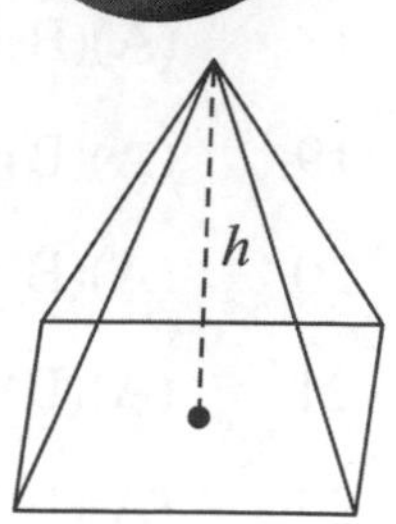

(5) Special Right Triangle

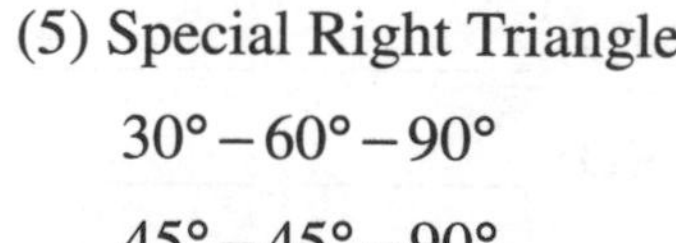

$30°-60°-90°$

$45°-45°-90°$

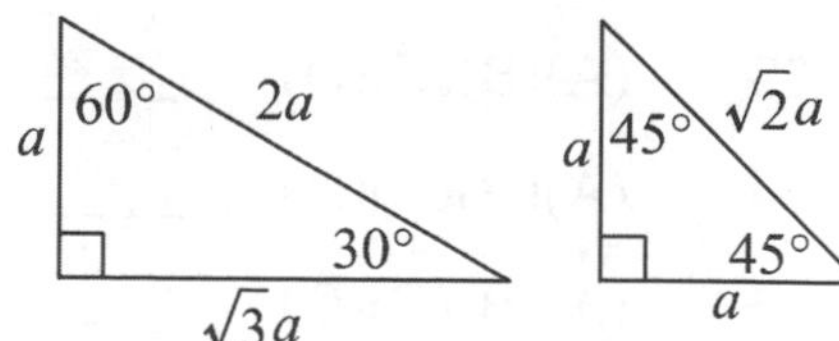

MODEL TEST 2
ANSWER SHEET

Question Number		Your Answers	Answer Key	Question Number		Your Answers	Answer Key
1	(A)(B)(C)(D)			26	(A)(B)(C)(D)		
2	(A)(B)(C)(D)			27	(A)(B)(C)(D)		
3	(A)(B)(C)(D)			28	(A)(B)(C)(D)		
4	(A)(B)(C)(D)			29	(A)(B)(C)(D)		
5	(A)(B)(C)(D)			30	(A)(B)(C)(D)		
6	(A)(B)(C)(D)			31	(A)(B)(C)(D)		
7	(A)(B)(C)(D)			32	(A)(B)(C)(D)		
8	(A)(B)(C)(D)			33	(A)(B)(C)(D)		
9	(A)(B)(C)(D)			34	(A)(B)(C)(D)		
10	(A)(B)(C)(D)			35	(A)(B)(C)(D)		
11	(A)(B)(C)(D)			36	(A)(B)(C)(D)		
12	(A)(B)(C)(D)			37	(A)(B)(C)(D)		
13	(A)(B)(C)(D)			38	(A)(B)(C)(D)		
14	(A)(B)(C)(D)			39	(A)(B)(C)(D)		
15	(A)(B)(C)(D)			40	(A)(B)(C)(D)		
16	(A)(B)(C)(D)			41	(A)(B)(C)(D)		
17	(A)(B)(C)(D)			42	(A)(B)(C)(D)		
18	(A)(B)(C)(D)			43	(A)(B)(C)(D)		
19	(A)(B)(C)(D)			44	(A)(B)(C)(D)		
20	(A)(B)(C)(D)			45	(A)(B)(C)(D)		
21	(A)(B)(C)(D)			46	(A)(B)(C)(D)		
22	(A)(B)(C)(D)			47	(A)(B)(C)(D)		
23	(A)(B)(C)(D)			48	(A)(B)(C)(D)		
24	(A)(B)(C)(D)			49	(A)(B)(C)(D)		
25	(A)(B)(C)(D)			50	(A)(B)(C)(D)		

CHAPTER 15
MODEL TEST 2
LEVEL 1

1. If $x<2$, which of the following is equivalent to $|x-5|$?

(A) $5+x$

(B) $5-x$

(C) $-x-5$

(D) $x-5$

2. If $f(x)=x^2+4x+5$, find $f^{-1}(f(x))$.

(A) x

(B) $\frac{1}{2}x^2$

(C) $25x^2-2x+3$

(D) $\frac{1}{2}x^2-x+1$

3. In the function $f(x)=\frac{2}{3}x+b$, b is a constant. If $f(-3)=9$, what is the value of $f(9)$?

(A) 7

(B) 9

(C) 12

(D) 17

4. $\begin{cases} xy=8 \\ y(2+x)=5 \end{cases}$

What is value of x?

(A) -5.33

(B) -2.18

(C) 2.18

(D) 5.33

5. The half-life of a radioactive substance is 5,000 years. (half-life is the time it takes for a radioactive substance to lose half of its radiation)

How much quantity is left after 25,000 years?

(A) $\frac{1}{4}$

(B) $\frac{1}{8}$

(C) $\frac{1}{16}$

(D) $\frac{1}{32}$

6.

4, 3, -1, -4, -3, 1, 4, 3, -1, -4, -3, 1, 4, ……

The sequence is shown above, the first term is 4 and the second term is 3. Each term after the second term can be obtained by subtractive the previous term from the term before it, what is the value of the 75th term ?

(A) -1

(B) -4

(C) -3

(D) 1

7. $\begin{cases} 3x+y=15 \\ 2x+3y=24 \end{cases}$

If (x, y) is a solution to the system equations above, which of the following is the value of $x+y$?

(A) -3

(B) -6

(C) -9

(D) 9

8. $y=t(x+1)(x-2)$ is a parabola with $t<0$, what is the maximum value of y in terms of t?

(A) $-\frac{9}{4}t$

(B) $-\frac{3}{2}t$

(C) $\frac{9}{4}t$

(D) $\frac{3}{2}t$

9. In the quadratic equation of $y=k(x+2)(x-3)$ where k is a non-zero constant. The graph of the equation is a parabola with vertex (a,b) in the xy-plane. What is the value of b?

(A) $-\frac{25}{4}k$

(B) $\frac{25}{4}k$

(C) $-\frac{1}{2}k$

(D) $\frac{1}{2}$

10. Which of the following equations has no solution ?

(A) $-3(x-2)^2+5=0$

(B) $-3(x-2)(x+5)=0$

(C) $3(x-2)^2+5=0$

(D) $3(x+3)(x+2)=0$

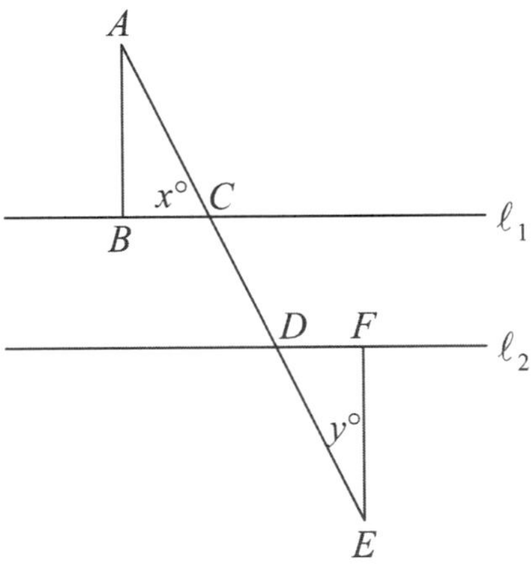

11. In the graph above, $\ell_1 \parallel \ell_2$, $\angle ABC = \angle DFE = 90°$, If $\sin x° = 0.5$, what is the value of $\cos y°$?

(A) 0.32

(B) 0.42

(C) 0.5

(D) 0.62

12. If a lag is a log, then a lag is a leg is true. Which of the following statement must be true?

(A) If a lag is not a log, then a lag is a leg.

(B) If a lag is not a log, then a lag is not a leg.

(C) If a leg is not a lag, then a log is not a lag.

(D) If a log is not a log, then a lag is a leg.

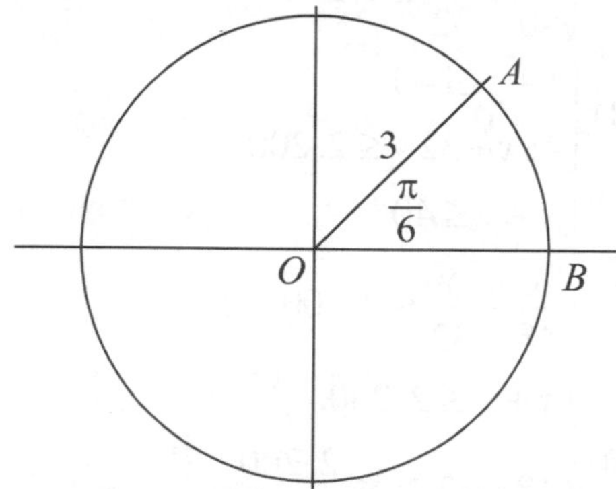

13. In the figure above, circle 0 with radius 3 and $\angle AOB = \dfrac{\pi}{6}$, what is the length of arc $\overset{\frown}{AB}$?

(A) $\dfrac{\pi}{2}$

(B) $\dfrac{3}{2}\pi$

(C) $\dfrac{\pi}{3}\pi$

(D) $\dfrac{2}{3}\pi$

14. $\begin{cases} ax+by=12 \\ 3x+7y=20 \end{cases}$

The system equations above has an infinite number of many solutions, what is the value of $a+b$?

(A) 3

(B) 6

(C) 10

(D) 12

15. A webside provides videos for customers to download. The cost for downloading a video is $\$7.99+tax\,(8\%)$, streaming a video charges $\$3.99+ tax\,(8\%)$. If one video was downloaded d times and streamed s times. What is the revenue of this video ?

(A) $(7.99\times d+3.99\times s)\times(1.08)$

(B) $(7.99\times0.08+3.99\times0.08)(s+d)$

(C) $(7.99+3.99)(s+d)\times(1.08)$

(D) $(7.99+3.99)(s+d)\times0.08$

16. A foreman of a certain production line wants to know the quality of his products. He randomly inspects 8 out of every 60 products. If the production line produce 15,000 products a day, how many of the products could be inspected per day?

(A) 200

(B) 260

(C) 300

(D) 360

17. Find the vertex of the parabola given by $f(x)=2x^2-8x+1$.

(A) $(2,-7)$

(B) $(-2, 7)$

(C) $(4, -7)$

(D) $(4, 7)$

18. Given $T_{n+2} = 2T_n + T_{n+1}$, where T_n represents the term at nth position. If $T_1 = 3$, $T_2 = 6$, what is the value of T_5 ?

(A) 32

(B) 38

(C) 42

(D) 48

19. If the graph of $2x + y + 3 = 0$ is perpendicular to the graph of $ax - 2y - 3 = 0$, what is the value of a?

(A) 1

(B) 2

(C) 3

(D) 4

20. Faith has 5 gallons of a 32% alcohol solution. How many quarts of an 80% alcohol solution should be added to Faith's original alcohol solution to create a 55% alcohol solution ?

(A) 18.4

(B) 24.6

(C) 32.2

(D) 44.3

21. Noah wants to pack the apples from his orchard for transportation. There are 40 boxes, each of the big box can hold 48 apples and each of the small box can hold 32 apples. And the truck can hold up to 2,200 apples. Which of the following can express such a system?

(A) $\begin{cases} x + y \le 40 \\ \dfrac{x}{80} + \dfrac{y}{32} \le 2,200 \end{cases}$

(B) $\begin{cases} x + y \le 40 \\ 48x + 32y \le 2,200 \end{cases}$

(C) $\begin{cases} x + y \le 40 \\ \dfrac{x}{48} + \dfrac{y}{32} \le 2,200 \end{cases}$

(D) $\begin{cases} x + y \le 2,200 \\ 48x + 32y \le \dfrac{2,200}{x + y} \end{cases}$

22. If $\theta = \sin^{-1}\dfrac{5}{13}$, What is the value of $\cos\theta$?

(A) $\dfrac{7}{13}$

(B) $\dfrac{8}{13}$

(C) $\dfrac{10}{13}$

(D) $\dfrac{12}{13}$

23. If $f(\sqrt{2x}) = 4x^2 + 2x + 3$, what is the value of $f(3)$?

(A) 26

(B) 43

(C) 65

(D) 93

24. AGE IN A LOCAL CHESS CLUB

6	6	8	9
10	10	12	12
12	14	14	15
16	16	17	18
18	18	18	18
42	48	56	58

What is the value of IQR?

(A) 6

(B) 7

(C) 8

(D) 9

25. If it is known to have two defective bulbs in a box of 10, two bulbs are drawn at random one at a time, what is the probability that none of the bulbs is defective?

(A) $\frac{28}{45}$

(B) $\frac{32}{50}$

(C) $\frac{46}{62}$

(D) $\frac{52}{68}$

26. The mean score of 12 students of a math class was 86. When two new students were enrolled, the mean score increased to 88. What was the average of these two new students?

(A) 87

(B) 90

(C) 94

(D) 100

27. Given function $f(x)$, which of the following functions corresponds to a horizontal compression, a reflection about the y-axis, and a downward shift of $f(x)$?

(A) $f(-\frac{1}{2}x)-3$

(B) $-f(2x)-3$

(C) $f(-2x)-3$

(D) $f(-\frac{1}{2}x)-3$

28. The graph of a parabola intercept the x-axis at $x=-6$ and $x=10$. If $f(4) = f(k)$, which of the following could be the value of k ?

(A) 0

(B) 2

(C) 4

(D) 6

29. After an international meeting all paticipants shook hands with each other. In total, there were 36 handshakes exchanged. How many participants were there at the meeting?

(A) 6
(B) 7
(C) 8
(D) 9

30. The speed of light is about 3.0×10^5 kilometers per second. If the circumference of the earth is about 4.0×10^4 $(2\times3.14\times6{,}371)$ km, how many times does the light circle the earth in one second? (Roughly)
(A) 25 circles
(B) 15 circles
(C) 7.5 circles
(D) 3.75 circles

31. If $2i$ is a zero of the function $p(x)$, which of the following must be a factor of $p(x)$?
(A) x^2+4
(B) x^2-4
(C) x^2+x+4
(D) x^2+x-4

32. In order to clear the invertory, a clothing store offers a 50% discount of the original price for all of their clothes. Allison is an employee of this store and enjoys another 20% discount for all goods. In addition, she can use a \$10 discount voucher. If a leather coat was originally priced at \$178.99, how much does Allison need to pay for this coat? (Rounded to cent)
(A) 58.60
(B) 61.60
(C) 63.25
(D) 66.55

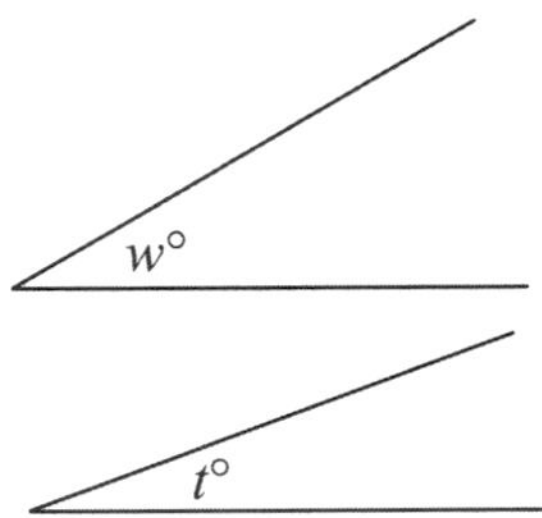

33. In the figures above $w°$ and $t°$ are two acute angles of a right triangle and $\sin w° = \cos t°$. If $w=3a-12$, $t=6a-18$, what is the value of a? (rounded to tenth)

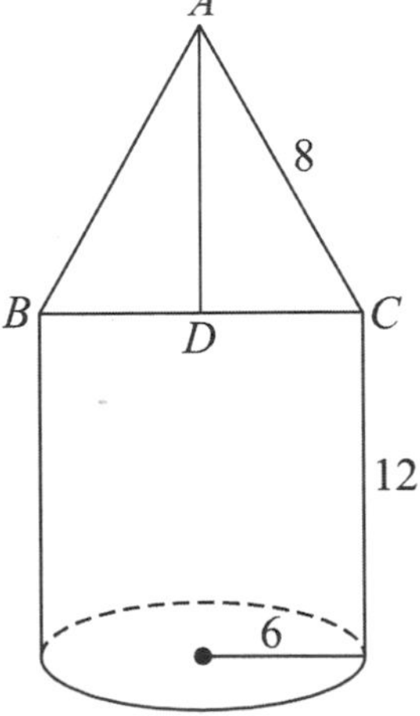

34. In the figure above, $\triangle ABC$ is a right circular cone on top of a cylinder. If the slant height of the cone is $AC=8$, and

the height of the cylinder is 12, and the radius of the base is 6, what is the total area exclude the base area ? (Rounded hundredth)

(A) 382.25

(B) 389.15

(C) 418.35

(D) 485.65

35. The magnitude of an earthquake is commonly expressed by the Richter scale. When Richter scale increases by 1 level, the earth quake intensity measured in wave amplitude increases by approximately 10 times, and the energy released intensity increases approximately 32 times. Suppose that there are two earthquakes with a difference of 1,000 times in intensity, how many times the difference in energy released?

36. All handbags sold at Store A are over \$100. Which of the following statements must be true?

(A) Handbags under \$100 are not sold at store A.

(B) All bags over \$100 are sold here.

(C) Among all merchandise sold by Store A, the ones over \$100 must be hand bags.

(D) All merchandise sold at Store A are over \$100.

37. If $3x-15\sqrt{x}-11=7$ is true, what is the value of the expression $12\sqrt{x}+5$?

(A) 55

(B) 60

(C) 77

(D) 90

38. If the equation of a parabola is given by $y+1=(x-1)^2$, what is the coordinates of its focus?

(A) $(1,\frac{3}{4})$

(B) $(1,-\frac{3}{4})$

(C) $(-1,\frac{3}{4})$

(D) $(-1,-\frac{3}{4})$

39. The formula of body mass index is (BMI)

$$\text{BMI}=\frac{W}{(H)^2}$$

W is the weight unit represented kilo gram, and H is height units represented by meters.

If measured by pound for weight and inches for height, the formula would be $\text{BMI}=k\cdot\frac{W}{(H)^2}$ what is the value fo k? (rounded to tenths)

40. If Katy's weight is 132 lbs and height is 5' 8". The normal BMI, is between 18.5 and 24, is Katy's BMI in the normal range ?

(A) Less than standard

(B) At standard

(C) Overweight

(D) None of the above

Box-and-Whisker Plot

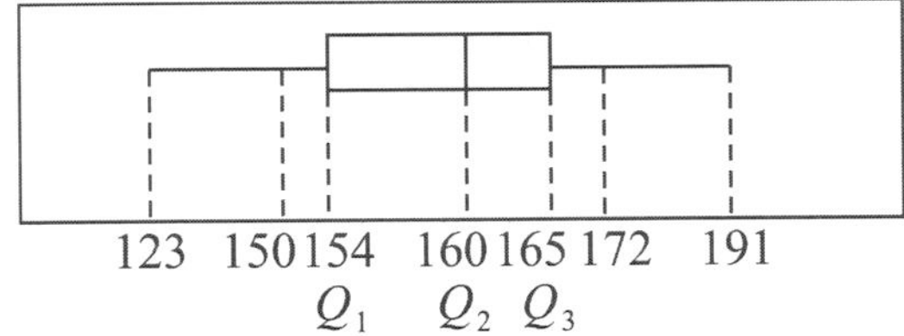

weight (pounds)

41. The figure above shows the Box-and-Whisker plot of the body weights of Chloe's classmates. If the percentile of 22% is at weight of 150 pounds, and the 82% is at weight of 172 pounds, which of the following ranges include the most number of students ?

(A) 123~154 pounds

(B) 150~160 pounds

(C) 160~172 pounds

(D) 165~191 pounds

42. There are three different alcohol solutions. The first solution is a 3-ounce fluid with 30% alcohol, the second soltion is 5 ounces with 50% alcohol, the third solution is a 3 ounces of 40% alcohol. If Caroline first mixes the 30% and 50% solutions together and then pours all the mixture into a 40% solution then what is the alcohol percentage in the final solution?

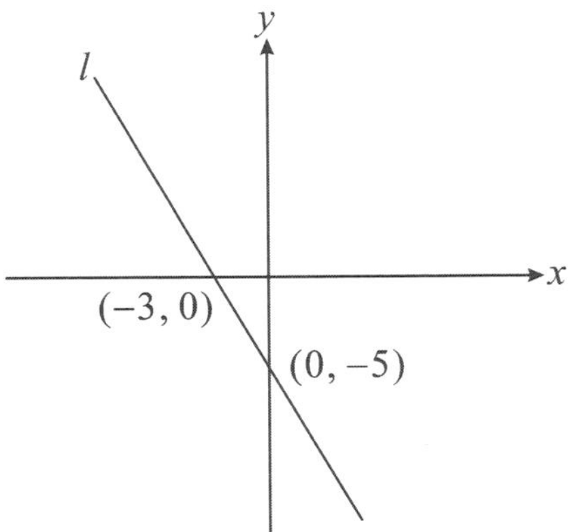

43. Line ℓ is shown above, if line m (not shown) is the result of translating line ℓ left by 4 units and up by 7 units, which of the following is the fuction of line m?

(A) $y=\frac{5}{3}x+2$

(B) $y=-\frac{5}{3}x-2$

(C) $y=-\frac{5}{3}x+\frac{26}{3}$

(D) $y=-\frac{5}{3}x-\frac{14}{3}$

44. If $f(x)=x^2-x-1$, $g(x+1)=f(x)$, and $r(x-1)=g(x+1)$. What is the remainder of $3g(x)+2r(x)$ divided by $x-1$?

(A) 0

(B) -1

(C) −2

(D) 2

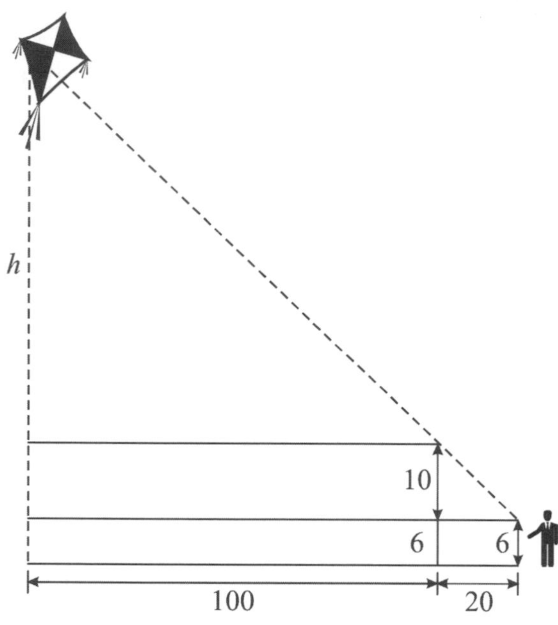

Note: Figure not drawn to scale

45. In the figure above, Shawn wanted to measure the height of the kite. If Shawn is 6 ft tall, and the height of the pole is 16 ft, what is the height of the kite?

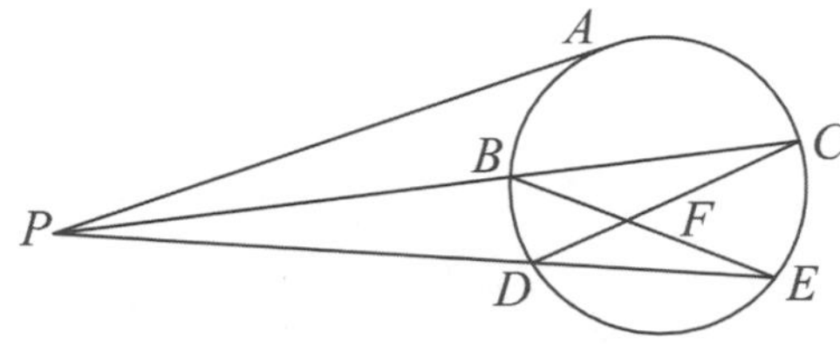

Note: Figure not drawn to scale

46. In the figure above, $\overline{PA}$ is tangent to the circle at point A, $\overline{PC}$ and $\overline{PE}$ intersected the circle at point B, C, D, and E respectively $\overline{BE}$ and $\overline{DC}$ intersected at point F, and $\overline{BF}:\overline{DF}$ is 5 : 3. If $PC=9$, $PE=7$ what is the length of $\overline{PA}$?

(A) $\sqrt{21}$

(B) $\sqrt{43}$

(C) 8

(D) $\sqrt{92}$

47. Ken bought a house for \$250,000 with a down payment of \$50,000 and a bank loan of \$200,000 with 6.8% APR (annual percentage rate). The monthly installment payment is \$1,301 for 360 periods in 30 years.

If Ken only wants to pay monthly interest and pay off all the principle at the end of 30 years. How much more does Ken have to pay ?

M (fixed monthly pay)

$$=\frac{p\cdot r\cdot(1+r)^n}{(1+r)^n-1}$$

$P=$ principle,

$r=$ interest rate per month

n = loan period

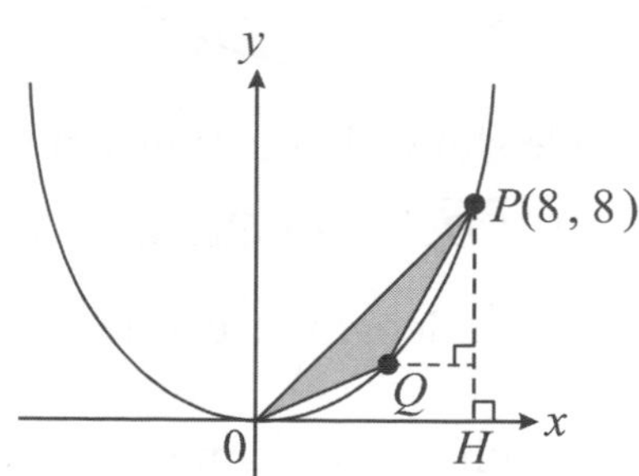

Note: Figure not drawn to scale

48. The figure above shows a parabola passes through origin O; point $P(8, 8)$ and point Q lie on the parabola. If the area of $\triangle OPQ$ is 8, what are the coordinates of point Q ?

(A) (2, 4)

(B) (4, 2)

(C) (2, 2)

(D) (4, 4)

49. If $|7-\frac{k}{3}|-2<5$, which of the following is the solution set ?

(A) $42>k$ or $k>0$

(B) $k>-42$ or $k<0$

(C) $42>k>0$

(D) $-42<k<0$

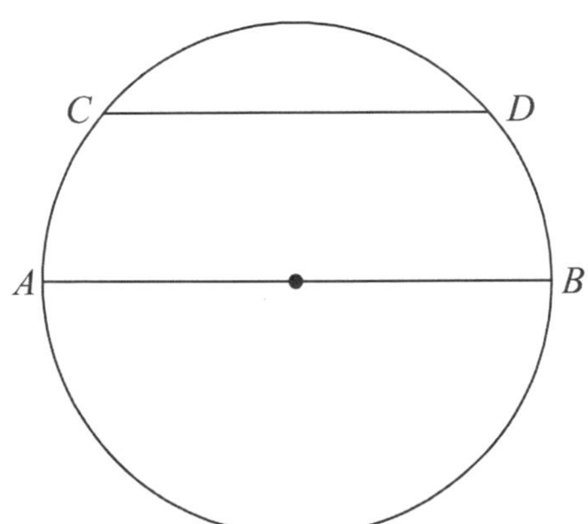

50. The circle shown has a radius of r inches. If chord $\overline{CD}$ is paralleled to diameter $\overline{AB}$, and the length of chord $\overline{CD}$ is $\frac{2}{3}$ of the length of the diameter, what is the distance in inches between $\overline{AB}$ and $\overline{CD}$ in terms of r?

(A) 0.75 r

(B) 0.86 r

(C) 0.89 r

(D) 0.92 r

MODEL TEST 2
ANSWER KEY

1. (B)
2. (A)
3. (D)
4. (A)
5. (D)
6. (A)
7. (D)
8. (A)
9. (A)
10. (C)
11. (C)
12. (C)
13. (A)
14. (B)
15. (A)
16. (A)
17. (A)
18. (D)
19. (A)
20. (A)
21. (B)
22. (D)
23. (D)
24. (B)
25. (A)
26. (D)
27. (C)
28. (A)
29. (D)
30. (C)
31. (A)
32. (B)
33. 13.3
34. (D)
35. 32,768
36. (A)
37. (C)
38. (B)
39. 714.28
40. (B)
41. (C)
42. 42%
43. (D)
44. (B)
45. 66
46. (A)
47. 138,617
48. (B)
49. (C)
50. (A)

MODEL TEST 2
ANSWERS and EXPLANATIONS

1. (B)

When $x-5>0$, $|x-5|=x-5$

When $x-5<0$,

$|x-5|=-(x-5)=-x+5$

Answer is (B)

2. (A)

$f^{-1}(f(x))=x$

Answer is (A)

3. (D)

$f(x)=\frac{2}{3}x+b$,

$f(-3)=\frac{2}{3}(-3)+b=-2+b$

Since $-2+b=9 \quad b=11$

The function is $f(x)=\frac{2}{3}x+11$

So, $f(9)=\frac{2}{3}(9)+11=17$

Answer is (D)

4. (A)

From $y(2+x)=5$, $2y+xy=5$

Since $xy=8$, so $2y+8=5$, $y=-\frac{3}{2}$.

$-\frac{3}{2}(2+x)=5$

$-3-\frac{3}{2}x=5$

So, $\frac{3}{2}x=-8$, $x=\frac{-16}{3}=-5.33$

Answer is (A)

5. (D)

$$Q=Q_1\times(\frac{1}{2})^{\frac{25,000}{5,000}}=Q_1(\frac{1}{2})^5=Q_1\cdot\frac{1}{32}$$

Answer is (D)

6. (A)

repeating for every 6 terms,

$75\div 6=12\cdots 3$

the third place of the repeating group should be (-1)

Answer is (A).

7. (D)

$$\begin{cases}3x+y=15\cdots①\\2x+3y=24\cdots②\end{cases}$$

Elimination of y ②−3①

$-7x=-21$, $x=3$

Substitute $x=3$ into ①

$3\times3+y=15$, $y=6$

$x+y=3+6=9$

Answer is (D)

8. (A)

$y=t(x^2-x-2)$

$=t[x^2-x+(\frac{1}{2})^2]-\frac{1}{4}t-2t$

$=t(x-\frac{1}{2})^2-\frac{9}{4}t$

Since $t<0$, the maximum value of y is

$-\frac{9}{4}t$

Answer is (A)

9. (A)

$y = k(x+2)(x-3)$

$= k(x^2 - x - 6)$

$= k[x^2 - x + (\frac{1}{2})^2 - (\frac{1}{2})^2 - 6]$

$= k[(x - \frac{1}{2})^2 - 6\frac{1}{4}]$

$= k(x - \frac{1}{2})^2 - \frac{25}{4}k$

Vertex is at $(\frac{1}{2}, -\frac{25}{4}k)$

$a = \frac{1}{2}$, $b = -\frac{25}{4}k$

Answer is (A)

10. (C)

$3(x-2)^2 + 5 > 0$

Answer is (C)

11. (C)

$\ell_1 \parallel \ell_2$, $\angle ACB = \angle FDE = 90° - y°$

$\sin \angle ACB = \sin x° = \sin \angle FDE = \cos y°$

$\cos y° = \sin x° = 0.5$

Answer is (C)

12. (C)

Only the contrapositive statement is equivant to the original statement.

Anwer is (C)

13. (A)

The circumference of the circle is $2\pi r$

$\Rightarrow 2 \times 3\pi = 6\pi$

So, $\frac{\overparen{AB}}{\frac{\pi}{6}} = \frac{6\pi}{2\pi}$ $\quad \therefore \overparen{AB} = \frac{6}{2} \times \frac{\pi}{6} = \frac{\pi}{2}$

Answer is (A)

14. (B)

$\frac{a}{3} = \frac{b}{7} = \frac{12}{20} = \frac{3}{5}$

So, $a = \frac{9}{5}$, $b = \frac{21}{5}$

$a + b = \frac{9}{5} + \frac{21}{5} = \frac{30}{5} = 6$

Answer is (B)

15. (A)

Downloading a video costs $7.99 plus 8% tax $\Rightarrow$ $7.99 \times (1.08)$ streaming a video costs $3.99 plus 8% tax $\Rightarrow$ $3.99 \times (1.08)$. Downloading d times costs. $7.99 \times (1.08) \times d$, streaming s time costs $3.99 \times (1.08) \times s$

$\therefore$ Total revenue

$= (1.08)(7.99d + 3.99s)$

Answer is (A)

16. (A)

It is a proportional problem. So, $\frac{8}{60} = \frac{x}{1500}$, $x = 200$

Answer is (A)

17. (A)

$2x^2-8x+1 \Rightarrow 2(x^2-4x+4)-8+1$

$\Rightarrow 2(x-2)^2-7$

So, the vertex is at $(2,-7)$.

Answer is (A)

18. (D)

$T_{n+2}=2T_n+T_{n+1}$

So, $T_3=2T_1+T_2=2\times3+6=12$

$T_4=2T_2+T_3=2\times6+12=24$

$T_5=2T_3+T_4=2\times12+24=48$

Answer is (D)

19. (A)

Two lines are perpendicular to each other

$m_1\times m_2=-1$

$2x+y+3=0 \Rightarrow y=-2x-3$

So, $m_1=-2$

Line $ax-2y-3=0 \Rightarrow 2y=ax-3$

$y=\frac{a}{2}x-\frac{3}{2} \quad m_2=\frac{a}{2}$

$m_1\times m_2=-2\times(\frac{a}{2})=-1$

Since perpendicular to each other

$-a=-1,\ a=1$

Answer is (A)

20. (A)

Step 1. draw a figure to help to understand the question

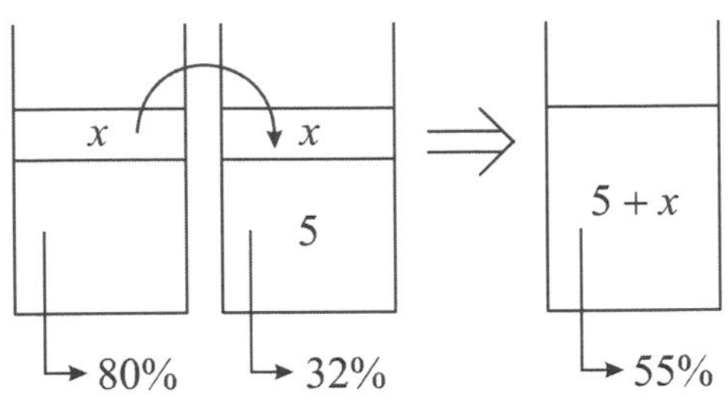

Step 2. Total Alcohol $=5\times0.32+x\cdot0.8$

Total solution $=5+x$

$0.55=\frac{5\times0.32+0.8x}{5+x}$

$5(0.55)+0.55x=5\times0.32+0.8x$

$0.8x-0.55x=5(0.55)-5\times0.32=1.15$

$\therefore\ x=\frac{1.15}{0.25}=4.6$ (gallons)

Step 3. 1 gallon = 4 quarts,

$4.6\times4=18.4$ quarts

Answer is (A)

21. (B)

Let the number of big boxes be x and small boxes y then $x+y=40$

The total number of apples are 2,200, that means $48x+32y\le 2,200$

So, the system equations

$\begin{cases} x+y\le 40 \\ 48x+32y\le 2,200 \end{cases}$

Answer is (B)

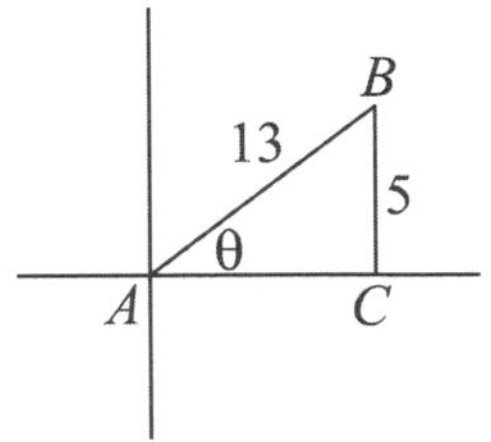

22. (D)

Let $\theta = \sin^{-1}\frac{5}{13}$, $\sin\theta = \frac{5}{13}$

draw a right triangle

$\overline{AC}^2 = 13^2 - 5^2 = 144$

$\therefore\ \overline{AC} = 12$

$\cos\theta = \frac{\overline{AC}}{\overline{AB}} = \frac{12}{13}$

Answer is (D)

23. (D)

Let $\sqrt{2x} = x'$, $2x = x'^2$, $x = \frac{x'^2}{2}$

$f(x') = 4(\frac{x'^2}{2})^2 + 2(\frac{x'^2}{2}) + 3$

$= x'^4 + x'^2 + 3$

$f(x) = x^4 + x^2 + 3$

so, $f(3) = (3)^4 + (3)^2 + 3$

$= 81 + 9 + 3 = 93$

Other solution:

$\sqrt{2x} = 3$, $2x = 9$, $x = \frac{9}{2}$

$4x^2 + 2x + 3 = 4 \cdot \frac{81}{4} + 2 \cdot \frac{9}{2} + 3$

$= 81 + 9 + 3 = 93$

Answer is (D)

24. (B)

For Q_1: $4\times 6 = 24$, $24\times 0.25 = 6$

The value of $Q_1 = \frac{10+12}{2} = 11$

Q_3 at $24\times 0.75 = 18$

The value of $Q_3 = \frac{18+18}{2} = 18$

$\text{IQR} = Q_3 - Q_1 = 18 - 11 = 7$

Answer is (B)

25. (A)

Pick two bulbs consecutively

$P(\text{first non-defective}) = \frac{8}{10} = \frac{4}{5}$

$P(\text{second non-defective}) = \frac{7}{9}$

$P(\text{both non-defective}) = \frac{4}{5} \times \frac{7}{9} = \frac{28}{45}$

Answer is (A)

26. (D)

Mean $= 86$

Total $= 86\times 12 = 1,032$

Two new students joined

Mean $= 88$

Total $= 88\times 14 = 1,232$

$1232 - 1032 = 200$

$200 \div 2 = 100$

Answer is (D)

27. (C)

(1) horizontal compression

$f(x) \rightarrow f(2x)$

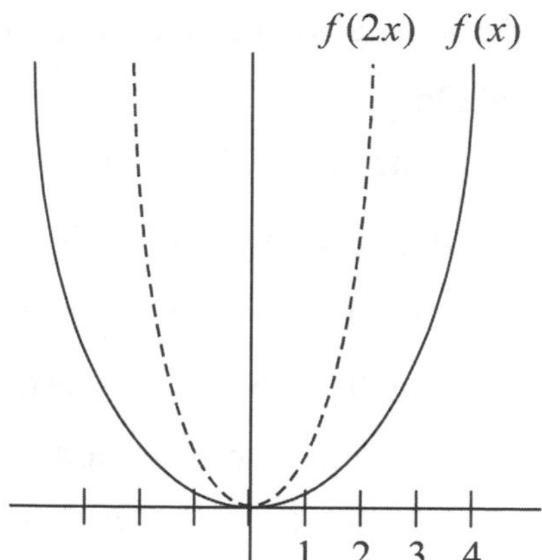

(2) reflection over the y-axis

$y = f(2) = 4$

$y = f(-2) = 4$, $f(2) = f(-2)$

$f(x) = f(-x)$

So, $f(2x) \to f(-2x)$

(3) down shift

$f(x) \to f(x) - k$

So, $f(-2x) - k$

Answer is $f(-2x) - 3 \Rightarrow$ (C)

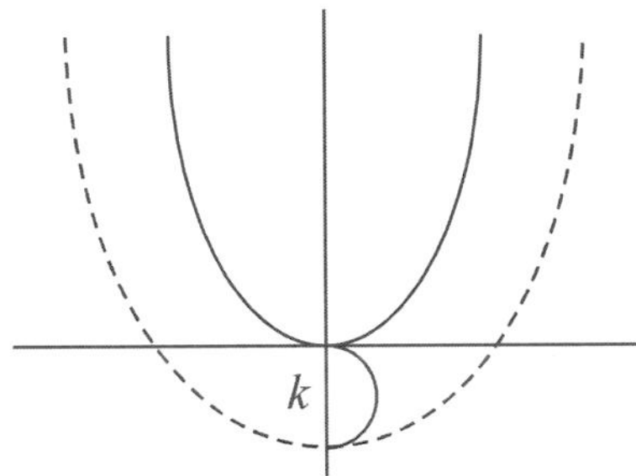

28.(A)

(1) The midpoint of two x-intercept is the x-coordinate of vertex $x = \dfrac{-6+10}{2} = 2$. It is also the symmetry of the parabola.

(2) $f(4)$ means, when $x = 4$, the y-coordinate of the point lies on the parabola. It is 2 units on the right of the symmetry. So, Its corresponding point 2 units on the left of the symmetry that is $2 - 2 = 0$, y-axis.

(3) $f(4) = f(0)$, $k = 0$

Answer is (A)

29. (D)

Method 1. Suppose there are n participants, then we have C_2^n handshakes

$$\Rightarrow \frac{n(n-1)(n-2)!}{(n-2)!2!} = \frac{n(n-1)}{2}$$

$$\frac{n^2 - n}{2} = 36, \quad n^2 - n - 72 = 0$$

$$(n-9)(n+8) = 0$$

$n = 9$ or $n = -8$ (false)

There are 9 participants.

Answer is (D)

30. (C)

The speed of light $= 3.0 \times 10^5$ km/sec.

The circumference of the earth

$= 4.0 \times 10^4$

Number of circles

$$= \frac{3.0 \times 10^5}{4.0 \times 10^4} = \frac{3}{4} \times 10 = 7.5$$

Answer is (C)

31. (A)

Set $p(x) = (x^2 + bx + c)Q(x)$

$a + bi$ is a zero, $a - bi$ is also a zero,

$0 + 2i$ is a zero, $0 - 2i$ is also a zero,

$c = (-2i)(+2i) = 4$

$b = (-2i) + (+2i) = 0$

$p(x) = (x^2 + 4)\,Q(x)$

Answer is (A)

32. (B)

$S(\text{coat}) = 178.99 \times 0.5 \times (1-0.2) - 10$

$= 61.60$

Answer is (B)

33. (13.3)

$\sin(w°) = \cos(90° - w) = \cos(t°)$

$90 - w = t = 6a - 18$

$90 - (3a - 12) = 6a - 18$

$90 + 12 - 3a = 6a - 18$

$9a = 120 \quad a = 13.3$

34. (D)

The area of the cone $\pi 8^2 \times \frac{60}{360} = 33.49$

The area of the cylinder $2 \cdot \pi(6) \cdot 12 = \pi$

$\times 12 \times 12 = 452.16$

Total area $= 452.16 + 33.49 = 485.65$

Answer is (D)

35. (32,768)

$E(r) = I \cdot 32^r, \quad A(r) = M \cdot 10^r$

r is the Richter scale, M and I are constants.

$\frac{A(r_1)}{A(r_2)} = \frac{M10^{r_1}}{M10^{r_2}} = 10^{(r_1 - r_2)}, \quad 1{,}000 = 10^3$

$r_1 - r_2 = 3$

$\frac{E(r_1)}{E(r_2)} = \frac{I \cdot 32^{r_1}}{I \cdot 32^{r_2}} = 32^{(r_1 - r_2)}$

$32^{n - r_2} = 32^3 = 32{,}768$

36. (A)

Only contrapositive statement is logically equivalent to the original statement. Therefore the answer is (A) Not over $100, not sold at store A.

37. (C)

$3x - 15\sqrt{x} - 11 = 7$

$\Rightarrow 3x - 15\sqrt{x} - 18 = 0$

$\Rightarrow (3\sqrt{x} + 3)(\sqrt{x} - 6) = 0$

$3\sqrt{x} + 3 = 0, \ \sqrt{x} = -1$ (undefined)

$\sqrt{x} - 6 = 0, \ \sqrt{x} = 6$

$12 \times 6 + 5 = 77$

Answer is (C)

38. (B)

$y + 1 = (x-1)^2$

$y = (x-1)^2 - 1$

$\frac{1}{4p} = 1, \ p = \frac{1}{4}$, vertex is at $(1, -1)$

Since $a = 1$, positive, concave upwards

focus is at $(1, -1 + \frac{1}{4}) \Rightarrow (1, -\frac{3}{4})$

Answer is (B)

39. (714.28)

1 kg = 2.2 pounds,

1 meter = 39.37 in

$\frac{w}{(H)^2} = \frac{1}{1^2} = 1$

1 BMI $= \frac{2.2}{(39.37)^2} = 0.0014 \text{ lb/in}^2$

$$\text{BMI} = \frac{x \text{ lb/in}^2}{0.0014 \text{ lb/in}^2} = x \cdot \frac{1}{0.0014}$$

$$\frac{1}{0.0014} = k$$

Therefore $k = \dfrac{1}{0.0014} = 714.28$

40. (B)

$$\text{BMI} = \frac{132}{(68)^2} \times 714.28 = 20.39$$

She is at standard.

Answer is (B)

41. (A) 123~154 has 25%

(B) 150~160 $50 - 22 = 28(\%)$

(C) 160~172 $82 - 50 = 32(\%)$

(D) 165~191 has 25%

Answer is (C)

42. **Step 1.** Understand the question

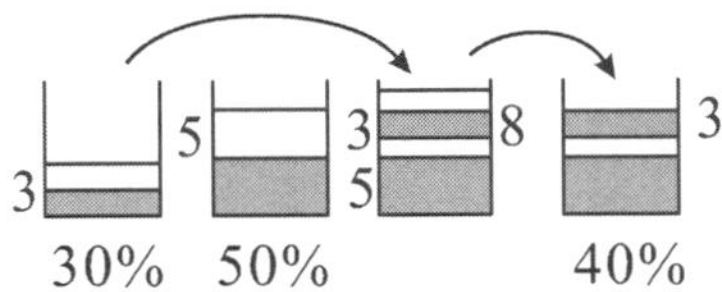

Step 2. Analyzing the problem

Alcohol in part 1: $3 \times 0.3 + 5 \times 0.5 = 3.4$

Alcohol in part 2: $3 \times 0.4 = 1.2$

Step 3.

$$\frac{3.4 + 1.2}{3 + 5 + 3} = \frac{4.6}{11} = 0.418 \doteq 0.42 \Rightarrow 42\%$$

43. (D)

$f(x) = mx + b$

According to the graph $b = -5$

$y = mx - 5$ passes through $(-3, 0)$

$\therefore\ 0 = -3m - 5,\ m = -\dfrac{5}{3}$

$l: y = -\dfrac{5}{3}x - 5$

Shift left 4 units $y = -\dfrac{5}{3}(x + 4) - 5$

Translate up 7 units

$$\begin{aligned} y &= -\frac{5}{3}(x+4) - 5 + 7 \\ &= -\frac{5}{3}(x+4) + 2 \\ &= -\frac{5}{3}x - \frac{20}{3} + 2 \\ &= -\frac{5}{3}x - \frac{14}{3} \end{aligned}$$

Answer is (D)

44. (B)

Sol 1. Since $g(x+1) = f(x) = x^2 - x - 1$,

So, $g(x') = (x'-1)^2 - (x'-1) - 1$

$= x'^2 - 2x' + 1 - x' + 1 - 1$

$= x'^2 - 3x' + 1$

$g(x) = x^2 - 3x + 1$

$r(x-1) = g(x+1) = x^2 - x - 1$

$r(x') = (x'+1)^2 - (x'+1) - 1$

$= x'^2 + 2x' + 1 - x' - 1 - 1$

$= x'^2 + x' - 1$

$r(x) = x^2 + x - 1$

So, $3g(x) + 2r(x)$

$\Rightarrow 3(x^2 - 3x + 1) + 2(x^2 + x - 1)$

$= 3x^2 - 9x + 3 + 2x^2 + 2x - 2$

$= 5x^2 - 7x + 1$

divided by $x - 1$

$\Rightarrow 5(1)^2 - 7(1) + 1 = -1$

Sol 2. $g(1)=g(0+1)=f(0)=-1$

$r(1)=r(2-1)=g(2+1)$

$=f(2)=2^2-2-1=1$

$3g(1)+2r(1)=3(-1)+2(1)=-1$

Answer is (B)

45. (66)

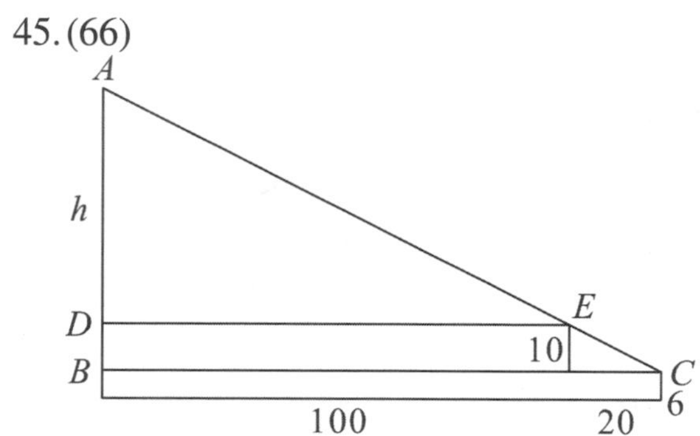

$\triangle ABC \sim \triangle ADE$

$\dfrac{AB}{BC}=\dfrac{AD}{DE}$

$\dfrac{h+10}{100+20}=\dfrac{h}{100}$

$100h+20h=100h+1000$

$20h=1000$

$h=50$

Hight of the kite $=50+6+10=66$

46. (A)

Analysis:

(1) $PA^2=PC\times PB$ or $PA^2=PE\times PD$

PC and PE are given, what is PB, or PD ?

(2) Through $\triangle DFE \sim \triangle BFC$, we can get the length of $\overline{BC}$ and $\overline{DE}$. Indirectly, and also we got the length of $\overline{PB}$ and $\overline{PD}$.

(3) Then $\overline{PA}^2=PC\times PB$

Operation:

(1) $\triangle BFC \sim \triangle DFE$ (a. a. a.)

(2) $\dfrac{BF}{DF}=\dfrac{BC}{DE}=\dfrac{5}{3}$, Let $BC=5k$, $DE=3k$

(3) $PC\times PB=PE\times PD$

$9\times(9-5k)=7\times(7-3k)$

$81-45k=49-21k$

$24k=32$, $k=\dfrac{4}{3}$

(4) $BC=5\times\dfrac{4}{3}=\dfrac{20}{3}$, $DE=3\times\dfrac{4}{3}=4$

(5) $PA^2=PC\times PB=9\times(9-\dfrac{20}{3})=21$

$PA=\sqrt{21}$

Answer is (A)

47. 138,617

$I=(200,000\times\dfrac{0.068}{12})\times360=408,000$

The formula for the mortilization mortgage is M (fixed monthly pay) $=\dfrac{pr(1+r)^n}{(1+r)^n-1}$

where P = principle, r = interest rate per month n = loan period

$$M=\frac{200,000\times\frac{0.068}{12}(1+\frac{0.068}{12})^{360}}{(1+\frac{0.068}{12})^{360}-1}$$

$=1303.85$

$1303.84\times360=469,386.14$

$408,000+200,000-469,386.14$

$=138,613.86$

48. (B)

(1) The area of $\triangle OPQ = \triangle OPH - \triangle OQH - \triangle QPH$

(2) Let the function of the parabola be $y = ax^2$

(3) Since point Q lies on the parabola , so, Let the coordinate of point Q be (x, ax^2); also point P also lies on the parabola $8 = a \times 8^2$, $a = \frac{1}{8}$

(4) The area of $\triangle OQH = \frac{1}{2} \times 8 \times (\frac{1}{8}x^2)$

The area of $\triangle QPH = \frac{1}{2} \times (8 - x) \times 8$

(5) $\triangle OPQ = \triangle OPH - \triangle OQH - \triangle QPH$

$$8 = \frac{1}{2} \times 8 \times 8 - \frac{1}{2} \times 8 \times \frac{x^2}{8} - \frac{1}{2} \times 8 \times (8 - x)$$

$$= \frac{1}{2} \times 8 \times 8 - \frac{1}{2} \times 8 \times \frac{x^2}{8} - \frac{1}{2} \times 8 \times 8 + \frac{1}{2} \times (8x)$$

$$= -\frac{1}{2}x^2 + \frac{8}{2}x = -\frac{x^2}{2} + 4x$$

$$x^2 - 8x + 16 = 0$$

$$(x - 4)^2 = 0$$

$$x = 4$$

(6) The coordinates of point Q are (4, 2)

Answer is (B)

49. (C)

$$|7 - \frac{k}{3}| - 2 < 5$$

$$\Rightarrow |7 - \frac{k}{3}| < 7$$

$$\Rightarrow -7 < 7 - \frac{k}{3} < 7$$

$$\Rightarrow -14 < -\frac{k}{3} < 0$$

$$\Rightarrow -42 < -k < 0$$

$$\Rightarrow 42 > k > 0$$

Answer is (C).

50. (A)

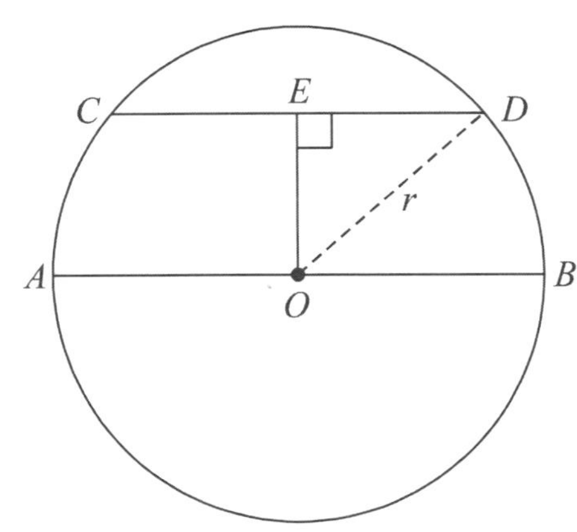

$\frac{\overline{CD}}{\overline{AB}} = \frac{2}{3}$ $\quad CD = \frac{2}{3} \cdot 2r = \frac{4}{3}r$

$\therefore\ ED = \frac{1}{2} \cdot \frac{4}{3}r = \frac{2}{3}r$

$OE^2 = r^2 - (\frac{2}{3}r)^2 = r^2(1 - \frac{4}{9}) = \frac{5}{9}r^2$

$\therefore\ OE = \frac{\sqrt{5}}{3}r = 0.75r$

Answer is (A).

CHAPTER 15
MODEL TEST 3
LEVEL 1

DIRECTION:

Fill in the correct or the best answer in the proper position on the answer sheet. You may use any available space for scratch work.

NOTES:

1. Calculator is permitted.
2. All figures are not drawn to scale.
3. All numbers are real, unless stated otherwise.
4. Some reference information:

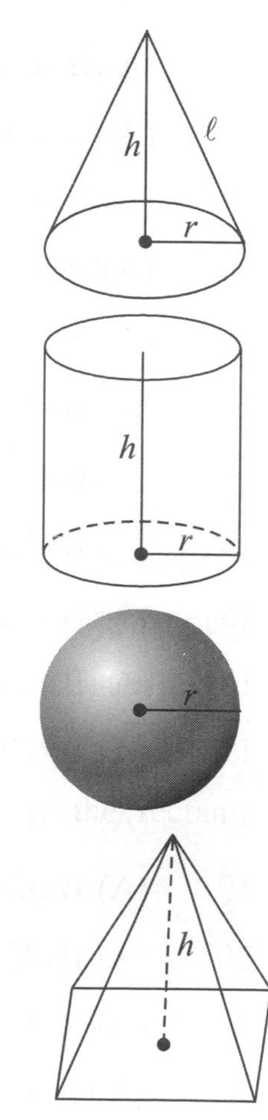

(1) Right Circular Cone

$V = \frac{1}{3}\pi r^2 \cdot h$, $A(\text{lateral surface}) = \frac{1}{2}c\ell = \pi r \cdot \ell$

(2) Cylinder

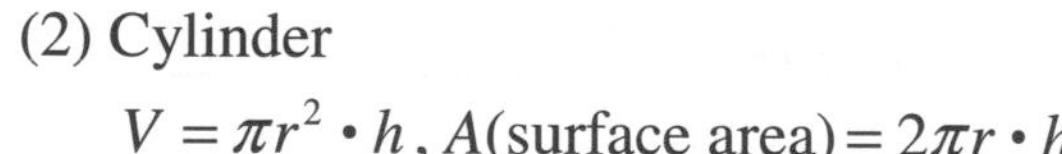

(3) Sphere

$V = \frac{4}{3}\pi r^3 \qquad A = 4\pi r^2$

(4) Pyramid

$V = \frac{1}{3}Bh$ (B: area of base)

(5) Special Right Triangle

$30° - 60° - 90°$

$45° - 45° - 90°$

a 60° 2a 30° $\sqrt{3}a$

a 45° $\sqrt{2}a$ 45° a

MODEL TEST 3
ANSWER SHEET

Question Number		Your Answers	Answer Key	Question Number		Your Answers	Answer Key
1	(A)(B)(C)(D)			26	(A)(B)(C)(D)		
2	(A)(B)(C)(D)			27	(A)(B)(C)(D)		
3	(A)(B)(C)(D)			28	(A)(B)(C)(D)		
4	(A)(B)(C)(D)			29	(A)(B)(C)(D)		
5	(A)(B)(C)(D)			30	(A)(B)(C)(D)		
6	(A)(B)(C)(D)			31	(A)(B)(C)(D)		
7	(A)(B)(C)(D)			32	(A)(B)(C)(D)		
8	(A)(B)(C)(D)			33	(A)(B)(C)(D)		
9	(A)(B)(C)(D)			34	(A)(B)(C)(D)		
10	(A)(B)(C)(D)			35	(A)(B)(C)(D)		
11	(A)(B)(C)(D)			36	(A)(B)(C)(D)		
12	(A)(B)(C)(D)			37	(A)(B)(C)(D)		
13	(A)(B)(C)(D)			38	(A)(B)(C)(D)		
14	(A)(B)(C)(D)			39	(A)(B)(C)(D)		
15	(A)(B)(C)(D)			40	(A)(B)(C)(D)		
16	(A)(B)(C)(D)			41	(A)(B)(C)(D)		
17	(A)(B)(C)(D)			42	(A)(B)(C)(D)		
18	(A)(B)(C)(D)			43	(A)(B)(C)(D)		
19	(A)(B)(C)(D)			44	(A)(B)(C)(D)		
20	(A)(B)(C)(D)			45	(A)(B)(C)(D)		
21	(A)(B)(C)(D)			46	(A)(B)(C)(D)		
22	(A)(B)(C)(D)			47	(A)(B)(C)(D)		
23	(A)(B)(C)(D)			48	(A)(B)(C)(D)		
24	(A)(B)(C)(D)			49	(A)(B)(C)(D)		
25	(A)(B)(C)(D)			50	(A)(B)(C)(D)		

MODEL CHAPTER 15
MODEL TEST 3
LEVEL 1

1. If $x+5=2(3-2x)$, then $x=$

(A) 0.2

(B) 0.4

(C) 0.6

(D) 0.8

2. If $3=k^3$, $9k=$

(A) k^6

(B) k^7

(C) k^9

(D) $9\sqrt{3}$

3. Which of the following lines is perpendicular to the line $2x+3y=8$

(A) $3x-2y=24$

(B) $3x+2y=24$

(C) $4x-6y=24$

(D) $4x+6y=24$

4. If the complex number $z=3-4i$, what is the value of $|z+3|$?

(A) $\sqrt{17}$

(B) $\sqrt{26}$

(C) $2\sqrt{13}$

(D) $3\sqrt{17}$

5. The base of the pyramid khufu in Egypt is a square, each side is 230 meters long. The height of the pyramid is 146.59 meters. Due to the age of the tower, many stones on the top and sides of the pyramid have fallen off. The height of the pyramid is only 136.5 meters and the bottom side is only 227 meters. What percentage of the original volume is the volume now?

(A) 97%

(B) 90%

(C) 87%

(D) 80%

6. If the probability of event A is $P(A)$, and event B is $P(B)$, and if $P(A\cap B)=\frac{1}{6}$, $P(B')=\frac{2}{5}$, $P(A\cup B)=\frac{2}{3}$, what is $P(A)$?

(A) $\frac{2}{15}$

(B) $\frac{4}{15}$

(C) $\frac{7}{30}$

(D) $\frac{13}{30}$

7. 2, 1, 5, 3, 8, 5, 11, 7, 14, ?

What is the next number in the above sequence ?

(A) 3

(B) 6

(C) 9

(D) 11

8. If $\cos x° = \sin(3x-30)°$, what is the value of $\tan(2x)°$?

(A) $\frac{\sqrt{3}}{3}$

(B) $1+\sqrt{3}$

(C) $\sqrt{3}$

(D) $\sqrt{3}-1$

9. There are three boxes A, B, and C they, contain the same number of balls of various colors. The black balls account for $\frac{1}{4}$, $\frac{1}{3}$, and $\frac{5}{12}$ of the total number of balls in box A, B, and C, respectively, Chloe pours all the balls in box B and box C into box A and picks one ball from box A at random, what is the probability that the ball is not the black ball ?

(A) $\frac{1}{3}$

(B) $\frac{2}{3}$

(C) $\frac{3}{5}$

(D) $\frac{4}{5}$

10. If $x^2-13=a(x^2-1)+b(x+1)+c(x-1)^2$ for all real number of x, where a, b, and c are constants and not equal to zeroes, what is the value of $a+b+c$?

(A) −9

(B) −1

(C) −5

(D) 5

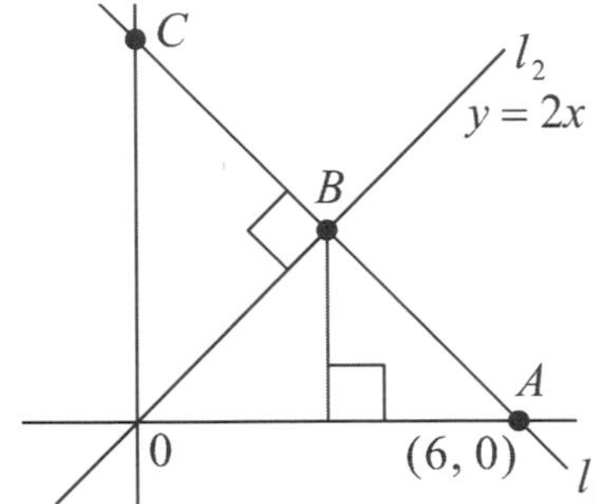

Note: Figure not drawn to scale

11. In the figure above, $l_2 : y=2x$ and $l_2 \perp l_1$, the x-intercept is at (6, 0). What is the area of $\triangle OBA$?

(A) $\frac{15}{4}$

(B) $\frac{36}{5}$

(C) 8

(D) $\frac{22}{7}$

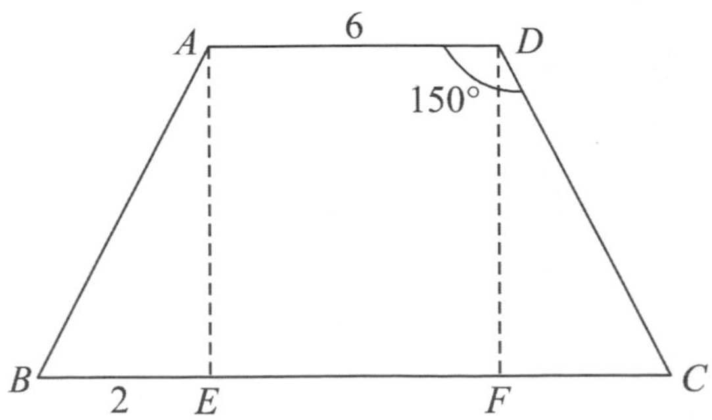

12. What is the area of the isosceles trapezoid above?

(A) 8.84

(B) 9.24

(C) 10.32

(D) 11.89

13. What is the domain of

$$f(x)=\frac{\sqrt{x-2}}{\sqrt{x^2-2x-8}}$$

(A) $-2<x<4$

(B) $x\geq 2$

(C) $x>4$

(D) $x<-2$ or $x>4$

14. $\begin{cases}\frac{2}{3}x+\frac{1}{6}y=15\\ kx-2y=18\end{cases}$

If the system of linear equations above has no solution and k is a constant, what is the value of k?

(A) -4

(B) -8

(C) 4

(D) 8

15.

	Male	Female	Total
Age ≥ 16	$\frac{4}{12}$		$\frac{40}{60}$
Age < 16		A	
Total	$\frac{4}{9}$		1

In Tina's school the fractions of male students and female students are shown as above table. If there are 360 students, what is the number of female students with the age under 16 years old ?

(A) 60

(B) 80

(C) 100

(D) 120

16. If the parabola $y=ax^2+bx+c$ passes through the points $(1,0)$, $(2,0)$, and $(0,-1)$, what is the value of $a+2b+3c$?

(A) $\frac{3}{2}$

(B) $-\frac{1}{2}$

(C) $\frac{2}{3}$

(D) $-\frac{2}{3}$

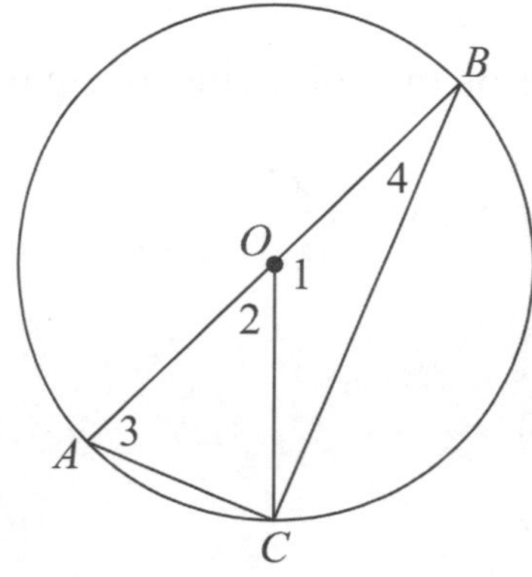

17. The figure above is a circle 0, which of the following is the measurement of $\angle 1$?

(A) $m\angle 3+m\angle 4$

(B) $2m\angle 3$

(C) $\frac{m\angle 3+m\angle 4}{2}$

(D) $2m\angle 2$

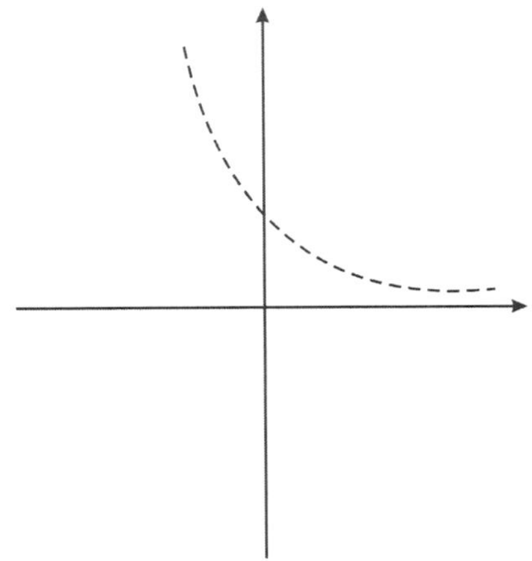

18. Which of the following best describes the type of relationship shown in the scatter-plot above?

(A) Exponential, positive

(B) Exponential, negative

(C) Linear, positive

(D) Linear, negative

19. Which of the following is equivalent to the expression $\{\frac{x^{\frac{1}{3}}}{x^{-3}}\}^2$?

(A) $x^{\frac{20}{3}}$

(B) $x^{\frac{3}{20}}$

(C) $x^{-\frac{1}{9}}$

(D) $x^{\frac{1}{9}}$

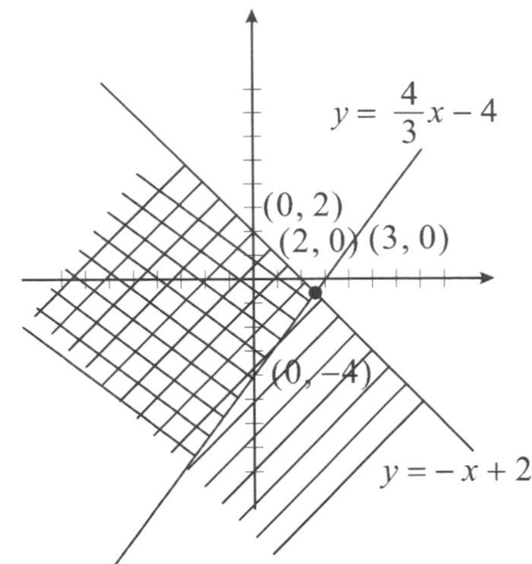

20. $\begin{cases} y>\frac{4}{3}x-4 \\ y<-x+2 \end{cases}$

The figure above shows the solution set for the given system of inequalities above. If (p, q) is a solution to the system inequalities, which of the following could be the greatest possible integer value of p?

(A) −1

(B) 1

(C) 2

(D) 3

21. If $\frac{1}{2}\leq 3-\frac{k}{4}\leq\frac{6}{5}$, what is the minimum possible value of k?

(A) 6.6

(B) 7.2

(C) 8.4

(D) 9.6

22. $\frac{x}{x+1}-\frac{1}{x}=1$

What is one possible solution of the equation above?

(A) $-\frac{1}{2}$

(B) $\frac{1}{2}$

(C) $-\frac{1}{3}$

(D) $\frac{1}{3}$

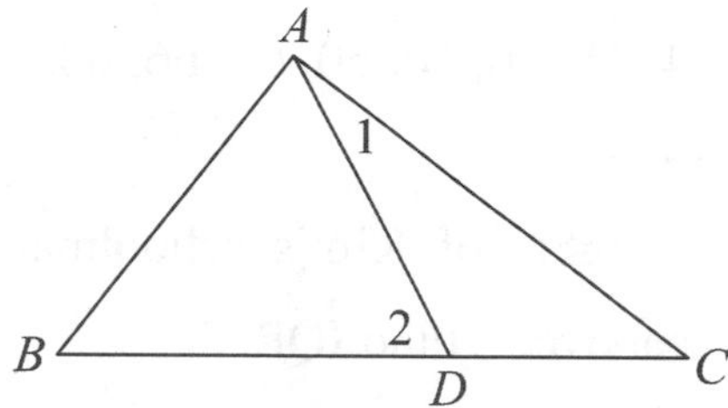

Note: Figure not drawn to scale

23. In the figure above, $\triangle ABC$ is an isosceles triangle with $\overline{AB}=\overline{AC}$. D is a point lies on $\overline{BC}$, if $AB=8\sqrt{3}$, $\angle 1=\angle ACD$ and $\angle 2=60°$, what is length of $\overline{BC}$?

24. If the function g is defined by $g(x)=-12+\frac{x^2}{3}$ and if $g(3k)=5k$, what is one possible value of k?

(A) 1

(B) 2

(C) 3

(D) 4

25.

2, 3, 5, 8, 13, 21, □

In the list above, what is the value of the 7th term?

(A) 29

(B) 34

(C) 41

(D) 49

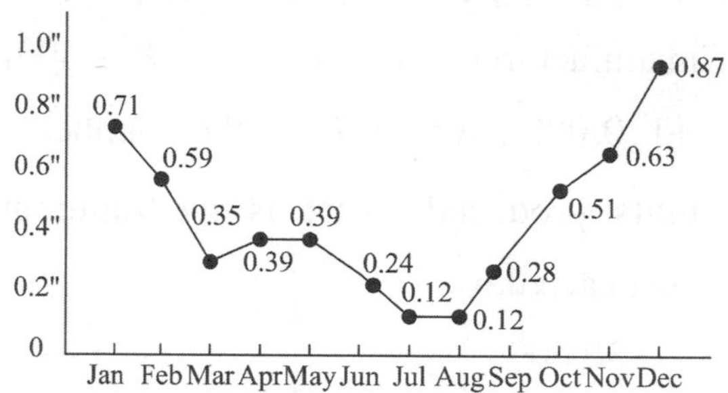

Note: Figure not drawn to sacle

26. The graph of the average rainfall in SS-City is shown above. What is the value of the median ?

(A) 0.28

(B) 0.35

(C) 0.39

(D) 0.51

27. What is value of the IQR ?

(A) 0.08

(B) 0.22

(C) 0.28

(D) 0.35

28. Are there any outlier(s)?

(A) Lower outlier is at 0.12 and no upper outlier

(B) Upper outlier is at 0.87 and no lower outlier

(C) Lover outlier is at 0.12 and upper outlier at 0.87

(D) None

29. If the daily production costs of a manufacturer are $C(P) = P^2 - 600P + 120,000$, where P is the number of units produced, what is the minimum cost each day ?

(A) 30,000

(B) 60,000

(C) 81,000

(D) 108,000

30. A biologist is doing research on a certain kind of bacteria. She put about 5,000 microbes in a experimental dish. For the first 3 days, the number of microbes in the dish doubled every 6 hours. If n represents the number of microbes after h hours, which of the following equations can best model this situation during 3-day period?

(A) $n = \frac{h}{6} \times 5,000$

(B) $n = 5,000(2)^{\frac{h}{6}}$

(C) $n = 6h \times 5,000$

(D) $n = 5,000(2)^{6h}$

31. If $(3^9)^{(3^6)} = 3^{(3^x)}$, what is the value of x?

(A) 6

(B) 8

(C) 10

(D) 15

32.

36, 36, 38, 42, 42, 46, 48, 48, 48, 52, 53, 54, 54, 55, 56, 60, 60, 64, 66, 68, 70, 72, 76, 80, 98

The weights of Kim's schoolmates are listed above. Find IQR.

(A) 12

(B) 14

(C) 16

(D) 18

33. If the graph of the function $f(x)$ passes through the point $(-2, 4)$, which of the following points does the graph of $-f(x+2)+4$ pass through?

(A) $(-2, 4)$

(B) $(-4, 4)$

(C) $(-4, -4)$

(D) $(-4, 0)$

34. If it is sunny tomorrow, I will go fishing. Which of the following statements is equivalent to the above statement.

(A) If it is not sunny tomorrow, I won't go fishing.

(B) If I go fishing tomorrow, it is a sunny day.

(C) If I do not go fishing tomorrow, it is not a sunny day.

(D) I will go fishing tomorrow anyway.

35. What is the range of the function $f(x) = |x+1|$ for the domain of $-2 \le x \le 2$?

(A) $2 \le y \le 3$

(B) $1 \le y \le 3$

(C) $0 \le y \le 3$

(D) $-2 \le y \le 3$

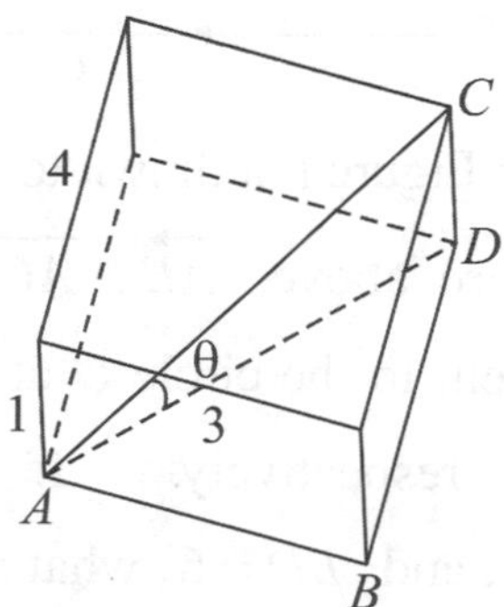

Note: Figure not drawn to scale

36. Figure above shows the rectangular solid with dimension 1, 3, and 4. If $\overline{AC}$ is the diagonal of the solid, which of the following could be the value of the θ which formed by the diagonal $\overline{AC}$ and the line $\overline{AD}$?

(A) 11.48°

(B) 23.32°

(C) 32.12°

(D) 38.08°

37. If $g(x) = -4x^2 + 5x - 8$, what is the maximum value of $g(x)$?

(A) −8.32

(B) −6.44

(C) 3.12

(D) 8.32

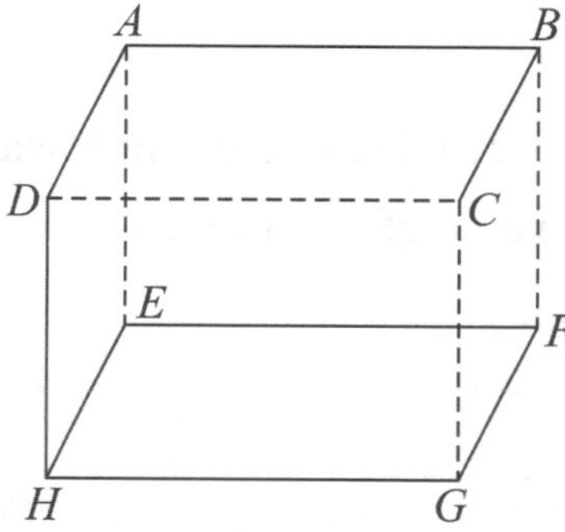

Note: Figure not drawn to scale

38. In the figure above shows a rectangular solid. If the area of □$ABFE$ is 48, □$CBFG$ is 36, and □$ABCD$ is 48. What is the volume of the rectangular solid ?

(A) 288

(B) 332

(C) 418

(D) 482

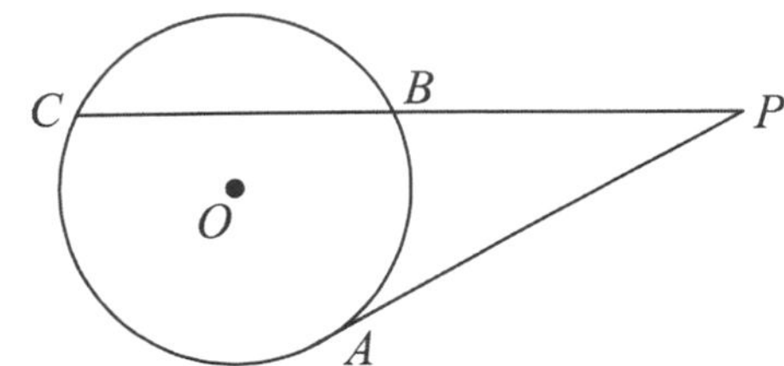

39. In the circle $\odot$ above, $\overline{PA}$ is tangent to $\odot$ at A, $\overline{PC}$ intersects $\odot$ at B and C. If $\overline{PA} = x-2$, $\overline{PC} = x+2$, and $\overline{BC} = 6$, what is the length of $\overline{PA}$?

(A) 2

(B) 3

(C) 4

(D) 5

40. If $x < -2$, which of the following could be the value of $|1-|x+1||$?

(A) $-x-2$

(B) x

(C) $x+2$

(D) $-x-1$

41. Suppose the graph of $f(x) = -3x^2$ is translated 4 units down and 3 units left. If the resulting graph represents the graph of $r(x)$, what is the value of $r(1.3)$?

(A) -50.31

(B) -52.23

(C) -54.12

(D) -59.47

42. Which of the following is the solution set for $x(x+1)(x-2) > 0$

(A) $2 < x$

(B) $-1 < x < 0$

(C) $\{-1 < x < 0\} \cap \{x < 2\}$

(D) $\{-1 < x < 0\} \cup \{2 < x\}$

43. The standard deviation of a data set is 6. If subtract 3 from each data in the data set, what is the new standard deviation of the new data set ?

(A) 3

(B) 4

(C) 5

(D) 6

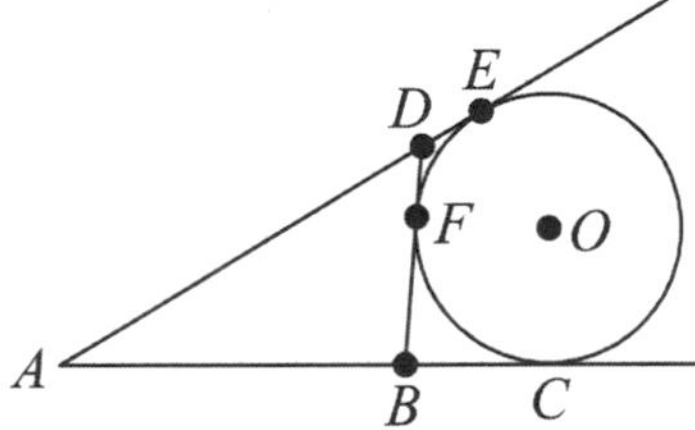

Note: Figure not drawn to scale

44. The figure above, $\overleftrightarrow{AE}$, $\overline{AC}$, and $\overline{BD}$ are tangent to the circle O at point C, E, and F, respectively. If $AD = 14$, $AB = 11$, and $BD = 6$, what is the length of $AE + AC$?

(A) 13.5

(B) 15.5

(C) 31

(D) 34.5

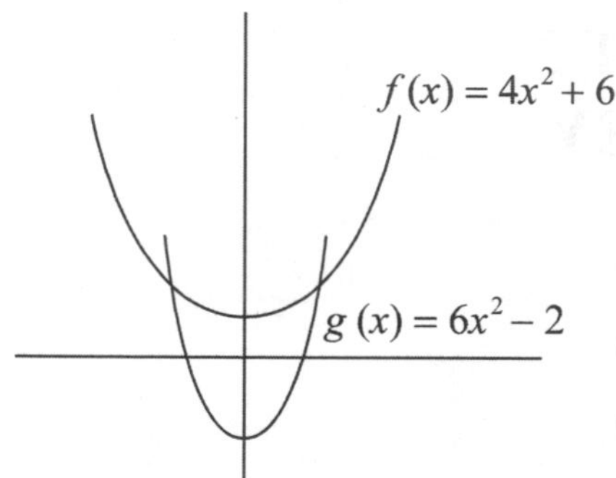

45. The parabola $f(x)=4x^2+6$ and the parabola $g(x)=6x^2-2$ are graphed in the xy-plane shown above, intersect at point (p, q) and $(-p, q)$, what is the value p?

(A) 0

(B) 1

(C) 2

(D) 3

46. What is the sum of the geometric series $12+6+3+\frac{3}{2}+\frac{3}{4}+\cdots$

(A) 24

(B) 36

(C) 48

(D) infinite

47. If $f(x)=x^2+x-3$, $g(x+2)=f(x)$, and $p(x+1)=g(x-1)$, what is the remainder of $2g(x)-3p(x)$ divided by $x+1$?

(A) 25

(B) 35

(C) –45

(D) –55

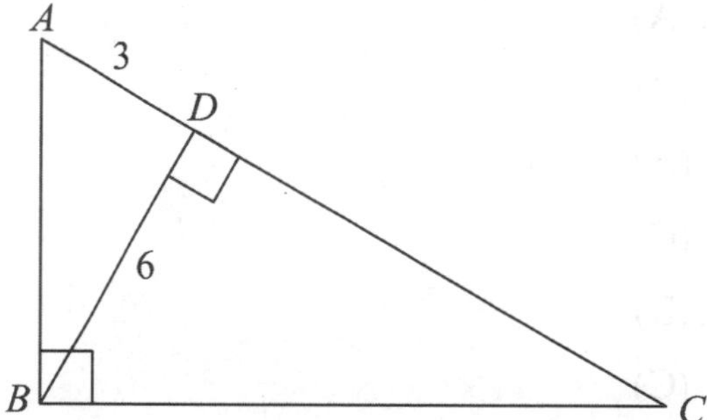

Note: Figure not drawn to scale

48. In the figure above, $\triangle ABC$ is a right triangle. $AD=3$, $BD=6$. What is the length of $\overline{AC}$?

(A) 10

(B) 12

(C) 15

(D) 18

49. If a complex number is $z=8-3i$, then $|z-3|=$

(A) $\sqrt{11}$

(B) $\sqrt{18}$

(C) $\sqrt{26}$

(D) $\sqrt{34}$

50. If the point $(0, 3)$, $(3, k)$, and $(6, 5)$ are collinear, what is the value of k?

(A) 2

(B) 3

(C) 4

(D) 5

MODEL TEST 3
ANSWER KEY

1. (A)
2. (B)
3. (A)
4. (C)
5. (B)
6. (C)
7. (C)
8. (C)
9. (B)
10. (A)
11. (B)
12. (B)
13. (C)
14. (B)
15. (B)
16. (B)
17. (B)
18. (B)
19. (A)
20. (C)
21. (B)
22. (A)
23. (24)
24. (C)
25. (B)
26. (C)
27. (D)
28. (D)
29. (A)
30. (B)
31. (B)
32. (D)
33. (D)
34. (C)
35. (C)
36. (A)
37. (B)
38. (A)
39. (C)
40. (A)
41. (D)
42. (D)
43. (D)
44. (C)
45. (C)
46. (A)
47. (C)
48. (C)
49. (D)
50. (C)

MODEL TEST 3
ANSWERS and EXPLANATIONS

1. (A)

$x+5=2(3-2x)=6-4x$

$5x=1,\ \ x=\frac{1}{5}$

Answer is (A)

2. (B)

$3^2=k^6=9$

$9k=k^6\cdot k=k^7$

Answer is (B)

3. (A)

$2x+3y=8\ \Rightarrow\ 3y=-2x+8$

$y=-\frac{2}{3}x+\frac{8}{3}$

$m=-\frac{2}{3}$

If they are perpendicular to each other,

$m_1\times m=-1$

$\therefore\ \ m_1=\frac{-1}{-\frac{2}{3}}=\frac{3}{2}$

only $3x-2y=24$

has the slope of $\frac{3}{2}$

Answer is (A)

4. (C)

$z=3-4i,\ \ z+3=3-4i+3=6-4i$

$|z+3|=\sqrt{6^2+4^2}=\sqrt{52}=2\sqrt{13}$

Answer is (C)

5. (B)

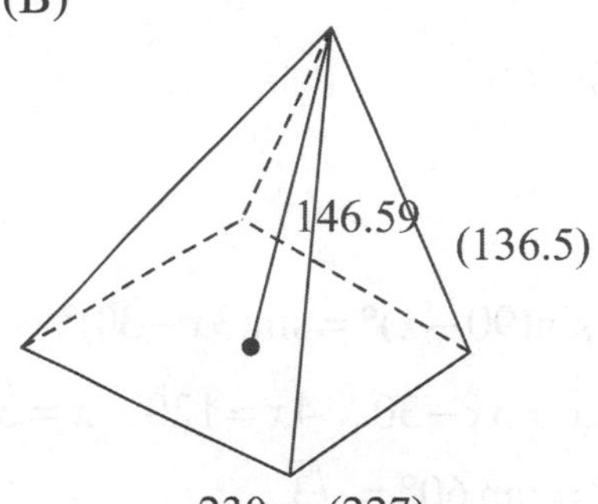

Original volume $=(230)^2\times\frac{1}{3}\times 146.59$

$=2{,}584{,}870\ \text{m}^3$

Volume (now) $=(227)^2\times(136.5)\times\frac{1}{3}$

$=2{,}344{,}569.5$

$\frac{v_{now}}{v_0}=\frac{2{,}344{,}569.5}{2{,}584{,}870}=0.9070$

about 90%

Answer is (B)

6. (C)

Since

$P(A\cup B)=P(A)+P(B)-P(A\cap B)$

$\frac{2}{3}=P(A)+[1-P(B')]-P(A\cap B)$

$=P(A)+\frac{3}{5}-\frac{1}{6}$

$P(A)=\frac{2}{3}-\frac{13}{30}=\frac{7}{30}$

Answer is (C)

7. (C)

$$\overset{2}{\frown}\ \overset{2}{\frown}\ \overset{2}{\frown}\ \overset{2}{\frown}$$
$$2,\ 1,\ 5,\ 3,\ 8,\ 5,\ 11,\ 7,\ 14,\ \boxed{9}$$
$$\underset{3}{\smile}\ \underset{3}{\smile}\ \underset{3}{\smile}\ \underset{3}{\smile}$$

Answer is (C)

8. (C)

$\cos x° = \sin(90 - x)° = \sin(3x - 30)°$

$\therefore\ 90 - x = 3x - 30,\ 4x = 120,\ x = 30°$

$\tan(2x)° = \tan 60° = \sqrt{3}$

Answer is (C)

9. (B)

	x	x	x
black balls	$\frac{1}{4}x$	$\frac{1}{3}x$	$\frac{5}{12}x$

$$P(\text{not black balls}) = \frac{\frac{3}{4}x + \frac{2}{3}x + \frac{7}{12}x}{x + x + x}$$

$$= \frac{\frac{24}{12}x}{3x} = \frac{2}{3}$$

Answer is (B)

10. (A)

$x^2 - 13 = a(x^2 + 1) + b(x + 1) + c(x - 1)^2$

When $x = 1$ $\quad -12 = a(1 + 1) + b(1 + 1)$,

$-6 = a + b$

When $x = -1$ $\quad -12 = c(-1 - 1)^2 = 4c$,

$c = -3$

So, $a + b + c = -6 - 3 = -9$

Answer is (A)

11. (B)

$l_1 \perp l_2,\ m_1 \times 2 = -1,\ m_1 = -\frac{1}{2}$

So, $l_1 : y = -\frac{1}{2}x + b$, substitute (6, 0) into the equation $0 = -\frac{1}{2}(6) + b,\ b = 3$,

$y = -\frac{1}{2}x + 3$, for l_1 intercept l_2 at point B

$\therefore\ -\frac{1}{2}x + 3 = 2x,\ \frac{5}{2}x = 3,\ x = \frac{6}{5}$,

$y = -\frac{1}{2}(\frac{6}{5}) + 3 = -\frac{3}{5} + 3 = \frac{12}{5}$

The coordinates of B are $(\frac{6}{5}, \frac{12}{5})$

The area of $\triangle OBA = \frac{1}{2} \times \frac{12}{5} \times 6 = \frac{36}{5}$

Answer is (B)

12. (B)

$\overline{AD} \parallel \overline{BC},\ \angle DCF = 180° - 150° = 30°$

$\frac{DF}{2} = \tan 30°$,

$DF = 2 \cdot \tan 30° = 2 \times 0.577 = 1.154$

Area of $ABCD$

$= \frac{\text{top side} + \text{bottom side}}{2} \times h$

$= \frac{6 + (2 + 6 + 2)}{2} \times 1.154$

$= 8 \times 1.154 = 9.23$

Answer is (B)

13. (C)

1. $\sqrt{x - 2} \geq 0,\ x \geq 2$
2. $\sqrt{x^2 - 2x - 8} \Rightarrow \sqrt{(x - 4)(x + 2)} > 0$

$(x-4)(x+2)>0$, $x>4$ or $x<-2$

3. $\{x\geq 2\}\cap\{x>4\} \Rightarrow \{x>4\}$

or $\{x\geq 2\}\cap\{x<-2\} \Rightarrow \phi$

4. The domain of the function is $\{x>4\}$

Answer is (C)

14. (B)

When the coefficients are proportional, then the system equations have no solution

$$\frac{\frac{2}{3}}{k}=\frac{\frac{1}{6}}{-2} \quad \therefore\ k=\frac{\frac{2}{3}}{\frac{1}{6}}\times(-2)=-8$$

Answer is (B)

15. (B)

$$1-\frac{4}{9}=\frac{5}{9}$$

$$\frac{40}{60}-\frac{4}{12}=\frac{1}{3}$$

$$A=\frac{5}{9}-\frac{5}{15}=\frac{25-15}{45}=\frac{10}{45}=\frac{2}{9}$$

$$\frac{2}{9}\times 360=80$$

Answer is (B)

16. (B)

$y=ax^2+bx+c$

passes (1, 0)

$0=a+b+c$ (1)

passes (2, 0)

$0=4a+2b+c$ (2)

passes (0, −1)

$-1=0+0+c$, $c=-1$

(2) − (1)×2 $\quad 2a-c=0 \quad 2a=-1$

$$a=-\frac{1}{2}$$

substitute ①

$$0=\frac{-1}{2}+b-1 \quad \therefore\ b=\frac{3}{2}$$

$\therefore\ a+2b+3c$

$$-\frac{1}{2}+2\times\frac{3}{2}+3(-1)=-\frac{1}{2}+3-3=-\frac{1}{2}$$

Answer is (B)

17. (B)

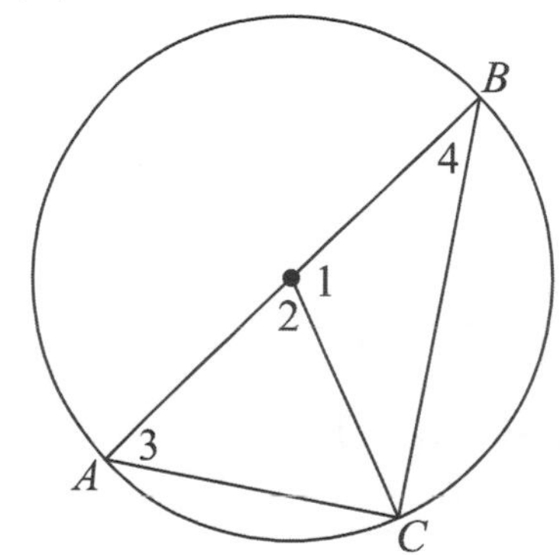

$m\angle 1=\overset{\frown}{BC}$

$m\angle 3=\frac{1}{2}\overset{\frown}{BC}$

$\therefore\ m\angle 1=2m\angle 3$

Answer is (B)

18. (B)

When x-coordinate varies bigger, y-coordinate decrease exponentially so, the realtionship between x and y is exponentially negative.

Answer is (B)

19. (A)

$$[\frac{x^{\frac{1}{3}}}{x^{-3}}]^2 = \frac{x^{\frac{2}{3}}}{x^{-6}} = x^{\frac{2}{3}-(-6)} = x^{\frac{2}{3}+6} = x^{\frac{20}{3}}$$

Answer is (A)

20. (C)

$$-x+2=\frac{4}{3}x-4,\ \frac{7}{3}x=6$$

$$x=\frac{18}{7}=2.57,$$

The maximum value of x is 2.57, so,

The maximum integer is 2.

Answer is (C)

21. (B)

From $\frac{1}{2}\le 3-\frac{k}{4}\le\frac{6}{5}$

$$\frac{1}{2}-3\le-\frac{k}{4}\ \Rightarrow\ -\frac{5}{2}\le-\frac{k}{4}$$

$$\Rightarrow\ -10\le-k\ \Rightarrow\ 10\ge k$$

From $3-\frac{k}{4}\le\frac{6}{5}$

$$-\frac{k}{4}\le\frac{6}{5}-3\ \Rightarrow\ -\frac{k}{4}\le\frac{-9}{5}$$

$$\Rightarrow\ -k\le-\frac{36}{5}\ \Rightarrow\ k\ge\frac{36}{5}$$

Therefore $\frac{36}{5}\le k\le 10$

The minimum value of k is $\frac{36}{5}$ or 7.2

Answer is (B)

22. (A)

$$\frac{x}{x+1}-\frac{1}{x}=1$$

$$\frac{x^2-x-1}{(x+1)\cdot x}=1$$

$$\frac{x^2-x-1}{x^2+x}=1$$

$$x^2-x-1=x^2+x$$

$$2x=-1,\ x=-\frac{1}{2}$$

Answer is (A)

23. (24)

Analysis: $\overline{BC}=\overline{BD}+\overline{DC}$

$\overline{AB}=\overline{AC}=8\sqrt{3}$

$m\angle 2=60°=\angle 1+\angle C=2\angle C$

$m\angle C=\frac{1}{2}\times 60°=30°$

$\angle B=\angle C=30°,\ \angle 2=60°$

Therefore, $m\angle BAD=90°$

$\because\ \frac{AB}{BD}=\frac{8\sqrt{3}}{x}=\sin 60°=\frac{\sqrt{3}}{2}$

$\therefore\ x=16$

then $\overline{BC}=BD+DC$

Operation:

(1) $\angle 2=\angle 1+\angle C,\ 60°=2\angle C,$

$\angle C=30°=\angle B$

(2) $\angle BAD=\angle B+\angle 2=30°+60°=90°$

(3) $\frac{AB}{BD}=\cos 30°=\frac{\sqrt{3}}{2}=\frac{8\sqrt{3}}{BD}$

(4) $BD=16,\ AD=16\cdot\sin 30°=8$

(5) $AD=DC=8,\ BD+DC=16+8$

$=24$

24. (C)

$$g(x)=-12+\frac{x^2}{3}$$

$$g(3k) = -12 + \frac{(3k)^2}{3} = -12 + \frac{9k^2}{3}$$
$$= -12 + 3k^2 = 5k$$
$$3k^2 - 5k - 12 = 0$$
$$(3k+4)(k-3) = 0$$
$$k = -\frac{4}{3} \quad \text{or} \quad k = 3$$

Answer is (C)

25. (B)

Since $5 = 2+3$, $8 = 3+5$

$13 = 5+8$, $21 = 8+13$

So, $13+21 = 34$, $T_7 = 13+21 = 34$

Answer is (B)

26. (C)

The data list 0.12, 0.12, 0.24, 0.28, 0.35, 0.39, 0.39, 0.51, 0.59, 0.63, 0.71, 0.87, there are 12 data. The median is at the middle position. Because there are even number of items in the list, median at $\frac{12}{2} = 6$ So, the value $= \frac{0.39+0.39}{2}$

$= 0.39$

Answer is (C)

27. (D)

Q_1 at $12 \times 0.25 = 3$ (position value)

$$Q_1 = \frac{0.24+0.28}{2} = 0.26$$

Q_3 at $12 \times 0.75 = 9$

$$Q_3 = \frac{0.59+0.63}{2} = 0.61$$

$\text{IQR} = Q_3 - Q_1 = 0.61 - 0.26 = 0.35$

Answer is (D)

28. (D)

Check Lower outlier $0.26 - 1.5 \times 0.35$

$= -0.265$

no lower outlier

Check upper outlier $0.61 + 1.5 \times 0.35$

$= 1.135 > 0.87$

no upper outlier

Answer is (D)

29. (A)

$$C(P) = P^2 - 600P + 120{,}000$$
$$= P^2 - 600P + (300)^2 - (300)^2 + 120{,}000$$
$$= (P-300)^2 - 90{,}000 + 120{,}000$$
$$= (P-300)^2 + 30{,}000$$

It is a figure of *a* parabola with vertex at (300, 30,000), and concave upwards. So, the minimum cost is 30,000 daily.

Answer is (A)

30. (B)

The number of bacteria is doubled for every 6 hours.

So, $n = 2^{\frac{n}{6}}$

model should be

$n = 5{,}000 \times 2^{\frac{h}{6}}$

Answer is (B)

31. (B)

$(3^9)^{(3^6)} = 3^{3^x}$

$3^{9\times3^6} = 3^{3^x}$

So, $9\times3^6 = 3^x$

$3\times3\times3^6 = 3^x$

$3^8 = 3^x$

$x = 8$

Answer is (B)

32. (D)

Q_1 at $25\times0.25 = 6.25$, Q_1 at 7th

$Q_1 = 48$

Q_3 at $25\times0.75 = 18.75$, Q_3 at 19th

$Q_3 = 66$

$IQR = Q_3 - Q_1 = 66 - 48 = 18$

Answer is (D)

33. (D)

Firstly, we have to know how the function $f(x)$ change to $-f(x+2)+4$.

(1) $f(x) \to f(x+2)$ shift left for 2 units $(-2, 4) \to (-4, 4)$

(2) $f(x+2) \to -f(x+2)$ flipped point $(-4, 4)$ changed to $(-4, -4)$

(3) $-f(x+2) \to -f(x+2)+4$ raised upward 4 units, point $(-4, -4)$ changed to $(-4, 0)$

Answer is (D)

34. (C)

If a statement is true the logically equivalent statement is contrapositive statement is also true.

The statement

If sunny, then go fishing.

The contrapositive statement is. If I don't go fishing, it is not a sunny day.

Answer is (C)

35. (C)

From algebraic, substitute $-2 \le x \le 2$ in to $f(x) = |x+1|$

$\Rightarrow |-2+1| \le y \le |2+1| \Rightarrow 1 \le y \le 3$

But when $x = -1$, then $|-1+1| = |0| = 0$ is the smallest, so, $0 \le y \le 3$

Answer is (C)

From Graph: The range of $f(x) = |x+1|$ is $0 \le y \le 3$

Answer is (C)

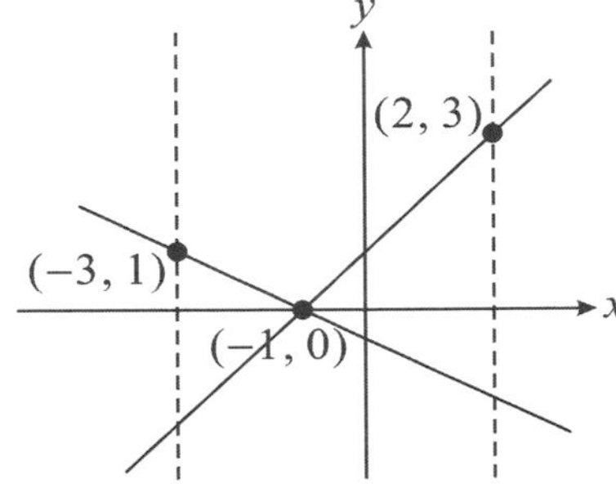

36. (A)

$AD^2 = AB^2 + BD^2$,

$AC^2 = AD^2 + CD^2 = 3^2 + 4^2 + 1^2$

$AC = \sqrt{3^2 + 4^2 + 1^2} = \sqrt{26}$,

$AD = \sqrt{3^2 + 4^2} = 5$

$\cos\theta = \dfrac{5}{\sqrt{26}} = 0.98$

$\theta = \cos^{-1} 0.98 \doteq 11.48°$

Answer is (A)

37. (B)

$g(x) = -4x^2 + 5x - 8$

$= -4[x^2 - \frac{5}{4}x + (\frac{5}{8})^2] + 4 \cdot (\frac{5}{8})^2 - 8$

$= -4(x - \frac{5}{8})^2 - \frac{103}{16}$

Vertex at $(\frac{5}{8}, -\frac{103}{16}) \Rightarrow (0.625, -6.438)$

Because the graph of $g(x)$ concave downwards, the maximum value of $g(x)$ is the y coordinate of the vertex.

Answer is (B)

38. (A)

area of $\square ABFE = l \cdot h = 48$

area of $\square CBFG = h \cdot w = 36$

area of $\square ABCD = l \cdot w = 48$

$\therefore\ (l \cdot h)(h \cdot w)(l \cdot w) = (lwh)^2$

$= (V)^2 = (48 \times 36 \times 48) = (82{,}944)$

$V = \sqrt{82{,}944} = 288$

Answer is (A)

39. (C)

Since $PA^2 = PB \times PC$

$PA^2 = (x-2)^2$,

$PB = PC - BC = (x+2) - 6 = x - 4$

$(x-2)^2 = (x-4) \times (x+2)$

$x^2 - 4x + 4 = x^2 - 2x - 8$

$-2x = -12$, $x = 6$

$PA = x - 2 = 6 - 2 = 4$

Answer is (C)

40. (A)

Since $x < -2$,

so, $||1 - |x+1|| = |1 - (-(x+1))|$

$= |1 + (x+1)| = |x+2| = -(x+2)$

$= -x - 2$

Answer is (A)

41. (D)

(1) $f(x) = -3x^2$ 4 units down

$f(x) - 4 \Rightarrow -3x^2 - 4$

(2) 3 units left $-3(x+3)^2 - 4$

$r(x) = -3(x+3)^2 - 4$

$\therefore\ r(1.3) = -3(1.3+3)^2 - 4$

$= -3(4.3)^2 - 4$

$= (-3)(18.49) - 4 = -59.47$

Answer is (D)

42. (D)

$x(x+1)(x-2) > 0$

(1)
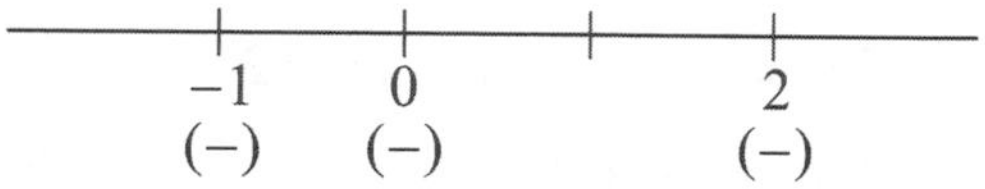

When $x < -1$ $x(x+1)(x-2) < 0$

(−) (−) (−) < 0

No solution

(2) When $-1 < x < 0$ $x(x+1)(x-2) > 0$

(−) (+) (−) > 0

(3) When $0 < x < 2$ $x(x+1)(x-2) < 0$

(+) (+) (−) < 0

(4) When $2<x$ $x(x+1)(x-2)>0$

$(+)\ (+)\quad (+)>0$

The solution sets are $-1<x<0$ and $2<x$

$\{-1<x<0\}\cup\{2<x\}$

Answer is (D)

43.All data values have been subtracted by 3, the standard deviation remains unchanged.

Answer is (D)

44.(C)

Analysis:

(1) $AC=AE$, $BC=BF$, $DF=DE$

(2) $AC+AE=2AC$

$=AB+BF+DF+AD$

$=11+6+14=31$

Answer is (C)

45.(C)

The y-coordinate of $f(x)$ and $g(x)$ is the same. Let $4x^2+6=6x^2-2$

$2x^2=8$, $x^2=4$, $x=\pm 2$

Answer is (C)

46.(A)

$12+6+3+\frac{3}{2}+\frac{3}{4}+\cdots\cdots$

$r=\frac{1}{2}\quad \frac{1}{2}\quad \frac{1}{2}\quad \frac{1}{2}$

$a=12$, $r=\frac{1}{2}$

$$S=\frac{12}{1-\frac{1}{2}}=\frac{12}{\frac{1}{2}}=24$$

Answer is (A)

47.(C)

1. Since $g(x+2)=f(x)=x^2+x-3$

So, $g(x')=(x'-2)^2+(x'-2)-3$

$=x'^2-4x'+4+x'-2-3$

$=x'^2-3x'-1$

$g(x)=x^2-3x-1$

Since $p(x+1)=g(x-1)$

$=(x-1)^2-3(x-1)-1$

$=x^2-2x+1-3x+3-1$

$=x^2-5x+3$

So, $p(x')=(x'-1)^2-5(x'-1)+3$

$=x'^2-2x'+1-5x'+5+3$

$=x'^2-7x'+9$

$p(x)=x^2-7x+9$

So, $2g(x)-3p(x)$ divided by $x+1$ is $2g(-1)-3p(-1)$

$=2(1+3-1)-3(1+7+9)$

$=6-51=-45$

(2)Other solution:

$g(-1)=g(-3+2)$

$=f(-3)=(-3)^2+(-3)-3$

$=9-3-3=3$

$p(-1)=p(-2+1)=g(-2-1)$

$=g(-3)=g(-5+2)$

$=f(-5)=(-5)^2+(-5)-3$

$=25-5-3=17$

$2g(-1)-3p(-1)=2\times 3-3\times 17=-45$

48. (C)

$\triangle ABD \sim \triangle BDC$

$\therefore \ \frac{AD}{BD}=\frac{BD}{CD} \quad \overline{BD}^2 = AD\times CD$

$6^2 = 3\times CD \quad \therefore \ CD=\frac{36}{3}=12$

$AC = CD + AD = 12+3=15$

Answer is (C)

49. (D)

$|z-3|=|8-3i-3|=|5-3i|$

$=\sqrt{25+9}=\sqrt{34}$

Answer is (D)

50. (C)

Collinear means all three points lie on the same line let the function of the line be $y=mx+b$

substitute $(0, 3)$ and $(6, 5)$ into the equation.

$3=b$, $5=6m+3$, $6m=2$, $m=\frac{1}{3}$

The function of the line $y=\frac{1}{3}x+3$

substitute $(3, k)$ to the function

$k=\frac{1}{3}(3)+3=1+3=4$

Answer is (C)

CHAPTER 16
MODEL TEST 4
LEVEL 2

DIRECTION:

Fill in the correct or the best answer in the proper position on the answer sheet. You may use any available space for scratch work.

NOTES:

1. Calculator is permitted.
2. All figures are not drawn to scale.
3. All numbers are real, unless stated otherwise.
4. Some reference information:

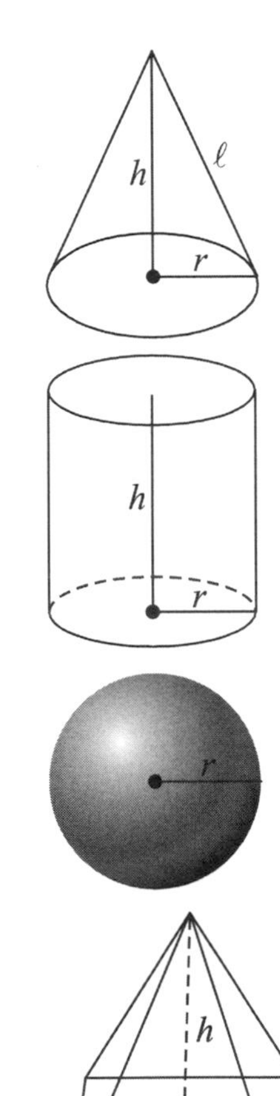

(1) Right Circular Cone

$V = \frac{1}{3}\pi r^2 \cdot h$, $A(\text{lateral surface}) = \frac{1}{2}c\ell = \pi r \cdot \ell$

(2) Cylinder

$V = \pi r^2 \cdot h$, $A(\text{surface area}) = 2\pi r \cdot h$

(3) Sphere

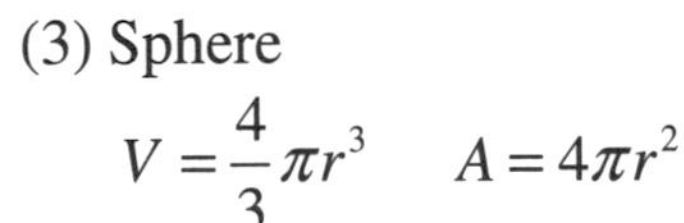

$V = \frac{4}{3}\pi r^3 \qquad A = 4\pi r^2$

(4) Pyramid

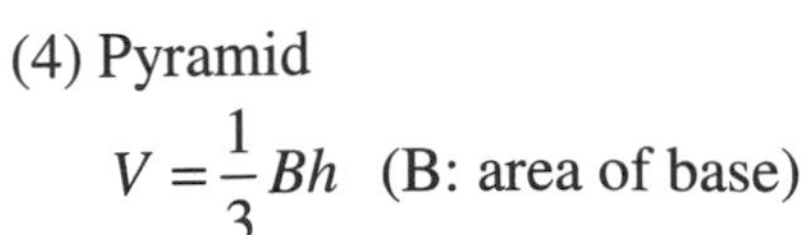

$V = \frac{1}{3}Bh$ (B: area of base)

(5) Special Right Triangle

$30° - 60° - 90°$

$45° - 45° - 90°$

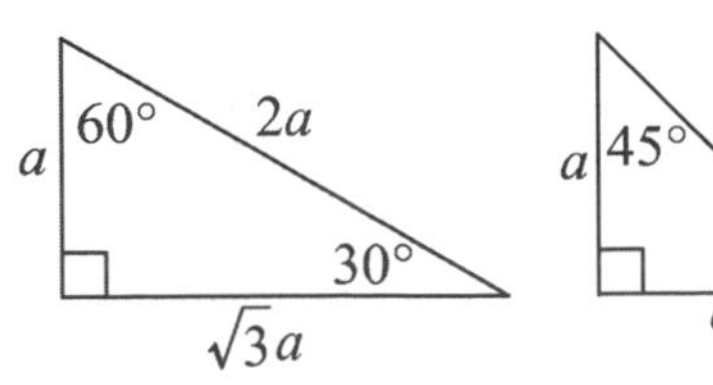

MODEL TEST 4
ANSWER SHEET

Question Number		Your Answers	Answer Key
1	(A)(B)(C)(D)	______	______
2	(A)(B)(C)(D)	______	______
3	(A)(B)(C)(D)	______	______
4	(A)(B)(C)(D)	______	______
5	(A)(B)(C)(D)	______	______
6	(A)(B)(C)(D)	______	______
7	(A)(B)(C)(D)	______	______
8	(A)(B)(C)(D)	______	______
9	(A)(B)(C)(D)	______	______
10	(A)(B)(C)(D)	______	______
11	(A)(B)(C)(D)	______	______
12	(A)(B)(C)(D)	______	______
13	(A)(B)(C)(D)	______	______
14	(A)(B)(C)(D)	______	______
15	(A)(B)(C)(D)	______	______
16	(A)(B)(C)(D)	______	______
17	(A)(B)(C)(D)	______	______
18	(A)(B)(C)(D)	______	______
19	(A)(B)(C)(D)	______	______
20	(A)(B)(C)(D)	______	______
21	(A)(B)(C)(D)	______	______
22	(A)(B)(C)(D)	______	______
23	(A)(B)(C)(D)	______	______
24	(A)(B)(C)(D)	______	______
25	(A)(B)(C)(D)	______	______

Question Number		Your Answers	Answer Key
26	(A)(B)(C)(D)	______	______
27	(A)(B)(C)(D)	______	______
28	(A)(B)(C)(D)	______	______
29	(A)(B)(C)(D)	______	______
30	(A)(B)(C)(D)	______	______
31	(A)(B)(C)(D)	______	______
32	(A)(B)(C)(D)	______	______
33	(A)(B)(C)(D)	______	______
34	(A)(B)(C)(D)	______	______
35	(A)(B)(C)(D)	______	______
36	(A)(B)(C)(D)	______	______
37	(A)(B)(C)(D)	______	______
38	(A)(B)(C)(D)	______	______
39	(A)(B)(C)(D)	______	______
40	(A)(B)(C)(D)	______	______
41	(A)(B)(C)(D)	______	______
42	(A)(B)(C)(D)	______	______
43	(A)(B)(C)(D)	______	______
44	(A)(B)(C)(D)	______	______
45	(A)(B)(C)(D)	______	______
46	(A)(B)(C)(D)	______	______
47	(A)(B)(C)(D)	______	______
48	(A)(B)(C)(D)	______	______
49	(A)(B)(C)(D)	______	______
50	(A)(B)(C)(D)	______	______

CHAPTER 16
MODEL TEST 4
LEVEL 2

1. What is the radius of the circle $x^2-4x+y^2+2y=20$?

(A) 3

(B) 4

(C) 5

(D) 6

2. What is the range of $f(x)=\sqrt{9-x^2}$

(A) $y\geq 3$

(B) $y\leq 3$

(C) $-3\leq y\leq 3$

(D) $0\leq y\leq 3$

3. Which of the following is the solution set of $|x^2-2|<2$?

(A) $0<x<2$

(B) $\{x<0\}\cup\{x>2\}$

(C) $\{x\neq 0\}\cup\{x>2\}$

(D) $\{x\neq 0\}\cap\{-2<x<2\}$

4. $\lim\limits_{x\to\infty}\dfrac{1}{\sqrt{n^2-n}+n}=$

(A) 0

(B) 1

(C) 2

(D) 3

(E) Infinite

5. If $y=f(\dfrac{2x}{x+1})=\dfrac{2}{3}x^2+x-1$, what is the value of $f(3)$?

(A) -3

(B) -2

(C) 3

(D) 2

6. The graph of the curve represented by $\begin{cases} x=\cos\theta \\ y=\sin\theta \end{cases}$ is

(A) a line

(B) an ellipse

(C) a circle

(D) a parabola

7. If $2-\sqrt{3}i$ is the root of the equation $ax^2+10x+c=0$, which of the following is the value of c ?

(A) 3

(B) 5

(C) 7

(D) 9

8. What is the equation of the horizontal asymptote of the function $f(x)=\dfrac{(2x-1)(x+4)}{(x+4)^2}$

(A) $y=-2$

(B) $y=-3$

(C) $y=2$

(D) $y=\frac{1}{2}$

9. If $\log_2(\log_3(\log_4^x))=1$, what is the value of x ?

(A) 6128

(B) 4096

(C) 16,384

(D) 262,144

10. $\sum_{n=1}^{\infty}(\frac{1}{3})^n+(\frac{1}{4})^n+(\frac{1}{5})^n$

(A) $\frac{3}{14}$

(B) $\frac{5}{12}$

(C) $\frac{11}{12}$

(D) $\frac{13}{12}$

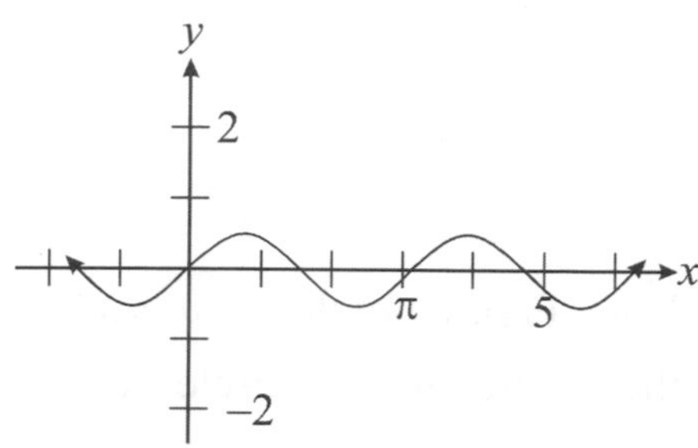

Note: Figure not drawn to scale

11. Which of the following equations has the graph shown above?

(A) $y=2\cos 2x$

(B) $y=\sin 2x$

(C) $y=\frac{1}{2}\sin x$

(D) $y=\frac{1}{2}\sin 2x$

12. Which of the following is the graph of $x^2-\frac{2}{3}x+y^2+y-4=0$

(A) an ellipse

(B) a parabola

(C) a circle

(D) a hyperbola

13. $\sqrt[4]{\sqrt[3]{\sqrt{x}}}=$

(A) $x^{\frac{1}{9}}$

(B) $x^{\frac{1}{12}}$

(C) $x^{\frac{1}{20}}$

(D) $x^{\frac{1}{24}}$

14. $\lim_{x\to 2}\frac{(2x^2-x-6)}{x^2-4}$

(A) $-\frac{7}{4}$

(B) $\frac{7}{4}$

(C) -2

(D) 2

15. If $f(x)=2x^2+1$, where $x\geq 0$, then what value could $f(f^{-1}(x))$ be ?

(A) $-\frac{1}{2}$

(B) $\frac{1}{2}$

(C) $\frac{1}{x}$

(D) x

16. In how many ways can 3 black balls and 4 white balls be selected from a box which contains 7 black and 6 white balls?

(A) 365

(B) 390

(C) 445

(D) 525

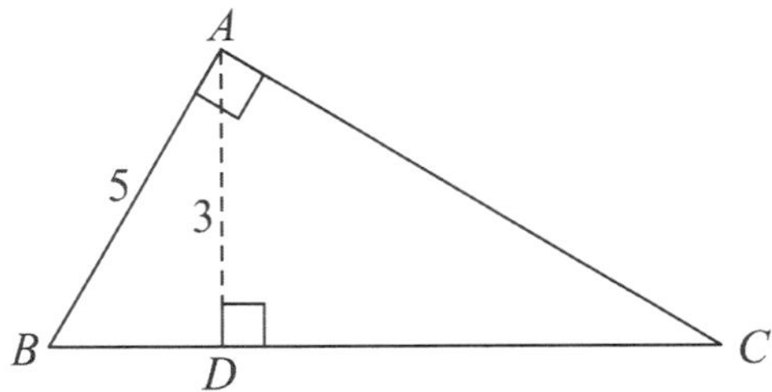

Note: Figure not drawn to scale

17. In the figure above, $\overline{AD} \perp \overline{BC}$, $\triangle ABC$ is a right triangle. The length of $\overline{AD} = 3$, $\overline{AB} = 5$ what is the area of $\triangle ABC$

(A) $\frac{75}{4}$

(B) $\frac{75}{8}$

(C) $\frac{65}{4}$

(D) $\frac{65}{8}$

18. Which of the following is the y-intercept of $y = \left|\sqrt{2}\sec 2(x + \frac{\pi}{6})\right|$?

(A) $2\sqrt{2}$

(B) $\sqrt{2}$

(C) $\sqrt{3}$

(D) $2\sqrt{3}$

19. If the probability that the defective of a light bulb is $\frac{1}{8}$, what is the probability that a package of 8 light bulbs has exactly two defective bulbs?

(A) 0.20

(B) 0.28

(C) 0.33

(D) 0.42

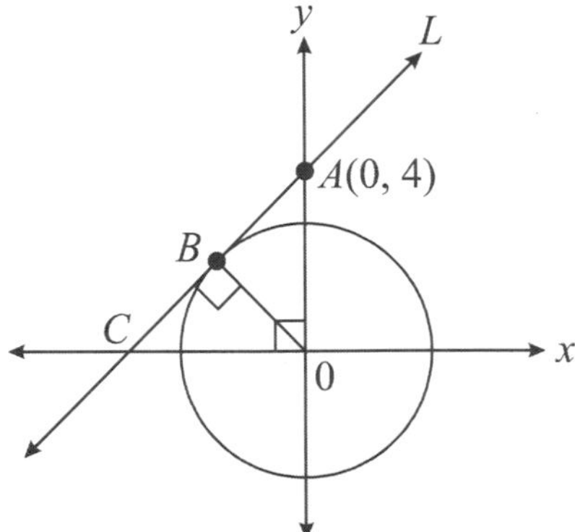

Note: Figure not drawn to scale

20. In the figure above, circle $x^2 + y^2 = 4$ with the center at the origin, and line L is tangent to the circle at point B and passes through the point (0, 4), what is the function of line L ?

(A) $y = \sqrt{3}x + 4$

(B) $y = -\sqrt{3}x + 4$

(C) $y = \sqrt{2}x + 2$

(D) $y = -\sqrt{x} + 2$

21. Which of the following quadratic equations has no solution?

(A) $3(x+4)^2 - 5 = 0$

(B) $-2(x+4)(x+3) = 0$

(C) $5(x-2)^2 + 4 = 0$

(D) $2(x-4)(x-3) = 0$

22. If f is a function defined over the set of all real numbers and $f(x+1) = 2x^2 + 7x + 6$, which of the following functions defines $f(x)$?

(A) $f(x) = 3x^2 + 2x + 1$

(B) $f(x) = -3x^2 + 2x + 1$

(C) $f(x) = 2x^2 + 3x + 1$

(D) $f(x) = 2x^2 - 3x + 1$

23. If the matrix $A = \begin{bmatrix} -1 & 2 \\ 0 & 3 \end{bmatrix}$, $B = \begin{bmatrix} 1 & 2 \\ -3 & 1 \end{bmatrix}$, Fine the determinant of $A + B$?

(A) −12

(B) 12

(C) −8

(D) 8

24. If point (1, 2) lies on the graph of the inverse of $f(x) = 3x^3 + 2x + k$, what is the value of k?

(A) −27

(B) −16

(C) 16

(D) 27

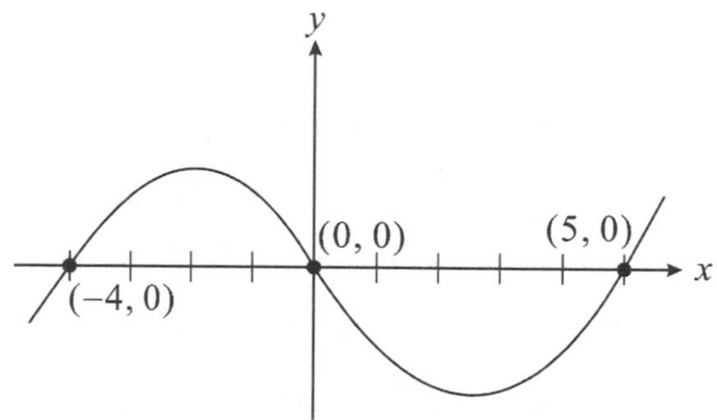

Note: Figure not drawn to scale

25. The graph above shows the function $f(x) = ax^3 + bx^2 + cx + d$, which of the following must be true?

(A) $b > 0$

(B) $b < 0$

(C) $a > 0$

(D) $a < 0$

26. If n is a integer, what is the remainder when $2x^{(n+2)} - 3x^{(2n+3)} + 5x^{(2n+1)} - 7$ is divided $x - 1$?

(A) −1

(B) −2

(C) −3

(D) −4

27. If $\sin\frac{\theta}{2}=\frac{1}{2}$, what is the value of $\cos\theta$?

(A) $-\frac{1}{2}$

(B) $\frac{1}{2}$

(C) $\frac{\sqrt{2}}{2}$

(D) $-\frac{\sqrt{2}}{2}$

28. There are 6 identical balls in the box, numbered 1, 2, 3, 4, 5, 6. If four balls are taken out of the box (the balls taken out are no longer put back), what is the probability that the sum of number of the balls taken out is odd?

(A) $\frac{1}{5}$

(B) $\frac{4}{15}$

(C) $\frac{8}{15}$

(D) $\frac{16}{36}$

29. If $\frac{\sin 2x}{1-\cos 2x}=\frac{1}{2}$, $0°\le x\le 180°$, what is the value of x?

(A) 25

(B) 45

(C) 64

(D) 84

30. There are 7 carriages on a light rail in a particular city, if we would like to paint one animal image on each carriage, and there are two cougars, three seahawks, and two huskies to choose from randomly, what is the probability of having a cougar image painted for the first carriage ?

(A) $\frac{1}{4}$

(B) $\frac{2}{7}$

(C) $\frac{1}{3}$

(D) $\frac{3}{7}$

31. What is the maximum value of $4\sin\theta\cos\theta$?

(A) 1

(B) 1.8

(C) 2

(D) 2.4

32. Which of the following is the graph of the polar equation $r=\frac{2}{2\sin\theta+3\cos\theta}$?

(A) *a* point

(B) *a* line

(C) *a* parabola

(D) *a* ellipse

33. If $f(x)=\log_3(x^3-54)$ and $f(p(1))=4$, which of the following could be $p(x)$?

(A) $p(x)=x^3+x^2-x+1$

(B) $p(x)=3x^3-2x^2+x+3$

(C) $p(x)=\sin(\frac{\pi}{2}x)+2$

(D) $p(x)=\cos(\frac{\pi}{2}x)+2$

34. If a coin is tossed four times, what is the probalility that exactly three heads will appears?

(A) $\frac{1}{2}$

(B) $\frac{1}{3}$

(C) $\frac{1}{4}$

(D) $\frac{1}{5}$

35. If $\log_{\sqrt{5}}^{x}=20$, what is the value of $\log_{5}^{x^3}$?

(A) 15

(B) 20

(C) 25

(D) 30

36. $\frac{1+i}{3-i}+\frac{2-i}{1-i}=$

(A) $\frac{17}{10}+\frac{9}{10}i$

(B) $\frac{12}{15}+\frac{7}{15}i$

(C) $3+5i$

(D) $5+3i$

37. What is the period of the graph of $y=-2\sin(4\pi x-\frac{\pi}{2})-1$

(A) $\frac{1}{2}$

(B) $\frac{1}{3}$

(C) $\frac{1}{4}$

(D) $\frac{1}{6}$

38. Find the coordinates of perpendicular point of L_1 and L_2 . L_1 passes through point (2, 1, 1) and is perpendicular to line $l_2: \frac{x-2}{-1}=\frac{y+1}{2}=\frac{z-1}{1}$.

39. If $\lim_{x\to-1}\frac{x^2+ax+b}{x+1}=7$

Where a and b are constants, what is the value of a?

(A) 6

(B) 7

(C) 8

(D) 9

40. If $\frac{x^2-4x-5}{x^2}<0$, which of the following is the solution of the inequality ?

(A) $x<5$

(B) $-1 < x < 5$

(C) $-1 < x < 5$ or $x \neq 0$

(D) $-1 < x < 0$ and $0 < x < 5$

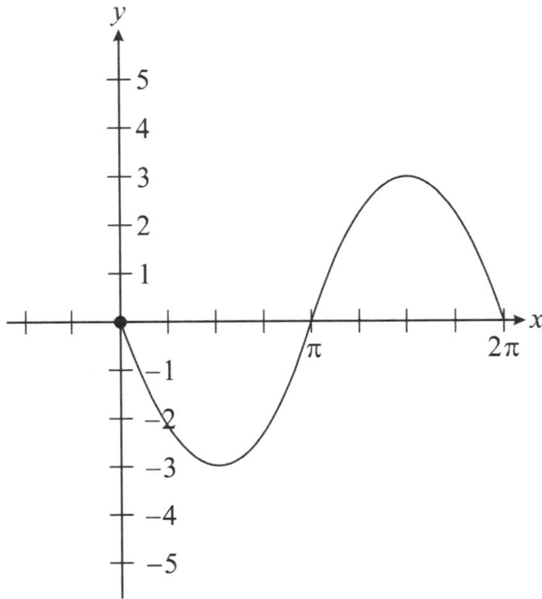

Note: Figure not drawn to scale

41. The figure above represents a portion of the graph, which of the following functions could represent the graph?

(A) $f(x) = 3\cos(x + \pi)$

(B) $f(x) = \cos 2x$

(C) $f(x) = 3\cos x$

(D) $f(x) = -3\sin x$

42. In triangle $\triangle ABC$, $a = 8$, $b = 8\sqrt{3}$, and $\angle A = 30°$, what is the measure of $\angle B$?

(A) $30°$

(B) $60°$

(C) $30°$ or $90°$

(D) $60°$ or $90°$

43. If $\log_3^5 = a$ and $\log_3^7 = b$, then which of the following could be equivalent to $\log_3 \sqrt{140}$?

(A) $\log_3^4 + a + b$

(B) $7a + 3b$

(C) $\frac{1}{2}(\log_3^4 + a + b)$

(D) $2(\log_3^4 + a + b)$

44. What are the asymptotes of the hyperbola whose equation is $9x^2 - 96y^2 = 288$?

(A) $y = 0.31x$

(B) $y = 0.44x$

(C) $y = 0.56x$

(D) $y = 0.73x$

45. If the polar equation $r = \frac{3}{3 - 2\sin\theta}$ which of the following represents its graph?

(A) Circle

(B) Ellipse

(C) Hyperbola

(D) Parabola

46. If $|x+2| + 3|x-1| - |x+1| = 4$, where x is *a* real number, what is one possible value of x?

(A) $-\frac{2}{3}$

(B) 0

(C) $\frac{2}{3}$

(D) $\frac{4}{3}$

47. $\lim_{n\to\infty}\frac{2^n-2^{n-1}}{2^n-1}$

(A) 0

(B) $\frac{1}{2}$

(C) $-\frac{1}{3}$

(D) $\frac{1}{5}$

48. If the first three terms in a geometric sequence whose terms are all positive are x, 18, $7x+12$, which of the following could be the value of 7th term?

(A) 2,864

(B) 3,212

(C) 3,908

(D) 4,374

49. $\lim_{x\to 2}\frac{x^2+x+a}{x-2}=5$, what is the value of a ?

(A) 2

(B) 3

(C) 5

(D) 6

50. In $x-y-z$ space,

There are 2 parallel lines

$L_1:\frac{x}{2}=\frac{y+1}{1}=\frac{z+1}{2}$, and

$L_2:\frac{x-1}{-1}=\frac{y}{1}=\frac{z-1}{3}$,

What is the distance between these two lines?

(A) 1.2

(B) 2.58

(C) 3.54

(D) 4.08

MODEL TEST 4
ANSWER KEY

1. (C)
2. (D)
3. (D)
4. (D)
5. (D)
6. (C)
7. (C)
8. (C)
9. (D)
10. (D)
11. (D)
12. (C)
13. (D)
14. (B)
15. (D)
16. (D)
17. (B)
18. (A)
19. (A)
20. (A)
21. (C)
22. (C)
23. (B)
24. (A)
25. (C)
26. (C)
27. (B)
28. (A)
29. (C)
30. (B)
31. (C)
32. (B)
33. (C)
34. (C)
35. (D)
36. (A)
37. (A)
38. $(\frac{4}{3}, \frac{1}{3}, \frac{5}{3})$
39. (D)
40. (D)
41. (D)
42. (B)
43. (C)
44. (A)
45. (B)
46. (B)
47. (B)
48. (D)
49. (C)
50. (C)

MODEL TEST 4
ANSWERS and EXPLANATIONS

1. (C)

$x^2-4x+y^2+2y=20$

$x^2-4x+4+y^2+2y+1-4-1-20=0$

$(x-2)^2+(y+1)^2=25$

radius is 5

Answer is (C)

2. (D)

$f(x)=\sqrt{9-x^2}$

The domain of $f(x)$

$9-x^2\ge 0$

$9\ge x^2$

$-3\le x\le 3$

When $x=0$, $y=3$

y can not be negative, so minimum value is zero, maximum value is $+3$,

The range of y $0\le y\le 3$

Answer is (D)

3. (D)

$-2<x^2-2<2$

$\Rightarrow\ 0<x^2<4$

$\Rightarrow\ \{0<x^2\}\cap\{x^2<4\}$

$\Rightarrow\ \{x\ne 0\}\cap\{-2<x<2\}$

Answer is (D)

4. (D)

$\lim\limits_{n\to\infty}\dfrac{1}{\sqrt{n^2-n}+n}$

$=\lim\limits_{n\to\infty}\dfrac{1\times(\sqrt{n^2-n}-n)}{(\sqrt{n^2-n}+n)(\sqrt{n^2-n}-n)}$

$=\lim\limits_{n\to\infty}\dfrac{\sqrt{n^2-n}-n}{n^2-n-n^2}=\lim\limits_{n\to\infty}\dfrac{\sqrt{n^2-n}-n}{-n}$

$=\lim\limits_{n\to\infty}\dfrac{\sqrt{n^2-n}}{-n}+1$

$=\lim\limits_{n\to\infty}\dfrac{\sqrt{1-\frac{1}{n}}}{-1}+1=-1+1=0$

Answer is (A)

5. (D)

Let $\dfrac{2x}{x+1}=3$, $3x+3=2x$, $x=-3$

$f(3)=\dfrac{2}{3}(-3)^2-3-1=2$

Answer is (D)

Other solution:

Let $t=\dfrac{2x}{x+1}$, $tx+t-2x=0$

$x(t-2)+t=0$, $x=\dfrac{-t}{t-2}$

So, $f(t)=\dfrac{2}{3}(\dfrac{-t}{t-2})^2+\dfrac{-t}{t-2}-1$

$f(3)=\dfrac{2}{3}(\dfrac{-3}{3-2})^2+\dfrac{-3}{3-2}-1$

$=6-3-1=2$

6. (C)

$x=\cos\theta$

$x^2=\cos^2\theta$

$y=\sin\theta$

$y^2=\sin^2\theta$

$x^2+y^2=\cos^2\theta+\sin^2\theta=1$

Answer is (C)

7. (C)

Since $2-\sqrt{3}i$ is one root of the equation the other conjugate root must be $2+\sqrt{3}i$

So, $c=(2-\sqrt{3}i)(2+\sqrt{3}i)=[4-(\sqrt{3}i)^2]$

$=4-3i^2=4+3=7$

Answer is (C)

8. (C)

$$f(x)=\frac{(2x-1)(x+4)}{(x+4)^2}=\frac{2x-1}{x+4}$$

$$\lim_{x\to\infty}f(x)=\lim_{x\to\infty}\frac{2x-1}{x+4}$$

$$=\lim_{x\to\infty}\frac{2-\frac{1}{x}}{1+\frac{4}{x}}=\frac{2}{1}=2$$

When $x\to\infty$, $y\to2$

horizontal asymptote $y=2$

Answer is (C)

9. (D)

$2^1=\log_3(\log_4^x)$

$3^2=\log_4^x$

$4^9=x$, $x=262{,}144$

Answer is (D)

10. (D)

$$\sum_{n=1}^{\infty}(\frac{1}{3})^n+\sum_{n=1}^{\infty}(\frac{1}{4})^n+\sum_{n=1}^{\infty}(\frac{1}{5})^n$$

$$=[\frac{1}{3}+(\frac{1}{3})^2+(\frac{1}{3})^3+\cdots+(\frac{1}{3})^n]$$

$$+[(\frac{1}{4})+(\frac{1}{4})^2+\cdots+(\frac{1}{4})^n]$$

$$+[(\frac{1}{5})+(\frac{1}{5})^2+\cdots+(\frac{1}{5})^n]$$

$$=\frac{\frac{1}{3}}{1-\frac{1}{3}}+\frac{\frac{1}{4}}{1-\frac{1}{4}}+\frac{\frac{1}{5}}{1-\frac{1}{5}}=\frac{\frac{1}{3}}{\frac{2}{3}}+\frac{\frac{1}{4}}{\frac{3}{4}}+\frac{\frac{1}{5}}{\frac{4}{5}}$$

$$=\frac{1}{2}+\frac{1}{3}+\frac{1}{4}=\frac{12+8+6}{24}=\frac{26}{24}=\frac{13}{12}$$

Answer is (D)

11. (D)

The graph shows:

Amplitude $=\frac{1}{2}$

period $=\frac{2\pi}{2}=\pi$

It is $y=\frac{1}{2}\sin 2x$

Answer is (D)

12. (C)

$$x^2-\frac{2}{3}x+y^2+y-4=0$$

$$(x^2-\frac{2}{3}x+\frac{1}{9})-\frac{1}{9}+y^2+y+\frac{1}{4}-\frac{1}{4}-4=0$$

$$(x-\frac{1}{3})^2+(y+\frac{1}{2})^2-\frac{1}{9}-\frac{1}{4}-4=0$$

$$(x-\frac{1}{3})^2+(y+\frac{1}{2})^2-\frac{4+9}{36}-4=0$$

$$(x-\frac{1}{3})^2+(y+\frac{1}{2})^2=\frac{157}{36}$$

$(x-\frac{1}{3})^2+(y+\frac{1}{2})^2=(\frac{\sqrt{157}}{6})^2$

It is a cicle with center at $(\frac{1}{3},-\frac{1}{2})$ and radius of $\frac{\sqrt{157}}{6}$

Answer is (C)

13. (D)

$\sqrt[4]{\sqrt[3]{\sqrt{x}}}=[(x^{\frac{1}{2}})^{\frac{1}{3}}]^{\frac{1}{4}}=x^{\frac{1}{2}\times\frac{1}{3}\times\frac{1}{4}}=x^{\frac{1}{24}}$

Answer is (D)

14. (B)

$\lim_{x\to 2}\frac{2x^2-x-6}{x^2-4}$

$=\lim_{x\to 2}\frac{(2x+3)(x-2)}{(x+2)(x-2)}$

$=\lim_{x\to 2}\frac{2x+3}{x+2}$

$=\frac{7}{4}$

Answer is (B)

15. (D)

Since $f(f^{-1}(x))=x$

Therefore, answer is (D)

16. (D)

$C_3^7\cdot C_4^6=\frac{7\cdot 6\cdot 5\cdot 4!}{4!\cdot 3!}\times\frac{6\cdot 5\cdot 4!}{2!\cdot 4!}=525$

Answer is (D)

17. (B)

$BD^2=5^2-3^2=25-9=16$

$BD=4$

$\triangle ABD\sim\triangle ABC$

$\frac{AB}{BC}=\frac{BD}{AB}\quad\therefore\ BC=\frac{25}{4}$

$\triangle ABC=\frac{1}{2}\times 3\times\frac{25}{4}=\frac{75}{8}$

Answer is (B)

18. (A)

When $x=0$.

We can find the y-intercept

$y=|\sqrt{2}\sec 2(x+\frac{\pi}{6})|$

$y=|\sqrt{2}\sec 2(0+\frac{\pi}{6})|$ (When $x=0$)

$=|\sqrt{2}\sec(2\times\frac{\pi}{6})|=|\sqrt{2}\sec\frac{\pi}{3}|$

$=\sqrt{2}\cdot\frac{1}{\cos\frac{\pi}{3}}$

$=\frac{\sqrt{2}}{\cos\frac{\pi}{3}}=\frac{\sqrt{2}}{\frac{1}{2}}=2\sqrt{2}$

Answer is (A)

19. (A)

$C_2^8(\frac{1}{8})^2(\frac{7}{8})^6=\frac{8\cdot 7\cdot 6!}{6!2!}\frac{7^6}{8^8}$

$=28\cdot\frac{117{,}649}{16{,}777{,}216}=0.20$

Answer is (A)

20. (A)

(1) Let the function of line L be $y=mx+b$ (Slope-intercept form), so,

it is $y=mx+4$

(2) m is $\tan(\angle ACO)$, Let's take the redius of the circle as a bridge, so, connect $\overline{BO}$.

(3) $m\angle ABO=90°$, $\sin(\angle BAO)=\dfrac{BO}{AO}$

$=\dfrac{2}{4}=\dfrac{1}{2}$, $m\angle BAO=30°$

So, $m\angle ACO=60°$, $\tan 60°=\sqrt{3}$

$=m$

(4) The function of line L $y=\sqrt{3}x+4$

Answer is (A)

21. (C)

(A) $3(x+4)^2-5=0$

$3(x+4)^2=5$

$(x+4)^2=\dfrac{5}{3}$

$x+4=\sqrt{\dfrac{5}{3}}$

$x=\sqrt{\dfrac{5}{3}}-4$

(B) $-2(x+4)(x+3)=0$

$x=-4$ or $x=-3$

(C) $5(x-2)^2+4=0$

$5(x-2)^2=-4$

$(x-2)^2=-\dfrac{4}{5}$

It is impossible

So, This equation has no solution

Answer is (C)

22. (C)

$f(x+1)=2x^2+7x+6$

Set $x'=x+1$ $\therefore$ $x=x'-1$

$f(x')=2(x'-1)^2+7(x'-1)+6$

$=2(x'^2-2x'+1)+7x'-7+6$

$=2x'^2-4x'+2+7x'-7+6$

$=2x'^2+3x'+1$

$\therefore$ $f(x)=2x^2+3x+1$

Answer is (C)

23. (B)

$A+B=\begin{bmatrix}-1 & 2\\ 0 & 3\end{bmatrix}+\begin{bmatrix}1 & 2\\ -3 & 1\end{bmatrix}$

$=\begin{bmatrix}-1+1 & 2+2\\ 0-3 & 3+1\end{bmatrix}=\begin{bmatrix}0 & 4\\ -3 & 4\end{bmatrix}$

determinant $=(0\times4)-(-3)\times4=12$

Answer is (B)

24. (A)

Since $P(1, 2)$ lies on the graph of the inverse of $f(x)$, then the inverse of the point must lie on the $f(x)$.

The inverse of the point (1, 2)

$\Rightarrow$ (y, x) $\Rightarrow (2, 1)$

lie on $f(x)=3x^3+2x+k$

$1=3(2)^3+2(2)+k$

$\therefore$ $k=-27$

Answer is (A)

25. (C)

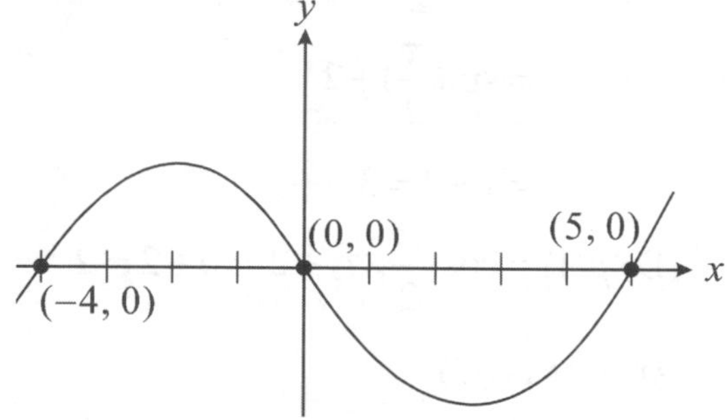

The highest power is odd with left downward right upward, it must be $a > 0$.

Answer is (C)

26. (C)

$2x^{(n+2)} - 3x^{(2n+3)} + 5x^{(2n+1)} - 7$ is divided by $x - 1$

$f(1) = 2\times(1)^{n+2} - 3(1)^{2n+3} + 5(1)^{2n+1} - 7$

$= 2 - 3 + 5 - 7 = -3$

Answer is (C)

27. (B)

Since $\sin\frac{\theta}{2} = \pm\sqrt{\frac{1-\cos\theta}{2}}$

Therefore $\pm\sqrt{\frac{1-\cos\theta}{2}} = \frac{1}{2}$

$\frac{1-\cos\theta}{2} = \frac{1}{4}$, $1-\cos\theta = \frac{1}{2}$

$\cos\theta = 1 - \frac{1}{2} = \frac{1}{2}$

Answer is (B)

28. (A)

Total outcomes $= C_4^6 = 30$

Desire outcomes:

(3 odd + 1 even) + (1 odd + 3 even)

$C_3^3 \times C_1^3 + C_1^3 \times C_3^3 = 1\times3 + 3\times1 = 6$

$P = \frac{6}{30} = \frac{1}{5}$

Answer is (A)

29. (C)

$$\frac{\sin 2x}{1-\cos 2x} = \frac{2\sin x\cos x}{1-(\cos^2 x - \sin^2 x)}$$

$$= \frac{2\sin x\cos x}{1-\cos^2 x + \sin^2 x}$$

$$= \frac{2\sin x\cos x}{\sin^2 x + \sin^2 x}$$

$$= \frac{2\sin x\cos x}{2\sin^2 x} = \frac{\cos x}{\sin x}$$

$$= \cot x = \frac{1}{2}$$

$\therefore\ x = \cot^{-1}\frac{1}{2} = 63.43 \doteq 64$

Answer is (C)

30. (B)

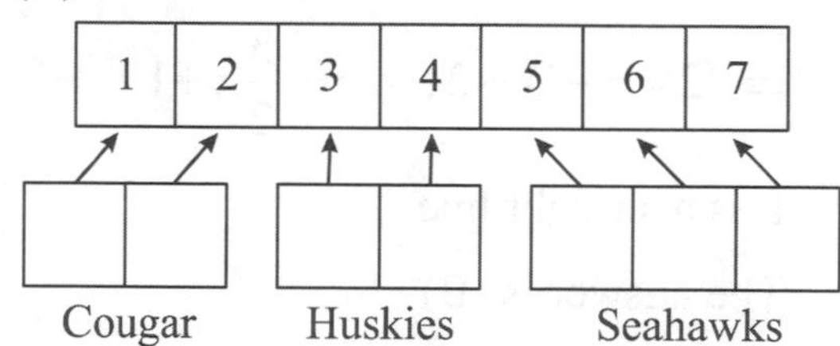

It is repeating permutation total ways of painting $= \frac{7!}{2!2!3!}$ paint one cougar on the first carriage, there are $\frac{6!}{2!\cdot 3!}$

So, P(cougar on the first carriage)

$$=\frac{\frac{6!}{2!3!}}{\frac{7!}{2!2!3!}}=\frac{60}{210}=\frac{2}{7}$$

Answer is (B)

31. (C)

Since $4\sin\theta\cos\theta=2\cdot 2\sin\theta\cos\theta$

$=2\cdot\sin 2\theta$

The maximum value of $\sin 2\theta$ is '1', so, the maximum value of $2\sin 2\theta=2\times 1$

$=2$

Answer is (C)

32. (B)

$$r=\frac{2}{2\sin\theta+3\cos\theta}$$

$\Rightarrow\ 2r\sin\theta+3r\cos\theta=2$

$\Rightarrow\ (2)\cdot(r)(\frac{y}{r})+(3)(r)(\frac{x}{r})=2$

$\Rightarrow\ 2y+3x=2$

$\Rightarrow\ 2y=-3x+2,\ \ y=-\frac{3}{2}x+1$

It is a straight line

The answer is (B)

33. (C)

Let $p(1)$ be k, then

$f(p(1))=\log_3(k^3-54)=4$

$k^3-54=3^4=81$

$k^3=27,\ k=3\ \Rightarrow\ p(1)=3$

(A) $p(1)=1+1-1+1=2$

(B) $p(1)=3(1)^3-2(1)^2+1+3=5$

(C) $p(1)=\sin(\frac{\pi}{2}\times 1)+2$

$=\sin(\frac{\pi}{2})+2$

$=1+2=3$

(D) $p(1)=\cos(\frac{\pi}{2}\times 1)+2=0+2=2$

Answer is (C)

34. (C)

$$C_3^4(\frac{1}{2})^3\cdot(\frac{1}{2})=\frac{4\cdot 3!}{1!\cdot 3!}\cdot(\frac{1}{8})(\frac{1}{2})$$

$$=4\cdot\frac{1}{16}=\frac{1}{4}$$

Answer is (C)

35. (D)

$$\log_{\sqrt{5}}^{x}=\frac{\log_5^x}{\log_5^{\sqrt{5}}}=\frac{\log_5^x}{\frac{1}{2}\log_5^5}=20$$

$\therefore\ \log_5^x=10,$

$\log_5^{x^3}=3\log_5^x=3\times 10=30$

Answer is (D)

36. (A)

$$\frac{1+i}{3-i}+\frac{2-i}{1-i}$$

$$=\frac{(1+i)(1-i)+(2-i)(3-i)}{(3-i)(1-i)}$$

$$=\frac{(1-i^2)+(6-5i+i^2)}{3-4i+i^2}$$

$$=\frac{1-i^2+6-5i+i^2}{2-4i}=\frac{7-5i}{2-4i}$$

$$=\frac{(7-5i)(2+4i)}{(2-4i)(2+4i)}=\frac{14-10i+28i-20i^2}{2^2-(4i)^2}$$

$=\dfrac{14+18i+20}{4-16i^2}=\dfrac{34+18i}{4+16}$

$=\dfrac{34+18i}{20}=\dfrac{17+9i}{10}$

Answer is (A)

37. (A)

$y=-2\sin(4\pi x-\dfrac{\pi}{2})-1$ period of sin is 2π, so, $\dfrac{2\pi}{4\pi}=\dfrac{1}{2}$

Answer is (A)

38. Find the coordinates of perpendicular point.

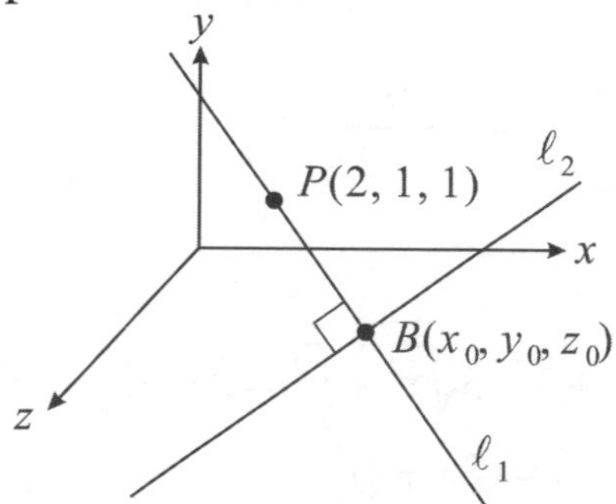

Assume perpendicular point at $B(x_0, y_0, z_0)$ which lies on l_2,

The equation of l_2:

$x_0=-t+2$, $y_0=2t-1$, $z_0=t+1$

$\overrightarrow{PB}=(x_0-2, y_0-1, z_0-1)$

$\overrightarrow{PB}=(-t+2-2, 2t-1-1, t+1-1)$

$=(-t, 2t-2, t)$

Since $\overrightarrow{PB}\perp l_2$, Therefore $\overrightarrow{PB}\cdot l_2=0$

$\overrightarrow{PB}\cdot l_2=((-t)(-1)+(2t-2)\cdot(2)+t(1))$

$=(t+4t-4+t)=6t-4=0$

$t=\dfrac{2}{3}$

$x_0=-\dfrac{2}{3}+2=\dfrac{4}{3}$

$y_0=2\cdot\dfrac{2}{3}-1=\dfrac{1}{3}$, $z_0=\dfrac{2}{3}+1=\dfrac{5}{3}$

The intersected point at $(\dfrac{4}{3},\dfrac{1}{3},\dfrac{5}{3})$

39. (D)

Since: $\lim\limits_{x\to-1}\dfrac{x^2+ax+b}{x+1}=7$

So, $f(x)=x^2+ax+b$ must have the factor of $(x+1)$.

$f(-1)=(-1)^2-a+b=0$, $a=1+b$

$x^2+ax+b=x^2+(1+b)x+b$

$=x^2+bx+x+b$

$=x(x+b)+(x+b)$

$=(x+1)(x+b)$

$\lim\limits_{x\to-1}\dfrac{x^2+ax+b}{x+1}=\lim\limits_{x\to-1}\dfrac{(x+1)(x+b)}{x+1}$

$=-1+b=7$

then $x=-1$, $b=8$, $a=1+8=9$

Answer is (D)

40. (D)

$\dfrac{x^2-4x-5}{x^2}=x^2(x^2-4x-5)<0$

$x^2-4x-5<0$

$(x+1)(x-5)<0$

$-1<x<5$ and $x\neq0$

So, $-1<x<0$ and $0<x<5$

Answer is (D)

41. (D)

It shows the sine curve, $y = \sin x$, reflected over the x-axis with an amplitude of 3. So, the graph is $f(x) = -3\sin x$

Answer is (D)

42. (B)

From Law of sine:

$$\frac{8}{\sin A} = \frac{8\sqrt{3}}{\sin B}$$

$$\frac{8}{\sin 30°} = \frac{8\sqrt{3}}{\sin B}$$

$$\frac{8}{\frac{1}{2}} = \frac{8\sqrt{3}}{\sin B}$$

Therefore $\sin B = \dfrac{\frac{1}{2} \bullet 8\sqrt{3}}{8} = \dfrac{\sqrt{3}}{2}$

$\therefore \; B = 60°$

Answer is (B)

43. (C)

$$\log_3 \sqrt{140} = \log_3 140^{\frac{1}{2}} = \frac{1}{2} \bullet \log_3 140$$

$$= \frac{1}{2}(\log_3 4 \times 35)$$

$$= \frac{1}{2}(\log_3^4 + \log_3^{35})$$

$$= \frac{1}{2}(\log_3^4 + \log_3^5 + \log_3^7)$$

$$= \frac{1}{2}(\log_3^4 + a + b)$$

Answer is (C)

44. (A)

Hyperbola: $9x^2 - 96y^2 = 288$

$$\Rightarrow \frac{x^2}{32} - \frac{y^2}{3} = 1$$

$a = \sqrt{32}$, $b = \sqrt{3}$

Therefore the asymptotes

$$= \pm\frac{b}{a}x = \pm\frac{\sqrt{3}}{\sqrt{32}}x = 0.31x$$

Answer is (A)

45. (B)

Since $y = r\sin\theta$

$$r = \frac{3}{3 - 2\sin\theta} = \frac{3}{3 - 2(\frac{y}{r})} = \frac{3}{3 - \frac{2y}{r}}$$

$$= \frac{3}{\frac{3r - 2y}{r}} = \frac{3r}{3r - 2y}$$

$$3r - 2y = 3$$

$$3r = 2y + 3,\; 3\sqrt{x^2 + y^2} = 2y + 3$$

$$9(x^2 + y^2) = (2y + 3)^2$$

$$9x^2 + 5y^2 - 12y - 9 = 0$$

$$\Rightarrow 9x^2 + 5(y - \frac{6}{5})^2 = \frac{81}{5}$$

$$\frac{9x^2}{\frac{81}{5}} + \frac{5(y - \frac{6}{5})^2}{\frac{81}{5}} = 1$$

$$\Rightarrow \frac{x^2}{(\frac{3}{\sqrt{5}})^2} + \frac{(y - \frac{6}{5})^2}{(\frac{9}{5})^2} = 1$$

It is an ellipse.

Answer is (B)

46. (B)

(1) if $x \geq 2$

$x+2+3x-3-x-1=4$

$3x=6$, $x=2$ (True, but there is no this answer)

(2) if $1 \leq x < 2$

$(x+2)+3(x-1)-(x+1)=4$

$x+2+3x-3-x-1=4$

$3x=6$, $x=2$ (false) $(x<2)$

(3) if $0 \leq x < 1$

$x+2-3(x-1)-(x+1)=4$

$x+2-3x+3-x-1=4$

$3x=0$

$x=0$ (True)

(4) if $-1 \leq x < 0$

$x+2+[-3(x-1)]-x-1=4$

$x+2-3x+3-x-1=4$ (false)

$x=0$

(5) if $-2 \leq x < -1$

$x+2+3[-(x-1)]-[-(x+1)]=4$

$x+2-3x+3+x+1=4$

$x=2$ (false)

(6) $x<-2$

$-(x+2)-3(x-1)+(x+1)=4$

$-x-2-3x+3+x+1-4=0$

$x=-\frac{2}{3}$ (false)

Only (3) stands.

Answer is (B)

47. (B)

$$\lim_{n\to\infty}\frac{2^n-2^{n-1}}{2^n-1}=\lim_{n\to\infty}\frac{\frac{2^n}{2^n}-\frac{1}{2}}{\frac{2^n}{2^n}-\frac{1}{2^n}}$$

$$=\lim_{n\to\infty}\frac{1-\frac{1}{2}}{1-\frac{1}{2^n}}=\frac{\frac{1}{2}}{1}=\frac{1}{2}$$

Answer is (B)

48. (D)

$\frac{18}{x}=\frac{7x+12}{18}$,

$18^2=x\cdot(7x+12)=7x^2+12x$

$\therefore\ 7x^2+12x-18^2=0$

$7x^2+12x-324=0$

$(x-6)(7x+54)=0$

$x=6$, or $x=-\frac{54}{7}$ (false)

$x=6$, $r=\frac{18}{6}=3$

The 7th term $=ar^{7-1}=6\cdot3^6=4,374$

Answer is (D)

49. (C)

Since $\lim_{x\to2}\frac{x^2+x+a}{x-2}=5$,

Numerator must have a factor of $x-2$.

So, $2^2+2+a=0$, $a=6$

$$\lim_{x\to2}\frac{x^2+x-a}{x-2}$$

$$=\lim_{x\to2}\frac{(x^2+x-6)}{x-2}$$

$$=\lim_{x\to2}\frac{(x-2)(x+3)}{x-2}$$

$= \lim_{x \to 2}(x+3) = 2+3 = 5$

Answer is (C)

50. (C)

1. Set point $p(4,1,3)$ lies on the line l_1.

   Because point p satisfies the equation $\frac{x}{2} = \frac{y+1}{1} = \frac{z+1}{2}$

2. Set an arbitrary point $B(x, y, z)$ lies on the ℓ_2.

   So, $x = -t+1$

   $y = t$

   $z = 3t+1$

3. $\overrightarrow{PB} = (-t+1-4, t-1, 3t+1-3)$

   $= (-t-3, t-1, 3t-2)$

4. If $\overrightarrow{PB} \perp \overrightarrow{\ell_2}$, then the point of intersection is the perpendicular point.

   $\overrightarrow{PB} \cdot \overrightarrow{\ell_2}$

   $= (-t-3, t-1, 3t-2) \cdot (-1, 1, 3)$

   $= (-t-3)(-1) + (t-1)(1) + (3t-2)(3)$

   $= t+3+t-1+9t-6$

   $= 11t-4$

   When $\overrightarrow{PB} \perp \overrightarrow{\ell_2}$, $11t-4=0$

   $t = \frac{4}{11}$

   Therefore the perpendicular point at

   $(-\frac{4}{11}+1, \frac{4}{11}, (3)\frac{4}{11}+1)$

   $\Rightarrow (\frac{7}{11}, \frac{4}{11}, \frac{23}{11})$

   The distance between point P and point B is also the distance between two lines.

$$D = \sqrt{(4-\frac{7}{11})^2 + (1-\frac{4}{11})^2 + (3-\frac{23}{11})^2}$$

$$= \sqrt{(\frac{37}{11})^2 + (\frac{7}{11})^2 + (\frac{10}{11})^2}$$

$$= \frac{\sqrt{1518}}{11}$$

$$= 3.54$$

Answer is (C)

CHAPTER 16
MODEL TEST 5
LEVEL 2

DIRECTION:

Fill in the correct or the best answer in the proper position on the answer sheet. You may use any available space for scratch work.

NOTES:

1. Calculator is permitted.
2. All figures are not drawn to scale.
3. All numbers are real, unless stated otherwise.
4. Some reference information:

(1) Right Circular Cone

$V = \frac{1}{3}\pi r^2 \cdot h$, $A(\text{lateral surface}) = \frac{1}{2}c\ell = \pi r \cdot \ell$

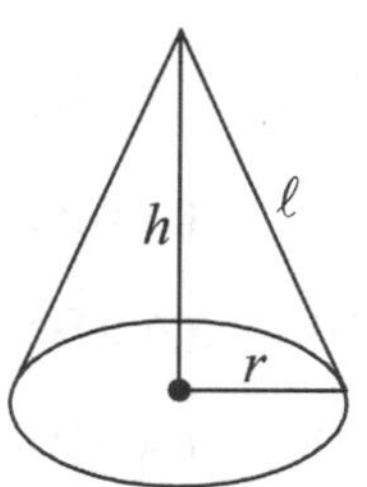

(2) Cylinder

$V = \pi r^2 \cdot h$, $A(\text{surface area}) = 2\pi r \cdot h$

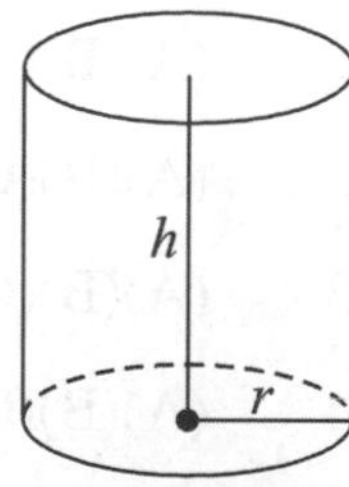

(3) Sphere

$V = \frac{4}{3}\pi r^3 \quad A = 4\pi r^2$

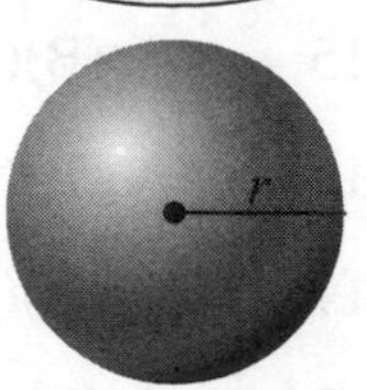

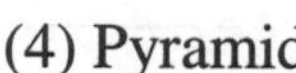
(4) Pyramid

$V = \frac{1}{3}Bh$ (B: area of base)

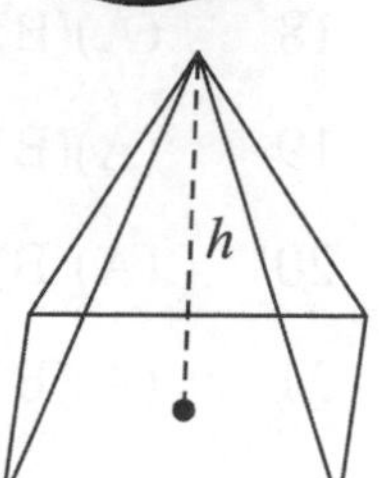

(5) Special Right Triangle

$30° - 60° - 90°$

$45° - 45° - 90°$

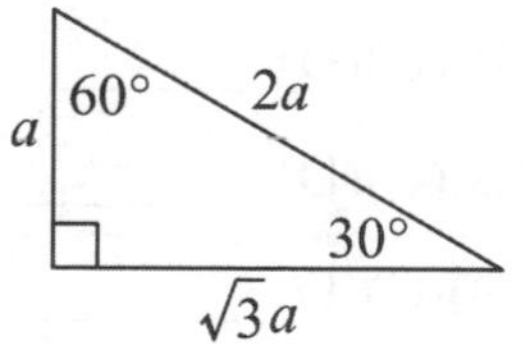

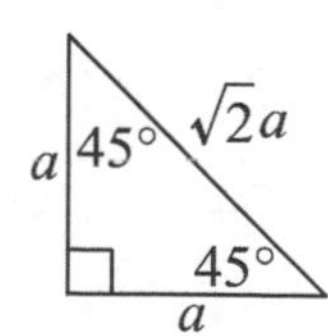

MODEL TEST 5
ANSWER SHEET

Question Number		Your Answers	Answer Key	Question Number		Your Answers	Answer Key
1	(A)(B)(C)(D)			26	(A)(B)(C)(D)		
2	(A)(B)(C)(D)			27	(A)(B)(C)(D)		
3	(A)(B)(C)(D)			28	(A)(B)(C)(D)		
4	(A)(B)(C)(D)			29	(A)(B)(C)(D)		
5	(A)(B)(C)(D)			30	(A)(B)(C)(D)		
6	(A)(B)(C)(D)			31	(A)(B)(C)(D)		
7	(A)(B)(C)(D)			32	(A)(B)(C)(D)		
8	(A)(B)(C)(D)			33	(A)(B)(C)(D)		
9	(A)(B)(C)(D)			34	(A)(B)(C)(D)		
10	(A)(B)(C)(D)			35	(A)(B)(C)(D)		
11	(A)(B)(C)(D)			36	(A)(B)(C)(D)		
12	(A)(B)(C)(D)			37	(A)(B)(C)(D)		
13	(A)(B)(C)(D)			38	(A)(B)(C)(D)		
14	(A)(B)(C)(D)			39	(A)(B)(C)(D)		
15	(A)(B)(C)(D)			40	(A)(B)(C)(D)		
16	(A)(B)(C)(D)			41	(A)(B)(C)(D)		
17	(A)(B)(C)(D)			42	(A)(B)(C)(D)		
18	(A)(B)(C)(D)			43	(A)(B)(C)(D)		
19	(A)(B)(C)(D)			44	(A)(B)(C)(D)		
20	(A)(B)(C)(D)			45	(A)(B)(C)(D)		
21	(A)(B)(C)(D)			46	(A)(B)(C)(D)		
22	(A)(B)(C)(D)			47	(A)(B)(C)(D)		
23	(A)(B)(C)(D)			48	(A)(B)(C)(D)		
24	(A)(B)(C)(D)			49	(A)(B)(C)(D)		
25	(A)(B)(C)(D)			50	(A)(B)(C)(D)		

CHAPTER 16
MODEL TEST 5
LEVEL 2

1. $\sin(\tan^{-1}\frac{1}{2}) =$, $0 < \theta < \frac{\pi}{2}$

(A) 0.32

(B) 0.37

(C) 0.41

(D) 0.45

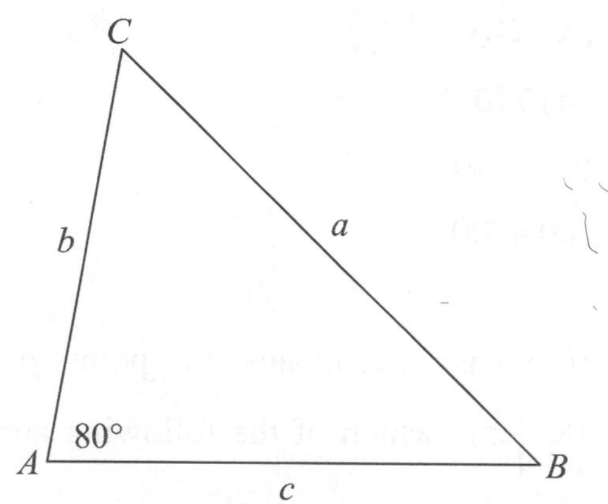

Note: Figure not drawn to scale

2. In the figure above $\angle A = 80°$, $a = 6$, $b = 4$, what is the value of $\angle B =$?

(A) 41°

(B) 45°

(C) 47°

(D) 51°

3. If vector $\overrightarrow{a} = (-1, 2)$ and $\overrightarrow{b} = (2, -3)$, what is the value of $|2\overrightarrow{a} + 3\overrightarrow{b}|$?

(A) $\sqrt{41}$

(B) $2\sqrt{3}$

(C) $\sqrt{52}$

(D) $3\sqrt{2}$

4. Which of the following is an odd function ?

(A) $f(x) = x^2 + 2$

(B) $f(x) = x^3 - 5$

(C) $f(x) = x^2 + \cos x$

(D) $f(x) = x^3 + 2x$

5. Evaluate $(1+i)^{12}$

(A) −64

(B) −32

(C) −12

(D) 12

6. If $f(x) = \sin x$ and $g(x) = 3x - 1$, which of the following could be an even function?

I. $f(x) \cdot g(x)$

II. $f(g(x))$

III. $g(f(x))$

(A) I only

(B) II only

(C) III only

(D) None of them

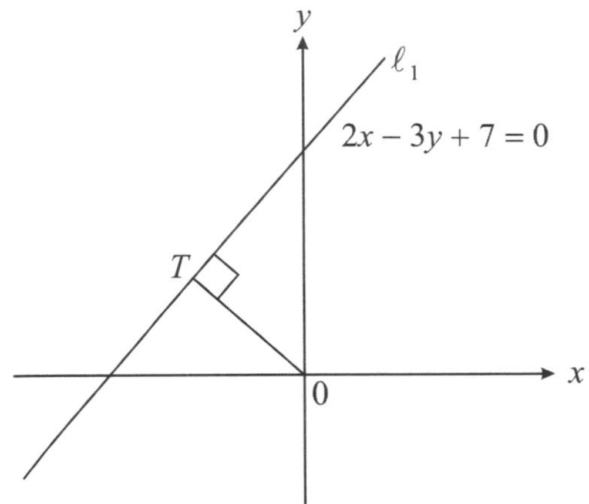

Note: Figure not drawn to scale

7. In the figure above, $\overline{OT}$ is perpendicular to line $2x-3y+7=0$, what is the length of $\overline{OT}$?

(A) 1.94

(B) 3.12

(C) 3.87

(D) 4.28

8. If matrix A has dimension $p\times q$, matrix B has dimension $q\times m$, and matrix C has dimension $m\times n$, which of the following must be true?

(A) AC exist

(B) CA exist

(C) ACB exist

(D) ABC exist

9. For the parametric equations $x=3\sin\frac{\theta}{2}$ and $y=6\cos\frac{\theta}{2}$, which of the following represents the graph of the points ?

(A) Line

(B) Circle

(C) Parabola

(D) Ellipse

10. A team of 5 students is to be selected randomly from 5 boys and 7 girls to represent their school to compete in the state-wide math championship.

If the team has at least one boy and one girl. How many ways to do the selection?

(A) 420

(B) 770

(C) 1440

(D) 4200

11. If polar coordinates of point p are $(8, \frac{3}{4}\pi)$, which of the following are the rectangular coordinates of point p ?

(A) $(-4\sqrt{2}, 4\sqrt{2})$

(B) $(4\sqrt{2}, -4\sqrt{2})$

(C) $(4\sqrt{2}, 4\sqrt{2})$

(D) $(-4\sqrt{2}, -4\sqrt{2})$

12. If $f(x)=3\ln x+2$ and $g(x)=e^x$, what is the value of $f(g(2))$?

(A) 4

(B) 6

(C) 8

(D) 2

13. What is the maximum value of $f(x)=2\cos^2 x+2\sin x+5$?

(A) 5

(B) 5.5

(C) 7

(D) 7.5

14. Which of the following is a horizontal asymptote to the function $f(x)=\dfrac{2x^4-5x^3+7x^2-x+2}{3x^4-2}$?

(A) $-\dfrac{1}{3}$

(B) $\dfrac{2}{3}$

(C) $\dfrac{1}{3}$

(D) $-\dfrac{2}{3}$

15. If $\sin\theta-\cos\theta=\dfrac{1}{3}$, then $\tan\theta+\cot\theta=$

(A) $\dfrac{1}{3}$

(B) $\dfrac{1}{4}$

(C) $\dfrac{9}{4}$

(D) $\dfrac{4}{9}$

16. What is the period of the function $g(x)=-3+\cos(\dfrac{2x-\pi}{2})$

(A) π

(B) 2π

(C) 3π

(D) 4π

17. If the ellipse $x^2-6x+2y^2-4y+6=k$ has a major axis of 8, where k is a costant, what is the value of k ?

(A) 4

(B) 8

(C) 9

(D) 11

18. Which of the following is the equation of the inverse of $f(x)=5^{x+1}$?

(A) $y=\log_5^{(x+1)}$

(B) $y=\log_5^{\frac{x}{5}}$

(C) $y=\log_5^{(x-1)}$

(D) $y=\log_5^{x}$

19. What is the domain of the function defined by $g(x)=\dfrac{\sqrt{2x-8}}{x^2+5x-6}$?

(A) $x\ge 4$ or $x\le -1$

(B) $x\ge 4$

(C) $1\le x\le 4$

(D) $x\ge 4$ and $-6<x<1$

20. If $f(x)=\log_2^{(x+1)}+3$, which of the following is $f^{-1}(x)$?

(A) $f^{-1}(x)=2^{x-3}-1$

(B) $f^{-1}(x)=2^{x-3}+1$

(C) $f^{-1}(x) = 2^{x-1} + 3$

(D) $f^{-1}(x) = 2^{x-1} - 3$

21. $(1+i)^{20} =$

(A) $2^{20} + i$

(B) $-2^{20} - i$

(C) 2^{10}

(D) -2^{10}

22. If the hyperbola $4x^2 - 9y^2 - 8x - 18y - 10 = k$ has a transverse axis of 4, where k is a constant, what is the value of k ?

(A) 5

(B) 7

(C) 9

(D) 11

23. If the graph of the rational function $f(x) = \dfrac{x^2 + ax - 6}{x - 2}$ does not have vertical asymptotes, what could be the value of a ?

(A) -1

(B) 1

(C) -2

(D) 2

24. The equation $xy - x - 3 = 0$ can be expressed by a set of parametric equations. If $y(t) = 3t - 2$, which of the following can be expressed by $x(t)$?

(A) $t - 3$

(B) $-t + 3$

(C) $\dfrac{3t+1}{3t-2}$

(D) $\dfrac{3t-2}{3t+1}$

25. Line L_1 passes through point (2, 1, 1) and is perpendicular to $L_2: \dfrac{x-2}{-1} = \dfrac{y+1}{2} = \dfrac{z-1}{1}$, what are the coordinates of the perpendicular point ?

(A) $A(1, 1, 2)$

(B) $A(-1, 1, 2)$

(C) $A(\frac{4}{3}, \frac{1}{3}, \frac{5}{3})$

(D) $A(-\frac{4}{3}, \frac{1}{3}, \frac{5}{3})$

26. Which of the following is the solution set of $\dfrac{x+1}{x} < 0$?

(A) $x < 0$

(B) $x < 1$

(C) $-1 < x < 0$

(D) $0 < x < 1$

27. If $g(x) = x^2 - 1$, where $x \geq 1$, Find $(gog^{-1})(x) \Rightarrow g(g^{-1}(x))$.

(A) x

(B) $\sqrt{x}$

(C) x^2-1

(D) $\sqrt{x^2+1}$

28. Given that there are 3 black balls, x white balls, and 1 red ball in the bag. Now take two balls from the bag. If the probability of the two balls being the same colour is $\frac{2}{7}$, how many white balls are in the bag ?

(A) 2

(B) 3

(C) 4

(D) 5

29. If $(\frac{1}{3})^{2a}=(\frac{1}{2})^{b}$, what is the value of $\frac{b}{a}$?

(A) 1.86

(B) 2.37

(C) 3.17

(D) 4.23

30. If $\lim\limits_{x\to -1}\frac{x^2+ax+b}{x+1}=3$, where a and b are constants, what is the value of a ?

(A) 2

(B) 3

(C) 4

(D) 5

31. Matrix $A=\begin{bmatrix}1 & 3\\ 2 & 4\end{bmatrix}$ and matrix $B=\begin{bmatrix}a\\ b\end{bmatrix}$. If $AB=\begin{bmatrix}1\\ 0\end{bmatrix}$, which of the following could be the value of a?

(A) -3

(B) -2

(C) -1

(D) 0

32. What is the value of $(1+i)^{60}$?

(A) -2^{30}

(B) 2^{30}

(C) $-20^{30}i$

(D) $20^{30}i$

33. What is the distance between point $p(1,-2,1)$ and line $L:\frac{x-1}{1}=\frac{y+2}{2}=\frac{z+1}{-1}$?

(A) 1.12

(B) 1.83

(C) 3.13

(D) 4.25

34. Which of the following is the smallest positive x-intercept for the graph of $y=4\sin(3x+\frac{2}{3}\pi)$?

(A) $\frac{1}{9}\pi$

(B) $\frac{1}{4}\pi$

(C) $\frac{1}{3}\pi$

(D) $\frac{2}{3}\pi$

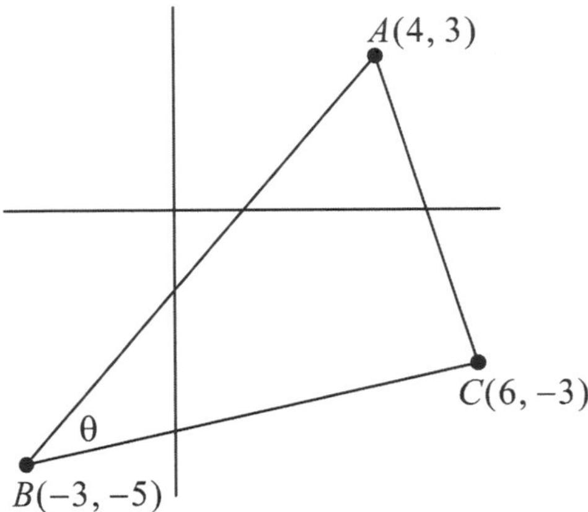

Note: Figure not drawn to scale

35. For the triangle shown above, what is the area of $\triangle ABC$?

(A) 29.14

(B) 43.22

(C) 50.18

(D) 58.34

36. If $f(x)=3$ for all real numbers x, then $f(x+2)+f(x-2)=$

(A) 0

(B) 3

(C) 4

(D) 6

37. Solve $\begin{cases} 3x-2y+z=3 \\ 2x+3y+5z=-11 \\ 2x-5y+3z=1 \end{cases}$

(A) $x=2$, $y=1$, $z=-1$

(B) $x=-1$, $y=1$, $z=2$

(C) $x=1$, $y=-1$, $z=-2$

(D) $x=-2$, $y=1$, $z=-1$

38. If $A=\begin{bmatrix} 2 & 1 \\ 0 & 3 \end{bmatrix}$, $B=\begin{bmatrix} -1 & 1 \\ 0 & 2 \end{bmatrix}$, What is the value of $(A+B)^{-1}$?

(A) $\frac{1}{5}\begin{bmatrix} 2 & 0 \\ -1 & 3 \end{bmatrix}$

(B) $\frac{1}{5}\begin{bmatrix} 1 & 2 \\ 0 & 5 \end{bmatrix}$

(C) $\frac{1}{5}\begin{bmatrix} 5 & -2 \\ 0 & 1 \end{bmatrix}$

(D) $\frac{1}{5}\begin{bmatrix} 3 & -1 \\ 0 & 2 \end{bmatrix}$

39. If $|px+1|\le q$, and $-4\le x\le 8$, what is the value of $p+q$?

(A) $\frac{5}{2}$

(B) $\frac{5}{3}$

(C) $\frac{3}{5}$

(D) $\frac{2}{5}$

40. What is the maximum value of $y=8\sin\theta\cos\theta$?

(A) 4

(B) 6

(C) 8

(D) 16

41. Throw 8 coins. What is the probability that eractly come up 4 heads ?

(A) 27.3%

(B) 31.8%

(C) 38.1%

(D) 44.2%

42. If $BC=\begin{bmatrix}3 & 2\\3 & 1\end{bmatrix}$, $BD=\begin{bmatrix}0 & -1\\3 & -2\end{bmatrix}$ and $C+D=\begin{bmatrix}-3 & 1\\3 & 0\end{bmatrix}$, which of the following is the matrix for C ?

(A) $\begin{bmatrix}1 & 2\\-1 & 1\end{bmatrix}$

(B) $\begin{bmatrix}-2 & 1\\1 & -1\end{bmatrix}$

(C) $\begin{bmatrix}-1 & 0\\2 & 1\end{bmatrix}$

(D) $\begin{bmatrix}-1 & 2\\1 & 0\end{bmatrix}$

43. 8, s, t, 24

In the list above s and t are positive numbers. The first three numbers form a geometric sequence, and the last three numbers form an arithmetic sequence. What is the value of $s+t$?

(A) 26

(B) 30

(C) 34

(D) 40

44. What is the y-coordinate of one focus of the ellipse $9x^2-36x+16y^2+64y=44$

(A) 2.65

(B) 3.18

(C) 4.65

(D) 5.32

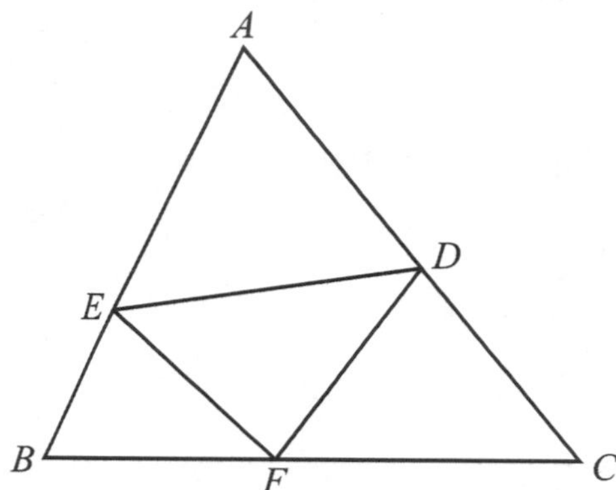

Note: Figure not drawn to scale

45. In the figure above, $\triangle ABC$ is an equilateral triangle, E, D lie on $\overline{AB}$ and $\overline{AC}$, respectively, if $AB=14$ and the area of parallelogram $AEFD$ is $24\sqrt{3}$, what is the length of $\overline{ED}$?

(A) $4\sqrt{3}$

(B) $3\sqrt{5}$

(C) $2\sqrt{13}$

(D) $5\sqrt{2}$

46. If $3.25^x=4.05^y$, what is the value of $\frac{x}{y}$?

(A) 1.19

(B) 2.16

(C) 2.89

(D) 3.25

47. If $P(A \cap B) = \frac{1}{4}$, $P(B') = 0.5$, $P(A \cup B) = \frac{1}{2}$, what is $P(A)$? (The probability of event A)

(A) $\frac{1}{4}$

(B) $\frac{1}{3}$

(C) $\frac{1}{2}$

(D) $\frac{2}{3}$

48. If $2\sin^2\theta - 5\cos^2\theta = 3\sin\theta\cos\theta$, $\frac{\pi}{2} < \theta < \pi$, then $\sin 2\theta + \cos 2\theta =$

(hint: $\sin 2\theta = \frac{2\tan\theta}{1+\tan^2\theta}$, $\cos 2\theta = \frac{1-\tan^2\theta}{1+\tan^2\theta}$)

(A) 1

(B) -1

(C) $\frac{2}{5}$

(D) $\frac{2}{3}$

49. If z is a complex number, and $z\bar{z} + 2(z+\bar{z}) - 12 = 0$, which of the following is the graph of the equation ?

(A) point

(B) straight line

(C) circle

(D) ellipse

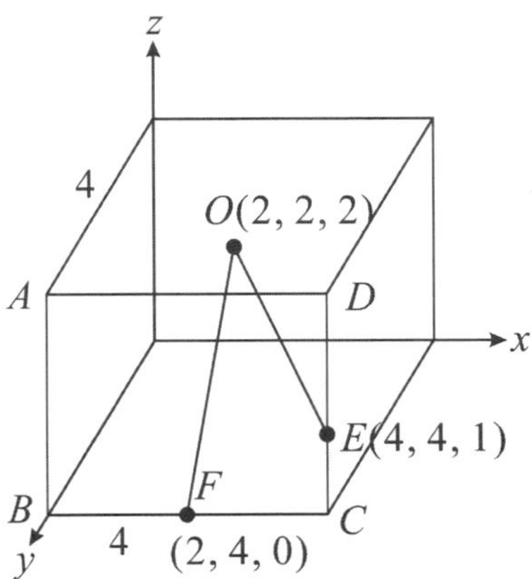

Note: Figure not drawn to scale

50. The figure above shows a cube and $\square ABCD$ is a face of the cube. O is the center of the cube with coordinates $(2, 2, 2)$, $E(4, 4, 1)$, and $F(2, 4, 0)$ lie on $\overline{DC}$ and $\overline{BC}$ respectively, what is the value of $\cos\angle EOF$?

(Round to Hundredth)

(A) 0.71

(B) 0.62

(C) 0.53

(D) 0.44

MODEL TEST 5
ANSWER KEY

1. (D)
2. (A)
3. (A)
4. (D)
5. (A)
6. (D)
7. (A)
8. (D)
9. (D)
10. (B)
11. (A)
12. (C)
13. (D)
14. (B)
15. (C)
16. (B)
17. (D)
18. (B)
19. (B)
20. (A)
21. (D)
22. (D)
23. (B)
24. (C)
25. (C)
26. (C)
27. (A)
28. (B)
29. (C)
30. (D)
31. (B)
32. (A)
33. (B)
34. (A)
35. (A)
36. (D)
37. (C)
38. (C)
39. (A)
40. (A)
41. (A)
42. (C)
43. (B)
44. (C)
45. (C)
46. (A)
47. (A)
48. (B)
49. (C)
50. (A)

MODEL TEST 5
ANSWERS and EXPLANATIONS

1. (D)

Assume $\tan\theta = \frac{1}{2}$

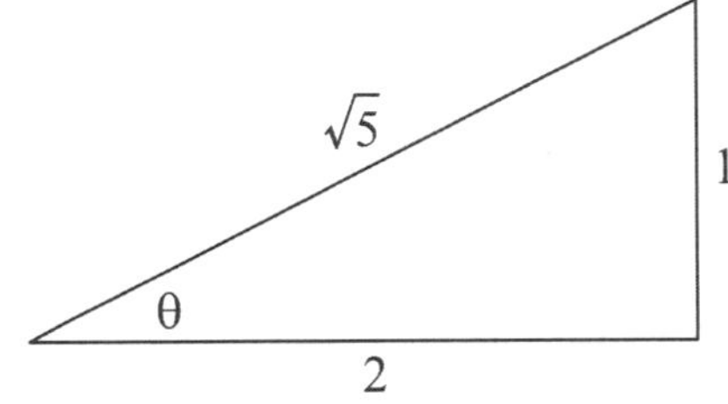

$\sin\theta = \frac{1}{\sqrt{5}} = \frac{\sqrt{5}}{5} = 0.45$

Answer is (D)

2. (A)

$\frac{a}{\sin 80°} = \frac{b}{\sin B}$

$\frac{6}{\sin 80°} = \frac{4}{\sin B}$

$\therefore\ \sin B = \frac{4}{6} \times \sin 80° = \frac{4}{6} \times 0.98 = 0.65$

$\angle B = \sin^{-1} 0.65 \doteq 41°$

Answer is (A)

3. (A)

$\overrightarrow{a} = (-1, 2)$, $2\overrightarrow{a} = (-2, 4)$

$\overrightarrow{b} = (2, -3)$, $3\overrightarrow{b} = (6, -9)$

$|2\overrightarrow{a} + 3\overrightarrow{b}| = |(-2+6), (4-9)|$

$= |4, -5| = \sqrt{4^2 + (-5)^2}$

$= \sqrt{16+25} = \sqrt{41}$

Answer is (A)

4. (D)

A. $f(x) = x^2 + 2$, $f(-x) = (-x)^2 + 2$

$= x^2 + 2$

$\therefore\ f(x) = f(-x)$ even function

B. $f(x) = x^3 - 5$, $f(-x) = (-x)^3 - 5$

$= -x^3 - 5 = -(x^3 + 5)$

$f(x) \neq f(-x)$, $f(x) \neq -f(-x)$

neither even nor odd function

C. $f(x) = x^2 + \cos x$, $f(-x) = (-x)^2$

$+ \cos(-x) = x^2 + \cos x$

$f(x) = f(-x)$, even function

D. $f(x) = x^3 + 2x$, $f(-x) = (-x)^3$

$+2(-x) = -x^3 - 2x = -(x^3 + 2x)$

$\therefore\ f(-x) = -f(x)$ odd function.

Answer is (D)

5. (A)

Since $(1+i)^2 = 1 + 2i + i^2 = 2i$

So, $(1+i)^{12} = [(1+i)^2]^6 = (2i)^6 = 2^6 i^6$

$= 64 \cdot (-1) = -64$

Answer is (A)

6. (D)

$f(x) = \sin x$ and $g(x) = 3x - 1$

I. $f(x) \cdot g(x) = (\sin x)(3x-1)$

$= 3x \cdot \sin x - \sin x$

$f(-x) \cdot g(-x)$

$= 3(-x) \cdot \sin(-x) - \sin(-x)$

$= -3(x)(-\sin x) + \sin x$

$= 3x \cdot \sin x + \sin x$

Not an even function

II. $f(g(x)) = \sin(3x-1)$

$f(g(-x)) = \sin[3(-x)-1]$

$= \sin(-3x-1)$

Not even function

III. $g(f(x)) = 3(\sin x) - 1$

$g(f(-x)) = 3(\sin(-x)] - 1$

$= 3(-\sin x) - 1$

$= -3\sin x - 1$

Not even function

Answer is (D)

7. (A)

$2x - 3y + 7 = 0$

$3y = 2x + 7$, $y = \frac{2}{3}x + \frac{7}{3}$

$m_1 = \frac{2}{3}$

$\overline{OT} \perp l_1$, $m_2 \times \frac{2}{3} = -1$

$\therefore\ m_2 = -\frac{3}{2}$

The function of $l_2 : y = m_2x + b$

$y = -\frac{3}{2}x + b$, since it passes origin,

$b = 0$

$y = -\frac{3}{2}x$

Since the y-coodinate is the same.

$\frac{2}{3}x + \frac{7}{3} = -\frac{3}{2}x$, $\frac{2}{3}x + \frac{3}{2}x = -\frac{7}{3}$

$\frac{13}{6}x = -\frac{7}{3}$

$x = -\frac{14}{13}$

$y = (-\frac{3}{2}) \times (-\frac{14}{13}) = \frac{42}{26} = \frac{21}{13}$

$OT = \sqrt{(-\frac{14}{13})^2 + (\frac{21}{13})^2} = 1.94$

Answer is (A)

8. (D)

$A_{p\times q}$, $B_{q\times m}$, $C_{m\times n}$

(A) AC $\quad q \neq m$ not exist

(B) CA $\quad n \neq p$ not exist

(C) ACB $\quad q \neq m$ not exist

(D) ABC $\quad q = q$, $m = m$ exist

Answer is (D)

9. (D)

$x = 3\sin\frac{\theta}{2}$, $\sin\frac{\theta}{2} = \frac{x}{3}$, $y = 6\cos\frac{\theta}{2}$,

$\cos\frac{\theta}{2} = \frac{y}{6}$

Since $\sin^2\frac{\theta}{2} + \cos^2\frac{\theta}{2} = 1 = (\frac{x}{3})^2 + (\frac{y}{6})^2$

$\frac{x^2}{9} + \frac{y^2}{36} = 1$

It is the graph of ellipse

Answer is (D)

10. (B)

boys 5 girls 7 → 5

$C_1^5 \cdot C_4^7 + C_2^5 \cdot C_3^7 + C_3^5 \cdot C_2^7 + C_4^5 \cdot C_1^7$

$$=5\times\frac{7\cdot6\cdot5\cdot4!}{3!\cdot4!}+\frac{5\cdot4\cdot3!}{3!\cdot2!}$$

$$\times\frac{7\cdot6\cdot5\cdot4!}{4!\cdot3!}+\frac{5\cdot4\cdot3!}{2!\cdot3!}$$

$$\times\frac{7\cdot6\cdot5!}{5!\cdot2!}+\frac{5\cdot4!}{1!\cdot4!}\times7$$

$$=5\times35+10\times35+10\times21+5\times7$$

$$=770$$

Answer is (B)

11. (A)

$$x=r\cos\theta=8\cos\frac{3}{4}\pi$$

$$=8\cdot(-\frac{\sqrt{2}}{2})=-4\sqrt{2},$$

$$y=r\sin\theta=8\sin\frac{3}{4}\pi=8\cdot(\frac{\sqrt{2}}{2})=4\sqrt{2}$$

Answer is (A)

12. (C)

$f(x)=3\ln^{x}+2,\quad g(x)=e^{x}$

$f(g(x))=3\ln^{e^{x}}+2$

$f(g(2))=3\times2\ln^{e}+2=3\times2+2=8$

Answer is (C)

13. (D)

$$2\cos^2x+2\sin x+5$$

$$=2(1-\sin^2x)+2\sin x+5$$

$$=2-2\sin^2x+2\sin x+5$$

$$=-2(\sin^2x-\sin x)+7$$

$$=-2[(\sin^2x-\sin x+\frac{1}{4})-\frac{1}{4}]+7$$

$$=-2(\sin x-\frac{1}{2})^2+\frac{15}{2}$$

When $\sin x=\frac{1}{2}$, $y=\frac{15}{2}$ is the maximum value

Answer is (D)

14. (B)

$$f(x)=\frac{2x^4-5x^3+7x^2-x+2}{3x^4-2}$$

Horizontal asymptote

$$\lim_{x\to\infty}\frac{2x^4-5x^3+7x^2-x+2}{3x^4-2}$$

$$=\lim_{x\to\infty}\frac{2-5\frac{1}{x}+7\frac{1}{x^2}-\frac{1}{x^3}+\frac{2}{x^4}}{3-\frac{2}{x^4}}$$

$=\frac{2}{3}$ Horizontal asymptote

$$y=\frac{2}{3}$$

15. (C)

$$\sin\theta-\cos\theta=\frac{1}{3}$$

$$\frac{\sin\theta}{\cos\theta}-\frac{\cos\theta}{\cos\theta}=\frac{1}{3\cos\theta}$$

$$\tan\theta-1=\frac{1}{3\cos\theta}$$

$$\tan\theta+\cot\theta$$

$$=\frac{\sin\theta}{\cos\theta}+\frac{\cos\theta}{\sin\theta}$$

$$=\frac{\sin^2\theta+\cos^2\theta}{\sin\theta\cos\theta}$$

$$=\frac{1}{\sin\theta\cos\theta}$$

Since $(\sin\theta-\cos\theta)^2=\sin^2\theta+\cos^2\theta$

$-2\sin\theta\cos\theta=1-2\sin\theta\cos\theta=(\frac{1}{3})^2$

$$2\sin\theta\cos\theta = 1-\frac{1}{9} = \frac{8}{9}$$

$$\sin\theta\cos\theta = \frac{4}{9}$$

$$\therefore\ \tan\theta+\cot\theta = \frac{1}{\sin\theta\cos\theta} = \frac{1}{\frac{4}{9}} = \frac{9}{4}$$

Answer is (C)

16. (B)

$$g(x) = -3+\cos(\frac{2x-\pi}{2})$$

$$= -3+\cos(x-\frac{\pi}{2})$$

The period of $\cos$ is 2π

$\therefore\ \frac{2\pi}{1} = 2\pi$ is the period of $g(x)$

Answer is (B)

17. (D)

$$x^2-6x+2y^2-4y+6 = k$$

$$\Rightarrow\ x^2-6x+9-9+2(y^2-2y+1)$$

$$-2+6 = k$$

$$\Rightarrow\ (x-3)^2+2(y-1)^2-5 = k$$

$$\Rightarrow\ (x-3)^2+2(y-1)^2 = k+5$$

$$\Rightarrow\ \frac{(x-3)^2}{k+5}+\frac{(y-1)^2}{\frac{k+5}{2}} = 1$$

So, $a^2 = k+5 = (\frac{8}{2})^2$

$\therefore\ k = 16-5 = 11$

Answer is (D)

18. (B)

Let $y = 5^{x+1}\ \Rightarrow\ \log_5^y = x+1$

$\Rightarrow\ \log_5^y - 1 = x$

Exchange x and y

$$y = \log_5^x - 1 = \log_5^x - \log_5^5 = \log_5^{\frac{x}{5}}$$

Answer is (B)

19. (B)

$$\text{domain of } g(x) = \frac{\sqrt{2x-8}}{x^2+5x-6} = \frac{\sqrt{2x-8}}{(x+6)(x-1)}$$

(1) For denominator

$x+6 \neq 0$, $x \neq -6$

$x-1 \neq 0$, $x \neq 1$

(2) For numerator

$\sqrt{2x-8} \geq 0$, $2x \geq 8$, $x \geq 4$

(3) Domain $x = \{x : x \geq 4\}$

Answer is (B)

20. (A)

$$y = f(x) = \log_2^{(x+1)}+3$$

$$y-3 = \log_2^{(x+1)}$$

$$2^{y-3} = x+1$$

$x = 2^{y-3}-1$ Exchange x and y

Then $y = 2^{x-3}-1\ \Rightarrow\ f^{-1}(x)$

Answer is (A)

21. (D)

$$(1+i)^{20} = (\sqrt{2}(\frac{1}{\sqrt{2}}+i\frac{1}{\sqrt{2}})]^{20}$$

$$= [\sqrt{2}(\cos\frac{\pi}{4}+i\sin\frac{\pi}{4})]^{20}$$

(Since $\sin\theta$ and $\cos\theta$ are

positive, $0 \le \theta \le \frac{\pi}{2}$)

(De moivre's formula)

$= (\sqrt{2})^{20}(\cos\frac{20\pi}{4} + i\sin\frac{20\pi}{4})$

$= 2^{10}(\cos 5\pi + i\sin 5\pi)$

$= 2^{10}(\cos\pi + i\sin\pi)$

$= 2^{10}(-1 + 0i)$

$= -2^{10}$

other solution:

$(1+i)^{20} = [(1+i)^2]^{10} = (1+2i+i^2)^{10}$

$= (1+2i-1)^{10}$

$= (2i)^{10} = 2^{10} \cdot i^{10}$

$= 2^{10} \cdot (-1) = -2^{10}$

Answer is (D)

22. (D)

$4x^2 - 9y^2 - 8x - 18y - 10 = k$

$\Rightarrow 4(x^2 - 2x + 1) - 4 - 9(y^2 + 2y + 1)$

$+9 - 10 = k$

$\Rightarrow 4(x-1)^2 - 9(y+1)^2 + 5 - 10 = k$

$\Rightarrow 4(x-1)^2 - 9(y+1)^2 = k + 5$

$\Rightarrow \frac{(x-1)^2}{\frac{k+5}{4}} - \frac{(y+1)^2}{\frac{k+5}{9}} = 1$

Since $a^2 = \frac{k+5}{4}$

$\therefore a = \frac{\sqrt{k+5}}{2}$

The length of transverse axis is $2a = 4$

Therefore, $\sqrt{k+5} = 4$, $k + 5 = 16$,

$k = 11$

Answer is (D)

23. (B)

If $f(x)$ does not have vertical asymptote, $x^2 + ax - 6$ must have the factor of $x - 2$

So, $(2)^2 + 2a - 6 = 0$

$4 + 2a - 6 = 0$

$2a = 2$

$a = 1$

Answer is (B)

24. (C)

$xy - x - 3 = 0$, since $y = 3t - 2$

So, $x(3t-2) - (3t-2) - 3 = 0$

$3xt - 2x - 3t + 2 - 3 = 0$

$x(3t-2) - 3t - 1 = 0$

$x(3t-2) = 3t + 1$

$x = \frac{3t+1}{3t-2}$

Answer is (C)

25. (C)

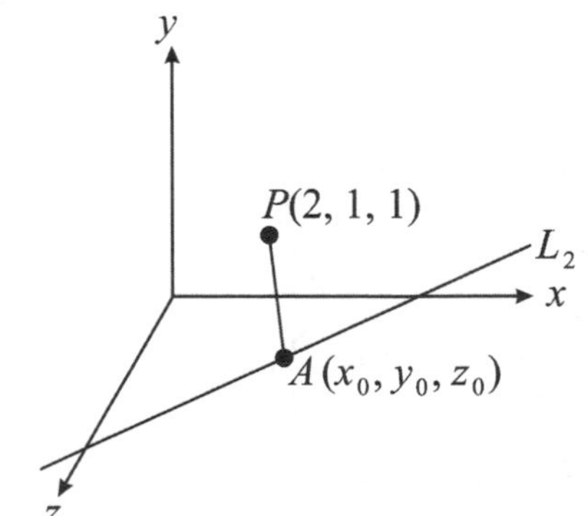

Set the perpendicular point is $A(x_0, y_0, z_0)$

$\overrightarrow{PA} = (x_0 - 2, y_0 - 1, z_0 - 1)$

Since point A lies on the line

$\frac{x-2}{-1}=\frac{y+1}{2}=\frac{z-1}{1}$

So, $x_0=-t+2$, $y_0=2t-1$, and

$z_0=t+1$

$\overrightarrow{PA}=(-t+2-2, 2t-1-1, t+1-1)$

$=(-t, 2t-2, t)$

Since $\overrightarrow{PA}\perp L_2$,

so, $\overrightarrow{PA}\cdot\overrightarrow{L_2}$

$=[(-t)\cdot(-1)+(2t-2)\cdot 2+(t)(1)]$

$=t+4t-4+t=0$

$6t=4$, $t=\frac{2}{3}$

$x_0=-\frac{2}{3}+2=\frac{4}{3}$, $y_0=2\cdot\frac{2}{3}-1=\frac{1}{3}$,

$z_0=\frac{2}{3}+1=\frac{5}{3}$

$A(\frac{4}{3},\frac{1}{3},\frac{5}{3})$

26. (C)

$\frac{x+1}{x}<0$

(1) When $x<-1$, $\frac{x+1}{x}>0$ (false)

(2) When $-1<x<0$, $\frac{x+1}{x}<0$ (true)

(3) When $0<x<1$, $\frac{x+1}{x}>0$ (false)

(4) When $1<x$, $\frac{x+1}{x}>0$ (false)

So, only (2) get $\frac{x+1}{x}<0$

(C) is the correct answer

Other solution: Multiply x^2 to both sides, $x^2+x<0$, $-1<x<0$.

Answer is (C)

27. (A)

$g(x)=x^2-1$

$y=x^2-1$, $x^2=y+1$, $x=\sqrt{y+1}$

Exchange $y=\sqrt{x+1}=g^{-1}(x)$

$g(g^{-1}(x))=(\sqrt{x+1})^2-1=x+1-1=x$

Since $f(f^{-1}(x))$ is always equivalent to x.

Answer is (A)

28. (B)

B W R

3 x 1

The probability of 2 balls being the same colour

$=\frac{C_2^3+C_2^x}{C_2^{3+x+1}}$

$=\frac{\frac{3\cdot 2!}{(3-2)!2!}+\frac{x(x-1)(x-2)!}{(x-2)!\cdot 2!}}{\frac{(x+4)(x+3)(x+4-2)!}{(x+4-2)!2!}}$

$=\frac{\frac{6+x^2-x}{2}}{\frac{(x+4)(x+3)}{2}}=\frac{x^2-x+6}{x^2+7x+12}=\frac{2}{7}$

$7x^2-7x+42=2x^2+14x+24$

$5x^2-21x+18=0$

$(5x-6)(x-3)=0$

$x=\frac{6}{5}$ (not integer), $x=3$

There are 3 white balls in the bag.

Answer is (B)

29. (C)

$$\left(\frac{1}{3}\right)^{2a} = \left(\frac{1}{2}\right)^{b}$$

$$\left(\frac{1}{3^2}\right)^{a} = \left(\frac{1}{2}\right)^{b}$$

$$\left(\frac{1}{9}\right)^{a} = \left(\frac{1}{2}\right)^{b}$$

$$\left(\frac{1}{9}\right) = \left(\frac{1}{2}\right)^{\frac{b}{a}}$$

$$\log^{\frac{1}{9}} = \frac{b}{a}\log^{\frac{1}{2}}$$

$$\frac{b}{a} = \frac{\log\frac{1}{9}}{\log\frac{1}{2}} = 3.17$$

Answer is (C)

30. (D)

When $x \to -1$, $f(x) = \dfrac{x^2+ax+b}{x+1} \to$ undefined unless x^2+ax+b has the factor of $x+1$. Since $\lim\limits_{x\to -1} = \dfrac{x^2+ax+b}{x+1} = 3$, it means it has a finite limit value of 3. So that when $x=-1$.

$(-1)^2 - a + b = 0$, $a = b+1$

$$x^2+ax+b$$
$$= x^2+(b+1)x+b$$
$$= x^2+bx+x+b$$
$$= x(x+1)+b(x+1)$$
$$= (x+1)(x+b)$$

$$\lim_{x\to -1}\frac{x^2+ax+b}{x+1} = \lim_{x\to -1}\frac{\cancel{(x+1)}(x+b)}{\cancel{x+1}}$$

$$= -1+b = 3$$

$\therefore\ b=4,\ a=5$

Answer is (D)

31. (B)

$$A = \begin{bmatrix} 1 & 3 \\ 2 & 4 \end{bmatrix},\quad B = \begin{bmatrix} a \\ b \end{bmatrix},\quad AB = \begin{bmatrix} 1 \\ 0 \end{bmatrix}$$

$$AB = \begin{bmatrix} 1 & 3 \\ 2 & 4 \end{bmatrix}\begin{bmatrix} a \\ b \end{bmatrix} = \begin{bmatrix} a+3b \\ 2a+4b \end{bmatrix} = \begin{bmatrix} 1 \\ 0 \end{bmatrix}$$

$\therefore\ a+3b=1,\ 2a+4b=0$

$\therefore\ b=1,\ a=-2$

Answer is (B)

32. (A)

$\sqrt{2}$ 1 Q 1

$r = \sqrt{1^2+1^2} = \sqrt{2}$, $a=1$, $b=1$, so, in quadrant I, $a+bi = r(\cos\theta + i\sin\theta)$

So, $(1+i)^{60} = (\sqrt{2})^{60}\left(\cos\frac{\pi}{4} + i\sin\frac{\pi}{4}\right)^{60}$

$$= (\sqrt{2})^{60}\left(\cos\frac{60\pi}{4} + i\sin\frac{60\pi}{4}\right)$$
$$= (\sqrt{2})^{60}(\cos 15\pi + i\sin 15\pi)$$
$$= (\sqrt{2})^{60}(\cos\pi + i\sin\pi)$$
$$= (\sqrt{2})^{60}(-1 + i\times 0) = -2^{30}$$

Other solution:

$(1+i)^2 = (1+2i+i^2) = 2i$

$(1+i)^{60} = [(1+i)^2]^{30} = (2i)^{30} = (2i)^{28+2}$

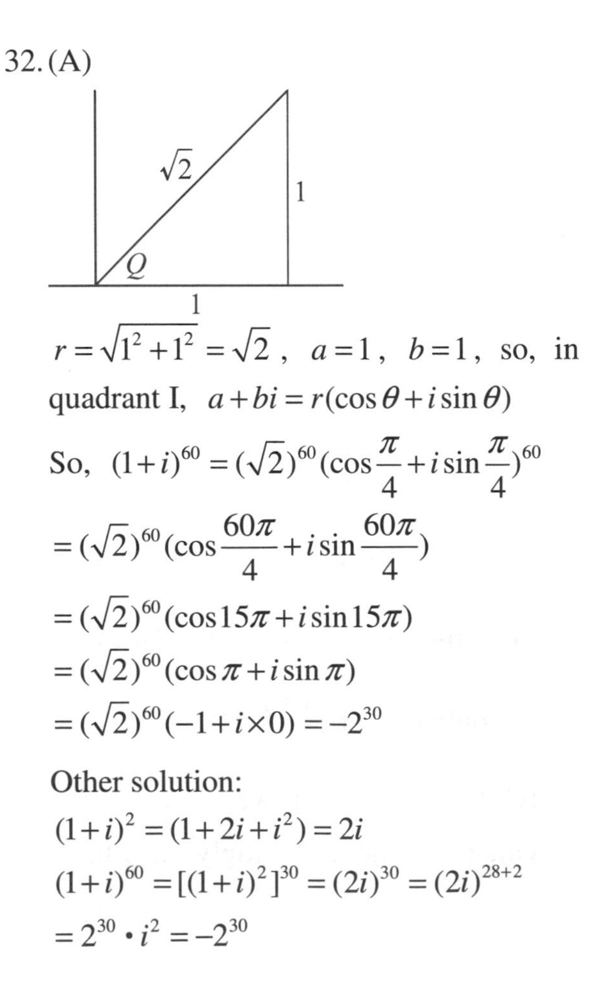

$= 2^{30}\bullet i^2 = -2^{30}$

33. (B)

Set an arbitrary point $B(x, y, z)$ to lie on the straight line $\frac{x-1}{1}=\frac{y+2}{2}=\frac{z+1}{-1}$

$x=t+1,\ y=2t-2,\ z=-t-1$

$\overrightarrow{PB}=(t+1-1, 2t-2+2, -t-1-1)$

$=(t, 2t, -t-2)$

$\overrightarrow{PB}\cdot\overrightarrow{l}=t+4t+t+2=t+4t+t+2$

$=6t+2=0,\ t=-\frac{1}{3}$

$B: x=\frac{-1}{3}+1=\frac{2}{3}$,

$y=2\cdot(-\frac{1}{3})-2=-\frac{2}{3}-2=-\frac{8}{3}$,

$z=\frac{1}{3}-1=-\frac{2}{3}$

$D=\sqrt{(1-\frac{2}{3})^2+(-2+\frac{8}{3})^2+(1+\frac{2}{3})^2}$

$=\sqrt{(\frac{1}{3})^2+(\frac{2}{3})^2+(\frac{5}{3})^2}$

$=\frac{1}{3}\sqrt{1+4+25}=\frac{1}{3}\sqrt{30}\doteq 1.83$

Answer is (B)

34. (A)

When $y=0$, the graph of sin curve yields x-intercept. So that, $3x+\frac{2}{3}\pi=0,\ \pi,\ \cdots$

When $3x+\frac{2}{3}\pi=0,\ x=-\frac{2}{9}\pi$ (negative)

When $3x+\frac{2}{3}\pi=\pi,\ x=\frac{\pi-\frac{2}{3}\pi}{3}=\frac{1}{9}\pi$

So, the smallest positive x-intercept is at

$x=\frac{\pi}{9}$

Answer is (A)

35. (A)

Let $\overrightarrow{BA}=\overrightarrow{a}$, and $\overrightarrow{BC}=\overrightarrow{b}$

$\triangle ABC$

$=\frac{1}{2}|\overrightarrow{a}||\overrightarrow{b}|\sin\theta$

$=\frac{1}{2}|\overrightarrow{a}||\overrightarrow{b}|\sqrt{1-\cos^2\theta}$

$=\frac{1}{2}\sqrt{|\overrightarrow{a}|^2|\overrightarrow{b}|^2-|\overrightarrow{a}|^2|\overrightarrow{b}|^2\cos^2\theta}$

$=\frac{1}{2}\sqrt{|\overrightarrow{a}|^2|\overrightarrow{b}|^2-(\overrightarrow{a}\cdot\overrightarrow{b})^2}$

$\overrightarrow{a}=(4-(-3), 3-(-5))=(7,8)$,

$\overrightarrow{b}=(6-(-3), -3-(-5))=(9, 2)$

$|\overrightarrow{a}|^2=(\sqrt{7^2+8^2})^2=113$,

$|\overrightarrow{b}|^2=(\sqrt{9^2+2^2})^2=85$

$(\overrightarrow{a}\cdot\overrightarrow{b})^2=(63+16)^2=6241$

$\therefore\ \triangle ABC=\frac{1}{2}\sqrt{113\times 85-6241}$

$=29$

Answer is (A)

36. (D)

$f(x)=3$, it means that $y=f(x)$ is always 3. The graph of $y=3$ is a straight line. Therefore, $f(x-2)$ $=f(x+2)=3$

$f(x-2)+f(x+2)=3+3=6$

Answer is (D)

37. (C)

Use Cramer's Rule.

$$D=\begin{vmatrix} 3 & -2 & 1 \\ 2 & 3 & 5 \\ 2 & -5 & 3 \end{vmatrix}$$

$\Rightarrow$ 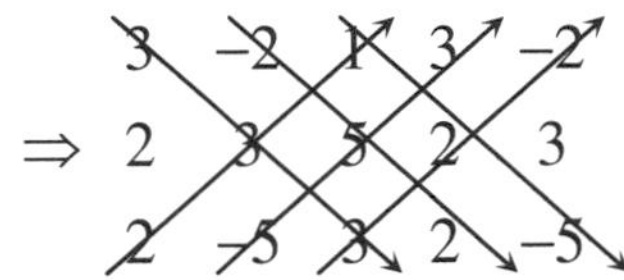

$=27-20-10-6+75+12=78$

$$D_1=\begin{vmatrix} 3 & -2 & 1 \\ -11 & 3 & 5 \\ 1 & -5 & 3 \end{vmatrix}$$

$\Rightarrow$ 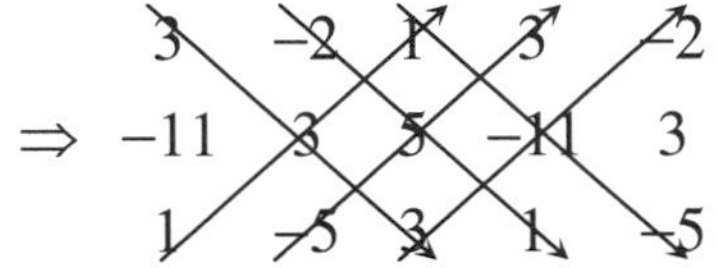

$=27-10+55-3+75-66=78$

$$D_2=\begin{vmatrix} 3 & 3 & 1 \\ 2 & -11 & 5 \\ 2 & 1 & 3 \end{vmatrix}$$

$\Rightarrow$ 

$=-99+30+2+22-15-18=-78$

$$D_3=\begin{vmatrix} 3 & -2 & 3 \\ 2 & 3 & -11 \\ 2 & -5 & 1 \end{vmatrix}$$

$\Rightarrow$ 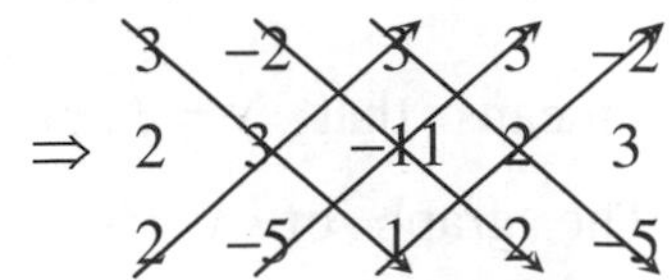

$=9+44-30-18-165+4=-156$

$$x=\frac{D_1}{D}=\frac{78}{78}=1$$

$$y=\frac{D_2}{D}=\frac{-78}{78}=-1$$

$$z=\frac{D_3}{D}=\frac{-156}{78}=-2$$

Answer is (C)

38. (C)

$$A+B=\begin{bmatrix} 2 & 1 \\ 0 & 3 \end{bmatrix}+\begin{bmatrix} -1 & 1 \\ 0 & 2 \end{bmatrix}$$

$$=\begin{bmatrix} 2-1 & 1+1 \\ 0+0 & 3+2 \end{bmatrix}=\begin{bmatrix} 1 & 2 \\ 0 & 5 \end{bmatrix}$$

$$\therefore\ (A+B)^{-1}=\frac{1}{5-0}\begin{bmatrix} 5 & -2 \\ 0 & 1 \end{bmatrix}$$

$$=\frac{1}{5}\begin{bmatrix} 5 & -2 \\ 0 & 1 \end{bmatrix}$$

Answer is (C)

39. (A)

If $-4\le x\le 8$, then the midpoint is

$\dfrac{8-(-4)}{2}=6$, $8-6=2$

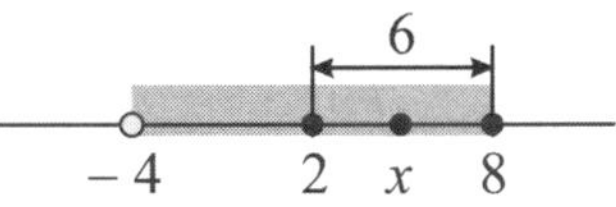

$|x-2|\le 6$ change it to the form of $|px+1|$

$|x-2|\le 6$

$|-\frac{1}{2}|\times|x-2|\le|-\frac{1}{2}|\times 6$

$|-\frac{1}{2}x+1|\le 3$

so, $p=-\frac{1}{2}$, $q=3$

$p+q=\frac{5}{2}$

Answer is (A)

40. (A)

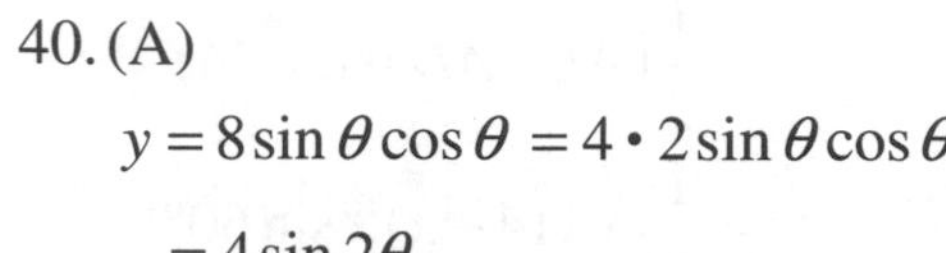

$y = 8\sin\theta\cos\theta = 4 \cdot 2\sin\theta\cos\theta$

$= 4\sin 2\theta$

So, the maximum value of y is 4.

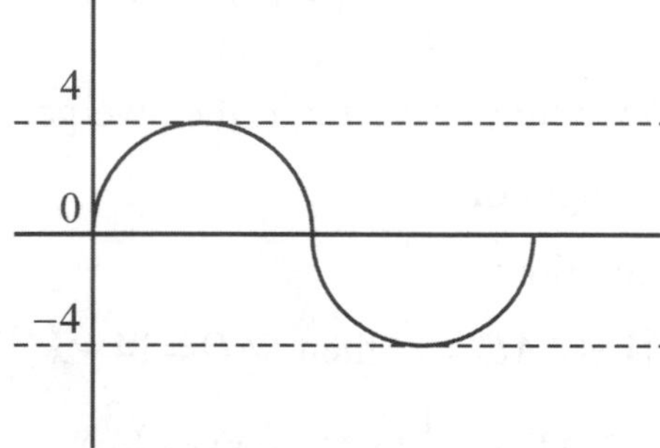

Answer is (A)

41. (A)

For a single coin, the probability of coming up heads is $\frac{1}{2}$, and tails is also $\frac{1}{2}$. So, P(4 heads) $= C_4^8 (\frac{1}{2})^4 (\frac{1}{2})^4$

$= \frac{8 \cdot 7 \cdot 6 \cdot 5 \cdot 4!}{4! \cdot 4!} \cdot \frac{1}{2^8}$

$= \frac{70}{256} = \frac{35}{128} = 0.273$

Answer is (A)

42. (C)

Since $BC + BD = B(C+D)$

$(C+D)^{-1} = \frac{1}{-3}\begin{bmatrix} 0 & -1 \\ -3 & -3 \end{bmatrix}$

$BC + BD = \begin{bmatrix} 3 & 2 \\ 3 & 1 \end{bmatrix} + \begin{bmatrix} 0 & -1 \\ 3 & -2 \end{bmatrix}$

$= \begin{bmatrix} 3 & 1 \\ 6 & -1 \end{bmatrix} = B(C+D)$

Since $B(C+D)(C+D)^{-1} = B$

$= \begin{bmatrix} 3 & 1 \\ 6 & -1 \end{bmatrix} \cdot (\frac{-1}{3}) \begin{bmatrix} 0 & -1 \\ -3 & -3 \end{bmatrix}$

$= (-\frac{1}{3}) \begin{bmatrix} -3 & -6 \\ 3 & -3 \end{bmatrix} = \begin{bmatrix} 1 & 2 \\ -1 & 1 \end{bmatrix}$

Since $B^{-1}(BC) = C = (B^{-1})\begin{bmatrix} 3 & 2 \\ 3 & 1 \end{bmatrix}$

$B = \begin{bmatrix} 1 & 2 \\ -1 & 1 \end{bmatrix}$, $B^{-1} = \frac{1}{3}\begin{bmatrix} 1 & -2 \\ 1 & 1 \end{bmatrix}$

$\therefore\ C = \frac{1}{3}\begin{bmatrix} 1 & -2 \\ 1 & 1 \end{bmatrix} \cdot \begin{bmatrix} 3 & 2 \\ 3 & 1 \end{bmatrix}$

$= \frac{1}{3}\begin{bmatrix} -3 & 0 \\ 6 & 3 \end{bmatrix} = \begin{bmatrix} -1 & 0 \\ 2 & 1 \end{bmatrix}$

Answer is (C)

43. (B)

Since 8, s, t are in geometric sequence

so, $t = (\frac{s}{8}) \cdot s = \frac{s^2}{8}$

Since, s, t, 24 are in arithmetic sequence,

So, $24 - t = t - s$

$\therefore\ 2t = 24 + s$

Substitute $2(\frac{s^2}{8}) = 24 + s$

$\frac{s^2}{4} = 24 + s$

$s^2 - 4s - 96 = 0$

$(s-12)(s+8) = 0$

$s = 12$ or $s = -8$ (false, s is positive)

$2t = 24 + 12 = 36$, $t = 18$

$t + s = 18 + 12 = 30$

Answer is (B)

44. (C)

$9x^2-36x+16y^2+64y-44=0$

$9(x^2-4x+4)+16(y^2+4y+4)-36-64$

$-44=0$

$9(x-2)^2+16(y+2)^2-144=0$

$\frac{(x-2)^2}{16}+\frac{(y+2)^2}{9}=1$

$\frac{(x-2)^2}{4^2}+\frac{(y+2)^2}{3^2}=1$

$a=4,\ \ b=3$

$c^2=a^2-b^2=4^2-3^2=7$

Center is at $(2,-2)$

focus is $2+\sqrt{7}=2+2.65=4.65$

or $2-\sqrt{7}=-0.65$

Answer is (C)

45. (C)

1. Analysis:

(1) Given $m\angle A=60°$, $\cos 60°=\frac{1}{2}$

$ED^2=AE^2+AD^2$

$-2(AE)(AD)\cos 60°$

$=AE^2+AD^2-(AE)(AD)$

(2) $AE=DF=DC$

($m\angle FDC=m\angle A=60°$, so $\triangle DFC$ is an equilateral triangle), if $AE=x$, then $AD=14-x$ ($DC=x$)

(3) Given $\square AEFD=24\sqrt{3}$, hidden information $\triangle AED=\frac{1}{2}\square AEFD$

$=12\sqrt{3}$

(4) $\triangle AED=12\sqrt{3}$

$=\frac{1}{2}(AE)(AD)\sin(\angle A)$

$=\frac{1}{2}(x)(14-x)\times\sin 60°$

$=\frac{1}{2}(14x-x^2)\times\frac{\sqrt{3}}{2}$

(5) we may solve for x, then we get the length of $\overline{ED}$

2. Operation:

(1) Let $AE=x$, then $AD=14-x$

$\triangle AED=\frac{1}{2}(x)(14-x)\times(\frac{\sqrt{3}}{2})$

$=\frac{\sqrt{3}}{4}(14x-x^2)=\frac{1}{2}\times 24\sqrt{3}$

$\therefore\ 14x-x^2=48$

$x^2-14x+48=0$

$(x-8)(x-6)=0$

$x=8$ or $x=6$

(2) $ED^2=8^2+(14-8)^2-2(8)(6)(\frac{1}{2})$

$=100-48=52$

$\therefore\ ED=\sqrt{52}=2\sqrt{13}$

Answer is (C)

46. (A)

$3.25^x=4.05^y$

$x\log^{3.25}=y\bullet\log^{4.05}$

$\frac{x}{y}=\frac{\log^{4.05}}{\log^{3.25}}=\frac{0.607}{0.512}=1.19$

Answer is (A)

47. (A)

Since

$(P\cup A)=P(A)+P(B)-P(A\cap B)$

$$\frac{1}{2} = P(A) + [1 - P(B')] - P(A \cap B)$$
$$= P(A) + (1 - 0.5) - \frac{1}{4}$$
$$= P(A) + 0.5 - 0.25$$
$$= P(A) + 0.25$$
$$\therefore \quad P(A) = 0.5 - 0.25 = 0.25$$

Answer is (A)

48. (B)

$$2\sin^2\theta - 5\cos^2\theta = 3\sin\theta\cos\theta$$
$$\Rightarrow \quad 2\sin^2\theta - 3\sin\theta\cos\theta - 5\cos^2\theta = 0$$

both sides divided by $\cos^2\theta$

yield, $2\frac{\sin^2\theta}{\cos^2\theta} - 3\frac{\sin\theta}{\cos\theta} - 5 = 0$

$$2\tan^2\theta - 3\tan\theta - 5 = 0$$
$$(2\tan\theta - 5)(\tan\theta + 1) = 0$$
$$\tan\theta = \frac{5}{2}$$

or, $\tan\theta = -1$

Since θ is in quadrant II, so

$\tan\theta = -1$, $\theta = 135°$

$$\sin 2\theta + \cos 2\theta = \frac{2\tan\theta}{1+\tan^2\theta} + \frac{1-\tan^2\theta}{1+\tan^2\theta}$$
$$= \frac{2(-1)}{1+1} + \frac{1-1}{1+1} = -1$$

Answer is (B)

49. (C)

Let $z = x + yi$, then $\bar{z} = x - yi$

$$z\bar{z} + 2(z + \bar{z}) - 12 = (x + yi)(x - yi)$$
$$+2[(x + yi) + (x - yi)] - 12$$
$$= x^2 + y^2 + 4x - 12$$
$$= x^2 + 4x + 4 + y^2 - 4 - 12$$
$$= (x+2)^2 + y^2 - 16 = 0$$
$$\therefore \quad (x+2)^2 + y^2 = 16 = 4^2$$

It is a circle

Answer is (C)

50. (A)

Since $\cos\theta = \dfrac{\overrightarrow{OE} \cdot \overrightarrow{OF}}{|\overrightarrow{OE}||\overrightarrow{OF}|}$

$$\overrightarrow{OE} = (4-2, 4-2, 1-2) = (2, 2, -1)$$
$$\overrightarrow{OF} = (2-2, 4-2, 0-2) = (0, 2, -2)$$
$$\cos\theta = \frac{0+4+2}{(\sqrt{9})(\sqrt{8})} = \frac{6}{3\sqrt{8}} = 0.71$$

Answer is (A)

MATH REVIEW

著　者：馮建中
出版者：馮建中
發行者：馮建中
代理經銷：白象文化事業有限公司
地址：401 台中市東區和平街228巷44號
電話：04-22208589
印刷者：迪生設計印刷公司
台中市國光路103號
ISBN 978-626-01-1260-8(平裝)
發行日：中華民國112年5月 初版
定　價：700元